Hormone Metabolism and Signaling in Plants

Hormone Metabolism and Signaling in Plants

Jiayang Li
Chuanyou Li
Steven M Smith

ACADEMIC PRESS

An imprint of Elsevier

Academic Press is an imprint of Elsevier
125 London Wall, London EC2Y 5AS, United Kingdom
525 B Street, Suite 1800, San Diego, CA 92101-4495, United States
50 Hampshire Street, 5th Floor, Cambridge, MA 02139, United States
The Boulevard, Langford Lane, Kidlington, Oxford OX5 1GB, United Kingdom

Notices
Knowledge and best practice in this field are constantly changing. As new research and
experience broaden our understanding, changes in research methods, professional practices, or
medical treatment may become necessary.

Practitioners and researchers must always rely on their own experience and knowledge in
evaluating and using any information, methods, compounds, or experiments described herein.
In using such information or methods they should be mindful of their own safety and the
safety of others, including parties for whom they have a professional responsibility.

To the fullest extent of the law, neither the Publisher nor the authors, contributors, or editors,
assume any liability for any injury and/or damage to persons or property as a matter of
products liability, negligence or otherwise, or from any use or operation of any methods,
products, instructions, or ideas contained in the material herein.

Library of Congress Cataloging-in-Publication Data
A catalog record for this book is available from the Library of Congress

British Library Cataloguing-in-Publication Data
A catalogue record for this book is available from the British Library

ISBN: 978-0-12-811562-6

For information on all Academic Press publications
visit our website at https://www.elsevier.com/books-and-journals

Working together
to grow libraries in
developing countries

www.elsevier.com • www.bookaid.org

Publisher: Glyn Jones
Acquisition Editor: Glyn Jones
Editorial Project Manager: Charlotte Rowley
Production Project Manager: Poulouse Joseph
Cover Designer: Mark Rogers

Typeset by TNQ Books and Journals

Contents

List of contributors

Bin Cai Institute of Botany, Chinese Academy of Sciences, Beijing, China

Zhi Juan Cheng Shandong Agricultural University, Taian, China

Yi Chen Key Laboratory of Analytical Chemistry for Living Biosystems, Institute of Chemistry, Chinese Academy of Sciences, Beijing, China

Jinfang Chu National Center for Plant Gene Research (Beijing), Institute of Genetics and Developmental Biology, Chinese Academy of Sciences, Beijing, China

Shuang Fang National Center for Plant Gene Research (Beijing), Institute of Genetics and Developmental Biology, Chinese Academy of Sciences, Beijing, China

Jian Feng State Key Laboratory of Plant Genomics, Institute of Genetics and Developmental Biology, Chinese Academy of Science, Beijing, China

Xiangdong Fu Institute of Genetics and Developmental Biology, Chinese Academy of Sciences, Beijing, China

Xiuhua Gao Institute of Genetics and Developmental Biology, Chinese Academy of Sciences, Beijing, China

Zhizhong Gong China Agricultural University, Beijing, China

Hongwei Guo South University of Science and Technology of China, Shenzhen, China; Peking University, Beijing, China

Zhenpeng Guo Key Laboratory of Analytical Chemistry for Living Biosystems, Institute of Chemistry, Chinese Academy of Sciences, Beijing, China

Dongdong Hao South University of Science and Technology of China, Shenzhen, China; Peking University, Beijing, China

Zuhua He Shanghai Institutes for Biological Sciences, Chinese Academy of Sciences, Shanghai, China

Yuxin Hu Institute of Botany, Chinese Academy of Sciences, Beijing, China

Zhaoyun Jiang Peking University, Beijing, China

Jing Bo Jin Institute of Botany, Chinese Academy of Sciences, Beijing, China

Chuanyou Li Institute of Genetics and Developmental Biology, Chinese Academy of Sciences, Beijing, China

Jia Li Lanzhou University, Lanzhou, China

Jiayang Li Institute of Genetics and Developmental Biology, Chinese Academy of Sciences, Beijing, China

Jigang Li China Agricultural University, Beijing, China

Lin Li Institute of Genetics and Developmental Biology, Chinese Academy of Sciences, Beijing, China

Chun-Ming Liu Institute of Botany, Chinese Academy of Sciences, Beijing, China; Institute of Crop Science, Chinese Academy of Agricultural Sciences, Beijing, China

Biao Ma Institute of Genetics and Developmental Biology, Chinese Academy of Sciences, Beijing, China

Li-Jia Qu Peking University, Beijing, China

Shi-Chao Ren Institute of Botany, Chinese Academy of Sciences, Beijing, China

Baoshuan Shang Institute of Botany, Chinese Academy of Sciences, Beijing, China

Yiting Shi State Key Laboratory of Plant Physiology and Biochemistry, China Agricultural University, Beijing, China

Steven M. Smith Institute of Genetics and Developmental Biology, Chinese Academy of Sciences, Beijing, China; University of Tasmania, Hobart, Australia

Xiu-Fen Song Institute of Botany, Chinese Academy of Sciences, Beijing, China

Yi Su Hunan Agricultural University, Changsha, China

Xiangzhong Sun South University of Science and Technology of China, Shenzhen, China; Peking University, Beijing, China

Bing Wang Institute of Genetics and Developmental Biology, Chinese Academy of Sciences, Beijing, China

Haijiao Wang Huazhong Agricultural University, Wuhan, China

Ruozhong Wang Hunan Agricultural University, Changsha, China

Xuelu Wang Huazhong Agricultural University, Wuhan, China

Yonghong Wang Institute of Genetics and Developmental Biology, Chinese Academy of Sciences, Beijing, China

Zhuoyun Wei Lanzhou University, Lanzhou, China

Yaorong Wu Institute of Genetics and Developmental Biology, Chinese Academy of Sciences, Beijing, China

Langtao Xiao Hunan Agricultural University, Changsha, China

Shitou Xia Hunan Agricultural University, Changsha, China

Daoxin Xie Tsinghua University, Beijing, China

Qi Xie Institute of Genetics and Developmental Biology, Chinese Academy of Sciences, Beijing, China

Peiyong Xin National Center for Plant Gene Research (Beijing), Institute of Genetics and Developmental Biology, Chinese Academy of Sciences, Beijing, China

Chun Yan Tsinghua University, Beijing, China

Shuhua Yang State Key Laboratory of Plant Physiology and Biochemistry, China Agricultural University, Beijing, China

Qingzhe Zhai Institute of Genetics and Developmental Biology, Chinese Academy of Sciences, Beijing, China

Jin-Song Zhang Institute of Genetics and Developmental Biology, Chinese Academy of Sciences, Beijing, China

Xian Sheng Zhang Shandong Agricultural University, Taian, China

Yingying Zhang Shanghai Institutes for Biological Sciences, Chinese Academy of Sciences, Shanghai, China

Jian-Min Zhou Institute of Genetics and Developmental Biology, Chinese Academy of Sciences, Beijing, China

Jianru Zuo State Key Laboratory of Plant Genomics, Institute of Genetics and Developmental Biology, Chinese Academy of Science, Beijing, China

About the editors

Jiayang Li

Dr. Jiayang Li was awarded the Degree of Bachelor of Agronomy from Anhui Agricultural University in 1982, the Degree of Master of Science from the Institute of Genetics and Developmental Biology (IGDB) of the Chinese Academy of Sciences (CAS) in 1984, and PhD in Biology from Brandeis University, United States, in 1991. After completing his postdoctoral research in the Boyce Thompson Institute at Cornell University, Dr. Li was recruited as a Professor of plant molecular genetics by IGDB in 1994. Dr. Li's laboratory is mainly interested in the molecular genetics of higher plant development, with a focus on the biosynthesis and action of plant hormones, including auxin, brassinosteroid, and strigolactones. Dr. Li has made seminal contributions to establishing forward genetics approaches to understand rice growth and to improve rice yield and quality through rational design, with his achievements receiving worldwide attention from scientific websites and public media. His achievements have also been widely recognized by scientists, and his publications have been highlighted in commentaries and cited by review articles, making him a Thomson-Reuters Highly Cited Researcher in the field of Plant and Animal Sciences in 2014. Dr. Li served as the Director General of IGDB from 1999 to 2004, Vice President of CAS from 2004 to 2011, and President of the Chinese Academy of Agricultural Sciences and Vice Minister of Agriculture for the People's Republic of China from 2011 to 2016. Dr. Li was elected a Member of the Chinese Academy of Sciences in 2001, a Fellow of The World Academy of Sciences in 2004, a Foreign Associate of the US National Academy of Sciences in 2011, a Member of the German Academy of Sciences Leopoldina in 2012, a Foreign Member of the European Molecular Biology Organization in 2013, and a Foreign Member of the Royal Society in 2015.

Chuanyou Li

Dr. Chuanyou Li received his PhD in Genetics at the Institute of Genetics, Chinese Academy of Sciences (CAS) in 1999. He did his postdoc training from 1999 to 2003 in the DOE-Plant Research Laboratory at Michigan State University. Since 2003, he serves as a professor and group leader at the State Key Laboratory of Plant Genomics, Institute of Genetics and Developmental Biology, CAS. Research in the Chuanyou Li laboratory is aimed at understanding the action mechanisms of jasmonate, which plays vital roles in regulating plant immunity and a wide range of developmental processes. Dr. Li and his colleagues found that the long-distance mobile signal in regulating systemic plant immunity is jasmonate, rather than the peptide systemin. He led more than 60 Chinese scientists to take part in the International Tomato Genome

Consortium and successfully decoded the genome of tomato, a unique model system for plant immunity and fruit biology. He connected the PLETHORA stem cell transcription factor pathway to jasmonate signaling and illustrated a molecular framework for jasmonate-induced regulation of root growth through interaction with the growth hormone auxin. His laboratory has a long-term focus on the transcriptional mechanism of MYC2, a basic helix-loop-helix protein that regulates diverse aspects of jasmonate responses. He found that the turnover of MYC2 stimulates its transcription activity, revealing an "activation by destruction" mechanism to regulate plant stress response and adaptive growth. He also linked MYC2 to the MED25 subunit of the Mediator complex in the transcription machinery. He has published more than 80 research papers in journals such as *Nature*, *Nature Genetics*, *PNAS*, *Plant Cell*, and *PLoS Genetics*, which received more than 4000 citations. Dr. Li serves as editor for several international journals including *Molecular Plant*, *Plant Molecular Biology*, and *Annals of Botany*.

Steven M Smith

Steven Smith began his scientific career as a technician in the Botany Department at Rothamsted Experimental Station in the United Kingdom, where he conducted bioassays of auxins, cytokinins, and gibberellins. He later completed a Master's degree studying cytokinin action in tissue culture cells at Indiana University, United States, under the mentorship of Carlos O. Miller, the discoverer of kinetin and of organogenesis mediated by auxin and cytokinin. After his PhD at Warwick University, United Kingdom, and postdoctoral studies at CSIRO in Canberra, Australia, during which time he conducted research on the biosynthesis of chloroplast proteins, he was a lecturer in molecular biology at the Edinburgh University, United Kingdom, for 20 years, studying plant metabolism and development. He was awarded an Australian Research Council Federation Fellowship and moved to the University of Western Australia in 2005 where he was a founding member of the Australian Research Council Centre of Excellence in Plant Energy Biology. There he discovered the mode of action of karrikins and made important contributions to research on strigolactones, brassinosteroids, and auxins. He has been Visiting Professor at the Institute of Genetics and Developmental Biology, Chinese Academy of Sciences, since 2013. He took up a new position at the University of Tasmania in 2015 and was recognized as a Thomson-Reuters Highly Cited Researcher in Plant and Animal Sciences in 2016.

Foreword

The first plant hormones were discovered around the middle of the 20th century, including auxin, gibberellic acid, ethylene, cytokinin, and abscisic acid. Subsequently several more have been discovered, including brassinosteroids, salicylic acid, jasmonates, and, most recently, strigolactones. In addition, there are many peptides that behave as hormones.

It is now recognized that plant hormones play central roles in every aspect of plant growth, development, and physiology. They often operate in concert to orchestrate complex developmental programs and to integrate plant responses to multiple environmental factors. These modes of action underpin the very high level of plasticity observed in plants and hence contribute to the extraordinary evolutionary success of plant life on earth.

For humans the impact of plant hormones is profound. Through domestication and selection, humans bred crop varieties that we now know carry particular hormone-related traits that provide many benefits, including increased yields of food and other products. The domestication of grasses to create cereals and corn provides an excellent example. In the 1960s and 1970s, the breeding of dwarf varieties of wheat and rice increased cereal and grain production dramatically as a direct result of the selection of mutants compromised in their response to gibberellin. This was described as the "green revolution." In the 1950s and 1960s, the discovery that auxin and cytokinin could be used to drive undifferentiated plant cells to regenerate into whole plants triggered another plant biotechnology revolution, which led to the ability of making transgenic plants. Today, the rapid increase in our understanding of plant hormone biology offers the opportunity for continued advances in agriculture.

This new book edited by Jiayang Li, Chuanyou Li, and Steven M Smith reflects not only the biological importance of plant hormones but also their strategic importance in future plant and food production. All the authors are based in China, and have been participants in a major 8-year program funded by the National Natural Science Foundation of China. The Chinese Ministry of Science and Technology and Ministry of Agriculture have also invested heavily in new programs and key laboratories to support such fundamental research in China. This investment by the Chinese Government is an acknowledgment of the importance of plant hormone research. Most of the senior authors have gained previous research experience in North America, Europe, and other Asia-Pacific countries, before returning to China. This book is therefore rooted in the international research community and is aimed at the international community of scientists and students.

The hormones discussed are organic chemicals that function exclusively as signals in the co-ordination of plant growth, development, and physiology, and defense. This book does not directly discuss chemical signals that are generated by cells as a result of their core metabolic activities, such as reactive oxygen species and nitric oxide. Nor does it provide any specific focus on inorganic ions such as calcium and nitrate, which, although providing essential signaling functions in plants, have a wide range of other roles. A third group of signaling chemicals that are not discussed in detail are metabolites such as sugars, amino acids, and polyamines because again, although they may have vital signaling functions and some are transported between tissues and organs of the plant, they have other primary functions.

Although numerous hormones have been discovered, there are very likely many more, as discussed by Smith et al. (Chapter 1). For example, the full range of peptides that contribute to plant development and physiology has not yet been fully appreciated, as discussed by Song et al. (Chapter 11). There is also an increasing awareness of the importance of microRNA molecules in signaling and control mechanisms, and these are discussed in the context of hormone signaling throughout the book.

There have been tens of thousands of research papers published on the subject of plant hormones, and hundreds of reviews. There are also excellent texts on plant hormones, which provide important perspectives and foundations. The focus of this book is on the biosynthesis and signaling mechanisms of the major hormones, and it discusses the latest research in a rapidly advancing field. Spectacular advances in technology, from increasingly sophisticated "omics" approaches to live imaging from subcellular to organismal scales, have transformed our understanding of plant hormone biology. The authors of this book have contributed to this transformation.

Each of nine hormone types is considered in a separate chapter (Chapters 2–10), with the first introductory chapter providing an overview of hormone functions (Smith et al., Chapter 1). Peptide hormones are also discussed in detail (Song et al., Chapter 11). One further chapter provides a discussion specifically of hormone function in stem cells, because of their critical importance in plant growth and development (Cheng et al., Chapter 12). Two further chapters provide overviews of assay and imaging systems for plant hormones using advanced biological principles (Su et al., Chapter 13) and analytical chemistry methods (Fang et al., Chapter 14). The main focus of the discussion is on molecular mechanisms and relates extensively to research in the reference plant *Arabidopsis thaliana*, and increasingly in rice. This work provides the essential foundation for future research directed at diverse crop species, and more widely at understanding how hormones can be deployed to address the challenge of increasing the production of food and plant products in a changing environment.

This book provides a valuable reference work for those contributing to this important endeavor.

Ottoline Leyser
Sainsbury Laboratory
Cambridge University, UK
January 2017

Hormone function in plants

1

Steven M. Smith[1,2], Chuanyou Li[1], Jiayang Li[1]
[1]Institute of Genetics and Developmental Biology, Chinese Academy of Sciences, Beijing, China; [2]University of Tasmania, Hobart, Australia

Summary

There are many endogenous signaling and regulatory molecules which can influence the growth, development and physiology of plants. Hormones are produced specifically for signaling. They are often transported from sites of synthesis to distant sites of action and they operate at very low concentrations. In contrast, some other chemicals may provide signals, but it may not be their main function or activity, such as primary metabolites, reactive oxygen species (ROS) and inorganic ions, and they often act locally within individual cells. The hormone family includes auxins, cytokinins (CK), gibberellins (GA), abscisic acid (ABA), ethylene (ETH), brassinosteroids (BR), strigolactones (SL), salicylic acid (SA), jasmonates (JA), and peptides. They are synthesized from common metabolic precursors, but use specialized pathways, and their production is very strictly controlled, both spatially and temporally. All hormones influence multiple aspects of plant function, and they influence the synthesis and actions of each other. The interactions between hormones, environmental signals, and developmental programs are so complex that the description and modeling of the whole system is very challenging. Some hormone receptors are membrane anchored (CK, ETH, BR, and peptides) while others are soluble (auxin, GA, ABA, SA, jasmonic acid (JA), and SL). Co-receptor complexes are formed during perception of hormones including auxin (with IAA/AUX transcriptional repressor proteins), JA (with JAZ transcriptional repressor proteins), and ABA (with phosphoprotein phosphatase PPC2). Hormone perception can lead to signal transduction through protein phosphorylation cascades (e.g., ABA, CK, BR, and peptides). Other hormone-receptor complexes trigger interaction with F-box proteins and ubiquitination enzymes that target proteins such as transcriptional repressors for degradation by the 26S proteasome (e.g., auxin, GA, SA, JA, and SL). Such signaling changes protein activities and gene transcription, with consequent changes to plant development and physiology. The effects of hormones are so profound that through breeding and agrochemical approaches in the 20th century, they gave us high-yielding, nutritious and resilient crops. In the 21st century we look to plant hormones to help meet the increasing demand for food production under ever-more challenging environmental conditions.

1.1 The nature of hormones

1.1.1 Hormones and signals

Since the discovery of the auxin indole-3-acetic acid in the 1930s, many endogenous signaling and regulatory molecules have been discovered, and more are yet to be discovered. There is great diversity among this array of signaling molecules, which

Hormone Metabolism and Signaling in Plants. http://dx.doi.org/10.1016/B978-0-12-811562-6.00001-3

includes small organic compounds, gases and volatiles, inorganic ions, oligosaccharides, peptides, and RNAs. Some of these substances are produced and act within individual cells, others pass between cells, others are transported between tissues and organs to regulate processes at the whole-plant level and yet others may be released into the environment to influence neighboring plants and other organisms.

These observations raise the important question of how we define plant hormones. It is not easy to define them, and there is no absolute definition. The hormone concept in plants was borrowed from that of animals. A classical and simple definition of a hormone in animals is an organic chemical produced in one organ and transported at very low concentrations to other sites in the animal to regulate specific processes in target tissues. Examples include adrenaline, thyroxine, growth hormone (somatotropin) and the steroidal sex hormones, testosterone and estrogen. There are several aspects of such a definition that do not fit so well for plants. Firstly it is more difficult to identify discrete source organs and specific target tissues. Many such chemical signals are produced throughout the plant and can act locally, as well as distally. Some signaling substances have multiple functions, such as sugars which serve as a source of carbon and energy in addition to signaling, or calcium ions which serve signaling, enzymatic, and structural roles. Some signals such as reactive oxygen species (ROS) and nitric oxide (NO) are probably produced in all cells as an inevitable consequence of conducting metabolic activities in an oxygen-containing environment. Sugars and some other metabolic signals operate at high (mM) concentrations while others such as calcium ions operate at low (μM) concentrations, while hormones typically function at very low levels (nM to pM).

For the present discussion and for the content of this book (Li et al., 2017a), we focus on nine hormones which are all small endogenous organic signaling chemicals, plus a 10th group comprising small peptides involved in the control of plant development (Fig. 1.1). Each of these hormones is discussed individually in more detail in subsequent chapters in this book. They are auxin (Qu et al., 2017), cytokinins (CK) (Feng et al., 2017), gibberellic acid (GA) and gibberellins (Gao et al., 2017), abscisic acid (ABA) (Li et al., 2017b), ethylene (ETH) (Hao et al., 2017), jasmonic acid (JA) and jasmonates (Zhai et al., 2017), salicylic acid (SA) (Jin et al., 2017), brassinosteroids (BR) (Wang et al., 2017a), strigolactones (SL) (Wang et al., 2017b), and peptides (Song et al., 2017). The structures of these hormones (Fig. 1.1) show that they are small organic chemicals, ranging in size from the 2-carbon ETH (C_2H_4) to BR with approximately 25 carbons.

The classical growth hormones are sometimes considered to be auxin, CK, GA, ABA, and ETH, because they have been known and understood for the longest period of time and they have profound effects on plant physiology and development (Davies, 2010). In the past, ABA was considered to be primarily a growth inhibitor and a stress hormone because it acts during exposure of plants to abiotic stress such as water deficit, but it is now known to be also required for many growth and developmental processes (Li et al., 2017b). Similarly JA and SA are often considered to be stress or defense hormones because they act when plants are challenged by pests, pathogens, and environmental stresses. However, it is clear that both JA and SA are also involved in other aspects of plant function (Zhai et al., 2017; Jin et al., 2017). Although BR

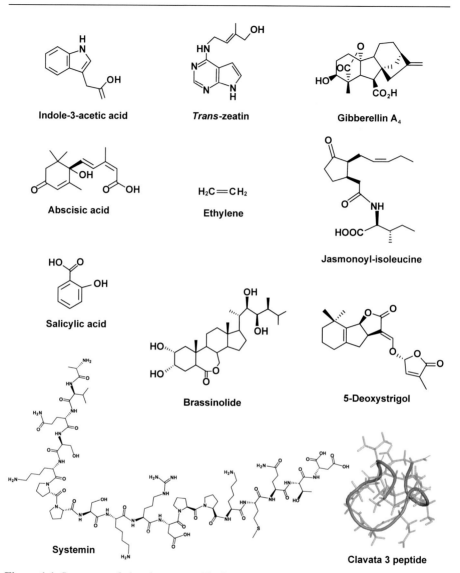

Figure 1.1 Structures of plant hormones. The hormones are: the auxin indole-3-acetic acid (IAA); the cytokinin (CK) *trans*-zeatin; Gibberellin A$_4$ which is an oxidized form of gibberellic acid (GA) also denoted as GA$_3$; Abscisic acid (ABA); Ethylene (ETH); Jasmonyl-isoleucine (JA-Ile) which is a conjugate of jasmonic acid (JA); Salicylic acid (SA); Brassinolide (BL), the first brassinosteroid (BR) to be studied; 5-Deoxystrigol (5DS), a precursor of strigol, both active strigolactones (SL); Systemin, a peptide first identified in tomato leaves following wounding and insect attack; Clavata 3 peptide (CLV3), here represented as a three-dimensional skeletal model.

and SL have been known for several decades, their essential roles in plant growth and development have become recognized now, with SL being identified as a multifunctional plant hormone only between 2000 and 2010 (Wang et al., 2017a,b). Peptide hormones were first recognized for their role in defense against pests but have since been found to participate in symbiotic interactions and plant development (Song et al., 2017). It is now clear that the concept of separate "growth" hormones and "stress" hormones does not hold up, because they all have multiple and diverse roles.

1.1.2 Discovering and uncovering

1.1.2.1 The process of discovery

The discovery of auxin was built upon founding observations of phototropic bending in oat coleoptiles, made by Charles Darwin (1880), and recorded in his book *The Power of Movement in Plants*. He provided evidence for a transmissible substance that moves asymmetrically down the coleoptile from the tip and in so doing, controls coleoptile elongation and bending. This response provided the basis for a bioassay which enabled the isolation of this growth substance and its identification as indole-3-acetic acid, which became known as auxin ("to grow") (Qu et al., 2017).

At about the same time, a role for ETH was recognized, firstly because the ETH in gas street lights was seen to affect plant growth (Doubt, 1917) and it was soon discovered that plants also produce ETH (Gane, 1934). Soon after that, in 1935, GA was discovered in the pathogenic fungus *Gibberella fujikuroi* which causes the "foolish seedling" disease in rice. The GA produced by the fungus causes elongation of the rice shoot. Later in the 1950s, it was found that plants produce their own GAs (Takahashi, 1998; Gao et al., 2017). Also in the 1950s, CK was discovered by Miller and Skoog in the United States while searching for factors that could stimulate plant cells to undergo sustained division in vitro (Amasino, 2005). They discovered that herring sperm DNA or yeast extract could achieve this, but only if the samples were old or had been autoclaved (Miller et al., 1955). This led to the discovery of kinetin, the first CK, and soon afterward zeatin, a naturally occurring CK was discovered in maize (Miller, 1961; Letham and Miller, 1965; Feng et al., 2017).

In the early 1960s UK researchers isolated a factor from sycamore leaves which induced bud dormancy, while a team in the United States discovered a factor which induced abscission in cotton fruits. The factors were initially named "dormin" and "abscisin" and were soon found to be identical and renamed abscisic acid (Li et al., 2017b). Although SA was discovered in the 19th century in willow (*Salix* sp.) trees, it was not until a century later that its role as a plant hormone was recognized by its ability to induce virus resistance in tobacco plants (White, 1979). Also in the 1970s, studies of the growth-promoting activity of a substance in pollen of *Brassica napus* L. led to the discovery of BR. The activity stimulated elongation of bean internodes and it was named "brassin" (Mitchell et al., 1970). The structure was determined after purification of 4 mg from 227 kg of pollen (Grove et al., 1979).

The first isolation of JA came from fractionation of compounds from the fragrant oil of jasmine flowers (*Jasminum grandiflorum*) (Demole et al., 1962), and subsequently jasmonates were found in other plant species. They were initially shown to inhibit seedling growth and induce senescence (Zhai et al., 2017). The first peptide hormone described was systemin, which was found to initiate defense responses in tomato leaves when the plants were attacked by insects or mechanical wounding. Importantly the signal was found to be transmitted systemically through the plant (Pearce et al., 1991). Now such peptides are known to participate in the control of plant development and symbioses with bacteria and fungi (reviewed by Song et al., 2017).

The discovery of SLs was made in root extracts of cotton plants by virtue of their ability to stimulate germination of seeds of parasitic plants of the Orobanchaceae, such as the witchweed *Striga hermonthica* (Cook et al., 1966). Active compounds were lactones, hence giving rise to the generic name of strigolactones. The benefit of SL secretion from the roots of host plants was realized when it was revealed that SLs exuded from the roots stimulated the development of hyphae of arbuscular mycorrhiza (AM) symbiotic fungi (Akiyama et al., 2005). However it was not until 2008 that SLs were recognized as plant hormones (Gomez-Roldan et al., 2008; Umehara et al., 2008).

1.1.2.2 Uncovering new hormones

An appreciation of the process of hormone discovery is important because there are still hormones and other signaling molecules to be discovered.

One intriguing example is a signal produced in Arabidopsis *bypass* mutants. In the *bypass* mutant, a signal is transmitted from roots to the shoot where it inhibits shoot growth and embryogenesis. The signal acts by disrupting CK signaling, expression of the gene coding for the WUSCHEL (WUS) transcription factor, and shoot apical meristem (SAM) maintenance (Lee et al., 2012, 2016). The normal function of the *BYPASS* gene is apparently to prevent the production of this signal. Its identity is unknown, but it appears to be a low-molecular weight metabolite rather than a macromolecule (Adhikari et al., 2013).

Other unidentified signals are produced from carotenoids, independently of ABA and SL biosynthesis. In one case a carotenoid-derived signal is required to establish a pre-pattern of lateral root formation in Arabidopsis (Van Norman et al., 2014). The signal is transmissible between cells, but its identity is unknown. In another study, a carotenoid-derived signal is required for leaf development, cell differentiation, and the expression of chloroplast and nuclear genes (Avendaño-Vázqueza et al., 2014). Its identity is unknown, and it is not known if it is transmissible between cells and tissues; so, it is unclear whether it could be considered to be a new hormone.

Another example arises from the discovery of karrikins, compounds produced by burning vegetation in wildfires, which can stimulate seed germination and seedling photomorphogenesis (Nelson et al., 2011; Flematti et al., 2013). Karrikins are butenolides similar to SL, and their perception in Arabidopsis requires the protein KARRIKIN INSENSITIVE2 (KAI2) which is a homolog of the SL receptor DWARF14 (Waters et al., 2012). However KAI2 does not recognize naturally occurring SLs and the endogenous substrate or ligand for KAI2 is produced independently of the biosynthetic pathway of known SLs (Waters et al., 2015). The *kai2* mutant has defects in seed germination, seedling development, and leaf shape, suggesting that the endogenous karrikin-like signal could be a new hormone.

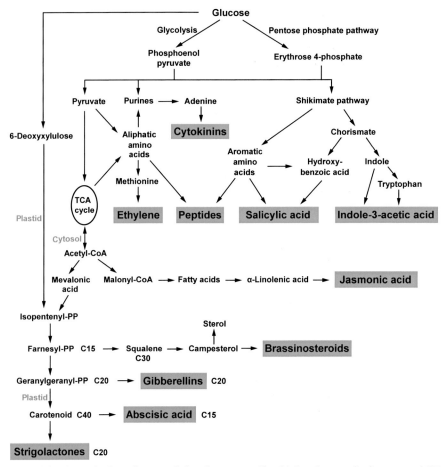

Figure 1.2 Biosynthetic pathways of plant hormones. Cytokinins also require isopentenyl-PP for the prenylation of adenine. Salicylic acid can be made via two different pathways, as can indole-3-acetic acid. Isopentenyl-PP is produced by two separate pathways, one in the plastid and one in the cytosol. Abscisic acid and strigolactones are made by oxidative cleavage of carotenoids. *CoA*, coenzyme A; *PP*, pyrophosphate; *TCA*, tricarboxylic acid cycle.

1.1.3   Making and breaking

1.1.3.1   Biosynthetic pathways

The pathways of biosynthesis of the known plant hormones have been elucidated in great detail, although some gaps in our knowledge still remain. In most cases it seems that the hormones are derived from metabolic pathways or products that have other functions (Fig. 1.2). Thus, IAA and SA are derived mainly from tryptophan and phenylalanine respectively, although alternative pathways exist for each via the precursors indole and chorismate, respectively. Meanwhile ETH is derived from methionine, CKs from adenine, and JA from lipoxygenase-catalyzed oxygenation of α-linolenic acid in the plastid. The other hormones are all terpenoids. Both ABA and SL are derived by oxidative cleavage of C-40 carotenoids in the plastid to produce different fragments which are further metabolized to the active hormones. The sterol campesterol which is quite abundant in plant membranes is the precursor for BR, while GA is derived from C-20 geranylgeranyl pyrophosphate, a ubiquitous diterpenoid precursor (Fig. 1.2).

A major recurring question is the extent to which different cell types synthesize the different plant hormones, and to what extent cells or tissues are dependent upon other cells for a supply of hormones. Expression of genes for hormone biosynthesis is detected in most tissues to variable extents, and hormone biosynthesis can be triggered by treatments such as light, wounding, or senescence. Furthermore the pluripotent nature of plant cells and the plasticity of plant development indicate that many or most cells have the potential to synthesize the full range of hormones. Nevertheless, we know that many developmental processes depend on the synthesis of hormones in one organ and transport to another for their action. The production of auxin in the SAM and its polar transport down the stem repressesing lateral shoot growth is a prime example (Qu et al., 2017). Another example is the repression of lateral shoot growth by SL transported from root to shoot (Wang et al., 2017b). Biosynthesis of ABA apparently occurs in the vascular tissues and the guard cells in shoots, whereas all tissues in seeds are thought to be involved in ABA biosynthesis (Li et al., 2017b).

1.1.3.2 Structural diversity

Except in the cases of ETH, SA, and ABA, all the other hormones are produced in numerous different forms. Auxin, CK, GA, JA, BR, SL, and peptide hormones all exist in dozens of different structural forms, many of which are active. This potentially provides a degree of specificity and selectivity, so that different chemical forms of a hormone can trigger different responses. For example, chlorinated IAA seems to be important in legume seeds while the carboxyl methyl ester form of IAA is apparently involved in leaf curling in Arabidopsis (Qu et al., 2017; Li et al., 2008; Qin et al., 2005). Conjugation of JAs with different amino acids provides a range of compounds with different activities (Zhai et al., 2017).

Another crucial aspect of structural diversity is the conversion of hormones between active and inactive forms. This provides a way to control the activity of a hormone in ways that may not be readily achieved through the simple control of rates of biosynthesis and breakdown of the active hormone. Thus, chemically modified forms of a hormone can provide a reservoir of inactive compound that can be rapidly converted to the active form for a rapid response. Such a mechanism also potentially provides a means of buffering against major changes in the amount of active hormone. Other roles of chemical modification might be to increase hormone stability or to facilitate transport within or between cells.

Glycosylation is a common form of chemical modification, which occurs with auxin, CK, GA, ABA, BR, SA, and JA. The active form of auxin is free IAA but most of the IAA in a plant tissue can be conjugated with ester-linked sugars or amide-linked amino acids (Qu et al., 2017). Commonly CKs occur as inactive nucleosides and nucleotides, which can be activated by hydrolysis of the phosphate and sugar moieties to give the active nucleobases (Feng et al., 2017). The most common ABA conjugate is the physiologically inactive glucosyl ester (ABA-GE), which is produced by glucosyl transferase activity. It can be hydrolyzed by β-glucosidase to release free ABA (Li et al., 2017b). It is suggested that such modification of active hormones allows the plant to rapidly and precisely control the amount of active hormone.

1.1.3.3 Degradation of hormones

The amount of active hormone is clearly vitally important to the effect that it has on target cells, and degrading a hormone to render it inactive is presumably just as important as synthesizing it. The balance between rates of synthesis and breakdown can determine the amount of active hormone. The rate at which a hormone is broken down might also determine the time during which the hormone remains active. For some hormone actions we might imagine that a very rapid response is required over a period of minutes, such as during wounding or rapid fluctuation in light quality. For other actions we could imagine that sustained hormone signaling over a period of hours or days is required, such as during wound callus formation, seed filling, or maintenance of meristem identity (Cheng et al., 2017). An exception to these principles lies with ETH, because there is no known breakdown pathway, so its amount is determined by the relative rates of biosynthesis and dissipation by diffusion.

For some hormones we have a good understanding of their pathways and control of breakdown. This is the case for ABA which is inactivated by oxidation via phaseic acid and dihydrophaseic acid (Li et al., 2017b). Modulation of CK levels is achieved by degradation by CK oxidase/dehydrogenase enzymes and this can have a profound effect on rice grain development (Feng et al., 2017). Inactivation of GA is also critical for controlling endogenous GA levels, involving 2β-hydroxylation catalyzed by GA 2-oxidase (GA2ox) enzymes. In Arabidopsis , a cytochrome P450 hydroxylates brassinolide (BL) into inactive 26-OH-BL (Wang et al., 2017a). Auxin is inactivated by oxidation to 2-oxoindole-3-acetic acid by the product of the Arabidopsis *DIOXYGENASE FOR AUXIN OXIDATION 1* (At*DAO1*) gene (Porco et al., 2016). Surprisingly, loss-of-function and gain-of-function genes did not greatly alter the amount of IAA in Arabidopsis plants, but did alter the amounts of 2-oxoindole-3-acetic acid and of IAA conjugates, indicating that there are other mechanisms that maintain IAA at suitable levels for normal growth (Porco et al., 2016).

For most hormones however we have little knowledge of their breakdown. An interesting example is that of ETH because it is a simple gas, and potentially is dissipated by diffusion rather than by degradation. If so, it is unclear if or how the amount of ETH can be regulated precisely. Another unusual example is provided by SL. Signaling by SL involves its destruction by the receptor protein DWARF14, which is a serine hydrolase enzyme (Wang et al., 2017b). DWARF14 attacks the butenolide moiety of SL which becomes covalently attached at the active site of the protein, while the residual tricyclic structure is released as an inactive degradation product (Yao et al., 2016). The destruction of SL by its receptor means that each hormone molecule can be used only once. We do not know how many times other hormone molecules are used before they become degraded.

Thus, fine control of rates of biosynthesis, breakdown, modification, and transport cannot only determine the amount and hence the magnitude of hormone action, but also the precise location and the duration of action. Understanding such complexity at the single cell level is a challenge, even for a single hormone, so elevating the complexity by considering multiple hormones and other signals becomes even more challenging. Progress toward visualizing the amounts and actions of hormones at the single cell level are discussed by Su et al. (2017).

1.1.4 Measuring and visualizing

1.1.4.1 Identification and quantification

Bioassays have been absolutely essential for the discovery of plant hormones, and continue to be vital today for many applications (Su et al., 2017). However, one of the major reasons that we have made great progress in our understanding of plant hormones has been the advances in chemical analysis, especially separation science and mass spectrometry. Hormones are present at extremely low concentrations in plant materials and the chemical composition of plant extracts is extremely complex. The difference in abundance of hormones and major metabolites such as sucrose can be more than 10 orders of magnitude (Chu et al., 2017; Su et al., 2017). The complexity of the matrix can interfere with attempts to separate and analyze hormones. Progress has been achieved with the use of simple "clean-up" or enrichment methods such as partitioning between aqueous and organic solvents, but now the use of solid phase separation methods have become easy and powerful. This is possible because the compounds in a complex extract have a wide range of different chemical properties. Some hormones such as SL, GA, BR are quite hydrophobic while others such as auxin and CK and more hydrophilic. A range of different cartridges can be purchased for use in such enrichment steps. Thereafter, liquid chromatography (LC) and gas chromatography (GC) instruments have become increasingly effective and efficient, and chemical modification agents have been developed and optimized to improve the separation and detection of target compounds.

Mass spectrometers have become increasingly sensitive, high resolution and fast. The use of "triple quadrapole" instruments allows multiple target ions to be monitored simultaneously, and "ion trap" instruments allow a chosen ion to be captured and fragmented for structural analysis. Coupled directly with LC or GC instruments, the separation, identification, and quantification of hormones can now be achieved very effectively, although it still requires specialized technicians. Methods have advanced such that simultaneous analysis of several hormones in a single extract is now possible (Chu et al., 2017). This takes us closer to the goal of characterizing all hormones in a single biological sample.

1.1.4.2 Visualizing hormone action

Reporter genes have played a vital role in characterizing hormone action in cells and tissues. Genes encoding a β-glucuronidase (GUS) and green fluorescent protein (GFP) are very powerful when linked to a promoter sequence from a hormone biosynthesis or hormone response gene. For the visualization of auxin levels or auxin activity, two widely used reporter genes are *DR5-GUS* and *DR5-GFP*, which combine multiple copies of an auxin response element linked to the core 35S promoter, and respond sensitively to auxins (Ulmasov et al., 1997; Benková et al., 2003). Transgenic plants containing a *DR5-GUS* gene are treated with the chromogenic GUS substrate and blue staining of cells reports the presence of auxin. Disadvantages of GUS are that it is a destructive assay which kills the cells and diffusion of the blue product of GUS reduces resolution, whereas fluorescent proteins such as GFP are nondestructive and have greater resolution and sensitivity. Such reporters act downstream of auxin signaling and the reporter proteins are stable, so they are not good at reporting rapid changes in hormone.

An alternative approach exploits the fact that signaling of some hormones leads to degradation of certain target proteins. An auxin "degron" system was constructed which encodes the auxin-binding domain of co-repressor protein IAA28 fused in frame to the yellow fluorescent protein (YFP) variant "VENUS," and expressed using the 35S promoter. The use of VENUS, which has a fast oxidation of its chromophore, was necessary because such IAA proteins exhibit rapid turnover and are difficult to visualize using GFP tagging. This degron reporter allowed screening of auxin distribution in transgenic plants (Brunoud et al., 2012; Zhou et al., 2011). Another example is provided by DELLA transcriptional repressor proteins which are degraded in response to GAs (Gao et al., 2017). A gene encoding a DELLA-GFP translational fusion protein expressed from the *DELLA* promoter has been used to monitor cellular GA changes in Arabidopsis since perception of GA leads to the degradation of the DELLA-GFP fusion (reviewed in Sun, 2011). A similar "degron" approach has used a JAS9 transcriptional repressor linked to the VENUS protein to quantify JA dynamics in response to stress with high spatiotemporal sensitivity (Larrieu et al., 2015). The abundance of JAS9-VENUS is dependent on bioactive JA isoforms, the COI1 co-receptor, and a functional Jas motif and proteasome activity (Zhai et al., 2017). The results demonstrate the value of developing such quantitative degron systems to provide high-resolution spatiotemporal data about phytohormone distribution and signaling in plants.

A very commonly employed method to determine protein–protein interactions *in planta* is the use of bimolecular fluorescence complementation (BiFC) assays in which N-terminal and C-terminal portions of a YFP are fused separately to two proteins that potentially interact. The interaction of those proteins can lead to close association of the two portions of the YFP such that upon excitation with light of the appropriate wavelength, fluorescence (or Förster) resonance energy transfer (FRET) occurs between the two halves of the YFP protein to produce a fluorescence signal. For example, to test auxin-dependent interactions between auxin response factor (ARF) and Aux/IAA proteins, different combinations were co-expressed in onion epidermal cells. Evidence was obtained for interaction between specific pairs of proteins in the nucleus of transformed cells (Piya et al., 2014). In another example, interaction between the ABA receptor and phosphoprotein phosphatase 2C (PP2C)-type phosphatases was monitored by BiFC to directly monitor cellular ABA dynamics in response to environmental cues. This system provides a FRET sensor with high affinities for ABA, and was used to map tissue ABA dynamics with high spatial and temporal resolution (Waadt et al., 2014). More direct biosensors methods are now being developed to detect hormones, which use a range of systems including FRET, surface plasmon resonance, electrochemical, and optical (luminescence) signals (Su et al., 2017).

1.1.5 Transport and communication

1.1.5.1 Functions of transporters

The spatial separation of sites of synthesis and action of hormones is central to their role in controlling plant development and in mediating communication between cells. Thus the sites of synthesis, storage, and action of hormones can all be different, thus requiring transport between these sites. We need to understand transport processes within cells, between cells, and between organs, including loading into and unloading from the vascular system. Most information is available for auxin because auxin transport plays a central role in the establishment of polarity and the relationship between cells in differentiation and development (Baluska et al., 2003).

Auxin transport can occur by passive diffusion through the plasma membrane depending on the protonation state of IAA, and by active transport (Qu et al., 2017). There are at least three different types of auxin transporter including PIN-formed (PIN) proteins (a class of anion carriers/channels), ATP-Binding Cassette subfamily B (ABCB) transport proteins, and a family of influx carriers/channels named AUX1 and Like-AUX1 (LAX). A possible fourth group is the PIN-LIKES (PILS) proteins showing only 10%–18% sequence identity with PINs and comprising seven members in *Arabidopsis thaliana*. The PIN proteins are responsible for auxin efflux from internal compartments such as the endoplasmic reticulum (ER), and from the cell. Crucially, the asymmetric distribution of PIN proteins in the apical, basal, or lateral plasma membrane can establish polar auxin transport (PAT) pathways or can influence one neighboring cell in preference to another. Similarly the AUX1/LAX influx carriers can determine the amount or direction of auxin uptake by a cell. In contrast ABCB transporters have a non-polar distribution but are proposed to regulate the amount of auxin in the cell available for PAT. One root-specific ABCB transporter is involved in auxin transport during root gravitropic bending and lateral root formation (Qu et al., 2017).

Similar to auxins, there are at least three types of CK transporter. These are purine permeases (PUP), equilibrative nucleoside transporters (ENTs), and ATP-binding cassette (ABC) transporters (Feng et al., 2017). They are represented by multiple members in each family, have different substrate specificities, and occur in different cells. The importance of ABC transporters is further illustrated by two such proteins in Arabidopsis. One (AtABCG25) functions as an ABA efflux transporter in vascular tissues, while another (AtABCG40) functions as an ABA influx transporter in guard cells (Kang et al., 2010; Kuromori et al., 2010). Such ABC proteins also have roles in intracellular transport, such as the Arabidopsis "comatose" protein, which not only imports fatty acids into peroxisomes but also benzoylates and precursors of JA and auxin (Bussell et al., 2014).

The nitrate or peptide transporter family (NPF) proteins have been found to transport not only nitrate and peptides but also auxin, ABA, GA, and JA (Chiba et al., 2015;

Kanno et al., 2012; Krouk et al., 2010; Saito et al., 2015). The AtNPF3.1 protein is targeted to the plasma membrane of root endodermis cells that accumulate bioactive GAs (Tal et al., 2016). Another NPF protein was identified as a glucosinolate transporter (GTR1) but can also transport GA_3 (Saito et al., 2015). The Arabidopsis nitrate (NO_3^-) transporter NRT1.2 was later identified as ABA-importing transporter 1 (AIT1), and shown to be an influx transporter of ABA, as well as NO_3^- (Kanno et al., 2012).

1.1.5.2 Long-distance transport

Long-distance transport of hormones coordinates growth such as in the imposition of apical dominance by auxin transported from the SAM, and in tropic responses such as phototropism and gravitropism. Transport of hormones between root and shoot establishes communication between organs and regulates source-sink relationships. Long-distance transport also provides systemic signaling systems such as during pest attack or symbiosis (Jin et al., 2017; Song et al., 2017).

The best understood transport system is that of auxin. The asymmetric distribution of PIN proteins in a cell establishes polarity by delivering auxin unidirectionally to a neighboring cell, which in turn can deliver auxin to the next cell. The polar distribution of PIN proteins is a self-reinforcing process which establishes a file of cells for the delivery of auxin from apex to base. The establishment of such a flow of auxin is described as canalization (de Jong et al., 2014; Qu et al., 2017).

Transport between root and shoot can be demonstrated by grafting experiments with hormone biosynthesis mutants and by biochemical methods using isotopically labeled compounds. Such experiments have demonstrated the translocation of GAs from sites of synthesis to the tissues and organs that require GAs for growth and development (Regnault et al., 2016). For example, leaf-derived GA_1 and GA_{20} are mobile signals that induce internode elongation, cambial activity, and fiber differentiation (Dayan et al., 2012). However, GA_{12} is also mobile within the vegetative tissues of the plant but it is not transferred to developing seeds (Regnault et al., 2016). Root–shoot transport of CK, SL, BR, and ABA has also been demonstrated in such ways.

Long-distance transport is achieved by phloem and xylem tissues, and hormones can be found in xylem and phloem sap. Such transport is highly specific. For example, xylem sap contains CKs comprising predominantly *trans*-zeatin ribosides, whereas phloem sap contains *cis*-zeatin ribosides and isopentenyl adenine, consistent with the idea that they are transported in opposite directions between shoot and root (Bishopp et al., 2011; Kudo et al., 2010). Such transport therefore requires specific loading and unloading. In Arabidopsis, ABC transporter AtABCG14 is involved in loading of *trans*-zeatin type CKs into the xylem for transport to shoots (Ko et al., 2014; Zhang et al., 2014). On the other hand, PUPs are primarily expressed in the phloem, and it is suggested that they participate in the uptake of CKs from xylem sap during long-distance transport. In Arabidopsis, AtPUP1 and AtPUP2 have been shown to transport adenine and nucleobase CKs whereas translocation of CK ribosides appears to be mediated by ENTs. In rice, an ENT protein participates in the selective transport of CK nucleosides into vascular tissues (Feng et al., 2017).

Pathogen infection induces transmission of systemic signals from the site of attack to other parts of the plant to establish systemic acquired resistance (SAR). Although local and systemic induction of SA biosynthesis is required for establishment of SAR, SA itself is not the mobile signal (Gaffney et al., 1993). Potential SAR mobile signals include methyl salicylic acid (Fu and Dong, 2013). Insect attack and other forms of mechanical wounding activate plant defense responses both locally at the site of the wounding and systemically in undamaged parts throughout the plant. A wealth of evidence indicates that systemin and JA work together in the same signaling pathway to promote systemic defense responses. Grafting experiments with tomato mutants defective in JA biosynthesis or signaling support a hypothesis that systemin acts at or near the site of wounding to amplify the production of JA, which in turn acts as a long-distance mobile signal to activate the systemic defense response (Li et al., 2002). While systemin is thought not to be a long-distance mobile signal, other peptide hormones are transported. Thus in *Lotus japonicus* CLE-RS2 produced in roots is found in xylem sap collected from shoots, implying that CLE-RSs may provide a long-distance mobile signal in the regulation of the initial step of nodulation (Okamoto et al., 2013).

1.2 Mechanisms of hormone action

1.2.1 Perception and transduction

1.2.1.1 Receptors and co-receptors

Some hormone receptors are membrane anchored (receptors for CK, ETH, BR, and peptides) while others are soluble (receptors for auxin, GA, ABA, JA, SA, and SL) (Fig. 1.3). Receptors for CK, BR, and peptides are located to the plasma membrane but the BR receptor BRI1 may undergo cycling through the endomembrane system. In addition, while CK receptor CK1 is in the plasma membrane, other CK receptors (e.g., Arabidopsis Histidine Kinases, AHKs) are present in the endomembrane system. The ETH receptor ETR1 is localized in the ER. Soluble receptors may be cytosolic (ABA, GA, SA, and SL) or nuclear (auxin and JA), or may migrate between cytosol and nucleus (GA, SA and SL).

The ABA receptor proteins PYR1 (PYR1) and PYR1-Like (PYL) are cytosolic and interact with cytosolic PPC2 to regulate the phosphorylation status of other proteins including protein kinase SUCROSE NON-FERMENTING KINASE 2 (SnRK2) (Li et al., 2017b). The SA receptor NPR1 (NONEXPRESSER OF PATHOGENESIS-RELATED PROTEIN 1) binds to SA and migrates to the nucleus where it acts as a co-activator of transcription factors such as TGA3 (TGACG sequence-specific binding protein) to activate defense genes (Jin et al., 2017). The amount of NPR1 is affected by its direct interaction with other SA receptors such that it is degraded in local tissues leading to the hypersensitive cell death response (HR) but accumulates in systemic tissues, suppressing the HR response. Receptors for GA and SL are members of a large family of α/β-fold hydrolases, although the GA receptor does not have hydrolase activity (Gao et al., 2017; Wang et al., 2017b). In Arabidopsis, binding of GA and SL to their respective receptors GID1

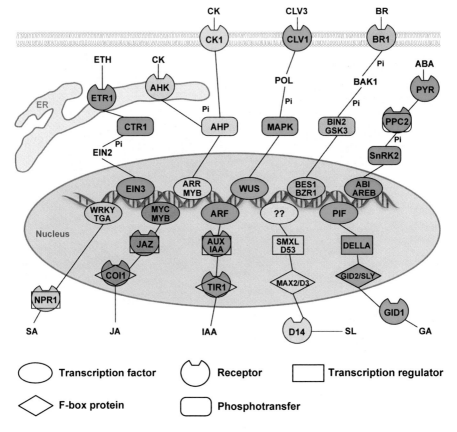

Figure 1.3 Receptors and signaling. The receptors for eight hormones and for one peptide (CLV3) are indicated, with an indication of their location in the cell. Membrane-bound receptors can potentially be located in more than one membrane as a result of cycling through the endomembrane system. Soluble receptors such as NPR1, D14, and GID1 translocate between cytosol and nucleus. Proteins involved in signaling from ETH, CK, CLV3, BR, ABA also translocate from cytosol to nucleus. *ABI*, ABA-insensitive; *AHK1*, Arabidopsis histidine kinase; *AHP*, Arabidopsis histidine phosphotransfer protein; *AREB*, ABA response element binding protein; *ARF*, auxin response factor; *ARR*, Arabidopsis response regulator; *AUX/IAA*, auxin/IAA-inducible protein; *BAK1*, BRI1-associated receptor kinase 1; *BES1*, bri1-EMS-suppressor 1; *BIN2*, BR-insensitive 2; *BRI*, brassinolide-insensitive 1; *BZR1*, brassinozole-resistant 1; *CK1*, cytokinin 1; *CLV1*, Clavata 1; *COI*, coronatine-insensitive 1; *CTR1*, constitutive triple response 1; *D14*, Dwarf14; *D3*, dwarf 3; *D53*, dwarf 53; *DELLA*, protein containing DELLA (Asp-Glu-Leu-Leu-Ala) motif; *EIN2*, ethylene-insensitive 2; *EIN3*, ethylene-insensitive 3; *ER*, endoplasmic reticulum; *ETR1*, ethylene-resistant 1; *GID1*, GA-insensitive Dwarf 1; *GID2*, GA-insensitive dwarf 2; *GSK3*, glycogen synthase kinase 3; *JAZ*, jasmonate-ZIM domain; *MAPK*, mitogen-activated protein kinase; *MAX2*, more axillary growth 2; *MYB*, myeloblastosis-related transcription factor; *MYC*, myelocytomatosis oncogene-like; *NPR1*, nonexpresser of pathogenesis-related protein 1; *PIF*, phytochrome-interacting factor; *POL*, POLTERGEIST; *PPC2*, phosphoprotein phosphatase type C2; *PYR*, pyrabactin resistance; *SLY*, sleepy; SMXL, suppressor of MAX2-like; *SnRK2*, SNF1-related protein kinase; *TGA*, TGACG sequence-specific binding protein; *TIR1*, transport inhibitor response 1; *WRKY*, protein containing WRKY (Trp-Arg-Lys-Tyr) motif; *WUS*, WUSCHEL. Hormones ETH, CK, CLV3, BR, ABA, GA, SL, IAA, JA, and SA are defined in Figure 1 legend.

(GA-INSENSITIVE DWARF 1) and D14 (DWARF 14) triggers conformational changes which lead to their interaction with F-box proteins SLY1 (SLEEPY 1) and MAX2 (MORE AXILLARY GROWTH2) respectively, and recruitment into Skp-Cullin-F-box (SCF) complexes. Auxin (IAA) and JA (Isoleucine-JA) on the other hand, bind directly to F-box proteins TIR1 (TRANSPORT INHIBITOR RESPONSE 1) and CORONATINE INSENSITIVE 1 (COI1) respectively, leading to formation of SCF complexes and recruitment of transcriptional co-repressors for degradation (Qu et al., 2017; Zhai et al., 2017).

In the case of auxin, a co-receptor complex is formed. The binding of auxin to TIR1 provides a surface which includes the auxin, to which AUX/IAA proteins bind. The auxin is considered to act as a "molecular glue" since it provides part of the binding surface for AUX/IAA. Similarly, perception of JA also involves the formation of a co-receptor complex consisting of COI1, JAZ, and an inositol pentakisphosphate. Structural and pharmacological studies revealed that COI1 contains an open pocket that recognizes the bioactive hormone JA-Ile. Binding of JA-Ile requires a bipartite JAZ degron sequence consisting of a conserved α-helix for COI1 docking and a loop region to trap the hormone. Another critical component of the JA-Ile co-receptor complex is inositol pentakisphosphate, which interacts with both COI1 and JAZ adjacent to the ligand (Sheard et al., 2010). The concept of a co-receptor complex is arguably subtle. The binding of other hormones to their receptors also triggers formation of specific protein complexes, but in these cases the hormone is contained within the receptor protein, and so does not form part of the contact between proteins in the complex. For example, the binding of ABA to PYR1 triggers formation of the complex with PPC2, but ABA is contained within the PYR1 structure. The perception of SL is unlike any other hormone because the receptor D14 is a serine hydrolase which attacks the SL molecule and creates a covalently attached intermediate (CLIM), which remains at the active site and triggers a conformational change in D14 enabling interaction with other proteins including F-box protein MAX2 (Yao et al., 2016).

Receptors for peptide hormones and BR are trans-membrane leucine-rich-repeat receptor kinases (LRR-RKs) with kinase domains on the cytoplasmic side of the membrane. Peptide hormones such as CLAVATA 3 (CLV3) and other CLV3/Endosperm surrounding region (CLE) peptides bind to receptors such as CLV1 and CLV2, which leads to the cytoplasmic kinase domain phosphorylating target proteins (Song et al., 2017). When BRs bind to receptor BRI1, the cytoplasmic kinase domain phosphorylates negative regulator BRI1 KINASE INHIBITOR 1 (BKI1), from which it then dissociates, leading BRI1 to phosphorylate BRI1-ASSOCIATED RECEPTOR KINASE 1 (BAK1) and trigger signaling (Wang et al., 2017a).

Receptors for ETH and CK are similar to bacterial two-component system regulators, which consist of a conserved histidine kinase (HK) and a response regulator protein (RR). Phosphotransfer from HK to RR results in the activation of RR and the generation of signaling output (West and Stock, 2001). In Arabidopsis, upon binding CK, the receptors are auto-phosphorylated at a highly conserved His residue. The phosphoryl group is subsequently transferred to phosphotransfer proteins and then to RR proteins (Feng et al., 2017). In ETH signaling, the binding of ETH to its receptors inactivates CONSTITUTIVE TRIPLE RESPONSE 1 (CTR1) (Hao et al., 2017) and in CK signaling the binding of CK to a receptor such as CK1 or CYTOKININ RESPONSE 1 (CRE1) triggers phosphorylation of Histidine Phosphotransfer (HP) proteins.

There are two broad modes of signaling downstream of perception. One is a phosphorylation cascade or relay that leads to changes in activity of downstream targets such as transcriptional regulators. The other is the targeted destruction of proteins such as transcriptional regulators through ubiquitination and proteolysis by the 26S proteasomal pathway. Both mechanisms operate in all hormone systems but one tends to be primary mode of action.

1.2.1.2   Protein phosphorylation

Perception of ABA, CK, ETH, BR, and peptides leads to signal transduction mainly through changes in protein phosphorylation. Several classes of protein kinases have been identified as important regulators of ABA signaling in addition to SnRK2. These include SnRK3s which are also known as CBL-interacting protein kinases (CIPKs), calcium-dependent protein kinases (CDPKs), and members of the mitogen-activated protein kinase (MAPK) family. A MAPK module typically includes combinations of at least three kinases including MAPK kinase (MAPKK) and MAPKK kinase (MAPKKK) (de Zelicourt et al., 2016). The activation of MAPKs by ABA has been reported in barley, maize, pea, and rice. The downstream targets for phosphorylation are transcriptional regulators such as ABA-insensitive (ABI) proteins and transcription factors such as ABA response element binding factors (AREBs). The ABI3 protein displays high sequence similarity to maize VIVIPAROUS 1 (VP1) both of which belong to the B3-domain transcription factor family while ABI4 is a member of the APETELLA2/ ETHYLENE RESPONSE FACTOR (AP2/ERF) family (Li et al., 2017b).

The SnRK2 kinase is also involved in BR signaling since it can phosphorylate the GLYCOGEN SYNTHASE KINASE 3 (GSK3) protein, also known as BR-INSENSITIVE 2 (BIN2), yet GSK3/BIN2 is also the prime target for phosphorylation by BAK1. The targets for GSK3/BIN2 are transcriptional regulators BRASSINOZOLE RESISTANT 1 (BZR1) and BRI1 EMS SUPPRESSOR 1 (BES1) (Wang et al., 2017a). Thus ABA and BR signaling can potentially influence each other through the activity of SnRK2s. Further, cross talk might be possible between ABA and CLV-type peptides since they both signal through MAPK modules. The initial step in CLV3-mediated CLV1 signaling is activation of a Rho-like GTPase (ROP), which in turn phosphorylates MAPK (Song et al., 2017). Thus cross talk between peptides, ABA, and BR signaling is possible through the involvement of common protein kinases.

In CK signaling, a two-component system has been elucidated in Arabidopsis involving a phospho-relay that sequentially transfers phosphoryl groups from the receptor, for example, CK1 or CRE1, to Arabidopsis Histidine Phosphotransfer (AHP) proteins. These AHPs can then phosphorylate transcriptional regulators such as Arabidopsis Response Regulators (ARRs). Type B-ARRs are MYB-type transcription factors that activate expression of other genes including type A-ARRs (Feng et al., 2017). Phosphorylation is also important in ETH signaling in which ETH perception leads to dephosphorylation of an ER-localized protein ETHYLENE INSENSITIVE 2 (EIN2). This leads to cleavage of EIN2 and release of a fragment that either participates in translational repression in cytoplasmic processing body (P-body), or migrates to the nucleus to activate master transcriptional regulators EIN3 and EIL1 (EIN3-LIKE 1) followed by the induction of multiple downstream genes (Hao et al., 2017).

Phosphorylation also plays a key role in ETH biosynthesis since ACC synthase has MAPK and CDPK phosphorylation sites which stabilize the protein when phosphorylated, whereas phosphoprotein phosphatases PP2C and PP2A can reduce the phosphorylation level of ACC synthase and promote its degradation (Agnieszka et al., 2014). The SA receptor NPR1 is normally phosphorylated at Ser55 and Ser59, which prevents SUMO3 modification and hence its interaction with transcription repressor WRKY70 (Jin et al., 2017). Thus expression of defense genes is downregulated. Stimulation by SA leads to dephosphorylation of NPR1 at Ser55 and Ser59, and conjugation with SUMO3, which promotes interaction between NPR1 and transcription factor TGA3 to activate expression of defense genes. Conjugation of NPR1 with SUMO3 also induces phosphorylation of NPR1 at Ser11 and Ser15, facilitating Cullin3-based ubiquitin ligase-dependent degradation of NPR1. The SA-induced rapid degradation of NPR1 is required for full activation of defense gene expression and establishment of SAR.

1.2.1.3   Proteasomal degradation of transcriptional regulators

The second major mechanism by which hormone signaling brings about a response is through the targeting of transcriptional regulators for degradation via ubiquitination and proteolysis by the 26S proteasome. This applies particularly to auxin, GA, JA, and SL, although such protein degradation also operates in the control systems of other hormones. In each of these four cases hormone binding to its cognate receptor leads to formation of complexes with F-box proteins which are substrate recognition subunits of SCF ubiquitin ligases. The complex then binds to target transcriptional regulators which are ubiquitinated and degraded.

The targets for TIR1 and closely related F-box proteins are Aux/IAA repressor proteins of which there are 29 in *Arabidopsis thaliana*. These interact in pairwise combinations with up to 23 ARF transcription factors and also recruit TOPLESS (TPL) and TPL-RELATED (TPR) co-repressors, leading to chromatin inactivation and silencing of ARF target genes (Wang and Estelle, 2014). Thus TIR1-targeted destruction of AUX/IAA proteins relieves repression of many genes in tissue-specific ways to control different aspects of growth and differentiation (Qu et al., 2017).

Signaling of GA operates in a similar way, such that in Arabidopsis the GA-GID1 complex recruits F-box protein GID2 and targets DELLA proteins for destruction. The DELLA proteins belong to a subfamily of the plant-specific GRAS transcriptional regulators whose name derives from GAI, RGA, and SCARECROW. The DELLA proteins lack a DNA-binding domain and there is no evidence of direct DNA binding. Instead they act as repressors of transcription factors and repressors of growth. They operate separately or cooperatively with each other at different developmental stages to regulate plant growth. Their GA-stimulated destruction by ubiquitination and proteolysis relieves inhibition, which is observed as GA-stimulated plant growth. They include examples in wheat (Rht1), maize (d8), barley (SLN1), and rice (SLR1) (Gao et al., 2017).

In JA signaling, the targets for proteolysis are JAZ proteins, of which there are 13 members in *A. thaliana*. These are transcriptional repressors that belong to the plant-specific TIFY family, defined by the presence of a TIF[F/Y]XG motif. The JAZ proteins bind and repress the activity of MYC2, a bHLH-type transcription factor which binds to the G-box (CACGTG) and differentially regulates two branches of JA-responsive genes involved in responses to wounding and pathogens (Zhai et al., 2017). Other targets of JAZ proteins include several MYC and MYB proteins and also bHLH proteins that act as transcriptional repressors. Thus JA-induced targeting of JAZ proteins for proteolysis leads to many diverse changes in gene expression and different responses (Zhai et al., 2017).

The mechanisms operating in signaling of SL follows a similar pattern (Wang et al., 2017b). The formation of the D14-CLIM product triggers binding to the F-box protein D3 (in rice) or MAX2 (in *A. thaliana*) and recruitment of proteins of the D53/SMXL family, of which there are nine and eight members respectively. Proteolysis of D53 in rice inhibits growth of tillers while destruction of SMXL6, 7, and 8 in *A. thaliana* inhibits lateral shoot growth. Thus D53/SMXL proteins promote such growth, in contrast to DELLA proteins which inhibit growth. Evidence suggests that D53/SMXL proteins are transcriptional repressors and can recruit TPL and TPR proteins into repressor complexes. The target genes of SL signaling are still under investigation. It is also possible that D53/SMXL proteins have other functions such as in protein–protein interactions in protein trafficking during auxin function (Wang et al., 2017b).

Proteasomal degradation also plays a major role in response to ETH since in the absence of ETH, EIN3 and EIL1 proteins are rapidly degraded by an SCF complex containing F-box proteins EIN3-BINDING PROTEIN 1 (EBF1) and EBF2. Two other F-box proteins EIN2 TARGETING PROTEIN 1 (ETP1) and ETP2 interact with EIN2 and control its level (Hao et al., 2017).

1.2.2 Homeostasis and fail-safe

In the early period of hormone research before any of the molecular components were identified, much emphasis was placed on measuring the amounts of hormones in plant tissues using bioassays, and treating plants with exogenous hormones to study their effects. In the 1980s new focus of attention became the hypothesis that the amounts of hormones were not as important as the sensitivity of plant tissues to plant hormones (Bradford and Trewavas, 1994). Now that we have a deeper understanding of the molecular mechanism in play, it is clear that responses to hormones depend on a multitude of factors, including input from developmental programs and environmental signals.

One recurring theme is usually referred to as hormone homeostasis. This refers to the fact that a hormone is observed to repress its own biosynthesis, particularly at the level of expression of biosynthesis genes. Similarly, in mutants that are insensitive to a hormone or have reduced hormone level, expression of the biosynthesis genes is increased. These effects on gene expression have been observed for essentially all the hormones, at least for some genes of hormone biosynthesis. Another way in which hormone levels can be maintained within certain limits is through the formation and breakdown of hormone conjugates. These observations are interpreted to indicate a form of control of the amount of hormone, or homeostasis. Homeostasis in general is a type of buffering which prevents the amount of a substance or a condition from changing greatly.

This does not seem to fit with the hypothesis that plant hormone signaling depends on precise changes in the amount and distribution of a particular hormone. Instead, the broad control of hormone biosynthesis seen at the level of gene expression is more likely to represent a type of "fail-safe" mechanism, which ensures that there is enough hormone but not too much. This is consistent with the observation that loss-of-function and gain-of-function *AtDAO1* genes did not greatly alter the amount of IAA in Arabidopsis plants, indicating that such "fail-safe" mechanisms are at play (Porco et al., 2016). The homeostasis concept does not fit well either, with the observation that there can apparently be very large variations in the amounts or activities of particular hormones in specific cells. For example the DR5-GUS reporter protein reveals wide variation in the level of auxin "activity" in different cells of any given tissue (Su et al., 2017). However this activity might reflect differences in responsiveness to IAA rather than levels of IAA.

Thus precise signaling actions are likely determined by several specific mechanisms such as cell-specific gene expression, hormone transport, hormone activation, and the activity of receptors. If so, homeostasis could be regarded as a fail-safe mechanism that does not contribute directly to hormonal regulation of plant growth and development.

Another dimension to this concept is that almost all hormones will apparently affect the biosynthesis or activity of several others, and most of the hormones can influence almost every aspect of plant development and function. This is referred to as "crosstalk" and presents a picture of a highly complex 4-dimensional network of interactions. This is discussed below and discussed in terms of targets of hormone action rather than networks of interactions.

1.3 Biological functions of hormones

1.3.1 Cell growth and differentiation

Hormones participate in every aspect of plant growth, development, and function, throughout the entire life cycle. Every cell is the product of hormonal activity, and all living cells are responsive to hormones. Auxin and CK are apparently essential for cell expansion, division and differentiation, because to our knowledge there are no mutants that completely lack auxin or CK (Qu et al., 2017; Feng et al., 2017).

The role of auxin in cell expansion was discovered through studies of the phototropic bending in coleoptiles in which asymmetric distribution of auxin promotes preferential elongation on one side. The growth effect of auxin in promoting cell expansion was later observed in cultured callus cells of tobacco, but without cell division. This led to the discovery of kinetin which could stimulate cell division so that the combination of auxin and CK could sustain prolonged growth and division of callus tissue (Miller et al., 1955). Subsequent experiments to optimize the amounts of auxin and CK immediately led to the discovery that in the appropriate ratios they could induce differentiation of callus to produce either roots or shoots (Skoog and Miller, 1957). With suitable manipulations of amounts of auxin and CK, a whole plant could be regenerated from undifferentiated callus tissue. Some cultured somatic cells were later shown to be capable of undergoing embryogenesis in vitro to produce somatic embryos and ultimately whole plants. These are among the most profound discoveries in plant biology because they illustrate the totipotency of plant cells, which has underpinned the development of the discipline of plant biotechnology.

It soon became apparent that the induction of crown-gall disease by *Agrobacterium tumefaciens* was the result of transfer into plant cells, of bacterial genes directing the biosynthesis of auxin and CK, with the consequent formation of callus tissue. By removing these auxin and CK biosynthesis genes, the bacterium became unable to induce crown-gall disease but was still capable of gene transfer to plant cells. Thus *Agrobacterium*-based gene vectors were invented and are the most widely used method of gene delivery today (Păcurar et al., 2011).

The process of embryogenesis, whether from somatic cells or zygote, depends crucially on auxin, CK, and peptide signaling. In Arabidopsis, four rounds of cell division generate a 16-cell globular embryo within which the identity and fate of each cell is already determined. Polarity then becomes established partly through the formation of auxin gradients and differential cell–cell communication. Subsequent asymmetric division and cell expansion leads to the morphogenesis of torpedo- and heart-shaped embryos (Song et al., 2017). The WUS homeodomain transcription factor is essential for embryogenesis and for the formation and maintenance of meristems. Expression of the *WUS* gene is already detected in 16-cell embryos, and *wus* mutations disrupt embryogenesis and meristem formation. The *WUS* gene is activated by CK signaling via type B ARRs (Song et al., 2017; Feng et al., 2017). Furthermore, mutants in auxin biosynthesis, transport, and response are defective in embryogenesis (Wójcikowska et al., 2013; Wang et al., 2015). Early embryo development is also regulated by members of the CLE peptide family (Song et al., 2017). The CLE8 peptide acts upstream of the WUS-RELATED HOMEOBOX (WOX) transcription factor WOX8 and mutations affecting CLE8 or WOX8 result in abnormal cell divisions in both embryo and suspensor (Fiume and Fletcher, 2012).

The CLE-WOX module, interacting with auxin and CK, provides a common unit in the regulation of the three major meristems, namely SAM, root apical meristem (RAM), and cambium. This regulatory system establishes cell layers (L1, L2, and L3) and maintains stem cell identity through a balance between cell division and cell differentiation. This is achieved by spatial and functional antagonism between auxin and CK signaling, while the CLV3-WUS pathway forms a feedback regulation loop to regulate stem cell maintenance and cell differentiation. In parallel with WUS, another homeodomain transcription factor SHOOT MERISTEMLESS (STM) also functions in stem cell maintenance. In addition, GA, BR, and ETH function to drive the patterning and growth of the SAM (Cheng et al., 2017).

While the SAM, RAM, and cambium are maintained throughout growth and development, the plasticity of plants is manifested by the switching of certain types of somatic cells into stem cell niches resembling those in the SAM and the RAM. These include lateral shoot and root meristems, the outgrowth of which determines plant architecture and is regulated by hormones in response to environmental factors. Another example is the wound response in which wound-induced hormone concentrations promote fate transition of competent cells, allowing the regeneration of new apical meristems. A mutually exclusive distribution pattern of auxin and CK provides positional information to reestablish the proper pattern by inducing the spatial expression of key SAM regulators (Cheng et al., 2017).

Hormones and peptides control many other aspects of cell growth and differentiation throughout development, such as the differentiation of xylem and phloem cells, trichomes, and stomatal guard cells. For example, several *EPIDERMAL PATTERNING FACTOR* (*EPF*) genes have been identified which encode peptides regulating guard cell number and distribution (Song et al., 2017). The *EPF1* gene is expressed in progenitor guard cells and controls stomatal patterning by regulating asymmetric divisions of guard mother cells. Mutation of *EPF1* results in clustering of guard cells. The stomagen (EPFL9) peptide is a 45-amino acid cysteine-rich peptide which is expressed in mesophyll cells rather than epidermal cells, and acts in a non-cell autonomous manner to regulate stomata patterning. In BR-deficient mutants, clustered stomata are observed, while BR treatment can reduce stomatal density. Such negative regulation of stomatal development by BR requires BIN2 but details of interactions with other hormones are unclear (Wang et al., 2017a).

1.3.2   Size and form

At the level of the whole plant, several hormones have a profound impact on growth and morphology. Leaf expansion, stem elongation, and shoot branching are all controlled by auxin, CK, SL, GA, and BR. Mutants lacking GA, BR, or SL are dwarf and are compromised in some aspects of development, but BR and SL mutants are still capable of reproduction while GA and JA mutants tend to produce infertile flowers. Both GA and BR promote cell expansion, whereas SL mutants are dwarf because the mutant invests resources in the growth of multiple secondary shoots at the expense of the primary shoot.

A classic function of auxin is in the imposition of apical dominance (de Jong et al., 2014; Qu et al., 2017). The auxin produced in the shoot apex is transported basipetally down the main stem in dicotyledonous plants such as pea or Arabidopsis. The PAT stream is mediated by PIN proteins and has the effect of repressing the outgrowth of lateral shoots, since decapitation of the plant releases the lateral buds from inhibition, while application of IAA to the cut apex can restore the inhibition. It is proposed that the PAT stream in the main stem prevents lateral buds from establishing their own PAT stream out into the main stem. Meanwhile auxin inhibits biosynthesis of CK, which normally promotes bud outgrowth. The hypothesis is that auxin in the PATS acts in part by downregulating the synthesis of CK, restricting its availability to axillary buds, thereby repressing their growth. Furthermore, auxin induces the biosynthesis of SL in roots, which is transported to the shoot and inhibits the outgrowth of lateral buds by an unknown mechanism (de Jong et al., 2014).

Mutants defective in SL biosynthesis or sensing have increased numbers of side shoots in dicotyledonous plants, or increased numbers of tillers in cereals. The extent of shoot branching or number of secondary shoots determines the shape of the plant and in crops this can play a major role in determining the effectiveness of the plant in light interception. It also determines the number of inflorescences and hence seed number and yield (Wang et al., 2017b). Collectively these hormones determine the architecture of the shoot.

The growth and architecture of the root system is also controlled by hormones. Indeed auxin is known as a rooting hormone and is used in horticulture to induce rooting of explants. Auxin is produced at the root apex and transported basipetally to direct cell expansion and differentiation. In response to gravity and mechano-stimulation the asymmetric distribution of auxin can lead to root curvature and hence directional growth. It also plays a role in triggering formation of lateral roots. The control of lateral root formation and outgrowth is also determined by CK and SL (Jung and McCouch, 2013). It has been reported that ABA produced in the shoot is required for root growth (McAdam et al., 2016) and that ETH regulates root elongation and root hair development (Chang et al., 2013).

1.3.3   Flowering and reproduction

The photoperiodic control of the transition from vegetative growth to flowering involves transmission of a signal from the leaves to the shoot meristems. The primary signal was considered to be a putative flowering hormone and was given the name "florigen." The hypothetical florigen is now believed to be a mobile protein, FLOWERING LOCUS T (FT) (Andrés et al., 2015). However, hormones including GA and ETH can promote flowering and increase the number of flowers while JA signaling can delay flowering. The JA response is mediated by interaction of a subset of JAZ proteins with transcription factors which negatively regulate the expression of *FT* (Zhai et al., 2015).

Peptide hormones play a profound role in floral development by controlling organ development in the floral meristem. Thus mutants defective in CLAVATA3 peptide or its receptor CLAVATA1 have enlarged floral meristems giving rise to malformed flowers with increased carpel number and club-shaped siliques (Song et al., 2017).

Stamen development in Arabidopsis requires JA, GA, and BR. Signaling by JA targets JAZ proteins that repress R2R3-MYB transcription factors MYB21 and MYB24, which regulate stamen development (Zhai et al., 2017). Synthesis of bioactive GAs in stamens is not only necessary for their development, but the GA is also transported to other tissues such as petals to support their growth (Hu et al., 2008). Stamens and pollen are also a rich source of BR which is essential for male fertility. It has been found that BES1 can bind to the promoter regions of genes encoding transcription factors essential for anther and pollen development. Mutants lacking BR show reduced length of stamen filaments resulting in the deposition of pollen on the ovary wall rather than on the stigmatic surface (Azpiroz et al., 1998). More importantly, BR mutants show reduced pollen number, efficiency of release, and viability (Ye et al., 2010). Thus male sterility is a common feature of BR mutants.

Ovule initiation and development also require BR signaling. In BR-deficient and BR-insensitive mutants, the siliques are shorter and thinner, while ovule and seed numbers are significantly reduced. In contrast the siliques are longer and thicker in the BR signalling-enhanced mutant *bzr1*-1D. It has been found that BR signaling regulates expression of genes involved in ovule and seed development through regulating gene expression by BZR1 (Huang et al., 2013).

Seed development is strongly influenced by hormone signaling. For example, in rice, a mutant which is deficient in IAA due to the absence of an IAA-glucose hydrolase is delayed in the transition from the syncytial to the cellular phase of endosperm development. The result is that the grains are longer than normal and have a greater mass, producing a greater grain yield per plant (Ishimaru et al., 2013). Another mutation that disrupts a CK oxidase results in increased CK which also leads to modified panicle structure, increased grain mass, and greater yield (Ashikari et al., 2005). Fruit and seed development are also influenced by JA and ETH. The role of ETH in fruit ripening is best known in climacteric fruits such as tomatoes, melons, and strawberries in which a burst of ETH production precedes the ripening phase (Chang et al., 2013). Mutants defective in ETH production or response are impaired in ripening, whereas treatment with exogenous ETH can accelerate ripening. The use of ETH in post-harvest ripening is well-known in commercial production and transportation of fruits.

1.3.4   Biotic and abiotic interactions

Plants interact with a wide range of other organisms, both beneficial and harmful. Symbiotic relationships forming nitrogen fixing root nodules or AM involve participation of several hormones. Nodulation is promoted by auxin, GA, SL, BR, CK, ABA, and JA, depending on the experimental system under investigation. Auxin, GA, SL, and ABA can also promote AM formation. However ETH and SA generally have inhibitory effects on such symbioses, potentially reflecting their role in defense against pathogens (Foo et al., 2014, 2016).

Of major importance in the regulation of nodulation are peptide hormones, which act in long-distance signaling to regulate the number of nodules. The number of root nodules that develop is controlled by the shoot, presumably to balance the supply and demand of resources such as sugars. This control system is referred to as autoregulation of nodulation (AON) and is mediated by mobile signals. Mutants of *L. japonicus* with excess nodules were found to be defective in a protein homologous to the Arabidopsis peptide receptor CLV1, thus implicating CLV3-like peptides in the control of nodule number (Nishimura et al., 2002; Song et al., 2013). Such peptides are believed to be transported between root and shoot to coordinate nodule number with shoot development in several legumes (Song et al., 2017).

The formation of AM is promoted by exudation of SL into the rhizosphere. The SL stimulates branching of the fungal hyphae, which then go on to colonize the root and form the arbuscules which facilitate the transfer of nitrate and phosphate from the fungus to the host plant in return for a supply of carbon (Akiyama et al., 2005). The fungal hyphae can apparently detect and respond to SL but the mechanism is unknown. In a new development it has been discovered that the formation of AM in rice depends on a protein homologous to the Arabidopsis karrikin receptor protein, KAI2 (Gutjahr et al., 2015; Waters et al., 2014). This discovery implies that the unidentified endogenous karrikin-like signal in plants plays an essential role in AM development.

Exudation of SL from roots has led to the exploitation of this signal by parasitic plants of the *Orobanchaceae* family (witchweeds and broomrapes). Seeds of such parasites, for example, *Striga* species, remain dormant in the soil until stimulated to germinate by SL exuded from the roots of a potential host plant. The seedling then attaches to the roots of the host and establishes a parasitic relationship which depletes the host of resources (Smith, 2014).

Defense responses against biotrophic pathogens and the establishment of SAR involve SA as a crucial player, whereas JA and ETH tend to be involved in responses to necrotrophic pathogens (Bari and Jones, 2009). Other biotic interactions of major significance are those of pests, including insects and other herbivores. Such pests induce responses through wounding and chemical signals which trigger host defense systems particularly comprising peptide hormones such as systemin (Song et al., 2017) and JA (Zhai et al., 2017).

Abiotic effectors of plant hormone signaling include light, water, temperature, salinity, and oxygen. Of particular importance is ABA which is involved in responses to drought, salt, and cold (Li et al., 2017b). Responses to drought, cold, salinity, and flooding also involve ETH signaling (Chang et al., 2013). Nevertheless, other plant hormones clearly play important roles in responses to abiotic stress, including BR (Che et al., 2010) and SL (Ha et al., 2014). It is a mistake to think in terms of separate growth hormones and stress hormones because they all have multiple functions and operate in concerted ways to mediate plant growth and development under a range of environmental conditions.

1.4 Integration of hormonal activities

1.4.1 Synergy and antagonism

The interactive nature of auxin and CK signaling in meristem formation and organogenesis has been outlined above. Such interaction is arguably antagonistic at the level of individual cells, but the outcome implies synergy or cooperativity since it generates a functionally differentiated meristem. This interaction is further integrated with peptide signaling in a cell-specific manner (Song et al., 2017). Normal SAM function also requires a balance of CK and GA signaling. Relatively high CK and low GA signaling is required for effective SAM development, but the later stages of cell maturation and elongation require low CK and high GA signaling (Jasinski et al., 2005; Sakamoto et al., 2001). Thus we see a complex set of interactions between hormones in both time and space.

Another much studied relationship is that between GA and ABA in the control of seed germination. Dormancy is induced by ABA, whereas germination is promoted by GA, so the balance between them can determine whether a seed will germinate (Gao et al., 2017). During seed development and maturation, ABA accumulates, whereas during imbibition of the mature seed, ABA is broken down via phaseic acid, while at the same time GA biosynthesis is activated, leading to germination. The balance between ABA and GA is very important to control germination. Premature germination in cereals can lead to the problem of pre-harvest sprouting on the mother plant, especially during wet weather. On the other hand, too much dormancy can lead to slow or erratic germination after harvest, which is a problem for malting. Drought during seed maturation will elevate ABA level and increase seed dormancy. It is also reported that ETH can stimulate seed germination and that BR antagonizes ABA during germination in Arabidopsis (Hu and Yu, 2014.)

Antagonism is observed between GA and JA in plant development and defense. While GA promotes hypocotyl and root elongation, JA can have the opposite effect. In contrast JA promotes defense responses and GA signaling can compromise defense responses. These observations potentially underlie the prioritization of defense overgrowth when plants face environmental stresses and pathogen attacks (Davière and Achard, 2016). They could also imply that the selection of GA-deficient dwarf cereals might have provided an added benefit of enhanced defenses.

There are many other examples of antagonism and synergy in plant development. Senescence and abscission tend to be accelerated by ETH and ABA but delayed by CK and SL. Secondary shoot growth is repressed by auxin and SL, but promoted by CK.

1.4.2   Nodes and hubs

The many examples of synergy and antagonism are commonly discussed in terms of hormonal "crosstalk." The magnitude of the challenge is reflected by the fact that each hormone typically influences the expression of hundreds or thousands of genes, as well as affects multiple developmental processes. For example, in Arabidopsis seedlings 10% of genes (nearly 3000) display ABA-regulated expression (Nemhauser et al., 2006). Similarly BR regulates more than 1000 genes, which is reflected in the binding of transcription factor BZR1 to more than 950 BR-regulated target genes (Sun et al., 2010) and BES1 binding to more than 1600 target genes (Yu et al., 2011). Furthermore, over 1000 genes are directly targeted by EIN3 during a time-course of ETH treatment (Chang et al., 2013). Frequently the same gene is influenced by more than one hormone, but the magnitude or direction of the response may vary in different tissues or at different stages of development. More importantly, each hormone can potentially influence the biosynthesis and responsiveness of other hormones.

Metabolism of GA is regulated by ABA, auxin, ETH, BR, and CK. Auxin changes the expression levels of the genes encoding enzymes of GA metabolism. In pea, auxin affects the expression of GA biosynthesis genes *PsGA3ox1* and *PsGA2ox1* (O'Neill and Ross, 2002). The auxin regulation of *PsGA2ox1* is partly dependent on DELLA, while *PsGA2ox2* is upregulated by auxin in a DELLA-independent manner (O'Neill et al., 2010). In Arabidopsis, Aux/IAA and ARF proteins directly upregulate the expression of *AtGA20ox* and *AtGA2ox* (Frigerio et al., 2006). The auxin transport inhibitors 1-N-naphthylpthalamic acid and 1-naphthoxyacetic acid (NOA) elevate the expression of At*GA20ox1* in shoots but not in roots, and only at specific developmental stages (Desgagne-Penix and Sponsel, 2008). Such findings suggest that auxin differentially regulates the expression of GA biosynthesis genes in pea and Arabidopsis and that developmental regulation and organ-, tissue-, and cell-specific regulation override both auxin and GA metabolic regulation, ensuring the appropriate temporal and spatial levels of GA. Auxin also negatively regulates CK biosynthesis in pea stems by repressing the expression of the *PsIPT* gene (Tanaka et al., 2006). The transcript levels of Arabidopsis *CYP735A1* and *CYP735A2* genes involved in the biosynthesis of *trans*-zeatin are downregulated by auxin or ABA in roots but are upregulated by CK suggesting a modulation of CK biosynthesis by ABA and auxin (Takei et al., 2004).

In another example, BR has been proposed as a master regulator of GA biosynthesis (Tong et al., 2014; Unterholzner et al., 2015). Accumulation of GA is promoted by BR-enhanced expression of GA metabolism genes, stimulating cell elongation and plant growth. Exogenous BR treatment and overexpression of BR biosynthesis gene *DWARF4* increase expression of several *GA20ox* genes in Arabidopsis (Lilley et al., 2013). However the view that BR is a master regulator of GA biosynthesis has been challenged (Ross and Quittenden, 2016), emphasizing the difficulty of interpreting interactions between different hormones.

This complex network of interactions has been one of the most challenging aspects of hormone action, but gradually we are starting to understand how it works. Instead of considering very many separate interactions in which all hormones seem to influence the synthesis and activity of others, the emerging picture is of "hubs" or "nodes" particularly centered on key transcriptional regulators (Fig. 1.4). It is emerging that DELLA proteins are particularly important because they interact with numerous other transcriptional regulators and so influence their activity, which in turn regulate the biosynthesis or signaling of other hormones (Gao et al., 2017). While DELLA proteins were discovered in relation to GA action, they also interact directly with JAZ, BZR, EIN3, and ARF proteins and so influence JA, BR, ETH, and auxin signaling, respectively (Fig. 1.4).

One way for DELLA proteins to regulate expression of target genes is to interact with DNA-binding domains of transcription factors and so inhibit their DNA-binding activity, thus acting as repressors of target gene expression. Another way is to interact with transcription factors in such a way as to act as transcriptional co-activators or co-repressors (Yoshida et al., 2014). For example, JA and GA antagonize each other in regulating seedling growth and resistance to pathogens, via interaction between JAZ and DELLA proteins. Interaction of DELLA proteins with JAZ proteins relieves inhibition of MYC2 and hence enhances JA-mediated inhibition of root growth, while GA-induced degradation of DELLA proteins frees JAZ proteins to bind to MYC2 and suppress JA signaling (Hong et al., 2012; Hou et al., 2010). The binding of JAZ proteins to DELLA proteins in turn disrupts DELLA binding to PHYTOCHROME INTERACTING FACTOR 3 (PIF3), which enhances GA-mediated hypocotyl elongation (Yang et al., 2012) (Fig. 1.4).

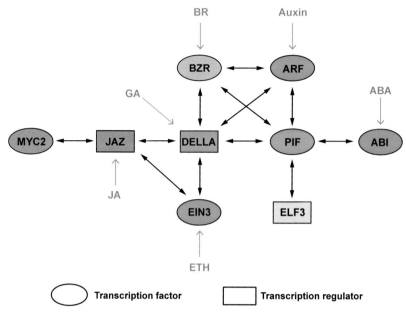

Figure 1.4 Nodes or hubs mediating hormonal interactions. Hormonal "crosstalk" explained in terms of a limited number of key transcriptional regulators which act as "hubs" or "nodes" to mediate the effects of more than one hormone on developmental and physiological processes. Each hormone is known to signal (shown by *blue arrows*) directly via one of these transcriptional regulators, but physical interaction between these proteins (shown by *black arrows*) influences signaling by other hormones. This representation is based on Arabidopsis transcriptional regulators some of which are specific individual proteins (EIN3, MYC2, and ELF3), while others represent families of proteins (JAZ, DELLA, BZR, ARF, PIF, and ABI). These transcriptional regulators also control the expression of hormone biosynthesis genes, and thus mediate hormonal regulation of hormone biosynthesis (crosstalk and homeostasis).

It is further shown that DELLA proteins interact with BR-responsive transcription factor BZR1 to inhibit the binding of BZR1 to its target promoters (Gallego-Bartolome et al., 2012). Thus GA-mediated breakdown of DELLA proteins is expected to activate BR-responsive genes, but BR also induces expression of genes for DELLA proteins (Lilley et al., 2013), implying the operation of a feedback mechanism. Furthermore, studies in rice show that BR promotes GA biosynthesis through binding of BZR1 to the promoters of GA biosynthetic genes (Tong et al., 2014). Thus the cross talk between BR and GA signaling can be explained in large part by interactions between DELLA and BZR proteins.

Similarly DELLA and JAZ proteins physically interact with EIN3 and related proteins to repress their transcriptional activity, thus providing a means for cross talk between GA, ETH, and JA (Fig. 1.4). Indeed there is evidence that EIN3 facilitates JA signaling, potentially because it sequesters JAZ repressors away from MYC2. Another manifestation of this interplay is that JA and ETH synergistically regulate plant resistance to necrotrophic pathogens through the JA-mediated destruction of JAZ proteins in concert with ETH-mediated induction of EIN3 activity (Zhai et al., 2017).

Other central players in hormonal cross talk are ABI transcription factors (Li et al., 2017b). Three key proteins in Arabidopsis, ABI3, ABI4, and ABI5, have different but overlapping roles in ABA-mediated responses in plants. They also mediate responses to light, sugars, salt, and cellular metabolism. Direct interaction between ABI5 and PIF1 (Oh et al., 2009) points to a mechanism for cross talk not only between ABA and light but also between ABA and all other hormones via PIF proteins (Fig. 1.4). Reports show that ABI4 mediates cross talk between ABA and CK during inhibition of lateral root development (Shkolnik-Inbar and Bar-Zvi, 2010), between ABA and JA in plant defense (Kerchev et al., 2011), and between ABA and GA in seed dormancy (Shu et al., 2016). Thus direct interaction of ABI proteins with other transcriptional regulators is likely to be discovered in hormonal cross talk.

In summary, "crosstalk" is probably best explained in terms of certain key regulators or "hubs," with DELLA proteins providing a clear example. Different isoforms of each key regulator with slight differences in their activity or binding specificity can be expressed at different times and at different places in the plant to achieve specific patterns of development or response to environmental factors. It is also important to recognize that interactions and mechanisms discovered in *Arabidopsis* will not necessarily apply to rice and other plants. Indeed such differences are inevitable because all plants are different.

1.5 Hormones and crops

The "green revolution" of the last century was founded on the intensification of agriculture using inputs from fossil fuels including mechanization, fertilizers, and irrigation, combined with crop genetics to produce varieties that respond to such inputs. At the center of this revolution was the creation of dwarf and semi-dwarf cereals that invest more of their available resources into grain and less into shoot elongation. The short-stem phenotype not only provides improved resource allocation but creates sturdy plants that resist wind damage. The dwarfing genes that underpin these developments including *Reduced height* (*Rht*) genes in wheat and *semi-dwarf* (*sd*) genes in rice were later discovered to be GA biosynthesis and response genes (Gao et al., 2017). In rice the *sd1* gene encodes GA 20-oxidase. For this reason GA can be referred to as the "green revolution hormone." However there are many other aspects of hormone function that have contributed to improvements in crop productivity. Changes to shoot architecture have produced varieties with optimized leaf number, shape, and disposition, resulting in greatly improved light interception, and varieties with improved ear or panicle structure. The Quantitative Trait Locus (QTL) *HTD1* (High Tillers and Dwarf) is a mutant of a carotenoid cleavage dioxygenase involved in SL biosynthesis. The *Gn1a* (Grain number) gene which encodes CK oxidase increases spikelet number and hence grain number and yield (Ashikari et al., 2005). The *TGW6* (Thousand Grain

Weight) gene encodes an IAA-glucose hydrolase and has increased grain size and yield. Strategies for the generation of high-yielding "super rice" to help meet future food demands include rational genetic design in which these hormone-based QTLs play a major role (Qian et al., 2016).

Future crop improvements will include the need for greater water use efficiency and abiotic stress tolerance. These traits depend on the coordinated activities of plant hormones. Tolerance to drought, cold, and salinity depends on modulation by ABA (Li et al., 2017b) and other hormones including SL (Wang et al., 2017b). Tolerance to flooding can be imparted by the *SNORKEL* and *SUBMERGENCE* (*SUB*) genes encoding transcription factors which mediate ETH signaling (Hao et al., 2017). Nutrient use efficiency is likely to be improved through changes to root architecture and physiology and potentially by exploitation of symbioses with rhizobia and mycorrhizae. Tolerance to pests and pathogens also depend on defense hormone signaling systems including SA, JA, and peptides. Thus hormone signaling underpins all aspects of crop improvement.

Since all hormones seem to influence the activities of others, there may be inevitable trade-offs and unintended consequences resulting from changes to hormone signaling. For example, there is a trade-off between seed dormancy and seed germination traits, which are governed partly by relative levels of GA and ABA signaling. Dormancy is required to prevent pre-harvest sprouting in cereals, but too much dormancy is undesirable for the seed and malting industries. The "trade-off" between disease resistance and growth has been recognized as a key factor to understand and control. While selection of desired phenotypes in crops has generally been very productive, rational design of plant hormone systems will require a much greater level of understanding, and will be crop-specific and context-specific. With inevitable climatic changes and diminishing availability of good arable land, a vital element in the breeding of new varieties to meet the requirements of the next generation will be plant hormones.

Abbreviations

ABA	Abscisic acid
ABA-GE	ABA glucose ester
ABC	ATP-binding cassette
ABI	ABA-insensitive
ACC	1-Aminocyclopropane-1-carboxylic acid
AHP	Arabidopsis histidine phosphotransfer protein
AIT1	ABA-importing transporter 1
AM	Arbuscular mycorrhiza
AON	Autoregulation of nodulation
AP2	APETELLA2
AREB	ABA response element binding factor
ARF	Auxin response factor
ARR	Arabidopsis response regulator

AtDAO	*Arabidopsis thaliana DIOXYGENASE FOR AUXIN OXIDATION 1*
AUX/IAA	Auxin/IAA inducible protein
AUX1	Auxin transporter
BAK1	BRI1-ASSOCIATED RECEPTOR KINASE 1
BES1	BRI1 EMS SUPPRESSOR 1
bHLH	Basic helix-loop-helix protein
BiFC	Bimolecular fluorescence complementation
BIN2	BR-INSENSITIVE 2
BKI1	BRI1 KINASE INHIBITOR 1
BR	Brassinosteroid
BRI1	Brassinosteroid-insensitive 1
BZR1	BRASINOZOLE RESISTANT 1
CBL	Calcineurin B-like
CDPK	Calcium-dependent protein kinase
CIPK	CBL-interacting protein kinase
CK	Cytokinin
CLE	CLV3/Endosperm surrounding region
CLIM	Covalently attached intermediate
CLV	CLAVATA
COI1	CORONATINE INSENSITIVE 1
CRE1	CYTOKININ RESPONSE 1
CTR1	CONSTITUTIVE TRIPLE RESPONSE 1
CYP	Cytochrome P450
D14	DWARF14
EBF	EIN3-BINDING PROTEIN
DELLA protein	Protein containing DELLA (Asp-Glu-Leu-Leu-Ala) motif
EIN2	ETHYLENE INSENSITIVE 2
EIL	EIN3-LIKE
ENT	Equilibrative nucleoside transporter
EPR	EXPRESSER OF PATHOGENESIS-RELATED PROTEIN 1
EPF	EPIDERMAL PATTERNING FACTOR
ERF	ETHYLENE RESPONSE FACTOR
ETH	Ethylene
ETP	EIN2 TARGETING PROTEIN
FRET	Fluorescence (or Förster) resonance energy transfer
FT	FLOWERING LOCUS T
GA	Gibberellic acid
GA2ox	GA 2-oxidase
GC	Gas chromatography
GFP	Green fluorescent protein
GID	GA-insensitive dwarf
GRAS	GAI, RGA, and SCARECROW
GSK3	GLYCOGEN SYNTHASE KINASE 3
GUS	β-Glucuronidase
HK	Histidine kinase
HP	Histidine phosphotransfer protein

HR	Hypersensitive cell death response
HTD	High tillers and dwarf
JA	Jasmonic acid
JAM	JASMONATE-ASSOCIATED MYC2-LIKE
JAS	Jasmonate domain
JAZ	JASMONATE ZIM DOMAIN
KAI	KARRIKIN INSENSITIVE
LAX	Like-AUX1
LC	Liquid chromatography
LRR-RK	Leucine-rich-repeat receptor kinase
MAPK	Mitogen-Activated protein kinase
MAX	MORE AXILLARY GROWTH
MS	Mass spectrometer/spectrometry
MYB	Myeloblastosis viral oncogene homolog
MYC	Myelocytomatosis oncogene like
NO	Nitric oxide
NPF	Nitrate or peptide transporter family
NPR	Nonexpresser of PR genes
NRT	Nitrate transporter
PAT	Polar auxin transport
PIF	PHYTOCHROME INTERACTING FACTOR
PILS	PIN-LIKES
PIN	PINOID
PP2C2	Phosphoprotein phosphatase type C2
PUP	Purine permease
PYR	Pyrabactin
PYL	PYR-like
QTL	Quantitative trait locus
RAM	Root apical meristem
Rht	Reduced height
ROP	Rho GTPase-related protein
ROS	Reactive oxygen species
RR	Response regulator protein
SA	Salicylic acid
SAM	Shoot apical meristem
SAR	Systemic acquired resistance
SCF	Skp-Cullin-F-box
sd	Semi-dwarf
SL	Strigolactone
SLN1	SLENDER 1 (barley)
SLR1	SLENDER 1 (rice)
SLY	SLEEPY
SnRK2	SUCROSE NON-FERMENTING KINASE 2
STM	SHOOT MERISTEMLESS
SUB	SUBMERGENCE
SUMO3	Small ubiquitin-like modifier
TGA	TGACG sequence-specific binding protein
TIFY	TIF[F/Y]XG motif

TIR1	TRANSPORT INHIBITOR RESPONSE 1
TPL	TOPLESS
TPR	TPL-RELATED
VP1	VIVIPAROUS 1
WOX	WUS-RELATED HOMEOBOX
WRKY	Protein containing WRKY (Trp-Arg-Lys-Tyr) motif
WUS	WUSCHEL
YFP	Yellow fluorescent protein

Acknowledgments

The authors sincerely thank Dr Lin Li for the preparation of figures and proof reading the manuscript. The research of authors JL and CL was supported by grants from the National Natural Science Foundation of China, the Ministry of Science and Technology of China, and the Chinese Academy of Sciences, and SMS was supported by the High-End Program of Foreign Experts (2013T1S0013).

References

Adhikari, E., Lee, D., Giavalisco, P., et al., 2013. Long-distance signalling in bypass1 mutants: bioassay development reveals the bps signal to be a metabolite. Mol. Plant 6, 164–173.

Agnieszka, L., Agata, C., Anna, K.M., et al., 2014. *Arabidopsis* protein phosphatase 2C ABI1 interacts with Type I ACC synthases and is involved in the regulation of ozone-induced ethylene biosynthesis. Mol. Plant 7, 960–976.

Akiyama, K., Matsuzaki, K., Hayashi, H., 2005. Plant sesquiterpenes induce hyphal branching in arbuscular mycorrhizal fungi. Nature 435, 824–827.

Amasino, R., 2005. 1955: kinetin arrives: the 50th anniversary of a new plant hormone. Plant Physiol. 138, 1177–1184.

Andrés, F., Romera-Branchat, M., Martínez-Gallegos, R., et al., 2015. Floral induction in *Arabidopsis* by FLOWERING LOCUS T requires direct repression of BLADE-ON-PETIOLE genes by the homeodomain protein PENNYWISE. Plant Physiol. 169, 2187–2199.

Ashikari, M., Sakakibara, H., Lin, S., et al., 2005. Cytokinin oxidase regulates rice grain production. Science 309, 741–745.

Avendaño-Vázqueza, A.O., Cordobaa, E., Llamas, E., et al., 2014. An uncharacterized apocarotenoid-derived signal generated in ζ-carotene desaturase mutants regulates leaf development and the expression of chloroplast and nuclear genes in *Arabidopsis*. Plant Cell 26, 2524–2537.

Azpiroz, R., Wu, Y., LoCascio, J., et al., 1998. An *Arabidopsis* brassinosteroid-dependent mutant is blocked in cell elongation. Plant Cell 10, 219–230.

Baluska, F., Wojtaszek, P., Volkmann, D., et al., 2003. The architecture of polarized cell growth: the unique status of elongating plant cells. Bioessays 25, 569–576.

Bari, R., Jones, J.D., 2009. Role of plant hormones in plant defense responses. Plant Mol. Biol. 69, 473–488.

Benková, E., Michniewicz, M., Sauer, M., et al., 2003. Local, efflux-dependent auxin gradients as a common module for plant organ formation. Cell 115, 591–602.

Bishopp, A., Lehesranta, S., Vaten, A., et al., 2011. Phloem-transported cytokinin regulates polar auxin transport and maintains vascular pattern in the root meristem. Curr. Biol. 21, 927–932.

Bradford, K., Trewavas, A., 1994. Sensitivity thresholds and variable time scales in plant hormone action. Plant Physiol. 105, 1029–1036.

Brunoud, G., Wells, D., Oliva, M., et al., 2012. A novel sensor to map auxin response and distribution at high spatio-temporal resolution. Nature 482, 103–106.

Bussell, J., Reichelt, M., Wiszniewski, A., et al., 2014. Peroxisomal ATP-binding cassette transporter COMATOSE and the multifunctional protein abnormal INFLORESCENCE MERISTEM are required for the production of benzoylated metabolites in Arabidopsis seeds. Plant Physiol. 164, 48–54.

Chang, K., Zhong, S., Weirauch, M., et al., 2013. Temporal transcriptional response to ethylene gas drives growth hormone cross-regulation in Arabidopsis. eLife 2, e00675.

Che, P., Bussell, J., Zhou, W., et al., 2010. Signalling from the endoplasmic reticulum activates brassinosteroid signalling and promotes acclimation to stress in Arabidopsis. Sci. Signal. 3, ra69.

Cheng, Z., Shang, B., Zhang, X., et al., 2017. Plant hormones and stem cells. In: Li, J., Li, C., Smith, S.M. (Eds.), Hormone Metabolism and Signaling in Plants, Academic Press, United States of America, 405–423.

Chiba, Y., Shimizu, T., Miyakawa, S., et al., 2015. Identification of Arabidopsis thaliana NRT1/PTR FAMILY (NPF) proteins capable of transporting plant hormones. J. Plant Res. 128, 679–686.

Chu, J., Fang, S., Xin, P., et al., 2017. Quantitative analysis of plant hormones based on LC-MS/MS. In: Li, J., Li, C., Smith, S.M. (Eds.), Hormone Metabolism and Signalling in Plants, Academic Press, United States of America, 471–526.

Cook, C., Whichard, L., Turner, B., et al., 1966. Germination of witchweed (Striga lutea Lour.): isolation and properties of a potent stimulant. Science 154, 1189–1190.

Dayan, J., Voronin, N., Gong, F., et al., 2012. Leaf-induced gibberellin signalling is essential for internode elongation, cambial activity, and fiber differentiation in tobacco stems. Plant Cell 24, 66–79.

Darwin, C.R., 1880. The Power of Movement in Plants. John Murray, London.

Davière, J., Achard, P., 2016. A pivotal role of DELLAs in regulating multiple hormone signals. Mol. Plant 9, 10–20.

Davies, P. (Ed.), 2010. Plant Hormones. Springer, New York.

de Jong, M., George, G., Ongaro, V., et al., 2014. Auxin and strigolactone signalling are required for modulation of Arabidopsis shoot branching by nitrogen supply. Plant Physiol. 166, 384–395.

de Zelicourt, A., Colcombet, J., Hirt, H., 2016. The role of MAPK modules and ABA during abiotic stress signalling. Trends Plant Sci. 21, 677–685.

Demole, E., Lederer, E., Mercier, D., 1962. Isolement et détermination De La structure Du Jasmonate De Méthyle, constituant odorant caractéristique de lessence de jasmin. Helv. Chim. Acta 45, 675–685.

Desgagne-Penix, I., Sponsel, V., 2008. Expression of gibberellin 20-oxidase1 (AtGA20ox1) in Arabidopsis seedlings with altered auxin status is regulated at multiple levels. J. Exp. Bot. 59, 2057–2070.

Doubt, S., 1917. The response of plants to illuminating gas. Bot. Gaz. 63, 209–224.

Feng, J., Shi, Y., Yang, S., 2017. Cytokinins. In: Li, J., Li, C., Smith, S.M. (Eds.), Hormone Metabolism and Signalling in Plants. Academic Press, United States of America. 77–97.

Fiume, E., Fletcher, J., 2012. Regulation of Arabidopsis embryo and endosperm development by the polypeptide signalling molecule CLE8. Plant Cell 24, 1000–1012.

Flematti, G., Waters, M., Adrian, S., et al., 2013. Karrikin and cyanohydrin smoke signals provide clues to new endogenous plant signalling compounds. Mol. Plants 6, 29–37.

Foo, E., Ferguson, B., Reid, J.B., 2014. Common and divergent roles of plant hormones in nodulation and arbuscular mycorrhizal symbioses. Plant Signal. Behav. 9, e29593.

Foo, E., McAdam, E.L., Weller, J.L., et al., 2016. Interactions between ethylene, gibberellins, and brassinosteroids in the development of rhizobial and mycorrhizal symbioses of pea. J. Exp. Bot. 67, 2413–2424.

Frigerio, M., Alabadi, D., Perez-Gomez, J., et al., 2006. Transcriptional regulation of gibberellin metabolism genes by auxin signalling in *Arabidopsis*. Plant Physiol. 142, 553–563.

Fu, Z., Dong, X., 2013. Systemic acquired resistance: turning local infection into global defense. Annu. Rev. Plant Biol. 64, 839–863.

Gaffney, T., Friedrich, L., Vernooij, B., et al., 1993. Requirement of salicylic acid for the induction of systemic acquired resistance. Science 261, 754–756.

Gallego-Bartolome, J., Minguet, E., Grau-Enguix, F., et al., 2012. Molecular mechanism for the interaction between gibberellin and brassinosteroid signalling pathways in *Arabidopsis*. Proc. Natl. Acad. Sci. U.S.A. 109, 13446–13451.

Gane, R., 1934. Production of ethylene by some fruits. Nature 134, 1008.

Gao, X., Zhang, Y., He, Z., 2017. Gibberillins. In: Li, J., Li, C., Smith, S.M. (Eds.), Hormone Metabolism and Signalling in Plants. Academic Press, United States of America. 107–146.

Gomez-Roldan, V., Fermas, S., Brewer, P., et al., 2008. Strigolactone inhibition of shoot branching. Nature 455, 189–194.

Grove, M., Spencer, G., Rohwedder, W., et al., 1979. Brassinolide, a plant growth-promoting steroid isolated from *Brassica napus* pollen. Nature 281, 216–217.

Gutjahr, C., Gobbato, E., Choi, J., et al., 2015. Rice perception of symbiotic arbuscular mycorrhizal fungi requires the karrikin receptor complex. Science 350, 1521–1524.

Ha, C.V., Leyva-González, M.A., Osakabe, Y., et al., 2014. Positive regulatory role of strigolactone in plant responses to drought and salt stress. Proc. Natl. Acad. Sci. U.S.A. 111, 851–856.

Hao, D., Sun, X., Ma, B., et al., 2017. Ethlyene. In: Li, J., Li, C., Smith, S.M. (Eds.), Hormone Metabolism and Signalling in Plants. Academic Press, United States of America. 203–234.

Hong, G., Xue, X., Mao, Y., et al., 2012. *Arabidopsis* MYC2 interacts with DELLA proteins in regulating sesquiterpene synthase gene expression. Plant Cell 24, 2635–2648.

Hou, X., Lee, L., Xia, K., et al., 2010. DELLAs modulate jasmonate signalling via competitive binding to JAZs. Dev. Cell 19, 884–894.

Hu, J., Mitchum, M., Barnaby, N., et al., 2008. Potential sites of bioactive gibberellin production during reproductive growth in *Arabidopsis*. Plant Cell 20, 320–336.

Hu, Y., Yu, D., 2014. BRASSINOSTEROID INSENSITIVE2 interacts with ABSCISIC ACID INSENSITIVE5 to mediate the antagonism of brassinosteroids to abscisic acid during seed germination in *Arabidopsis*. Plant Cell 26, 4394–4408.

Huang, H., Jiang, W., Hu, Y., et al., 2013. BR signal influences *Arabidopsis* ovule and seed number through regulating related genes expression by BZR1. Mol. Plant 6, 456–469.

Ishimaru, K., Hirotsu, N., Madoka, Y., et al., 2013. Loss of function of the IAA-glucose hydrolase gene TGW6 enhances rice grain weight and increases yield. Nat. Genet. 45, 707–711.

Jasinski, S., Piazza, P., Craft, J., et al., 2005. KNOX action in *Arabidopsis* is mediated by coordinate regulation of cytokinin and gibberellin activities. Curr. Biol. 15, 1560–1565.

Jin, J., Cai, B., Zhou, J., 2017. Salicylic acid. In: Li, J., Li, C., Smith, S.M. (Eds.), Hormone Metabolism and Signalling in Plants. Academic Press, United States of America. 273–285.

Jung, J., McCouch, S., 2013. Getting to the roots of it: genetic and hormonal control of root architecture. Front. Plant Sci. 4, 186.

Kang, J., Hwang, J., Lee, M., et al., 2010. PDR-type ABC transporter mediates cellular uptake of the phytohormone abscisic acid. Proc. Natl. Acad. Sci. U.S.A. 107, 2355–2360.

Kanno, Y., Hanada, A., Chiba, Y., et al., 2012. Identification of an abscisic acid transporter by functional screening using the receptor complex as a sensor. Proc. Natl. Acad. Sci. U.S.A. 109, 9653–9658.

Kerchev, P., Pellny, T., Vivancos, P., et al., 2011. The transcription factor ABI4 is required for the ascorbic acid-dependent regulation of growth and regulation of jasmonate-dependent defense signalling pathways in *Arabidopsis*. Plant Cell 23, 3319–3334.

Ko, D., Kang, J., Kiba, T., et al., 2014. *Arabidopsis* ABCG14 is essential for the root-to-shoot translocation of cytokinin. Proc. Natl. Acad. Sci. U.S.A. 111, 7150–7155.

Krouk, G., Lacombe, B., Bielach, A., et al., 2010. Nitrate-regulated auxin transport by NRT1.1 defines a mechanism for nutrient sensing in plants. Dev. Cell 18, 927–937.

Kudo, T., Kiba, T., Sakakibara, H., 2010. Metabolism and long-distance translocation of cytokinins. J. Integr. Plant Biol. 52, 53–60.

Kuromori, T., Miyaji, T., Yabuuchi, H., et al., 2010. ABC transporter AtABCG25 is involved in abscisic acid transport and responses. Proc. Natl. Acad. Sci. U.S.A. 107, 2361–2366.

Larrieu, A., Champion, A., Legrand, J., et al., 2015. A fluorescent hormone biosensor reveals the dynamics of jasmonate signalling in plants. Nat. Commun. 16, 6043.

Lee, D., Parrott, D., Adhikari, E., et al., 2016. The mobile bypass signal arrests shoot growth by disrupting shoot apical meristem- maintenance, cytokinin signalling, and WUS transcription factor expression. Plant Physiol. 171, 2178–2190.

Lee, D., Van Norman, J., Murphy, C., et al., 2012. In the absence of BYPASS1-related gene function, the bps signal disrupts embryogenesis by an auxin-independent mechanism. Development 139, 805–815.

Letham, D., Miller, C., 1965. Identity of kinetin-like factors from *Zea mays*. Plant Cell Physiol. 6, 355–359.

Li, J., Li, C., Smith, S.M. (Eds.), 2017a. Hormone Metabolism and Signalling in Plants. Academic Press, United States of America.

Li, J., Wu, Y., Xie, Q., et al., 2017b. Abscisic acid. In: Li, J., Li, C., Smith, S.M. (Eds.), Hormone Metabolism and Signaling in Plants. Academic Press, United States of America, 161–189.

Li, L., Li, C., Lee, G., et al., 2002. Distinct roles for jasmonate synthesis and action in the systemic wound response of tomato. Proc. Natl. Acad. Sci. U.S.A. 99, 6416–6421.

Li, L., Hou, X., Tsuge, T., et al., 2008. The possible action mechanisms of indole-3-acetic acid methyl ester in *Arabidopsis*. Plant Cell Rep. 27, 575–584.

Lilley, J., Gan, Y., Graham, I., et al., 2013. The effects of DELLAs on growth change with developmental stage and brassinosteroid levels. Plant J. 76, 165–173.

McAdam, S., Brodribb, T., Ross, J., 2016. Shoot-derived abscisic acid promotes root growth. Plant Cell Environ. 39, 652–659.

Miller, C., Skoog, F., Von Saltza, M., et al., 1955. Kinetin, a cell division factor from deoxyribonucleic acid. J. Am. Chem. Soc. 77, 1392.

Miller, C., 1961. A kinetin-like compound in maize. Proc. Natl. Acad. Sci. U.S.A. 47, 170–174.

Mitchell, J., Mandava, N., Worley, J., et al., 1970. Brassins-a new family of plant hormones from rape pollen. Nature 225, 1065–1066.

Nelson, D., Scaffidi, A., Dun, E., et al., 2011. F-box protein MAX2 has dual roles in karrikin and strigolactone signalling in *Arabidopsis thaliana*. Proc. Natl. Acad. Sci. U.S.A. 108, 8897–8902.

Nemhauser, J., Hong, F., Chory, J., 2006. Different plant hormones regulate similar processes through largely nonoverlapping transcriptional responses. Cell 126, 467–475.

Nishimura, R., Hayashi, M., Wu, G., et al., 2002. HAR1 mediates systemic regulation of symbiotic organ development. Nature 420, 426–429.

Van Norman, J., Zhang, J., Cazzonelli, C., et al., 2014. Periodic root branching in *Arabidopsis* requires synthesis of an uncharacterized carotenoid derivative. Proc. Natl. Acad. Sci. U.S.A. 111, E1300–E1309.

Oh, E., Kang, H., Yamaguchi, S., et al., 2009. Genome-wide analysis of genes targeted by PHYTOCHROME INTERACTING FACTOR 3-LIKE5 during seed germination in *Arabidopsis*. Plant Cell 21, 403–419.

Okamoto, S., Shinohara, H., Mori, T., et al., 2013. Root-derived CLE glycopeptides control nodulation by direct binding to HAR1 receptor kinase. Nat. Commun. 4, 2191.

O'Neill, D., Davidson, S., Clarke, V., et al., 2010. Regulation of the gibberellin pathway by auxin and DELLA proteins. Planta 232, 1141–1149.

O'Neill, D., Ross, J., 2002. Auxin regulation of the gibberellin pathway in pea. Plant Physiol. 130, 1974–1982.

Păcurar, D., Thordal-Christensen, H., Păcurar, M., et al., 2011. *Agrobacterium tumefaciens*: from crown gall tumors to genetic transformation. Physiol. Mol. Plant Pathol. 76, 76–81.

Pearce, G., Strydom, D., Johnson, S., et al., 1991. A polypeptide from tomato leaves induces wound-inducible inhibitor proteins. Science 253, 895–898.

Piya, S., Shrestha, S.K., Binder, B., et al., 2014. Protein-protein interaction and gene co-expression maps of ARFs and Aux/IAAs in *Arabidopsis*. Front. Plant Sci. 5, 744.

Porco, S., Pencik, A., Rashed, A., et al., 2016. Dioxygenase-encoding AtDAO1 gene controls IAA oxidation and homeostasis in *Arabidopsis*. Proc. Natl. Acad. Sci. U.S.A. 113, 11016–11021.

Qian, Q., Guo, L., Smith, S.M., et al., 2016. Breeding high-yield superior quality hybrid super rice by rational design. Natl. Sci. Rev. 3, 283–294.

Qin, G., Gu, H., Zhao, Y., et al., 2005. An indole-3-acetic acid carboxyl methyltransferase regulates *Arabidopsis* leaf development. Plant Cell 17, 2693–2704.

Qu, L., Jiang, Z., Li, J., 2017. Auxin. In: Li, J., Li, C., Smith, S.M. (Eds.), Hormone Metabolism and Signalling in Plants. Academic Press, United States of America. 39–65.

Regnault, T., Daviere, J., Achard, P., 2016. Long-distance transport of endogenous gibberellins in *Arabidopsis*. Plant Signal. Behav. 11, e1110661.

Ross, J., Quittenden, L., 2016. Interactions between brassinosteroids and gibberellins: synthesis or signalling? Plant Cell 28, 829–832.

Saito, H., Oikawa, T., Hamamoto, S., et al., 2015. The jasmonate-responsive GTR1 transporter is required for gibberellin-mediated stamen development in *Arabidopsis*. Nat. Commun. 6, 6095.

Sakamoto, T., Kamiya, N., Ueguchi-Tanaka, M., et al., 2001. KNOX homeodomain protein directly suppresses the expression of a gibberellin biosynthetic gene in the tobacco shoot apical meristem. Genes Dev. 15, 581–590.

Sheard, L., Tan, X., Mao, H., et al., 2010. Jasmonate perception by inositol-phosphate-potentiated COI1-JAZ co-receptor. Nature 468, 400–405.

Shkolnik-Inbar, D., Bar-Zvi, D., 2010. ABI4 mediates abscisic acid and cytokinin inhibition of lateral root formation by reducing polar auxin transport in *Arabidopsis*. Plant Cell 22, 3560–3573.

Shu, K., Chen, Q., Wu, Y., et al., 2016. ABI4 mediates antagonistic effects of abscisic acid and gibberellins at transcript and protein levels. Plant J. 85, 348–361.

Skoog, F., Miller, C., 1957. Chemical regulation of growth and organ formation in plant tissue cultures in vitro. Symp. Soc. Exp. Biol. 11, 118–131.

Smith, S.M., 2014. What are strigolactones and why are they important to plants and soil microbes? BMC Biol. 12, 19.

Song, X.F., Guo, P., Ren, S.C., et al., 2013. Antagonistic peptide technology for functional dissection of CLV3/ESR genes in Arabidopsis. Plant Physiol. 161, 1076–1085.

Song, X., Ren, S., Liu, C., 2017. Peptide hormones. In: Li, J., Li, C., Smith, S.M. (Eds.), Hormone Metabolism and Signalling in Plants. Academic Press, United States of America. 361–395.

Su, Xia, Wang, R., et al., 2017. Phytohormonal quantification based on biological principles. In: Li, J., Li, C., Smith, S.M. (Eds.), Hormone Metabolism and Signalling in Plants. Academic Press, United States of America. 431–459.

Sun, T.P., 2011. The molecular mechanism and evolution of the GA-GID1-DELLA signalling module in plants. Curr. Biol. 21, R338–R345.

Sun, Y., Fan, X., Cao, D., et al., 2010. Integration of brassinosteroid signal transduction with the transcription network for plant growth regulation in *Arabidopsis*. Dev. Cell 19, 765–777.

Takahashi, N., 1998. Discovery of gibberellin. In: Kung, S., Yang, S. (Eds.), Discoveries in Plant Biology. World Scientific Publishing Co., Singapore, pp. 17–32.

Takei, K., Yamaya, T., Sakakibara, H., 2004. *Arabidopsis* CYP735A1 and CYP735A2 encode cytokinin hydroxylases that catalyze the biosynthesis of trans-Zeatin. J. Biol. Chem. 279, 41866–41871.

Tal, I., Zhang, Y., Jorgensen, M., et al., 2016. The *Arabidopsis* NPF3 protein is a GA transporter. Nat. Commun. 7, 11486.

Tanaka, M., Takei, K., Kojima, M., et al., 2006. Auxin controls local cytokinin biosynthesis in the nodal stem in apical dominance. Plant J. 45, 1028–1036.

Tong, H., Xiao, Y., Liu, D., et al., 2014. Brassinosteroid regulates cell elongation by modulating gibberellin metabolism in rice. Plant Cell 26, 4376–4393.

Ulmasov, T., Murfett, J., Hagen, G., et al., 1997. Aux/IAA proteins repress expression of reporter genes containing natural and highly active synthetic auxin response elements. Plant Cell 9, 1963–1971.

Umehara, M., Hanada, A., Yoshida, S., et al., 2008. Inhibition of shoot branching by new terpenoid plant hormones. Nature 455, 195–200.

Unterholzner, S., Rozhon, W., Papacek, M., et al., 2015. Brassinosteroids are master regulators of gibberellin biosynthesis in *Arabidopsis*. Plant Cell 27, 2261–2272.

Waadt, R., Hitomi, K., Nishimura, N., et al., 2014. FRET-based reporters for the direct visualization of abscisic acid concentration changes and distribution in *Arabidopsis*. Elife 3, e01739.

Wang, B., Chu, J., Yu, T., et al., 2015. Tryptophan-independent auxin biosynthesis contributes to early embryogenesis in *Arabidopsis*. Proc. Natl. Acad. Sci. U.S.A. 112, 4821–4826.

Wang, H., Wei, Z., Li, J., et al., 2017a. Brassinosteroids. In: Li, J., Li, C., Smith, S.M. (Eds.), Hormone Metabolism and Signalling in Plants. Academic Press, United States of America. 291–316.

Wang, B., Wang, Y., Li, J., 2017b. Strigolactgones. In: Li, J., Li, C., Smith, S.M. (Eds.), Hormone Metabolism and Signalling in Plants. Academic Press, United States of America. 327–350.

Wang, R., Estelle, M., 2014. Diversity and specificity: auxin perception and signalling through the TIR1/AFB pathway. Curr. Opin. Plant Biol. 21, 51–58.

Waters, M., Nelson, D., Scaffidi, A., et al., 2012. Specialisation within the DWARF14 protein family confers distinct responses to karrikins and strigolactones in *Arabidopsis*. Development 139, 1285–1295.

Waters, M., Scaffidi, A., Moulin, S., et al., 2015. A *Selaginella moellendorffii* ortholog of KARRIKIN INSENSITIVE2 functions in *Arabidopsis* development but cannot mediate responses to karrikins or strigolactones. Plant Cell 27, 1925–1944.

Waters, M., Scaffidi, A., Sun, Y., et al., 2014. The karrikin response system of *Arabidopsis*. Plant J. 79, 623–631.

West, A., Stock, A., 2001. Histidine kinases and response regulator proteins in two-component signalling systems. Trends Biochem. Sci. 26, 369–376.

White, R., 1979. Acetylsalicylic-acid (aspirin) induces resistance to tobacco mosaic virus in tobacco. Virology 99, 410–412.

Wójcikowska, B., Jaskóła, K., Gąsiorek, P., et al., 2013. LEAFY COTYLEDON2 (LEC2) promotes embryogenic induction in somatic tissues of *Arabidopsis*, via YUCCA-mediated auxin biosynthesis. Planta 238, 425–440.

Yang, D., Yao, J., Mei, C., et al., 2012. Plant hormone jasmonate prioritizes defense over growth by interfering with gibberellin signalling cascade. Proc. Natl. Acad. Sci. U.S.A. 109, E1192–E1200.

Yao, R., Ming, Z., Yan, L., et al., 2016. DWARF14 is a non-canonical hormone receptor for strigolactone. Nature 536, 469–473.

Ye, Q., Zhu, W., Li, L., et al., 2010. Brassinosteroids control male fertility by regulating the expression of key genes involved in *Arabidopsis* anther and pollen development. Proc. Natl. Acad. Sci. U.S.A. 107, 6100–6105.

Yoshida, H., Hirano, K., Sato, T., et al., 2014. DELLA protein functions as a transcriptional activator through the DNA binding of the indeterminate domain family proteins. Proc. Natl. Acad. Sci. U.S.A. 111, 7861–7866.

Yu, X., Li, L., Zola, J., et al., 2011. A brassinosteroid transcriptional network revealed by genome-wide identification of BESI target genes in *Arabidopsis thaliana*. Plant J. 65, 634–646.

Zhai, Q., Yan, C., Li, L., 2017. Jasmonates. In: Li, J., Li, C., Smith, S.M. (Eds.), Hormone Metabolism and Signalling in Plants. Academic Press, United States of America. 243–263.

Zhai, Q., Zhang, X., Wu, F., et al., 2015. Transcriptional mechanism of jasmonate receptor COI1-mediated delay of flowering time in *Arabidopsis*. Plant Cell 27, 2814–2828.

Zhang, K., Novak, O., Wei, Z., et al., 2014. *Arabidopsis* ABCG14 protein controls the acropetal translocation of root-synthesized cytokinins. Nat. Commun. 5, 3274.

Zhou, R., Benavente, L., Stepanova, A., et al., 2011. A recombineering-based gene tagging system for *Arabidopsis*. Plant J. 66, 712–723.

Auxins

2

Zhaoyun Jiang[1], Jiayang Li[2], Li-Jia Qu[1]
[1]Peking University, Beijing, China; [2]Institute of Genetics and Developmental
Biology, Chinese Academy of Sciences, Beijing, China

Summary

The plant hormone auxin is involved in the regulation of plant growth and development. Indole-3-acetic acid (IAA) was the first plant hormone identified and is the predominant form of auxin. Auxin is a small organic acid yet influences cell expansion and division, cell elongation and differentiation, and a variety of physiological responses, thus significantly affecting the final shape and function of cells and tissues in plants. Extensive genetic studies have revealed the pathways of auxin production, auxin transport, and auxin signaling. Auxin is mainly derived from the same biosynthetic pathway as tryptophan, and also from a tryptophan-independent pathway. It is produced particularly in shoot and root meristems and is transported long distance in the vasculature to other parts of the plant. Of particular significance is the role of auxin transporters known as PIN (PINOID) proteins that create auxin gradients within and between cells and help to establish long-distance polar transport pathways. Auxin signaling involves binding to F-box protein TIR1/AFB (TRANSPORT INHIBITOR RESPONSE/AUXIN SIGNALING F-BOX) together with a regulatory protein of the AUX/IAA family, forming a coreceptor complex. This leads to ubiquitination and degradation of the AUX/IAA protein with the release of ARF (AUXIN TRANSCRIPTION FACTOR) transcription factors, resulting in consequent changes in expression of target genes. These processes influence almost every aspect of plant development and growth. A comprehensive understanding of auxin regulation will facilitate the development of effective agricultural practices using auxin. In this chapter, we focus on auxin metabolism, auxin transport, and auxin signaling.

2.1   Discovery and functions of auxins

Auxin has fascinated and puzzled plant scientists for more than 100 years. In 1880, botanist Charles Darwin and his son published their book, *The Power of Movement in Plants*, proposing that a substance transduced from the tip of the oat coleoptile to the lower portion induces bending toward directional light (Darwin, 1880). This proposal by Darwin and his son was one of the first scientific descriptions of the action of auxin. Several closely related hormones are known collectively as auxin or auxins. The chemical structure of the most common plant auxin, indole-3-acetic acid (IAA), was determined in the 1930s. The name auxin is derived from the Greek word "auxien," which means "to grow" (Teale et al., 2006). Growth inhibition of the primary root, stimulation of lateral root initiation, and adventitious rooting were some of the first biological processes determined to be influenced by auxin. Auxins and their synthetic cousins have been used to boost plant growth and kill weeds. Excessively high

Hormone Metabolism and Signaling in Plants. http://dx.doi.org/10.1016/B978-0-12-811562-6.00002-5

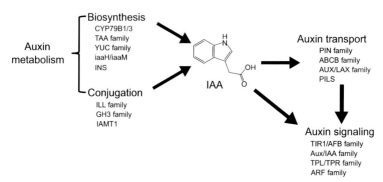

Figure 2.1 A schematic model of auxin metabolism, transport, and signal transduction in plants.

concentrations of auxins are deadly to plants; the herbicide 2,4-dichlorophenoxyacetic acid (2,4-D) is a synthetic auxin, whereas the defoliant "Agent Orange" contains a combination of synthetic auxins (Vogel, 2006). The physiological effects of auxins are used in bioassays for the identification of endogenous and artificial compounds exerting auxin functions.

This chapter summarizes recent research on auxin metabolism, auxin transport, and auxin signaling (Fig. 2.1).

2.2 Auxin metabolism

2.2.1 Structures of natural and synthetic auxins

Many small molecules, when supplied exogenously, induce an auxin response. These compounds include naturally occurring active auxin, such as IAA, 4-chloroindole-3-acetic acid (4-Cl-IAA), and phenylacetic acid (PAA); naturally occurring inactive auxin precursors, such as indole-3-pyruvic acid (IPA), indole acetamine (IAM), indole-3-acetaldoxime (IAOx), indole-3-acetonitrile (IAN), and indole-3-acetaldehyde (IAAld); and naturally occurring auxin storage forms, such as indole-3-butyric acid (IBA), methyl-IAA (MeIAA), and auxins conjugated to amino acids or sugars. In addition, synthetic compounds such as 2,4-D, 1-naphthaleneacetic acid (NAA), 3,6-dichloro-2-methoxybenzoic acid (dicamba), and 4-amino-3,5,6-trichloropicolinic acid (picloram) induce an auxin response (Korasick et al., 2013). These forms of auxin act cooperatively to control levels of active auxin in plants; however, IAA is the predominant form of auxin in plants, and studies on auxin have mainly focused on IAA.

There are two forms of IAA in plants: free IAA and conjugated IAA. Free IAA is the active form of auxin, but it comprises only up to 25% of total IAA, depending on the tissue and plant species under investigation (Ludwig-Muller, 2011). Conjugated IAA is considered to be a storage form or intermediate destined for degradation. Conjugation of IAA is one of the mechanisms through which plants regulate free auxin levels based on a variety of developmental and environmental factors (Woodward and

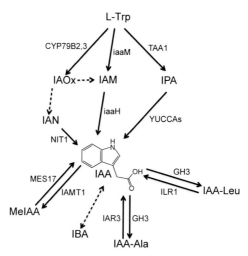

Figure 2.2 Main pathways for Trp-dependent indole-3-acetic acid (IAA) biosynthesis and conjugation.

Bartel, 2005). The cellular pool of auxin consists of free IAA, precursors of IAA, and conjugated forms of IAA, the latter two of which can contribute to free IAA. Therefore there are two ways to produce free IAA: de novo biosynthesis and release from IAA conjugates.

2.2.2 Pathways for de novo auxin biosynthesis

2.2.2.1 The precursor role of Trp

IAA is believed to be synthesized mainly from precursors generated via the shikimate pathway, which produces precursors for the biosynthesis of several indole compounds, aromatic amino acids (L-Trp, L-Phe, and L-Tyr), alkaloids, and other aromatic metabolites, lignins, and flavonoids. Therefore the shikimate pathway is an essential pathway for primary and secondary metabolism in plants. Multiple pathways contribute to the complexity of auxin biosynthesis (Fig. 2.2 and see Chapter 3). Parallel Trp-dependent and Trp-independent pathways function in different organs, developmental stages, and environmental conditions. The various routes of auxin biosynthesis are independently and differentially regulated to form a metabolic network capable of dynamic changes, maintaining auxin homeostasis and supplying auxin to meet local demand. Therefore identifying the predominant pathway of auxin biosynthesis and combining data from various species are challenging problems (Kriechbaumer et al., 2016; Tivendale et al., 2014).

For many years, Trp has been proposed as a primary precursor for IAA biosynthesis based on biochemical evidence, e.g., feeding of isotope-labeled Trp to plants resulted in the formation of isotope-labeled IAA (Sugawara et al., 2009). Several Trp-dependent auxin biosynthetic pathways contribute to IAA production, including the IAOx pathway, the IAM pathway, and the IPA pathway. Trp-independent auxin biosynthesis was proposed as a major route of IAA biosynthesis in maize and Arabidopsis based on metabolite quantification and feeding experiments in Trp biosynthesis mutants (Normanly et al., 1993; Wright et al., 1991). Studies suggest that the Trp-independent pathway is not a major route for IAA biosynthesis, but a cytosol-localized indole synthase in this pathway plays an important role in the production of IAA precursor (Wang et al., 2015).

2.2.2.2   The IAOx pathway

It has been suggested that the IAOx pathway functions only in crucifers (Sugawara et al., 2009). However, IAN, a downstream intermediate, has also been detected in maize coleoptiles (Bak et al., 1998). Cytochrome P450 enzymes CYP79B2 and CYP79B3 convert Trp to IAOx (Hull et al., 2000; Mikkelsen et al., 2000; Zhao et al., 2002) (Fig. 2.2). IAOx is largely used to produce defense compounds such as glucosinolates or camalexins, but it is also used to produce IAA. Overexpression of *CYP79B2* results in increased levels of indole glucosinolates (IGs), IAN, and free IAA, accompanied by long hypocotyls and epinastic cotyledons (Korasick et al., 2013). Conversely, the *cyp79b2 cyp79b3* double mutant displays decreased IAOx, IAN, IAM, and free IAA under normal and elevated temperatures, suggesting that both IAN and IAM are downstream intermediates of the IAOx pathway. In addition, *cyp79b2 cyp79b3* displays slightly shorter petioles and smaller leaves, consistent with roles for IAOx-derived auxin in driving these processes. Although IAN, an intermediate downstream from IAOx, was previously thought to be downstream of the glucosinolate pathway, *superroot1* mutants, defective in glucosinolate biosynthesis, display normal IAN levels, consistent with IAN being produced independently of glucosinolates; however, the enzymatic steps between IAOx and IAN are yet to be identified. IAN can be converted to active IAA through the activities of the NIT1 family of nitrolases (Korasick et al., 2013).

A previous study demonstrated that CYP79A2, a homolog of CYP79B2 and CYP79B3 in Arabidopsis, can catalyze the conversion of Phe to phenyl acetaldoxime (PAOx) with narrow substrate specificity in vitro (Wittstock and Halkier, 2000). Overexpression of *CYP79A2* significantly increases phenyl glucosinolate levels in Arabidopsis. Moreover, endogenous glucobrassicin (GB) levels are markedly reduced (<0.01%) in *cyp79b2 cyp79b3* double mutants (Zhao et al., 2002). Taken together, these results suggest that indolic glucosinolates (IGs) are largely produced by CYP79A61, could produce both PAOx and IAOx in heterologous systems, and might contribute to plant defense and auxin formation (Irmisch et al., 2015). These findings suggest that the distribution of the IAOx pathway in the plant kingdom might be wider than is currently thought. Thus further investigation of IAOx-dependent IAA biosynthesis in maize and similar systems is crucial.

2.2.2.3   The IPA pathway

This pathway appears to be the main pathway through which free IAA is produced and is the only pathway in which every step from Trp to IAA has been identified.

Conversion of Trp to IAA via the IPA pathway is a two-step process: the TRYPTOPHAN AMINOTRANSFERASE OF ARABIDOPSIS (TAA) family of Trp aminotransferases converts Trp to IPA, whereas the YUCCA (YUC) family of flavin monooxygenases converts IPA to IAA (Fig. 2.2). Mutation of the *TAA1* gene, also known as *SHADE AVOIDANCE3, WEAK ETHYLENE INSENSITIVE8,* or *TRANSPORT INHIBITOR RESPONSE2 (TAA1/SAV3/WEI8/TIR2)*, the product of which has been shown to convert Trp to IPA, results in decreased abundance of free IAA (Stepanova et al., 2008; Tao et al., 2008). The double mutant *wei8-1 tar2-1*, defective in both *TAA1* and the related gene *TRYPTOPHAN AMINOTRANSFERASE RELATED2 (TAR2)*, displays altered meristem function and floral phenotypes suggestive of decreased auxin abundance (Stepanova et al., 2008). In addition, *tir2* mutants defective in *TAA1* display impaired temperature-dependent hypocotyl elongation, gravitropism, root hair formation, and lateral root development (Yamada et al., 2009). In comparison with wild-type plants, the *wei8-1 tar2-1* double mutant accumulates less IPA and less IAA, whereas TAA1 overexpression lines accumulate more IPA, consistent with roles for IAA enzymes in converting Trp to IPA in an auxin biosynthetic pathway (Mashiguchi et al., 2011). YUC enzymes convert IPA to IAA. Overexpression of many YUC family members results in auxin overproduction phenotypes. Higher order (multiple) *yuc* mutants display decreased activity of the auxin-inducible *DR5-GUS* reporter gene (Cheng et al., 2006). They also have defects in floral patterning and vascular formation; the *yuc1 yuc4 yuc10 yuc11* quadruple mutant does not develop a hypocotyl or root meristem (Korasick et al., 2013). The *yuc1 yuc2 yuc4 yuc6* quadruple mutants hyperaccumulate IPA, whereas YUC6 overexpression lines hypoaccumulate IPA (Mashiguchi et al., 2011), consistent with roles for YUC family members in converting IPA to IAA. Because overexpression of YUC family members results in auxin overproduction phenotypes and TAA1 overexpression lines resemble wild-type plants, YUC activity is likely the rate-limiting step of the IPA pathway (Korasick et al., 2013). Interestingly, IAAld has been identified in several plant species, is hypothesized to be an auxin precursor, and is included as an intermediate in many proposed auxin biosynthetic pathways (Woodward and Bartel, 2005). In addition, IAAld application results in increased free IAA abundance, consistent with direct conversion of IAAld to IAA *in planta*. It has been suggested that *ARABIDOPSIS ALDEHYDE OXIDASE1 (AAO1)* is required for conversion of IAAld into IAA (Bottcher et al., 2014; Koiwai et al., 2000; Seo et al., 1998), but this proposal has been questioned. The *AAO1* gene belongs to the *aldehyde oxidase (AO)* gene family with four members in Arabidopsis. Members of the AO family require a molybdenum cofactor sulfurase, encoded by *ABA DEFICIENT 3 (ABA3)*, for enzyme activity (Bittner et al., 2001; Xiong et al., 2001). However, an *aba3* mutant that fails to produce the molybdenum cofactor required for AAO activity displays no obvious auxin-related defects and does not hyperaccumulate IAAld, suggesting that AAO members do not contribute to the regulation of auxin levels or regulation of IAAld-to-IAA conversion. At the same time, IAAld is not hypothesized to be an intermediate in proposed auxin biosynthetic pathways, despite its natural occurrence and despite *in planta* conversion of supplied IAAld to IAA. Therefore IAAld is an orphan intermediate in the currently proposed IAA biosynthetic pathways (Woodward and Bartel, 2005).

One study found that YUC-mediated auxin biogenesis is required for cell fate transition occurring during de novo root organogenesis in Arabidopsis (Chen et al., 2016); the authors measured the concentration of IAA to provide direct evidence of auxin production after leaf explant detachment, a process involving YUC-mediated auxin biosynthesis. Further analysis showed that YUC1 and YUC4 act relatively quickly (within 4 h) in response to wounding after detachment under both light and dark conditions and promote auxin biosynthesis in both mesophyll and competent cells, whereas YUC5, YUC8, and YUC9 primarily respond in dark conditions. In addition, YUC2 and YUC6 contribute to rooting by providing a basal auxin level in the leaf. This study showed that *YUC* genes have different roles in de novo root organogenesis from leaf explants in response to multiple signals (Chen et al., 2016).

2.2.2.4   The IAM pathway

This pathway for IAA synthesis is extensively studied in bacteria, and *iaaM* and *iaaH* genes from *Agrobacterium tumefaciens* have been used for transgenic overproduction of IAA in several plant species. The *iaaM* and *iaaH* genes encode tryptophan 2-monooxygenase (iaaM) and indole-3-acetamide hydrolase (iaaH), respectively, which produce IAA from Trp via IAM in a two-step reaction, in which iaaM catalyzes the conversion of Trp to IAM, and iaaH converts IAM to IAA (Fig. 2.2). IAM is present in many plant species, including Arabidopsis, maize, rice, and tobacco (Novak et al., 2012; Sugawara et al., 2009). IAM hydrolases (AtAMI, NtAMI), which convert IAM to IAA, have been isolated from Arabidopsis and tobacco (Ljung, 2013; Nemoto et al., 2009; Pollmann et al., 2006). It has been proposed that plants also produce IAA from the IAM pathway in a manner similar to that of auxin-producing bacteria (Mano and Nemoto, 2012; Pollmann et al., 2003). The presence of IAM in maize and rice appears to be significant since they do not produce IAOx (Kasahara, 2015; Korasick et al., 2013). In addition, IAA abundance is increased in lines overexpressing *iaaM*, leading to high-auxin phenotypes (Kasahara, 2015), suggesting that plants are capable of converting IAM to IAA in vivo. However, no direct evidence has been found for the biological significance of AMI1 in IAA biosynthesis using *atami1* mutants or plants overexpressing *AtAMI1*. In addition, the enzymes that catalyze the initial step of the IAM pathway have not been identified in plants. Although bioinformatics studies suggested that members of the AMI1 family are widely distributed in higher plants, currently available evidence is insufficient to indicate that the IAM pathway is a major route of IAA biosynthesis in plants (Kasahara, 2015).

2.2.2.5   The Trp-independent pathway

For many years, Trp has been proposed as a primary precursor for IAA biosynthesis based on biochemical evidence, e.g., feeding of isotope-labeled Trp to plants could result in the formation of isotope-labeled IAA. TAA and CYP79B, which metabolize Trp in IAA biosynthesis, were identified in later studies. In contrast, the Trp-independent pathway was proposed as a major route of IAA biosynthesis in maize and Arabidopsis based on metabolite quantification and feeding experiments in Trp biosynthesis mutants (Normanly et al., 1993). Endogenous IAA abundance was remarkably higher in Trp biosynthesis mutants in comparison with that of wild-type plants when IAA and its metabolites (e.g., IAA-amino acid conjugates and IAA-glucoside) were quantified altogether as IAA following strong base treatment (alkaline hydrolysis of IAA metabolites). In addition, when ^{15}N-labeled anthranilate was fed to an Arabidopsis Trp biosynthesis defective mutant, *trp2-1*, levels of ^{15}N incorporation were higher for IAA rather than for Trp (Normanly et al., 1993). Taken together, these biochemical findings have led researchers to posit that IAA is mainly produced from Trp precursors, e.g., indole (IND) and indole-3-glycerol phosphate (IGP), rather than from Trp. Furthermore, IGP was later proposed as a branch-point intermediate in the Trp-independent pathway in Arabidopsis because IAA levels were significantly reduced in IGP synthase-antisense transgenic plants. IAA abundance was remarkably increased in Trp biosynthesis mutants *trp3-1* and *trp2-1* when it was quantified using the alkaline hydrolysis-based method (Ouyang et al., 2000). However, it was suggested that the Trp-independent pathway was an artifact by demonstrating that IGP and GB levels were dramatically increased in *trp3-1* mutants, whereas a remarkable increase in apparent IAA abundance was the result of degradation of IGP and GB during alkaline hydrolysis-based quantification (Muller and Weiler, 2000).

It has been reported that cytosol-localized indole synthase (INS), a close paralog of plastid-localized Trp synthase α-subunit encoded by *TRP3*, functions in the Trp-independent pathway in Arabidopsis (Wang et al., 2015). This study demonstrated that INS produces IND as an initial precursor of IAA in the cytosol, although the underlying pathway from IND to IAA has not been solved. Endogenous IAA levels are slightly decreased in seedlings of *ins* mutants, whereas levels of IAA are increased in seedlings of *trp3-1* and *trp2-1* mutants. Interestingly, *ins* mutants do not display major developmental defects in adult plants. However, in seedlings, impaired hypocotyl elongation induced by high temperature is observed in *ins-1* mutants, and an additive effect in *wei8-1* mutants is observed. It is proposed that INS functions in a pathway parallel to TAA1, contributing to the control of hypocotyl elongation. More importantly, INS plays an important role in the establishment of the apical-basal pattern during early embryogenesis, demonstrating that Trp-dependent and Trp-independent IAA biosynthetic pathways are coordinated to regulate embryogenesis of higher plants. Further genetic and biochemical investigations of the Trp-independent pathway are needed to understand the role of this pathway in IAA biosynthesis (Kasahara, 2015).

2.2.3 IAA conjugation

2.2.3.1 Functions and types of conjugation

In addition to being biosynthesized de novo, free IAA can also be made available by releasing IAA from its conjugated forms, or from indole butyric acid (IBA) (Woodward and Bartel, 2005). In fact, the majority of IAA in plants exists in conjugated forms, which are proposed to serve primarily as a storage pool. The main forms of auxin conjugates in higher plants include ester-linked simple and complex carbohydrate conjugates, amide-linked amino acid conjugates, and amide-linked peptide and protein conjugates (Ludwig-Muller, 2011). Auxin conjugates are generally considered inactive; any observed auxin activity is attributed to active auxin released by conjugate hydrolysis (Ludwig-Muller, 2011; Woodward and Bartel, 2005). Interestingly, the composition of IAA conjugates varies between plant species. For example, the major conjugate auxin form in maize kernels is ester-linked sugars (Normanly et al., 1995), whereas Arabidopsis and many other dicots primarily store IAA as amide-linked amino acid conjugates (Bajguz and Piotrowska, 2009).

IAA can be conjugated via ester bonds with simple alcohols and sugars such as *myo*-inositol and *myo*-inositol-sugars. In addition, IAA can also be conjugated with amino acids, peptides, or proteins via amide bonds. Free IAA is produced when IAA conjugates are hydrolyzed. Hydrolysis of conjugates provides plants with a means of modulating free IAA levels that may be quicker or more flexible than de novo biosynthesis. Ester-linked conjugates are found in endosperm tissues of monocots and dicots, whereas amide-linked IAA-1-amino acid conjugates (IAA-aa) predominate in mature dicot seeds and vegetative tissues of most light-grown plants, including monocots and dicots. MeIAA is the carboxyl methyl ester form of IAA and might be involved in leaf flattening in Arabidopsis (Qin et al., 2005).

2.2.3.2 Amide-linked conjugates of IAA

Several amide-linked conjugates of IAA, known as IAA-aa conjugates, have been identified in plants, including IAA-leucine (IAA-Leu), IAA-alanine (IAA-Ala), IAA-aspartate (IAA-Asp), IAA-glutamate (IAA-Glu), and IAA-tryptophan (IAA-Trp) (Kowalczyk and Sandberg, 2001; Ostin et al., 1998). The occurrence of other conjugates such as IAA-valine (IAA-Val) and IAA-phenylalanine (IAA-Phe) has been postulated based on the identification of oxidative metabolites in Arabidopsis (Kai et al., 2007). These IAA conjugates function in storage, inactivation, or inhibition of auxin (Sanchez Carranza et al., 2016). IAA-aa conjugates have been implicated in a variety of biological processes. For example, IAA-Glu and IAA-Asp, which are readily synthesized upon incubation of plants with high concentrations of IAA, are considered precursors for auxin degradation (Staswick, 2009). IAA-Trp requires the auxin receptor TIR1 (TRANSPORT INHIBITOR RESPONSE1) for full activity, although its reactive mechanism is not yet elucidated (Staswick, 2009). Other conjugates such as IAA-Ala, IAA-Leu, IAA-Phe, and IAA-Val are biologically active and mimic the effect of free IAA in plant developmental responses including root and hypocotyl elongation inhibition (Sanchez Carranza et al., 2016).

Mutant screens based on reduced sensitivity to biologically active IAA-aa in root growth inhibition assays led to the identification of a specific group of amidohydrolases (Bartel and Fink, 1995; LeClere et al., 2002). The first mutant to be identified was *IAA-Leu-resistant1* (*ilr1*); the gene encodes an amidohydrolase with high affinity for IAA-Leu and IAA-Phe (Bartel and Fink, 1995). In Arabidopsis, the ILR1-like (ILL) family consists of seven members, one of which was independently identified as IAA-Ala RESISTANT3 (IAR3), and are ILR1, ILL1, ILL2, ILL3, IAR3 (ILL4), ILL5, and ILL6. The best characterized members of the ILL family are ILR1, ILL1, ILL2, and IAR3; for these enzymes, IAA-amino acid cleavage activity and substrate specificity have been revealed by in vitro enzymatic assays (LeClere et al., 2002;

Widemann et al., 2013). IAR3 and ILL2 show the highest catalytic activity with IAA-Ala as a substrate, whereas ILR1 is the most efficient at hydrolyzing IAA-Leu and IAA-Phe (Fig. 2.2). ILL3 and ILL6 show no activity or very little activity on IAA-aa in vitro (LeClere et al., 2002; Widemann et al., 2013). The *ILL5* gene is an apparent pseudogene. IAA-Asp and IAA-Glu are not efficiently hydrolyzed by any of the Arabidopsis amidohydrolases. Single mutants of the ILR1-like family such as *ilr1*, *ill2*, and *iar3*, display reduced sensitivity to biologically active IAA-aa conjugates, whereas plants overexpressing these hydrolases show higher sensitivity to certain IAA-aa conjugates (Davies et al., 1999). The triple mutant *ilr1 iar3 ill2* is less sensitive to IAA-Phe and IAA-Ala and essentially insensitive to IAA-Leu (Rampey et al., 2004). These genetic data correlate with hydrolysis rates for each IAA-aa conjugate reported from in vitro enzymatic assays, suggesting that the hydrolase activity of ILR1-like proteins is directly linked to the biological activity of IAA-aa conjugates (Sanchez Carranza et al., 2016). IAA-Leu was identified as an additional substrate for IAR3 (Sanchez Carranza et al., 2016) (Fig. 2.2). The IAR3, ILL2, and ILR1 proteins reside in the endoplasmic reticulum (ER), indicating that hydrolases in this compartment regulate amido-IAA hydrolysis, thus activating auxin signaling (Sanchez Carranza et al., 2016).

In one study, *IAR3* was identified as a new target of miR167a (Kinoshita et al., 2012). IAR3 was cleaved at the miR167a complementary site. Abundance of miR167a decreased under high osmotic stress, whereas *IAR3* mRNA abundance increased. IAR3 hydrolyzes an inactive form of auxin (IAA-Ala) and releases bioactive auxin (IAA), which influences root development. In contrast with wild-type plants, *iar3* mutants showed reduced IAA accumulation and did not display root architecture changes in response to high osmotic stress. Transgenic plants expressing a cleavage-resistant form of *IAR3* mRNA accumulated a significant amount of *IAR3* mRNA and showed enhanced lateral root development in comparison with that of transgenic plants expressing wild-type *IAR3*. Expression of an inducible noncoding RNA to sequester miR167a by target mimicry increased *IAR3* mRNA abundance, further confirming the inverse relationship between these molecules. Sequence comparison revealed that the miR167 target site on *IAR3* mRNA is conserved in evolutionarily distant plant species (Kinoshita et al., 2012). This study also showed that *IAR3* is necessary for drought tolerance.

Some functions of auxin conjugates have been elucidated using mutant Arabidopsis. However, phenotypes are usually not very clear because there are gene families for each amidohydrolase and auxin conjugate synthetase, as well as UDP-glucose transferases. Arabidopsis single mutants of various auxin conjugate hydrolase genes did not show a phenotype apart from lower sensitivity to defined auxin conjugates (Ludwig-Muller, 2011). Less IAA is formed when a hydrolase contributing to IAA transformation is deleted, reducing the inhibitory effect of free IAA. Triple mutants of Arabidopsis hydrolases, but not single or double mutants, show altered phenotypes when they are not cultivated on auxin conjugates (Rampey et al., 2004). Overexpression of hydrolases contributing to auxin conjugate formation also did not result in an apparent phenotype. Expression patterns of auxin conjugate hydrolases from *Arabidopsis thaliana*, *Arabidopsis suecica*, *Medicago truncatula*, and *Triticum aestivum* indicate their presence in growing tissues, specifically in seedlings (Ludwig-Muller, 2011).

In Arabidopsis, 20 amido synthases encoded by the large *Grethchen Hagen 3* (*GH3*) family of genes conjugate IAA, as well as some other plant hormones such as jasmonic acid and salicylic acid, with amino acids to form amide conjugates (Staswick et al., 2005). The GH3 family of auxin conjugate synthetases also plays many roles during development. Single mutants of *GH3* in Arabidopsis do not show apparent phenotypes, whereas overexpression of *GH3.2* (*yadakari1-D, ydk1-D*) and *GH3.6* (*dwarf in light 1-D, dfl1-D*) caused strong developmental phenotypes such as short hypocotyls (Nakazawa et al., 2001; Takase et al., 2004). Mutants with altered *GH3.9*, a class II auxin conjugate synthetase, displayed a shorter root phenotype and higher sensitivity to auxin-regulated root growth, indicating a role for GH3.9 in root development (Takase et al., 2004). A line overexpressing *GH3.5* (*gh3.5-1D*) showed an altered root phenotype and a curled leaf phenotype (Zhang et al., 2007). Despite their probable overlapping functions, individual members of the *GH3* family might have tissue-specific roles.

Arabidopsis has 19 *GH3* genes, *Oryza sativa* has 12 paralogs, whereas the moss *Physcomitrella patens* has only two *GH3* genes (Terol et al., 2006). In *P. patens*, single knockout mutants also did not show a phenotype; growth was retarded only when they were grown on high concentrations of IAA. However, double knockout *P. patens* mutants grow somewhat more slowly than wild-type moss, even without added IAA. In *O. sativa*, several *GH3* insertion mutants were investigated using rice retrotransposon insertion mutants; the phenotypes of the insertion mutants showed dwarfism, sterility, vivipary, and leaf withering. However, the specificity of the phenotypic changes to insertional mutagenesis of *OsGH3* genes must be demonstrated experimentally. Overexpression of rice *GH3.8* produces a retarded growth phenotype. A *GH3* gene from *Capsicum chinense* (*CcGH3*) was expressed in fruit when auxin levels were decreasing, but further experiments demonstrated that it was induced by endogenous ethylene (Liu et al., 2005). When overexpressed in tomato, *CcGH3* promoted ripening of ethylene-treated fruit. Therefore *CcGH3* might mediate the connection between auxin and ethylene signaling in ripening fruit.

Genes of the *GH3* family are thought to function in light-regulated development of Arabidopsis. Analysis of *YDK1* (*GH3.2*) has shown that expression of this gene is inhibited by blue and far-red light (Takase et al., 2004). Expression of the *DFL2* (*GH3.10*) gene was induced transiently after a red light pulse (Takase et al., 2003). Overexpression of *DFL2* caused a short hypocotyl phenotype when plants were grown under red and blue light. These results suggest that DFL2, a putative adenylase, acts downstream of the red light signaling pathway to regulate hypocotyl elongation. There are several possibilities as to how DFL2 regulates hypocotyl elongation. One possibility is that DFL2 might be involved in the adenylation of phytohormones even though activities are low in biochemical assays. Another possibility is that other proteins induced by light are required for the function of DFL2 that might in turn alter the activities of phytohormones, resulting in hypocotyl elongation in the light (Takase et al., 2003). However, for DFL2, no adenylation of IAA has been demonstrated, so it is not clear whether the observed phenotypes are connected to auxin metabolism (Ludwig-Muller, 2011). Furthermore, *GH3.5* (*WES1*) was found to be involved in shade-avoidance responses and to act downstream of phytochrome B. For the two *P. patens GH3* genes, no involvement in light-regulated growth was reported.

2.2.3.3   Indole butyric acid

IBA has long been used in agriculture to promote root initiation/growth from plant cuttings. Arabidopsis plants accumulate a detectable amount of IBA. However, little is known regarding the manner in which IBA is synthesized in plants. IBA inhibits primary root elongation and stimulates lateral root formation. Genetic screens for Arabidopsis mutants resistant to exogenous IBA have identified many loci (*ibr*, IBA resistant). The physiological roles of IBA-derived IAA are difficult to determine because the enzymes responsible for conversion of IBA to IAA (Fig. 2.2) may also participate in other pathways such as peroxisomal fatty acid metabolism. The identification of Arabidopsis IBA resistance (*ibr*) mutants showed that disruption of the *ENOYL-COA HYDRATASE2* (*ECH2*) gene impairs IBA responsiveness, but appears to leave sugar and fatty acid metabolism unchanged (Strader et al., 2010, 2011). Similarly, a peroxisomal mutant lacking a short chain dehydrogenase/reductase (SDRa) is an *ibr* mutant (Wiszniewski et al., 2009). Further analysis of *ech2* and other *ibr* mutants demonstrated that IBA-derived IAA plays an important role in root hair and cotyledon cell expansion (Strader et al., 2010, 2011).

2.2.3.4   Carboxyl methyl IAA

The carboxyl methyl form of IAA (MeIAA) is thought to be inactive in plants. IAA carboxyl methyltransferase 1 (IAMT1) catalyzes methyl ester formation to produce MeIAA (Fig. 2.2), a nonpolar molecule (Qin et al., 2005; Zhao et al., 2007, 2008). The Arabidopsis gain-of-function mutant *iamt1-D* displays a dramatic hyponastic leaf phenotype caused by increased expression of *IAMT1*, suggesting that methylation of IAA plays an important role in regulating leaf flattening and auxin homeostasis. Although both exogenous IAA and MeIAA inhibited primary root and hypocotyl elongation, MeIAA was much more potent than IAA in a hypocotyl elongation assay, indicating that IAA activity can be regulated by methylation. Expression of *IAMT1* was spatially and temporally regulated during the development of both rosette and cauline leaves. Changing expression patterns and/or levels of *IAMT1* often lead to dramatic leaf curling phenotypes. Reduced expression of *TCP* genes, which regulate leaf curvature, may partially account for the curly leaf phenotype of *iamt1-D* mutants. The identification of IAMT1 and elucidation of its role in Arabidopsis leaf development have broadened our understanding of auxin-regulated developmental processes (Li et al., 2008; Qin et al., 2005).

The salicylic acid-binding protein 2 (SABP2) in tobacco hydrolyzes methyl salicylate to salicylic acid (Forouhar et al., 2005). The homologues of tobacco SABP2 in Arabidopsis were identified as methyl esterases (MES). Biochemical assays of 15 MES proteins with various substrates revealed several candidate MeIAA esterases (MES1, MES2, MES3, MES7, MES9, MES16, MES17, and MES18) capable of hydrolyzing

MeIAA (Yang et al., 2008). Among the available T-DNA insertional mutants of these genes, only T-DNA insertional mutant *mes17* displays significantly decreased sensitivity of roots to MeIAA compared with that of wild type, while remaining as sensitive to free IAA as wild-type roots. Roots of Arabidopsis plants overexpressing *MES17* showed increased sensitivity to MeIAA, but not to IAA. In addition, *mes17* plants have longer hypocotyls and display increased expression of auxin-responsive gene DR5-GUS, suggesting perturbation of IAA homeostasis (Yang et al., 2008). These results suggest that MES17 converts MeIAA to IAA (Fig. 2.2).

Unlike other IAA conjugates, MeIAA is a nonpolar molecule that can diffuse rapidly and widely. Further investigation is required to determine whether this specific rapid-diffusion property plays an important role in specific tissues or biological processes.

2.2.4   Transcriptional control of auxin metabolism

Auxin biosynthesis is regulated at the transcriptional level by developmental signals and environmental cues. The specific expression patterns of *TAA* genes and *YUC* genes are determined by transcription factors that bind to the regulatory regions of the genes. Several groups of transcription factors have been identified by their ability to bind directly to the regulatory regions of *TAA* genes and/or *YUC* genes (Zhao, 2014). The SHORT INTENOTES/STYLISH (SHI/STY) family of transcription factors plays an important role in leaf and flower development by regulating expression of *YUC* genes (Eklund et al., 2010; Sohlberg et al., 2006). The STY1 protein binds directly to the short motif of ACTCTAC in the *YUC4* promoter, activating expression of *YUC4* (Eklund et al., 2010). It also binds to the promoter region of *YUC8* and activates *YUC8* expression (Eklund et al., 2010). The NGATHA (NGA1) family of transcription factors, which redundantly controls style development in a dosage-dependent manner, positively regulates *YUC2* and *YUC4* in the apical domain of Arabidopsis gynoecium (Alvarez et al., 2009; Trigueros et al., 2009); however, it is unclear whether such regulation by NGA1 transcription factors is direct or indirect. LEAFY COTYLEDON2 (LEC2), a central regulator of embryogenesis, binds directly to the promoters of *YUC2, YUC4,* and *YUC10* to activate expression of these *YUC* genes (Stone et al., 2008; Wojcikowska et al., 2013). In Arabidopsis, indeterminate domain (IDD) protein family members are important in plant development. The *IDD14, IDD15,* and *IDD16* genes are important regulators of lateral organ morphogenesis and gravitropism. It has been shown that IDD proteins directly target *YUC5* and *TAA1* (Cui et al., 2013). Interestingly, class III HD-Zip transcription factor REVOLUTA (REV) also directly binds to the promoters of *TAA1* and *YUC5*, suggesting that the functions of REV in shoot development and organ polarity are probably mediated by altering auxin biosynthesis (Brandt et al., 2012). PLETHORA transcription factors (PLT3, PLT5, and PLT7) control phyllotaxis in Arabidopsis by regulating expression of *YUC* genes (Pinon et al., 2013; Zhao, 2014).

Genes controlling auxin conjugate synthesis and hydrolysis are differentially regulated among tissues, during development, upon abiotic stresses, and in response to pathogen infection. The *GH3* family genes are part of the superfamily of auxin-inducible genes that also includes *Aux/IAA* genes (reviewed later). Expression of at least some *GH3* genes in Arabidopsis is mediated by auxin-responsive transcription factors (ARFs). ARFs are regulated by proteolytic inactivation of Aux/IAA repressors via the TIR signaling pathway (discussed later). ARFs are positive regulators of *GH3* gene transcription and increase the abundance of auxin amide conjugates. Arabidopsis has 23 ARFs, some of which bind to *GH3* promoters to regulate transcription. ARF8 was the first ARF shown to be capable of regulating *IAA* expression (Tian et al., 2004). Three *GH3* genes were downregulated in an *arf8-1* mutant and upregulated in *ARF8*-overexpression lines; the latter consistently showed reduced abundance of free auxin. These results suggest that ARF8 might positively regulate expression of *GH3*, resulting in the formation of IAA-amino acid conjugates.

2.3 Auxin transport

2.3.1 *The key role of auxin transport in plant development*

Auxin is unusual among phytohormones in that it is specifically and actively transported. Although rates of synthesis and conjugation are undoubtedly important contributors to the overall auxin status of plants, it is the fine concentration gradients across only a few cells that have powerful effects on plant development (Teale et al., 2006). These observations have made auxin transport one of the most studied topics in plant development.

Local biosynthesis, degradation, and conjugation contribute to the modulation of IAA homeostasis at the cellular level. Availability of free IAA inside the cell is also controlled by auxin transport via diverse transporter systems, and is essential for forming a local auxin gradient, maxima and minima. Auxin transport occurs in two distinct pathways: passive diffusion through the plasma membrane (PM) and active cell-to-cell transport, depending on the protonation state of IAA. IAA is a weak acid with a dissociation constant of $P_k = 4.8$. In a neutral or basic environment, IAA^- will be the most abundant form (99.4% ionized at pH 7.0), whereas in the acidic extracellular space protonated IAA (IAAH) is predominant (about 20% protonated at pH 5.5) (Estelle, 1998; Kramer and Bennett, 2006). IAAH can enter into the cell through the PM by passive diffusion or active transport by PM importers. Once inside the cytoplasm, which has neutral pH, IAA^- becomes the predominant form. IAA^- cannot freely diffuse out of the cell and must be actively transported by efflux carrier proteins (Balzan et al., 2014). The differential localization of transporters at specific sites on the PM creates a directional auxin flow that eventually establishes a polar auxin transport (PAT) stream through adjacent cells (Balzan et al., 2014).

Short-range and long-range polar cell-to-cell transport is critical in auxin's role as a ubiquitous chemical messenger (Baluska et al., 2003). Long-range transport (a few centimeters and more) is the part of the PAT system most accessible to experimentation and, not accidentally, it was the first that was discovered and subsequently extensively studied. Since the 1960s, when radioactive auxin became available, the following overall picture emerged. Auxin, which is synthesized in apical shoot meristems and leaf primordia, is transported in a basal direction, called basipetal, by specialized cells that reside in the vascular bundles of leaves and of vegetative and generative (inflorescence) stems (Boot et al., 2016). With the advent of molecular genetics, the main focus of PAT research shifted toward a search for genes and corresponding protein products involved in PAT. According to the current model, the direction of transport is governed by apical, basal, or sometimes lateral PM localization of members of the PIN-FORMED (PIN) auxin transport proteins, a class of putative auxin anion carriers/channels, as assumed in the chemiosmotic theory. Another class of proteins consists of the P-GLYCOPROTEIN (PGP), MULTIDRUG RESISTANCE (MDR), or ATP-BINDING CASSETTE SUBFAMILY B (ABCB) auxin transport proteins, which have nonpolar distribution. ABCBs have been proposed to regulate the amount of auxin in the cell available for PAT. A third class of auxin transport proteins belongs to the AUXIN RESISTANT1 (AUX1) and LIKE AUXIN RESISTANT1 (LAX) family of influx carriers/channels (AUX1/LAX proteins). In the weakly acidic apoplast, a portion of IAA is protonated and can pass through the nonpolar cell membrane by diffusion. Nevertheless, the larger proportion of extracellular IAA occurs as an anion and is imported in a manner facilitated by (AUX1/LAX) proteins. In Arabidopsis, this family consists of four genes, *AUX1*, *LAX1*, *LAX3*, and *LAX3*, which are all apolarly localized in the PM. It is assumed that these genes play a role in keeping IAA in transporting cells, thereby overcoming leakage of auxin from transport channels. Some newly described auxin transporters were also proposed recently, such as PIN-LIKES (PILS).

2.3.2   PIN proteins

The PIN proteins are the most studied family of auxin transporters in plants. There are eight members of the PIN family. The *PIN* genes encode integral membrane proteins with two conserved domains forming transmembrane helices, typically five at both the N and C termini, and a less conserved central hydrophilic loop of variable length (Ganguly et al., 2012; Krecek et al., 2009). In Arabidopsis, PIN1, PIN2, PIN3, PIN4, and PIN7 have a longer loop (ranging in size from 298 to 377 amino acid residues), PIN5 and PIN8 have a shorter loop (27–46 residues), and PIN6 contains an intermediate form (Ganguly et al., 2010; Krecek et al., 2009; Viaene et al., 2013). "Long" PINs are generally inserted into the PM (Fig. 2.3), whereas "short" PINs are located in the ER (Fig. 2.3) and are thought to contribute to intracellular auxin homeostasis (Cazzonelli et al., 2013; Mravec et al., 2009). It has been demonstrated that PIN5 is also localized to the PM, depending on cell type and developmental stage, and that PIN5, PIN6, and PIN8 function in polar cell-to-cell transport of auxin by regulating coordinated influx and efflux of IAA into and out of the ER (Balzan et al., 2014; Bender et al., 2013; Ganguly et al., 2014; Sawchuk et al., 2013).

Although all examined PIN proteins transport IAA, PIN family members differ in their ability to transport other auxinic compounds. For example, PIN2 and PIN7 efflux 2,4-D, whereas PIN1 does not display this ability (Yang and Murphy, 2009). In addition, PIN4 and PIN7 transport NAA (Petrasek et al., 2006), whereas PIN1 and PIN2 do not (Blakeslee et al., 2007). These differing substrate specificities may contribute to differences in whole-plant responses to various auxinic compounds.

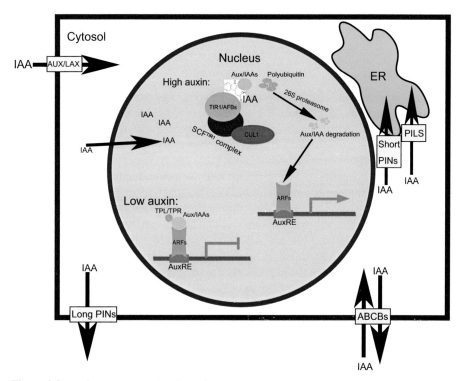

Figure 2.3 Auxin transport and main auxin signaling. Auxin transport: AUX/LAX proteins localize to different faces of the cell depending on the particular cell type, where they act to influx auxin from the apoplast into the cytoplasm. The long PIN proteins efflux auxin and establish auxin gradients. The short PIN proteins and PILS proteins localize to the endoplasmic reticulum (ER), where they efflux auxin from the cytoplasm into the ER lumen. The ABCB family of auxin transporters localize to the PM to efflux auxin outside the cell. Some members of the ABCB family have been shown to display both efflux and influx activity based on the cytoplasmic concentration of auxin. Auxin signaling: In general, auxin level is low, AUX/IAAs interact with ARFs, which specifically occupy auxin-responsive promoter elements (AuxREs) in numerous auxin-regulated genes. AUX/IAA proteins repress ARF function to inhibit the expression of downstream genes. When auxin level is higher, the TIR1/AFB F-box proteins, which participate in an SCF (Skp1-Cullin-F-box) E3 ubiquitin ligase, interact with Aux/IAA repressor proteins to form a coreceptor. The interaction leads to ubiquitination and consequent degradation of the Aux/IAA repressor proteins through the 26S proteasome, relieving repression of the ARF transcription factors and allowing for auxin-regulated gene transcription.

Unlike the "long" PIN proteins involved in polar transport of IAA, the "short" PIN proteins, PIN5, PIN6, and PIN8, localize to the ER to transport IAA from the cytoplasm into the ER. PIN5, PIN6, and PIN8 also transport NAA. At present, the specific role of auxin in the ER in regulating plant growth is unknown; however, it appears that transporting auxin into the ER may serve to alter cytoplasmic levels of free IAA. In addition to PIN5, PIN6, and PIN8, the PILS also facilitate auxin influx into the ER and may have roles in regulating the cytoplasmic pool of free auxin and auxin-amino acid conjugates (Barbez et al., 2012; Enders and Strader, 2015).

2.3.3   The ABC subfamily B transporters

The ABC subfamily of membrane proteins includes more than 100 different members in plants (Kang et al., 2011). ABCB includes homologs of the mammalian MDR/PGP proteins, several of which are involved in auxin transport. ABCB transporters are integral membrane proteins that actively transport chemically diverse substrates across the lipid bilayers of cellular membranes (Balzan et al., 2014). The core unit of a functional ABC transporter consists of four domains: two nucleotide-binding domains (NBDs) and two transmembrane domains (TMDs). The two NBDs of an ABC transporter unite to bind and hydrolyze ATP, providing the driving force for transport, whereas the TMDs are involved in substrate recognition and translocation across the membrane. Arabidopsis has 22 ABCBs. The first ABCBs characterized as functioning in IAA translocation were identified in Arabidopsis seedlings (Noh et al., 2001). ABCB1, ABCB4, ABCB14, ABCB15, ABCB19, and ABCB21 are associated with auxin transport (Fig. 2.3), although not exclusively (Cho and Cho, 2013). The most studied ABCBs are ABCB1, ABCB4, and ABCB19, which function in auxin-driven root development and require the activity of immunophilin TWISTED DWARF1 (TWD1), also known as FK506-binding protein 42 (FKBP42), to be correctly inserted at the PM (Wu et al., 2010). The first plant *MDR*-like gene cloned in Arabidopsis was *ABCB1/PGP1*. The ABCB1/PGP1 protein is localized at the PM in the root and shoot apices of seedlings (Noh et al., 2001). The first *pgp1* mutant to be studied displayed only a subtle phenotype, but mutation of a new allele designated *pgp1-2* produces plants with a shorter hypocotyl and dwarf phenotype under long-day conditions (Ye et al., 2013). ABCB19 functions together with ABCB1 in long-distance transport of auxin along the plant main axis in coordination with PIN1 and regulates root and cotyledon development, and also the tropic bending response in Arabidopsis (Bandyopadhyay et al., 2007; Christie et al., 2011; Lin and Wang, 2005; Nagashima et al., 2008). ABCB4 is a root-specific transporter involved in auxin transport during root gravitropic bending, root elongation, and lateral root formation (Cho et al., 2012; Santelia et al., 2005). ABCB21 encodes a protein that is the closest homolog to ABCB4 and is expressed in the aerial parts of the seedling and root pericycle cells. ABCB21 functions as a facultative importer/exporter that controls cellular auxin concentrations in a manner similar to ABCB14 (Kamimoto et al., 2012). It has been proposed that ABCB14 and ABCB15 participate in auxin transport (Kaneda et al., 2011).

2.3.4   Auxin influx carriers

The AUX1/LAX proteins are auxin influx carriers that are mostly responsible for auxin transport from the apoplast into the cytoplasm. The existence of auxin importers in plants was first demonstrated in a study of an Arabidopsis mutant *auxin insensitive 1* (*aux1*), which has a defective root gravitropic response. AUX1 belongs to a small gene family comprising four highly conserved proteins that share similarities with amino acid transporters: AtAUX1, AtLAX1, AtLAX2, and AtLAX3 (Peret et al., 2012). This family of four proteins regulates several developmental processes, including lateral root formation (AUX1 and LAX3) and cotyledon vascular patterning (LAX2) (Peret et al., 2012; Swarup et al., 2008). AUX1 and LAX1 act redundantly in regulating the phyllotactic pattern in Arabidopsis, although LAX2 is not expressed in the SAM L1 layer.

2.3.5   Other auxin transporters

The PIN-LIKES (PILS) proteins are the most recently characterized family of plant auxin transport proteins and include seven members in Arabidopsis. PILS proteins show low (10%–18%) sequence identity with PINs and are all capable of transporting auxin across the PM in heterologous systems (Barbez et al., 2012). PILS proteins regulate intracellular auxin accumulation at the ER (Fig. 2.3) and thus reduce the availability of free auxin that can reach the nucleus, possibly exerting a role in auxin signaling comparable with that of PIN5 (Barbez and Kleine-Vehn, 2013; Barbez et al., 2012). The PILS family is conserved throughout the plant lineage, having representatives in several taxa, including unicellular algae, in which *PIN* genes have not been found. These findings indicate that the *PILS* family could be evolutionarily older than the *PIN* (Feraru et al., 2012; Viaene et al., 2013). Six *PILS* have been identified in rice, whereas 10 have been identified in maize, 7 in sorghum, and 8 in *Brachypodium* (Feraru et al., 2012).

A putative MATE (multidrug and toxic compound extrusion) transporter, ADP1 (ALTERED DEVELOPMENT PROGRAM), was reported to play an essential role in regulating lateral organ outgrowth, and thus in maintaining normal Arabidopsis architecture. Elevated expression levels of *ADP1* resulted in accelerated plant growth and increased the numbers of axillary branches and flowers. Molecular and genetic evidence demonstrated that phenotypes of plants overexpressing *ADP1* were caused by reduction of local auxin levels in the meristematic regions. This study further discovered that this reduction was probably due to decreased levels of auxin biosynthesis in the local meristematic regions based on the measured reduction in IAA levels and gene expression data. Simultaneous inactivation of *ADP1* and its three closest homologs led to growth retardation, a relative reduction of the lateral organ number, and slightly elevated auxin abundance. This study indicated that ADP1-mediated regulation of local auxin abundance in meristematic regions is an essential contributor to plant architecture maintenance that functions by restraining the outgrowth of lateral organs (Li et al., 2014). Consistent with reduced local auxin biosynthesis, the protein abundance of PIN1, PIN3, and PIN7 was reduced in the *adp1*-D mutant without accompanying changes in transcription. In addition, subcellular analysis revealed that overexpression of ADP1 inhibited endocytosis of PIN proteins (Li et al., 2015). ADP1 is localized to a post-Golgi endomembrane compartment and acts upstream of, or coordinately with, YUCCAs in auxin biosynthesis. YUCCA6 was localized to a similar endomembrane compartment (Kim et al., 2007), suggesting that ADP1 may function in mobilization of IAA precursors to YUCCAs (for conversion to IAA) or movement of IAA out of endosomal compartments, although the latter scenario is less likely. The function of ADP1 may be homeostatic and involve reversible activity because NorM, a prokaryotic MATE crystallized from *Vibrio cholera*, may exhibit a conformational change on substrate binding (He et al., 2010). ADP1 may also be involved in auxin cellular homeostasis, which is maintained by PIN5 (Mravec et al., 2009) and PILS auxin transporters. Because pollen-specific PIN8 is the only auxin exporter reported in the ER (Ding et al., 2012), future studies should assess whether ADP1 can balance auxin homeostasis in the ER. The lack of successful ADP1 protein expression in multiple heterologous expression systems has prevented more detailed biochemical characterization (Li et al., 2014).

2.3.6 Inhibitors of auxin transport

PAT and vesicle cycling are inhibited by synthetic auxin transport inhibitors, including 1-N-naphthylphthalamic acid (NPA), counteracting the effect of auxin. One uncovered the underlying targets and mechanism involved in this process. Nuclear magnetic resonance was used to map the NPA-binding surface on Arabidopsis ABCB chaperone TWD1. The ACTIN7 (ACT7) protein was identified as a relevant, although likely indirect, TWD1 interactor. TWD1-dependent regulation of actin filament organization and the presence of efflux transporters at the membrane were demonstrated. The *act7* and *twd1* mutants share developmental and physiological phenotypes indicative of defects in auxin transport, which can be phenocopied by NPA treatment or chemical actin (de)stabilization. In addition, it was shown that TWD1 determines the downstream locations of auxin efflux transporters by adjusting actin filament debundling and dynamizing processes and mediating the action of NPA on the latter. This study also showed that this function is evolutionarily conserved because TWD1 expression in budding yeast alters polarization and cell polarity and provides NPA sensitivity (Zhu et al., 2016).

Plants transport the endogenous auxin precursor IBA in addition to active auxins (i.e., IAA, 2,4-D, and NAA). Early IBA transport assays relied on the ability of IBA to affect plant morphology distant from the site of application, raising the question of whether IBA itself or IBA-derived IAA moved through these tissues to create these morphological changes. Later studies using radiolabeled or heavy IBA demonstrated that IBA and/or IBA conjugates may travel long distances through plant tissues. Because IAA and IBA are chemically similar, one might hypothesize that these compounds are transported by the same mechanism. However, examined IAA carriers, including AUX1, PIN2, PIN7, ABCB19, do not transport IBA, suggesting that unique carriers are required for IBA transport (Enders and Strader, 2015; Strader and Bartel, 2011).

2.4 Auxin signaling

2.4.1 Auxin gradients and perception

Changes in auxin concentration are perceived by auxin receptors to initiate auxin signaling. Auxin-responsive gene expression relies on the TRANSPORT INHIBITOR

RESPONSE1/AUXIN SIGNALING F-BOX (TIR1/AFB) pathway to trigger expression of genes controlling auxin-regulated cell division, expansion, and differentiation. Auxin signaling through the TIR1/AFB pathway involves three major protein families: auxin-binding TIR1/AFB F-box proteins, AUXIN RESPONSE FACTOR (ARF) transcription factors, and AUXIN/INDOLE-3-ACETIC ACID INDUCIBLE (Aux/IAA) repressor proteins (Korasick et al., 2015; Powers and Strader, 2016; Wang and Estelle, 2014). The core parts of the auxin response apparatus are encoded by 6 *TIR1/AFB* genes, 29 *Aux/IAA* genes, 23 *ARF* genes, and 5 *TPL/TPR* genes in *A. thaliana* (Wang and Estelle, 2014) (Fig. 2.3). For each family, developmental regulation of cell type-specific mRNA expression at multiple levels, cellular control of protein abundance and activity, and functional diversification of protein domains provide a vast repertoire for combinatorial interactions between core components (Dinesh et al., 2016). Integration of these complicated processes can possibly account for appropriate interpretation of the context-specific information of auxin. Auxin distribution profiles in a field of cells may range from steep maxima to distinct minima. Such complex auxin gradients are often modified by internal and external cues and have been implicated in nonlinear regulation of numerous auxin-mediated processes relevant to the adaptation of plant form and function. Differential expression of Aux/IAA family members seems to play a significant role in tuning auxin responses because Aux/IAA family members determine the affinities of coreceptor pairs for auxin and its structural analogs. A broad range of auxin concentrations can be sensed by numerous TIR1/AFB:Aux/IAA coreceptor combinations, resulting in different Aux/IAA degradation rates. Aux/IAA repressors engage in sophisticated Aux/IAA:ARF interaction networks and are often products of early auxin genes, which establish robust negative feedback loops. Finally, ARF-dependent selection of downstream target genes is thought to confer specificity to auxin responses (Dinesh et al., 2016).

In general, when the auxin level is low, AUX/IAAs interact with ARFs (Kim et al., 1997; Vernoux et al., 2011), which specifically occupy auxin-responsive promoter elements (AuxREs) in numerous auxin-regulated genes (Guilfoyle and Hagen, 2007) (Fig. 2.3). AUX/IAA proteins repress ARF function in two ways: sequestering ARFs away from their target promoters (Farcot et al., 2015) and recruiting TOPLESS (TPL)/TPL-RELATED (TPR) corepressors, leading to chromatin inactivation and silencing of ARF target genes (Causier et al., 2012; Kagale and Rozwadowski, 2011; Long et al., 2006; Szemenyei et al., 2008). ARF repression by Aux/IAA-dependent recruitment of TPL/TPR corepressors is much studied. The N-terminal DI (domain I, the primary structures of most AUX/IAAs share four regions of sequence conservation known as domains I–IV) of most AUX/IAAs binds TPL/TPR proteins via its ETHYLENE-RESPONSIVE ELEMENT BINDING FACTOR-ASSOCIATED AMPHIPHILIC REPRESSION (EAR)-like motif (D/E-L-X-L-X-L), a prototypic repressor motif found in many plant transcriptional regulators (Causier et al., 2012; Kagale and Rozwadowski, 2011). When the auxin level is increased, TIR1/AFB F-box proteins, which are components of an SCF E3 ubiquitin ligase complex, interact with Aux/IAA repressor proteins to form a coreceptor, with auxin acting as the "molecular glue" (Chapman and Estelle, 2009; Kepinski and Leyser, 2005; Salehin et al., 2015). This interaction leads to ubiquitination and degradation of Aux/IAA repressor proteins through the 26S proteasome, relieving repression of ARF transcription factors and allowing auxin-regulated gene transcription (Salehin et al., 2015).

2.4.2   Regulation of TIR1/AFB stability

The SCF ubiquitin protein ligases (E3s) are composed of CULLIN (CUL1), the RING protein RBX1, SKP1 (Arabidopsis SKP1-like protein, ASK, in plants), and F-box protein1 (FBP1) (Deshaies and Joazeiro, 2009; Hua and Vierstra, 2011). Plant genomes encode large numbers of FBPs, the substrate recognition subunit of SCF ubiquitin ligases. Most eukaryotes have a large number of FBPs, ranging from 20 in budding yeast to approximately 700 in Arabidopsis (Gagne et al., 2002). The abundance of FBPs can be regulated at several levels (Yu et al., 2015). In particular, studies in mammals and yeast have shown that FBPs can be degraded by an autocatalytic mechanism once they are assembled into the SCF complex, particularly in the absence of substrate (Bosu and Kipreos, 2008; Galan and Peter, 1999). In Arabidopsis, SCF^{TIR1} functions in auxin perception and promotes degradation of Aux/IAA transcriptional repressors (Wang and Estelle, 2014) (Fig. 2.3). TIR1 belongs to a small family of FBPs, the AFB proteins, which consist of an F-box domain at the N-terminus and a large leucine-rich repeat domain that forms the auxin-binding pocket (Tan et al., 2007). In one study, the residues in the H1 helix of the F-box domain of TIR1 are required for binding to CUL1 and assembly into an SCF complex (Yu et al., 2015). A yeast two-hybrid system was used to identify novel mutations of TIR1 that altered its properties. The analysis of these mutants revealed that TIR1 associates with the CUL1 subunit of the SCF complex through the N-terminal H1 helix of the F-box domain. Mutations that untether TIR1 from CUL1 stabilize the FBP and cause auxin resistance and associated growth defects, probably by protecting TIR1 substrates from degradation. Based on these results, it was proposed that TIR1 is subject to autocatalytic degradation when assembled into an SCF complex. This study also showed that a key amino acid variation in the F-box domain of auxin signaling F-box AFB1 reduces its ability to form an SCF complex, resulting in increased AFB1 levels (Yu et al., 2015).

The TIR1 protein is unstable in seedlings, raising the possibility that changes in TIR1 stability may influence auxin signaling (Yu et al., 2015). Assembly and regulation of SCF complexes such as SCF^{TIR1} are highly dynamic processes that involve several proteins and protein complexes, including ubiquitin-like protein Nedd8 (that is, neddylation), the CONSTITUTIVE PHOTOMORPHOGENESIS 9 (COP9) signalosome (SCN) complex, and the cullin-associated Nedd8-disassociated (CAND1) protein (Chuang et al., 2004; del Pozo and Estelle, 1999; Duda et al., 2011; Enchev et al., 2012; Pierce et al., 2013; Schwechheimer et al., 2001). These proteins regulate the activity of the SCF complex and enable the exchange of substrate adapters as required during changing cellular conditions. As the Arabidopsis genome encodes hundreds of FBPs, the challenge of regulating SCF complex assembly is particularly acute. Indeed, the RUB1, CSN, and CAND1 have all been implicated in SCF$^{TIR1/AFB}$ function (Chuang et al., 2004; Gray et al., 1999). In addition to these factors, genetic studies show that a protein called SUPPRESSOR OF G2 ALLELE SKP1 (SGT1) is required for the auxin response (Gray et al., 2003). There are two SGT1 genes in Arabidopsis: *SGT1a* and *SGT1b*. A mutant allele of *SGT1b*, called *enhancer of tir1 auxin resistance 3* (*eta3*), was isolated as an enhancer of the *tir1-1* mutant, implicating SGT1 in the auxin response. SGT1 is a cochaperone of HEAT SHOCK FACTOR 90 (HSP90) and has been implicated in a variety of processes in eukaryotes. SGT1 was first identified in budding yeast and is required for assembly of kinetochores and activity of E3 ligase SCF^{Cdc4p} (Catlett and Kaplan, 2006; Zhang et al., 2008). SGT1 and HSP90 play critical roles in regulating abiotic and biotic stress responses in plants (Mittler et al., 2012; Xu et al., 2013). For example, they are required for normal function of NBD leucine-rich repeat (NLR) immune sensor proteins in plants and animals (Shirasu, 2009; Takahashi et al., 2003). However, their roles in ambient temperature sensing and signaling have not been investigated. One study showed that the HSP90-SGT1 chaperone system is required for the response of plants to an increase in ambient temperature (Wang et al., 2016). TIR1 is rapidly stabilized at 29°C; this change is HSP90 dependent and associated with increased seedling growth. Furthermore, HSP90 and SGT1 exist in a complex with TIR1 in plants. In addition, the mutant SGT1b protein encoded by the *eta3* gene displays reduced binding to TIR1, suggesting that auxin resistance in *eta3* mutants is directly related to TIR1 function. This study suggests that the HSP90-TIR1 module integrates environmental temperature and auxin signaling to regulate plant development and reveals a link between the molecular networks regulating the plant growth response to ambient temperature and the HSP-HSF module that plays a central role in heat sensing and signaling (Wang et al., 2016). The chaperone HSP90 is required for temperature-dependent growth and the auxin response. HSP90 and high temperature stabilize auxin coreceptors. TIR1 coimmunoprecipitates with HSP90 and SGT1. The mechanism of HSP90 action appears to be complex and is poorly understood. HSP90 has a very flexible and dynamic structure and cooperates with many cochaperones with diverse activities. Some cochaperones act to regulate the ATPase activity of HSP90, but this is not the case with SGT1. Extensive studies of plant NLR proteins led to a model in which SGT1 binds to both HSP90 and SGT1, possibly acting to stabilize a functional complex. However, RAR1 (*r*equired for M*la*12 *r*esistance) is not required for the auxin response, suggesting that SGT1 is sufficient to promote the interaction between TIR1 and HSP90. Alternatively, an unknown protein may serve the function of RAR1. Finally, the observation that HSP90 interacts directly with TIR1 in vitro raises the possibility that the inherent affinity of HSP90 for TIR1 might overcome the requirement for RAR1. In general, HSP90 is thought to act late in the protein folding pathway on unfolded and completely folded proteins (Karagoz and Rudiger, 2015; Saibil, 2013). Because most of the 18 LRRs in TIR1 contribute to the auxin-binding pocket, HSP90, in concert with SGT1, may have an important role in maintaining proper folding. Although SGT1 acts in concert with HSP90 to regulate TIR1 folding and/or stability in our current model, it is also possible that SGT1 has a function that is independent of HSP90 (Deng et al., 2016; Takahashi et al., 2003).

2.4.3   Control of the 26S proteasome

The ubiquitin/26S proteasome proteolytic pathway selectively removes regulatory proteins, providing efficient and rapid control of many cellular processes and playing a critical role in regulating various aspects of hormone signaling, development, and stress responses in plants. The proteasome is highly conserved, but little is known regarding the manner in which proteasome activity is regulated in mammals or plants. Bovine proteasome inhibitor 31 (PI31) (Ma et al., 1992) and its homologues in mice (Zaiss et al., 1999) and humans (McCutchen-Maloney et al., 2000) diminish the activity of the purified 26S proteasome and are necessary for sperm differentiation. In recent research, Arabidopsis PROTEASOME REGULATOR1 (PTRE1), which is homologous to human PI31, was identified (Yang et al., 2016). PTRE1 stimulates 26S proteasome activity and influences auxin-related processes during plant growth and development (Yang et al., 2016). It is thought that PTRE1 acts in concert with the TIR1-AFB pathway to buffer the degradation of Aux/IAA proteins and thus modulate expression of auxin-responsive genes in a precise manner. Knockout mutant *ptre1* plants are dwarfs and display developmental defects related to auxin, including small and curved leaves, altered shoot apical dominance, short siliques, and arrested embryogenesis, consistent with widespread expression of PTRE1. In addition, *ptre1* seedlings showed a defective phototropic response, suggesting that auxin signaling might be suppressed. Complementation of *ptre1* rescued the defective growth phenotype, confirming the role of PTRE1 in regulating plant growth (Yang et al., 2016). Moreover, *ptre1* mutants showed an altered response to auxin. In accordance with their auxin resistance phenotypes, *ptre1* mutants exhibited not only significantly reduced activation of DR5:GFP, but also reduced auxin-mediated gene expression (Kong et al., 2016). Thus PTRE1 is required for auxin-mediated suppression of 26S proteasome activity, which is responsible for Aux/IAA degradation in the auxin response pathway. These findings show that PTRE1 and TIR1 regulate auxin signaling (Yang et al., 2016). It is reported that cyclophilin-catalyzed peptidyl-prolyl isomerization of IAA11 promotes Aux/IAA degradation in rice, demonstrating that Aux/IAA degradation is regulated at multiple levels (Jing et al., 2015).

2.4.4 Recognition of auxin-responsive DNA elements

Auxin rapidly induces (2–30 min) primary response genes of three families known as *Aux/IAAs*, *GH3s*, and *SAURs* (Abel and Theologis, 1996). A comparison of several transcription profiling studies revealed that the early response to auxin (<30 min) is characterized by mostly upregulated mRNA expression (Delker et al., 2010; Paponov et al., 2008). Computational analyses of the genome-wide distribution of TGTCTC-type AuxREs showed a strong association with transcriptional start sites and proximal promoter regions of auxin-induced genes. In addition, this analysis identified several coupling elements forming composite AuxREs, including additional TGTCTC-type elements and the binding sites of bZIP and MYB transcription factors (Berendzen et al., 2012; Mironova et al., 2014).

Using multiple tandem copies of inverted TGTCTC repeats as a bait, the founding member of the Arabidopsis ARF family, ARF1, was selected in a yeast one-hybrid screen and shown to bind in vitro to a distinctly spaced palindromic TGTCTC motif (e.g., the ER7 element) (Ulmasov et al., 1997). Most ARF proteins contain three separable regions with specific functions: the conserved N-terminal DNA-binding domain (DBD), the variable middle region of biased amino acid composition determining either ARF activator function (Q-rich as in ARF5-8 and ARF19) or ARF repressor activity (S, P, L/G-rich), and the C-terminal protein–protein interaction domain shared by the ARF and Aux/IAA families (Dinesh et al., 2016). The core DBD region of ARFs is related to the plant-specific B3 domain (Yamasaki et al., 2013), but additional flanking residues are necessary for efficient binding to *AuxRE* (Ulmasov et al., 1999).

The high-resolution DBD crystal structures of ARF5, an activator, and ARF1, a phylogenetically distant repressor, were reported, allowing unprecedented insight into the mechanism underlying ARF:DNA interaction (Boer et al., 2014). Both ARF DBDs fold into three distinct subdomains. The regions at both flanks of the B3-type subdomain, which adopts a seven-stranded open β-barrel structure, fold into a second, structurally novel subdomain (DD) that forms a highly curved, "taco"-like five-stranded β-sheet. A distinct surface of the DD subdomain facilitates robust ARF dimerization via a network of hydrogen bonds and hydrophobic interaction (Dinesh et al., 2016).

Members of the ARF and Aux/IAA families interact directly via their homologous C-terminal regions. The importance of ARF:Aux/IAA interaction for ARF repression was demonstrated in transfection assays of Arabidopsis leaf mesophyll protoplasts with ARF proteins lacking the C-terminal interaction domain (resulting in constitutive high-auxin responses) (Guilfoyle, 2015; Krogan et al., 2012). The primary structures of most Aux/IAAs share four regions of sequence conservation known as domains I–IV (DI–DIV) (Overvoorde et al., 2005). The N-terminal DI region recruits TPL/TPR corepressors to target promoters (Li et al., 2011; Szemenyei et al., 2008), the degron motif of the central DII region interacts with TIR1/AFBs and is required for auxin-dependent coreceptor assembly (Salehin et al., 2015), and the C-terminal DIII/IV region mediates homotypic and heterotypic interactions within and between the Aux/IAA and ARF families (Vernoux et al., 2011). Although the N-terminal region of most Aux/IAAs (comprising regions DI and DII) is predicted to be intrinsically disordered to a large extent, a bioinformatics analysis of the C-terminal DII/IV regions of ARF and Aux/IAA family members indicated a single protein–protein interaction domain (Guilfoyle and Hagen, 2012). This domain is related in secondary structure to the more ancient Phox/Bem1p (PB1) domain and in tertiary structure to the globular ubiquitin-like β-grasp fold (Guilfoyle and Hagen, 2012). Canonical PB1 domains comprise two helices and a mixed five-stranded β-sheet and are classified into three types depending on the conservation and presence of oppositely charged and oriented surface patches. PB1 domains may expose a conserved acidic cluster (D-X-D/E-D-Xn-D) known as the octicosapeptide repeat, p40phox, Cdc24p, atypical PKC interaction domain (OPCA) motif (type I), a basic surface patch with an invariant lysine residue as its hallmark (type II), or both characteristic structural features (type I/II). Electrostatic interactions between the two different faces drive specific PB1 dimer formation between type I and type II PB1 domains or polymerization by directional front-to-back association of type I/II PB1 monomers (Dinesh et al., 2016; Sumimoto et al., 2007).

2.4.5 Auxin-binding protein ABP1

In early screens for auxin receptors, auxin-binding protein 1 (ABP1) was identified based on its ability to bind auxin with high affinity and soon became an important extracellular auxin receptor candidate based mainly on electrophysiological studies utilizing antibodies against ABP1 (Hertel et al., 1972; Lobler and Klambt, 1985). Electrophysiological studies showed rapid ABP1-mediated modulation of PM ion transport shortly after exposure to auxin. In subsequent research, the auxin-binding activity of ABP1 was characterized in detail by biochemical studies, while its protein structure, including the auxin-binding pocket, was revealed (Napier et al., 2002; Woo et al., 2002). Phylogenetic studies have shown that ABP1 homologues are present in the genomes of all plant species, with more than one copy present in the genomes of maize, rice, poplar, and *P. patens* (Michalko et al., 2016; Tromas et al., 2010). However, in part because of the predominant subcellular localization of ABP1 in the ER in maize, where the conditions for auxin binding are unfavorable, the biological importance of ABP1 as a PM auxin receptor has been a matter of debate (Habets and Offringa, 2015). These discussions were altered by the isolation of *abp1-c1* and *abp1-TD1*, two new *abp1* knockout alleles in Arabidopsis, which show no obvious phenotypes under standard growth conditions (Gao et al., 2015). The contradiction between this observation and the previously published embryo-lethal phenotypes of *adp1* mutants (Chen et al., 2001) has been clarified by showing that the embryonic lethality of *abp1-1* and *abp1-1s* was caused by disruption of tightly linked neighboring gene *BELAYA SMERT* (*BSM*), rather that knockout of ABP1 (Dai et al., 2015; Michalko et al., 2015). This correction and the demonstration of normal embryo development in *abp1* knockouts suggest that ABP1 plays no essential role in early Arabidopsis embryogenesis.

When knocking out a gene in Arabidopsis has no resulting phenotype, other proteins generally overlap functionally with the deleted gene. This explanation might also apply to ABP1; new *abp1* null mutants have no phenotype, and the old suspected mutant *adp1-1* had the embryonic lethality phenotype caused by the deletion of the neighboring *BSM*, as mentioned earlier. It is quite possible that there are some other proteins that functionally overlap with ABP1. Regarding SS12S (SS12S construct), SS12K (SS12K construct), and *abp1-AS* (p35S:AlcR>pAlcA:ABP1AS), they were raised from immunization and RNAi in a Col-0 background (Braun et al., 2008), and the antibody and antisense RNA raised against ABP1 may also inhibit other auxin-binding proteins with overlapping function with ABP1. Most recently, also in the Col-0 background, overexpression of ABP1 caused gain-of-function auxin-related phenotypes, which were masked by point mutations targeting the auxin-binding site of ABP1, confirming the important role of ABP1 in auxin-mediated processes (Grones et al., 2015). It is possible that the overexpressed and mutated form of ABP1 may occupy the binding site in SPIKE1 (SPK1) or TRANSFER MEMBRANE KINASE (TMK), preventing wild-type ABP1 from binding with them. After perceiving the auxin signal, ABP1 interacts, directly or indirectly, with TMK or SPK1, other unidentified proteins, which transfer the signal into the cell to the Rho of plants (ROP). ROPs interact with their effectors to regulate the endocytosis/exocytosis of the auxin efflux carrier PIN proteins to mediate PAT across the PM (Woo et al., 2002). Alternatively, mutated ABP1 may form complexes with wild-type ABP1 or functional homologues, thus masking their function, because ABP1 has been known to form dimers to perform its function (Woo et al., 2002). These findings indicate that other proteins may functionally overlap with ABP1 in Arabidopsis.

2.4.6 Other progress in understanding auxin signaling

It was found that some communications exist between target of rapamycin (TOR) and auxin signaling in plants (Deng et al., 2016). After yeast FKBP12 was introduced into DR5::GUS homozygous plants, the DR5/FKBP12 transgenic plants were treated with rapamycin or KU63794 (a new inhibitor of TOR). Staining for β-galactosidase (GUS) activity in the treated plants showed that the auxin content of the root tips was decreased in comparison with that of the control plants. The DR5/FKBP12 plants lost sensitivity to auxin after treatment with rapamycin. Auxin-defective phenotypes, including short primary roots, fewer lateral roots, and loss of gravitropism, occurred in DR5/BP12 plants when seedlings were treated with rapamycin and KU63794. These findings show that the combination of rapamycin and KU63794 significantly inhibits TOR and auxin signaling in DR5/FKBP12 plants, demonstrating that TOR is essential for auxin signaling transduction in Arabidopsis (Deng et al., 2016).

2.5 Summary points

1. The plant hormone auxin is involved in the regulation of plant growth and development, influencing cell expansion and division, cell elongation and differentiation, and a variety of physiological responses, thus significantly affecting the final shape and function of cells and tissues in all higher plants. IAA is the predominant form of auxin in plants, and most auxin studies have been mainly focused on IAA.

2. There are two forms of IAA in plants: free IAA and conjugated IAA. Free IAA is the active form of auxin, but it comprises only up to 25% of total IAA, depending on the tissue and plant species under investigation. There are two ways to produce free IAA: de novo biosynthesis and release from IAA conjugates.

3. Several Trp-dependent auxin biosynthetic pathways contribute to IAA production, including the IAOx pathway, the IAM pathway, and the IPA pathway. Trp-independent auxin biosynthesis was proposed as a major route of IAA biosynthesis in maize and Arabidopsis based on metabolite quantification and feeding experiments in Trp biosynthesis mutants.

4. IAA conjugates include MeIAA, and IAA conjugated to amino acids or sugars. Conjugated IAA and IBA are considered to be the storage forms or intermediates destined for transport and/or degradation. IAA conjugation is one of the mechanisms through which plants regulate free auxin levels.

5. Auxin transport occurs in two distinct pathways, passive diffusion through the PM and active cell-to-cell transport, depending on the protonation state of IAA. The direction of transport is governed by apical, basal, or sometimes lateral PM localization of the PIN auxin transport proteins. Another class of proteins consists of PGP, MDR, and ABCB auxin transport proteins, which have nonpolar distribution. A third class of auxin transport proteins belongs to the AUX1/LAX family of influx carriers/channels. Some newly described auxin transporters were also proposed recently, such as PILS.

6. Changes in the auxin concentration are perceived by auxin receptors to initiate auxin signaling. Auxin-responsive gene expression relies on the TIR1/AFB pathway to trigger expression of genes controlling auxin-regulated cell division, expansion, and differentiation.

2.6   Future perspectives

1. There are many IAA conjugates known in plants now. The characterization of the functions and pathways of these IAA conjugates may be important for fully understanding auxin regulation mechanisms.
2. Auxin transporters help to form local gradients in auxin concentration. There may be other mechanisms contributing to auxin concentration gradients. There is evidence to suggest that some IAA modification might help IAA to form local concentration gradients to regulate plant responses to environmental cues, for instance, gravity. Further study on the roles of local auxin transport in plant development and plant responses to environmental stimuli will help to broaden and deepen our understanding of auxin function.
3. IAA receptors (TIR1 and AFBs) perceive IAA to trigger auxin signaling. IAA is the predominant and active form of auxin in plants. Whether other forms of auxins are active or share the same signaling pathway with IAA needs further study.
4. The TIR1/AFB-dependent pathway of auxin signaling has been characterized to some extent, but transcriptional control of this pathway is not well understood. This aspect needs more investigation.
5. Studies of plant hormone signaling pathways are well advanced in Arabidopsis. Future studies should focus on understanding how these different hormone signaling pathways are integrated to regulate and control the whole life span of plants.

Abbreviations

2,4-D	2,4-Dichlorophenoxyaceticacid
AAO1	ARABIDOPSIS ALDEHYDE OXIDASE1
ABA3	ABA DEFICIENT 3
ABP1	Auxin-binding protein 1
ADP1	ALTERED DEVELOPMENT PROGRAM
ARF	AUXIN RESPONSE FACTOR
ASK	Arabidopsis SKP1-like protein
Aux/IAA	AUXIN/INDOLE-2-ACETIC ACID INDUCIBLE
AUX1/LAX	AUX1/LIKE-AUX1
BSM	BELAYA SMERT
CUL1	CULLIN
EAR	ETHYLENE-RESPONSIVE ELEMENT BINDING FACTOR-ASSOCIATED AMPHIPHILIC REPRESSION
ECH2	ENOYL-COA HYDRATASE2
eta3	enhancer of tir1 auxin resistance 3
FBPs	F-box proteins
GH3	Grethchen Hagen 3
HSP90	HEAT SHOCK FACTOR 90
IAA	Indole-3-acetic acid
iaaH	INDOLE-3-ACETAMIDE HYDROLASE
iaaM	TRYPTOPHAN 2-MONOOXY-GENASE
IAMT1	IAA carboxyl methyltransferase 1

IAR3	IAA-Ala RESISTANT3
IBA	Indole butyric acid
IDD	Indeterminate domain
IGP	Indole-3-glycerol phosphate
IGS	IGP synthase
ILL	ILR1-like
ILR1	IAA-Leu-resistant1
IND	Indole
INS	Indole synthase
LEC2	LEAFY COTYLEDON2
MES	Methyl esterases
Nedd8	RELATED TO UBIQUITIN in Arabidopsis
NGA1	NGATHA
NPA	1-N-naphthylphthalamic acid
PB1	Phox/Bem1p
PGP, MDR, and ABCB	P-GLYCOPROTEIN, MULTIDRUG RESISTANCE, and ATP-BINDING CASSETTE SUBFAMILY B
PILS	PIN-LIKES
PIN	PIN-FORMED
PLT	PLETHORA transcription factors
PTRE1	PROTEASOME REGULATOR1
REV	REVOLUTA
SGT1	SUPPRESSOR OF G2 ALLELE SKP1
SHI/STY	SHORT INTENOTES/STYLISH
TAA	TRYPTOPHAN AMINOTRANSFERASE OF ARABIDOPSIS
TAA1/SAV3/ WEI8/TIR2	TAA1/SHADE AVOIDANCE3/WEAK ETHYLENE INSENSITIVE8/TRANSPORT INHIBITOR RESPONSE2
TIR1/AFB	TRANSPORT INHIBITOR RESPONSE/AUXIN SIGNALING F-BOX
TPL	TOPLESS
TPR	TPL-RELATED
TWD1	TWISTED DWARF1
YUC	YUCCA

Acknowledgments

We are grateful to Prof. Yunde Zhao for his valuable comments and suggestions. This work is supported by Natural Science Foundation of China (Grant No 31230006 and 90717003).

References

Abel, S., Theologis, A., 1996. Early genes and auxin action. Plant Physiol. 111, 9–17.

Alvarez, J.P., Goldshmidt, A., Efroni, I., Bowman, J.L., Eshed, Y., 2009. The *NGATHA* distal organ development genes are essential for style specification in *Arabidopsis*. Plant Cell 21, 1373–1393.

Bajguz, A., Piotrowska, A., 2009. Conjugates of auxin and cytokinin. Phytochemistry 70, 957–969.

Bak, S., Nielsen, H.L., Halkier, B.A., 1998. The presence of CYP79 homologues in glucosino-late-producing plants shows evolutionary conservation of the enzymes in the conversion of amino acid to aldoxime in the biosynthesis of cyanogenic glucosides and glucosinolates. Plant Mol. Biol. 38, 725–734.

Baluska, F., Samaj, J., Menzel, D., 2003. Polar transport of auxin: carrier-mediated flux across the plasma membrane or neurotransmitter-like secretion? Trends Cell Biol. 13, 282–285.

Balzan, S., Johal, G.S., Carraro, N., 2014. The role of auxin transporters in monocots develop-ment. Front. Plant Sci. 5, 393.

Bandyopadhyay, A., Blakeslee, J.J., Lee, O.R., Mravec, J., Sauer, M., Titapiwatanakun, B., Makam, S.N., Bouchard, R., Geisler, M., Martinoia, E., et al., 2007. Interactions of PIN and PGP auxin transport mechanisms. Biochem. Soc. Trans. 35, 137–141.

Barbez, E., Kleine-Vehn, J., 2013. *Divide Et Impera*—cellular auxin compartmentalization. Curr. Opin. Plant Biol. 16, 78–84.

Barbez, E., Kubes, M., Rolcik, J., Beziat, C., Pencik, A., Wang, B., Rosquete, M.R., Zhu, J., Dobrev, P.I., Lee, Y., et al., 2012. A novel putative auxin carrier family regulates intracel-lular auxin homeostasis in plants. Nature 485, 119–122.

Bartel, B., Fink, G.R., 1995. ILR1, an amidohydrolase that releases active indole-3-acetic acid from conjugates. Science 268, 1745–1748.

Bender, R.L., Fekete, M.L., Klinkenberg, P.M., Hampton, M., Bauer, B., Malecha, M., Lindgren, K., M, J.A., Perera, M.A., Nikolau, B.J., et al., 2013. *PIN6* is required for nectary auxin response and short stamen development. Plant J. 74, 893–904.

Berendzen, K.W., Weiste, C., Wanke, D., Kilian, J., Harter, K., Droge-Laser, W., 2012. Bioinformatic cis-element analyses performed in *Arabidopsis* and rice disclose bZIP- and MYB-related binding sites as potential AuxRE-coupling elements in auxin-mediated tran-scription. BMC Plant Biol. 12, 125.

Bittner, F., Oreb, M., Mendel, R.R., 2001. ABA3 is a molybdenum cofactor sulfurase required for activation of aldehyde oxidase and xanthine dehydrogenase in *Arabidopsis thaliana*. J. Biol. Chem. 276, 40381–40384.

Blakeslee, J.J., Bandyopadhyay, A., Lee, O.R., Mravec, J., Titapiwatanakun, B., Sauer, M., Makam, S.N., Cheng, Y., Bouchard, R., Adamec, J., et al., 2007. Interactions among PIN-FORMED and P-glycoprotein auxin transporters in *Arabidopsis*. Plant Cell 19, 131–147.

Boer, D.R., Freire-Rios, A., van den Berg, W.A., Saaki, T., Manfield, I.W., Kepinski, S., Lopez-Vidrieo, I., Franco-Zorrilla, J.M., de Vries, S.C., Solano, R., et al., 2014. Structural basis for DNA binding specificity by the auxin-dependent ARF transcription factors. Cell 156, 577–589.

Boot, K.J., Hille, S.C., Libbenga, K.R., Peletier, L.A., van Spronsen, P.C., van Duijn, B., Offringa, R., 2016. Modelling the dynamics of polar auxin transport in inflorescence stems of *Arabidopsis thaliana*. J. Exp. Bot. 67, 649–666.

Bosu, D.R., Kipreos, E.T., 2008. Cullin-RING ubiquitin ligases: global regulation and activa-tion cycles. Cell Div. 3, 7.

Bottcher, C., Chapman, A., Fellermeier, F., Choudhary, M., Scheel, D., Glawischnig, E., 2014. The biosynthetic pathway of indole-3-carbaldehyde and indole-3-carboxylic acid deriva-tives in *Arabidopsis*. Plant Physiol. 165, 841–853.

Brandt, R., Salla-Martret, M., Bou-Torrent, J., Musielak, T., Stahl, M., Lanz, C., Ott, F., Schmid, M., Greb, T., Schwarz, M., et al., 2012. Genome-wide binding-site analysis of REVOLUTA reveals a link between leaf patterning and light-mediated growth responses. Plant J. 72, 31–42.

Braun, N., Wyrzykowska, J., Muller, P., David, K., Couch, D., Perrot-Rechenmann, C., Fleming, A.J., 2008. Conditional repression of AUXIN BINDING PROTEIN1 reveals that it coordinates cell division and cell expansion during postembryonic shoot development in *Arabidopsis* and tobacco. Plant Cell 20, 2746–2762.

Catlett, M.G., Kaplan, K.B., 2006. Sgt1p is a unique co-chaperone that acts as a client adaptor to link Hsp90 to Skp1p. J. Biol. Chem. 281, 33739–33748.

Causier, B., Ashworth, M., Guo, W., Davies, B., 2012. The TOPLESS interactome: a framework for gene repression in *Arabidopsis*. Plant Physiol. 158, 423–438.

Cazzonelli, C.I., Vanstraelen, M., Simon, S., Yin, K., Carron-Arthur, A., Nisar, N., Tarle, G., Cuttriss, A.J., Searle, I.R., Benkova, E., et al., 2013. Role of the Arabidopsis PIN6 auxin transporter in auxin homeostasis and auxin-mediated development. PLoS One 8, e70069.

Chapman, E.J., Estelle, M., 2009. Mechanism of auxin-regulated gene expression in plants. Annu. Rev. Genet. 43, 265–285.

Chen, J.G., Ullah, H., Young, J.C., Sussman, M.R., Jones, A.M., 2001. ABP1 is required for organized cell elongation and division in *Arabidopsis* embryogenesis. Genes Dev. 15, 902–911.

Chen, L., Tong, J., Xiao, L., Ruan, Y., Liu, J., Zeng, M., Huang, H., Wang, J.W., Xu, L., 2016. *YUCCA*-mediated auxin biogenesis is required for cell fate transition occurring during *de novo* root organogenesis in *Arabidopsis*. J. Exp. Bot.

Cheng, Y., Dai, X., Zhao, Y., 2006. Auxin biosynthesis by the YUCCA flavin monooxygenases controls the formation of floral organs and vascular tissues in *Arabidopsis*. Genes Dev. 20, 1790–1799.

Cho, M., Cho, H.T., 2013. The function of ABCB transporters in auxin transport. Plant Signal. Behav. 8, e22990.

Cho, M., Lee, Z.W., Cho, H.T., 2012. ATP-binding cassette B4, an auxin-efflux transporter, stably associates with the plasma membrane and shows distinctive intracellular trafficking from that of PIN-FORMED proteins. Plant Physiol. 159, 642–654.

Christie, J.M., Yang, H., Richter, G.L., Sullivan, S., Thomson, C.E., Lin, J., Titapiwatanakun, B., Ennis, M., Kaiserli, E., Lee, O.R., et al., 2011. phot1 inhibition of ABCB19 primes lateral auxin fluxes in the shoot apex required for phototropism. PLoS Biol. 9, e1001076.

Chuang, H.W., Zhang, W., Gray, W.M., 2004. Arabidopsis ETA2, an apparent ortholog of the human cullin-interacting protein CAND1, is required for auxin responses mediated by the SCF^{TIR1} ubiquitin ligase. Plant Cell 16, 1883–1897.

Cui, D., Zhao, J., Jing, Y., Fan, M., Liu, J., Wang, Z., Xin, W., Hu, Y., 2013. The *Arabidopsis* IDD14, IDD15, and IDD16 cooperatively regulate lateral organ morphogenesis and gravitropism by promoting auxin biosynthesis and transport. PLoS Genet. 9, e1003759.

Dai, X., Zhang, Y., Zhang, D., Chen, J., Gao, X., Estelle, M., Zhao, Y., 2015. Embryonic lethality of *Arabidopsis abp1-1* is caused by deletion of the adjacent BSM gene. Nat. Plants 1.

Darwin, C., 1880. The Power of Movement in Plants. John Murray, London.

Davies, R.T., Goetz, D.H., Lasswell, J., Anderson, M.N., Bartel, B., 1999. IAR3 encodes an auxin conjugate hydrolase from *Arabidopsis*. Plant Cell 11, 365–376.

del Pozo, J.C., Estelle, M., 1999. The *Arabidopsis* cullin AtCUL1 is modified by the ubiquitin-related protein RUB1. Proc. Natl. Acad. Sci. U.S.A. 96, 15342–15347.

Delker, C., Poschl, Y., Raschke, A., Ullrich, K., Ettingshausen, S., Hauptmann, V., Grosse, I., Quint, M., 2010. Natural variation of transcriptional auxin response networks in *Arabidopsis thaliana*. Plant Cell 22, 2184–2200.

Deng, K., Yu, L., Zheng, X., Zhang, K., Wang, W., Dong, P., Zhang, J., Ren, M., 2016. Target of rapamycin is a key player for auxin signaling transduction in *Arabidopsis*. Front. Plant Sci. 7, 291.

Deshaies, R.J., Joazeiro, C.A., 2009. RING domain E3 ubiquitin ligases. Annu. Rev. Biochem. 78, 399–434.

Dinesh, D.C., Villalobos, L.I., Abel, S., 2016. Structural biology of nuclear auxin action. Trends Plant Sci. 21, 302–316.

Ding, Z., Wang, B., Moreno, I., Duplakova, N., Simon, S., Carraro, N., Reemmer, J., Pencik, A., Chen, X., Tejos, R., et al., 2012. ER-localized auxin transporter PIN8 regulates auxin homeostasis and male gametophyte development in *Arabidopsis*. Nat. Commun. 3, 941.

Duda, D.M., Scott, D.C., Calabrese, M.F., Zimmerman, E.S., Zheng, N., Schulman, B.A., 2011. Structural regulation of cullin-RING ubiquitin ligase complexes. Curr. Opin. Struct. Biol. 21, 257–264.

Eklund, D.M., Staldal, V., Valsecchi, I., Cierlik, I., Eriksson, C., Hiratsu, K., Ohme-Takagi, M., Sundstrom, J.F., Thelander, M., Ezcurra, I., et al., 2010. The *Arabidopsis thaliana* STYLISH1 protein acts as a transcriptional activator regulating auxin biosynthesis. Plant Cell 22, 349–363.

Enchev, R.I., Scott, D.C., da Fonseca, P.C., Schreiber, A., Monda, J.K., Schulman, B.A., Peter, M., Morris, E.P., 2012. Structural basis for a reciprocal regulation between SCF and CSN. Cell Rep. 2, 616–627.

Enders, T.A., Strader, L.C., 2015. Auxin activity: past, present, and future. Am. J. Bot. 102, 180–196.

Estelle, M., 1998. Polar auxin transport. New support for an old model. Plant Cell 10, 1775–1778.

Farcot, E., Lavedrine, C., Vernoux, T., 2015. A modular analysis of the auxin signalling network. PLoS One 10, e0122231.

Feraru, E., Vosolsobe, S., Feraru, M.I., Petrasek, J., Kleine-Vehn, J., 2012. Evolution and structural diversification of PILS putative auxin carriers in plants. Front. Plant Sci. 3, 227.

Forouhar, F., Yang, Y., Kumar, D., Chen, Y., Fridman, E., Park, S.W., Chiang, Y., Acton, T.B., Montelione, G.T., Pichersky, E., et al., 2005. Structural and biochemical studies identify tobacco SABP2 as a methyl salicylate esterase and implicate it in plant innate immunity. Proc. Natl. Acad. Sci. U.S.A. 102, 1773–1778.

Gagne, J.M., Downes, B.P., Shiu, S.H., Durski, A.M., Vierstra, R.D., 2002. The F-box subunit of the SCF E3 complex is encoded by a diverse superfamily of genes in *Arabidopsis*. Proc. Natl. Acad. Sci. U.S.A. 99, 11519–11524.

Galan, J.M., Peter, M., 1999. Ubiquitin-dependent degradation of multiple F-box proteins by an autocatalytic mechanism. Proc. Natl. Acad. Sci. U.S.A. 96, 9124–9129.

Ganguly, A., Lee, S.H., Cho, H.T., 2012. Functional identification of the phosphorylation sites of Arabidopsis PIN-FORMED3 for its subcellular localization and biological role. Plant J. 71, 810–823.

Ganguly, A., Lee, S.H., Cho, M., Lee, O.R., Yoo, H., Cho, H.T., 2010. Differential auxin-transporting activities of PIN-FORMED proteins in *Arabidopsis* root hair cells. Plant Physiol. 153, 1046–1061.

Ganguly, A., Park, M., Kesawat, M.S., Cho, H.T., 2014. Functional analysis of the hydrophilic loop in intracellular trafficking of *Arabidopsis* PIN-FORMED proteins. Plant Cell 26, 1570–1585.

Gao, Y., Zhang, Y., Zhang, D., Dai, X., Estelle, M., Zhao, Y., 2015. Auxin binding protein 1 (ABP1) is not required for either auxin signaling or *Arabidopsis* development. Proc. Natl. Acad. Sci. U.S.A. 112, 2275–2280.

Gray, W.M., del Pozo, J.C., Walker, L., Hobbie, L., Risseeuw, E., Banks, T., Crosby, W.L., Yang, M., Ma, H., Estelle, M., 1999. Identification of an SCF ubiquitin-ligase complex required for auxin response in *Arabidopsis thaliana*. Genes Dev. 13, 1678–1691.

Gray, W.M., Muskett, P.R., Chuang, H.W., Parker, J.E., 2003. Arabidopsis SGT1b is required for SCF[TIR1]-mediated auxin response. Plant Cell 15, 1310–1319.

Grones, P., Chen, X., Simon, S., Kaufmann, W.A., De Rycke, R., Nodzynski, T., Zazimalova, E., Friml, J., 2015. Auxin-binding pocket of ABP1 is crucial for its gain-of-function cellular and developmental roles. J. Exp. Bot. 66, 5055–5065.

Guilfoyle, T.J., 2015. The PB1 domain in auxin response factor and Aux/IAA proteins: a versatile protein interaction module in the auxin response. Plant Cell 27, 33–43.

Guilfoyle, T.J., Hagen, G., 2007. Auxin response factors. Curr. Opin. Plant Biol. 10, 453–460.

Guilfoyle, T.J., Hagen, G., 2012. Getting a grasp on domain III/IV responsible for auxin response factor–IAA protein interactions. Plant Sci. 190, 82–88.

Habets, M.E., Offringa, R., 2015. Auxin binding protein 1: a red herring after all? Mol. Plant 8, 1131–1134.

He, X., Szewczyk, P., Karyakin, A., Evin, M., Hong, W.X., Zhang, Q., Chang, G., 2010. Structure of a cation-bound multidrug and toxic compound extrusion transporter. Nature 467, 991–994.

Hertel, R., Thomson, K.S., Russo, V.E., 1972. In vitro auxin binding to particulate cell fractions from corn coleoptiles. Planta 107, 325–340.

Hua, Z., Vierstra, R.D., 2011. The cullin-RING ubiquitin-protein ligases. Annu. Rev. Plant Biol. 62, 299–334.

Hull, A.K., Vij, R., Celenza, J.L., 2000. Arabidopsis cytochrome P450s that catalyze the first step of tryptophan-dependent indole-3-acetic acid biosynthesis. Proc. Natl. Acad. Sci. U.S.A. 97, 2379–2384.

Irmisch, S., Zeltner, P., Handrick, V., Gershenzon, J., Kollner, T.G., 2015. The maize cytochrome P450 CYP79A61 produces phenylacetaldoxime and indole-3-acetaldoxime in heterologous systems and might contribute to plant defense and auxin formation. BMC Plant Biol. 15, 128.

Jing, H., Yang, X., Zhang, J., Liu, X., Zheng, H., Dong, G., Nian, J., Feng, J., Xia, B., Qian, Q., et al., 2015. Peptidyl-prolyl isomerization targets rice Aux/IAAs for proteasomal degradation during auxin signalling. Nat. Commun. 6, 7395.

Kagale, S., Rozwadowski, K., 2011. EAR motif-mediated transcriptional repression in plants: an underlying mechanism for epigenetic regulation of gene expression. Epigenetics 6, 141–146.

Kai, K., Horita, J., Wakasa, K., Miyagawa, H., 2007. Three oxidative metabolites of indole-3-acetic acid from Arabidopsis thaliana. Phytochemistry 68, 1651–1663.

Kamimoto, Y., Terasaka, K., Hamamoto, M., Takanashi, K., Fukuda, S., Shitan, N., Sugiyama, A., Suzuki, H., Shibata, D., Wang, B., et al., 2012. Arabidopsis ABCB21 is a facultative auxin importer/exporter regulated by cytoplasmic auxin concentration. Plant Cell Physiol. 53, 2090–2100.

Kaneda, M., Schuetz, M., Lin, B.S., Chanis, C., Hamberger, B., Western, T.L., Ehlting, J., Samuels, A.L., 2011. ABC transporters coordinately expressed during lignification of Arabidopsis stems include a set of ABCBs associated with auxin transport. J. Exp. Bot. 62, 2063–2077.

Kang, J., Park, J., Choi, H., Burla, B., Kretzschmar, T., Lee, Y., Martinoia, E., 2011. Plant ABC transporters. Arabidopsis Book 9, e0153.

Karagoz, G.E., Rudiger, S.G., 2015. Hsp90 interaction with clients. Trends Biochem. Sci. 40, 117–125.

Kasahara, H., 2015. Current aspects of auxin biosynthesis in plants. Biosci. Biotechnol. Biochem. 80, 34–42.

Kepinski, S., Leyser, O., 2005. The Arabidopsis F-box protein TIR1 is an auxin receptor. Nature 435, 446–451.

Kim, H.B., Lee, H., Oh, C.J., Lee, N.H., An, C.S., 2007. Expression of EuNOD-ARP1 encoding auxin-repressed protein homolog is upregulated by auxin and localized to the fixation zone in root nodules of Elaeagnus umbellata. Mol. Cells 23, 115–121.

Kim, J., Harter, K., Theologis, A., 1997. Protein–protein interactions among the Aux/IAA proteins. Proc. Natl. Acad. Sci. U.S.A. 94, 11786–11791.

Kinoshita, N., Wang, H., Kasahara, H., Liu, J., Macpherson, C., Machida, Y., Kamiya, Y., Hannah, M.A., Chua, N.H., 2012. *IAA-Ala Resistant3*, an evolutionarily conserved target of miR167, mediates *Arabidopsis* root architecture changes during high osmotic stress. Plant Cell 24, 3590–3602.

Koiwai, H., Akaba, S., Seo, M., Komano, T., Koshiba, T., 2000. Functional expression of two *Arabidopsis* aldehyde oxidases in the yeast *Pichia pastoris*. J. Biochem. 127, 659–664.

Kong, X., Zhang, L., Ding, Z., 2016. 26S Proteasome: Hunter and Prey in auxin signaling. Trends Plant Sci. 21, 546–548.

Korasick, D.A., Enders, T.A., Strader, L.C., 2013. Auxin biosynthesis and storage forms. J. Exp. Bot. 64, 2541–2555.

Korasick, D.A., Jez, J.M., Strader, L.C., 2015. Refining the nuclear auxin response pathway through structural biology. Curr. Opin. Plant Biol. 27, 22–28.

Kowalczyk, M., Sandberg, G., 2001. Quantitative analysis of indole-3-acetic acid metabolites in *Arabidopsis*. Plant Physiol. 127, 1845–1853.

Kramer, E.M., Bennett, M.J., 2006. Auxin transport: a field in flux. Trends Plant Sci. 11, 382–386.

Krecek, P., Skupa, P., Libus, J., Naramoto, S., Tejos, R., Friml, J., Zazimalova, E., 2009. The PIN-FORMED (PIN) protein family of auxin transporters. Genome Biol. 10, 249.

Kriechbaumer, V., Botchway, S.W., Hawes, C., 2016. Localization and interactions between Arabidopsis auxin biosynthetic enzymes in the TAA/YUC-dependent pathway. J. Exp. Bot. 67, 4195–4207.

Krogan, N.T., Ckurshumova, W., Marcos, D., Caragea, A.E., Berleth, T., 2012. Deletion of MP/ARF5 domains III and IV reveals a requirement for Aux/IAA regulation in Arabidopsis leaf vascular patterning. New Phytol. 194, 391–401.

LeClere, S., Tellez, R., Rampey, R.A., Matsuda, S.P., Bartel, B., 2002. Characterization of a family of IAA-amino acid conjugate hydrolases from *Arabidopsis*. J. Biol. Chem. 277, 20446–20452.

Li, H., Tiwari, S.B., Hagen, G., Guilfoyle, T.J., 2011. Identical amino acid substitutions in the repression domain of auxin/indole-3-acetic acid proteins have contrasting effects on auxin signaling. Plant Physiol. 155, 1252–1263.

Li, J., Li, R., Jiang, Z., Gu, H., Qu, L.-J., 2015. ADP1 affects abundance and endocytosis of PIN-FORMED proteins in *Arabidopsis*. Plant Signal. Behav. 10, e973811.

Li, L., Hou, X., Tsuge, T., Ding, M., Aoyama, T., Oka, A., Gu, H., Zhao, Y., Qu, L.-J., 2008. The possible action mechanisms of indole-3-acetic acid methyl ester in *Arabidopsis*. Plant Cell Rep. 27, 575–584.

Li, R., Li, J., Li, S., Qin, G., Novak, O., Pencik, A., Ljung, K., Aoyama, T., Liu, J., Murphy, A., et al., 2014. ADP1 affects plant architecture by regulating local auxin biosynthesis. PLoS Genet. 10, e1003954.

Lin, R., Wang, H., 2005. Two homologous ATP-binding cassette transporter proteins, AtMDR1 and AtPGP1, regulate *Arabidopsis* photomorphogenesis and root development by mediating polar auxin transport. Plant Physiol. 138, 949–964.

Liu, K., Kang, B.C., Jiang, H., Moore, S.L., Li, H., Watkins, C.B., Setter, T.L., Jahn, M.M., 2005. A *GH3*-like gene, *CcGH3*, isolated from *Capsicum chinense* L. fruit is regulated by auxin and ethylene. Plant Mol. Biol. 58, 447–464.

Ljung, K., 2013. Auxin metabolism and homeostasis during plant development. Development 140, 943–950.

Lobler, M., Klambt, D., 1985. Auxin-binding protein from coleoptile membranes of corn (*Zea mays* L.). II. Localization of a putative auxin receptor. J. Biol. Chem. 260, 9854–9859.

Long, J.A., Ohno, C., Smith, Z.R., Meyerowitz, E.M., 2006. TOPLESS regulates apical embry-onic fate in *Arabidopsis*. Science 312, 1520–1523.

Ludwig-Muller, J., 2011. Auxin conjugates: their role for plant development and in the evolu-tion of land plants. J. Exp. Bot. 62, 1757–1773.

Ma, C.P., Slaughter, C.A., DeMartino, G.N., 1992. Identification, purification, and character-ization of a protein activator (PA28) of the 20 S proteasome (macropain). J. Biol. Chem. 267, 10515–10523.

Mano, Y., Nemoto, K., 2012. The pathway of auxin biosynthesis in plants. J. Exp. Bot. 63, 2853–2872.

Mashiguchi, K., Tanaka, K., Sakai, T., Sugawara, S., Kawaide, H., Natsume, M., Hanada, A., Yaeno, T., Shirasu, K., Yao, H., et al., 2011. The main auxin biosynthesis pathway in *Arabidopsis*. Proc. Natl. Acad. Sci. U.S.A. 108, 18512–18517.

McCutchen-Maloney, S.L., Matsuda, K., Shimbara, N., Binns, D.D., Tanaka, K., Slaughter, C.A., DeMartino, G.N., 2000. cDNA cloning, expression, and functional characterization of PI31, a proline-rich inhibitor of the proteasome. J. Biol. Chem. 275, 18557–18565.

Michalko, J., Dravecka, M., Bollenbach, T., Friml, J., 2015. Embryo-lethal phenotypes in early *abp1* mutants are due to disruption of the neighboring *BSM* gene. F1000Res. 4, 1104.

Michalko, J., Glanc, M., Perrot-Rechenmann, C., Friml, J., 2016. Strong morphological defects in conditional *Arabidopsis abp1* knock-down mutants generated in absence of functional ABP1 protein. F1000Res. 5, 86.

Mikkelsen, M.D., Hansen, C.H., Wittstock, U., Halkier, B.A., 2000. Cytochrome P450 CYP79B2 from *Arabidopsis* catalyzes the conversion of tryptophan to indole-3-acetaldoxime, a pre-cursor of indole glucosinolates and indole-3-acetic acid. J. Biol. Chem. 275, 33712–33717.

Mironova, V.V., Omelyanchuk, N.A., Wiebe, D.S., Levitsky, V.G., 2014. Computational analy-sis of auxin responsive elements in the *Arabidopsis thaliana* L. genome. BMC Genomics 15 (Suppl. 12), S4.

Mittler, R., Finka, A., Goloubinoff, P., 2012. How do plants feel the heat? Trends Biochem. Sci. 37, 118–125.

Mravec, J., Skupa, P., Bailly, A., Hoyerova, K., Krecek, P., Bielach, A., Petrasek, J., Zhang, J., Gaykova, V., Stierhof, Y.D., et al., 2009. Subcellular homeostasis of phytohormone auxin is mediated by the ER-localized PIN5 transporter. Nature 459, 1136–1140.

Muller, A., Weiler, E.W., 2000. Indolic constituents and indole-3-acetic acid biosynthesis in the wild-type and a tryptophan auxotroph mutant of *Arabidopsis thaliana*. Planta 211, 855–863.

Nagashima, A., Uehara, Y., Sakai, T., 2008. The ABC subfamily B auxin transporter AtABCB19 is involved in the inhibitory effects of *N*-1-naphthyphthalamic acid on the phototropic and gravitropic responses of *Arabidopsis* hypocotyls. Plant Cell Physiol. 49, 1250–1255.

Nakazawa, M., Yabe, N., Ichikawa, T., Yamamoto, Y.Y., Yoshizumi, T., Hasunuma, K., Matsui, M., 2001. *DFL1*, an auxin-responsive *GH3* gene homologue, negatively regulates shoot cell elongation and lateral root formation, and positively regulates the light response of hypocotyl length. Plant J. 25, 213–221.

Napier, R.M., David, K.M., Perrot-Rechenmann, C., 2002. A short history of auxin-binding proteins. Plant Mol. Biol. 49, 339–348.

Nemoto, K., Hara, M., Suzuki, M., Seki, H., Muranaka, T., Mano, Y., 2009. The *NtAMI1* gene functions in cell division of tobacco BY-2 cells in the presence of indole-3-acetamide. FEBS Lett. 583, 487–492.

Noh, B., Murphy, A.S., Spalding, E.P., 2001. Multidrug resistance-like genes of *Arabidopsis* required for auxin transport and auxin-mediated development. Plant Cell 13, 2441–2454.

Normanly, J., Cohen, J.D., Fink, G.R., 1993. *Arabidopsis thaliana* auxotrophs reveal a tryptophan-independent biosynthetic pathway for indole-3-acetic acid. Proc. Natl. Acad. Sci. U.S.A. 90, 10355–10359.

Normanly, J., Slovin, J.P., Cohen, J.D., 1995. Rethinking auxin biosynthesis and metabolism. Plant Physiol. 107, 323–329.

Novak, O., Henykova, E., Sairanen, I., Kowalczyk, M., Pospisil, T., Ljung, K., 2012. Tissue-specific profiling of the *Arabidopsis thaliana* auxin metabolome. Plant J. 72, 523–536.

Ostin, A., Kowalyczk, M., Bhalerao, R.P., Sandberg, G., 1998. Metabolism of indole-3-acetic acid in *Arabidopsis*. Plant Physiol. 118, 285–296.

Ouyang, J., Shao, X., Li, J., 2000. Indole-3-glycerol phosphate, a branch point of indole-3-acetic acid biosynthesis from the tryptophan biosynthetic pathway in *Arabidopsis thaliana*. Plant J. 24, 327–333.

Overvoorde, P.J., Okushima, Y., Alonso, J.M., Chan, A., Chang, C., Ecker, J.R., Hughes, B., Liu, A., Onodera, C., Quach, H., et al., 2005. Functional genomic analysis of the *AUXIN/INDOLE-3-ACETIC ACID* gene family members in *Arabidopsis thaliana*. Plant Cell 17, 3282–3300.

Paponov, I.A., Paponov, M., Teale, W., Menges, M., Chakrabortee, S., Murray, J.A., Palme, K., 2008. Comprehensive transcriptome analysis of auxin responses in *Arabidopsis*. Mol. Plant 1, 321–337.

Peret, B., Swarup, K., Ferguson, A., Seth, M., Yang, Y., Dhondt, S., James, N., Casimiro, I., Perry, P., Syed, A., et al., 2012. *AUX/LAX* genes encode a family of auxin influx transporters that perform distinct functions during *Arabidopsis* development. Plant Cell 24, 2874–2885.

Petrasek, J., Mravec, J., Bouchard, R., Blakeslee, J.J., Abas, M., Seifertova, D., Wisniewska, J., Tadele, Z., Kubes, M., Covanova, M., et al., 2006. PIN proteins perform a rate-limiting function in cellular auxin efflux. Science 312, 914–918.

Pierce, N.W., Lee, J.E., Liu, X., Sweredoski, M.J., Graham, R.L., Larimore, E.A., Rome, M., Zheng, N., Clurman, B.E., Hess, S., et al., 2013. Cand1 promotes assembly of new SCF complexes through dynamic exchange of F box proteins. Cell 153, 206–215.

Pinon, V., Prasad, K., Grigg, S.P., Sanchez-Perez, G.F., Scheres, B., 2013. Local auxin biosynthesis regulation by PLETHORA transcription factors controls phyllotaxis in *Arabidopsis*. Proc. Natl. Acad. Sci. U.S.A. 110, 1107–1112.

Pollmann, S., Neu, D., Lehmann, T., Berkowitz, O., Schafer, T., Weiler, E.W., 2006. Subcellular localization and tissue specific expression of amidase 1 from *Arabidopsis thaliana*. Planta 224, 1241–1253.

Pollmann, S., Neu, D., Weiler, E.W., 2003. Molecular cloning and characterization of an amidase from *Arabidopsis thaliana* capable of converting indole-3-acetamide into the plant growth hormone, indole-3-acetic acid. Phytochemistry 62, 293–300.

Powers, S.K., Strader, L.C., 2016. Up in the air: untethered factors of auxin response. F1000Res. 5, 133.

Qin, G., Gu, H., Zhao, Y., Ma, Z., Shi, G., Yang, Y., Pichersky, E., Chen, H., Liu, M., Chen, Z., et al., 2005. An indole-3-acetic acid carboxyl methyltransferase regulates *Arabidopsis* leaf development. Plant Cell 17, 2693–2704.

Rampey, R.A., LeClere, S., Kowalczyk, M., Ljung, K., Sandberg, G., Bartel, B., 2004. A family of auxin-conjugate hydrolases that contributes to free indole-3-acetic acid levels during *Arabidopsis* germination. Plant Physiol. 135, 978–988.

Saibil, H., 2013. Chaperone machines for protein folding, unfolding and disaggregation. Nat. Rev. Mol. Cell Biol. 14, 630–642.

Salehin, M., Bagchi, R., Estelle, M., 2015. SCF^{TIR1}/^{AFB}-based auxin perception: mechanism and role in plant growth and development. Plant Cell 27, 9–19.

Sanchez Carranza, A.P., Singh, A., Steinberger, K., Panigrahi, K., Palme, K., Dovzhenko, A., Dal Bosco, C., 2016. Hydrolases of the ILR1-like family of *Arabidopsis thaliana* modulate auxin response by regulating auxin homeostasis in the endoplasmic reticulum. Sci. Rep. 6, 24212.

Santelia, D., Vincenzetti, V., Azzarello, E., Bovet, L., Fukao, Y., Duchtig, P., Mancuso, S., Martinoia, E., Geisler, M., 2005. MDR-like ABC transporter AtPGP4 is involved in aux-in-mediated lateral root and root hair development. FEBS Lett. 579, 5399–5406.

Sawchuk, M.G., Edgar, A., Scarpella, E., 2013. Patterning of leaf vein networks by convergent auxin transport pathways. PLoS Genet. 9, e1003294.

Schwechheimer, C., Serino, G., Callis, J., Crosby, W.L., Lyapina, S., Deshaies, R.J., Gray, W.M., Estelle, M., Deng, X.W., 2001. Interactions of the COP9 signalosome with the E3 ubiquitin ligase SCF[TIR1] in mediating auxin response. Science 292, 1379–1382.

Seo, M., Akaba, S., Oritani, T., Delarue, M., Bellini, C., Caboche, M., Koshiba, T., 1998. Higher activity of an aldehyde oxidase in the auxin-overproducing superroot1 mutant of *Arabidopsis thaliana*. Plant Physiol. 116, 687–693.

Shirasu, K., 2009. The HSP90-SGT1 chaperone complex for NLR immune sensors. Annu. Rev. Plant Biol. 60, 139–164.

Sohlberg, J.J., Myrenas, M., Kuusk, S., Lagercrantz, U., Kowalczyk, M., Sandberg, G., Sundberg, E., 2006. STY1 regulates auxin homeostasis and affects apical-basal patterning of the *Arabidopsis* gynoecium. Plant J. 47, 112–123.

Staswick, P.E., 2009. The tryptophan conjugates of jasmonic and indole-3-acetic acids are endogenous auxin inhibitors. Plant Physiol. 150, 1310–1321.

Staswick, P.E., Serban, B., Rowe, M., Tiryaki, I., Maldonado, M.T., Maldonado, M.C., Suza, W., 2005. Characterization of an Arabidopsis enzyme family that conjugates amino acids to indole-3-acetic acid. Plant Cell 17, 616–627.

Stepanova, A.N., Robertson-Hoyt, J., Yun, J., Benavente, L.M., Xie, D.Y., Dolezal, K., Schlereth, A., Jurgens, G., Alonso, J.M., 2008. TAA1-mediated auxin biosynthesis is essential for hormone crosstalk and plant development. Cell 133, 177–191.

Stone, S.L., Braybrook, S.A., Paula, S.L., Kwong, L.W., Meuser, J., Pelletier, J., Hsieh, T.F., Fischer, R.L., Goldberg, R.B., Harada, J.J., 2008. *Arabidopsis* LEAFY COTYLEDON2 induces maturation traits and auxin activity: implications for somatic embryogenesis. Proc. Natl. Acad. Sci. U.S.A. 105, 3151–3156.

Strader, L.C., Bartel, B., 2011. Transport and metabolism of the endogenous auxin precursor indole-3-butyric acid. Mol. Plant 4, 477–486.

Strader, L.C., Culler, A.H., Cohen, J.D., Bartel, B., 2010. Conversion of endogenous indole-3-butyric acid to indole-3-acetic acid drives cell expansion in *Arabidopsis* seedlings. Plant Physiol. 153, 1577–1586.

Strader, L.C., Wheeler, D.L., Christensen, S.E., Berens, J.C., Cohen, J.D., Rampey, R.A., Bartel, B., 2011. Multiple facets of Arabidopsis seedling development require indole-3-butyric acid-derived auxin. Plant Cell 23, 984–999.

Sugawara, S., Hishiyama, S., Jikumaru, Y., Hanada, A., Nishimura, T., Koshiba, T., Zhao, Y., Kamiya, Y., Kasahara, H., 2009. Biochemical analyses of indole-3-acetaldoxime-dependent auxin biosynthesis in *Arabidopsis*. Proc. Natl. Acad. Sci. U.S.A. 106, 5430–5435.

Sumimoto, H., Kamakura, S., Ito, T., 2007. Structure and function of the PB1 domain, a protein interaction module conserved in animals, fungi, amoebas, and plants. Sci. STKE 2007, re6.

Swarup, K., Benkova, E., Swarup, R., Casimiro, I., Peret, B., Yang, Y., Parry, G., Nielsen, E., De Smet, I., Vanneste, S., et al., 2008. The auxin influx carrier LAX3 promotes lateral root emergence. Nat. Cell Biol. 10, 946–954.

Szemenyei, H., Hannon, M., Long, J.A., 2008. TOPLESS mediates auxin-dependent transcriptional repression during *Arabidopsis* embryogenesis. Science 319, 1384–1386.

Takahashi, A., Casais, C., Ichimura, K., Shirasu, K., 2003. HSP90 interacts with RAR1 and SGT1 and is essential for RPS2-mediated disease resistance in *Arabidopsis*. Proc. Natl. Acad. Sci. U.S.A. 100, 11777–11782.

Takase, T., Nakazawa, M., Ishikawa, A., Kawashima, M., Ichikawa, T., Takahashi, N., Shimada, H., Manabe, K., Matsui, M., 2004. *ydk1-D*, an auxin-responsive *GH3* mutant that is involved in hypocotyl and root elongation. Plant J. 37, 471–483.

Takase, T., Nakazawa, M., Ishikawa, A., Manabe, K., Matsui, M., 2003. *DFL2*, a new member of the *Arabidopsis GH3* gene family, is involved in red light-specific hypocotyl elongation. Plant Cell Physiol. 44, 1071–1080.

Tan, X., Calderon-Villalobos, L.I., Sharon, M., Zheng, C., Robinson, C.V., Estelle, M., Zheng, N., 2007. Mechanism of auxin perception by the TIR1 ubiquitin ligase. Nature 446, 640–645.

Tao, Y., Ferrer, J.L., Ljung, K., Pojer, F., Hong, F., Long, J.A., Li, L., Moreno, J.E., Bowman, M.E., Ivans, L.J., et al., 2008. Rapid synthesis of auxin via a new tryptophan-dependent pathway is required for shade avoidance in plants. Cell 133, 164–176.

Teale, W.D., Paponov, I.A., Palme, K., 2006. Auxin in action: signalling, transport and the control of plant growth and development. Nat. Rev. Mol. Cell Biol. 7, 847–859.

Terol, J., Domingo, C., Talon, M., 2006. The *GH3* family in plants: genome wide analysis in rice and evolutionary history based on EST analysis. Gene 371, 279–290.

Tian, C.E., Muto, H., Higuchi, K., Matamura, T., Tatematsu, K., Koshiba, T., Yamamoto, K.T., 2004. Disruption and overexpression of *auxin response factor 8* gene of *Arabidopsis* affect hypocotyl elongation and root growth habit, indicating its possible involvement in auxin homeostasis in light condition. Plant J. 40, 333–343.

Tivendale, N.D., Ross, J.J., Cohen, J.D., 2014. The shifting paradigms of auxin biosynthesis. Trends Plant Sci. 19, 44–51.

Trigueros, M., Navarrete-Gomez, M., Sato, S., Christensen, S.K., Pelaz, S., Weigel, D., Yanofsky, M.F., Ferrandiz, C., 2009. The *NGATHA* genes direct style development in the *Arabidopsis* gynoecium. Plant Cell 21, 1394–1409.

Tromas, A., Paponov, I., Perrot-Rechenmann, C., 2010. AUXIN BINDING PROTEIN 1: functional and evolutionary aspects. Trends Plant Sci. 15, 436–446.

Ulmasov, T., Hagen, G., Guilfoyle, T.J., 1997. ARF1, a transcription factor that binds to auxin response elements. Science 276, 1865–1868.

Ulmasov, T., Hagen, G., Guilfoyle, T.J., 1999. Dimerization and DNA binding of auxin response factors. Plant J. 19, 309–319.

Vernoux, T., Brunoud, G., Farcot, E., Morin, V., Van den Daele, H., Legrand, J., Oliva, M., Das, P., Larrieu, A., Wells, D., et al., 2011. The auxin signalling network translates dynamic input into robust patterning at the shoot apex. Mol. Syst. Biol. 7, 508.

Viaene, T., Delwiche, C.F., Rensing, S.A., Friml, J., 2013. Origin and evolution of PIN auxin transporters in the green lineage. Trends Plant Sci. 18, 5–10.

Vogel, G., 2006. Plant science. Auxin begins to give up its secrets. Science 313, 1230–1231.

Wang, B., Chu, J., Yu, T., Xu, Q., Sun, X., Yuan, J., Xiong, G., Wang, G., Wang, Y., Li, J., 2015. Tryptophan-independent auxin biosynthesis contributes to early embryogenesis in *Arabidopsis*. Proc. Natl. Acad. Sci. U.S.A. 112, 4821–4826.

Wang, R., Estelle, M., 2014. Diversity and specificity: auxin perception and signaling through the TIR1/AFB pathway. Curr. Opin. Plant Biol. 21, 51–58.

Wang, R., Zhang, Y., Kieffer, M., Yu, H., Kepinski, S., Estelle, M., 2016. HSP90 regulates temperature-dependent seedling growth in *Arabidopsis* by stabilizing the auxin co-receptor F-box protein TIR1. Nat. Commun. 7, 10269.

Widemann, E., Miesch, L., Lugan, R., Holder, E., Heinrich, C., Aubert, Y., Miesch, M., Pinot, F., Heitz, T., 2013. The amidohydrolases IAR3 and ILL6 contribute to jasmonoyl-isoleucine hormone turnover and generate 12-hydroxyjasmonic acid upon wounding in *Arabidopsis* leaves. J. Biol. Chem. 288, 31701–31714.

Wiszniewski, A., Zhou, W., Smith, S., Bussell, J., 2009. Arabidopsis genes encoding a peroxiso-mal oxidoreductase-like protein and an acyl-CoA synthetase-like protein that are required for responses to pro-auxins. Plant Mol. Biol. 69, 503–515.

Wittstock, U., Halkier, B.A., 2000. Cytochrome P450 CYP79A2 from *Arabidopsis thaliana* L. Catalyzes the conversion of L-phenylalanine to phenylacetaldoxime in the biosynthesis of benzylglucosinolate. J. Biol. Chem. 275, 14659–14666.

Wojcikowska, B., Jaskola, K., Gasiorek, P., Meus, M., Nowak, K., Gaj, M.D., 2013. LEAFY COTYLEDON2 (LEC2) promotes embryogenic induction in somatic tissues of *Arabidopsis*, via YUCCA-mediated auxin biosynthesis. Planta 238, 425–440.

Woo, E.J., Marshall, J., Bauly, J., Chen, J.G., Venis, M., Napier, R.M., Pickersgill, R.W., 2002. Crystal structure of auxin-binding protein 1 in complex with auxin. EMBO J. 21, 2877–2885.

Woodward, A.W., Bartel, B., 2005. Auxin: regulation, action, and interaction. Ann. Bot. 95, 707–735.

Wright, A.D., Sampson, M.B., Neuffer, M.G., Michalczuk, L., Slovin, J.P., Cohen, J.D., 1991. Indole-3-acetic acid biosynthesis in the mutant maize orange pericarp, a tryptophan auxo-troph. Science 254, 998–1000.

Wu, G., Otegui, M.S., Spalding, E.P., 2010. The ER-localized TWD1 immunophilin is neces-sary for localization of multidrug resistance-like proteins required for polar auxin transport in *Arabidopsis* roots. Plant Cell 22, 3295–3304.

Xiong, L., Ishitani, M., Lee, H., Zhu, J.K., 2001. The *Arabidopsis LOS5/ABA3* locus encodes a molybdenum cofactor sulfurase and modulates cold stress- and osmotic stress-responsive gene expression. Plant Cell 13, 2063–2083.

Xu, J., Xue, C., Xue, D., Zhao, J., Gai, J., Guo, N., Xing, H., 2013. Overexpression of *GmHsp90s*, a *heat shock protein 90 (Hsp90)* gene family cloning from soybean, decrease damage of abiotic stresses in *Arabidopsis thaliana*. PLoS One 8, e69810.

Yamada, M., Greenham, K., Prigge, M.J., Jensen, P.J., Estelle, M., 2009. The *TRANSPORT INHIBITOR RESPONSE2* gene is required for auxin synthesis and diverse aspects of plant development. Plant Physiol. 151, 168–179.

Yamasaki, K., Kigawa, T., Seki, M., Shinozaki, K., Yokoyama, S., 2013. DNA-binding domains of plant-specific transcription factors: structure, function, and evolution. Trends Plant Sci. 18, 267–276.

Yang, B.J., Han, X.X., Yin, L.L., Xing, M.Q., Xu, Z.H., Xue, H.W., 2016. *Arabidopsis* PROTEASOME REGULATOR1 is required for auxin-mediated suppression of protea-some activity and regulates auxin signalling. Nat. Commun. 7, 11388.

Yang, H., Murphy, A.S., 2009. Functional expression and characterization of Arabidopsis ABCB, AUX 1 and PIN auxin transporters in *Schizosaccharomyces pombe*. Plant J. 59, 179–191.

Yang, Y., Xu, R., Ma, C.J., Vlot, A.C., Klessig, D.F., Pichersky, E., 2008. Inactive methyl indole-3-acetic acid ester can be hydrolyzed and activated by several esterases belonging to the AtMES esterase family of *Arabidopsis*. Plant Physiol. 147, 1034–1045.

Ye, L., Liu, L., Xing, A., Kang, D., 2013. Characterization of a dwarf mutant allele of Arabidopsis MDR-like ABC transporter *AtPGP1* gene. Biochem. Biophys. Res. Commun. 441, 782–786.

Yu, H., Zhang, Y., Moss, B.L., Bargmann, B.O., Wang, R., Prigge, M., Nemhauser, J.L., Estelle, M., 2015. Untethering the TIR1 auxin receptor from the SCF complex increases its stabil-ity and inhibits auxin response. Nat. Plants 1.

Zaiss, D.M., Standera, S., Holzhutter, H., Kloetzel, P., Sijts, A.J., 1999. The proteasome inhibitor PI31 competes with PA28 for binding to 20S proteasomes. FEBS Lett. 457, 333–338.

Zhang, M., Boter, M., Li, K., Kadota, Y., Panaretou, B., Prodromou, C., Shirasu, K., Pearl, L.H., 2008. Structural and functional coupling of Hsp90- and Sgt1-centred multi-protein complexes. EMBO J. 27, 2789–2798.

Zhang, Z., Li, Q., Li, Z., Staswick, P.E., Wang, M., Zhu, Y., He, Z., 2007. Dual regulation role of GH3.5 in salicylic acid and auxin signaling during *Arabidopsis–Pseudomonas syringae* interaction. Plant Physiol. 145, 450–464.

Zhao, N., Ferrer, J.L., Ross, J., Guan, J., Yang, Y., Pichersky, E., Noel, J.P., Chen, F., 2008. Structural, biochemical, and phylogenetic analyses suggest that indole-3-acetic acid methyltransferase is an evolutionarily ancient member of the SABATH family. Plant Physiol. 146, 455–467.

Zhao, N., Guan, J., Lin, H., Chen, F., 2007. Molecular cloning and biochemical characterization of indole-3-acetic acid methyltransferase from poplar. Phytochemistry 68, 1537–1544.

Zhao, Y., 2014. Auxin biosynthesis. Arabidopsis Book 12, e0173.

Zhao, Y., Hull, A.K., Gupta, N.R., Goss, K.A., Alonso, J., Ecker, J.R., Normanly, J., Chory, J., Celenza, J.L., 2002. Trp-dependent auxin biosynthesis in *Arabidopsis*: involvement of cytochrome P450s CYP79B2 and CYP79B3. Genes Dev. 16, 3100–3112.

Zhu, J., Bailly, A., Zwiewka, M., Sovero, V., Di Donato, M., Ge, P., Oehri, J., Aryal, B., Hao, P., Linnert, M., et al., 2016. TWISTED DWARF1 mediates the action of auxin transport inhibitors on actin cytoskeleton dynamics. Plant Cell 28, 930–948.

Cytokinins

3

Jian Feng[1], Yiting Shi[2], Shuhua Yang[2], Jianru Zuo[1]
[1]State Key Laboratory of Plant Genomics, Institute of Genetics and Developmental Biology, Chinese Academy of Science, Beijing, China; [2]State Key Laboratory of Plant Physiology and Biochemistry, China Agricultural University, Beijing, China

Summary

Cytokinin is one of the so-called classic plant growth phytohormones and functions to promote cell division and cell differentiation. Cytokinins are adenine derivatives that carry a variable side chain at the N^6 position of the purine. As a key growth-promoting phytohormone, cytokinin is involved in almost all aspects of plant growth and development. Cytokinin signaling is mediated by a two-component system involved in sequential transfer of phosphoryl groups from the receptors to downstream effectors. In Arabidopsis, upon binding cytokinin, the receptors, a small class of His kinases, are autophosphorylated at a highly conserved His residue. The phosphoryl group is subsequently transferred to PHOSPHOTRANSFER PROTEINS and then to RESPONSE REGULATORS, eventually activating the transcription of downstream effector genes, thereby turning on the signaling pathway. This chapter mainly summarizes recent progress and our current understanding on cytokinin metabolism, translocation, signal transduction, regulatory mechanisms, and physiological roles of cytokinin in plant growth and development.

3.1 Discovery and functions of cytokinins

Cytokinins are a class of plant growth hormones (phytohormones) that promote cell division and cell differentiation. Cytokinins are adenine derivatives that carry an isoprene-derived or an aromatic side chain at the N^6 position of the purine (Fig. 3.1). In the 1950s, Folke Skoog and colleagues isolated the first cytokinesis-promoting factor, kinetin, from autoclaved herring sperm DNA (Miller et al., 1955a, 1955b). Since then several other growth-promoting factors similar to kinetin have been identified from plants. The first naturally occurring cytokinin, *trans*-zeatin (*t*Z), was isolated from maize (*Zea mays* L.) endosperm (Miller, 1961). In the following years, researchers isolated compounds with cytokinin activities from many plant species (Mok and Mok, 2001). Currently, the most prevalent and most studied natural cytokinins are isopentenyladenine (iP) and *t*Z.

The discovery of cytokinins facilitated the development of methods for plant tissue culture and also had a profound effect on studies of plant biology. Cytokinins participate in regulating various processes in plant growth and development, including female gamete and embryo development, seed germination, vascular development, shoot apical meristem development, photomorphogenesis, leaf senescence, and floral development, as well as regulating adaptive responses to environmental stresses (Schmülling, 2002). In the past decades, the discovery of many enzymes that control

Hormone Metabolism and Signaling in Plants. http://dx.doi.org/10.1016/B978-0-12-811562-6.00003-7

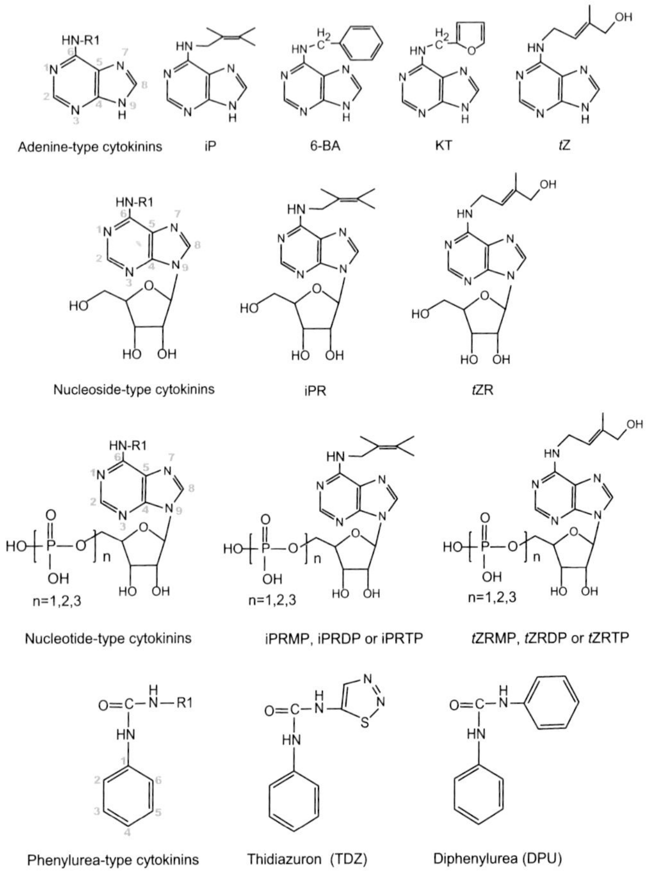

Figure 3.1 Structures of representative active cytokinin species. *6-BA*, 6-benzyl adenine; *DPU*, diphenylurea; *iP*, N^6-(Δ^2-isopentenyl)-adenine; *iPR*, iP riboside; *iPRDP*, iP riboside 5′-diphosphate; *iPRMP*, iP riboside 5′-monophosphate; *iPRTP*, iP riboside 5′-triphosphate; *KT*, kinetin; *TDZ*, thidiazuron; *tPRDP*, *trans*-zeatin-diphosphate; *tPRTP*, *trans*-zeatin-triphosphate; *tZ*, *trans*-zeatin; *tZR*, *trans*-zeatin riboside; *tZRMP*, *trans*-zeatin-riboside 5′-monophosphate.

the modification and activity of cytokinins has shed new light on the basic molecular mechanisms of cytokinin biosynthesis (Sakakibara, 2006; Werner and Schmülling, 2009; Zürcher and Müller, 2016).

3.2 Structures and types of cytokinins

In plants, the naturally occurring cytokinins are adenine derivatives. Based on the structure of the N^6-side chain, cytokinins are classified into isoprenoid cytokinins and aromatic cytokinins (Martin et al., 2001; Mok and Mok, 2001). The isoprenoid cytokinins have an isopentenyl side chain and include N^6-(Δ^2-isopentenyl)-adenine (iP) and zeatin. For zeatin, the *trans*- and *cis*-configurations of the hydroxylated side chain produce *t*Z and *cis*-zeatin (*c*Z), respectively. Aromatic cytokinins have a benzyl or hydroxybenzyl group at the N^6-position, such as kinetin, N^6-benzyl adenine (6-BA), and topolin. The amount of aromatic cytokinin is relatively low in plant tissues. The highly active endogenous plant aromatic cytokinins include *ortho*-topolin (*o*T), *meta*-topolin (*m*T), and the methoxy derivatives of 6-BA (Strnad, 1997; Tarkowska et al., 2003). Topolins are derivatives formed by substitution of a benzyl ring on the N^6-position, and the methoxytopolins are found to be the biologically active metabolites of 6-benzyl adenine (6-BA). Cytokinin derivatives in plants generally occur as nucleobases (free bases) and their conjugates with various moieties, including nucleosides, nucleotides (Fig. 3.1), and glycosides (*O*- and *N*-glycosides). In most cases, the nucleobases function as the active form and the nucleosides are inactive forms, which serve as a reservoir for cytokinin storage. In addition to the naturally occurring cytokinins, a number of nonnatural cytokinins have been synthesized, including phenylurea-type *N*, *N'*-diphenylurea, thidiazuron, adenine-type kinetin, and 6-BA (Fig. 3.1).

The activities of cytokinins vary widely depending on the plant species, tissues, and developmental stages. For example, iP and *t*Z are the major cytokinin derivatives in Arabidopsis and a moss (*Funaria hygrometrica*), whereas a high abundance of bioactive *c*Z has been observed in maize and rice (*Oryza sativa*) (Kudo et al., 2012; Mok and Mok, 2001; Spiess, 1975; Veach et al., 2003). Cytokinins with different structures or modifications have different specificities for cytokinin receptors. For example, experiments employing ^{3}H-labeled cytokinins provided direct evidence that most cytokinin receptors have a high affinity for the free bases, but low affinity for the riboside derivatives of cytokinins (Yamada et al., 2001). An Arabidopsis cytokinin receptor, AHK3, is sensitive to *t*Z and *trans*-zeatin riboside monophosphate (*t*ZRMP) (Spichal et al., 2004). However, a maize cytokinin receptor, ZmHK1, can bind *c*Z, whereas another maize cytokinin receptor, ZmHK2, recognizes *trans*-zeatin riboside (*t*ZR) and *t*Z, suggesting that each cytokinin receptor has a preference for a specific ligand (Yonekura-Sakakibara et al., 2004).

3.3 Cytokinin synthesis, metabolism and transport

3.3.1 Cytokinins biosynthesis

3.3.1.1 Cytokinins biosynthetic pathways

Plants synthesize isoprenoid cytokinins by two pathways: the primary or de novo synthesis pathway and the tRNA degradation pathway. Most cytokinins are produced by the de novo pathway (Haberer and Joseph, 2002). The first step of de novo synthesis produces a cytokinin precursor, iP riboside 5′-tri-, di-, or monophosphate (iPRTP, iPRDP, or iPRMP), by transferring the prenyl moiety from dimethylallyl diphosphate (DMAPP) to the N^6 position of ATP, ADP, or AMP (Taya et al., 1978). This reaction is the rate-limiting step of isoprenoid cytokinin biosynthesis and is catalyzed by adenylate isopentenyltransferase (IPT), an enzyme that was initially identified in the slime mold *Dictyostelium discoideum*. The first *IPT* gene, named *Tmr*, was characterized from the Ti (tumor-inducing) plasmid of the crown gall-forming bacterium *Agrobacterium tumefaciens*. The *Tmr* gene is present in the transferred DNA (T-DNA) of the Ti plasmid, which integrates into the genome of host plants to induce tumor formation (Akiyoshi et al., 1984; Barry et al., 1984). Another *IPT* gene, *Tzs*, is present in the virulence region of the nopaline-type Ti-plasmid of *A. tumefaciens* (Akiyoshi et al., 1984; Barry et al., 1984).

The Arabidopsis genome contains nine *IPT* genes (*AtIPT1* to *AtIPT9*). *AtIPT1* and *AtIPT3* through *AtIPT8* belong to the ATP/ADP IPT clade (Kakimoto, 2001, 2003a; Takei et al., 2001). Overexpression of *AtIPT4* and *AtIPT8* (also named *PGA22*) resulted in increased iP levels and a typical cytokinin response in the transgenic plant lines (Kakimoto, 2001; Sun et al., 2003). An *ipt1, 3, 5, 7* quadruple mutant showed retarded growth and decreased levels of iP, *t*Z, and their derivatives, suggesting that IPTs are major contributors to cytokinin biosynthesis (Kakimoto, 2001; Miyawaki et al., 2006; Sun et al., 2003). Similar activities for IPTs are also exhibited in petunia (*Petunia hybrida*) and hops (*Humulus lupulus* L.), indicating that the IPT-catalyzed de novo synthesis pathway is an evolutionarily conserved mechanism in plants (Sakano et al., 2004; Zubko et al., 2002).

Arabidopsis IPTs preferentially use ADP or ATP rather than AMP for the first step in cytokinin biosynthesis (Kakimoto, 2001). Therefore, most of the natural iP ribosides are iPRDP and iPRTP, indicating that iPRMP mostly derives from iPRDP and iPRTP (Fig. 3.2). From iPRMP, further hydrolytic reactions then release iP riboside (iPR) and the free base iP. Biochemical studies have revealed that there are two possible pathways for *t*Z biosynthesis, iPRMP-dependent and iPRMP-independent pathways. The first step of the iPRMP-dependent pathway of *t*Z synthesis is catalyzed by a cytochrome P450 monooxygenase CYP735A (Chen and Leisner, 1984; Takei et al., 2001, 2004b). In Arabidopsis, CYP735A1 and CYP735A2 preferentially utilize iP nucleotides but not the nucleoside or free-base forms to produce *t*Z nucleotides (Takei et al., 2004b) (Fig. 3.2). In addition, an iPRMP-independent pathway for *t*Z biosynthesis has been proposed, in which zeatin is synthesized from AMP and side chain precursor to produce zeatin by IPT (Åstot et al., 2000).

Cytokinins can also be produced from recycled tRNAs. The first step of this pathway is catalyzed by tRNA isopentenyltransferase (tRNA-IPT), an enzyme that transfers an isopentenyl unit from dimethylallyl diphosphate (DMAPP) to the N^6 of the nucleotide adjacent to the 3′-end of the anticodon in the tRNA. Two Arabidopsis genes (*AtIPT2* and *AtIPT9*) and two orthologous genes in rice (*OsIPT9* and *OsIPT10*) encode tRNA IPTs (Miyawaki et al., 2006; Sakamoto et al., 2006). A loss-of-function mutant of both *AtIPT2* and *AtIPT9* led to significantly reduced *c*ZR and *c*ZRMP contents, suggesting that tRNA degradation is the main source of *c*Z-type cytokinins (Miyawaki et al., 2006). Thus, iP- and *t*Z-type cytokinins are biosynthesized through IPTs acting on ATP and ADP, whereas *c*Z-type cytokinins are biosynthesized through the action of tRNA IPTs (Fig. 3.2).

3.3.1.2 Activation of cytokinins

Two pathways have been proposed for the activation of cytokinins from the inactive riboside forms: the one-step reaction pathway (also known as the direct activation pathway) and the two-step reaction pathway (Chen, 1997; Kurakawa et al., 2007). In the two-step reaction pathway, which is considered to be the major pathway for activation of cytokinins, nucleotides are hydrolyzed to nucleobases (iP and *t*Z) firstly by cleavage of the phosphate moiety and then by cleavage of the sugar moiety (Chen and

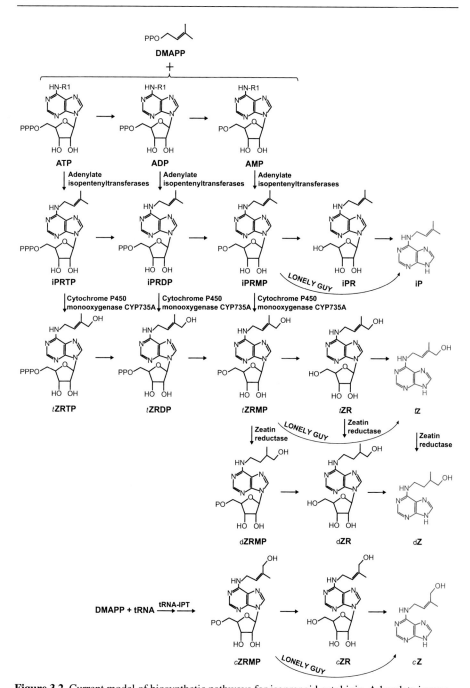

Figure 3.2 Current model of biosynthetic pathways for isoprenoid cytokinin. Adenylate isopentenyltransferases (IPTs) preferentially use ADP or ATP for the first step in cytokinin biosynthesis to form iPRTP or iPRDP. These cytokinin nucleotides are converted into the corresponding *t*Z-nucleotides, *t*ZRTP and *t*PRDP, by cytochrome P450 monooxygenase CYP735A. iPRMP and *t*ZRTP are converted to iP and *t*Z by cytokinin riboside 5′-monophosphate phosphoribohydrolase (LONELY GUY, LOG). *c*Z-type cytokinins are biosynthesized by tRNA IPTs.

Kristopeit, 1981). Work in moss (*Physcomitrella patens*) and maize identified a nucleoside N-ribohydrolase (NRH) that catalyzes the second step of the reaction (Kopecna et al., 2013), the hydrolysis of iPR to iP. The protein LONELY GUY (OsLOG) catalyzes the one-step pathway in rice. OsLOG functions as a cytokinin riboside 5′-monophosphate phosphoribohydrolase that directly converts a cytokinin nucleotide to a free-base form by release of a ribose 5′-monophosphate (Kurakawa et al., 2007). In Arabidopsis, nine *LOG* genes (*LOG1* to *LOG9*) have been identified as *OsLOG* homologs (Chang et al., 2015; Kuroha et al., 2009), suggesting that the *LOG*-dependent one-step pathway is conserved in both monocot and dicot plants (Fig. 3.2).

3.3.1.3   Sites of cytokinins biosynthesis

Early studies suggested that cytokinin biosynthesis mainly occurs in roots, but recent studies show that cytokinin biosynthesis occurs throughout the plant. For example, *AtIPT* genes are expressed in several tissues of shoots, including leaves, stems, flowers, and siliques. *AtIPT1* gene is predominantly expressed in the vascular stele of roots, leaf axils, ovules, and immature seeds; *AtIPT3* is expressed in the phloem companion cells; *AtIPT4* and *AtIPT8* are expressed in immature seeds, with the highest expression in the chalazal endosperm. *AtIPT5* is expressed in the lateral root primordia, root-cap columella, the upper parts of young inflorescences, and fruit abscission zones. *AtIPT7* is expressed in the endodermis of the root elongation zone, trichomes on young leaves, and pollen tubes (Miyawaki et al., 2004). These findings indicate that expression of *AtIPT* gene is spatially differentiated and biosynthesis of cytokinins occurs at various sites in plants. Similar to *IPT* genes, *LOG* genes also show differential expression in various tissues during plant development (Kuroha et al., 2009), suggesting that the distribution of different types of cytokinins varies among plant tissues.

Xylem sap contains predominantly *t*Z-type ribosides, whereas phloem sap contains iP and *c*Z-type ribosides (Hirose et al., 2008). Consistent with this observation that *t*Z-type cytokinins are dominant cytokinins in xylem, *CYP735A* gene responsible for hydroxylation of the isopentenyl side chain are mainly expressed in roots (Beveridge et al., 1997; Takei et al., 2004b).

3.3.1.4   Regulation of cytokinins biosynthesis

The transcription of cytokinin biosynthetic genes is modulated by multiple factors, including hormones and macronutrients. In Arabidopsis, cytokinins promote cell division by antagonizing auxin. Auxin induces the expression of *AtIPT5* and *AtIPT7*, whereas cytokinins repress the expression of *AtIPT1*, *AtIPT3*, *AtIPT5*, and *AtIPT7* in shoot meristem (Cheng et al., 2013b). Auxin also negatively regulates cytokinin biosynthesis in the nodal stem by repressing the expression of the *PsIPT* gene in pea (*Pisum sativum* L.) (Tanaka et al., 2006). The transcript levels of *CYP735A1* and *CYP735A2* in roots are upregulated by cytokinins but downregulated by auxin or abscisic acid (ABA) (Takei et al., 2004b), suggesting a modulation of cytokinin biosynthesis by auxin and ABA.

Macronutrients also influence cytokinin biosynthesis. Nitrate promotes the accumulation of various cytokinins in maize and *t*Z-type cytokinins in the root of Arabidopsis (Takei et al., 2002). This accumulation of cytokinins probably results from the induction of *AtIPT3* expression, as the nitrate-dependent accumulation of cytokinins was greatly reduced in an *ipt3* null mutant (Takei et al., 2004a). Other macronutrients, including sulfate and phosphate, also regulate *AtIPT3* expression (Takei et al., 2004a).

In addition to transcriptional regulation, posttranslational modifications also modulate cytokinin biosynthesis. For instance, AtIPT3 contains typical CaaX boxes, which are short cysteine-containing motifs recognized by a farnesyl transferase. The farnesylation directs the subcellular localization of AtIPT3 (Galichet et al., 2008). Nonfarnesylated AtIPT3 is mainly localized in plastids, but farnesylated AtIPT3 is found in the nucleus and cytoplasm.

3.3.2 Cytokinins degradation

Along with biosynthesis and activation, degradation of cytokinin plays an important role in modulating cytokinin levels in plants. Endogenous cytokinins with an unsaturated side chain (iP, *cZ*, *tZ*, etc.) are irreversibly degraded by cytokinin oxidase/dehydrogenase (CKX) enzymes. These are oxidoreductases containing a flavin adenine dinucleotide (FAD)-binding motif. They selectively cleave the unsaturated N^6-side chain from iP, *cZ*, *tZ*, and their ribosides (Armstrong, 1994; Paces et al., 1971; Whitty and Hall, 1974). A CKX enzyme was first purified from corn kernels where it has an important function in fine-tuning local cytokinin concentration (Houba-Hérin et al., 1999; Morris et al., 1999). The maize genome has 13 *ZmCKX* genes, which are expressed in grains, roots, leaves, and immature ears (Bilyeu et al., 2001; Massonneau et al., 2004; Vyroubalová et al., 2009). Among these enzymes, the activities of ZmCKX1 and ZmCKX3 have been demonstrated in yeast (Houba-Hérin et al., 1999; Massonneau et al., 2004; Morris et al., 1999). Arabidopsis has seven *CKX* genes (*AtCKX1* to *AtCKX7*). *AtCKX1* is mainly expressed in the shoot apex, young floral tissues, and the root-hypocotyl junction; *AtCKX2* is expressed in the shoot apex and stipules. *AtCKX4* is predominantly expressed in trichomes, stipules, and root caps. *AtCKX5* expression is mainly confined to the edges of the youngest emerging leaves and in shoot and root meristems, whereas *AtCKX6* is expressed in the vasculature (Werner et al., 2003). Thus, *CKXs* are expressed in various types of cells and organs and probably have overlapping functions in certain plant tissues. Interestingly, AtCKX1 and AtCKX3 localize in the plant vacuole, whereas AtCKX2 localizes in the endoplasmic reticulum, implying that subcellular compartmentalization of AtCKXs is important for their functional specificity. Alteration of expression of *CKXs* results in strong phenotypes in many plant species. For example, overexpression of *AtCKX1* to *AtCKX4* using the 35S promoter in Arabidopsis resulted in reduced cytokinin levels and cytokinin-deficient phenotypes in shoots and roots (Werner et al., 2003). Overexpression of the orchid (*Dendrobium* Sonia) *DSCKX1* gene in Arabidopsis also led to cytokinin-deficient developmental phenotypes due to decreased cytokinin levels (Yang et al., 2002, 2003).

Overexpression of *CKX* genes can cause significantly increased seed size in crops, suggesting potential applications for cytokinins in improving crop yields. In rice, an important quantitative trait locus *GN1A* controls grain yield and the causal gene encodes a cytokinin oxidase (OsCKX2) (Ashikari et al., 2005). Transgenic plants expressing antisense constructs of *OsCKX2* have elevated cytokinin levels in the shoot apical meristem, which led to an increase in inflorescence meristems and in the number of reproductive organs, resulting in enhanced grain productivity in rice (Ashikari et al., 2005). *AtCKX3* and *AtCKX5* are mainly expressed in meristems and a *ckx3 ckx5* double mutant formed larger inflorescences, thereby leading to an increase of approximately 55% (per plant) in seed yield compared with the wild type (Bartrina et al., 2011). In tobacco and Arabidopsis, root-specific overexpression of *CKX* genes significantly promotes the growth and development of roots, which in turn enhanced drought tolerance and nutrient absorption (Werner et al., 2003). Thus, *CKXs* are important targets for molecular approaches to improve crop yields.

3.3.3   Modification of cytokinins

3.3.3.1   The functions of modification

Modification of active cytokinins allows the plant to efficiently control the amount of active cytokinin forms. Usually, cytokinin modification occurs either on the side chain or on the adenine moiety by which cytokinins convert between free bases and the riboside form. This conversion is crucial for the activity of cytokinins, through which the active nucleobase cytokinin can be rapidly converted into the inactive form in plants.

3.3.3.2   Modification of the adenine moiety

Modifications of the adenine moiety include phosphorylation, dephosphorylation, *N*-glucosylation, and *N*-alanylation. In glucosylation, a sugar moiety is attached to the N^3, N^7, and N^9 positions of the purine ring (*N*-glucosylation) or at the hydroxyl group of the zeatin side chain (*O*-glucosylation). In Arabidopsis, two closely related genes, *UGT76C1* and *UGT76C2*, encoding *N*-glucosyltransferases, bring about *N*-glucosylation at the N^7 and N^9 positions of cytokinins. Overexpression of *UGT76C* increased the abundance of *N*-glucosylated cytokinins and reduced the sensitivity to exogenous *t*Z (Hou et al., 2004; Wang et al., 2011, 2013). To date, the *N*-glucosyl hydrolases have not been identified; therefore this mechanism is thought to inactivate excess cytokinins.

3.3.3.3   Modification of the cytokinins side chain

O-glucosylation is a major modification that occurs during *t*Z metabolism, and is catalyzed by zeatin *O*-glucosyltransferase (Dixon et al., 1989; Veach et al., 2003). Natural zeatin in plant tissues is composed of *t*Z, *c*Z, and dZ. *O*-glucosylation of *t*Z, *c*Z, and dZ side chains can convert them to storage forms. *O*-glucosyltransferase genes for *t*Z have been cloned in *Phaseolus lunatus*, maize, rice, and Arabidopsis (Jin et al., 2013; Kudo et al., 2012; Martin et al., 2001, 1999; Veach et al., 2003). Transgenic plants overexpressing Arabidopsis *O*-glucosyltransferase gene *UGT85A1* had significantly increased levels of *t*Z-*O*-glucoside (*t*ZOG) (Jin et al., 2013). Inactive *t*ZOG serves as the storage and transport form of zeatin (Armstrong, 1994), which can be activated by β-glucosidase to release active zeatin (Brzobohaty et al., 1993).

3.3.4   Cytokinins transport

Studies have shown that different types of cytokinins can be transported in different ways. Using reciprocal grafting with *ipt1, 3, 5, 7* mutants and wild-type Arabidopsis plants, iP appeared to be mainly transported from shoots to roots, whereas *t*Z might be mainly transported from roots to shoots (Bishopp et al., 2011b; Kudo et al., 2010). Analysis of the *cyp735a1 cyp735a2* double mutant phenotype demonstrated that root to shoot transport of *t*Z is essential for shoot development (Kiba et al., 2013). Thus, cytokinins must be transported to target cells through an efficient transport system (Burkle et al., 2003). Three types of proteins function in the transport of cytokinins: purine permeases (PUPs), nucleoside transporters (equilibrative nucleoside transporters, ENTs), and ATP-binding cassette (ABC) transporters.

Two Arabidopsis genes, *AtPUP1* and *AtPUP2*, encode transporters that localize to the plasma membrane and mediate cytokinin uptake into cell (Gillissen et al., 2000). *AtPUPs* are primarily expressed in phloem, and it is suggested that PUPs participate in the uptake of cytokinins from xylem sap during long-distance transport. AtPUP1 and AtPUP2 have been shown to transport adenine and nucleobase cytokinin derivatives in yeast and Arabidopsis cells (Burkle et al., 2003). However, the Arabidopsis genome has more than 20 *PUP* genes and further study is required to determine their individual functions.

In higher plants, the translocation of cytokinin ribosides appears to be mediated by ENTs. Arabidopsis AtENT8 (also known as SOI33) and AtENT3 mediate iPR uptake in yeast (Chen et al., 2006; Sun et al., 2005). In rice, OsENT2 participates in the selective transport of cytokinin nucleosides into vascular tissues (Hirose et al., 2005).

Studies by two groups demonstrated that the Arabidopsis ABC transporter subfamily G14 (AtABCG14) is involved in loading of *t*Z-type cytokinins into the xylem for transport from roots to shoots (Ko et al., 2014; Zhang et al., 2014). Expression of *AtABCG14* is observed in the pericycle and stellar cells of root. Loss of *AtABCG14* function severely impaired the translocation of *t*Z-type cytokinins from roots to shoots, thus affecting plant growth and development (Ko et al., 2014; Zhang et al., 2014).

3.4 Cytokinin perception and signal transduction

3.4.1 Discovery of cytokinin signaling systems

Cytokinins were discovered in the 1950s (Miller et al., 1955b). Early studies mainly focused on the physiological roles and metabolism of cytokinin. Until the late 1990s, our understanding of cytokinin signal transduction remained limited in spite of many years of effort, including the use of genetic, biochemical, and physiological approaches. It was proposed that calmodulin and a G-protein-linked receptor are involved in cytokinin signaling (Elliott, 1983; Plakidou-Dymock et al., 1998). However, no direct evidence supports the involvement of these factors in cytokinin signaling. In addition, extensive genetic screening of Arabidopsis mutant populations has aimed to identify mutants with altered responses to externally applied cytokinins. However, in contrast to the successes in elucidating other phytohormone pathways, such as the auxin and ethylene pathways, this analogous approach did not return useful results for the cytokinin pathway. In fact, numerous attempts have been unable to identify a mutation that mainly or specifically affects the cytokinin signal transduction pathway. One of the reasons for this failure was that exogenous cytokinins evoke the ethylene response, mainly by stimulating ethylene biosynthesis. Thus, most of the previously identified cytokinin-insensitive mutants affect the ethylene pathway (Su and Howell, 1992; Vogel et al., 1998). In addition, components of the cytokinin signaling pathway are highly redundant. These issues have made it difficult to investigate cytokinin signaling via genetic approaches.

In the mid-1990s, Japanese scientist Tatsuo Kakimoto and colleagues carried out genetic screens to identify Arabidopsis mutants with constitutive or impaired cytokinin responses, including rapid cell proliferation and shoot formation in tissue culture in the absence of cytokinin. An important breakthrough was made with the identification of *CYTOKININ INDEPENDENT 1* (*CKI1*) and *CYTOKININ RESPONSE 1* (*CRE1*) as positive regulators of cytokinin signaling (Inoue et al., 2001; Kakimoto, 1996). An additional breakthrough was the identification of a class of cytokinin-inducible genes, known as type-A Arabidopsis response regulators (ARRs) (Brandstatter and Kieber, 1998; Imamura et al., 1998). Because of these discoveries, further studies were able to elucidate the cytokinin signal transduction cascade, which we describe in the following section.

3.4.2   Outline of cytokinin signal transduction

Cytokinin signaling (summarized in Fig. 3.3) is mediated by a two-component system involving a phosphorelay that functions by sequential transfer of phosphoryl groups from receptors to downstream components (Hwang and Sheen, 2001; Hwang et al., 2012; Schaller et al., 2011; Sheen, 2002; To and Kieber, 2008). The Arabidopsis cytokinin receptor family is composed of three histidine kinases (AHKs): AHK2, AHK3, and AHK4, the latter also named CRE1 or WOODENLEG (WOL). Upon binding of cytokinin, the receptor is activated by autophosphorylation at a conserved histidine residue within the kinase domain, which is subsequently transferred to a conserved aspartate residue on the receiver domain of the same protein. The phosphoryl groups are then transferred from the receptor to the conserved His residue in Arabidopsis HISTIDINE PHOSPHOTRANSFER PROTEINS (AHPs). The phosphorylated AHPs may relocate from the cytosol to the nucleus by unknown mechanisms and transfer the phosphoryl groups to downstream components, including Arabidopsis type-B and type-A response regulators (ARRs). Type-B ARRs are a group of MYB-class transcription factors that are activated upon receiving a phosphoryl group and then directly regulate the expression of type-A *ARR* genes and other downstream target genes, which bring about responses to cytokinins. In turn, type-A ARR proteins negatively regulate phosphorelay activity, thereby forming a feedback regulatory loop. Among the three cytokinin receptors, AHK4/CRE1/WOL possesses kinase activity that is stimulated upon the binding of cytokinin, but in the absence of cytokinin it functions as a phosphatase, acting on AHPs. Therefore, cytokinin signaling is mediated by a bidirectional phosphorelay that is regulated by the cytokinin receptors and several negative regulators (Hwang and Sheen, 2001; Hwang et al., 2012; Inoue et al., 2001; Kakimoto, 2003b; Kieber, 2002; Punwani et al., 2010; Sheen, 2002).

In Arabidopsis, the cytokinin signaling pathway is composed of 3 AHK receptors, 6 AHPs, 10 type-A ARRs, and 11 type-B ARRs. In each gene family, different

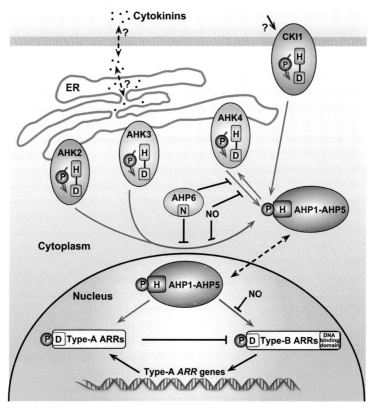

Figure 3.3 Core steps of the cytokinin signaling pathway. The cytokinin signaling cascade is initiated by cytokinin binding to the cytokinin receptors AHK2, AHK3, and AHK4 within the lumen of the endoplasmic reticulum (ER). After binding to cytokinins, the cytokinin receptors are autophosphorylated at conserved histidine residues in the kinase domain. The phosphate groups are then transferred to the conserved aspartic acid residues in the receiver domain of the receptors, and then are likely transferred to the histidine residues of AHP1–AHP5 in the cytoplasm. The phosphorylated AHPs translocate into the nucleus by an unknown mechanism and transfer the phosphate groups to the conserved aspartic acid residues in the receiver domains of type-A ARRs or type-B ARRs. In the absence of cytokinins, AHK4/CRE1/WOL removes phosphate groups from AHPs. CKI1 also mediates phosphorylation of AHPs in a cytokinin-independent manner. The stability of type-A ARR proteins may be regulated by phosphorylation. The phosphorylated type-B ARRs activate the expression of downstream genes, regulating plant growth and development. Type-B ARRs activate the expression of type-A *ARR*s, and type-A ARRs, in turn, act to repress the activity of type-B ARRs by a negative feedback mechanism. P denotes the phosphate group; H and D indicate histidine and aspartic acid; black solid circles indicate cytokinins.

members exert both redundant and specific functions, which may explain the difficulty in identifying a recessive mutation specific to the cytokinin signaling pathway. Genome sequencing and phylogenetic analyses indicated that two-component system elements, which mediate cytokinin responses, are present in the monocot land plant rice as well as in early diverging land plants *Marchantia polymorpha* and *P. patens*. Further studies in these species may help us better understand the evolution of cytokinin signaling (Choi et al., 2012; Du et al., 2007; Gruhn et al., 2014; Pils and Heyl, 2009; Schaller et al., 2007; Tsai et al., 2012; von Schwartzenberg et al., 2016).

3.4.3   Histidine kinases

3.4.3.1   Introduction to cytokinin receptors

The *Arabidopsis thaliana* genome encodes six histidine kinases, three of which, AHK2, AHK3, and AHK4/CRE1/WOL, comprise the cytokinin receptor family (Fig. 3.4). CRE1 is the first cytokinin receptor identified from the "shoot formation assay". The callus tissues of *cre1* mutants were less sensitive to cytokinin, indicating that *CRE1* is a critical positive regulator of cytokinin signaling (Inoue et al., 2001). The *cre1* mutant was also allelic to the *woodenleg* (*wol*) mutation, which was originally identified as a mutant with fewer cells and lacking phloem in the root vasculature (Mähönen et al., 2000).

Similar two-component signaling pathways are also present in bacteria and yeast; thus *Escherichia coli* and fission yeast *Schizosaccharomyces pombe* are used as important systems to study the functions of the components of Arabidopsis cytokinin signaling (Laub and Goulian, 2007; Schaller et al., 2011). In fission yeast, *SLN1* encodes an osmosensing histidine kinase. The *sln1Δ* mutant is lethal under high-osmolality

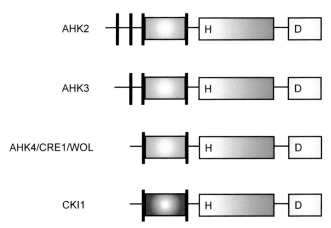

Figure 3.4 Structures of cytokinin-related histidine kinases in Arabidopsis. Black rectangle, transmembrane segment; green rectangle, the extracellular CHASE domain; purple rectangle, the extracellular domain of CKI1; pale-blue rectangle, the kinase domain; yellow rectangle, the receiver domain. H and D indicate conserved histidine and aspartic acid residues, respectively.

conditions. Expression of *CRE1* in yeast in the presence of cytokinin suppresses the lethality of the *sln1Δ* mutant. Notably, *tZ* is effective in this yeast system, but *cZ* and other phytohormones (auxin, ABA, and gibberellin) are ineffective (Inoue et al., 2001). Similar experiments with AHK4, as well as with AHK2 and AHK3, gave similar results using the fission yeast and bacterial phosphorelay system (Inoue et al., 2001; Suzuki et al., 2001; Yamada et al., 2001). These results are consistent with CRE1/AHK4, AHK2, and AHK3 being cytokinin-binding receptors.

Arabidopsis cytokinin receptors share a similar structural organization, including transmembrane domains, cytokinin-binding sites, the kinase domain, and the receiver domain (Fig. 3.4). Subcellular localization studies indicated that the cytokinin receptors are predominately localized in the membranes of the endoplasmic reticulum (ER; see Fig. 3.3), suggesting that ER plays important roles in cytokinin signal transduction (Caesar et al., 2011; Lomin et al., 2011; Wulfetange et al., 2011).

3.4.3.2 Cytokinin recognition by receptors

In Arabidopsis cytokinin receptors, the cytokinin-binding sites (also known as CHASE domains) are located between transmembrane domains. Biochemical and genetic approaches have demonstrated that AHK receptors can specifically bind bioactive cytokinins. Biochemical studies in fission yeast and *E. coli* suggested that cytokinins directly bind to the three AHK receptors through the CHASE domains (Inoue et al., 2001; Suzuki et al., 2001; Yamada et al., 2001). In the *wol* mutant, a single amino acid residue substitution occurred within the CHASE domain of AHK4/CRE1/WOL (T281I) resulting in the loss of cytokinin-binding activity of the mutant WOL protein (Mähönen et al., 2000, 2006b). Cytokinin receptors exhibit different affinities for various cytokinin species. In Arabidopsis, AHK3 and AHK4 bind to *tZ* and iP, but differ significantly in their binding to other cytokinins. Unlike AHK4, AHK3 recognizes ribosides and ribotides as well as *cZ* and dZ, although with low sensitivity (Spichal et al., 2004; Yamada et al., 2001). Similar to AHK4, the CHASE domain of AHK2 also shows a high affinity for iP and *tZ* (Stolz et al., 2011). In maize, ZmHK1 and ZmHK3 are more responsive to free-base cytokinins (iP, *cZ*, and *tZ*), whereas ZmHK2 tends to respond with greater sensitivity to *tZ* and its riboside form (Yonekura-Sakakibara et al., 2004). In addition, the binding of cytokinins to receptors is pH dependent (Lomin et al., 2015; Romanov et al., 2006). The ligand-binding activities of the three AHK receptors and of ZmHK1 are reduced at low pH. ZmHK1 shows particular sensitivity to pH, suggesting that some cytokinin receptors may also function as pH sensors in the lumen of ER (Lomin et al., 2015).

To understand the mechanism of AHK4 activation by cytokinin, Hothorn et al. (2011) determined the crystal structure of the CHASE domain of AHK4 (residues 126–395). The N-terminus of the AHK4 CHASE domain folds into a long stalk followed by two PAS-like domains. The ligand-binding pocket of AHK4 is occupied by iP. The T278I mutation in AHK4 (the *wol* allele) most likely restricts the overall size of the binding pocket and blocks its cytokinin-binding activity, consistent with the associated loss-of-function phenotype of the *wol* mutant (Hothorn et al., 2011).

3.4.3.3 Mechanism of activation and inactivation of the cytokinin receptor AHK4

The AHK4/CRE1/WOL protein is a much-studied member of the cytokinin receptor family. Most *ahk4* mutations do not result in obvious growth or developmental defects, but the *wol* allele causes a short-root phenotype and xylem differentiation defects, which is similar to the *ahk2 ahk3 ahk4* triple mutant phenotype. Genetic analyses suggest that *wol* is a dominant-negative mutation (Mähönen et al., 2000). Biochemical studies indicate that AHK4 exhibits both kinase and phosphatase activities. In the presence of cytokinin, AHK4 showed kinase activity, which initiated autophosphorylation leading to the transfer of the phosphoryl group to AHPs; in the absence of cytokinin, AHK4 showed phosphatase activity that dephosphorylated AHPs. The wol mutant protein lacks cytokinin binding activity and is "locked" in the phosphatase form. As a consequence, the *wol* mutation blocks the phosphotransfer from the cytokinin receptors AHK2 and AHK3 to downstream targets, phenocopying the *ahk2 ahk3 ahk4* triple mutants. Unlike AHK4, AHK2 and AHK3 have either no or very low phosphatase activity or such phosphatase activity is overcome by kinase activity in the absence of cytokinins (Inoue et al., 2001; Mähönen et al., 2006b).

3.4.3.4 Specific and redundant functions of cytokinin receptors

Under normal growth conditions, the *ahk* single mutants and the *ahk2 cre1* or *ahk3 cre1* double mutants do not show significant developmental defects, indicating that the cytokinin receptors have redundant functions in plant growth and development. In the presence of exogenous cytokinin, the *ahk3* and *ahk4* single mutants exhibited a cytokinin-insensitive phenotype. Although the *ahk2* mutants responded normally to cytokinin, the calli of the *ahk2 ahk3* double mutants exhibited a strong cytokinin-insensitive phenotype, suggesting a positive role of *AHK2* in cytokinin signaling (Higuchi et al., 2004; Inoue et al., 2001; Nishimura et al., 2004). Interestingly, AHK3 was reported to be specifically involved in cytokinin-regulated leaf senescence through phosphorylation of ARR2 (Kim et al., 2006). The *ahk4* mutants showed cytokinin-insensitive root growth, whereas the root growth of the *ahk2* and *ahk3* single mutants and the *ahk2 ahk3* double mutants showed similar cytokinin responses to that of wild type, indicating a specific role of *AHK4* in root growth. The *ahk2 ahk3* double mutants developed an enhanced root system through fast growth of the primary root and more prolific lateral roots, suggesting that cytokinin plays a negative role in regulating root growth (Riefler et al., 2006). The *ahk2 ahk3* double mutants had semidwarf shoots with smaller leaves and shorter inflorescence stems (Nishimura et al., 2004), suggesting that cytokinin regulates shoot development mainly through *AHK2* and *AHK3*. The *ahk2 ahk3 ahk4* triple mutants exhibited serious developmental defects as well as a loss of cytokinin responses (Bencivenga et al., 2012; Cheng et al., 2013a; Higuchi et al., 2004; Nishimura et al., 2004; Riefler et al., 2006). Taken together, these results indicate that cytokinin is essential for plant growth and development, in which cytokinin receptors play both redundant and specific roles.

3.4.3.5   Histidine kinase CKI1

In addition to the three cytokinin receptors described earlier, an additional *AHK* gene, *CYTOKININ INDEPENDENT 1* (*CKI1*), has been implicated in cytokinin signaling (Kakimoto, 1996). The *CKI1* gene encodes a receptor-like protein histidine kinase (HK) containing the putative transmembrane domains, a histidine kinase domain, and a receiver domain (Fig. 3.4). CKI1 is incapable of binding cytokinin because it lacks the CHASE domain (Yamada et al., 2001). In Arabidopsis protoplasts, transient overexpression of *CKI1* constitutively induced the expression of the cytokinin-responsive gene *ARR6*. Ectopic expression of *CKI1* induced cytokinin responses (Hwang and Sheen, 2001). CKI1 can phosphorylate or dephosphorylate AHPs, suggesting that CKI1 may play an important role in cytokinin signaling (Mähönen et al., 2006b).

CKI1 may be activated by unknown signals and activates cytokinin signaling via a hormone-independent mechanism (Fig. 3.3). Loss-of-function mutations in *CKI1* caused abnormal development of female gametophytes (Deng et al., 2010; Hejátko et al., 2003; Pischke et al., 2002). However, male gametogenesis and vegetative growth were not affected in the *cki1* loss-of-function mutants, suggesting that *CKI1* specifically regulates female gametogenesis. Genetic studies indicate that *CKI1* acts upstream of *AHP* and that *CKI1-AHP*-dependent signaling is essential for the development of female gametophytes (Deng et al., 2010).

3.4.4   Histidine phosphotransfer proteins

3.4.4.1   AHP1 to AHP5 function as positive regulators of cytokinin signaling

The *Arabidopsis thaliana* genome has five *AHP* genes that encode histidine phosphotransfer (HPt) proteins (AHP1 to AHP5) containing the conserved histidine residue for activity; Arabidopsis also has a pseudo-HPt (AHP6) lacking the histidine phosphorylation site. When expressed in *Saccharomyces cerevisiae*, AHP1 exhibits in vivo activity that can rescue a mutation of the yeast *YPD1* gene, which encodes a typical HPt protein involved in osmosensing signal transduction (Suzuki et al., 1998). Using an in vitro phosphotransfer system, researchers demonstrated that AHPs can accept a phosphoryl group from AHK4/CRE1/WOL and then transfer the phosphoryl groups to the receiver domains of type-A and type-B ARRs (Imamura et al., 2001; Suzuki et al., 1998, 2002, 2001). Multiple interactions among AHKs, AHPs, and ARRs were detected in yeast two-hybrid assays (Dortay et al., 2008, 2006).

The AHPs function as positive regulators of cytokinin signaling, acting redundantly, consistent with their high degree of sequence similarity. The *ahp* single mutants did not display an obvious morphology change or altered cytokinin sensitivity in root and hypocotyls elongation assays. The *ahp1 ahp2 ahp3* triple mutants showed reduced sensitivity to cytokinin. The *ahp2 ahp3 ahp5* triple mutants have a short-root phenotype, impaired vascular development, and reduced inhibition of hypocotyl elongation in response to cytokinin. The combination of *ahp* mutants into the quintuple mutant, *ahp1 ahp2-1 ahp3 ahp4 ahp5*, which includes the weak *ahp2-1* allele, showed various abnormalities similar to those found in the cytokinin receptor triple mutants (Hutchison et al., 2006). The stronger *ahp* quintuple mutant, *ahp1 ahp2-2 ahp3 ahp4 ahp5*, which includes an *ahp2-2* null mutant allele, exhibited severe defects in megagametogenesis similar to *cki1* and type-B *arr* mutants (see later discussion). The *ahk* triple mutants also showed defects in early female gametophyte development, suggesting that cytokinin signaling is essential for female gametophyte development, in which both *CKI1* and cytokinin receptors activate cytokinin signaling via AHPs (Cheng et al., 2013a; Deng et al., 2010; Kinoshita-Tsujimura and Kakimoto, 2011). In Arabidopsis and rice, AHPs are localized in both the cytosol and nucleus in the absence or presence of exogenous cytokinin (Punwani et al., 2010; Tsai et al., 2012). The mechanisms regulating active transport of AHPs between cytosol and nucleus remain unknown.

3.4.4.2   AHP6 inhibits cytokinin signaling

The *ahp6* mutation was identified in a genetic screen for suppressors of the dominant-negative *wol* mutant. The *ahp6* mutation partially restored the *wol* defects in vascular development and conferred increased cytokinin responses in the adventitious root formation assay, indicating that *AHP6* interacts with *CRE1/AHK4/WOL* to regulate cytokinin signaling. Expression of *AHP6* is downregulated by cytokinin treatment. However, AHP6 is predicted to be a nonfunctional HPt protein because it lacks the conserved histidine residue that is present in other functional AHPs (AHP1 to AHP5) and that is required for phosphotransfer. Indeed, biochemical studies suggest that AHP6 does not function as a phosphotransfer protein. Moreover, AHP6 can inhibit phosphotransfer from the kinase domain to the receiver domain of the SLN1 histidine kinase and from AHP1 to ARR1. The evidence suggests that AHP6 acts as an inhibitor of cytokinin signaling by interacting with the phosphorelay machinery (Mähönen et al., 2006a) (Fig. 3.3).

One study reported that cytokinins promote the bisymmetric distribution of PIN-FORMED (PIN) proteins, which function as auxin efflux transporters, resulting in low auxin responses in procambial cells. A high amount of auxin induces the expression of the gene encoding cytokinin signaling inhibitor AHP6, establishing a feedback loop between auxin and cytokinin that can specify vascular patterns (Bishopp et al., 2011a). AHP6 is also involved in regulating phyllotactic patterns. Intracellular movement of AHP6 in shoot apical meristems generates differential cytokinin signaling, which is required for the proper timing of organ initiation (Besnard et al., 2014).

3.4.4.3   Nitric oxide regulates cytokinin signaling

The cytokinin pathway actively interacts with other signaling pathways. One such example is the involvement of nitric oxide (NO) in the regulation of cytokinin signaling. NO is an important signaling molecule that regulates diverse physiological and pathological processes (Besson-Bard et al., 2008). A major bioactivity of NO is to modify proteins by *S*-nitrosylation, a redox-based posttranslational modification resulting from covalent addition of an NO group to the reactive Cys thiol of a protein to form S-nitrosothiol (Astier et al., 2011; Hess et al., 2005). The AHP proteins were found to be S-nitrosylated and S-nitrosylation of AHP1 at Cys 115 repressed its phosphorylation and subsequent transfer of the phosphoryl group to ARR1 (Fig. 3.3), suggesting that redox signaling and cytokinin signaling coordinate plant growth and development (Feng et al., 2013). Interestingly, direct chemical interaction between cytokinin and NO has been reported. The authors suggest that cytokinin may act as an NO scavenger to protect against NO-induced stress (Liu et al., 2013).

3.4.5   Response regulators

3.4.5.1   Type-B ARR transcription factors mediate cytokinin-regulated gene expression

Type-B ARRs act as transcription factors mediating the primary cytokinin transcriptional responses (Fig. 3.3). The *Arabidopsis thaliana* genome has 11 type-B *ARR*s falling into three subfamilies based on phylogenetic analyses. The type-B ARRs belong to the MYB family of transcription factors containing a receiver domain at the N-terminus. There are three conserved amino acids in the receiver domain, two aspartic acid (D) residues and a lysine (K), and it is hence also named the DDK domain. The second aspartic acid residue in the DDK domain can accept a phosphoryl group. The C-terminal extension of type-B ARRs contains a plant-specific DNA-binding domain, referred to as the GARP domain (named after maize GOLDEN2, Arabidopsis ARRs, *Chlamydomonas* PSR1, and Arabidopsis PHR1) (D'Agostino et al., 2000; Sakai et al., 2000). Expression of genes encoding type-B *ARR*s does not appear to be responsive to cytokinin. However, the stabilities of type-B ARR proteins are regulated by cytokinins, mediated by the 26S proteasome degradation machinery (Kim et al., 2013a,b; Kurepa et al., 2014).

Type-B ARR proteins are activated upon receiving phosphoryl groups and directly binding to the *cis*-regulatory elements in the promoters of target genes, which include genes encoding type-A ARRs. The type-B ARR1 binds a common DNA target sequence 5′-(A/G)GAT(T/C)-3′. This motif is significantly enriched in the promoter regions of type-A *ARR* genes (Hosoda et al., 2002; Hwang and Sheen, 2001; Imamura et al., 2001; Mähönen et al., 2006a; Sakai et al., 2000, 2001; Taniguchi et al., 2007). Based on this core binding motif, a synthetic promoter was designed for real-time monitoring of cytokinin responses in vivo (Müller and Sheen, 2008).

The DDK domain is proposed to inhibit the activity of the C-terminal transcription activation domain. Transgenic plants overexpressing individual type-B *ARR*s (*ARR1*, *ARR11*, *ARR14*, *ARR18*, and *ARR21*) that lack DDK domains displayed constitutive cytokinin responses (Imamura et al., 2003; Liang et al., 2012; Sakai et al., 2001; Tajima et al., 2004; Veerabagu et al., 2012).

No obvious phenotype was observed in type-B *arr* single mutants. Higher-order mutants displayed reduced cytokinin sensitivity in diverse cytokinin response assays and developmental abnormalities. Similar to the cytokinin receptor triple mutants, the *arr1 arr10 arr12* triple mutants showed severe developmental defects and almost complete insensitivity to exogenously applied cytokinin, suggesting that *ARR1*, *ARR10*, and *ARR12* play important roles in mediating cytokinin responses (Argyros et al., 2008; Ishida et al., 2008; Mason et al., 2005; Yokoyama et al., 2007).

As transcription factors, type-B ARRs regulate plant growth and development through the activation of target genes. Both ARR1 and ARR12 activate *SHY2/IAA3*, which represses auxin signaling. Activation of *SHY2/IAA3* negatively regulates the auxin transporter *PIN* genes causing a redistribution of auxin and promoting cell differentiation (Dello Ioio et al., 2008). Gibberellins selectively repress the expression of *ARR1* mediated by DELLA proteins via unknown mechanisms, regulating cell division and differentiation in root apical meristems (Moubayidin et al., 2010).

3.4.5.2   Type-A ARRs function as negative regulators of cytokinin signaling

The *Arabidopsis thaliana* genome encodes 10 typical type-A ARRs (ARR3, ARR4, ARR5, ARR6, ARR7, ARR8, ARR9, ARR15, ARR16, and ARR17) containing an N-terminal phosphoryl receiver domain and a short C-terminal variable extension. Phylogenetic analysis indicated that the 10 type-A ARRs fall into five pairs with similar amino acid sequences (D'Agostino et al., 2000; D'Agostino and Kieber, 1999). Most of the type-A ARRs localize in the nucleus except ARR3 and ARR16, which localize in the cytosol (Dortay et al., 2008; Kiba et al., 2002).

Expression of type-A *ARR* genes is rapidly induced in response to cytokinin treatment, indicating that type-A *ARRs* are primary response genes in cytokinin signaling (Fig. 3.3). Their induction occurs in the absence of de novo protein synthesis and their promoters typically contain multiple type-B ARR-binding sites (Brandstatter and Kieber, 1998; Imamura et al., 1998).

Cytokinin not only stimulates transcription of type-A *ARR*s, but also regulates ARR protein stability, which plays an important role in negatively regulating cytokinin signaling. Multiple type-A ARR proteins were stabilized in the presence of exogenous cytokinin in a phosphorylation-dependent manner. In addition, the accumulation of a subset of type-A ARRs was increased by MG132, a specific inhibitor of the 26S proteasome. These results suggest that the stability of type-A ARRs is regulated by both cytokinin and proteasome degradation pathways (Lee et al., 2008; Ren et al., 2009; To et al., 2007).

Similar to proteins in other two-component systems, type-A ARRs functionally overlap in the regulation of cytokinin signaling. Single type-A *arr* mutants showed a wild-type morphological phenotype and normal responses to cytokinin inhibition of primary root growth and lateral root formation. Higher-order mutants displayed increasing sensitivities to cytokinin treatment, based on various cytokinin response assays (Kiba et al., 2003; To et al., 2004). Overexpression of several type-A *ARRs* caused reduced cytokinin sensitivity in root elongation and an early-senescence or early-flowering phenotype, suggesting that type-A ARRs have negative regulatory roles in cytokinin signal transduction (Kiba et al., 2003; Lee et al., 2008, 2007; Osakabe et al., 2002; Ren et al., 2009).

3.4.6   Other regulators of cytokinin signaling

A subset of *APETALA2* genes, the *CYTOKININ RESPONSE FACTORS* (*CRFs*), was reported to be upregulated by cytokinin in a type-B ARR-dependent manner. In addition, CRF proteins rapidly relocalized to the nucleus in response to cytokinin treatment and this relocalization depended on both AHKs and AHPs, but was independent of either type-A or type-B ARRs, suggesting that CRFs act in parallel with type-B ARRs to mediate cytokinin signaling (Rashotte et al., 2006). Analyses of loss-of-function *crf* mutants indicated that CRF proteins function redundantly to modulate shoot and root growth and promote leaf senescence. Furthermore, CRF proteins mediated the expression of a set of cytokinin-responsive genes that overlap with type-B ARR targets (Raines et al., 2016; Rashotte et al., 2006). Studies indicate that CRF proteins transcriptionally control the auxin transporter *PIN* genes via specific *PIN CYTOKININ RESPONSE ELEMENT* (*PCRE*) domains. Removal of this *PCRE* element effectively uncoupled expression of *PIN* from CRF-mediated cytokinin regulation and compromised cytokinin sensitivity (Šimášková et al., 2015). These results suggest that CRF proteins may fine-tune cytokinin–auxin cross talk by regulating auxin transport in root growth and development.

Arabidopsis *FUMONISIN B1-RESISTANT12* (*FBR12*), which encodes the eukaryotic translation initiation factor 5A, was reported to be involved in cytokinin signaling (Ren et al., 2013). The *fbr12* mutants displayed defective protoxylem development and reduced sensitivity to exogenous cytokinin in primary root growth, shoot formation, and reporter gene expression. Double-mutant analyses indicated that *FBR12* genetically interacts with cytokinin receptor *CRE1* and downstream *AHP* genes. Moreover, FBR12 forms a protein complex with CRE1 and AHP1. Cytokinin treatment reduced the interactions of CRE1-FBR12 and AHP1-FBR12, but enhanced the interaction between AHP1 and CRE1, suggesting that cytokinin regulates CRE1-AHP1-FBR12 complex formation. Interestingly, the *fbr12* mutant phenotype was partially suppressed by the *ahp6* mutation and the expression of *AHP6* was enhanced in the *fbr12* mutants, suggesting that the transcription of *AHP6* is regulated by FBR12 (Ren et al., 2013). These results suggest that FBR12 functions as a key regulator of protoxylem development by modulating cytokinin signaling (Fig. 3.3).

3.5   Summary points

1. A subset of genes encoding key enzymes in cytokinin metabolism and modification has been identified in plants.
2. The shoot-to-root transport system of cytokinin involves various types of influx and efflux transporters in the phloem or xylem of whole plants.
3. Cytokinin signaling is mediated by a two-component system. A key feature of the two-component system is the sequential transfer of phosphoryl groups, through which the signal is amplified and transduced.
4. This phosphoryl-transfer mechanism is unique to cytokinin signaling, not present in any other phytohormone signaling pathways thus far identified in higher plants. Compared with other phytohormone signaling pathways, the scheme of the two-component cytokinin signaling system appears to be relatively simple.

5. The cytokinin pathway actively interacts with other signaling pathways in diverse biological processes. These advances are not covered in this chapter but interested readers are referred to several comprehensive reviews (Chandler and Werr, 2015; Ha et al., 2012; Hwang et al., 2012; Kieber and Schaller, 2014; Perilli et al., 2010; Schaller et al., 2015, 2014; Su and Zhang, 2014).

3.6 Future perspectives

1. The identification of unknown enzymes involved in the critical steps of cytokinin metabolism, such as cytokinin nucleosidases/nucleotidases, zeatin isomerization, as well as modification enzymes, will contribute to our understanding of the control of cytokinin concentration in plants.
2. To better understand the regulation of the cytokinin catabolic pathway, it will be important to learn how the transcription of cytokinin biosynthetic genes is controlled by external factors, including nutrition, phytohormone, and environmental signals.
3. Improvements in cytokinin analytical systems are required to quantify cytokinin production during plant development and in response to environmental stimuli.
4. The identification is needed of new regulatory components, the inactivation mechanism of the pathway, the precise function of type-A ARR proteins, and the genome-wide downstream targets of type-B ARR and CRF transcription factors.
5. It will be of great interest to understand the molecular mechanisms of the regulatory roles of cytokinin in plant growth and development, especially in the context of interacting with other phytohormone signaling pathways.
6. The latter aspect is particularly attractive for the dissection of agronomically important traits in crops, as demonstrated by studies on cytokinin-regulated panicle development (Ashikari et al., 2005; Li et al., 2013) and coregulation of shoot branching by cytokinin and strigolactone (Domagalska and Leyser, 2011; Janssen et al., 2014), which makes the cytokinin pathway a promising target for a new wave of Green Revolution.

Abbreviations

6-BA	N^6-benzyl adenine
ABA	Abscisic acid
ABCG14	ABC transporter subfamily G14
ABC transporter	ATP-binding cassette transporter
AHK	Arabidopsis HISTIDINE KINASE
AHP	Arabidopsis HISTIDINE PHOSPHOTRANSFER PROTEIN
ADP	Adenosine diphosphate
AMP	Adenosine monophosphate
ATP	Adenosine triphosphate
ARR	Arabidopsis response regulator
CK	Cytokinin
CKI1	CYTOKININ INDEPENDENT 1
CKX	Cytokinin oxidase/dehydrogenase
CRE1	CYTOKININ RESPONSE 1
CRF	CYTOKININ RESPONSE FACTOR
CZ	cis-Zeatin

D	Aspartic acid
DMAPP	Dimethylallyl diphosphate
DPU	N'-diphenylurea
ENT	Equilibrative nucleoside transporters
ER	Endoplasmic reticulum
FAD	Flavin adenine dinucleotide
FBR12	FUMONISIN B1-RESISTANT12
HPt	Histidine phosphotransfer
IP	N^6-(Δ^2-isopentenyl)-adenine
iPR	iP riboside
iPRDP	iP riboside 5'-diphosphate
iPRMP	iP riboside 5'-monophosphate
iPRTP	iP riboside 5'-triphosphate
IPT	Isopentenyltransferase
K	Lysine
LOG	LONELY GUY
mT	$meta$-topolin
NO	Nitric oxide
oT	$ortho$-topolin
P450	Cytochrome P450 monooxygenase
PCRE	PIN CYTOKININ RESPONSE ELEMENT
PIN	PIN-FORMED
PUP	Purine permeases
Ti	Tumor-inducing
TDZ	Thidiazuron
tZ	$trans$-zeatin
tZOG	tZ-O-glucoside
tZR	$trans$-zeatin riboside
tZRMP	$trans$-zeatin riboside monophosphate
UGT76C1/2	UDP-GLUCOSYL TRANSFERASE 76C1/2
WOL	WOODENLEG

Acknowledgments

We thank Ms. Juli Peng for preparing figures and editing literature; Dr. Qian Qian for discussion. This work is supported by grants from the National Natural Science Foundation of China to J.Z. (grant nos. 90817107 and 91217302) and S.Y. (grant nos. 90817007 and 91417310).

References

Akiyoshi, D.E., Klee, H., Amasino, R.M., et al., 1984. T-DNA of *Agrobacterium tumefaciens* encodes an enzyme of cytokinin biosynthesis. Proc. Natl. Acad. Sci. U.S.A. 81, 5994–5998.

Argyros, R.D., Mathews, D.E., Chiang, Y.H., et al., 2008. Type B response regulators of *Arabidopsis* play key roles in cytokinin signaling and plant development. Plant Cell 20, 2102–2116.

Armstrong, D.J., 1994. Cytokinin oxidase and the regulation of cytokinin degradation. In: Mok, D.W.S., Mok, M.C. (Eds.), Cytokinins: Chemistry, Activity, and Function. CRC Press, Boca Raton, pp. 139–154.

Ashikari, M., Sakakibara, H., Lin, S., et al., 2005. Cytokinin oxidase regulates rice grain production. Science 309, 741–745.

Astier, J., Rasul, S., Koen, E., et al., 2011. S-nitrosylation: an emerging post-translational protein modification in plants. Plant Sci. 181, 527–533.

Åstot, C., Dolezal, K., Nordström, A., et al., 2000. An alternative cytokinin biosynthesis pathway. Proc. Natl. Acad. Sci. U.S.A. 97, 14778–14783.

Barry, G.F., Rogers, S.G., Fraley, R.T., et al., 1984. Identification of a cloned cytokinin biosynthetic gene. Proc. Natl. Acad. Sci. U.S.A. 81, 4776–4780.

Bartrina, I., Otto, E., Strnad, M., et al., 2011. Cytokinin regulates the activity of reproductive meristems, flower organ size, ovule formation, and thus seed yield in *Arabidopsis thaliana*. Plant Cell 23, 69–80.

Bencivenga, S., Simonini, S., Benková, E., et al., 2012. The transcription factors BEL1 and SPL are required for cytokinin and auxin signaling during ovule development in *Arabidopsis*. Plant Cell 24, 2886–2897.

Besnard, F., Refahi, Y., Morin, V., et al., 2014. Cytokinin signalling inhibitory fields provide robustness to phyllotaxis. Nature 505, 417–421.

Besson-Bard, A., Pugin, A., Wendehenne, D., 2008. New insights into nitric oxide signaling in plants. Annu. Rev. Plant Biol. 59, 21–39.

Beveridge, C.A., Murfet, I.C., Kerhoas, L., Sotta, B., Miginiac, E., Rameau, C., 1997. The shoot controls zeatin riboside export from pea roots. Evidence from the branching mutant *rms4*. Plant J. 11, 339–345.

Bilyeu, K.D., Cole, J.L., Laskey, J.G., et al., 2001. Molecular and biochemical characterization of a cytokinin oxidase from maize. Plant Physiol. 125, 378–386.

Bishopp, A., Help, H., El-Showk, S., et al., 2011a. A mutually inhibitory interaction between auxin and cytokinin specifies vascular pattern in roots. Curr. Biol. 21, 917–926.

Bishopp, A., Lehesranta, S., Vaten, A., et al., 2011b. Phloem-transported cytokinin regulates polar auxin transport and maintains vascular pattern in the root meristem. Curr. Biol. 21, 927–932.

Brandstatter, I., Kieber, J.J., 1998. Two genes with similarity to bacterial response regulators are rapidly and specifically induced by cytokinin in *Arabidopsis*. Plant Cell 10, 1009–1019.

Brzobohaty, B., Moore, I., Kristoffersen, P., et al., 1993. Release of active cytokinin by a b-glucosidase localized to the maize root meristem. Science 262, 1051–1054.

Burkle, L., Cedzich, A., Dopke, C., et al., 2003. Transport of cytokinins mediated by purine transporters of the PUP family expressed in phloem, hydathodes, and pollen of *Arabidopsis*. Plant J. 34, 13–26.

Caesar, K., Thamm, A.M.K., Witthöft, J., et al., 2011. Evidence for the localization of the *Arabidopsis* cytokinin receptors AHK3 and AHK4 in the endoplasmic reticulum. J. Exp. Bot. 62, 5571–5580.

Chandler, J.W., Werr, W., 2015. Cytokinin–auxin crosstalk in cell type specification. Trends Plant Sci. 20, 291–300.

Chang, L., Ramireddy, E., Schmülling, T., 2015. Cytokinin as a positional cue regulating lateral root spacing in *Arabidopsis*. J. Exp. Bot. 66, 4759–4768.

Chen, C.M., 1997. Cytokinin biosynthesis and interconversion. Physiol. Plant 101, 665–673.

Chen, C.M., Kristopeit, S.M., 1981. Metabolism of cytokinin: deribosylation of cytokinin ribonucleoside by adenosine nucleosidase from wheat germ cells. Plant Physiol. 68, 1020–1023.

Chen, C.M., Leisner, S.M., 1984. Modification of cytokinins by cauliflower microsomal enzymes. Plant Physiol. 75, 442–446.

Chen, K.L., Xu, M.X., Li, G.Y., et al., 2006. Identification of AtENT3 as the main transporter for uridine uptake in *Arabidopsis* roots. Cell Res. 16, 377–388.

Cheng, C.Y., Mathews, D.E., Schaller, G.E., et al., 2013a. Cytokinin-dependent specification of the functional megaspore in the Arabidopsis female gametophyte. Plant J. 73, 929–940.

Cheng, Z.J., Wang, L., Sun, W., et al., 2013b. Pattern of auxin and cytokinin responses for shoot meristem induction results from the regulation of cytokinin biosynthesis by AUXIN RESPONSE FACTOR3. Plant Physiol. 161, 240–251.

Choi, J., Lee, J., Kim, K., et al., 2012. Functional identification of *OsHk6* as a homotypic cytokinin receptor in rice with preferential affinity for iP. Plant Cell Physiol. 53, 1334–1343.

D'Agostino, I.B., Deruère, J., Kieber, J.J., 2000. Characterization of the response of the *Arabidopsis* response regulator gene family to cytokinin. Plant Physiol. 124, 1706–1717.

D'Agostino, I.B., Kieber, J.J., 1999. Molecular mechanisms of cytokinin action. Curr. Opin. Plant Biol. 2, 359–364.

Dello Ioio, R., Nakamura, K., Moubayidin, L., et al., 2008. A genetic framework for the control of cell division and differentiation in the root meristem. Science 322, 1380–1384.

Deng, Y., Dong, H., Mu, J., et al., 2010. *Arabidopsis* histidine kinase CKI1 acts upstream of HISTIDINE PHOSPHOTRANSFER PROTEINS to regulate female gametophyte development and vegetative growth. Plant Cell 22, 1232–1248.

Dixon, S.C., Martin, R.C., Mok, M.C., et al., 1989. Zeatin glycosylation enzymes in *Phaseolus*: isolation of *O*-glucosyltransferase from *P. lunatus* and comparison to *O*-xylosyltransferase from *P. vulgaris*. Plant Physiol. 90, 1316–1321.

Domagalska, M.A., Leyser, O., 2011. Signal integration in the control of shoot branching. Nat. Rev. Mol. Cell Biol. 12, 211–221.

Dortay, H., Gruhn, N., Pfeifer, A., et al., 2008. Toward an interaction map of the two-component signaling pathway of *Arabidopsis thaliana*. J. Proteome Res. 7, 3649–3660.

Dortay, H., Mehnert, N., Bürkle, L., et al., 2006. Analysis of protein interactions within the cytokinin-signaling pathway of *Arabidopsis thaliana*. FEBS J. 273, 4631–4644.

Du, L., Jiao, F., Chu, J., et al., 2007. The two-component signal system in rice (*Oryza sativa* L.): a genome-wide study of cytokinin signal perception and transduction. Genomics 89, 697–707.

Elliott, D.C., 1983. Inhibition of cytokinin-regulated responses by calmodulin-binding compounds. Plant Physiol. 72, 215–218.

Feng, J., Wang, C., Chen, Q., et al., 2013. *S*-nitrosylation of phosphotransfer proteins represses cytokinin signaling. Nat. Commun. 4, 1529.

Galichet, A., Hoyerova, K., Kaminek, M., et al., 2008. Farnesylation directs AtIPT3 subcellular localization and modulates cytokinin biosynthesis in *Arabidopsis*. Plant Physiol. 146, 1155–1164.

Gillissen, B., Burkle, L., Andre, B., et al., 2000. A new family of high-affinity transporters for adenine, cytosine, and purine derivatives in *Arabidopsis*. Plant Cell 12, 291–300.

Gruhn, N., Halawa, M., Snel, B., et al., 2014. A subfamily of putative cytokinin receptors is revealed by an analysis of the evolution of the two-component signaling system of plants. Plant Physiol. 165, 227–237.

Ha, S., Vankova, R., Yamaguchi-Shinozaki, K., et al., 2012. Cytokinins: metabolism and function in plant adaptation to environmental stresses. Trends Plant Sci. 17, 172–179.

Haberer, G., Kieber, J.J., 2002. Cytokinins. New insights into a classic phytohormone. Plant Physiol. 128, 354–362.

Hejátko, J., Pernisová, M., Eneva, T., et al., 2003. The putative sensor histidine kinase CKI1 is involved in female gametophyte development in *Arabidopsis*. Mol. Genet. Genomics 269, 443–453.

Hess, D.T., Matsumoto, A., Kim, S.O., et al., 2005. Protein *S*-nitrosylation: purview and parameters. Nat. Rev. Mol. Cell Biol. 6, 150–166.

Higuchi, M., Pischke, M.S., Mähönen, A.P., et al., 2004. *In planta* functions of the *Arabidopsis* cytokinin receptor family. Proc. Natl. Acad. Sci. U.S.A. 101, 8821–8826.

Hirose, N., Makita, N., Yamaya, T., et al., 2005. Functional characterization and expression analysis of a gene, *OsENT2*, encoding an equilibrative nucleoside transporter in rice suggest a function in cytokinin transport. Plant Physiol. 138, 196–206.

Hirose, N., Takei, K., Kuroha, T., et al., 2008. Regulation of cytokinin biosynthesis, compartmentalization and translocation. J. Exp. Bot. 59, 75–83.

Hosoda, K., Imamura, A., Katoh, E., et al., 2002. Molecular structure of the GARP family of plant Myb-related DNA binding motifs of the *Arabidopsis* response regulators. Plant Cell 14, 2015–2029.

Hothorn, M., Dabi, T., Chory, J., 2011. Structural basis for cytokinin recognition by *Arabidopsis thaliana* histidine kinase 4. Nat. Chem. Biol. 7, 766–768.

Hou, B., Lim, E.K., Higgins, G.S., et al., 2004. *N*-glucosylation of cytokinins by glycosyltransferases of *Arabidopsis thaliana*. J. Biol. Chem. 279, 47822–47832.

Houba-Hérin, N., Pethe, C., d'Alayer, J., et al., 1999. Cytokinin oxidase from *Zea mays*: purification, cDNA cloning and expression in moss protoplasts. Plant J. 17, 615–626.

Hutchison, C.E., Li, J., Argueso, C., et al., 2006. The *Arabidopsis* histidine phosphotransfer proteins are redundant positive regulators of cytokinin signaling. Plant Cell 18, 3073–3087.

Hwang, I., Sheen, J., 2001. Two-component circuitry in *Arabidopsis* cytokinin signal transduction. Nature 413, 383–389.

Hwang, I., Sheen, J., Müller, B., 2012. Cytokinin signaling networks. Annu. Rev. Plant Biol. 63, 353–380.

Imamura, A., Hanaki, N., Umeda, H., et al., 1998. Response regulators implicated in His-to-Asp phosphotransfer signaling in *Arabidopsis*. Proc. Natl. Acad. Sci. U.S.A. 95, 2691–2696.

Imamura, A., Kiba, T., Tajima, Y., et al., 2003. *In vivo* and *in vitro* characterization of the ARR11 response regulator implicated in the His-to-Asp phosphorelay signal transduction in *Arabidopsis thaliana*. Plant Cell Physiol. 44, 122–131.

Imamura, A., Yoshino, Y., Mizuno, T., 2001. Cellular localization of the signaling components of *Arabidopsis* His-to-Asp phosphorelay. Biosci. Biotechnol. Biochem. 65, 2113–2117.

Inoue, T., Higuchi, M., Hashimoto, Y., et al., 2001. Identification of CRE1 as a cytokinin receptor from *Arabidopsis*. Nature 409, 1060–1063.

Ishida, K., Yamashino, T., Yokoyama, A., et al., 2008. Three type-B response regulators, ARR1, ARR10 and ARR12, play essential but redundant roles in cytokinin signal transduction throughout the life cycle of *Arabidopsis thaliana*. Plant Cell Physiol. 49, 47–57.

Janssen, B.J., Drummond, R.S.M., Snowden, K.C., 2014. Regulation of axillary shoot development. Curr. Opin. Plant Biol. 17, 28–35.

Jin, S.H., Ma, X.M., Kojima, M., et al., 2013. Overexpression of glucosyltransferase UGT85A1 influences *trans*-zeatin homeostasis and *trans*-zeatin responses likely through *O*-glucosylation. Planta 237, 991–999.

Kakimoto, T., 1996. CKI1, a histidine kinase homolog implicated in cytokinin signal transduction. Science 274, 982–985.

Kakimoto, T., 2001. Identification of plant cytokinin biosynthetic enzymes as dimethylallyl diphosphate: ATP/ADP isopentenyltransferases. Plant Cell Physiol. 42, 677–685.

Kakimoto, T., 2003a. Biosynthesis of cytokinins. J. Plant Res. 116, 233–239.

Kakimoto, T., 2003b. Perception and signal transduction of cytokinins. Annu. Rev. Plant Biol. 54, 605–627.

Kiba, T., Takei, K., Kojima, M., et al., 2013. Side-chain modification of cytokinins controls shoot growth in Arabidopsis. Dev. Cell 27, 452–461.

Kiba, T., Yamada, H., Mizuno, T., 2002. Characterization of the ARR15 and ARR16 response regulators with special reference to the cytokinin signaling pathway mediated by the AHK4 histidine kinase in roots of *Arabidopsis thaliana*. Plant Cell Physiol. 43, 1059–1066.

Kiba, T., Yamada, H., Sato, S., et al., 2003. The type-A response regulator, ARR15, acts as a negative regulator in the cytokinin-mediated signal transduction in *Arabidopsis thaliana*. Plant Cell Physiol. 44, 868–874.

Kieber, J.J., 2002. Tribute to Folke Skoog: recent advances in our understanding of cytokinin biology. J. Integr. Plant Biol. 21, 1–2.

Kieber, J.J., Schaller, G.E., 2014. Cytokinins. The Arabidopsis Book/Am. Soc. Plant Biol. 12, e0168.

Kim, H.J., Chiang, Y.H., Kieber, J.J., et al., 2013a. SCF^{KMD} controls cytokinin signaling by regulating the degradation of type-B response regulators. Proc. Natl. Acad. Sci. U.S.A. 110, 10028–10033.

Kim, H.J., Kieber, J.J., Schaller, G.E., 2013b. The rice F-box protein KISS ME DEADLY2 functions as a negative regulator of cytokinin signalling. Plant Signal. Behav. 8, e26434.

Kim, H.J., Ryu, H., Hong, S.H., et al., 2006. Cytokinin-mediated control of leaf longevity by AHK3 through phosphorylation of ARR2 in *Arabidopsis*. Proc. Natl. Acad. Sci. U.S.A. 103, 814–819.

Kinoshita-Tsujimura, K., Kakimoto, T., 2011. Cytokinin receptors in sporophytes are essential for male and female functions in *Arabidopsis thaliana*. Plant Signal. Behav. 6, 66–71.

Ko, D., Kang, J., Kiba, T., et al., 2014. *Arabidopsis* ABCG14 is essential for the root-to-shoot translocation of cytokinin. Proc. Natl. Acad. Sci. U.S.A. 111, 7150–7155.

Kopecna, M., Blaschke, H., Kopecny, D., et al., 2013. Structure and function of nucleoside hydrolases from *Physcomitrella patens* and maize catalyzing the hydrolysis of purine, pyrimidine, and cytokinin ribosides. Plant Physiol. 163, 1568–1583.

Kudo, T., Kiba, T., Sakakibara, H., 2010. Metabolism and long-distance translocation of cytokinins. J. Integr. Plant Biol. 52, 53–60.

Kudo, T., Makita, N., Kojima, M., et al., 2012. Cytokinin activity of *cis*-zeatin and phenotypic alterations induced by overexpression of putative *cis*-zeatin-*O*-glucosyltransferase in rice. Plant Physiol. 160, 319–331.

Kurakawa, T., Ueda, N., Maekawa, M., et al., 2007. Direct control of shoot meristem activity by a cytokinin-activating enzyme. Nature 445, 652–655.

Kurepa, J., Li, Y., Smalle, J.A., 2014. Cytokinin signaling stabilizes the response activator ARR1. Plant J. 78, 157–168.

Kuroha, T., Tokunaga, H., Kojima, M., et al., 2009. Functional analyses of LONELY GUY cytokinin-activating enzymes reveal the importance of the direct activation pathway in *Arabidopsis*. Plant Cell 21, 3152–3169.

Laub, M.T., Goulian, M., 2007. Specificity in two-component signal transduction pathways. Annu. Rev. Genet. 41, 121–145.

Lee, D.J., Kim, S., Ha, Y.M., et al., 2008. Phosphorylation of *Arabidopsis* response regulator 7 (ARR7) at the putative phospho-accepting site is required for ARR7 to act as a negative regulator of cytokinin signaling. Planta 227, 577–587.

Lee, D.J., Park, J.Y., Ku, S.J., et al., 2007. Genome-wide expression profiling of *ARABIDOPSIS RESPONSE REGULATOR 7* (*ARR7*) overexpression in cytokinin response. Mol. Genet. Genomics 277, 115–137.

Li, S., Zhao, B., Yuan, D., et al., 2013. Rice zinc finger protein DST enhances grain production through controlling Gn1a/OsCKX2 expression. Proc. Natl. Acad. Sci. U.S.A. 110, 3167–3172.

Liang, Y., Wang, X., Hong, S., et al., 2012. Deletion of the initial 45 residues of ARR18 induces cytokinin response in *Arabidopsis*. J. Genet. Genomics 39, 37–46.

Liu, W.Z., Kong, D.D., Gu, X.X., et al., 2013. Cytokinins can act as suppressors of nitric oxide in *Arabidopsis*. Proc. Natl. Acad. Sci. U.S.A. 110, 1548–1553.

Lomin, S.N., Krivosheev, D.M., Steklov, M.Y., et al., 2015. Plant membrane assays with cytokinin receptors underpin the unique role of free cytokinin bases as biologically active ligands. J. Exp. Bot. 66, 1851–1863.

Lomin, S.N., Yonekura-Sakakibara, K., Romanov, G.A., et al., 2011. Ligand-binding properties and subcellular localization of maize cytokinin receptors. J. Exp. Bot. 62, 5149–5159.

Mähönen, A.P., Bishopp, A., Higuchi, M., et al., 2006a. Cytokinin signaling and its inhibitor AHP6 regulate cell fate during vascular development. Science 311, 94–98.

Mähönen, A.P., Bonke, M., Kauppinen, L., et al., 2000. A novel two-component hybrid molecule regulates vascular morphogenesis of the *Arabidopsis* root. Gene Dev. 14, 2938–2943.

Mähönen, A.P., Higuchi, M., Törmäkangas, K., et al., 2006b. Cytokinins regulate a bidirectional phosphorelay network in *Arabidopsis*. Curr. Biol. 16, 1116–1122.

Müller, B., Sheen, J., 2008. Cytokinin and auxin interaction in root stem-cell specification during early embryogenesis. Nature 453, 1094–1097.

Martin, R.C., Mok, M.C., Habben, J.E., et al., 2001. A maize cytokinin gene encoding an *O*-glucosyltransferase specific to *cis*-zeatin. Proc. Natl. Acad. Sci. U.S.A. 98, 5922–5926.

Martin, R.C., Mok, M.C., Mok, D.W., 1999. Isolation of a cytokinin gene, *ZOG1*, encoding zeatin *O*-glucosyltransferase from *Phaseolus lunatus*. Proc. Natl. Acad. Sci. U.S.A. 96, 284–289.

Mason, M.G., Mathews, D.E., Argyros, D.A., et al., 2005. Multiple type-B response regulators mediate cytokinin signal transduction in *Arabidopsis*. Plant Cell 17, 3007–3018.

Massonneau, A., Houba-Herin, N., Pethe, C., et al., 2004. Maize cytokinin oxidase genes: differential expression and cloning of two new cDNAs. J. Exp. Bot. 55, 2549–2557.

Miller, C.O., 1961. A kinetin-like compound in maize. Proc. Natl. Acad. Sci. U.S.A. 47, 170–174.

Miller, C.O., Skoog, F., Okumura, F.S., et al., 1955a. Structure and synthesis of kinetin. J. Am. Chem. Soc. 78, 2662–2663.

Miller, C.O., Skoog, F., Von Saltza, M.H., et al., 1955b. J. Am. Chem. Soc. 77, 1392.

Miyawaki, K., Matsumoto-Kitano, M., Kakimoto, T., 2004. Expression of cytokinin biosynthetic isopentenyltransferase genes in *Arabidopsis*: tissue specificity and regulation by auxin, cytokinin, and nitrate. Plant J. 37, 128–138.

Miyawaki, K., Tarkowski, P., Matsumoto-Kitano, M., et al., 2006. Roles of *Arabidopsis* ATP/ADP isopentenyltransferases and tRNA isopentenyltransferases in cytokinin biosynthesis. Proc. Natl. Acad. Sci. U.S.A. 103, 16598–16603.

Mok, D.W., Mok, M.C., 2001. Cytokinin metabolism and action. Annu. Rev. Plant Biol. 52, 89–118.

Morris, R.O., Bilyeu, K.D., Laskey, J.G., et al., 1999. Isolation of a gene encoding a glycosylated cytokinin oxidase from maize. Biochem. Biophys. Res. Commun. 255, 328–333.

Moubayidin, L., Perilli, S., Dello Ioio, R., et al., 2010. The rate of cell differentiation controls the *Arabidopsis* root meristem growth phase. Curr. Biol. 20, 1138–1143.

Nishimura, C., Ohashi, Y., Sato, S., et al., 2004. Histidine kinase homologs that act as cytokinin receptors possess overlapping functions in the regulation of shoot and root growth in *Arabidopsis*. Plant Cell 16, 1365–1377.

Osakabe, Y., Miyata, S., Urao, T., et al., 2002. Overexpression of *Arabidopsis* response regulators, ARR4/ATRR1/IBC7 and ARR8/ATRR3, alters cytokinin responses differentially in the shoot and in callus formation. Biochem. Biophys. Res. Commun. 293, 806–815.

Paces, V., Werstiuk, E., Hall, R.H., 1971. Conversion of *N*-(delta-isopentenyl)adenosine to adenosine by enzyme activity in tobacco tissue. Plant Physiol. 48, 775–778.

Perilli, S., Moubayidin, L., Sabatini, S., 2010. The molecular basis of cytokinin function. Curr. Opin. Plant Biol. 13, 21–26.

Pils, B., Heyl, A., 2009. Unraveling the evolution of cytokinin signaling. Plant Physiol. 151, 782–791.

Pischke, M.S., Jones, L.G., Otsuga, D., et al., 2002. An *Arabidopsis* histidine kinase is essential for megagametogenesis. Proc. Natl. Acad. Sci. U.S.A. 99, 15800–15805.

Plakidou-Dymock, S., Dymock, D., Hooley, R., 1998. A higher plant seven-transmembrane receptor that influences sensitivity to cytokinins. Curr. Biol. 8, 315–324.

Punwani, J.A., Hutchison, C.E., Schaller, G.E., et al., 2010. The subcellular distribution of the *Arabidopsis* histidine phosphotransfer proteins is independent of cytokinin signaling. Plant J. 62, 473–482.

Raines, T., Shanks, C., Cheng, C.Y., et al., 2016. The cytokinin response factors modulate root and shoot growth and promote leaf senescence in *Arabidopsis*. Plant J. 85, 134–147.

Rashotte, A.M., Mason, M.G., Hutchison, C.E., et al., 2006. A subset of *Arabidopsis* AP2 transcription factors mediates cytokinin responses in concert with a two-component pathway. Proc. Natl. Acad. Sci. U.S.A. 103, 11081–11085.

Ren, B., Chen, Q., Hong, S., et al., 2013. The *Arabidopsis* eukaryotic translation initiation factor eIF5A-2 regulates root protoxylem development by modulating cytokinin signaling. Plant Cell 25, 3841–3857.

Ren, B., Liang, Y., Deng, Y., et al., 2009. Genome-wide comparative analysis of type-A *Arabidopsis response regulator* genes by overexpression studies reveals their diverse roles and regulatory mechanisms in cytokinin signaling. Cell Res. 19, 1178–1190.

Riefler, M., Novak, O., Strnad, M., et al., 2006. *Arabidopsis* cytokinin receptor mutants reveal functions in shoot growth, leaf senescence, seed size, germination, root development, and cytokinin metabolism. Plant Cell 18, 40–54.

Romanov, G.A., Lomin, S.N., Schmülling, T., 2006. Biochemical characteristics and ligand-binding properties of Arabidopsis cytokinin receptor AHK3 compared to CRE1/AHK4 as revealed by a direct binding assay. J. Exp. Bot. 57, 4051–4058.

Sakai, H., Aoyama, T., Oka, A., 2000. *Arabidopsis* ARR1 and ARR2 response regulators operate as transcriptional activators. Plant J. 24, 703–711.

Sakai, H., Honma, T., Aoyama, T., et al., 2001. ARR1, a transcription factor for genes immediately responsive to cytokinins. Science 294, 1519–1521.

Sakakibara, H., 2006. Cytokinins: activity, biosynthesis, and translocation. Annu. Rev. Plant Physiol. 57, 431–449.

Sakamoto, T., Sakakibara, H., Kojima, M., et al., 2006. Ectopic expression of KNOTTED1-like homeobox protein induces expression of cytokinin biosynthesis genes in rice. Plant Physiol. 142, 54–62.

Sakano, Y., Okada, Y., Matsunaga, A., et al., 2004. Molecular cloning, expression, and characterization of adenylate isopentenyltransferase from hop (*Humulus lupulus* L.). Phytochemistry 65, 2439–2446.

Schaller, G.E., Bishopp, A., Kieber, J.J., 2015. The yin-yang of hormones: cytokinin and auxin interactions in plant development. Plant Cell 27, 44–63.

Schaller, G.E., Doi, K., Hwang, I., et al., 2007. Nomenclature for two-component signaling elements of rice. Plant Physiol. 143, 555–557.

Schaller, G.E., Shiu, S.H., Armitage, J.P., 2011. Two-component systems and their co-option for eukaryotic signal transduction. Curr. Biol. 21, R320–R330.

Schaller, G.E., Street, I.H., Kieber, J.J., 2014. Cytokinin and the cell cycle. Curr. Opin. Plant Biol. 21, 7–15.

Schmülling, T., 2002. New insights into the functions of cytokinins in plant development. J. Plant Growth Regul. 21, 40–49.

Sheen, J., 2002. Phosphorelay and transcription control in cytokinin signal transduction. Science 296, 1650–1652.

Šimášková, M., O'Brien, J.A., Khan, M., et al., 2015. Cytokinin response factors regulate PIN-FORMED auxin transporters. Nat. Commun. 6, 8717.

Spichal, L., Rakova, N.Y., Riefler, M., et al., 2004. Two cytokinin receptors of *Arabidopsis thaliana*, CRE1/AHK4 and AHK3, differ in their ligand specificity in a bacterial assay. Plant Cell Physiol. 45, 1299–1305.

Spiess, L.D., 1975. Comparative activity of isomers of zeatin and ribosyl-zeatin on *Funaria hygrometrica*. Plant Physiol. 55, 583–585.

Stolz, A., Riefler, M., Lomin, S.N., et al., 2011. The specificity of cytokinin signalling in *Arabidopsis thaliana* is mediated by differing ligand affinities and expression profiles of the receptors. Plant J. 67, 157–168.

Strnad, M., 1997. The aromatic cytokinins. Physiol. Plant 101, 674–688.

Su, W., Howell, S.H., 1992. A single genetic locus, Ckr1, defines *Arabidopsis* mutants in which root growth is resistant to low concentrations of cytokinin. Plant Physiol. 99, 1569–1574.

Su, Y.H., Zhang, X.S., 2014. The hormonal control of regeneration in plants. In: Brigitte, G. (Ed.). Brigitte, G. (Ed.), Curr Top Dev Biol, vol. 108. Academic Press, pp. 35–69.

Sun, J., Hirose, N., Wang, X., Wen, P., Xue, L., Sakakibara, H., Zuo, J., 2005. *Arabidopsis SOI33/AtENT8* gene encodes a putative equilibrative nucleoside transporter that is involved in cytokinin transport in planta. J. Integr. Plant Biol. 47, 588–603.

Sun, J., Niu, Q.W., Tarkowski, P., et al., 2003. The Arabidopsis *AtIPT8/PGA22* gene encodes an isopentenyl transferase that is involved in de novo cytokinin biosynthesis. Plant Physiol. 131, 167–176.

Suzuki, T., Imamura, A., Ueguchi, C., et al., 1998. Histidine-containing phosphotransfer (HPt) signal transducers implicated in His-to-Asp phosphorelay in *Arabidopsis*. Plant Cell Physiol. 39, 1258–1268.

Suzuki, T., Ishikawa, K., Yamashino, T., et al., 2002. An *Arabidopsis* histidine-containing phosphotransfer (HPt) factor implicated in phosphorelay signal transduction: overexpression of AHP2 in plants results in hypersensitiveness to cytokinin. Plant Cell Physiol. 43, 123–129.

Suzuki, T., Miwa, K., Ishikawa, K., et al., 2001. The *Arabidopsis* sensor His-kinase, AHK4, can respond to cytokinins. Plant Cell Physiol. 42, 107–113.

Tajima, Y., Imamura, A., Kiba, T., et al., 2004. Comparative studies on the type-B response regulators revealing their distinctive properties in the His-to-Asp phosphorelay signal transduction of *Arabidopsis thaliana*. Plant Cell Physiol. 45, 28–39.

Takei, K., Sakakibara, H., Sugiyama, T., 2001. Identification of genes encoding adenylate isopentenyltransferase, a cytokinin biosynthesis enzyme, in *Arabidopsis thaliana*. J. Biol. Chem. 276, 26405–26410.

Takei, K., Takahashi, T., Sugiyama, T., et al., 2002. Multiple routes communicating nitrogen availability from roots to shoots: a signal transduction pathway mediated by cytokinin. J. Exp. Bot. 53, 971–977.

Takei, K., Ueda, N., Aoki, K., et al., 2004a. AtIPT3 is a key determinant of nitrate-dependent cytokinin biosynthesis in *Arabidopsis*. Plant Cell Physiol. 45, 1053–1062.

Takei, K., Yamaya, T., Sakakibara, H., 2004b. *Arabidopsis CYP735A1* and *CYP735A2* encode cytokinin hydroxylases that catalyze the biosynthesis of *trans*-Zeatin. J. Biol. Chem. 279, 41866–41872.

Tanaka, M., Takei, K., Kojima, M., Sakakibara, H., Mori, H., 2006. Auxin controls local cytokinin biosynthesis in the nodal stem in apical dominance. Plant J. 45, 1028–1036.

Taniguchi, M., Sasaki, N., Tsuge, T., et al., 2007. ARR1 directly activates cytokinin response genes that encode proteins with diverse regulatory functions. Plant Cell Physiol. 48, 263–277.

Tarkowska, D., Dolezal, K., Tarkowski, P., et al., 2003. Identification of new aromatic cytokinins in *Arabidopsis thaliana* and *Populus* x *canadensis* leaves by LC-(+) ESI-MS and capillary liquid chromatography/frit-fast atom bombardment mass spectrometry. Physiol. Plant 117, 579–590.

Taya, Y., Tanaka, Y., Nishimura, S., 1978. 5'-AMP is a direct precursor of cytokinin in *Dictyostelium discoideum*. Nature 271, 545–547.

To, J.P.C., Deruère, J., Maxwell, B.B., et al., 2007. Cytokinin regulates type-A *Arabidopsis* response regulator activity and protein stability via two-component phosphorelay. Plant Cell 19, 3901–3914.

To, J.P.C., Haberer, G., Ferreira, F.J., et al., 2004. Type-A *Arabidopsis* response regulators are partially redundant negative regulators of cytokinin signaling. Plant Cell 16, 658–671.

To, J.P.C., Kieber, J.J., 2008. Cytokinin signaling: two-components and more. Trends Plant Sci. 13, 85–92.

Tsai, Y.C., Weir, N.R., Hill, K., et al., 2012. Characterization of genes involved in cytokinin signaling and metabolism from rice. Plant Physiol. 158, 1666–1684.

Veach, Y.K., Martin, R.C., Mok, D.W., et al., 2003. *O*-glucosylation of *cis*-zeatin in maize. Characterization of genes, enzymes, and endogenous cytokinins. Plant Physiol. 131, 1374–1380.

Veerabagu, M., Elgass, K., Kirchler, T., et al., 2012. The *Arabidopsis* B-type response regulator 18 (ARR18) homomerizes and positively regulates cytokinin responses. Plant J. 72, 721–731.

Vogel, J.P., Schuerman, P., Woeste, K., et al., 1998. Isolation and characterization of *Arabidopsis* mutants defective in the induction of ethylene biosynthesis by cytokinin. Genetics 149, 417–427.

von Schwartzenberg, K., Lindner, A.C., Gruhn, N., et al., 2016. CHASE domain-containing receptors play an essential role in the cytokinin response of the moss *Physcomitrella patens*. J. Exp. Bot. 67, 667–679.

Vyroubalová, Š., Václavíková, K., Turečková, V., et al., 2009. Characterization of new maize genes putatively involved in cytokinin metabolism and their expression during osmotic stress in relation to cytokinin levels. Plant Physiol. 151, 433–447.

Wang, J., Ma, X.M., Kojima, M., et al., 2011. *N*-glucosyltransferase UGT76C2 is involved in cytokinin homeostasis and cytokinin response in *Arabidopsis thaliana*. Plant Cell Physiol. 52, 2200–2213.

Wang, J., Ma, X.M., Kojima, M., et al., 2013. Glucosyltransferase UGT76C1 finely modulates cytokinin responses via cytokinin *N*-glucosylation in *Arabidopsis thaliana*. Plant Physiol. Biochem. 65, 9–16.

Werner, T., Motyka, V., Laucou, V., Smets, R., Van Onckelen, H., Schmülling, T., 2003. Cytokinin-deficient transgenic Arabidopsis plants show multiple developmental alterations indicating opposite functions of cytokinins in the regulation of shoot and root meristem activity. Plant Cell 15, 2532–2550.

Werner, T., Schmülling, T., 2009. Cytokinin action in plant development. Curr. Opin. Plant Biol. 12, 527–538.

Whitty, C.D., Hall, R.H., 1974. A cytokinin oxidase in *Zea mays*. Can J. Biochem. 52, 789–799.

Wulfetange, K., Lomin, S.N., Romanov, G.A., et al., 2011. The cytokinin receptors of *Arabidopsis* are located mainly to the endoplasmic reticulum. Plant Physiol. 156, 1808–1818.

Yamada, H., Suzuki, T., Terada, K., et al., 2001. The *Arabidopsis* AHK4 histidine kinase is a cytokinin-binding receptor that transduces cytokinin signals across the membrane. Plant Cell Physiol. 42, 1017–1023.

Yang, S.H., Yu, H., Goh, C.J., 2002. Isolation and characterization of the orchid cytokinin oxidase *DSCKX1* promoter. J. Exp. Bot. 53, 1899–1907.

Yang, S.H., Yu, H., Goh, C.J., 2003. Functional characterisation of a cytokinin oxidase gene *DSCKX1* in *Dendrobium* orchid. Plant Mol. Biol. 51, 237–248.

Yokoyama, A., Yamashino, T., Amano, Y., et al., 2007. Type-B ARR transcription factors, ARR10 and ARR12, are implicated in cytokinin-mediated regulation of protoxylem differentiation in roots of *Arabidopsis thaliana*. Plant Cell Physiol. 48, 84–96.

Yonekura-Sakakibara, K., Kojima, M., Yamaya, T., et al., 2004. Molecular characterization of cytokinin-responsive histidine kinases in maize. Differential ligand preferences and response to *cis*-zeatin. Plant Physiol. 134, 1654–1661.

Zürcher, E., Müller, B., 2016. Cytokinin synthesis, signaling, and function-advances and new insights. Int. Rev. Cell Mol. Biol. 324, 1–38.

Zhang, K., Novak, O., Wei, Z., et al., 2014. Arabidopsis ABCG14 protein controls the acropetal translocation of root-synthesized cytokinins. Nat. Commun. 5, 3274.

Zubko, E., Adams, C.J., Macháèková, I., et al., 2002. Activation tagging identifies a gene from *Petunia hybrida* responsible for the production of active cytokinins in plants. Plant J. 29, 797–808.

Gibberellins

4

Xiuhua Gao[1], Yingying Zhang[2], Zuhua He[2], Xiangdong Fu[1]
[1]Institute of Genetics and Developmental Biology, Chinese Academy of Sciences, Beijing, China; [2]Shanghai Institutes for Biological Sciences, Chinese Academy of Sciences, Shanghai, China

Summary

Gibberellins (GAs), a class of diterpenoid phytohormones, produced by plants and some fungi play an important role in modulating diverse processes throughout plant growth and development. So far, up to 136 different gibberellin molecules have been discovered, only a few of which are bioactive, such as GA_1, GA_3, GA_4, and GA_7. Recent studies on GA biosynthesis, metabolism, transport, and signaling, as well as cross talk between GA and other plant hormones and environmental cues have achieved great progress along with the advancement of molecular genetics and functional genomics. Accumulating evidences suggest that the "de-repression" model makes it possible to explain signal transduction mechanisms in GA action. Bioactive GAs promote plant growth and development by promoting the degradation of the DELLA proteins, a family of nuclear growth repressors. The GA signal is perceived by the soluble receptor protein GIBBERELLIN INSENSITIVE DWARF1 (GID1) that undergoes a conformational change and then promotes GA-GID1-DELLA association with the Skp1-Cullin-F-box (SCF) E3 ubiquitin-ligase complex via the F-box protein (SLEEPY1 [SLY1] in Arabidopsis and GIBBERELLIN INSENSITIVE DWARF2 [GID2] in rice), thereby targeting the DELLA proteins for degradation via the 26S proteasome pathway. Evidence also shows that GAs act as mobile molecules that can pass through the plasma membrane for cell-to-cell transport. In this chapter, we focus on findings on GA biosynthesis, perception, and signal transduction pathways, highlighting how the evolutionary conserved GA-GID1-DELLA regulatory module is connected to developmental and environmental responses.

4.1 Functions of gibberellins

Gibberellins (GAs) are a large family of tetracyclic diterpenoid plant hormones that regulate many different aspects of plant growth and development through the entire life cycle of the plant, including promotion of cell division and elongation, seed germination, stem and hypocotyl elongation, root growth, and flowering induction (Daviere et al., 2008; Sun, 2011; Sun and Gubler, 2004; Vera-Sirera et al., 2016). In addition, GAs also regulate plant adaptation to biotic and abiotic stresses (Daviere and Achard, 2015; Yang et al., 2008, 2012).

The action of GA in promoting plant growth was first discovered in 1930s by studies of the rice *Bakanae* disease. The rice plants infected by the pathogenic fungus *Gibberella fujikuroi* exhibited excessive stem elongation such that they fell over easily (Silverstone and Sun, 2000). Later, the metabolite produced from the

Hormone Metabolism and Signaling in Plants. http://dx.doi.org/10.1016/B978-0-12-811562-6.00004-9

pathogenic fungus was identified as GA. There are at least 136 fully characterized GAs, named from GA_1 to GA_{136}, which have been identified from various bacteria, fungi, and plants (Hedden and Thomas, 2012; Silverstone and Sun, 2000). However, only a few of the GAs, such as GA_1, GA_3, GA_4, and GA_7, have biological activity as regulators of plant growth and development. The genetic evidence has revealed that GA_1 and GA_4 are major active GAs in most plant species although GA_3 has been identified in plants. Moreover, the bioactivity of GA_4 is stronger than that of GA_1 in both Arabidopsis and rice (Cowling et al., 1998; Magome et al., 2013; Nomura et al., 2013; Ueguchi-Tanaka et al., 2007). GA homeostasis is tightly feedback controlled by GA metabolism and signaling (Hedden and Kamiya, 1997).

The most remarkable discovery of GA functions is that GAs are critical to the "Green Revolution": the semidwarf mutant varieties of wheat and rice greatly improved grain productivity, which resulted from the *sd1* mutation in rice and *Rht1* mutation in wheat. The rice *SD1* (*SEMI-DWARF1*) gene encodes a key GA biosynthetic enzyme, OsGA20ox2, and the wheat *Rht1* gene encodes the GA repressor DELLA (Peng et al., 1999; Sasaki et al., 2002; Spielmeyer et al., 2002). Therefore, GA biosynthesis and signaling are critical to plant growth and development, and also of vital importance in agriculture. This chapter will summarize GA biosynthesis, transport, and signal transduction, and highlight the molecular mechanism of GA turning on the system to relieve DELLA repression and the function of DELLAs in integrating multiple signaling to control plant growth and development, as well as cross talk with other plant hormones and environmental responses.

4.2   Gibberellin biosynthesis, inactivation, transport and regulation

4.2.1   Gibberellin biosynthesis pathway

More than 136GAs have been identified so far (see http://www.plant-hormones. info/gibberellins.htm for details). Most GAs are biologically inactive, and are mainly intermediates and precursors in the biosynthesis or degradation of bioactive GAs (Hedden and Phillips, 2000; Hedden and Thomas, 2012; Serrani et al., 2007; Yamaguchi, 2008; Yamaguchi and Kamiya, 2000; Zi et al., 2014). Only a few GAs, including GA_1, GA_3, GA_4, and GA_7, exhibit biological activity toward various aspects of plant growth and development in the whole life cycle, such as seed germination, stem elongation, leaf expansion, flowering, and fruit and seed development (Hedden and Sponsel, 2015; Plackett et al., 2011; Sakamoto et al., 2004). These bioactive GAs were primarily identified from the model plants: the dicot Arabidopsis and the monocot rice.

The GA biosynthesis pathway has been well established through extensive biochemical and genetic studies in higher plants over a long period, which requires a set of enzymes, including terpene synthases (TPSs), cytochrome P450 monooxygenases (P450s), and 2-oxoglutarate-dependent dioxygenases (2ODDs). Many genes that encode these enzymes have been functionally identified and their regulation elucidated. These enzymatic reactions, as well as the production of bioactive GAs, are strictly and precisely controlled in both their timing and subcellular localization at the tissue and organ levels (Hedden and Thomas, 2012; Itoh et al., 1999; Phillips, 1998). Accordingly, GA biosynthesis can be generally divided into three stages in three subcellular compartments: plastids, the endomembrane system, and the cytosol (Fig. 4.1) (Han and Zhu, 2011; Hedden and Thomas, 2012; Yamaguchi, 2008).

The first stage of GA biosynthesis is the conversion of geranylgeranyl diphosphate (GGDP) to *ent*-kaurene, which occurs in a two-step process (Aach et al., 1997; Helliwell et al., 2001b; Smith et al., 1998; Sun and Kamiya, 1994). The initial step is the conversion of GGDP, a common C20 precursor for diterpenoids into *ent*-kaurene by two TPS enzymes, *ent*-copalyl diphosphate (*ent*-CDP) synthase (CPS) and *ent*-kaurene synthase (KS), in the plastids (Fig. 4.1). Loss-of-function mutations of *CPS* or *KS* in Arabidopsis and rice cause severe GA-deficient dwarf phenotypes, whereas overexpression of both *CPS* and *KS* greatly increases levels of the early intermediates *ent*-kaurene and *ent*-kaurenoic acid, with only slight increases in later metabolites (Sakamoto et al., 2004; Sun et al., 1992; Yamaguchi et al., 1998b). As a result, these overexpression lines have wild-type levels of bioactive GAs and do not exhibit any GA overdose phenotypes (Fleet et al., 2003; Otsuka et al., 2004). Therefore, plants have developed the ability to maintain GA homeostasis despite the accumulation of early intermediates. This GA homeostasis is likely attributed to the strict control of gene expression. While the *KS* gene seems to be constitutively expressed, *CPS* is expressed at low levels in a cell-specific manner during development and may thus act as a "gatekeeper" to control the flux of *ent*-kaurene metabolites in the early stage of the GA biosynthesis pathway (Fleet et al., 2003; Olszewski et al., 2002; Silverstone et al., 1997a). Interestingly, both CPS and KS are encoded by a single gene in most plant species; their expression is not under feedback regulation by the activity of the GA signaling pathway, unlike later GA biosynthesis genes which are usually feedback-regulated (Fleet et al., 2003; Koornneef and Van der Veen, 1980; Silverstone et al., 1997a). The *AtCPS* (also called *GA1*) gene was first isolated and identified from the *ga1* mutants, also known as "non-germinating GA-dwarfs" (Aach et al., 1997; Helliwell et al., 2001b; Sun and Kamiya, 1994; Sun et al., 1992). The Arabidopsis *ga1* mutants are male-sterile dwarfs whose phenotypes can be converted to the wild-type by repeated application of exogenous GA (Koornneef and Van der Veen, 1980). Using a *GUS* fusion reporter transgene, *AtCPS* was found to be expressed in the developing embryos in the provasculature, suggesting that GA might play an important role in vasculature development (Yamaguchi et al., 2001).

The rice genome has four *CPS*-like genes, of which *OsCPS3* is a pseudogene (Sakamoto et al., 2004). Interestingly, *OsCPS4* encodes a *syn*-copalyl diphosphate synthase that functions in phytoalexin but not GA biosynthesis (Toyomasu et al., 2014). Both *OsCPS1* and *OsCPS2* are located on chromosome 2. The OsCPS1 protein is essential for GA biosynthesis (Toyomasu et al., 2015). The rice loss-of-function mutation in *OsCPS1* exhibits GA-deficient dwarfism, indicating its critical

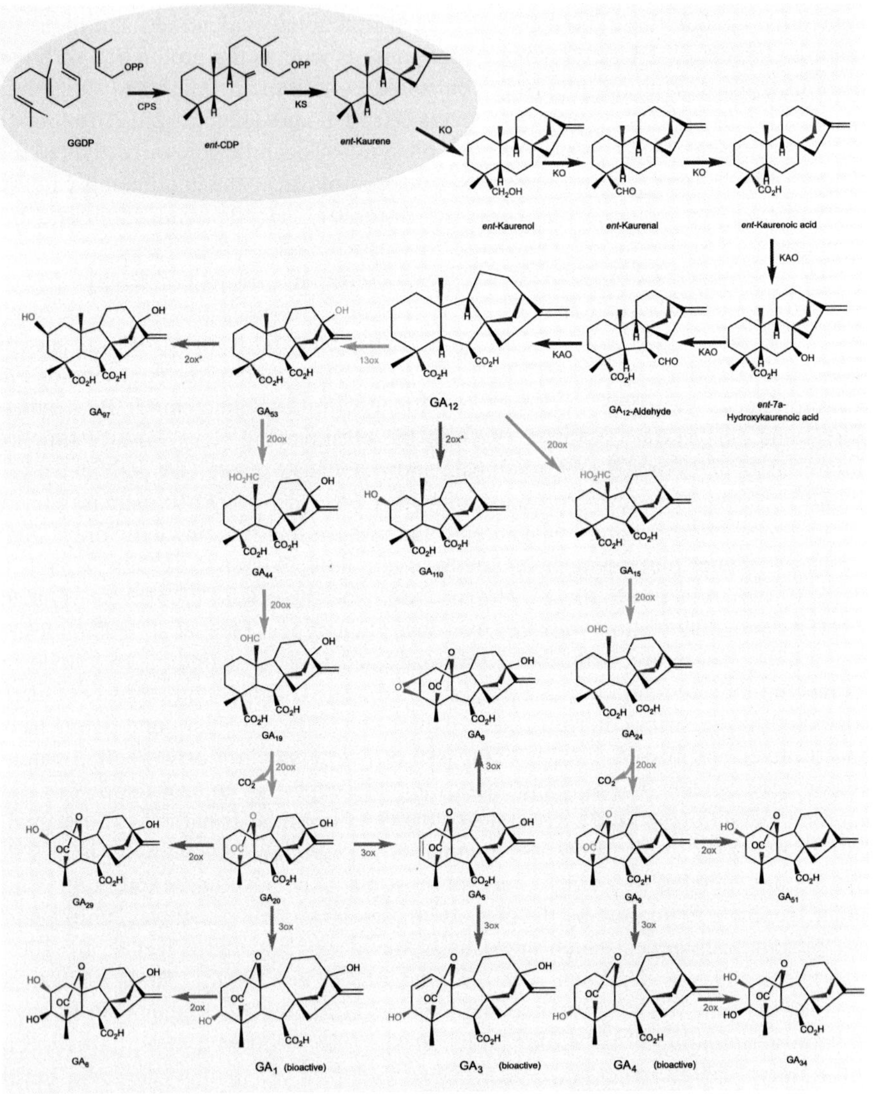

Figure 4.1 The principal pathways of gibberellin (GA) biosynthesis and deactivation by GA 2-oxidase in higher plants, showing subcellular compartmentalization between plastids, the endomembrane system, and cytoplasm. In each metabolic reaction, the modification is highlighted in color. 2ox, GA 2-oxidase (class I and II); 2ox*, GA 2-oxidase (class III); 3ox, GA 3-oxidase; 13ox, GA 13-oxidase; 20ox, GA 20-oxidase; CPS, *ent*-copalyl diphosphate synthase; *ent*-CDP, *ent*-copalyl diphosphate; GGDP, geranylgeranyl diphosphate; KAO, *ent*-kaurenoic acid oxidase; KO, *ent*-kaurene oxidase; KS, *ent*-kaurene synthase.

role in early GA biosynthesis in rice. In contrast, OsCPS2 functions in rice phytoalexin biosynthesis and cannot rescue *OsCPS1* loss-of-function mutants under the control of its native promoter. However, *OsCPS2* can complement the severe dwarf phenotype of the *Oscps1-1* mutant with a transgene driven by the *OsCPS1* promoter (Sakamoto et al., 2004; Toyomasu et al., 2015). Therefore, OsCPS1 and OsCPS2 have probably been subjected to sub-functionalization with overlapping but distinct enzymatic activities. Similarly, two pumpkin *CPS* genes, *CmCPS1* and *CmCPS2,,* display different expression patterns. Recombinant fusion CmCPS proteins have CPS activity in vitro. These *CPS* genes may fulfill their functions in GA biosynthesis at different developmental stages (Smith et al., 1998). Therefore, proper tissue or cell-specific expression of *CPS* genes is critical for their biological roles in the biosynthesis of GAs and other terpenoids; this also suggests that GAs act in specific cells.

The gene *AtKS* (also known as *GA2*) is also single copy in Arabidopsis, and was first identified as the homolog of the pumpkin *KS* gene with genetic confirmation by mutant complementation (Yamaguchi et al., 1996, 1998b). The *ga2-1* mutant, similar to the *ga1* mutants, is a severe GA-deficient mutant, with a non-germinating, extreme dwarf phenotype. By contrast, the rice genome contains a family of *KS-like (OsKSL)* genes, including a pseudogene *OsKSL3*, which are responsible for the biosynthesis of various diterpenoids (Tezuka et al., 2015; Xu et al., 2007). Only *OsKS1*, the closest homolog of Arabidopsis *AtKS*, functions in GA biosynthesis (Margis-Pinheiro et al., 2005; Sakamoto et al., 2004; Xu et al., 2007). Recently, the maize *ZmKS1* gene was also isolated by a map-based cloning experiment from the GA-deficient mutant *dwarf-5* (*d5*). The *ZmKS1* locus contains a tandem array of three *TPS* genes, *ZmTPS1, ZmKSL3,* and *ZmKSL5*. Only the ZmKSL3 protein serves as the KS for GA metabolism in maize (Fu et al., 2016). Therefore, the genetic control in the first step of GA biosynthesis is most likely conserved in diverse plants.

In the second stage of the GA biosynthesis pathway, the conversion of *ent*-kaurene to GA_{12} via stepwise oxidation is catalyzed by *ent*-kaurene oxidase (KO) and *ent*-kaurenoic acid oxidase (KAO). KO belongs to the CYP701A P450 subfamily, which initiates the first cytochrome P450-mediated step in GA biosynthesis in the endoplasmic reticulum (ER) and in the plastid envelope (Helliwell et al., 1998, 1999, 2001a; Ko et al., 2008; Morrone et al., 2010). The single AtKO protein is a multifunctional cytochrome P450 that catalyzes the sequential oxidation of *ent*-kaurene from $-CH_3 \rightarrow -CH_2OH \rightarrow -CHO \rightarrow -COOH$ on C-19 to produce the intermediates *ent*-kaurenol and *ent*-kaurenal and the end product *ent*-kaurenoic acid through three steps of reactions (Helliwell et al., 1998, 1999). The AtKO protein is encoded by the *GA3* gene in Arabidopsis. The loss-of-function mutant of *GA3* is also GA-deficient dwarf. The *AtKO* gene is expressed in all examined tissues, with the highest expression levels in inflorescence tissues (Helliwell et al., 1998), consistent with high levels of bioactive GAs during plant flowering. Subsequent studies have clarified the reaction sequence, enzymatic features, and substrate specificity of KO. In the rice genome, genes for five CYP701A subfamily members, *OsKO1* to *OsKO5*, are tandemly arrayed on chromosome 6 as a result of gene duplication (Sakamoto et al., 2004).

Notably, only OsKO2 (D35) exhibits KO activity and is required for GA biosynthesis (Itoh et al., 2004; Peters, 2006; Wang et al., 2012). Mutations in *OsKO2* also cause severe GA deficiency and dwarfism. The pea (*Pisum sativum*) homologous gene *PsKO1* (also known as *LH*) of the CYP701A subfamily was isolated and confirmed to have KO activity in GA biosynthesis. Similar to other plant *KO* genes, *PsKO1* is expressed in all tissues, including stems, roots, leaf, root, apical bud, and developing seeds (Davidson et al., 2004). Therefore, KO is also likely to be conserved across the plant kingdom, with a single functional gene in most plant species examined. Moreover, expression of the *KO* gene is not under feedback regulation by GAs (Hedden and Phillips, 2000; Olszewski et al., 2002); it also acts as a "gatekeeper" of early GA biosynthesis.

The next three steps of the GA biosynthesis pathway, from *ent*-kaurenoic acid, *ent*-7-hydroxykaurenoic acid, and GA_{12}-aldehyde to GA_{12}, are catalyzed by the action of KAO (Helliwell et al., 2001a). These reactions occur on the outer membranes of the plastid and require the transport of *ent*-kaurene from the organelles by an unknown mechanism. KAO belongs to the CYP88A subfamily that is localized in the ER. Two *KAO* genes in Arabidopsis, *AtKAO1* and *AtKAO2*, that encode CYP88A proteins are expressed in all tissues from germinating seeds to developing organs (Helliwell et al., 2001a,b; Regnault et al., 2014). The *Atkao1* and *Atkao2* single mutants are indistinguishable from wild-type plants. In comparison with wild-type plants, the *Atkao1 Atkao2* double mutant exhibits the typical GA-deficient phenotype, similar to that observed in the severely GA-deficient *ga1-3* mutants. Therefore, AtKAO1 and AtKAO2 are functionally redundant. Moreover, AtKAO1 and AtKAO2 exhibit equivalent enzymatic activities in GA catabolism in a yeast expression system (Helliwell et al., 2001a; Regnault et al., 2014). Similar to the other early GA biosynthesis genes, analysis of *AtKAO1* and *AtKAO2* expression patterns revealed that both genes are also mainly expressed in germinating seeds and young developing organs throughout plant development (Regnault et al., 2014). In contrast to the rice *CPS* and *KO* genes, the rice genome contains only a single *OsKAO* gene. The *Oskao* mutant also has a severe dwarf phenotype without flower or seed development with extremely low levels of intermediates and bioactive GA_1 (Sakamoto et al., 2004). Similar to Arabidopsis, pea has two *KAO* genes, *PsKAO1* (also known as *NA*) and *PsKAO2*. *PsKAO1* is expressed in the stem, apical bud, leaf, pod, and root. In these organs, GA levels are greatly reduced in the *na* mutant. *PsKAO2* is expressed only in seeds, and this expression pattern may explain the normal seed development and GA levels observed in *na* mutant plants, indicating that *PsKAO1* and *PsKAO2* subtly regulate plant growth through their complementary expression patterns (Davidson et al., 2003).

It has been widely recognized that the early steps in GA biosynthesis in higher plants are conserved before GA_{12}. By contrast, in the third stage of GA biosynthesis, GA_{12} and GA_{53} are converted to various GA intermediates and bioactive GAs by GA 20-oxidase (GA20ox) and GA 3-oxidase (GA3ox) through two parallel pathways. Both GA20ox and GA3ox are soluble 2-oxoglutarate-dependent dioxygenases (2ODDs) that are present in the cytosol (Chiang et al., 1995; Plackett et al., 2012; Reinecke et al., 2013; Sponsel et al., 1997; Xu et al., 1995). Therefore, GA_{12} lies at a branch-point in the GA biosynthesis pathway: the non-13-hydroxylation branch leads to the production of 13-H GAs, including GA_4, and the 13-hydroxylation branch leads to the biosynthesis of 13-OH GAs, including GA_1. The non-13-hydroxylation branch has been extensively studied, and all key enzymes have been identified. With a 3- or 4-step process, GA_{12} is converted into various GA intermediates and bioactive GA_4 through the non-13-hydroxyl step. GA_{12} is also a substrate for the 13-hydroxylation branch in the production of GA_{53} (13-OH GA_{12}), which is a precursor for GA_1 (Phillips, 1998; Yamaguchi, 2008). In contrast, the 13-hydroxylation branch has remained elusive, the enzyme (GA 13-oxidase, GA13ox) catalyzing the 13-hydroxylation step has not been identified, and the biological significance of the GA 13-hydroxylation reaction remained unknown for a long time. Recently, two rice cytochrome P450 genes, *CYP714B1* and *CYP714B2*, were found to encode GA13ox (Magome et al., 2013). Recombinant CYP714B1 and CYP714B2 can convert GA_{12} into GA_{53}. Consequently, the levels of 13-OH GAs, including GA_1, are decreased, whereas those of 13-H GAs, including GA_4, were increased in the *cyp714b1 cyp714b2* double mutant. GA_4 is more active than GA_1 in stem elongation in rice (Zhu et al., 2006). Interestingly, transgenic Arabidopsis plants overexpressing *CYP714B1* or *CYP714B2* are dwarfs due to the decreased levels of GA_4 in spite of the increased levels of GA_1, supporting the notion that GA_4 is more active than GA_1 in promoting plant growth (Magome et al., 2013). Therefore, CYP714B1 and CYP714B2 play important roles in plant growth and development through catalyzing GA 13-hydroxylation that fine-tunes GA homeostasis.

The GA20ox enzymes are encoded by multigene families that are responsible for the production of C_{19}-GAs using C_{20}-GAs as substrates. GA_9 and GA_{20} are precursors of bioactive GAs, which are converted by GA20ox via oxidation of C-20 to an aldehyde followed by the removal of this C atom and the formation of a lactone (Coles et al., 1999; Hedden and Phillips, 2000; Lange et al., 1994; Yamaguchi, 2008). Arabidopsis contains five paralogous *GA20ox* genes, *AtGA20ox1* to *AtGA20ox5*, while rice contains four paralogous *GA20ox* genes, *OsGA20ox1* to *OsGA20ox4*. A loss-of-function mutant of AtGA20ox1 has a reduced stem height but otherwise appears to be developmentally normal. Increased expression of the *AtGA20ox1* (*GA5*) gene in transgenic Arabidopsis plants causes an increase in GA levels and GA overdose phenotypes, indicating the GA20ox enzymes represent an important regulatory node and limit the production of bioactive GA in plants (Coles et al., 1999; Hedden and Phillips, 2000; Huang et al., 1998; Phillips et al., 1995; Plackett et al., 2012). Loss-of-function mutant studies of AtGA20ox1 and AtGA20ox2 identified partial functional redundancy between the two paralogs, which have partially overlapping expression patterns throughout development (Rieu et al., 2008b). Through systematic mutant analysis, AtGA20ox1, AtGA20ox2, and AtGA20ox3 were shown to have significant effects on floral organ and anther development, while AtGA20ox4 and AtGA20ox5 were shown to have very minor roles (Plackett et al., 2012). The *OsGA20ox2* (*SD1*) gene is well-known as the "Green Revolution" gene and is the most important gene employed in modern rice breeding (Ashikari et al., 2002; Monna et al., 2002; Qian et al., 2016; Spielmeyer et al., 2002). Semidwarf *sd1* mutant plants possess short, strong stalks and are more resistant to lodging than tall plants, resulting in greatly increased harvest index and hence grain productivity. Another gene, *OsGA20ox1*, has also been identified and shown to affect plant stature (Oikawa et al., 2004; Toyomasu et al., 1997). Combining QTL analysis with microarray expression profiling, OsGA20ox1 was shown to play a dominant role in increasing plant height and leaf sheath length at the initial growth stage (Abe et al., 2012; Yano et al., 2012). OsGA20ox1 and OsGA20ox2 control plant stature through coordinative regulation of bioactive GA levels (Ashikari et al., 2002; Oikawa et al., 2004).

The inactive precursors GA_9 and GA_{20} are hydroxylated by GA3ox enzymes to form the biologically active hormones GA_4 and GA_1 through 3β-hydroxylation (Yamaguchi, 2008). GA3ox enzymes are also encoded by multiple genes in all plant species. There are four GA3ox enzymes in Arabidopsis and two GA3ox enzymes in rice (Chiang et al., 1995; Hedden and Phillips, 2000; Itoh et al., 2001; Plackett et al., 2011; Yamaguchi et al., 1998a). The *AtGA3ox1* (*GA4*) gene was genetically identified by T-DNA tagging and expression in *Escherichia coli* (Chiang et al., 1995; Williams et al., 1998). Loss of *AtGA3ox1* results in a semidwarf phenotype. While *Atga3ox2* single mutants do not have any phenotype, the *Atga3ox1 Atga3ox2* double mutant has a smaller leaf diameter, a shorter phenotype, and a severe seed germination defect compared with the *Atga3ox1* single mutant (Mitchum et al., 2006). In rice, *OsGA3ox1* and *OsGA3ox2* (*D18*) have been functionally identified (Itoh et al., 2001). Different expression patterns of the two *OsGA3ox* genes were found in vegetative and reproductive organs: *OsGA3ox2* was broadly functional in vegetative and reproductive organs, whereas *OsGA3ox1* preferentially functioned in the reproductive organs (Itoh et al., 2001; Kaneko et al., 2003; Sakamoto et al., 2004). *ZmGA3ox2* is the causative gene for the maize *dwarf1* (*d1*) phenotype (Chen and Tan, 2015). While another maize putative GA3ox, ZmGA3ox1, does not have detectable enzymatic activity, indicating that ZmGA3ox2 (D1) provides the predominant GA3ox activity in maize. Therefore, different paralogous ODD enzymes have redundant functions in GA biosynthesis.

The production of bioactive GAs has been shown to be limited by the GA 20-oxidation step (Coles et al., 1999; Huang et al., 1998; Israelsson et al., 2004; Plackett et al., 2012). Whether GA 3-oxidation is a rate-limiting step in GA biosynthesis remains controversial. In hybrid aspen, transgenic *GA20ox* overexpression plants were found to produce higher bioactive GA levels, while transgenic *GA3ox* overexpression plants had increased 3β-hydroxylation activity but did not have significant increases in GA_1 and GA_4 levels nor changes in tree growth or morphology (Israelsson et al., 2004). Similarly, the overexpression of pea *PsGA3ox1* (*LE*) in tobacco plants only induced slight variations in phenotype and active GA_1 levels, while the slender phenotype of hybrid transgenic tobacco lines overexpressing *PsGA3ox1* was relatively similar to that of the *CcGA20ox* overexpression parental line (Gallego-Giraldo et al., 2008). However, overexpression of pumpkin *CmGA3ox1* in Arabidopsis resulted in a GA overdose phenotype with increased levels of endogenous GA_4 (Radi et al., 2006). The expression of *PsGA3ox1* in pea also affected GA homeostasis, catabolism, and gene expression, as well as GA_1 levels and plant phenotype (Reinecke et al., 2013).

4.2.2   Gibberellin inactivation pathway

Plants regulate the abundance and type of GA precisely in time and place, together with other plant hormones which regulate the growth and development and collectively respond to changes in their environment. In general, the majority of GAs are inactive, providing a means to rapidly regulate their bioactive levels. With this scenario, the life cycle of plants is controlled by precise GA levels through balancing active and inactive GAs.

Deactivation mechanisms and pathways to effectively regulate bioactive hormone levels are critical for proper plant growth and development. Several different GA inactivation pathways have been revealed that are critical for controlling endogenous GA levels. The most studied deactivation pathway is 2β-hydroxylation, which is catalyzed by a class of GA 2-oxidase (GA2ox) enzymes that are soluble 2-oxoglutarate-dependent dioxygenases (2ODDs) (Huang et al., 2010; Lo et al., 2008; Olszewski et al., 2002; Shan et al., 2014; Thomas et al., 1999; Yamaguchi and Kamiya, 2000). The GA2ox family was characterized through genetic, transgenic, and biochemical approaches. The overexpression of GA2ox enzymes in rice, Arabidopsis, and other plants resulted in similar dwarf phenotypes with reduced bioactive GA levels (Huang et al., 2010; Lee and Zeevaart, 2005; Lo et al., 2008; Sakamoto et al., 2001b; Schomburg et al., 2003; Shan et al., 2014). There are two classes of GA2ox based on their substrates: a larger class of C_{19}-GA2ox and a smaller class of C_{20}-GA2ox (Hedden and Phillips, 2000; Sakamoto et al., 2001b; Schomburg et al., 2003; Shan et al., 2014; Thomas et al., 1999). Different GA2ox enzymes convert bioactive GAs and their precursors as C_{20}-GAs and C_{19}-GA substrates to limit bioactive GA levels and regulate many stages of plant development. The C_{19}GA2ox can hydroxylate the C-2 of bioactive C_{19}-GAs (GA_1 and GA_4), as well as C_{19}-GA precursors, such as GA_9 and GA_{20}, to produce the catabolites, GA_8, GA_{34}, GA_{29}, and GA_{51}, respectively. The C_{20}GA2ox subgroup only acts on C_{20}-GA precursors, such as GA_{12} and GA_{53}, to form GA_{110} and GA_{97}, but not on C_{19}-GAs. The C_{20}GA2ox enzymes contain three unique conserved motifs that are absent in the class of C_{19}GA2ox (Lee and Zeevaart, 2005), which were found to be important for the activity of this class of GA2ox (Lo et al., 2008). The recombinant SoGA2ox1 (from *Spinacia oleracea*) has both C_{19}-GA and C_{20}-GA activities (Lee and Zeevaart, 2002). The GA2ox family can be divided into three classes on the basis of the phylogenetic relationships (Lee and Zeevaart, 2005). Members of classes I and II catabolize C_{19}-GAs, and class III members can only hydroxylate C_{20}-GAs. Arabidopsis has five $C_{19}GA2ox$ genes *AtGA2ox1*, *AtGA2ox2*, *AtGA2ox3*, *AtGA2ox4*, and *AtGA2ox6* (*AtGA2ox5*, is a pseudo gene) (Olszewski et al., 2002; Thomas et al., 1999; Yamaguchi and Kamiya, 2000) and two $C_{20}GA2ox$ genes (*AtGA2ox7* and *AtGA2ox8*) (Schomburg et al., 2003). C_{19}-GA2-oxidation limits bioactive GA content and regulates plant development at various stages during the plant life cycle: $C_{19}GA2ox$ enzymes prevent seed germination, delay vegetative and floral phase transitions, limit the number of flowers produced per inflorescence, and suppress elongation of the pistil prior to fertilization (Rieu et al., 2008a). Genetic analysis revealed that $C_{19}GA2ox$ enzymes participate in a major GA inactivation pathway that limits bioactive GA content and regulates plant growth and development (Lo et al., 2008; Rieu et al., 2008a). Increased expression of either *AtGA2ox7* or *AtGA2ox8*, two $C_{20}GA2oxs$, significantly reduced GA levels and caused dwarf phenotypes in Arabidopsis and tobacco. Double *Atga2ox7 Atga2ox8* mutants had higher levels of active GAs and displayed GA overdose phenotypes (Schomburg et al., 2003). In the rice genome, a total of 10 putative GA2oxs were identified, including seven $C_{19}GA2ox$ genes (*OsGA2ox1-4, OsGA2ox7-8,,* and *OsGA2ox10*) and three $C_{20}GA2ox$ genes (*OsGA2ox5, OsGA2ox6,* and *OsGA2ox9*), which are differentially regulated and act in concert or individually to control the homeostasis of GA levels (Lee and Zeevaart, 2005; Lo et al., 2008; Sakamoto et al., 2001b, 2004; Sasaki et al., 2003; Shan et al., 2014). In maize, 10 putative *GA2ox* genes (*ZmGA2ox1-ZmGA2ox10*) were also predicted with public maize databases (Song et al., 2011).

Remarkably, the cytochrome P450 monooxygenase CYP714 family represents a novel class of enzymes that maintains the balance of active and inactive GAs (Fig. 4.2). The *CYP714D1* gene (well-known as *ELONGATED UPPERMOST INTERNODE, EUI*) encodes the GA 16α,17-epoxidase in rice (Zhu et al., 2006), which catalyzes 16α,17-epoxidation of non-13-hydroxy GAs (GA_{12}, GA_9, and GA_4) into 16α,17-epoxy-GAs, but not of 13-hydroxy GAs (GA_{53}, GA_{20}, and GA_1). Then 16α,17-epoxides are converted to 16α,17-$[OH]_2$-GAs (GA 16,17-dihydrodiols) by EUI-related enzymes. The rice *eui* mutants accumulate extremely high levels of GA_4 (Zhu et al., 2006); while the ectopic expression of *EUI* under the control of the 35S promoter or the promoters of the rice GA biosynthesis genes *OsGA3ox2* and *OsGA20ox2* dramatically reduced GA levels, as well as plant morphology (Zhang et al., 2008; Zhu et al., 2006). These results suggest that EUI is a principal GA deactivation enzyme and 16α,17-epoxidation of GAs may be a general deactivation mechanism. GA 16,17-dihydrodiols have been found in diverse plants, and the EUI-mediated GA deactivation is a conserved mechanism in plants. However, the EUI immediate products should be epoxy-GAs (Zhu et al., 2006), which should be further catalyzed into $(OH)_2$-GAs by another unrecognized enzyme(s) that can hydrolyze epoxy-GAs. Arabidopsis has two CYP714 members, CYP714A1 and CYP714A2, named as ELA1 (EUI-like) and ELA2 respectively, which have redundant but also distinct roles in GA metabolism (Zhang et al., 2011a). ELA1 can catalyze the conversion of GA_{12} to 16-carboxy GA_{12} (16-carboxy-16β,17-dihydro GA_{12}), which is a major GA_{12} metabolite (Nomura et al., 2013). Functional analysis shows that ELA2 may play a role in the inactivation of non-13-hydroxy GAs such as GA_{12} and GA_9 by C16-carboxylation or a similar oxidation (Nomura et al., 2013). ELA2 likely acts as a GA13-oxidase or GA 12α-oxidase of GAs or GA precursors (such as *ent*-kaurenoic acid; *ent*-7α-hydroxy kaurenoic acid, GA_{12}-7-aldehyde, and GA_{12}) in plants, depending on the substrates (Nomura et al., 2013). Intriguingly, 12α-hydroxy GA_{12} (GA_{111}) is produced as a major product and 13-hydroxy GA_{12} (GA_{53}) as a minor product, when GA_{12} is used as a substrate for CYP714A2 (Nomura et al., 2013). Ectopic expression of *PtCYP714A3*, an *EUI* homolog in *Populus trichocarpa*, in the rice *eui* mutant could rescue the rice excessive-shoot-growth phenotype. The overexpression of *PtCYP714A3* in rice led to a semidwarf phenotype and reduced endogenous bioactive GA levels. The results indicate that CYP714A3 is likely also to be involved in

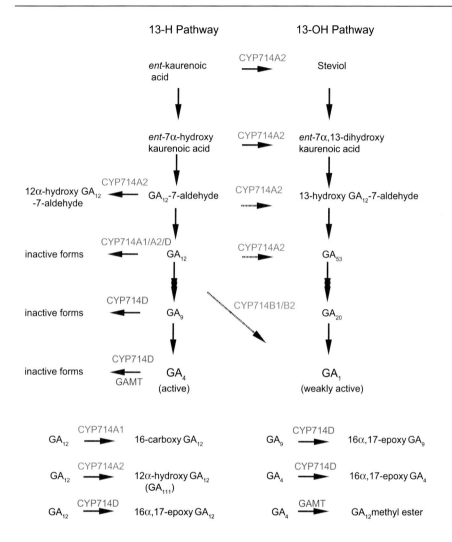

Figure 4.2 Different members of CYP714 family play important roles in GA biosynthesis and deactivation pathways in higher plants, while GA was deactivated by gibberellin methyl-transferases (GAMTs) through methylation in Arabidopsis. Arrows of 714A2 indicate strong (*solid*) or weak (*dotted*) activities of CYP714A2 to the synthesis of respective GAs.

the GA deactivation; however, the enzymatic functions of CYP714A3 have not yet been identified (Wang et al., 2016).

Two GA methyltransferases, GAMT1 and GAMT2, catalyze the methylation of active GAs to generate inactive GA methyl esters in Arabidopsis seeds (Fig. 4.2) (Varbanova et al., 2007). Siliques of the *gamt1 gamt2* double mutant accumulate high levels of active GAs. In contrast, overexpression of *AtGAMT1* reduces the level of the major bioactive GA_4, resulting in typical GA deficiency semidwarf phenotypes and increased tolerance to drought stress in transgenic plants (Nir et al., 2014; Varbanova et al., 2007). Thus, the methylation of GAs is part of the mechanism that regulates the levels of active GAs in plants. However, whether the methylation of GAs is a common deactivation reaction in higher plant species remains to be determined.

In conclusion, the GA metabolism pathway, which includes GA biosynthesis and catabolism, is a complex process that is regulated by multiple genes through the entire life cycle of plant growth and development. The uncertainties regarding the mechanisms of the GA metabolism pathway require further study.

4.2.3   Regulation of GA biosynthesis and metabolism

4.2.3.1   Feedback regulation of GA homeostasis

GAs play diverse roles in plant development. GA homeostasis must be fine-tuned with both GA biosynthesis and catabolism. Regulation of GA biosynthesis and catabolism occurs at a number of steps under different conditions (Colebrook et al., 2014; Hirose et al., 2013; Weiss and Ori, 2007; Yamaguchi, 2008). Therefore, the levels of bioactive GAs are maintained via feedback and feedforward regulation of GA metabolism (Hedden and Phillips, 2000), including regulation of the transcription of core GA signaling components, such as the GA receptor GID1 and repressor DELLA (Griffiths et al., 2006; Hedden and Thomas, 2012).

Unlike later GA biosynthesis genes, *CPS* genes appear not to be regulated by a feedback mechanism (Silverstone et al., 1997a). GA dynamics are mainly targeted to 2ODDs in the GA metabolism pathway to establish homeostasis. Negative feedback regulation of *AtGA20ox1* expression has been demonstrated by application of bioactive GAs or a high endogenous bioactive GA concentration (Xu et al., 1999). Consistent with this model, GA20ox enzymes in Arabidopsis and rice are highly elevated in GA-deficient mutants (Phillips et al., 1995; Zhang et al., 2008; Zhu et al., 2006). The *AtGA3ox1* gene, but not the *AtGA3ox2* gene, is under negative feedback regulation during Arabidopsis seed germination (Yamaguchi et al., 1998a). Bioactive GAs regulate *AtGA3ox1* transcript abundance in a dose-dependent manner, and downregulation of *AtGA3ox1* is not triggered by the immediate precursor GA_9 (Cowling et al., 1998). *AtGA20ox1* and *AtGA3ox1* genes were highly upregulated in both *CYP714A1* and *CYP714A2* overexpression plants which have lower endogenous GA_4 levels (Nomura et al., 2013). Furthermore, the feedback regulation of *AtGA3ox1* is modified by AGF1 (AT-hook protein of GA feedback regulation) to maintain GA homeostasis by binding to the *cis*-acting sequence of the *AtGA3ox1* promoter (Matsushita et al., 2007). More specifically, the expression of *AtGA3ox1* is repressed by the DOF transcription factor DAG1 (DOF AFFECTING GERMINATION1), which acts downstream of PIL5 (PHYTOCHROME-INTERACTING FACTOR 3-LIKE 5) in the light-mediated seed germination pathway (Boccaccini et al., 2014). Similarly in rice, expression of *OsGA3ox1*, *OsGA3ox2,* and *OsGA20ox2*, whose proteins catalyze formation of active GAs, was shown to be under negative feedback regulation in *eui* mutants that accumulate high levels of bioactive GAs (Zhang et al., 2008; Zhu et al., 2006). In contrast, the expression of the GA deactivation genes *AtGA2ox1*, *AtGA2ox2*, *OsGA2ox1*, and *OsGA2ox3* is upregulated upon GA_3 application or high endogenous bioactive GA levels (Sasaki et al., 2003; Thomas et al., 1999; Zhang et al., 2008; Zhu et al., 2006). In addition, *OsGA20ox2*, *OsGA20ox4*, and *OsGA3ox2* were upregulated in the *OsGA2ox6* overexpression mutant (Huang et al., 2010).

As the central GA signaling components, the GA receptor GID1, DELLA proteins, and the F-box proteins SLY1 in Arabidopsis or GID2 in rice, also function in GA homeostatic regulation (Boccaccini et al., 2014; Dill and Sun, 2001; Fukazawa et al., 2014; Sasaki et al., 2003; Ueguchi-Tanaka et al., 2007; Zentella et al., 2007). DELLA proteins affect GA homeostasis by direct feedback regulation of the GA biosynthesis genes and receptor genes (Zentella et al., 2007). Expression of *AtGA3ox1* is downregulated in seeds of the DELLA double mutant *gai-t6 rga28* in Arabidopsis (Oh et al., 2007). The expression of the *OsGA20ox2* (*SD1*) gene is upregulated in the rice *gid1* and *gid2* mutants, and the levels of bioactive GA_1 are highly elevated (Sasaki et al., 2003; Ueguchi-Tanaka et al., 2005). Moreover, the DELLA binding transcription factor GAF1 (GAI-ASSOCIATED FACTOR1) regulates *AtGA20ox2* and *AtGID1b* and is thereby involved in feedback regulation of GA biosynthesis (Fukazawa et al., 2014). YAB1, a member of the YABBY family of C2C2 zinc finger transcription factors, as a mediator of feedback inhibition of GA biosynthesis, regulates *OsGA3ox2* and *OsGA2ox3* gene expression in rice (Dai et al., 2007). The precise sites and timing of GA biosynthesis and responses must be controlled for proper plant growth and development. GAs are usually present at low concentrations (0.1–100 ng/g fresh weight) in most vegetative and floral tissues, consistent with activities of metabolic enzymes (Hedden and Phillips, 2000). In general, the highest levels of bioactive GAs are found in actively growing organs, such as expanding internodes and developing flowers (Hu et al., 2008; Silverstone et al., 1997a; Zhang et al., 2011a, 2008; Zhu et al., 2006). Therefore, different GA metabolism and signaling genes are expressed with different temporal and spatial patterns.

The synthesis of *ent*-kaurene is an early regulatory step of GA biosynthesis controlling the flow of metabolites into the pathway. The *CPS* gene is expressed with low levels in rapidly growing tissues and vascular elements of expanded leaves, which might act as a source of GAs or their precursors (Fleet et al., 2003; Silverstone et al., 1997a). However, the expression of *KS* is not tissue specific and is much higher than that of *CPS* (Smith et al., 1998; Yamaguchi et al., 1996, 1998b). Moreover, *AtCPS* is expressed in the provasculature of the embryonic axis in germinating seeds and *AtKO* and *AtGA3ox* genes are expressed in the cortex and endodermis (Yamaguchi et al., 2001), suggesting that expression of GA biosynthetic genes closely correlates with the sites of GA response. The intermediate *ent*-kaurene is relatively volatile and has been found to be released into the external environment, which might contribute to reducing the flux through the GA biosynthesis pathway in the *AtCPS* overexpression line (Fleet et al., 2003; Otsuka et al., 2004). Synthesis of *ent*-kaurene in spinach increases under long-day conditions, and the GA content also rises (Zeevaart et al., 1993).

While CPS activity may regulate flux through the early steps of the GA biosynthesis pathway, the concentration of bioactive GAs is more precisely determined by the activity of the 2ODDs, GA20ox, and GA3ox enzymes. The different paralogous *2ODD* genes show distinct but overlapping expression patterns, and as such contribute unequally to different developmental processes. In Arabidopsis, each *AtGA3ox* gene exhibits a unique organ-specific expression pattern, suggesting distinct physiological roles played by individual *AtGA3ox* members during development (Hu et al., 2008; Mitchum et al., 2006).

Differential expression of *AtGA2ox* genes during the plant life cycle is also associated with seed germination and development of seedlings, flowers, and secondary shoots. Therefore, they also act in concert or individually to control active GA levels during plant development (Aach et al., 1997; Silverstone et al., 1997a; Thomas et al., 1999; Zhu et al., 2006). Several AtGA2ox enzymes were downregulated in correlation with rapid seed germination when increased GA levels are necessary for seed germination (Lo et al., 2008). In rice, *OsGA20ox2* and *OsGA3ox2* are predominantly expressed in the flowers (Kaneko et al., 2003), where *OsGA2ox6* is also highly expressed (Huang et al., 2010), probably establishing a feedforward regulatory loop of bioactive GA abundance during the procreative period. Similar precise regulation is also adapted by the novel rice GA deactivation enzyme EUI and the Arabidopsis homologs CYP714A1 and CYP714A2 with strong tissue and developmental specificity (Zhang et al., 2008; Zhu et al., 2006).

4.2.3.2   Environmental regulation of GA metabolism

Many environmental responses are regulated through GA abundance, and GA metabolism is regulated by environmental signals, such as light, temperature, water, and nutrient status, as well as by other abiotic and biotic stresses. Light is one of the major environmental factors that affects plant growth and development (Kamiya and García-Martínez, 1999). GA metabolism is sensitive to changes in light quantity, quality, or duration, which may result in increased or decreased GA content. The photoreceptor phytochrome regulates transcript levels of GA20ox and GA3ox enzymes in germinating lettuce (*Lactuca sativa*) (Toyomasu et al., 1998) and Arabidopsis seeds (Yamaguchi et al., 1998a). Expression of *LsGA3ox1* was dramatically increased and *LsGA20ox2* expression was decreased by red light (R) treatment, while *LsGA3ox2* and *LsGA20ox2* expression were unaffected in seeds under any light conditions. These results suggest that GA_1 content increases in lettuce seeds by inducing *LsGA3ox1* expression via phytochrome action (Toyomasu et al., 1998). The expression of *AtGA3ox1* and *AtGA3ox2* genes is relatively high in germinating seeds under continuous white light and is elevated by red light, which may result in an increase in the biosynthesis of active GAs to promote seed germination (Yamaguchi et al., 1998a). Therefore, light-regulated expression of *GA3ox* genes via phytochrome action may be a common mechanism in plant species whose germination is dependent on a light stimulus. Cryptochromes can also regulate *GA20ox* and *GA2ox* expression in Arabidopsis seedlings (Achard et al., 2007b; Zhao et al., 2007) and rice seedlings (Hirose et al., 2012, 2013). High levels of *AtGA20ox1* and low levels of *AtGA2ox* genes expression were found in Arabidopsis seedling hypocotyls grown in the dark. Conversely, Arabidopsis seedling hypocotyls contained low *AtGA20ox1* and high *AtGA2ox1* transcripts when grown in continuous light. These studies indicate that the biosynthesis and accumulation of bioactive GAs are critical to regulate the length of the seedling hypocotyls (Achard et al., 2007b). Blue light affects bioactive GA level through regulating the expression of *AtGA2ox1*, *AtGA20ox1*, and *AtGA3ox1*, which requires cryptochromes (Zhao et al., 2007). In rice seedlings, phytochromes mediate the repression of GA biosynthesis capacity, including the repression of two *GA20ox* genes, *OsGA20ox2* and *OsGA20ox4*, and a *GA3ox* gene, *OsGA3ox2*, and the induction of four *GA2ox* genes, *OsGA2ox4* to *OsGA2ox7* (Hirose et al., 2012). A further study indicated that independent photoreceptors in rice separately but cooperatively mediate GA metabolism, for example, blue light sensed by cryptochrome 1 (cry1a and cry1b) induced the expression of the four *GA2ox* genes (*OsGA2ox4–OsGA2ox7*) (Hirose et al., 2013). Phytochrome B, together with auxiliary action of phytochrome A, mediates the repression of *GA20ox* genes (*OsGA20ox2* and *OsGA20ox4*). These independent effects cumulatively reduce active GA contents, leading to the suppression of leaf sheath elongation.

The circadian clock directly influences GA biosynthesis and signaling. Thus *AtGA20ox1*, but not *AtGA3ox1*, is activated under long-day (LD) conditions when Arabidopsis plants were transferred from short-day (SD) to LD conditions (Xu et al., 1997). Similarly, the expression level of *GA20ox* is higher under LD conditions than under SD conditions in spinach (Wu et al., 1996). Therefore, stem elongation under LD conditions is at least partly due to increased expression of *GA20ox* with photoperiodic responses. In Arabidopsis, PIL5 is a negative regulator of phytochrome-dependent seed germination. PIL5 represses the expression of some GA biosynthesis genes including *AtGA3ox1* and *AtGA3ox2* and activates the transcription of a GA catabolism gene (*AtGA2ox2*) through directly regulating the expression of *GAI* (*Gibberellin-Insensitive*) and *RGA* (*REPRESSOR* of *ga1-3*) (Oh et al., 2006, 2007). Furthermore, EARLY FLOWERING 3 (ELF3) is fundamental to core circadian clock activity and to time-of-day-specific regulation of the *PHYTOCHROME-INTERACTING FACTOR 4* (*PIF4*) and *PIF5* genes. Moreover, ELF3 negatively regulates *AtGA20ox1* and *AtGA20ox2* expression, which depends strongly on the redundant activities of PIF4 and PIF5 (Filo et al., 2015).

Endogenous GA levels are also regulated by cold, salt, and high temperature stimuli through GA biosynthesis pathway enzymes and GA deactivation enzymes (Colebrook et al., 2014; Hedden and Phillips, 2000; Toh et al., 2008). For example, low temperatures activate GA biosynthesis during seed imbibition. The *AtGA3ox1* gene, but not other *GA3ox* genes (*AtGA3ox2, AtGA3ox3,* and *AtGA3ox4*), is induced by cold temperature in dark-imbibed seeds, and *Atga3ox1* mutant seeds are defective in both cold-stimulated GA_4 accumulation and germination, compared to wild-type seeds (Yamauchi et al., 2004). The embryonic regulators LEAFY COTYLEDON2 (LEC2) and FUSCA3 (FUS3) are involved in multiple aspects of Arabidopsis seed development, including negative regulation of GA biosynthesis or action. Transcript levels of *AtGA3ox1* and *AtGA20ox1* were reduced by transient FUS3 protein expression (Gazzarrini et al., 2004). Moreover, FUS3 protein physically interacts with two RY elements (CATGCATG) present in the *AtGA3ox2* promoter and represses *AtGA3ox2* expression in the epidermis of the developing embryo (Curaba et al., 2004).

Levels of bioactive GAs are increased in immature seeds of *lec2* and *fus3* mutants, and *AtGA3ox2* expression is ectopically activated in *lec2* and *fus3* mutant embryos. The MADS domain transcriptional regulator AGAMOUS-Like15 (AGL15) promotes somatic embryogenesis by binding DNA and regulating gene expression. *AtGA2ox6* is a direct target of AGL15 determined in vitro and in vivo in both Arabidopsis and *Brassica*, and is activated in seeds that overexpress *AGL15* (Wang et al., 2004). Thus, AGL15 plays a role in lowering GA content through upregulation of a GA deactivation gene during embryogenesis (Wang et al., 2002, 2004). Furthermore, *LEC2* and *FUS3* were identified to be direct target genes of AGL15 (Zheng et al., 2009), to negatively regulate bioactive GA levels (Curaba et al., 2004; Gazzarrini et al., 2004). Such GA regulatory loops should act through all developmental stages.

Cold treatment also induced a transient increase in *AtGA2ox1* transcript levels (Achard et al., 2008). In Arabidopsis, six *AtGA2ox* genes, including *AtGA2ox1*, *AtGA2ox2*, *AtGA2ox4*, *AtGA2ox6*, *AtGA2ox7*, and *AtGA2ox8*, were upregulated under high-salinity stress; thus, endogenous GA levels could be reduced as a result of the induction of GA2ox leading to repression of growth as an adaptation to stress (Magome et al., 2008). The salinity-responsive *DWARF AND DELAYED FLOWERING 1* (*DDF1*) gene encodes an AP2 (Apetala2) transcription factor of the DREB1 (dehydration-responsive element binding protein)/CBF (C-repeat binding factor) subfamily. Overexpression of the *DDF1* gene causes dwarfism mainly by reducing the levels of bioactive GA in transgenic Arabidopsis (Magome et al., 2004). Subsequently, *AtGA2ox7* is strongly upregulated in transgenic plants overexpressing *DDF1*, which likely promotes the expression of *AtGA2ox7* directly upon exposure to salt stress (Magome et al., 2008). A zinc finger transcription factor gene from *Chrysanthemum morifolium*, *CmBBX24*, is induced by dehydration, low temperature (4°C), and salt conditions (Yang et al., 2014). The *CmBBX24* overexpression or RNAi lines showed clear changes in the expression of *CmGA20ox* and *CmGA3ox*, whereas the expression of other tested *CmGA2ox* genes was not altered. Therefore, the change in abiotic stress tolerance in *CmBBX24* RNAi lines may be related to the influence of increased levels of bioactive GAs through expression of GA biosynthesis genes (Yang et al., 2014). GAs also influence drought tolerance in plants (Colebrook et al., 2014). Overexpression lines of *AtGA20ox1* or *Atga2ox* quintuple mutants with high GA content presented decreased drought tolerance, while *Atga20ox1 Atga20ox2*, *Atga3ox1 Atga3ox2* double mutants, or *Atga20ox1 Atga20ox2 Atga20ox3* triple mutant with low GA levels or GA deficiency presented dramatically increased drought tolerance compared to the wild-type Arabidopsis (Colebrook et al., 2014).

The GA-deactivating enzyme EUI is also involved in response to biotic and abiotic stresses through the regulation of active GAs in rice and Arabidopsis (Yang et al., 2008, 2013). The *eui* mutants accumulate exceptionally large amounts of biologically active GAs, while the level of GA_4 is dramatically reduced in *EUI* overexpression transgenic plants (Zhu et al., 2006). Interestingly, *eui* mutants or *EUI* overexpression compromises or increases, respectively, resistance to bacterial blight and rice blast. Exogenous application of GA_3 and an inhibitor of GA synthesis (uniconazole) could increase disease susceptibility and resistance, respectively. Similarly, GA application also decreased the disease resistance. Therefore, the change in resistance is attributed to GA levels (Yang et al., 2008). Consistent with this, the GA metabolism genes *OsGA20ox2* and *OsGA2ox1* were downregulated during pathogen challenge. Therefore, GAs play a negative role in rice basal disease resistance (Yang et al., 2008).

The *AtGA3ox1* gene plays an essential role during seed imbibition and is induced by cold temperatures in dark-imbibed seeds, suggesting that the GA biosynthesis pathway and the cellular distribution of bioactive GAs are altered during seed imbibition under low temperature conditions (Yamauchi et al., 2004). The zeaxanthin epoxidase gene *ZEP* (also known *ABA1*) and three 9-*cis*-epoxycarotenoid dioxygenase genes *NCED2*, *NCED5*, and *NCED9*, which are the enzymes in ABA (abscisic acid) biosynthesis, are all upregulated and ABA levels are elevated under high temperatures in Arabidopsis seeds. Genes encoding GA20ox (*AtGA20ox1*, *AtGA20ox2*, and *AtGA20ox3*) and GA3ox (*AtGA3ox1* and *AtGA3ox2*) are suppressed at high temperatures, leading to low levels of bioactive GAs in imbibed seeds (Toh et al., 2008). In addition, *AtGA3ox1* and *AtGA3ox2* were severely repressed at high temperature in wild-type seeds, but their expression was completely de-repressed in ABA-deficient mutant seeds at high temperature. Simultaneously, the negative GA regulators SPINDLY (SPY) and RGA-LIKE 2 (RGL2) are enhanced to suppress GA signaling pathway (Toh et al., 2008). High temperatures stimulate ABA synthesis and repress GA synthesis and signaling through the action of ABA in Arabidopsis seeds. The *CBF1* (*DREB1b*) gene is a transcriptional activator transiently induced by cold, which modulates GA metabolism via upregulation of *AtGA2ox3* and *AtGA2ox6* gene transcript levels and decreases in bioactive GAs (Achard et al., 2008).

4.2.3.3   Regulation of GA biosynthesis and inactivation by other hormones

Interactions among different plant hormones affect overlapping processes in plant growth and development. In this section, we focus on the regulation of GA metabolism by ABA, auxin, ethylene, brassinosteroid (BR), and cytokinin (CK). The antagonistic roles played by GA and ABA regulate numerous developmental processes (Razem et al., 2006; Weiss and Ori, 2007). In Arabidopsis seeds higher amounts of ABA accumulated in GA-deficient *ga1-3* mutant seeds (Oh et al., 2007), whereas GA biosynthesis is increased in ABA-deficient *aba2* mutant seeds via activation of *AtGA3ox1* and *AtGA3ox2* expression (Seo et al., 2006). Furthermore, activation of GA biosynthesis genes is also observed in the *aba2-2* mutant during seed development (Seo et al., 2006). ABA reduces bioactive GA levels by decreasing *AtGA20ox1* and increasing *AtGA2ox6* transcript levels (Zentella et al., 2007). In addition, *aba2-2* mutants show significant expression of GA biosynthesis genes and repression of *SPY* expression even at high temperatures. The resistance to thermal inhibition of germination in *aba2-1* seeds is suppressed by a GA biosynthesis inhibitor, paclobutrazol (PAC) (Toh et al., 2008). Therefore, ABA plays an important role in the suppression of GA biosynthesis. The rice *eui* mutant with a higher level of bioactive GA was less sensitive to ABA, while the *EUI* overexpression lines of Arabidopsis and rice, with a GA phenotype, were hypersensitive to ABA (Yang et al., 2013). This evidence suggests that GAs play a role in the abiotic stress response probably through modulating the ABA signaling pathway. It was found that endogenous and exogenous ABA could induce GA deactivation through upregulating *EUI* gene expression (Yaish et al., 2010). Overexpression of *OsAP2-39* results in yield decrease in transgenic rice lines, which upregulates *EUI*, while ABA application induces the expression of *EUI* and suppresses the expression of *OsAP2-39*. The expression of both *EUI* and the ABA biosynthesis gene *OsNCED-1* is directly controlled by *OsAP2-39*, illustrating a mechanism that leads to homeostasis and balance of these hormones (Yaish et al., 2010).

The regulation of GA biosynthesis by auxin has been attributed to changes in the expression levels of the genes encoding GA biosynthesis or deactivation enzymes. In pea, GA_1 content is regulated by auxin, which affects the expression of *PsGA3ox1* in the expanding shoot (Ross et al., 2000). Exogenous auxin treatment also affects the expression of *PsGA3ox1* and *PsGA2ox1* genes in a dose-dependent manner (O'Neill and Ross, 2002). In Arabidopsis, the AUXIN/indole-3-acetic acid (Aux/IAA) and auxin response factor proteins (see auxin chapter in this book for details) directly regulate the expression of GA metabolism genes through upregulating the expression of *AtGA20ox* and *AtGA2ox* (Frigerio et al., 2006). The auxin regulation of *PsGA2ox1* is partly dependent on DELLA, while *PsGA2ox2* was upregulated by auxin in a DELLA-independent manner (O'Neill et al., 2010). This finding suggests that auxin differentially regulates the expression of GA biosynthesis genes involved in pea and Arabidopsis plants. Auxin also positively regulates GA biosynthesis with respect to the regulation of cell expansion and tissue differentiation, and the regulation of GA metabolism genes by auxin is tissue specific (Frigerio et al., 2006). The auxin transport inhibitors (ATIs) 1-N-naphthylthalamic acid (NPA) and 1-naphthoxyacetic acid (NOA) elevate the expression of *AtGA20ox1* in shoot tissues, but not in root tissues, and only at certain developmental stages (Desgagne-Penix and Sponsel, 2008). These observations suggest that developmental regulation and organ-, tissue-, and cell-specific regulation override both auxin and GA metabolic regulation, ensuring the appropriate temporal and spatial levels of GAs.

Similar to GA, BR is a major hormone that regulates plant cell elongation. BR regulates the biosynthesis of GA to promote plant growth (Unterholzner et al., 2015). Exogenous BR treatment and overexpression of BR biosynthesis gene *DWARF4* (*DWF4*) increase expression of several *GA20ox* genes (*AtGA20ox1, AtGA20ox2,* and *AtGA20ox5*) in Arabidopsis (Lilley et al., 2013). It has also been found that BR regulates cell elongation by modulating GA metabolism in rice (Tong et al., 2014). Under physiological conditions, BR promotes GA accumulation by regulating the expression of GA metabolism genes to stimulate cell elongation. BR greatly induces the expression of *OsGA3ox2* (*D18*), one GA biosynthesis gene, leading to increased levels of GA_1 (Tong et al., 2014). The expression of the five *GA20ox* genes and *AtGA3ox1* is decreased in the Arabidopsis BR biosynthesis mutant *cyp90a1* (also named *cpd*) and in BR signaling mutants such as *bri1-1*. Moreover, *AtGA20ox1* can rescue many of the developmental defects of BR mutants, and the transcription factors BES1 and BZR1 functioning in BR signaling likely mediate this GA-mediated response through binding the promoters of *AtGA20ox1* and other GA biosynthesis genes (Unterholzner et al., 2015). Therefore, cross talk between BR and GA occurs not only at the level of signals but also at the level of hormone biosynthesis (Hofmann, 2015; Tong and Chu, 2016; Unterholzner et al., 2016).

The role of ethylene in GA metabolism is less clear. Ethylene represses GA biosynthesis or suppresses GA responses via DELLA stabilization (Weiss and Ori, 2007). In the *ctr1-1* mutant, which exhibits constitutive ethylene responses, the levels of GA_1 and GA_4 are substantially reduced and the expression levels of *AtGA3ox1* and *AtGA20ox1* genes are elevated through a negative feedback mechanism, suggesting that the ethylene signal is initially targeted to the GA metabolism pathway (Achard et al., 2007a). Ethylene also regulates GA metabolism during fruit set in tomato (Shinozaki et al., 2015). The ethylene-insensitive *Sletr1-1* mutant fruits have elevated levels of bioactive GAs, most likely through increasing transcription of *SlGA20ox3*, and repressing transcription of *SlGA2ox4* and *SlGA2ox5*. Therefore, ethylene plays a role in tomato fruit set by suppressing GA biosynthesis (Shinozaki et al., 2015). The AtERF11 protein is a member of the ETHYLENE RESPONSE FACTOR (ERF) subfamily VIII-B-1a of ERF/AP2 transcription factors in Arabidopsis. Overexpression of *AtERF11* resulted in elevated bioactive GA levels by upregulating expression of *AtGA3ox1* and *AtGA20ox* genes. AtERF11 also enhances GA signaling by antagonizing the function of DELLA proteins via direct protein–protein interactions (Zhou et al., 2016). Therefore, AtERF11 plays a dual role in promoting internode elongation by inhibiting ethylene biosynthesis and activating GA biosynthesis and signaling.

Antagonistic effects are exerted by GA and CK in many developmental processes (Greenboim-Wainberg et al., 2005). A balance of CK and GA signals are required for normal SAM (shoot apical meristem) function (Jasinski et al., 2005; Sakamoto et al., 2001a). Whereas SAM development requires high CK and low GA levels/signals, the later stages of cell maturation and elongation require the opposite: low CK and high GA levels. CK can also regulate GA levels by directly inhibiting GA biosynthesis and indirectly promoting GA deactivation (Weiss and Ori, 2007). KNOX (Class-I KNOTTED1-like homeobox) proteins are key regulators in the SAM. KNOX activity is mediated by both GA and CK (Jasinski et al., 2005). CK and KNOX both induce the expression of *GA2ox* at the SAM probably to block bioactive GAs (Jasinski et al., 2005). The potato StBEL5 (BEL1-like transcription factor) and POTH1 (potato homeobox1) mediate developmental processes by regulating CK and GA levels. Overexpression of either gene increases tuber yields by lowering GA levels and increasing CK levels, which can directly bind to the promoter of *StGA20ox1* (Chen et al., 2004).

4.2.4   Gibberellin transport

Plants produce and accumulate appropriate levels of bioactive GAs to ensure normal growth. The fine-tuning of gene expression in GA biosynthesis and metabolism pathways co-ordinately control the levels of GAs (Hedden and Thomas, 2012). In addition, studies suggest the existence of local and long-distance GA transport in plants (Dayan et al., 2012; Eriksson et al., 2006; Katsumi et al., 1983; Proebsting et al., 1992; Ragni et al., 2011; Regnault et al., 2015; Shani et al., 2013; Tal et al., 2016). Biochemical and micrografting experiments have demonstrated the translocation of GAs from synthetic sites to the tissues and organs that require GAs for growth and development (Regnault et al., 2016).

Studies in pea provided the first evidence for the transport of GAs from root to shoot, by employing grafting experiments in a GA biosynthesis mutant *na* mutant that blocks *ent*-7α-hydroxykaurenoic acid conversion to GA_{12}-aldehyde, resulting in reduced levels of active GAs within the shoot (Ingram and Reid, 1987). After application of $[^2H, ^3H]$-labeled GA_1, GA_{19}, and GA_{20} to the *Na* rootstocks (the donor) in *na* scions grafted to *Na* rootstocks (*na/Na* grafts), the labeled GAs could be detected in the *na* scions (the receiver) (Proebsting et al., 1992). Moreover in *na/Na* grafts, GA_1 concentration of the *na* scions was normal, GA_{20} content increased, but GA_{19} was hardly translocated to shoot apices of the *na* scions, suggesting that GA_{20} was the major transported GA in peas (Proebsting et al., 1992). In the grass *Lolium temulentum*, GA_5 was shown to be transported from the leaf to the shoot apex (King et al., 2001). In Arabidopsis, GA_4 application to a single leaf significantly decreased the total number of leaves formed before flowering of the wild-type and induced flowering in the GA-deficient *ga1-13* mutant, suggesting that GA_4 likely acts as a mobile GA from the leaf to the shoot apex (Eriksson et al., 2006). In addition, de novo synthesis of bioactive GAs is not only necessary for stamen development, but also is transported to nearby tissues, such as petals, to support their growth (Hu et al., 2008). Further micrografting experiments between the *ga1-3* mutant and the wild-type L*er* plants showed that the L*er* scions could restore hypocotyl xylem expansion in the *ga1-3* rootstocks, whereas impaired GA signaling did not affect xylem expansion systemically in L*er/ga1-3* grafts (L*er* scions grafted to *ga1-3* rootstocks). Thus, the mobility of the shoot-derived GAs contributes to regulate hypocotyl xylem expansion (Ragni et al., 2011). Similarly, leaf-derived GA_1 and GA_{20} are mobile signals that induce GA-promoting internode elongation, cambial activity, and fiber differentiation in tobacco stems (Dayan et al., 2012). The GA precursor GA_{12} is also a long-distance mobile GA signal through the vascular system. The shoot-to-root translocation of GA_{12} induces degradation of the DELLA proteins in roots (Regnault et al., 2015). Although endogenous GA_{12} easily moves throughout the plant and promotes the growth of recipient tissues and organs, the plant-produced GAs fail to compensate for the germination defects of progeny seeds in the GA-deficient *ga1-3* mutant, suggesting that endogenous GAs are not transmitted to offspring in Arabidopsis (Regnault et al., 2016).

The fluorescently labeled GA compounds (GA_3-Fl and GA_4-Fl) were shown to accumulate in the endodermis of the root elongation zone after application to Arabidopsis roots (Shani et al., 2013), which is consistent with previous studies that the endodermis is the major GA-responsive tissue in the roots (Heo et al., 2011; Shani et al., 2013; Ubeda-Tomas et al., 2008, 2009; Zhang et al., 2011b). Two transcription factors TEMPRANILLO1 (TEM1) and TEM2 negatively regulate trichome initiation from the mesophyll cells beneath the epidermis. Surprisingly, GA_3-Fl accumulation in the mesophyll of rosette leaves in the Arabidopsis *tem1-1 tem2-2* mutants was increased and distributed throughout a much larger leaf area in comparison with wild-type plants (Matías-Hernández et al., 2016). Indeed, TEM is known to inhibit GA biosynthesis, and also represses the expression of several GA transporters, including NITRATE TRANSPORTER1/PEPTIDE TRANSPORTER FAMILY (NPF) NPF2.3, NPF2.10, and NPF3.1 (Jiao, 2016). Therefore, TEM is essential for GA distribution.

Accumulating evidence suggests that the movement of GA across membranes does not occur by simple diffusion but requires transporter proteins that are strictly regulated during plant growth and development. The NPF family proteins were initially identified as nitrate or peptide transporters (Léran et al., 2014; Tsay et al., 2007), and were later found to also transport auxin, ABA, GA, and/or JA (jasmonic acid) hormones (Chiba et al., 2015; Kanno et al., 2012; Krouk et al., 2010; Saito et al., 2015). NPF3.1 has been proven to be a unique GA transporter in plants (Tal et al., 2016). The AtNPF3.1 protein is targeted to the plasma membrane of root endodermis cells that accumulate bioactive GAs. Interestingly, expression of *AtNPF3.1* is repressed by GA treatments, suggesting a feedback regulation. Another nitrate/peptide transporter GTR1 (glucosinolate transporter1, also known as NPF2.10) was identified as the high-affinity, proton-dependent glucosinolate-specific transporter (Nour-Eldin et al., 2012). Interestingly, GTR1 can also transport GA_3. Consistent with this observation, the *gtr1* mutants exhibit severely impaired filament elongation and anther dehiscence (Saito et al., 2015). Therefore, levels of bioactive GAs in special tissues or cells are determined not only by local GA biosynthesis and catabolism but also by GA translocation through a GA transporter.

4.3 Gibberellin perception and signaling

4.3.1 GID1 protein: a soluble GA receptor

To identify GA receptors, mutants altered in GA responses in Arabidopsis were selected and analyzed, but no GA receptor mutants were identified, perhaps because of functional redundancy within GA signaling components in Arabidopsis. However, a breakthrough was achieved with the discovery of the GID1 protein as a GA receptor in rice (Ueguchi-Tanaka et al., 2005).

In order to investigate the GA signaling molecular mechanism, the rice *gid1* mutant was identified, showing the severe dwarf phenotype with wide, dark green leaf blades. Genetic analyses of the *gid1* mutant revealed that the dwarf phenotype was inherited in a recessive manner. The *gid1* mutant must be maintained as a heterozygote because of absence of fertile flowers, and it appears to be completely insensitive to GA (Sasaki et al., 2003; Ueguchi-Tanaka et al., 2005). Analysis of *gid1* mutants with the well-characterized α-amylase test for GA response in seeds showed that even when the mutants were treated with a very high level of bioactive GA, production of α-amylase was still undetectable in *gid1-1* aleurone layers. Additionally, the endogenous accumulation of GA_1 in 1-month-old *gid1-1* shoots was up to 100-fold higher than that in wild-type plants. The *GID1* gene encodes a previously uncharacterized protein with similarity to hormone-sensitive lipases (HSLs). The GA-binding kinetic analysis revealed that the half-time for both association and dissociation between GST-GID1 and 16,17-dihydro-GA_4 was within 5 min., and the fusion protein was shown to have high-affinity only for bioactive GAs. The GA-binding kinetics may be critical for soluble receptors, because the sensitivity of the system to subtle intracellular GA concentrations alterations results in profound and compounding effects on gene regulation (Ueguchi-Tanaka et al., 2005). The GID1-GFP fusion protein was primarily localized in nuclei, with a fainter cytosolic signal, and uniconazole or GA_3 treatment did not change the subcellular localization. Importantly, GID1 bound to the rice DELLA protein SLENDER RICE1 (SLR1) in a bioactive GA-dependent manner (Ueguchi-Tanaka et al., 2005). Further experiments demonstrated that the DELLA and VHYNP domains of SLR1 are required for the GID1-SLR1 interaction (Ueguchi-Tanaka et al., 2007).

The Arabidopsis genome encodes three GA receptors, AtGID1a, AtGID1b, and AtGID1c (Nakajima et al., 2006). Bioactive GA-dependent interaction between AtGID1a and RGA was obtained by co-immunoprecipitation analysis. Expression of Arabidopsis *AtGID1a*, *AtGID1b*, or *AtGID1c* genes could rescue the rice *gid1-1* dwarf phenotype (Nakajima et al., 2006). Similar to rice, the AtGID1a-GFP fusion protein was predominantly localized to the nucleus, and to the cytoplasm, and this localization was not altered by treatments with GA or with the GA biosynthesis inhibitor PAC (Fleck and Harberd, 2002; Silverstone et al., 2001; Ueguchi-Tanaka et al., 2005; Willige et al., 2007). Cytosolic AtGID1a also plays an important role in initiating GA signaling and responses. Expression AtGID1a-GFP fusion protein with either a nuclear export signal (NES) or a nuclear localization signal (NLS) in the Arabidopsis *Atgid1a Atgid1c* mutant, both subcellular localizations of AtGID1a-GFP fusion protein were able to restore GA responsiveness. A possible explanation is that activated cytosolic AtGID1a interacts with DELLA proteins before they enter the nucleus (Livne and Weiss, 2014). It is known that both the DELLA and VHYNP domains of RGA were necessary for AtGID1a-RGA interaction (Griffiths et al., 2006), whereas the DELLA domain of GAI alone can mediate GA-dependent AtGID1a-GAI interactions and that the presence of the adjacent VHYNP domain does not contribute to enhance this interaction. Yeast three-hybrid assays revealed that GID1 induces the interaction between SLY1 and RGA (or interaction between GID2 and SLR1) in a GA-dependent manner, providing an explanation of how GA-mediated degradation of DELLA proteins is achieved (Griffiths et al., 2006; Hirano et al., 2010). Thus the DELLA domain acts as a receiver domain for GA-GID1 (Willige et al., 2007).

In Arabidopsis, *AtGID1a*, *AtGID1b*, and *AtGID1c* genes are expressed in all tissues and have both overlapping and specific roles in growth and development (Griffiths et al., 2006; Iuchi et al., 2007). Although single *Atgid1* mutants developed normally, both *Atgid1a Atgid1c* and *Atgid1a Atgid1b* double mutants displayed reduced stem elongation and lower male fertility, and the *Atgid1a Atgid1b Atgid1c* triple mutant displayed severe growth defects, with a more severe dwarf phenotype than that of the GA-deficient *ga1-3* mutant. These genetic evidences suggest that AtGID1 isoforms are the GA receptors in Arabidopsis and have some degree of functional redundancy (Griffiths et al., 2006). Indeed, the *Atgid1a Atgid1b Atgid1c* triple mutant did not germinate readily and only started to grow when the seed coat was removed after imbibition. The *Atgid1a Atgid1b Atgid1c* triple mutant was insensitive to exogenous treatment of GA, and seedlings of this triple mutant were severe dwarfs that grew only a few millimeteres high after one month. By comparing *Atgid1* multiple mutants with *sly1 Atgid1* double mutants, roles of AtGID1a, AtGID1b, and AtGID1c in proteolytic and non-proteolytic GA signaling were demonstrated. Three alleles of AtGID1 were found to play different roles in germination, stem elongation, and fertility involving proteolytic and non-proteolytic GA signaling (Hauvermale et al., 2014).

Ala scanning experiments using conserved amino acid residues among the rice and three Arabidopsis GID1 proteins revealed that 12 blocks are essential for GA-binding activity and 13 blocks are important for GID1-SLR1 interaction (Ueguchi-Tanaka et al., 2007). The detailed crystal structure analysis of the GA receptor GID1 provided us with a better understanding of how GAs operate at the molecular level. The GA-binding site of GID1 protein corresponds to the active site of the HSL domain, and four helices at the N terminus and the central part of GID1 form a lid closure (Fig. 4.3). Both the lid and the binding pocket containing GA are necessary for the DELLA interaction. DELLA interacts with the GID1-GA complex at its N-terminal region, and as a result of DELLA binding, the GID1-GA complex is stabilized (Fig. 4.3) (Ueguchi-Tanaka et al., 2007). In 2008, two research groups independently demonstrated crystal structures of rice GID1 (OsGID1)-GA complex and the Arabidopsis GID1 (AtGID1a)-GA-DELLA complex (Murase et al., 2008; Shimada et al., 2008). The structures of the GA_3- and GA_4-AtGID1a-GAI complexes display a stocky structure with a globular AtGID1a that is bound on one side by the GAI-DELLA domain (Fig. 4.3). The AtGID1a protein is monomeric and is composed of one α/β core domain with an N-terminal extension that extends up the core surface toward the DELLA domain. The crystal structure studies provided the idea that bioactive GA is an allosteric inducer of AtGID1a, which causes conformational changes that allow

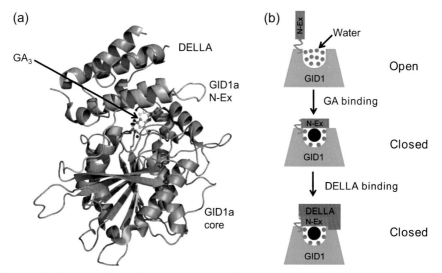

Figure 4.3 The GA-GID1-DELLA complex. (a) Crystal structure of the complex that contains GA_3, AtGID1a, and the DELLA domain of Arabidopsis GAI protein. Ribbon representation of the GA_3-GID1a-DELLA complex, with the DELLA domain (brown), the GID1a N-terminal extension (N-Ex, green) and the GID1a core domain (light blue). The GA_3 molecule is represented as a space-filling model. (b) A model for the GA-dependent GID1-DELLA interaction. GA binding induces a conformational change in the N-Ex of GID1 for DELLA binding.

the receptor to associate with DELLA proteins, but GA does not interact directly with DELLAs itself (Murase et al., 2008). This finding is consistent with the crystal structures of GA_4-OsGID1 and GA_3-OsGID1 (Shimada et al., 2008). Further Ala scanning experiments demonstrated that the conserved residues within plant GID1 proteins, but not among HSL proteins, are necessary for GA-binding activity, indicating that these residues have been recruited to establish a receptor for GA from the ancestral HSL structure (Shimada et al., 2008).

Thus, the GA perception mechanism differs from that of auxin, which serves as the "molecular glue" that brings together a substrate protein and an F-box protein without changing the structure of either protein or requiring the involvement of a third protein (Hedden, 2008; Tan et al., 2007). In contrast, the GA receptor can be activated by the allosteric effector GA to function as the "ubiquitination chaperone" that stimulates substrate recognition by the SCF complex (Lumba et al., 2010; Murase et al., 2008).

4.3.2   The evolutionary conservation of GA-GID1-DELLA module

Bioactive GAs promote growth and development throughout the life cycle of the flowering plant. Comparative studies suggest that the functional GA-GID1-DELLA module is highly conserved among vascular plants, but not in the bryophytes (Hirano et al., 2007; Sun, 2011; Yasumura et al., 2007). The proteins encoded by the *Selaginella moellendorffii* homologs of *GID1*, *DELLA*, and *GID2/SLY1* genes in vascular plants functioned well, in a manner similar to that in flowering plants, whereas the homologous proteins in *Physcomitrella patens* did not interact or functionally substitute for their flowering plant homologs (Hirano et al., 2007; Sun, 2011; Yasumura et al., 2007). The sequence alignment of the GID1 ortholog in *S. moellendorffii* shows 47% similarity to that of flowering plants, and includes an Arg265 residue which is important for GA binding (Vandenbussche et al., 2007). However, SmGID1 proteins showed lower binding affinity for bioactive GA_4 and low specificity for various other GAs when compared with GID1 of flowering plants. Studies have shown that this low affinity for GA_4 and relaxed specificity were caused by substitution of Ile133 by Leu or Val at the position that faces C2 of the *ent*-gibberellane skeleton (Ueguchi-Tanaka and Matsuoka, 2010). Taken together, these results indicate that GID1 evolved from a member of the HSLs and was further modified to have higher affinity and more strict selectivity for bioactive GAs by adapting the amino acids involved in GA binding in the course of step-by-step evolution (Shimada et al., 2008).

4.3.3   DELLA proteins: the key mediators of GA signaling

Through genetic analysis and biochemical studies, several key components in GA signaling have been identified. The DELLA proteins are characterized by their DELLA domain at the N terminus and are highly conserved in plants. They function genetically as growth repressors, and include examples in wheat (Rht1), maize (d8), barley (SLN1), rice (SLR1), grape (VvGAI), and tomato (PROCERA) (Boss and Thomas, 2002; Jasinski et al., 2008; Richards et al., 2001). The Arabidopsis DELLA protein family comprises GAI, RGA, RGA-LIKE 1 (RGL1), RGL2, and RGL3. The first identified DELLA protein was GAI, which was isolated from the Arabidopsis *gai-1* mutant (Peng and Harberd, 1993). The semidominant *gai-1* mutant is GA-insensitive, and contains a 51-bp in-frame deletion leading to a deletion of 17 amino acid residues in the DELLA domain of GAI (Peng et al., 1997). The *RGA* gene was identified based on its ability to suppress the dwarf phenotype of a GA-deficient *ga1-3* mutant (Silverstone et al., 1997b, 1998). The GAI and RGA proteins show 83% identity at the amino acid level, and the same deletion of 17 amino acid residues in the DELLA domain of RGA also produced dwarfism with similarity to *gai-1* (Dill et al., 2001). The other DELLA proteins, namely, RGL1, RGL2, and RGL3, were identified after Arabidopsis genome sequencing was completed (Lee et al., 2002).

The DELLA proteins belong to a subfamily of the plant-specific GRAS gene family of putative transcription factors whose name derives from its first three identified members, GAI, RGA, and SCARECROW (SCR). The DELLA proteins have a conserved C-terminal region and a more divergent N-terminal GA perception region (Fig. 4.4). The N-terminal region of the DELLA proteins differs from that of the rest of the GRAS family (Peng et al., 1997). In Arabidopsis, three specific motifs were identified in the N-terminal domain: the DELLA and VHYNP domains, which are responsible for the interaction between DELLA and GID1 proteins (Asano et al., 2009; Dill et al., 2001; Itoh et al., 2002; Peng et al., 1997; Silverstone et al., 2007), and polymeric Ser/Thr/Val motifs (poly S/T/V), which could be targets of phosphorylation or glycosylation. The C-terminal GRAS domain is characterized by two leucine heptad repeats (LHRs) which may mediate protein–protein interactions, a putative NLS (Dill et al., 2001; Silverstone et al., 2001) and three conserved motifs, VHIID, PFYRE and SAW, which mediate the secondary interactions with the GID1 and F-box proteins (Hirano et al., 2010).

Genetic analysis suggested that the DELLA proteins function as key components of GA signaling, since loss-of-function mutations of DELLA could suppress the phenotypes of the GA-deficient *ga1-3* mutant. The different Arabidopsis DELLA proteins have both overlapping and specific roles in the regulation of plant growth and development. The *rga* or *gai-t6* mutant could partially suppress defects of *ga1-3*; the *gai-t6 rga-24* double mutant could completely suppress the *ga1-3* phenotype, including the effects on leaf expansion, stem elongation, trichome initiation, flowering time, and apical dominance (Dill et al., 2001). Thus, RGA and GAI behave as the major

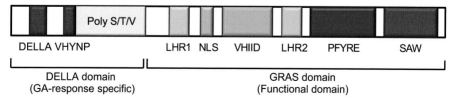

Figure 4.4 The domain structure of DELLA proteins. The unique N-terminal "DELLA domain" of the DELLA proteins contains two highly conserved motifs (named DELLA and VHYNP) and a Poly S/T/V region. The C-terminal region is GRAS domain, which contains two LHR, one NLS, and three conserved motifs, VHIID, PFYRE, and SAW. LHR, Leu heptad repeat; NLS, nuclear localization signal; Poly S/T/V, polymeric Ser and Thr and Val.

GA repressors during vegetative growth and floral induction, although, GAs are still required for seed germination and floral development in the *ga1-3 gai-t6 rga-24* triple mutant. However, the *rgl2* null mutant can suppress the germination defect of *ga1-3* (Lee et al., 2002). The *rgl2-1* mutant could germinate in the presence of the GA bio-synthesis inhibitor PAC, indicating that RGL2 has a key role in seed germination (Lee et al., 2002; Tyler et al., 2004). Furthermore, a combination of *rga*, *rgl1* and *rgl2* null mutations rescued the floral development defect and fertility of *ga1-3*, confirm-ing their roles in modulating floral development (Cheng et al., 2004). Regarding the biological function of RGL3, it was reported that RGL3 is essential for fully enhanc-ing JA and ethylene-mediated defense responses (Wild and Achard, 2013; Wild et al., 2012). These results indicate that DELLA proteins function separately and/or coop-eratively with each other at different developmental stages to regulate plant growth.

4.3.4   GA-promoted growth is dependent on degradation of DELLA proteins

Genetic studies suggest that SLY1 in Arabidopsis and its ortholog GID2 in rice are positive regulators of GA signaling. Both *SLY1* and *GID2* encode homologous F-box proteins and function as subunits of the SCF E3 ligase complex, which is required for GA-mediated degradation of DELLA proteins (Dill et al., 2004). The *sly1* null mutant fails to degrade DELLA proteins and exhibits GA-insensitive dwarf pheno-types (McGinnis et al., 2003). However, the *sly1-10* dwarf phenotype is suppressed in the *gai-t6 rga-24* double mutant (Dill et al., 2004; Fu et al., 2004). The direct inter-action between SLY1 or GID2 and DELLA proteins has been demonstrated using the yeast two-hybrid assay, and further co-immunoprecipitation analysis confirmed their roles in recruiting DELLA proteins and targeting them for degradation by SCF$^{SLY1/GID2}$ E3 ubiquitin-ligase proteins complex (Dill et al., 2004; Fu et al., 2004; Sasaki et al., 2003). As mentioned above, alterations to the DELLA domain as seen in *gai* mutants, render the mutant proteins resistant to GA-induced degradation, leading to a GA-insensitive dwarf phenotype (Fu et al., 2004). However, the GA-insensitive dwarf phenotype of *gai* mutants was suppressed by the gain-of-function mutant *gai revert-ant2-1* (*gar2-1*), which encodes a mutant version of SLY1 protein with increased affinity for DELLA proteins (Fu et al., 2004). In addition, the *gar2-1* mutant could suppress the effects of the *ga1-3* mutant throughout the plant life cycle. Thus, the affinity among components of DELLA-SCF$^{SLY1/GID2}$ proteins complex is important in the regulation of the accumulation of DELLA proteins (Fu et al., 2004).

Previous studies have shown that protein kinase inhibitors strongly inhibited GA-mediated degradation of barley SLN1 protein, but phosphorylation of DELLA proteins enhanced their interaction with the F-box component of SCF$^{SLY1/GID2}$ in Arabidopsis, barley, and rice (Fu et al., 2002, 2004; Gomi et al., 2004). However, the strength of the interaction between SLY1 and DELLA proteins is not GA-dependent (Fu et al., 2004; Itoh et al., 2005). Thus, phosphorylation of DELLA proteins may be essential for their biological functions, but might not be necessary for GA-mediated proteasome-dependent protein degradation. Indeed, Arabidopsis AtGID1b is co-immunoprecipitated with an FLAG-SLY1 fusion protein using an antibody which recognizes the FLAG epitope, supporting a model in which SLY1 regulates DELLA stability through an interaction with the GA-GID1-DELLA complex (Ariizumi et al., 2011). A study has shown that the E3 SUMO (small ubiquitin-like modifier) ligase AtSIZ1 positively regulates SLY1-mediated GA signaling through SLY1 sumoylation in Arabidopsis (Kim et al., 2015). The AtSIZ1 protein physically interacts with SLY1, in which Lys122 is the principal SUMO conjugation site. In the *Atsiz1-2* mutant, SLY1 was less abundant than that in wild-type plants, but the RGA protein was more abundant in the *Atsiz1-2* mutant than in wild-type plants. In addition, GA-promoted SLY1 sumoylation, and sumoylated SLY1 interacted with DELLA proteins, which in turn appears to regulate the degradation of the DELLA proteins. These findings reveal that sumoylation of F-box protein SLY1 is critical for GA-mediated degradation of DELLA proteins (Kim et al., 2015).

Arabidopsis has a homolog of *SLY1* known as *SNEEZY* (*SNE*) or *SLEEPY2* (*SLY2*), and the full-length *SLY1* and *SNE* genes have 55% DNA and 33% amino acid sequence homology (Ariizumi et al., 2011). No apparent phenotypic difference was observed in *sne* mutants when compared with wild-type plant in the vegetative phase, but loss-of-function mutations in *SNE* caused increased ABA sensitivity in seed germination. The *sly1 sne* double mutant exhibited a severe dwarf phenotype with a significant decrease in plant fertility when compared to the *sly1* mutant, indicating that SNE normally functions as a redundant positive regulator of GA signaling (Ariizumi and Steber, 2011). Furthermore, overexpression of *SNE* (*SLY2*) under the control of cauliflower mosaic virus 35S promoter could partially rescue the *sly1* dwarf phenotype, suggesting that SNE can functionally replace SLY1 (Fu et al., 2004; Strader et al., 2004). Interestingly, overexpression of *SLY1* in the *sly1* mutant resulted in the degradation of both RGA and RGL2 protein, whereas overexpression of *SNE* in the *sly1* mutant was associated with a decrease in RGA protein levels but not in RGL2 protein levels (Ariizumi et al., 2011). Although AtSIZ1 could interact with SLY1, AtSIZ1 did not interact with SLY2, indicating that AtSIZ1 mediates GA responses in a SLY1-dependent manner (Kim et al., 2015).

4.3.5   GA-GID1 mediated release of DELLA proteins repression

The GID1-mediated GA action in relieving DELLA protein repression has been demonstrated. GA perception is mediated by its receptor GID1, and the binding of bioactive GA to GID1 induces the formation of a GA-GID1-DELLA protein complex, which enhances their binding to the E3 ubiquitin-ligase SCF[SLY1/GID2] complex (Griffiths et al., 2006). This then results in the degradation of DELLA proteins by the 26S proteasome (Daviere et al., 2008; Gao et al., 2011; Hartweck, 2008; Jiang and Fu, 2007; Shimada et al., 2008; Xu et al., 2014), thereby relieving the growth suppression caused by DELLAs and resulting in GA-promoted growth and other GA responses (Fig. 4.5).

Although this proteolysis-dependent model can explain most GA responses, the *sly1* and *gid2* mutants accumulate more DELLA proteins but display less severe dwarf phenotypes than that of the *ga1-3* or *gid1* mutants. Thus, a proteolysis-independent model was proposed to explain these phenomena, and the inactivation may result directly from protein–protein interactions or indirectly through posttranslational modification (Ariizumi et al., 2008; Ueguchi-tanaka et al., 2008). Indeed, the deactivation of DELLA proteins can be accomplished by GA and GID1 alone in the presence of a

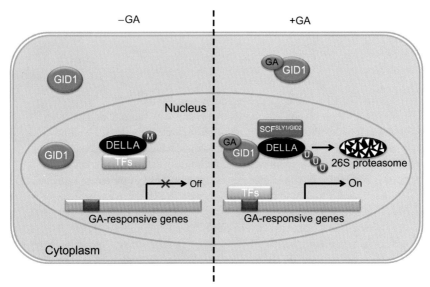

Figure 4.5 The de-repression regulatory model for DELLA-mediated GA signaling. The DELLA proteins act as the key repressors of GA signaling. In the absence of GA, the DELLA proteins restrain growth by sequestering transcription factors into inactive protein complexes. Conversely, GA promotes plant growth by the 26S proteasome-dependent degradation of DELLA proteins. The binding of GA to its receptor GID1 protein permits the interaction between GID1 and DELLA. The formation of the GA-GID1-DELLA complex enhances the interaction between DELLA and the F-box protein component of SCF$^{SLY1/GID2}$, resulting in polyubiquitination of DELLA and its targeting for degradation via the 26S proteasome pathway. The degradation of DELLA proteins releases transcription factors, which in turn activates the expression of GA-responsive genes. M, posttranslationally modified DELLA; U, ubiquitinated DELLA.

functional DELLA motif and does not require SLY1/GID2-mediated degradation of DELLA proteins (Ariizumi et al., 2008). Overexpression of *GID1* could relieve the strong *sly1-2* seed dormancy phenotype through increased GA-GID1-DELLA complex formation (Ariizumi et al., 2013). The DELLA proteins as transcriptional regulators can provide an alternative mechanism for the integration of developmental and environmental signals into plant development and defense responses. For example, GAF1 interacts with both DELLA proteins and co-repressor TOPLESS RELATED (TPR). GA converts the GAF1 complex from a transcriptional activator to a repressor via GA-mediated degradation of DELLA proteins (Fukazawa et al., 2015).

4.3.6   GA signal transduction requires the modification of DELLA proteins

Plant growth is repressed by DELLA proteins but their GA-induced degradation is quantitatively controlled by the level of endogenous GAs in a dose-dependent manner, leading to a stimulation of plant growth (Dill et al., 2001; Fu and Harberd, 2003).

It is known that DELLA proteins are phosphorylated in both Arabidopsis and rice (Fu et al., 2004; Gomi et al., 2004; Hussain et al., 2005, 2007; Itoh et al., 2002, 2005; Sasaki et al., 2003). However, whether the phosphorylation of DELLA proteins affects their function and/or stability has been unclear. Studies showed that dephosphorylation of serine/threonine was probably necessary for GA-mediated degradation of RGL2 in tobacco BY2 cells, and six mutations in conserved serine/threonine residues (RGL2^{S441D}, RGL2^{S542D}, RGL2^{T271E}, RGL2^{T319E}, RGL2^{T411E}, and RGL2^{T535E}) that mimicked a state of constitutive phosphorylation caused a resistance to GA-induced degradation (Hussain et al., 2005). Further experiments demonstrated that tyrosine phosphorylation might be a prerequisite for GA-induced degradation of RGL2 (Hussain et al., 2007). However, exogenous GA treatment induced accumulation of both the phosphorylated and non-phosphorylated SLR1 proteins with similar kinetics in callus of the *gid2* mutant, and GA-induced degradation of both forms of SLR1 proteins with similar half-life in rice wild-type callus cells, suggested that the phosphorylation of DELLA proteins is not essential for their GA-mediated degradation (Itoh et al., 2005). The rice SLR1 protein can be phosphorylated by EARLY FLOWERING1 (EL1) in vitro. GA-mediated degradation of SLR1-YFP fusion protein was significantly enhanced in the *el1* mutant when compared with the wild-type plants, suggesting that phosphorylation of DELLA protein may be essential for DELLA protein activity and its biological function in GA signaling (Dai and Xue, 2010).

Several studies have shown that microcystin-LR (inhibitor of PP1/PP2A-type phosphatase) markedly inhibited the degradation of maltose binding protein (MBP)-tagged RGA fusion protein in a cell-free proteasome assay, suggesting that the degradation of the DELLA proteins requires serine/threonine dephosphorylation activity (Wang et al., 2009). A study has shown that phosphatase TOPP4, which is a member of the protein phosphatase 1 (PP1) family in Arabidopsis, acts as a positive regulator in GA signaling and promotes the GA-mediated degradation of DELLA proteins through directly dephosphorylating DELLA proteins (Qin et al., 2014). To investigate the serine/threonine sites of DELLA proteins, six conserved serine/threonine sites of Arabidopsis RGA were substituted with alanine (RGA6A) or aspartic acid (RGA6D) to mimic the states of constitutive dephosphorylation and phosphorylation, respectively. The overexpression of de-phosphomimic RGA in wild-type plants caused DELLA protein-deficient phenotypes, whereas overexpression of phosphomimic RGA caused GA-deficient phenotypes with nondegradable RGA protein (Wang et al., 2014).

Modification by O-linked *N*-acetylglucosamine (O-GlcNAc) also plays an important role in GA signaling. The Arabidopsis *SPY* gene encodes an O-GlcNAc transferase that acts as a negative regulator of GA signaling via O-GlcNAcylation of DELLA proteins (Filardo et al., 2009; Robertson et al., 1998; Shimada et al., 2006; Swain et al., 2001). Loss-of-function mutations in *SPY* could rescue both GA-deficient *ga1-3* and GA-insensitive *gai* dwarf phenotypes, but do not lead to changes in the stability of DELLA proteins (Shimada et al., 2006; Silverstone et al., 2007). Studies have shown that SWI3C, the core component of Switch (SWI)/Sucrose Non-fermenting (SNF)-type chromatin-remodeling complexes (CRCs), could interact with both SPY and DELLA proteins, suggesting a potential role for DELLA and SPY in the modulation of GA responses through the regulation of chromatin architecture (Sarnowska et al., 2013).

Although Arabidopsis O-GlcNAc transferase SECRET AGENT (SEC) could modify the DELLA proteins, genetic analysis revealed that SEC and SPY play distinct roles in the regulation of GA signaling. SEC is a positive regulator, whereas SPY is a negative regulator of GA responses. The transient co-expression experiments in tobacco cells showed that RGA is O-GlcNAcylated by SEC but not by SPY. Moreover, O-GlcNAcylation of RGA inhibits RGA binding to its interactors PHYTOCHROME-INTERACTING FACTOR 3 (PIF3), PIF4, JA ZIM domain protein 1 (JAZ1), and BRASSINAZOLE-RESISTANT 1 (BZR1). These findings reveal that O-GlcNAcylation of DELLA proteins by SEC provides a fine-tuning mechanism for integration of developmental and environmental signals into GA responses (Zentella et al., 2016).

The posttranslational sumoylation of DELLA proteins suggests an additional mechanism for controlling their stability and activity. The DELLA proteins can be sumoylated at the conserved carboxyl-terminal region, and GID1 also contains a SUMO-interacting motif. The SUMO-conjugated DELLA proteins interact with a SUMO-interacting motif in GID1 in a GA-independent manner. The SUMO-conjugated DELLAs sequester GID1 and lead to accumulation of non-sumoylated DELLA proteins, resulting in restraint of growth which is beneficial under conditions of salinity stress (Conti et al., 2014).

4.3.7 GA-DELLA regulatory system in growth, developmental, and environmental responses

4.3.7.1 Multiple functions of DELLA proteins

Comparative studies have demonstrated that the GA-GID1-DELLA module is highly conserved among vascular plants, and GA-mediated degradation of DELLA proteins is quantitatively controlled by the level of endogenous GAs in a dose-dependent manner. The DELLA proteins lack a DNA-binding domain and there is no evidence of direct DNA binding. So, how do DELLA proteins regulate the expression level of the target genes? It has been shown that there are two ways for DELLA proteins to regulate expression of target genes. The first one is to interact with DNA-binding domains of transcription factors and inhibit their DNA-binding activity, and consequently repress target gene expression. The second one is to interact with other transcription factors and act as transcriptional co-activators or co-repressors to modulate the expression of downstream genes (Fig. 4.6) (Yoshida and Ueguchi-Tanaka, 2014). Accumulating evidence suggests that DELLA proteins are able to integrate multiple hormone and environmental signals to control plant growth and development by modulating stability and/or function of DELLA proteins (Fig. 4.7).

4.3.7.2 Antagonism of hormone signaling by interaction with DELLA proteins

It is well-known that GA and ABA play opposite roles in the regulation of seed dormancy and germination. Bioactive GA promotes seed germination, while ABA is involved in the establishment and maintenance of seed dormancy (Koornneef et al., 2002). Genetic analysis revealed that ABSCISIC ACID INSENSITIVE 3 (ABI3) and

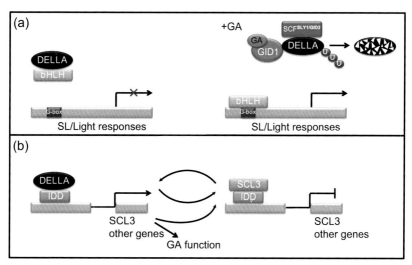

Figure 4.6 The DELLA proteins function as transcriptional repressors or activators (a) The DELLA proteins function as transcriptional repressors through physically interacting with transcription factor bHLH (e.g., PIFs and ALC) and exert negative effect by preventing its binding to the promoter of its target genes. In contrast, GA-mediated degradation of DELLA proteins releases bHLH, and consequently promotes the expression of its target genes. (b) The DELLA proteins function as transcriptional activators. DELLA proteins interact with IDD and activate the expression of the downstream genes, such as *SCL3*. The subsequently accumulated SCL3 protein causes an increase in SCL3/IDD complex, and decreases the formation of the DELLA/IDD complex and consequent suppression of downstream genes, including *SCL3* transcripts.

ABI5 are negative regulators of seed germination (Piskurewicz et al., 2008). Previous studies have shown that RGL2 represses Arabidopsis seed germination via the regulation of the transcriptional level of *XERICO*, a gene encoding a RING-H2 zinc finger protein, which promotes endogenous ABA biosynthesis. Moreover, both ABI3 and ABI5 could interact with DELLA proteins to form a complex, which binds to the promoters of *SOMNUS* and other high-temperature-inducible genes and subsequently activate their expression, thereby regulating seed germination in response to high temperature (Lim et al., 2013).

Studies have demonstrated that the cross talk between BR and GA signaling involves several aspects of plant growth and development. In Arabidopsis, the transcription factors BZR1 and BRASSINOSTEROID-INSENSITIVE 1 EMS-SUPPRESSOR1 (BES1) are positive regulators of BR signaling. Direct interaction between DELLA proteins and BZR1 blocks its ability to bind to the promoters of BR-regulated genes. Conversely, the GA-induced degradation of DELLA proteins releases transcription factors (Bai et al., 2012; Gallego-Bartolome et al., 2012; Li et al., 2012). Both DELLA and BZR1 proteins interact with PIF4 (Bernardo-Garcia et al., 2014; Oh et al., 2007) to form a DELLA-BZR1-PIF4 complex which enables plant growth to be regulated by GA, BR, and environmental factors acting through a core transcription network (Bai et al., 2012).

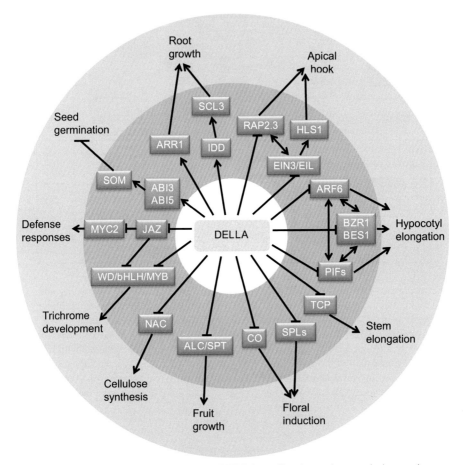

Figure 4.7 GA signal transduction requires DELLA-mediated protein–protein interactions. DELLA proteins can interact with multiple target proteins and consequently regulate multiple aspects of plant growth, development, and adaptation to stress.

Additionally, DELLA proteins interact with auxin response factor ARF6, which together with BZR1 and PIF4 can create a BZR1-ARF6-PIF-DELLA module, which provides a means to regulate Arabidopsis hypocotyl cell elongation (Oh et al., 2014).

There is cross talk between GA and ethylene signaling in the regulation of Arabidopsis apical hook development. Plant responses to ethylene are mediated by regulation of EIN3-Binding F-box protein 1 and 2 (EBF1/2)-dependent degradation of the ethylene-regulated transcription factors ETHYLENE INSENSITIVE3 (EIN3). Ethylene responses are regulated by DELLA proteins through interaction with the DNA-binding domains of EIN3 and EIN3-LIKE1 (EIL1) proteins, which in turn repress EIN3/EIL1-regulated *HOOKLESS1* (*HLS1*) expression and apical hook formation (An et al., 2012). A DELLA-interacting protein RELATED TO APETALA2.3 (RAP2.3) was found to be involved in the regulation of apical hook development.

The RAP2.3 protein belongs to the group VII ETHYLENE RESPONSE FACTOR of the APETALA2/ethylene responsive element binding protein superfamily, and the interaction with GAI impairs the DNA-binding activity of RAP2.3 to its target promoters (Marín-de la Rosa et al., 2014). Thus, the coordinated regulation of ethylene and GA signaling occurs through interaction between DELLA and EIN3/EIL1 proteins.

Cytokinins and GAs are known to exert antagonistic regulation of multiple developmental processes in plants (Weiss and Ori, 2007). Studies have shown that DELLA proteins interact with a transcription factor ARR1 (Arabidopsis response regulator 1), which is a key component of cytokinin signaling, and directly activate immediately responsive genes, which in turn regulate root meristem maintenance and skotomorphogenesis (Marín-de la Rosa et al., 2015; Moubayidin et al., 2010). In addition, GA represses the expression of *ARR1* at early stages of meristem development, and DELLA proteins are recruited by ARR1 to the promoters of cytokinin-regulated genes, where DELLA proteins act as transcriptional co-activators, thereby enhancing their expression (Marín-de la Rosa et al., 2015).

Studies have shown that plant-specific transcriptional factor TCP (TEOSINTE BRANCHED1/CYCLOIDEA/PROLIFERATING CELL FACTOR) is involved in strigolactone (SL) signal transduction pathways (Braun et al., 2012; Drummond et al., 2015; Guan et al., 2012; Rameau et al., 2015), which inhibit shoot branching and function during rhizospheric communication with symbiotic fungi and parasitic weeds. Indeed, DELLA proteins interact with class I TCP factors, such as TCP14 and TCP15, and block TCP function by binding to their DNA-recognition domain (Daviere et al., 2014; Resentini et al., 2015). The observation that DWARF14 (D14, a SL receptor) can physically interact in yeast with rice DELLA protein SLR1 suggests that DELLA proteins might mediate cross talk between SL and GA signaling (Nakamura et al., 2013). However, experimental evidence identifying the function of a D14-DELLA interaction has not yet been forthcoming.

4.3.7.3   GA signal transduction requires DELLA-mediated protein–protein interactions

The DELLA proteins also interact with transcription factor SQUAMOSA PROMOTER BINDING-LIKE (SPL), which promotes flowering in Arabidopsis by activating miR172 and MADS-box genes. The interaction between DELLA and SPL proteins interferes with SPL transcriptional activity, decreases the expression of miR172 in leaves and of MADS-box genes at the shoot apex under long-day conditions, which in turn delays floral induction (Yu et al., 2012). Flowering is promoted by GA in Arabidopsis by activating floral meristem identity genes such as *LEAFY* (*LFY*) and *SUPPRESSOR OF OVEREXPRESSION* CONSTANS1 (*SOC1*). The accumulation of DELLA proteins delays flowering under short-day conditions by repressing *LFY* and *SOC1* transcription (Achard et al., 2004, 2007a; Moon et al., 2003; Mutasa-Göttgens and Hedden, 2009). A new study showed that DELLA proteins physically interact with the critical flowering activator CONSTANS (CO), and the regulation of GA-promoting flowering in leaves under long-day conditions is mediated in part through DELLA-mediated repression of CO (Xu et al., 2016). Other

DELLA-interacting proteins include BOTRYTIS SUSCEPTIBLE1 INTERACTOR (BOI), BOI-RELATED GENE1 (BRG1), BRG2, and BRG3 which are collectively referred to as BOIs, and belong to the RING domain protein family. Using yeast two-hybrid screening, immunoprecipitation, and pull-down assays, BOIs were shown to interact with DELLAs and form a complex that binds to the promoter of GA-responsive genes and repress their responses, including seed germination, the juvenile-to-adult phase transition, and flowering time (Park et al., 2013).

In addition, the interaction between DELLA and ALCATRAZ (ALC) represses the expression of ALC target genes in the regulation of Arabidopsis fruit development (Arnaud et al., 2010). The in vivo interaction with the bHLH transcription factors PIF3 and PIF4 has been shown to affect the DNA-binding activity of the PIFs, resulting in an inhibition of PIF-mediated gene expression and hypocotyl elongation (de Lucas et al., 2008; Feng et al., 2008). The PIF homolog SPATULA (SPT) regulates seed dormancy and restrains cotyledon expansion and fruit growth (Fuentes et al., 2012; Josse et al., 2011). Interaction between SPT and DELLA proteins has been demonstrated in yeast two-hybrid experiments (Gallego-Bartolomé et al., 2010). Interestingly, SPT acts as a growth repressor in the regulation of GA-induced fruit development in a DELLA-independent manner (Fuentes et al., 2012). These findings indicate that DELLA proteins act as transcriptional repressors by preventing the binding of DELLA-associated transcription factors to their promoter region of target genes.

In addition to functioning as transcriptional repressors, DELLA proteins can function as transcriptional co-activators. In Arabidopsis, SCARECROW-LIKE 3 (SCL3) is a positive regulator of GA-induced root growth and development. Expression of *SCL3* is induced by DELLA proteins and repressed by GA treatment. Direct interaction of SCL3 with DELLA proteins promotes GA signaling by antagonizing the activity of DELLA proteins (Heo et al., 2011; Zhang et al., 2011b). The Arabidopsis genome contains 16 INDETERMINATE DOMAIN (IDD) proteins, which are characterized by a distinct arrangement of zinc (Zn) finger motifs (Colasanti et al., 2006). Several IDD proteins such as IDD1, IDD3, IDD5, IDD9, IDD10, and a homologous protein GAF1 interact directly with DELLA proteins, and regulate GA signaling in Arabidopsis (Feurtado et al., 2011; Fukazawa et al., 2014; Yoshida et al., 2014). The interaction between DELLA and IDD1 regulates seed dormancy (Feurtado et al., 2011). Indeed, IDD3, IDD5, IDD9, and IDD10 not only interact with DELLA but also with SCL3 protein. The three proteins DELLAs, SCL3, and IDDs constitute a "co-activator/co-repressor exchange regulation system" that fine-tunes GA feedback regulation (Yoshida et al., 2014; Yoshida and Ueguchi-Tanaka, 2014). In this model, DELLA proteins act as co-activators, and the interaction between DELLA and IDD proteins enhances expression of *SCL3*. Furthermore, the accumulation of SCL3 protein also acts as a co-repressor; an increased level of the SCL3-IDD protein complex can suppress its own expression (Yoshida et al., 2014; Yoshida and Ueguchi-Tanaka, 2014).

Studies have shown that cellulose synthesis is also regulated by the GA-DELLA regulatory system in rice. In the presence of GA, the degradation of SLR1 is induced, releasing the "top-layer" NAC transcription factors, which are required for secondary wall formation, enabling these factors to bind and upregulate the downstream target MYB61 and consequently enhancing the transcriptional levels of rice *CELLULOSE SYNTHASE* (*CESA*). In contrast, SLR1 interacts directly with NACs, which in turn inhibits an NAC-MYB-CESA signaling cascade. These findings reveal a conserved mechanism for the regulation of secondary wall cellulose synthesis in land plants (Huang et al., 2015).

Chromatin-remodeling is a crucial regulator of gene expression in eukaryotes (Ho and Crabtree, 2010; Jarillo et al., 2009). The negative regulator of the light signaling pathway known as PICKLE (PKL) encodes an ATP-dependent CHROMODOMAIN HELICASE-DNA BINDING3 (CHD3) type chromatin-remodeling factor of the SWITCH/SUCROSE NONFERMENTING (SWI/SNF) family (Jing and Lin, 2013; Jing et al., 2013). Previous studies have demonstrated that PKL (also known as EPP1) protein interacts directly with PIF3 and BZR1 to promote hypocotyl growth by repressing the H3K27me3 modification of cell elongation-related genes in Arabidopsis. However, DELLA proteins physically interact with PKL and block the PKL-PIF3 interaction, resulting in deactivation of these genes (Zhang et al., 2014). More interestingly, a study has revealed that GA regulates microtubule organization via an interaction between the DELLA proteins and the prefoldin complex (PFD), a co-chaperone required for tubulin folding (Locascio et al., 2013). Without GA, the DELLA-PFD interaction retained the localization of the complex in the nucleus, thus compromising α/β-tubulin heterodimer availability in the cytoplasm. Conversely, in the presence of GA, DELLA proteins are degraded, and the prefoldin complex stays in the cytoplasm and is functional (Locascio et al., 2013).

4.3.7.4 Integration of developmental and environmental signals by DELLA proteins

As sessile organisms, plants have evolved to survive adverse environmental conditions by adapting their pattern of growth and development to any environmental changes. Both light and GA regulate many important plant developmental processes. Light induces photomorphogenesis, leading to inhibition of hypocotyl growth, whereas GAs promote etiolated growth and increase hypocotyl elongation. Interaction of DELLA proteins with the bHLH transcription factors PIF3 and PIF4 blocks PIF transcriptional activity by binding to the DNA-recognition domain of these factors, resulting in inhibition of PIF-mediated expression of target genes and hypocotyl elongation (de Lucas et al., 2008; Feng et al., 2008). The GA-induced degradation of DELLA proteins releases PIFs and promotes PIF-activated gene expression. The interaction between DELLAs and PIFs integrates both GA and light signals to modulate photomorphogenesis.

JA is an important plant hormone involved in the regulation of plant development and stress responses, and JAZ proteins function as the key repressors of JA signaling. Studies have demonstrated antagonistic roles of GAs in JA-mediated plant development and defense against biotic or abiotic stress. The direct interaction between DELLA and JAZ inhibits JAZ activity, thereby enhancing the ability of MYC2 to regulate JA-mediated root growth inhibition and susceptibility to hemibiotrophic pathogen *Pst DC3000* (*Pseudomonas syringae* pv *tomato DC3000*) (Hou et al., 2010; Wild et al., 2012; Yang et al., 2012). Moreover, JAZ9 interrupts the RGA-PIF3 interaction and releases PIFs in the GA pathway to enhance hypocotyl elongation (Yang et al., 2012).

These findings reveal the important function of the core DELLA/JAZ/PIF complex in prioritizing plant defenses over growth when facing environmental stresses and pathogen attacks (Daviere and Achard, 2015).

In contrast to their antagonistic roles in modulating growth and defense, GA and JA signaling also synergistically regulate several developmental processes. Both DELLA and JAZ proteins interact with the WD-repeat/bHLH/MYB complex members of the bHLH transcription factors GL3/EGL3 and the MYB factor GL1 to mediate synergism between GA and JA signaling in regulating trichome development (Grebe, 2012; Qi et al., 2011, 2014). Moreover, JAZ and DELLA proteins interact with MYC2 and prevent MYC2 binding to the promoters of sesquiterpene synthase genes *TPS21* and *TPS11* (Hong et al., 2012). GA-induced degradation of DELLA proteins causes the upregulation of the two JA biosynthesis genes *DEFECTIVE IN ANTHER DEHISCENCE 1* (*DAD1*) and *LIPOXYNENASE 1* (*LOX1*), subsequently promoting JA biosynthesis and the expression of *MYB21*, *MYB24*, and *MYB57* (Cheng et al., 2009).

4.4   Summary points

- GA biosynthesis can be divided into three stages in three subcellular compartments comprising plastids, endomembranes, and cytosol, and the key synthesis enzymes have been identified.
- GA inactivation pathways involving covalent modification and conjugation have been revealed, which are important for controlling endogenous GA levels.
- GA is a mobile molecule that passes through the plasma membrane via specific transporter proteins for cell-to-cell transport, an essential process for GA response.
- The key components of GA signaling have been identified, and evolutionary conserved "Relief of Repression" regulatory model is demonstrated.
- DELLA proteins function as growth repressors and are modulated by phosphorylation, O-GlcNAcylation, and sumoylation.
- DELLA proteins integrated with other hormonal and environmental signaling molecules control plant growth and development.

4.5   Future perspectives

- GA plays an important role in the regulation of plant growth and stress adaptation. Studies have begun to uncover how GA promotes plant growth and development at the molecular level through GID1-mediated degradation of DELLA proteins. Accumulating evidences suggest that DELLA proteins modulate multiple hormonal and environmental signaling activities through transcriptional regulation and protein–protein interactions. However, several questions still remain to be resolved.
- The appropriate level of bioactive GAs is important for normal plant growth and development. Recent biochemical and micrografting studies have suggested that the movement of various GAs from production sites to the tissues and organs that require GAs contributes to the coordinated regulation of a diversity of processes associated with plant growth and stress adaptation. Although plant shoots and roots follow distinct developmental trajectories, GA

biosynthesis and transport are tightly coordinated to optimize whole plant GA-mediated responses in a fluctuating environment. However, the molecular basis of GA mobility and long-distance coordination is little understood.

- It is known that a plasma membrane-localized receptor may be required for GA-dependent induction of α-amylase production in the cells of cereal aleurone layer. Although soluble GA receptor GID1 protein is present in the cytoplasm, studies revealed that GA-GID1-DELLA protein complex is localized to the nucleus. It raises the possibility that the presence of membrane-localized co-receptor may be required for a GID1-dependent but DELLA-independent GA signaling pathway.

- The de-repression regulatory model suggests that GA promotes plant growth via the GA-mediated degradation of the DELLA proteins. However, studies suggest that a DELLA-independent signal transduction pathway may be involved in GA signaling. For example, Arabidopsis fruit growth is induced by GA in both *global* (lacking GAI, RGA, RGL1, RGL2 and RGL3) and *ga1 global* mutants, whereas exogenous GA treatment is unable to restore fruit growth in a *Atgid1a Atgid1b Atgid1c* triple mutant (Fuentes et al., 2012). The implication is that a DELLA-independent pathway is involved in the control of GA responses, although the components in DELLA-independent GA signaling have yet to be identified.

- GA-mediated plant growth does not always require the degradation of DELLA proteins. Previous studies showed that both *sly1* mutants accumulate more DELLA proteins but display less severe dwarf phenotypes than that of *ga1-3* or *Atgid1a Atgid1b Atgid1c* triple mutant, and the upregulation of *GID1* could rescue the *sly1* or *gid2* dwarf phenotypes without decreasing the accumulation levels of DELLA proteins (Ariizumi et al., 2008; Ueguchi-tanaka et al., 2008).

- It is proposed that modification by phosphorylation, O-GlcNAcylation, or sumoylation is critical for the stability and activity of DELLA proteins, although the molecular mechanism underlying this modification remains to be determined.

- It is known that DELLA proteins physically interact with various transcription factors that integrate multiple developmental and environmental signals into cell proliferation and differentiation through GA-mediated degradation of DELLA proteins. However, it is unclear how GA action is coordinated at the molecular level through dynamic regulation of DELLA-associated protein complex in response to developmental and/or environmental changes.

- Future studies should use a combined biochemical and systems biology approach to identify new components involved in GA responses and generate a robust framework for GA signaling that integrates multiple developmental and environmental signals to ensure normal plant growth and adaptation in a fluctuating environment.

Abbreviations

2ODD	2-Oxoglutarate-dependent dioxygenase
ABA	Abscisic acid
ABI3	ABSCISIC ACID INSENSITIVE 3
AGL15	AGAMOUS-Like15
ALC	ALCATRAZ
AP2	APETALA2
ARR1	Arabidopsis response regulator 1

ATI	Auxin transport inhibitor
BES1	BRASSINOSTEROID-INSENSITIVE 1 EMS-SUPPRESSOR1
BOI	BOTRYTIS SUSCEPTIBLE1 INTERACTOR
BRG1	BOI-RELATED GENE1
BZR1	BRASSINAZOLE-RESISTANT 1
CBF	C-repeat binding factor
CDP	Copalyl diphosphate
CESA	CELLULOSE SYNTHASE
CHD3	CHROMODOMAIN HELICASE-DNA BINDING3
CK	Cytokinin
CO	CONSTANS
CPS	*ent*-Copalyl diphosphate synthase
D1	DWARF1
D14	DWARF14
D18	OsGA3ox2
D35	OsKO2
DAD1	DEFECTIVE IN ANTHER DEHISCENCE 1
DAG1	DOF AFFECTING GERMINATION1
DDF1	DWARF AND DELAYED FLOWERING 1
DREB1	Dehydration-responsive element binding protein
DWF4	DWARF4
EBF1/2	EIN3-Binding F-box protein 1 and 2
EIL1	EIN3-LIKE1
EIN3	ETHYLENE INSENSITIVE3
ELF1	EARLY FLOWERING1
ELF3	EARLY FLOWERING3
ent-CDP	*ent*-Copalyl diphosphate
ERF	ETHYLENE RESPONSE FACTOR
EUI	ELONGATED UPPERMOST INTERNODE
FUS3	FUSCA3
GA	Gibberellin
GA1	AtCPS
GA2	AtKS
GA3	AtKO
GA4	AtGA3ox1
GA5	AtGA20ox1
GA2ox	GA 2-oxidase
GA3ox	GA 3-oxidase
GA13ox	GA 13-oxidase
GA20ox	GA 20-oxidase
GAF1	GAI-ASSOCIATED FACTOR1
GAI	Gibberellin-Insensitive
GAMT	Gibberellin methyltransferase
gar2-1	*gai revertant2-1*
GGDP	Geranylgeranyl diphosphate
GID1	GIBBERELLIN INSENSITIVE DWARF1
GID2	GIBBERELLIN INSENSITIVE DWARF2

GRAS	The name derives from its first three identified members, GAI, RGA, and SCARECROW (SCR)
GTR1	Glucosinolate transporter1
HLS1	HOOKLESS1
HSLs	Hormone-sensitive lipases
IAA	Indole-3-acetic acid
IDD	INDETERMINATE DOMAIN
JA	Jasmonic acid
JAZ1	JA ZIM domain protein 1
KAO	*ent*-Kaurenoic acid oxidase
KNOX	Class-I KNOTTED1-like homeobox
KO	*ent*-Kaurene oxidase
KS	*ent*-Kaurene synthase
LE	PsGA3ox1
LEC2	LEAFY COTYLEDON2
LFY	LEAFY
LH	PsKO1
LHRs	Leucine heptad repeats
LOX1	LIPOXYNENASE 1
MBP	Maltose binding protein
NA	PsKAO1
NCED	9-*Cis*-epoxycarotenoid dioxygenase
NES	Nuclear export signal
NLS	Nuclear localization signal
NOA	1-Naphthoxyacetic acid
NPA	1-N-naphthylthalamic acid
NPF	NITRATE TRANSPORTER1/PEPTIDE TRANSPORTER FAMILY
O-GlcNAc	O-linked *N*-acetylglucosamine
PAC	Paclobutrazol
PFD	Prefoldin complex
PIF3	PHYTOCHROME-INTERACTING FACTOR 3
PIF4	PHYTOCHROME-INTERACTING FACTOR 4
PIF5	PHYTOCHROME-INTERACTING FACTOR 5
PIL5	PHYTOCHROME-INTERACTING FACTOR 3-LIKE 5
PKL	PICKLE
POTH1	Potato homeobox 1
PP1	Protein phosphatase 1
Pst DC3000	*Pseudomonas syringae* pv *tomato DC3000*
P450	Cytochrome P450 monooxygenase
RAP2.3	RELATED TO APETALA2.3
RGA	REPRESSOR of *ga1-3*
RGL1	RGA-LIKE 1
RGL2	RGA-LIKE 2
SAM	Shoot apical meristem
SCF	Skp1-Cullin-F-box
SCL3	SCARECROW-LIKE 3
SCR	SCARECROW
SD1	SEMI-DWARF1

SEC	SECRET AGENT
SL	Strigolactone
SLR1	SLENDER RICE1
SLY1	SLEEPY1
SLY2	SLEEPY2
SNE	SNEEZY
SOC1	SUPPRESSOR OF OVEREXPRESSION CONSTANS1
SPL	SQUAMOSA PROMOTER BINDING-LIKE
SPT	SPATULA
SPY	SPINDLY
StBEL5	BEL1-like transcription factor in potato
SUMO	Small Ubiquitin-like Modifier
SWI/SNF	SWITCH/SUCROSE NONFERMENTING
SWI3C	The core component of Switch (SWI)/Sucrose Non-fermenting (SNF)-type chromatin-remodeling complexes (CRCs)
TCP	TEOSINTE BRANCHED1/CYCLOIDEA/PROLIFERATING CELL FACTOR
TEM1	TEMPRANILLO1
TPR	TOPLESS RELATED
TPS	Terpene synthase
YAB1	YABBY1
ZEP	Zeaxanthin epoxidase

References

Aach, H., Bode, H., Robinson, D., et al., 1997. *ent*-Kaurene synthase is located in proplastids of meristematic shoot tissues. Planta 202, 211–219.

Abe, A., Takagi, H., Fujibe, T., et al., 2012. OsGA20ox1, a candidate gene for a major QTL controlling seedling vigor in rice. Theor. Appl. Genet. 125, 647–657.

Achard, P., Baghour, M., Chapple, A., et al., 2007a. The plant stress hormone ethylene controls floral transition via DELLA-dependent regulation of floral meristem-identity genes. Proc. Natl. Acad. Sci. U.S.A. 104, 6484–6489.

Achard, P., Liao, L., Jiang, C., et al., 2007b. DELLAs contribute to plant photomorphogenesis. Plant Physiol. 143, 1163–1172.

Achard, P., Gong, F., Cheminant, S., et al., 2008. The cold-inducible CBF1 factor-dependent signaling pathway modulates the accumulation of the growth-repressing DELLA proteins via its effect on gibberellin metabolism. Plant Cell 20, 2117–2129.

Achard, P., Herr, A., Baulcombe, D.C., et al., 2004. Modulation of floral development by a gibberellin-regulated microRNA. Development 131, 3357–3365.

An, F., Zhang, X., Zhu, Z., et al., 2012. Coordinated regulation of apical hook development by gibberellins and ethylene in etiolated *Arabidopsis* seedlings. Cell Res. 22, 915–927.

Ariizumi, T., Hauvermale, A.L., Nelson, S.K., et al., 2013. Lifting DELLA repression of Arabidopsis seed germination by nonproteolytic gibberellin signaling. Plant Physiol. 162, 2125–2139.

Ariizumi, T., Lawrence, P.K., Steber, C.M., 2011. The role of two F-box proteins, SLEEPY1 and SNEEZY, in Arabidopsis gibberellin signaling. Plant Physiol. 155, 765–775.

Ariizumi, T., Murase, K., Sun, T.P., et al., 2008. Proteolysis-independent downregulation of DELLA repression in Arabidopsis by the gibberellin receptor GIBBERELLIN INSENSITIVE DWARF1. Plant Cell 20, 2447–2459.

Ariizumi, T., Steber, C.M., 2011. Mutations in the F-box gene *SNEEZY* result in decreased Arabidopsis GA signaling. Plant Signal. Behav. 6, 831–833.

Arnaud, N., Girin, T., Sorefan, K., et al., 2010. Gibberellins control fruit patterning in *Arabidopsis thaliana*. Genes Dev. 24, 2127–2132.

Asano, K., Hirano, K., Ueguchi-Tanaka, M., et al., 2009. Isolation and characterization of dominant dwarf mutants, *Slr1-d*, in rice. Mol. Genet. Genomics 281, 223–231.

Ashikari, M., Sasaki, A., Ueguchi-Tanaka, M., et al., 2002. Loss-of-function of a rice gibberellin biosynthetic gene, *GA20 oxidase* (*GA20ox-2*), led to the rice 'Green revolution'. Breed. Sci. 52, 143–150.

Bai, M.Y., Shang, J.X., Oh, E., et al., 2012. Brassinosteroid, gibberellin and phytochrome impinge on a common transcription module in Arabidopsis. Nat. Cell Biol. 14, 810–817.

Bernardo-Garcia, S., de Lucas, M., Martinez, C., et al., 2014. BR-dependent phosphorylation modulates PIF4 transcriptional activity and shapes diurnal hypocotyl growth. Genes Dev. 28, 1681–1694.

Boccaccini, A., Santopolo, S., Capauto, D., et al., 2014. The DOF protein DAG1 and the DELLA protein GAI cooperate in negatively regulating the *AtGA3ox1* gene. Mol. Plant 7, 1486–1489.

Boss, P.K., Thomas, M.R., 2002. Association of dwarfism and floral induction with a grape 'green revolution' mutation. Nature 416, 847–850.

Braun, N., de Saint Germain, A., Pillot, J.P., et al., 2012. The pea TCP transcription factor PsBRC1 acts downstream of strigolactones to control shoot branching. Plant Physiol. 158, 225–238.

Chen, H., Banerjee, A., Hannapel, D., 2004. The tandem complex of BEL and KNOX partners is required for transcriptional repression of ga20ox1. Plant J. 38, 276–284.

Chen, Y., Tan, B.C., 2015. New insight in the Gibberellin biosynthesis and signal transduction. Plant Signal. Behav. 10, e1000140.

Cheng, H., Qin, L., Lee, S., et al., 2004. Gibberellin regulates Arabidopsis floral development via suppression of DELLA protein function. Development 131, 1055–1064.

Cheng, H., Song, S., Xiao, L., et al., 2009. Gibberellin acts through jasmonate to control the expression of MYB21, MYB24, and MYB57 to promote stamen filament growth in Arabidopsis. PLoS Genet. 5, e1000440.

Chiang, H., Hwang, l, Goodma, H., 1995. Isolation of the *Arabidopsis* GA4 locus. Plant Cell Online 7, 195–201.

Chiba, Y., Shimizu, T., Miyakawa, S., et al., 2015. Identification of *Arabidopsis thaliana* NRT1/PTR FAMILY (NPF) proteins capable of transporting plant hormones. J. Plant Res. 128, 679–686.

Colasanti, J., Tremblay, R., Wong, A.Y., et al., 2006. The maize *INDETERMINATE1* flowering time regulator defines a highly conserved zinc finger protein family in higher plants. BMC Genomics 7, 158.

Colebrook, E.H., Thomas, S.G., Phillips, A.L., et al., 2014. The role of gibberellin signalling in plant responses to abiotic stress. J. Exp. Biol. 217, 67–75.

Coles, J., Phillips, A., Croker, S., et al., 1999. Modification of gibberellin production and plant development in Arabidopsis by sense and antisense expression of gibberellin 20-oxidase genes. Plant J. 17, 547–556.

Conti, L., Nelis, S., Zhang, C., et al., 2014. Small Ubiquitin-like Modifier protein SUMO enables plants to control growth independently of the phytohormone gibberellin. Dev. Cell 28, 102–110.

Cowling, R.J., Kamiya, Y., Seto, H., et al., 1998. Gibberellin dose-response regulation of *GA4* gene transcript levels in Arabidopsis. Plant Physiol. 117, 1195–1203.

Curaba, J., Moritz, T., Blervaque, R., et al., 2004. *AtGA3ox2*, a key gene responsible for bioactive gibberellin biosynthesis, is regulated during embryogenesis by *LEAFY COTYLEDON2* and *FUSCA3* in Arabidopsis. Plant Physiol. 136, 3660–3669.

Dai, C., Xue, H.W., 2010. Rice early flowering1, a CKI, phosphorylates DELLA protein SLR1 to negatively regulate gibberellin signalling. EMBO J. 29, 1916–1927.

Dai, M., Zhao, Y., Ma, Q., et al., 2007. The rice *YABBY1* gene is involved in the feedback regulation of gibberellin metabolism. Plant Physiol. 144, 121–133.

Davidson, S.E., Elliott, R.C., Helliwell, C.A., et al., 2003. The pea gene *NA* encodes *ent*-kaurenoic acid oxidase. Plant Physiol. 131, 335–344.

Davidson, S.E., Smith, J.J., Helliwell, C.A., et al., 2004. The pea gene *LH* encodes *ent*-kaurene oxidase. Plant Physiol. 134, 1123–1134.

Daviere, J.M., Achard, P., 2015. A pivotal role of DELLAs in regulating multiple hormone signals. Mol. Plant 9, 10–20.

Daviere, J.M., de Lucas, M., Prat, S., 2008. Transcription factor interaction: a central step in DELLA function. Curr. Opin. Genet. Dev. 18, 295–303.

Daviere, J.M., Wild, M., Regnault, T., et al., 2014. Class I TCP-DELLA interactions in inflorescence shoot apex determine plant height. Curr. Biol. 24, 1923–1928.

Dayan, J., Voronin, N., Gong, F., et al., 2012. Leaf-induced gibberellin signaling is essential for internode elongation, cambial activity, and fiber differentiation in tobacco stems. Plant Cell 24, 66–79.

de Lucas, M., Daviere, J.M., Rodriguez-Falcon, M., et al., 2008. A molecular framework for light and gibberellin control of cell elongation. Nature 451, 480–484.

Desgagne-Penix, I., Sponsel, V.M., 2008. Expression of *gibberellin 20-oxidase1* (*AtGA20ox1*) in Arabidopsis seedlings with altered auxin status is regulated at multiple levels. J. Exp. Bot. 59, 2057–2070.

Dill, A., Jung, H.S., Sun, T.P., 2001. The DELLA motif is essential for gibberellin-induced degradation of RGA. Proc. Natl. Acad. Sci. U.S.A. 98, 14162–14167.

Dill, A., Sun, T., 2001. Synergistic derepression of gibberellin signaling by removing RGA and GAI function in *Arabidopsis thaliana*. Genetics 159, 777–785.

Dill, A., Thomas, S.G., Hu, J., et al., 2004. The Arabidopsis F-box protein SLEEPY1 targets gibberellin signaling repressors for gibberellin-induced degradation. Plant Cell 16, 1392–1405.

Drummond, R.S., Janssen, B.J., Luo, Z., et al., 2015. Environmental control of branching in petunia. Plant Physiol. 168, 735–751.

Eriksson, S., Bohlenius, H., Moritz, T., et al., 2006. GA4 is the active gibberellin in the regulation of *LEAFY* transcription and *Arabidopsis* floral initiation. Plant Cell 18, 2172–2181.

Feng, S., Martinez, C., Gusmaroli, G., et al., 2008. Coordinated regulation of *Arabidopsis thaliana* development by light and gibberellins. Nature 451, 475–479.

Feurtado, J.A., Huang, D., Wicki-Stordeur, L., et al., 2011. The *Arabidopsis* C_2H_2 zinc finger INDETERMINATE DOMAIN1/ENHYDROUS promotes the transition to germination by regulating light and hormonal signaling during seed maturation. Plant Cell 23, 1772–1794.

Filardo, F., Robertson, M., Singh, D.P., et al., 2009. Functional analysis of HvSPY, a negative regulator of GA response, in barley aleurone cells and *Arabidopsis*. Planta 229, 523–537.

Filo, J., Wu, A., Eliason, E., et al., 2015. Gibberellin driven growth in elf3 mutants requires PIF4 and PIF5. Plant Signal. Behav. 10, e992707.

Fleck, B., Harberd, N.P., 2002. Evidence that the *Arabidopsis* nuclear gibberellin signalling protein GAI is not destabilised by gibberellin. Plant J. 32, 935–947.

Fleet, C.M., Yamaguchi, S., Hanada, A., et al., 2003. Overexpression of *AtCPS* and *AtKS* in *Arabidopsis* confers increased *ent*-kaurene production but no increase in bioactive gibberellins. Plant Physiol. 132, 830–839.

Frigerio, M., Alabadi, D., Perez-Gomez, J., et al., 2006. Transcriptional regulation of gibberellin metabolism genes by auxin signaling in *Arabidopsis*. Plant Physiol. 142, 553–563.

Fu, J., Ren, F., Lu, X., et al., 2016. A tandem array of *ent*-kaurene synthases in maize with roles in gibberellin and more specialized metabolism. Plant Physiol. 170, 742–751.

Fu, X., Harberd, N.P., 2003. Auxin promotes *Arabidopsis* root growth by modulating gibberellin response. Nature 421, 740–743.

Fu, X., Richards, D.E., Ait-Ali, T., et al., 2002. Gibberellin-mediated proteasome-dependent degradation of the barley DELLA protein SLN1 repressor. Plant Cell 14, 3191–3200.

Fu, X., Richards, D.E., Fleck, B., et al., 2004. The *Arabidopsis* mutant *sleepy1*[gar2-1] protein promotes plant growth by increasing the affinity of the SCF[SLY1] E3 ubiquitin ligase for DELLA protein substrates. Plant Cell 16, 1406–1418.

Fuentes, S., Ljung, K., Sorefan, K., et al., 2012. Fruit growth in *Arabidopsis* occurs via DELLA-dependent and DELLA-independent gibberellin responses. Plant Cell 24, 3982–3996.

Fukazawa, J., Ito, T., Kamiya, Y., et al., 2015. Binding of GID1 to DELLAs promotes dissociation of GAF1 from DELLA in GA dependent manner. Plant Signal. Behav. 10, e1052923.

Fukazawa, J., Teramura, H., Murakoshi, S., et al., 2014. DELLAs function as coactivators of GAI-associated FACTOR1 in regulation of gibberellin homeostasis and signaling in *Arabidopsis*. Plant Cell 26, 2920–2938.

Gallego-Bartolomé, J., Minguet, E.G., Marin, J.A., et al., 2010. Transcriptional diversification and functional conservation between DELLA proteins in Arabidopsis. Mol. Biol. Evol. 27, 1247–1256.

Gallego-Bartolome, J., Minguet, E.G., Grau-Enguix, F., et al., 2012. Molecular mechanism for the interaction between gibberellin and brassinosteroid signaling pathways in *Arabidopsis*. Proc. Natl. Acad. Sci. U.S.A. 109, 13446–13451.

Gallego-Giraldo, L., Ubeda-Tomas, S., Gisbert, C., et al., 2008. Gibberellin homeostasis in tobacco is regulated by gibberellin metabolism genes with different gibberellin sensitivity. Plant Cell Physiol. 49, 679–690.

Gao, X.H., Xiao, S.L., Yao, Q.F., et al., 2011. An updated GA signaling 'Relief of repression' regulatory model. Mol. Plant 4, 601–606.

Gazzarrini, S., Tsuchiya, Y., Lumba, S., et al., 2004. The transcription factor *FUSCA3* controls developmental timing in *Arabidopsis* through the hormones gibberellin and abscisic acid. Dev. Cell 7, 373–385.

Gomi, K., Sasaki, A., Itoh, H., et al., 2004. GID2, an F-box subunit of the SCF E3 complex, specifically interacts with phosphorylated SLR1 protein and regulates the gibberellin-dependent degradation of SLR1 in rice. Plant J. 37, 626–634.

Grebe, M., 2012. The patterning of epidermal hairs in *Arabidopsis* – updated. Curr. Opin. Plant Biol. 15, 31–37.

Greenboim-Wainberg, Y., Maymon, I., Borochov, R., et al., 2005. Cross talk between gibberellin and cytokinin: the *Arabidopsis* GA response inhibitor SPINDLY plays a positive role in cytokinin signaling. Plant Cell 17, 92–102.

Griffiths, J., Murase, K., Rieu, I., et al., 2006. Genetic characterization and functional analysis of the GID1 gibberellin receptors in *Arabidopsis*. Plant Cell 18, 3399–3414.

Guan, J.C., Koch, K.E., Suzuki, M., et al., 2012. Diverse roles of strigolactone signaling in maize architecture and the uncoupling of a branching-specific subnetwork. Plant Physiol. 160, 1303–1317.

Han, F., Zhu, B., 2011. Evolutionary analysis of three gibberellin oxidase genes in rice, *Arabidopsis*, and soybean. Gene 473, 23–35.

Hartweck, L.M., 2008. Gibberellin signaling. Planta 229, 1–13.

Hauvermale, A.L., Ariizumi, T., Steber, C.M., 2014. The roles of the GA receptors *GID1a*, *GID1b*, and *GID1c* in *sly1*-independent GA signaling. Plant Signal. Behav. 9, e28030.

Hedden, P., 2008. Plant biology: gibberellins close the lid. Nature 456, 455–456.

Hedden, P., Kamiya, Y., 1997. Gibberellin biosynthesis: enzymes, genes and their regulation. Annu. Rev. Plant Physiol. Plant Mol. Biol. 48, 431–460.

Hedden, P., Phillips, A., 2000. Gibberellin metabolism: new insights revealed by the genes. Trends Plant Sci. 5, 523–530.

Hedden, P., Sponsel, V., 2015. A century of gibberellin research. J. Plant Growth Regul. 34, 740–760.

Hedden, P., Thomas, S.G., 2012. Gibberellin biosynthesis and its regulation. Biochem. J. 444, 11–25.

Helliwell, C., Chandler, P., Poole, A., et al., 2001a. The CYP88A cytochrome P450, *ent*-kaurenoic acid oxidase, catalyzes three steps of the gibberellin biosynthesis pathway. Proc. Natl. Acad. Sci. U.S.A. 98, 2065–2070.

Helliwell, C., Sheldon, C., Olive, M., et al., 1998. Cloning of the *Arabidopsis ent*-kaurene oxidase gene *GA3*. Proc. Natl. Acad. Sci. U.S.A. 95, 9019–9024.

Helliwell, C.A., Poole, A., Peacock, W.J., et al., 1999. Arabidopsis *ent*-kaurene oxidase catalyzes three steps of gibberellin biosynthesis. Plant Physiol. 119, 507–510.

Helliwell, C.A., Sullivan, J.A., Mould, R.M., et al., 2001b. A plastid envelope location of *Arabidopsis ent*-kaurene oxidase links the plastid and endoplasmic reticulum steps of the gibberellin biosynthesis pathway. Plant J. 28, 201–208.

Heo, J.O., Chang, K.S., Kim, I.A., et al., 2011. Funneling of gibberellin signaling by the GRAS transcription regulator SCARECROW-LIKE 3 in the *Arabidopsis* root. Proc. Natl. Acad. Sci. U.S.A. 108, 2166–2171.

Hirano, K., Asano, K., Tsuji, H., et al., 2010. Characterization of the molecular mechanism underlying gibberellin perception complex formation in rice. Plant Cell 22, 2680–2696.

Hirano, K., Nakajima, M., Asano, K., et al., 2007. The GID1-mediated gibberellin perception mechanism is conserved in the Lycophyte *Selaginella moellendorffii* but not in the Bryophyte *Physcomitrella patens*. Plant Cell 19, 3058–3079.

Hirose, F., Inagaki, N., Hanada, A., et al., 2012. Cryptochrome and phytochrome cooperatively but independently reduce active gibberellin content in rice seedlings under light irradiation. Plant Cell Physiol. 53, 1570–1582.

Hirose, F., Inagaki, N., Takano, M., 2013. Differences and similarities in the photoregulation of gibberellin metabolism between rice and dicots. Plant Signal. Behav. 8, e23424.

Ho, L., Crabtree, G.R., 2010. Chromatin remodelling during development. Nature 463, 474–484.

Hofmann, N.R., 2015. Taking hormone crosstalk to a new level: brassinosteroids regulate gibberellin biosynthesis. Plant Cell 27.

Hong, G.J., Xue, X.Y., Mao, Y.B., et al., 2012. *Arabidopsis* MYC2 interacts with DELLA proteins in regulating sesquiterpene synthase gene expression. Plant Cell 24, 2635–2648.

Hou, X., Lee, L.Y., Xia, K., et al., 2010. DELLAs modulate jasmonate signaling via competitive binding to JAZs. Dev. Cell 19, 884–894.

Hu, J., Mitchum, M.G., Barnaby, N., et al., 2008. Potential sites of bioactive gibberellin production during reproductive growth in *Arabidopsis*. Plant Cell 20, 320–336.

Huang, D., Wang, S., Zhang, B., et al., 2015. A gibberellin-mediated DELLA-NAC signaling cascade regulates cellulose synthesis in rice. Plant Cell 27, 1681–1696.

Huang, J., Tang, D., Shen, Y., et al., 2010. Activation of gibberellin 2-oxidase 6 decreases active gibberellin levels and creates a dominant semi-dwarf phenotype in rice (*Oryza sativa* L.). J. Genet. Genomics 37, 23–36.

Huang, S., Raman, A., Ream, J., et al., 1998. Overexpression of 20-oxidase confers a gibberellin overproduction phenotype in *Arabidopsis*. Plant Physiol. Biochem. 118, 773–781.

Hussain, A., Cao, D., Cheng, H., et al., 2005. Identification of the conserved serine/threonine residues important for gibberellin-sensitivity of *Arabidopsis* RGL2 protein. Plant J. 44, 88–99.

Hussain, A., Cao, D., Peng, J., 2007. Identification of conserved tyrosine residues important for gibberellin sensitivity of *Arabidopsis* RGL2 protein. Planta 226, 475–483.

Ingram, T.J., Reid, J.B., 1987. Internode length in Pisum: gene *na* may block gibberellin synthesis between *ent*-7α-hydroxykaurenoic acid and gibberellin A_{12}-aldehyde. Plant Physiol. 83, 1048–1053.

Israelsson, M., Mellerowicz, E., Chono, M., et al., 2004. Cloning and overproduction of gibberellin 3-oxidase in hybrid aspen trees. Effects on gibberellin homeostasis and development. Plant Physiol. 135, 221–230.

Itoh, H., Sasaki, A., Ueguchi-Tanaka, M., et al., 2005. Dissection of the phosphorylation of rice DELLA protein, SLENDER RICE1. Plant Cell Physiol. 46, 1392–1399.

Itoh, H., Tanaka-Ueguchi, M., Kawaide, H., et al., 1999. The gene encoding tobacco gibberellin 3β-hydroxylase is expressed at the site of GA action during stem elongation and flower organ development. Plant J. 20, 15–24.

Itoh, H., Tatsumi, T., Sakamoto, T., et al., 2004. A rice semi-dwarf gene, *Tan-Ginbozu* (*D35*), encodes the gibberellin biosynthesis enzyme, *ent*-kaurene oxidase. Plant Mol. Biol. 54, 533–547.

Itoh, H., Ueguchi-Tanaka, M., Sato, Y., et al., 2002. The gibberellin signaling pathway is regulated by the appearance and disappearance of SLENDER RICE1 in nuclei. Plant Cell 14, 57–70.

Itoh, H., Ueguchi-Tanaka, M., Sentoku, N., et al., 2001. Cloning and functional analysis of two gibberellin 3β-hydroxylase genes that are differently expressed during the growth of rice. Proc. Natl. Acad. Sci. U.S.A. 98, 8909–8914.

Iuchi, S., Suzuki, H., Kim, Y.C., et al., 2007. Multiple loss-of-function of *Arabidopsis* gibberellin receptor AtGID1s completely shuts down a gibberellin signal. Plant J. 50, 958–966.

Jarillo, J.A., Pineiro, M., Cubas, P., et al., 2009. Chromatin remodeling in plant development. Int. J. Dev. Biol. 53, 1581–1596.

Jasinski, S., Piazza, P., Craft, J., et al., 2005. KNOX action in *Arabidopsis* is mediated by coordinate regulation of cytokinin and gibberellin activities. Curr. Biol. 15, 1560–1565.

Jasinski, S., Tattersall, A., Piazza, P., et al., 2008. PROCERA encodes a DELLA protein that mediates control of dissected leaf form in tomato. Plant J. 56, 603–612.

Jiang, C., Fu, X., 2007. GA action: turning on de-DELLA repressing signaling. Curr. Opin. Plant Biol. 10, 461–465.

Jiao, Y., 2016. Trichome formation: gibberellins on the move. Plant Physiol. 170, 1174–1175.

Jing, Y., Lin, R., 2013. PICKLE is a repressor in seedling de-etiolation pathway. Plant Signal. Behav. 8.

Jing, Y., Zhang, D., Wang, X., et al., 2013. *Arabidopsis* chromatin remodeling factor PICKLE interacts with transcription factor HY5 to regulate hypocotyl cell elongation. Plant Cell 25, 242–256.

Josse, E.M., Gan, Y., Bou-Torrent, J., et al., 2011. A DELLA in disguise: SPATULA restrains the growth of the developing *Arabidopsis* seedling. Plant Cell 23, 1337–1351.

Kamiya, Y., García-Martínez, J., 1999. Regulation of gibberellin biosynthesis by light. Curr. Opin. Plant Biol. 2, 398–403.

Kaneko, M., Itoh, H., Inukai, Y., et al., 2003. Where do gibberellin biosynthesis and gibberellin signaling occur in rice plants? Plant J. 35, 104–115.

Kanno, Y., Hanada, A., Chiba, Y., et al., 2012. Identification of an abscisic acid transporter by functional screening using the receptor complex as a sensor. Proc. Natl. Acad. Sci. U.S.A. 109, 9653–9658.

Katsumi, M., Foard, D., Phinney, B., 1983. Evidence for the translocation of gibberellin A_3 and gibberellin-like substances in grafts between normal, dwarf$_1$ and dwarf$_5$ seedlings of *Zea mays* L. Plant Cell Physiol. 24, 379–388.

Kim, S.I., Park, B.S., Kim do, Y., et al., 2015. E3 SUMO ligase AtSIZ1 positively regulates SLY1-mediated GA signalling and plant development. Biochem. J. 469, 299–314.

King, R.W., Moritz, T., Evans, L.T., et al., 2001. Long-day induction of flowering in *Lolium temulentum* involves sequential increases in specific gibberellins at the shoot apex. Plant Physiol. 127, 624–632.

Ko, K.W., Lin, F., Katsumata, T., et al., 2008. Functional identification of a rice *ent*-kaurene oxidase, OsKO2, using the *Pichia pastoris* expression system. Biosci. Biotechnol. Biochem. 72, 3285–3288.

Koornneef, M., Bentsink, L., Hilhorst, H., 2002. Seed dormancy and germination. Curr. Opin. Plant Biol. 5, 33–36.

Koornneef, M., Van der Veen, J., 1980. Induction and analysis of gibberellin sensitive mutants in *Arabidopsis thaliana* (L) Heyhn. Theor. Appl. Genet. 58, 257–263.

Krouk, G., Lacombe, B., Bielach, A., et al., 2010. Nitrate-regulated auxin transport by NRT1.1 defines a mechanism for nutrient sensing in plants. Dev. Cell 18, 927–937.

Léran, S., Varala, K., Boyer, J.C., et al., 2014. A unified nomenclature of NITRATE TRANSPORTER 1/PEPTIDE TRANSPORTER family members in plants. Trends Plant Sci. 19, 5–9.

Lange, T., Hedden, P., Graebe, J., 1994. Expression cloning of a gibberellin 20-oxidase, a multifunctional enzyme involved in gibberellin biosynthesis. Proc. Natl. Acad. Sci. U.S.A. 91, 8552–8556.

Lee, D.J., Zeevaart, J.A., 2002. Differential regulation of RNA levels of gibberellin dioxygenases by photoperiod in spinach. Plant Physiol. 130, 2085–2094.

Lee, D.J., Zeevaart, J.A., 2005. Molecular cloning of *GA 2-oxidase3* from spinach and its ectopic expression in *Nicotiana sylvestris*. Plant Physiol. 138, 243–254.

Lee, S., Cheng, H., King, K.E., et al., 2002. Gibberellin regulates *Arabidopsis* seed germination via *RGL2*, a *GAI/RGA*-like gene whose expression is up-regulated following imbibition. Genes Dev. 16, 646–658.

Li, Q.F., Wang, C., Jiang, L., et al., 2012. An interaction between BZR1 and DELLAs mediates direct signaling crosstalk between brassinosteroids and gibberellins in *Arabidopsis*. Sci. Signal. 5, ra72.

Lilley, J.L.S., Gan, Y., Graham, I.A., et al., 2013. The effects of DELLA s on growth change with developmental stage and brassinosteroid levels. Plant J. 76, 165–173.

Lim, S., Park, J., Lee, N., et al., 2013. ABA-INSENSITIVE3, ABA-INSENSITIVE5, and DELLAs interact to activate the expression of *SOMNUS* and other high-temperature-inducible genes in imbibed seeds in *Arabidopsis*. Plant Cell 25, 4863–4878.

Livne, S., Weiss, D., 2014. Cytosolic activity of the gibberellin receptor GIBBERELLIN INSENSITIVE DWARF1A. Plant Cell Physiol. 55, 1727–1733.

Lo, S.F., Yang, S.Y., Chen, K.T., et al., 2008. A novel class of gibberellin 2-oxidases control semidwarfism, tillering, and root development in rice. Plant Cell 20, 2603–2618.

Locascio, A., Blazquez, M.A., Alabadi, D., 2013. Dynamic regulation of cortical microtubule organization through prefoldin-DELLA interaction. Curr. Biol. 23, 804–809.

Lumba, S., Cutler, S., McCourt, P., 2010. Plant nuclear hormone receptors: a role for small molecules in protein–protein interactions. Annu. Rev. Cell Dev. Biol.

Magome, H., Nomura, T., Hanada, A., et al., 2013. *CYP714B1* and *CYP714B2* encode gibberellin 13-oxidases that reduce gibberellin activity in rice. Proc. Natl. Acad. Sci. U.S.A. 110, 1947–1952.

Magome, H., Yamaguchi, S., Hanada, A., et al., 2004. *Dwarf* and *delayed-flowering 1*, a novel *Arabidopsis* mutant deficient in gibberellin biosynthesis because of overexpression of a putative AP2 transcription factor. Plant J. 37, 720–729.

Magome, H., Yamaguchi, S., Hanada, A., et al., 2008. The DDF1 transcriptional activator upregulates expression of a gibberellin-deactivating gene, *GA2ox7*, under high-salinity stress in *Arabidopsis*. Plant J. 56, 613–626.

Marín-de la Rosa, N., Pfeiffer, A., Hill, K., et al., 2015. Genome wide binding site analysis reveals transcriptional coactivation of cytokinin-responsive genes by DELLA proteins. PLoS Genet. 11, e1005337.

Marín-de la Rosa, N., Sotillo, B., Miskolczi, P., et al., 2014. Large-scale identification of gibberellin-related transcription factors defines group VII ETHYLENE RESPONSE FACTORS as functional DELLA partners. Plant Physiol. 166, 1022–1032.

Margis-Pinheiro, M., Zhou, X.R., Zhu, Q.H., et al., 2005. Isolation and characterization of a *Ds*-tagged rice (*Oryza sativa* L.) GA-responsive dwarf mutant defective in an early step of the gibberellin biosynthesis pathway. Plant Cell Rep. 23, 819–833.

Matías-Hernández, L., Aguilar-Jaramillo, A.E., Osnato, M., et al., 2016. TEMPRANILLO reveals the mesophyll as crucial for epidermal trichome formation. Plant Physiol. 170, 1624–1639.

Matsushita, A., Furumoto, T., Ishida, S., et al., 2007. AGF1, an AT-hook protein, is necessary for the negative feedback of AtGA3ox1 encoding GA 3-oxidase. Plant Physiol. 143, 1152–1162.

McGinnis, K.M., Thomas, S.G., Soule, J.D., et al., 2003. The Arabidopsis *SLEEPY1* gene encodes a putative F-box subunit of an SCF E3 ubiquitin ligase. Plant Cell 15, 1120–1130.

Mitchum, M., Yamaguchi, S., Hanada, A., et al., 2006. Distinct and overlapping roles of two gibberellin 3-oxidases in *Arabidopsis* development. Plant J. 45, 804–818.

Monna, L., Kitazawa, N., Yoshino, R., et al., 2002. Positional cloning of rice semidwarfing gene, *sd-1*: rice "green revolution gene" encodes a mutant enzyme involved in gibberellin synthesis. DNA Res. 9, 11–17.

Moon, J., Suh, S.S., Lee, H., et al., 2003. The *SOC1* MADS-box gene integrates vernalization and gibberellin signals for flowering in *Arabidopsis*. Plant J. 35, 613–623.

Morrone, D., Chen, X., Coates, R.M., et al., 2010. Characterization of the kaurene oxidase CYP701A3, a multifunctional cytochrome P450 from gibberellin biosynthesis. Biochem. J. 431, 337–344.

Moubayidin, L., Perilli, S., Dello Ioio, R., et al., 2010. The rate of cell differentiation controls the *Arabidopsis* root meristem growth phase. Curr. Biol. 20, 1138–1143.

Murase, K., Hirano, Y., Sun, T.P., et al., 2008. Gibberellin-induced DELLA recognition by the gibberellin receptor GID1. Nature 456, 459–463.

Mutasa-Göttgens, E., Hedden, P., 2009. Gibberellin as a factor in floral regulatory networks. J. Exp. Bot. 60, 1979–1989.

Nakajima, M., Shimada, A., Takashi, Y., et al., 2006. Identification and characterization of *Arabidopsis* gibberellin receptors. Plant J. 46, 880–889.

Nakamura, H., Xue, Y.L., Miyakawa, T., et al., 2013. Molecular mechanism of strigolactone perception by DWARF14. Nat. Commun. 4, 2613.

Nir, I., Moshelion, M., Weiss, D., 2014. The *Arabidopsis* gibberellin methyl transferase 1 suppresses gibberellin activity, reduces whole-plant transpiration and promotes drought tolerance in transgenic tomato. Plant Cell Environ. 37, 113–123.

Nomura, T., Magome, H., Hanada, A., et al., 2013. Functional analysis of *Arabidopsis* CYP714A1 and CYP714A2 reveals that they are distinct gibberellin modification enzymes. Plant Cell Physiol. 54, 1837–1851.

Nour-Eldin, H.H., Andersen, T.G., Burow, M., et al., 2012. NRT/PTR transporters are essential for translocation of glucosinolate defence compounds to seeds. Nature 488, 531–534.

O'Neill, D.P., Davidson, S.E., Clarke, V.C., et al., 2010. Regulation of the gibberellin pathway by auxin and DELLA proteins. Planta 232, 1141–1149.

O'Neill, D.P., Ross, J.J., 2002. Auxin regulation of the gibberellin pathway in pea. Plant Physiol. 130, 1974–1982.

Oh, E., Yamaguchi, S., Hu, J., et al., 2007. PIL5, a phytochrome-interacting bHLH protein, regulates gibberellin responsiveness by binding directly to the *GAI* and *RGA* promoters in *Arabidopsis* seeds. Plant Cell 19, 1192–1208.

Oh, E., Yamaguchi, S., Kamiya, Y., et al., 2006. Light activates the degradation of PIL5 protein to promote seed germination through gibberellin in *Arabidopsis*. Plant J. 47, 124–139.

Oh, E., Zhu, J.Y., Bai, M.Y., et al., 2014. Cell elongation is regulated through a central circuit of interacting transcription factors in the *Arabidopsis* hypocotyl. Elife 3.

Oikawa, T., Koshioka, M., Kojima, K., et al., 2004. A role of OsGA20ox1, encoding an isoform of gibberellin 20-oxidase, for regulation of plant stature in rice. Plant Mol. Biol. 55, 687–700.

Olszewski, N., Sun, T-p, Gubler, F., 2002. Gibberellin signaling biosynthesis catabolism and response pathways. Plant Cell 14, S61–S80.

Otsuka, M., Kenmoku, H., Ogawa, M., et al., 2004. Emission of *ent*-kaurene, a diterpenoid hydrocarbon precursor for gibberellins, into the headspace from plants. Plant Cell Physiol. 45.

Park, J., Nguyen, K.T., Park, E., et al., 2013. DELLA proteins and their interacting RING Finger proteins repress gibberellin responses by binding to the promoters of a subset of gibberellin-responsive genes in *Arabidopsis*. Plant Cell 25, 927–943.

Peng, J., Carol, P., Richards, D.E., et al., 1997. The Arabidopsis *GAI* gene defines a signaling pathway that negatively regulates gibberellin responses. Genes Dev. 11, 3194–3205.

Peng, J., Harberd, N.P., 1993. Derivative alleles of the Arabidopsis gibberellin-insensitive (*gai*) mutation confer a wild-type phenotype. Plant Cell 5, 351–360.

Peng, J., Richards, D.E., Hartley, N.M., et al., 1999. 'Green revolution' genes encode mutant gibberellin response modulators. Nature 400, 256–261.

Peters, R.J., 2006. Uncovering the complex metabolic network underlying diterpenoid phytoalexin biosynthesis in rice and other cereal crop plants. Phytochemistry 67, 2307–2317.

Phillips, A., 1998. Gibberellins in *Arabidopsis*. Plant Physiol. Biochem. 36, 115–124.

Phillips, A., Ward, D., Uknes, S., et al., 1995. Isolation and expression of three gibberellin 20-oxidase cDNA clones from Arabidopsis. Plant Physiol. 108, 1049–1057.

Piskurewicz, U., Jikumaru, Y., Kinoshita, N., et al., 2008. The gibberellic acid signaling repressor RGL2 inhibits *Arabidopsis* seed germination by stimulating abscisic acid synthesis and ABI5 activity. Plant Cell 20, 2729–2745.

Plackett, A., Powers, S., Fernandez-Garcia, N., et al., 2012. Analysis of the developmental roles of the *Arabidopsis* gibberellin 20-oxidases demonstrates that *GA20ox1*, *-2*, and *-3* are the dominant paralogs. Plant Cell 24, 941–960.

Plackett, A., Thomas, S., Wilson, Z., et al., 2011. Gibberellin control of stamen development: a fertile field. Trends Plant Sci. 16, 568–578.

Proebsting, W.M., Hedden, P., Lewis, M.J., et al., 1992. Gibberellin concentration and transport in genetic lines of pea: effects of grafting. Plant Physiol. 100, 1354–1360.

Qi, T., Huang, H., Wu, D., et al., 2014. *Arabidopsis* DELLA and JAZ proteins bind the WD-repeat/bHLH/MYB complex to modulate gibberellin and jasmonate signaling synergy. Plant Cell 26, 1118–1133.

Qi, T., Song, S., Ren, Q., et al., 2011. The Jasmonate-ZIM-domain proteins interact with the WD-Repeat/bHLH/MYB complexes to regulate Jasmonate-mediated anthocyanin accumulation and trichome initiation in *Arabidopsis thaliana*. Plant Cell 23, 1795–1814.

Qian, Q., Guo, L., Smith, S.M., et al., 2016. Breeding high-yield superior-quality hybrid superrice by rational design. Natl. Sci. Rev. http://dx.doi.org/10.1093/nsr/nww1006.

Qin, Q., Wang, W., Guo, X., et al., 2014. Arabidopsis DELLA protein degradation is controlled by a type-one protein phosphatase, TOPP4. PLoS Genet. 10, e1004464.

Radi, A., Lange, T., Niki, T., et al., 2006. Ectopic expression of pumpkin gibberellin oxidases alters gibberellin biosynthesis and development of transgenic *Arabidopsis* plants. Plant Physiol. 140, 528–536.

Ragni, L., Nieminen, K., Pacheco-Villalobos, D., et al., 2011. Mobile gibberellin directly stimulates *Arabidopsis* hypocotyl xylem expansion. Plant Cell 23, 1322–1336.

Rameau, C., Bertheloot, J., Leduc, N., et al., 2015. Multiple pathways regulate shoot branching. Front. Plant Sci. 5, 741.

Razem, F., Baron, K., Hill, R., 2006. Turning on gibberellin and abscisic acid signaling. Curr. Opin. Plant Biol. 9, 454–459.

Regnault, T., Davière, J., Wild, M., et al., 2015. The gibberellin precursor GA12 acts as a long-distance growth signal in *Arabidopsis*. Nat. Plants 1, 15073.

Regnault, T., Daviere, J.M., Achard, P., 2016. Long-distance transport of endogenous gibberellins in *Arabidopsis*. Plant Signal. Behav. 11, e1110661.

Regnault, T., Daviere, J.M., Heintz, D., et al., 2014. The gibberellin biosynthetic genes AtKAO1 and AtKAO2 have overlapping roles throughout *Arabidopsis* development. Plant J. 80, 462–474.

Reinecke, D.M., Wickramarathna, A.D., Ozga, J.A., et al., 2013. Gibberellin 3-oxidase gene expression patterns influence gibberellin biosynthesis, growth, and development in pea. Plant Physiol. 163, 929–945.

Resentini, F., Felipo-Benavent, A., Colombo, L., et al., 2015. TCP14 and TCP15 mediate the promotion of seed germination by gibberellins in *Arabidopsis thaliana*. Mol. Plant 8, 482–485.

Richards, D.E., King, K.E., Ait-Ali, T., et al., 2001. How gibberellin regulates plant growth and development: a molecular genetic analysis of gibberellin signaling. Annu. Rev. Plant Physiol. Plant Mol. Biol. 52, 67–88.

Rieu, I., Eriksson, S., Powers, S.J., et al., 2008a. Genetic analysis reveals that C19-GA 2-oxidation is a major gibberellin inactivation pathway in *Arabidopsis*. Plant Cell 20, 2420–2436.

Rieu, I., Ruiz-Rivero, O., Fernandez-Garcia, N., et al., 2008b. The gibberellin biosynthetic genes *AtGA20ox1* and *AtGA20ox2* act, partially redundantly, to promote growth and development throughout the Arabidopsis life cycle. Plant J. 53, 488–504.

Robertson, M., Swain, S.M., Chandler, P.M., et al., 1998. Identification of a negative regulator of gibberellin action, HvSPY, in barley. Plant Cell 10, 995–1007.

Ross, J., O'Neill, D., Smith, J., et al., 2000. Evidence that auxin promotes gibberellin A1 biosynthesis in pea. Plant J. 21, 547–552.

Saito, H., Oikawa, T., Hamamoto, S., et al., 2015. The jasmonate-responsive GTR1 transporter is required for gibberellin-mediated stamen development in *Arabidopsis*. Nat. Commun. 6, 6095.

Sakamoto, T., Kamiya, N., Ueguchi-Tanaka, M., et al., 2001a. KNOX homeodomain protein directly suppresses the expression of a gibberellin biosynthetic gene in the tobacco shoot apical meristem. Genes Dev. 15, 581–590.

Sakamoto, T., Kobayashi, M., Itoh, H., et al., 2001b. Expression of a gibberellin 2-oxidase gene around the shoot apex is related to phase transition in rice. Plant Physiol. 125.

Sakamoto, T., Miura, K., Itoh, H., et al., 2004. An overview of gibberellin metabolism enzyme genes and their related mutants in rice. Plant Physiol. 134, 1642–1653.

Sarnowska, E.A., Rolicka, A.T., Bucior, E., et al., 2013. DELLA-interacting SWI3C core subunit of switch/sucrose nonfermenting chromatin remodeling complex modulates gibberellin responses and hormonal cross talk in *Arabidopsis*. Plant Physiol. 163, 305–317.

Sasaki, A., Ashikari, M., Ueguchi-Tanaka, M., et al., 2002. Green revolution: a mutant gibberellin-synthesis gene in rice. Nature 416, 701–702.

Sasaki, A., Itoh, H., Gomi, K., et al., 2003. Accumulation of phosphorylated repressor for gibberellin signaling in an F-box mutant. Science 299, 1896–1898.

Schomburg, F., Bizzell, C., Lee, D., et al., 2003. Overexpression of a novel class of gibberellin 2-oxidases decreases gibberellin levels and creates dwarf plants. Plant Cell 15, 151–163.

Seo, M., Hanada, A., Kuwahara, A., et al., 2006. Regulation of hormone metabolism in *Arabidopsis* seeds: phytochrome regulation of abscisic acid metabolism and abscisic acid regulation of gibberellin metabolism. Plant J. 48, 354–366.

Serrani, J.C., Sanjuan, R., Ruiz-Rivero, O., et al., 2007. Gibberellin regulation of fruit set and growth in tomato. Plant Physiol. 145, 246–257.

Shan, C., Mei, Z., Duan, J., et al., 2014. OsGA2ox5, a gibberellin metabolism enzyme, is involved in plant growth, the root gravity response and salt stress. PLoS One 9, e87110.

Shani, E., Weinstain, R., Zhang, Y., et al., 2013. Gibberellins accumulate in the elongating endodermal cells of Arabidopsis root. Proc. Natl. Acad. Sci. U.S.A. 110, 4834–4839.

Shimada, A., Ueguchi-Tanaka, M., Nakatsu, T., et al., 2008. Structural basis for gibberellin recognition by its receptor GID1. Nature 456, 520–523.

Shimada, A., Ueguchi-Tanaka, M., Sakamoto, T., et al., 2006. The rice *SPINDLY* gene functions as a negative regulator of gibberellin signaling by controlling the suppressive function of the DELLA protein, SLR1, and modulating brassinosteroid synthesis. Plant J. 48, 390–402.

Shinozaki, Y., Hao, S., Kojima, M., et al., 2015. Ethylene suppresses tomato (*Solanum lycopersicum*) fruit set through modification of gibberellin metabolism. Plant J. 83, 237–251.

Silverstone, A.L., Chang, C.W., Krol, E., et al., 1997a. Developmental regulation of the gibberellin biosynthetic gene GA1 in *Arabidopsis thaliana*. Plant J. 12, 9–19.

Silverstone, A.L., Mak, P.Y., Martinez, E.C., et al., 1997b. The new RGA locus encodes a negative regulator of gibberellin response in *Arabidopsis thaliana*. Genetics 146, 1087–1099.

Silverstone, A.L., Ciampaglio, C.N., Sun, T., 1998. The Arabidopsis *RGA* gene encodes a transcriptional regulator repressing the gibberellin signal transduction pathway. Plant Cell 10, 155–169.

Silverstone, A.L., Jung, H.S., Dill, A., et al., 2001. Repressing a repressor: gibberellin-induced rapid reduction of the RGA protein in *Arabidopsis*. Plant Cell 13, 1555–1566.

Silverstone, A.L., Sun, T., 2000. Gibberellins and the green revolution. Trends Plant Sci. 5, 1–2.

Silverstone, A.L., Tseng, T.S., Swain, S.M., et al., 2007. Functional analysis of SPINDLY in gibberellin signaling in *Arabidopsis*. Plant Physiol. 143, 987–1000.

Smith, M., Yamaguchi, S., Ait-Ali, T., et al., 1998. The first step of gibberellin biosynthesis in pumpkin is catalyzed by at least two copalyl diphosphate synthases encoded by differentially regulated genes. Plant Physiol. 118, 1411–1419.

Song, J., Guo, B., Song, F., et al., 2011. Genome-wide identification of gibberellins metabolic enzyme genes and expression profiling analysis during seed germination in maize. Gene 482, 34–42.

Spielmeyer, W., Ellis, M.H., Chandler, P.M., 2002. Semidwarf (*sd-1*), "green revolution" rice, contains a defective gibberellin 20-oxidase gene. Proc. Natl. Acad. Sci. U.S.A. 99, 9043–9048.

Sponsel, V., Schmidt, F., Porter, S., et al., 1997. Characterization of new gibberellin-responsive semidwarf mutants of Arabidopsis. Plant Physiol. 115, 1009–1020.

Strader, L.C., Ritchie, S., Soule, J.D., et al., 2004. Recessive-interfering mutations in the gibberellin signaling gene *SLEEPY1* are rescued by overexpression of its homologue, *SNEEZY*. Proc. Natl. Acad. Sci. U.S.A. 101, 12771–12776.

Sun, T.-P., Kamiya, Y., 1994. The Arabidopsis *GAI* locus encodes the cyclase *ent*-kaurene synthetase A of gibberellin biosynthesis. The Plant Cell 6, 1509–1518.

Sun, T., Goodman, H.M., Ausubel, F.M., 1992. Cloning the Arabidopsis *GA1* locus by genomic subtraction. Plant Cell 4, 119–128.

Sun, T.P., 2011. The molecular mechanism and evolution of the GA-GID1-DELLA signaling module in plants. Curr. Biol. 21, R338–R345.

Sun, T.P., Gubler, F., 2004. Molecular mechanism of gibberellin signaling in plants. Annu. Rev. Plant Biol. 55, 197–223.

Swain, S.M., Tseng, T.S., Olszewski, N.E., 2001. Altered expression of *SPINDLY* affects gibberellin response and plant development. Plant Physiol. 126, 1174–1185.

Tal, I., Zhang, Y., Jorgensen, M.E., et al., 2016. The *Arabidopsis* NPF3 protein is a GA transporter. Nat. Commun. 7, 11486.

Tan, X., Calderon-Villalobos, L.I., Sharon, M., et al., 2007. Mechanism of auxin perception by the TIR1 ubiquitin ligase. Nature 446, 640–645.

Tezuka, D., Ito, A., Mitsuhashi, W., et al., 2015. The rice *ent*-KAURENE SYNTHASE LIKE 2 encodes a functional *ent*-beyerene synthase. Biochem. Biophys. Res. Commun. 460, 766–771.

Thomas, S., Phillips, A., Hedden, P., 1999. Molecular cloning and functional expression of gibberellin 2-oxidases, multifunctional enzymes involved in gibberellin deactivation. Plant Biol. 96, 4698–4703.

Toh, S., Imamura, A., Watanabe, A., et al., 2008. High temperature-induced abscisic acid biosynthesis and its role in the inhibition of gibberellin action in Arabidopsis seeds. Plant Physiol. 146, 1368–1385.

Tong, H., Chu, C., 2016. REPLY: brassinosteroid regulates gibberellin synthesis to promote cell elongation in rice: critical comments on Ross and Quittenden's letter. Plant Cell 28, 833–835.

Tong, H., Xiao, Y., Liu, D., et al., 2014. Brassinosteroid regulates cell elongation by modulating gibberellin metabolism in rice. Plant Cell 26, 4376–4393.

Toyomasu, T., Kawaide, H., Mitsuhashi, W., et al., 1998. Phytochrome regulates gibberellin biosynthesis during germination of photoblastic lettuce Seeds. Plant Physiol. 118, 1517–1523.

Toyomasu, T., Kawaide, H., Sekimoto, H., et al., 1997. Cloning and characterization of a cDNA encoding gibberellin 20-oxidase from rice (*Oryza sativa*) seedlings. Physiol. Plant. 99, 111–118.

Toyomasu, T., Usui, M., Sugawara, C., et al., 2015. Transcripts of two *ent*-copalyl diphosphate synthase genes differentially localize in rice plants according to their distinct biological roles. J. Exp. Bot. 66, 369–376.

Toyomasu, T., Usui, M., Sugawara, C., et al., 2014. Reverse-genetic approach to verify physiological roles of rice phytoalexins: characterization of a knockdown mutant of OsCPS4 phytoalexin biosynthetic gene in rice. Physiologia Plant. 150, 55–62.

Tsay, Y.F., Chiu, C.C., Tsai, C.B., et al., 2007. Nitrate transporters and peptide transporters. FEBS Lett. 581, 2290–2300.

Tyler, L., Thomas, S.G., Hu, J., et al., 2004. DELLA proteins and gibberellin-regulated seed germination and floral development in *Arabidopsis*. Plant Physiol. 135, 1008–1019.

Ubeda-Tomas, S., Federici, F., Casimiro, I., et al., 2009. Gibberellin signaling in the endodermis controls *Arabidopsis* root meristem size. Curr. Biol. 19, 1194–1199.

Ubeda-Tomas, S., Swarup, R., Coates, J., et al., 2008. Root growth in *Arabidopsis* requires gibberellin/DELLA signalling in the endodermis. Nat. Cell Biol. 10, 625–628.

Ueguchi-Tanaka, M., Ashikari, M., Nakajima, M., et al., 2005. GIBBERELLIN INSENSITIVE DWARF1 encodes a soluble receptor for gibberellin. Nature 437, 693–698.

Ueguchi-tanaka, M., Hirano, K., Hasegawa, Y., et al., 2008. Release of the repressive activity of rice DELLA protein SLR1 by gibberellin does not require SLR1 degradation in the gid2 mutant. Plant Cell 20, 2437–2446.

Ueguchi-Tanaka, M., Matsuoka, M., 2010. The perception of gibberellins: clues from receptor structure. Curr. Opin. Plant Biol.

Ueguchi-Tanaka, M., Nakajima, M., Katoh, E., et al., 2007. Molecular interactions of a soluble gibberellin receptor, GID1, with a rice DELLA protein, SLR1, and gibberellin. Plant Cell 19, 2140–2155.

Unterholzner, S.J., Rozhon, W., Papacek, M., et al., 2015. Brassinosteroids are master regulators of gibberellin biosynthesis in *Arabidopsis*. Plant Cell 27, 2261–2272.

Unterholzner, S.J., Rozhon, W., Poppenberger, B., 2016. REPLY: interaction between brassinosteroids and gibberellins: synthesis or signaling? In *Arabidopsis*, both!. Plant Cell 28, 836–839.

Vandenbussche, F., Fierro, A.C., Wiedemann, G., et al., 2007. Evolutionary conservation of plant gibberellin signalling pathway components. BMC Plant Biol. 7, 65.

Varbanova, M., Yamaguchi, S., Yang, Y., et al., 2007. Methylation of gibberellins by *Arabidopsis* GAMT1 and GAMT2. Plant Cell 19, 32–45.

Vera-Sirera, F., Gomez, M.D., Perez-Amador, M.A., 2016. DELLA proteins, a group of GRAS transcription regulators that mediate gibberellin signaling. Plant Transcr. Factors 313–328.

Wang, C., Yang, Y., Wang, H., et al., 2016. Ectopic expression of a cytochrome P450 monooxygenase gene *PtCYP714A3* from *Populus trichocarpa* reduces shoot growth and improves tolerance to salt stress in transgenic rice. Plant Biotechnol. J. 14. http://dx.doi.org/10.1111/pbi.12544.

Wang, F., Zhu, D., Huang, X., et al., 2009. Biochemical insights on degradation of *Arabidopsis* DELLA proteins gained from a cell-free assay system. Plant Cell 21, 2378–2390.

Wang, H., Caruso, L.V., Downie, A.B., et al., 2004. The embryo MADS domain protein AGAMOUS-like 15 directly regulates expression of a gene encoding an enzyme involved in gibberellin metabolism. Plant Cell 16, 1206–1219.

Wang, H., Tang, W., Cong, Z., et al., 2002. A chromatin immunoprecipitation (ChIP) approach to isolate genes regulated by AGL15, a MADS domain protein that preferentially accumulates in embryos. Plant J. 32, 831–843.

Wang, Q., Hillwig, M.L., Wu, Y., et al., 2012. CYP701A8: a rice *ent*-kaurene oxidase paralog diverted to more specialized diterpenoid metabolism. Plant Physiol. 158, 1418–1425.

Wang, W., Zhang, J., Qin, Q., et al., 2014. The six conserved serine/threonine sites of REPRESSOR of *ga1-3* protein are important for its functionality and stability in gibberellin signaling in *Arabidopsis*. Planta 240, 763–779.

Weiss, D., Ori, N., 2007. Mechanisms of cross talk between gibberellin and other hormones. Plant Physiol. 144, 1240–1246.

Wild, M., Achard, P., 2013. The DELLA protein RGL3 positively contributes to jasmonate/ethylene defense responses. Plant Signal. Behav. 8, e23891.

Wild, M., Daviere, J.M., Cheminant, S., et al., 2012. The Arabidopsis DELLA RGA-LIKE3 is a direct target of MYC2 and modulates jasmonate signaling responses. Plant Cell 24, 3307–3319.

Williams, J., Phillips, A.L., Gaskin, P., et al., 1998. Function and substrate specificity of the gibberellin 3β-hydroxylase encoded by the Arabidopsis *GA4* gene. Plant Physiol. 117, 559–563.

Willige, B.C., Ghosh, S., Nill, C., et al., 2007. The DELLA domain of GA INSENSITIVE mediates the interaction with the GA INSENSITIVE DWARF1A gibberellin receptor of *Arabidopsis*. Plant Cell 19, 1209–1220.

Wu, K., Li, L., Gage, D.A., et al., 1996. Molecular cloning and photoperiod-regulated expression of gibberellin 20-oxidase from the long-day plant spinach. Plant Physiol. 110, 547–554.

Xu, F., Li, T., Xu, P.B., et al., 2016. DELLA proteins physically interact with CONSTANS to regulate flowering under long days in *Arabidopsis*. FEBS Lett. 590, 541–549.

Xu, H., Liu, Q., Yao, T., et al., 2014. Shedding light on integrative GA signaling. Curr. Opin. Plant Biol. 21, 89–95.

Xu, M., Wilderman, P.R., Morrone, D., et al., 2007. Functional characterization of the rice kaurene synthase-like gene family. Phytochemistry 68, 312–326.

Xu, Y., Gage, D., Zeevaart, J., 1997. Gibberellins and stem growth in *Arabidopsis thaliana* – effects of photoperiod on expression of the GA4 and GA5 loci. Plant Physiol. 114, 1471–1476.

Xu, Y., Li, L., Gage, D., et al., 1999. Feedback regulation of GA5 expression and metabolic engineering of gibberellin levels in *Arabidopsis*. The Plant Cell 11, 927–935.

Xu, Y., LI, L., Wu, K., et al., 1995. The GA5 locus of *Arabidopsis thaliana* encodes a multifunctional gibberellin 20-oxidase: molecular cloning and functional expression. Plant Biol. 92, 6640–6644.

Yaish, M.W., El-Kereamy, A., Zhu, T., et al., 2010. The APETALA-2-like transcription factor OsAP2-39 controls key interactions between abscisic acid and gibberellin in rice. PLoS Genet. 6, e1001098.

Yamaguchi, S., 2008. Gibberellin metabolism and its regulation. Annu. Rev. Plant Biol. 59, 225–251.

Yamaguchi, S., Kamiya, Y., 2000. Gibberellin biosynthesis: its regulation by endogenous and environmental signals. Plant Cell Physiol. 41, 251–257.

Yamaguchi, S., Kamiya, Y., Sun, T., 2001. Distinct cell-specific expression patterns of early and late gibberellin biosynthetic genes during Arabidopsis seed germination. Plant J. 28, 443–453.

Yamaguchi, S., Saito, T., Abe, H., et al., 1996. Molecular cloning and characterization of a cDNA encoding the gibberellin biosynthetic enzyme *ent*-kaurene synthase B from pumpkin (*Cucurbita maxima* L.). Plant J. 10, 203–213.

Yamaguchi, S., Smith, M.W., Brown, R.G.S., et al., 1998a. Phytochrome regulation and differential expression of gibberellin 3β-hydroxylase genes in germinating *Arabidopsis* seeds. Plant Cell 10, 2115–2126.

Yamaguchi, S., Sun, T., Kawaide, H., et al., 1998b. The GA2 locus of *Arabidopsis thaliana* encodes *ent*-kaurene synthase of gibberellin biosynthesis. Plant Physiol. 116, 1271–1278.

Yamauchi, Y., Ogawa, M., Kuwahara, A., et al., 2004. Activation of gibberellin biosynthesis and response pathways by low temperature during imbibition of *Arabidopsis thaliana* seeds. Plant Cell 16, 367–378.

Yang, D., Dong, W., Zhang, Y., et al., 2013. Gibberellins modulate abiotic stress tolerance in plants. Sci. Sin. 43, 1119–1126.

Yang, D., Li, Q., Deng, Y., et al., 2008. Altered disease development in the *eui* mutants and *Eui* overexpressors indicates that gibberellins negatively regulate rice basal disease resistance. Mol. Plant 1, 528–537.

Yang, D.L., Yao, J., Mei, C.S., et al., 2012. Plant hormone jasmonate prioritizes defense over growth by interfering with gibberellin signaling cascade. Proc. Natl. Acad. Sci. U.S.A. 109, E1192–E1200.

Yang, Y., Ma, C., Xu, Y., et al., 2014. A zinc finger protein regulates flowering time and abiotic stress tolerance in chrysanthemum by modulating gibberellin biosynthesis. Plant Cell 26, 2038–2054.

Yano, K., Takashi, T., Nagamatsu, S., et al., 2012. Efficacy of microarray profiling data combined with QTL mapping for the identification of a QTL gene controlling the initial growth rate in rice. Plant Cell Physiol. 53, 729–739.

Yasumura, Y., Crumpton-Taylor, M., Fuentes, S., et al., 2007. Step-by-step acquisition of the gibberellin-DELLA growth-regulatory mechanism during land-plant evolution. Curr. Biol. 17, 1225–1230.

Yoshida, H., Hirano, K., Sato, T., et al., 2014. DELLA protein functions as a transcriptional activator through the DNA binding of the indeterminate domain family proteins. Proc. Natl. Acad. Sci. U.S.A. 111, 7861–7866.

Yoshida, H., Ueguchi-Tanaka, M., 2014. DELLA and SCL3 balance gibberellin feedback regulation by utilizing INDETERMINATE DOMAIN proteins as transcriptional scaffolds. Plant Signal. Behav. 9, e29726.

Yu, S., Galvao, V.C., Zhang, Y.C., et al., 2012. Gibberellin regulates the *Arabidopsis* floral transition through miR156-targeted SQUAMOSA PROMOTER BINDING-LIKE transcription factors. Plant Cell 24, 3320–3332.

Zeevaart, J., Gage, D., Talon, M., 1993. Gibberellin A1 is required for stem elongation in spinach. Proc. Natl. Acad. Sci. U.S.A. 90, 7401–7405.

Zentella, R., Hu, J., Hsieh, W.P., et al., 2016. O-GlcNAcylation of master growth repressor DELLA by SECRET AGENT modulates multiple signaling pathways in *Arabidopsis*. Genes Dev. 30, 164–176.

Zentella, R., Zhang, Z.L., Park, M., et al., 2007. Global analysis of DELLA direct targets in early gibberellin signaling in *Arabidopsis*. Plant Cell 19, 3037–3057.

Zhang, D., Jing, Y., Jiang, Z., et al., 2014. The chromatin-remodeling factor PICKLE integrates brassinosteroid and gibberellin signaling during skotomorphogenic growth in *Arabidopsis*. Plant Cell 26, 2472–2485.

Zhang, Y., Zhang, B., Yan, D., et al., 2011a. Two *Arabidopsis* cytochrome P450 monooxygenases, CYP714A1 and CYP714A2, function redundantly in plant development through gibberellin deactivation. Plant J. 67, 342–353.

Zhang, Y., Zhu, Y., Peng, Y., et al., 2008. Gibberellin homeostasis and plant height control by EUI and a role for gibberellin in root gravity responses in rice. Cell Res. 18, 412–421.

Zhang, Z.L., Ogawa, M., Fleet, C.M., et al., 2011b. SCARECROW-LIKE 3 promotes gibberellin signaling by antagonizing master growth repressor DELLA in Arabidopsis. Proc. Natl. Acad. Sci. U.S.A. 108, 2160–2165.

Zhao, X., Yu, X., Foo, E., et al., 2007. A study of gibberellin homeostasis and cryptochrome-mediated blue light inhibition of hypocotyl elongation. Plant Physiol. 145, 106–118.

Zheng, Y., Ren, N., Wang, H., et al., 2009. Global identification of targets of the Arabidopsis MADS domain protein AGAMOUS-Like15. Plant Cell 21, 2563–2577.

Zhou, X., Zhang, Z.L., Park, J., et al., 2016. The ERF11 transcription factor promotes internode elongation by activating gibberellin biosynthesis and signaling. Plant Physiol. 171, 2760–2770.

Zhu, Y., Nomura, T., Xu, Y., et al., 2006. ELONGATED UPPERMOST INTERNODE encodes a cytochrome P450 monooxygenase that epoxidizes gibberellins in a novel deactivation reaction in rice. Plant Cell 18, 442–456.

Zi, J., Mafu, S., Peters, R.J., 2014. To gibberellins and beyond! Surveying the evolution of (di)terpenoid metabolism. Annu. Rev. Plant Biol. 65, 259–286.

Abscisic acid

5

Jigang Li[1], Yaorong Wu[2], Qi Xie[2], Zhizhong Gong[1]
[1]China Agricultural University, Beijing, China; [2]Institute of Genetics and
Developmental Biology, Chinese Academy of Sciences, Beijing, China

Summary

The classical plant hormone abscisic acid (ABA) was discovered over 50 years ago. ABA accumulates rapidly in plants in response to environmental stresses, such as drought, cold, or high salinity, and plays important roles in the adaptation to and survival of these stresses. This "stress hormone" also functions in many other processes throughout the plant life cycle, acting in embryo development and seed maturation, seed dormancy and germination, seedling establishment, vegetative development, root growth, stomatal movement, flowering, pathogen response, and senescence. It is transported in the vascular tissues to coordinate root and shoot development and function. Receptors for ABA have been identified as a family of soluble proteins, which upon binding ABA form coreceptor complexes with phosphoprotein phosphatase 2C (PP2C) phosphoprotein phosphatases. The resulting inhibition of activity of PP2C enzymes leads to changes in phosphorylation of protein kinases and transcription factors, to mediate the multiple effects of ABA. The elucidation of ABA perception mechanisms and the core components of the signal transduction mechanisms from ABA perception to downstream gene expression has expanded our understanding of the functions of ABA. This chapter summarizes our current understanding of the key components of ABA metabolism, transport, physiological functions, signal transduction, gene expression, and proteolysis.

5.1 Discovery and functions of abscisic acid

Abscisic acid (ABA), a classic plant hormone, was isolated multiple times in different studies. Researchers in the early 1950s isolated acidic compounds, referred to as β-inhibitors, from plants; they separated these compounds by paper chromatography and showed that β-inhibitors inhibit coleoptile elongation in oat. In the early 1960s, scientists in the United States isolated an abscission-accelerating compound from young cotton fruits, called "abscisin II." Simultaneously, UK researchers isolated a dormancy-inducing factor from sycamore leaves, called "dormin." The structure of abscisin II was determined in 1965 and dormin was subsequently shown to be chemically identical to abscisin II. This compound was renamed **abscisic acid** to reflect its supposed involvement in the abscission process and later work determined that the β-inhibitor is also ABA (reviewed in Cutler et al., 2010; Finkelstein, 2013).

ABA is a 15-carbon (C_{15}) compound that belongs to the terpenoid class of metabolites. The orientation of the carboxyl group at carbon 2 determines the *trans* or *cis* isomers of ABA. In addition, an asymmetric carbon atom at position 1′ in the ring determines the *S* (+) or *R* (−) enantiomers. The naturally occurring form is (*S*)-*cis*-ABA,

Hormone Metabolism and Signaling in Plants. http://dx.doi.org/10.1016/B978-0-12-811562-6.00005-0

whereas commercially available ABA is commonly a mixture of *S* and *R* forms in approximately equal amounts. Light can isomerize (*S*)-*cis*-ABA and (*R*)-*cis*-ABA to the biologically inactive forms (*S*)-2-*trans*-ABA and (*R*)-2-*trans*-ABA, respectively. Both (*S*)-*cis*-ABA and (*R*)-*cis*-ABA have strong activity in prolonged ABA responses, such as seed maturation; however, the (*S*)-*cis*-ABA has stronger activity in rapid ABA responses, such as stomatal closure (reviewed in Cutler et al., 2010; Finkelstein, 2013). Hereafter, the term ABA refers to (*S*)-*cis*-ABA.

Although ABA was originally thought to induce abscission, later work showed that ethylene, rather than ABA, regulates abscission. The presence of ABA in abscising organs reflects its roles in promoting senescence and/or stress responses, which precede abscission. ABA promotes abscission through ethylene (Cracker and Abeles, 1969). In Arabidopsis, the ABA-activated calcium-dependent protein kinases (CDPKs), CPK4 and CPK11, phosphorylate the N-termini of the ethylene biosynthetic enzyme 1–aminocyclopropane-1–carboxylate synthases (ACS). This phosphorylation stabilizes ACS and increases ethylene biosynthesis (Luo et al., 2014). However, there is a complicated interplay between ethylene and ABA in the control of plant growth and development (Ma et al., 2014). For example, in rice, ethylene inhibits root growth through ABA production, whereas ABA negatively regulates the ethylene response in the coleoptile by modulating *OsEIN2* expression (Ma et al., 2014).

Historically, ABA has been thought to be a growth inhibitor; however, young tissues have high ABA levels and ABA-deficient mutants are severely stunted, demonstrating that ABA is an important regulator of plant growth. ABA is a ubiquitous plant hormone and is synthesized in almost all cells that contain chloroplasts or amyloplasts. Also, the major components of ABA signaling appear to be conserved across land plants (Umezawa et al., 2010). ABA has been found in some bacteria, fungi, and a variety of metazoans ranging from sea sponges to humans. However, ABA functions as a crucial signaling molecule only in plants; in other organisms ABA is a secondary metabolic component.

Over the past three decades, a combination of molecular genetics, biochemical, and pharmacological studies have identified almost all of the enzymes in ABA biosynthesis and catabolism, with over 100 loci regulating ABA responses and thousands of genes that are regulated by ABA in various contexts (reviewed in Finkelstein, 2013). ABA is the central regulator of plant resistance to abiotic stresses such as drought, salinity, and low temperature. In addition, ABA regulates important aspects of plant growth and development, including embryo development and seed maturation, seed dormancy and germination, seedling establishment, vegetative development, root growth, stomatal movement, flowering, pathogen response, and senescence (reviewed in Finkelstein, 2013).

5.2   ABA metabolism

5.2.1   ABA biosynthesis

As with the other plant hormones, the extent of ABA responses depends on its concentration within a particular plant tissue and on the sensitivity of the tissue to ABA. ABA biosynthesis, catabolism, and transport all contribute to the concentration of active ABA.

In fungi, ABA is biosynthesized directly from **farnesyl diphosphate**; however, in plants, ABA is biosynthesized indirectly from **carotenoids** (reviewed in Nambara and Marion-Poll, 2005). In plants, the early steps of ABA biosynthesis take place in chloroplasts and other plastids and begin with the C_5 **isopentenyl diphosphate** (IPP), the biological isoprene unit that is also a precursor of cytokinins, gibberellins (GAs), strigolactones, and brassinosteroids. The addition of three IPP molecules to dimethylallyl diphosphate, a double-bond isomer of IPP, generates **geranylgeranyl diphosphate** (GGPP), a C_{20} precursor for several groups of plastidial isoprenoids including carotenoids (reviewed in Ruiz-Sola and Rodríguez-Concepción, 2012; Kirby and Keasling, 2009). The first committed step of carotenoid biosynthesis is the production of C_{40} **phytoene** from the condensation of two GGPP molecules (Fig. 5.1) catalyzed by the enzyme **phytoene synthase**. This step is considered to be the main bottleneck in the carotenoid pathway. Subsequent desaturation, isomerization, and cyclization steps lead to the production of either α- or β-carotene; however, only β-carotene is further metabolized into ABA via **zeaxanthin**. Production of zeaxanthin is catalyzed by **β-carotene hydroxylases** encoded by two homologous genes (*BCH1* and *BCH2*) in Arabidopsis.

Arabidopsis ABA-deficient mutants have defects in the biosynthetic steps downstream of zeaxanthin. For example, *aba1*, the first described ABA-deficient mutant of Arabidopsis, was isolated as a suppressor of *ga1*, a nongerminating GA-deficient mutant (Koornneef et al., 1982). The *ABA1* gene encodes **zeaxanthin epoxidase** (ZEP), the enzyme catalyzing the conversion of zeaxanthin to all-*trans*-**violaxanthin** (Rock and Zeevaart, 1991; Barrero et al., 2005). This reaction can be reversed by **violaxanthin de-epoxidase** to produce photoprotective zeaxanthin in response to a sudden increase in light intensity (Fig. 5.1). All-*trans*-violaxanthin is converted to **9′-*cis*-neoxanthin** or **9-*cis*-violaxanthin**. Arabidopsis ABA-deficient *aba4* mutants have reduced endogenous levels of 9′-*cis*-neoxanthin and all-*trans*-neoxanthin, but have increased levels of all-*trans*-violaxanthin and 9-*cis*-violaxanthin, leading to the conclusion that ABA4 functions as a neoxanthin synthase or as one of the components required for neoxanthin synthesis (North et al., 2007). To date, no mutants defective in the isomerization of all-*trans*-violaxanthin or all-*trans*-neoxanthin have been isolated.

The first committed step in the ABA biosynthetic pathway is the oxidative cleavage of 9′-*cis*-neoxanthin and/or 9-*cis*-violaxanthin, producing the first C_{15} intermediate, **xanthoxin**, a neutral growth inhibitor that has similar physiological properties to those of ABA (Fig. 5.1). This is a rate-limiting regulatory step and is catalyzed by **9-cis-epoxycarotenoid dioxygenase** (NCED) (Nambara and Marion-Poll, 2005). The NCED enzymes are encoded by multigene families in all species analyzed, and have nine potential members in Arabidopsis (Tan et al., 2003). The expression patterns of *NCED* genes vary in response to stress and developmental signals. For example, *AtNCED3* is induced for ABA production upon water stress, whereas *AtNCED6* and *AtNCED9* are induced for ABA production in seeds (Lefebvre et al., 2006; Frey et al., 2012). Moreover, the tissue localizations of different NCEDs differ, although they are all localized in plastids; AtNCED5 is bound to plastid membranes but AtNCED2, AtNCED3, and AtNCED6 have both soluble and membrane-bound forms (Tan et al., 2003).

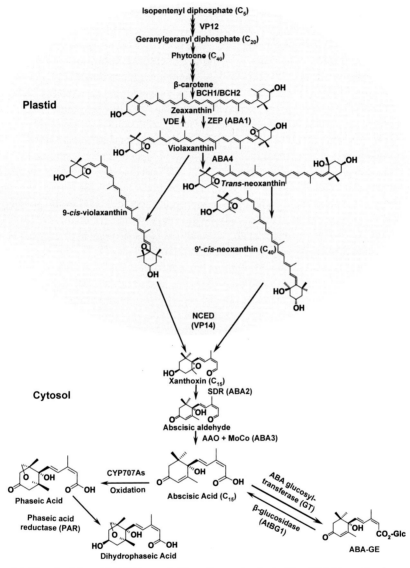

Figure 5.1 The abscisic acid (ABA) metabolic pathways in plants. The ABA biosynthesis, degradation, and conjugation pathways are shown in relation to the cellular compartments where these events occur. Biosynthesis steps blocked in the Arabidopsis mutants (*aba1*, *aba2*, *aba3*, and *aba4*) and maize mutants (*vp12* and *vp14*) are indicated.

Xanthoxin, which is synthesized in the plastids, moves to the cytoplasm where it is converted to ABA via oxidative steps involving the intermediate **abscisic aldehyde** (Fig. 5.1). Conversion of xanthoxin to abscisic aldehyde is catalyzed by a **short-chain dehydrogenase/reductase** (SDR)-like enzyme encoded by *ABA2* in Arabidopsis (Cheng et al., 2002b; Gonzalez-Guzman et al., 2002). The final step is catalyzed by a family of **abscisic aldehyde oxidases** (AAOs) that all require a molybdenum cofactor (MoCo). Arabidopsis *ABA3* encodes a sulfurase that produces a functional cofactor. The *aba3* mutant lacks all AAO activity and is unable to synthesize ABA (Bittner et al., 2001; Xiong et al., 2001).

Characterization of maize *viviparous* (*vp*) mutants, which show **precocious germination** of seeds while still attached to the plant, a phenomenon also known as **vivipary**, has contributed greatly to the understanding of the ABA biosynthetic pathway. The *vp* mutants are blocked at various points in the terpenoid and carotenoid biosynthetic pathway. For example, *vp12* is deficient in GGPP synthase and *VP14* was the first cloned *NCED* gene. More details regarding vivipary are discussed later.

5.2.2   ABA catabolism and storage

In addition to ABA biosynthesis, catabolism and conjugation function as major mechanisms for regulating ABA levels in vivo (reviewed in Nambara and Marion-Poll, 2005). ABA catabolism involves ABA hydroxylation and plants have three different ABA hydroxylation pathways (C-7′, C-8′, and C-9′), among which 8′-hydroxylation is thought to be the predominant ABA catabolic pathway (Cutler and Krochko, 1999). This step is catalyzed by 8′-hydroxylase and the resulting 8′-hydroxy-ABA is spontaneously rearranged to form **phaseic acid** (PA), which is further catabolized to dihydrophaseic acid (Fig. 5.1). ABA 8′-hydroxylase is a membrane-bound cytochrome P450 (CYP450) monooxygenase classified as **CYP707A**. Single or multiple mutants defective in *CYP450* genes contain higher endogenous levels of ABA and display enhanced stress resistance and increased seed dormancy (Kushiro et al., 2004; Saito et al., 2004; Okamoto et al., 2006).

ABA and its catabolites can also undergo conjugation reactions catalyzed by glucosyl transferases, with the most common conjugate being glucosyl ester (ABA-GE) (Fig. 5.1). ABA-GE is probably a deactivated product, as it is physiologically inactive. It is thought to be a storage or transport form of ABA and can be hydrolyzed by β-glucosidase (AtBG1 in Arabidopsis) to release free ABA (Lee et al., 2006; Xu et al., 2012).

5.3   ABA transport

In vegetative plant tissues, ABA biosynthesis is believed to take place in the vascular tissues and the guard cells, based on the expression patterns of the key enzymes of the ABA biosynthetic pathway, such as the enzymes catalyzing the last three steps of ABA biosynthesis (NCED, SDR, and AAO). By contrast, in seeds, all tissues are thought to be involved in ABA biosynthesis.

Long-distance ABA transport from roots to leaves has been extensively discussed in relation to stomatal closure in response to drought (reviewed in Boursiac et al., 2013). Stomata will close when a part of the root system is exposed to water stress, even if the water status of leaves remains unchanged. Moreover, the stomatal aperture correlates with the ABA concentration in xylem sap, suggesting that ABA synthesized in roots experiencing drought conditions could transit to the xylem sap and be transported to the leaves to induce stomatal closure. However, during water stress, ABA accumulates first in shoot vascular tissues and only later appears in roots and guard cells (Christmann et al., 2005). Reciprocal grafting experiments demonstrated that ABA synthesis in shoots was essential for the plant's response to root drying (Christmann et al., 2007). Moreover, studies challenged the need for ABA from the roots in stomatal regulation by showing that only ABA synthesized in guard cells triggered stomatal closure (Bauer et al., 2013). Furthermore, it has now been reported that ABA produced in the shoot is required for root growth (McAdam et al., 2016).

The molecular identification of ABA transporters emphasizes the importance of ABA transport for proper ABA signaling throughout the plant. The Arabidopsis ATP-BINDING CASSETTE G25 (AtABCG25) and AtABCG40 were identified as ABA transporters (Kang et al., 2010; Kuromori et al., 2010). AtABCG25 functions as an efflux transporter and AtABCG40 functions as an influx transporter of ABA. The promoter activity of *AtABCG25* was detected in vascular tissues, whereas that of *AtABCG40* was detected in guard cells, supporting the idea that ABA is transported from vascular tissues to guard cells (Kang et al., 2010; Kuromori et al., 2010). However, the single mutants of *atabcg25* and *atabcg40* did not show phenotypes typically observed in ABA-deficient mutants such as *aba1*, *aba2*, and *aba3*, suggesting that the ABA transport system might be highly complex and redundant (Kang et al., 2010; Kuromori et al., 2010). The Arabidopsis ABA-IMPORTING TRANSPORTER 1 (AIT1), previously characterized as a nitrate (NO_3^-) transporter (designated NRT1.2), was later shown to be an influx transporter of ABA as well as NO_3^- (Kanno et al., 2012). The *AIT1* gene is expressed in imbibed seeds and vascular tissues, and all three transporters localize at the plasma membrane. AIT1 is proposed to regulate stomatal aperture in inflorescence stems and may link nitrogen status to ABA signaling since it is also a NO_3^- transporter (Kanno et al., 2012).

5.4 ABA functions

5.4.1 Seed maturation and dormancy

From both an ecological and agricultural perspective, the plant life cycle can be considered to begin and end with the seed. A mature Arabidopsis seed consists of three major components, each with a distinct genotype: the **embryo** that will become the vegetative plant is **diploid** (having one maternal and one paternal genome equivalent); the **endosperm**, a single-cell layer surrounding the embryo that provides nourishment for embryo development, is **triploid** (having two maternal and one paternal genome equivalents); and the **testa** (seed coat), an outer layer of dead tissue, is strictly of maternal origin. Seed development comprises two major phases, **embryogenesis** and **seed maturation**. During embryogenesis, the single-celled zygote follows a defined pattern of cell division and differentiation to form the mature embryo. Seed maturation begins when the developing embryo ceases cell division and starts growing by cell enlargement as it begins to accumulate storage reserves. Finally, the embryos of so-called **orthodox seeds** dehydrate (losing up to 90% of their water) and become desiccation tolerant. As a consequence, their metabolic activity decreases and the seeds enter a **quiescent** state. In some cases the seeds become **dormant**, a state in which a viable seed fails to germinate under favorable conditions. Quiescent seeds will germinate upon rehydration; however, the germination of dormant seeds requires additional treatments or signals. In contrast to orthodox seeds, **recalcitrant seeds** have a high moisture content at maturity and are not desiccation tolerant (reviewed in Bentsink and Koornneef, 2008; Rajjou et al., 2012).

Typically, the ABA content in seeds is very low during early embryogenesis (i.e., during histodifferentiation and early pattern formation); however, ABA increases when the developing embryos transition to the maturation phase, usually peaking around midmaturation. ABA levels usually decline precipitously during late seed development, particularly during the maturation drying phase. However, another peak of ABA occurs in developing embryos prior to **desiccation** during later development of Arabidopsis seeds. Reciprocal crosses between the wild type and ABA-deficient mutants suggest that the first peak of ABA is maternally derived and is important to prevent precocious germination at the end of the cell division phase of embryogenesis (Karssen et al., 1983). However, de novo synthesis of ABA in developing embryos may also contribute to the first peak, with maternal ABA serving as a signal to trigger ABA biosynthesis in the embryo or endosperm. Consistent with this notion, ABA biosynthetic genes are actively expressed during mid-seed development (10 days after pollination) in the Arabidopsis embryo and endosperm (Xiong and Zhu, 2003; Lefebvre et al., 2006). The second peak of ABA accumulation depends on the synthesis of ABA in the embryo and is essential for the induction of dormancy (Karssen et al., 1983; Frey et al., 2004).

Genetic variations in the structure and/or pigmentation of the testa or the surrounding layers (such as the pericarp) lead to altered seed dormancy or **longevity** (the duration of seed viability in prolonged storage) in many species. In addition, when the immature embryo is removed from the testa and placed in culture medium before the onset of dormancy, it germinates precociously. However, the addition of ABA to the culture medium inhibits this precocious germination. These observations, together with the fact that the endogenous ABA level is high during mid- to late seed development, indicate that the maternal ABA located in testa and fruit tissues keeps developing embryos in their embryonic state.

Vivipary, also known as **preharvest sprouting**, occurs in some grain crops when they mature in wet weather and studies of various ABA mutants and transgenic lines support the involvement of ABA in the prevention of vivipary and the induction of dormancy. Several maize *viviparous* (*vp*) mutants have been identified; some of these are deficient in ABA biosynthesis and one (*vp1*) is insensitive to ABA. Vivipary in the ABA-deficient mutants can be partially prevented by treatment with exogenous ABA. Maize VP1 and its Arabidopsis ortholog, ABA INSENSITIVE 3 (ABI3), along with FUSCA 3 (FUS3) and LEAFY COTYLEDON 2 (LEC2), are closely related B3-domain family transcription factors, and LEC1 is a HAP3 subunit of the CCAAT-binding transcription factor also known as nuclear factor Y (NF-Y) (reviewed in Finkelstein, 2013). These four regulators play prominent roles in the overall control of seed maturation with apparently overlapping functions (Raz et al., 2001). Mutants deficient in any of these genes do not exhibit dormancy and often germinate precociously, and their seeds have phenotypes typical of the vegetative phase, rather than the reproductive phase, including reduced desiccation tolerance, and expression of germination-related genes. In addition, the expression of *LEC1*, *LEC2*, *ABI3*, and *FUS3* displays a temporal and spatial pattern during Arabidopsis seed development, and there is substantial cross-regulation among these genes (reviewed in Suzuki and McCarty, 2008; Finkelstein, 2013).

Arabidopsis ABA-deficient (*aba*) mutants are nondormant at maturity. Reciprocal crosses between wild-type and *aba* mutant plants showed that seeds exhibited dormancy only when the embryo itself could produce ABA during seed maturation. Exogenously applied or maternal ABA was not effective in inducing dormancy in the embryos of *aba* mutants. In addition to the ABA levels, sensitivity to ABA also plays an important role in the induction of dormancy. For example, *aba insensitive* (*abi*) mutants display varying reductions in seed dormancy, although the ABA levels in these seeds are higher than in the wild type throughout seed development. Therefore, although ABA is important for the inception of dormancy in developing seeds, high ABA levels are not required for the maintenance of dormancy during late maturation and desiccation of seeds, and there is no clear relationship between the ABA content of mature dry seeds (or grains) and the degree of dormancy.

ABA inhibits seed germination and GAs promote seed germination. Also, in most plants, the peak in ABA coincides with a decline in the level of GA. The importance of the ratio of these two hormones (**ABA:GA ratio**) was elegantly demonstrated by a genetic screen for suppressors of a GA-deficient (*ga-1*) mutant (Koornneef et al., 1982). The *ga-1* mutant seeds could not germinate in the absence of exogenous GA, and the seeds of mutant plants that had regained the ability to germinate likely carried a suppressor mutation. This screen led to the isolation of the first ABA-deficient mutants of Arabidopsis and about half of the suppressed plants were deficient in *ABA1*, encoding ZEP, which generates the epoxycarotenoid precursor of the ABA biosynthetic pathway. The suppressed seeds carrying *ga1* and *aba1* mutations could germinate because dormancy had not been induced in the seeds, so subsequent biosynthesis of GA was no longer required. Conversely, a screen for suppressors of the ABA-insensitive mutation *abi1-1* identified both GA-deficient and GA-resistant mutants (Steber et al., 1998). These studies illustrated that the ABA:GA ratio serves as the primary determinant of seed dormancy and germination. The amounts of GA and ABA present in a target tissue and the ability of the target tissue to detect and respond to each of the hormones determine the relative hormonal activities of ABA and GA.

During seed maturation, the water content of seeds gradually declines as water is replaced by the deposition of insoluble storage reserves, and **desiccation** (maturation drying) is usually the terminal event for orthodox seeds. Desiccation can severely damage membranes and other cellular components. For orthodox seeds, ABA induces **desiccation tolerance** during seed development, allowing their survival for long periods in a dry state. Desiccation tolerance depends in part on the ability of the antioxidant defense systems to scavenge the reactive oxygen species (ROS) generated during desiccation. Desiccation tolerance also involves the formation of an intracellular "glassy state" by a combination of sugars and **late embryogenesis abundant** (LEA) proteins. The LEA proteins become abundant during late seed maturation and are thought to act as chaperones to protect macromolecular structures against desiccation injury. ABA regulates the accumulation of storage proteins and desiccation protectants during embryogenesis. Exogenously applied ABA promotes the accumulation of storage proteins and LEA proteins in cultured embryos of many species. In addition, ABA treatment can induce the synthesis of some LEA proteins in vegetative tissues, indicating that ABA regulates the biosynthesis of some LEA proteins.

5.4.2   Seed germination

Seed germination, the resumption of growth of the quiescent or dormant embryo of the mature seed, begins with water uptake by the dry seeds and ends with the emergence of the embryonic axis, usually the radicle, from its surrounding tissues (reviewed in Bentsink and Koornneef, 2008; Rajjou et al., 2012). Because germination irreversibly commits a seed to grow into a seedling, plants have evolved sophisticated mechanisms to ensure that germination occurs only under optimal environmental conditions. Because the ABA:GA ratio plays a decisive role in maintaining seed dormancy, ABA homeostasis shifts during germination toward a catabolic state and ABA signals are downregulated, whereas GA homeostasis shifts to favor GA biosynthesis and GA signaling is activated. Therefore seed germination is often associated with a sharp decrease in the ABA:GA ratio.

Germination depends on several environmental factors, e.g., water, oxygen, temperature, and often light and nitrate as well. Of these, water is the most essential factor. The water content in mature, air-dried seeds ranges from 5% to 15%, well below the threshold required for fully active metabolism. The initial, rapid uptake of water by dry seeds is referred to as **imbibition**. A rapid decrease in endogenous ABA levels occurs in both dormant and nondormant seeds during early seed imbibition. In Arabidopsis, the expression of *CYP707A2*, which encodes an ABA 8′-hydroxylase and plays a key role in inactivating ABA during germination, rapidly increases after seed imbibition (Kushiro et al., 2004). However, several GA biosynthetic genes (such as *AtGA20ox3*, *AtGA3ox1*, and *AtGA3ox2*) are upregulated upon imbibition, whereas expression of all known *GA 2-oxidase* genes remains at a low level (Ogawa et al., 2003).

The importance of light in seed germination was first discovered by examination of the red (R)/far-red (FR) light-controlled reversible germination of lettuce (c.v. Grand Rapids) seeds (Borthwick et al., 1952). Red light promotes seed germination, whereas subsequent treatment with FR light reverses the R light-mediated induction of seed germination. The germination response of lettuce seeds repeatedly treated with R/FR cycles is determined by the last light treatment. **Phytochromes**, plant photoreceptors that sense R and FR light, are the primary sensors for light-regulated seed germination. Light induces the expression of GA biosynthetic genes (such as *AtGA3ox1* and *AtGA3ox2*) and an ABA catabolic gene (*CYP707A2*). Light also represses the expression of ABA biosynthetic genes (such as *ABA1*, *NCED6*, and *NCED9*) and GA catabolic genes (such as *GA2ox2*) (Yamaguchi et al., 1998; Seo et al., 2006; Oh et al.,

2007; Yamauchi et al., 2007). Light-controlled seed germination is mediated in part by a bHLH transcription factor, PHYTOCHROME-INTERACTING FACTOR 1 (PIF1; Oh et al., 2004, 2007, 2009).

Low temperature, or **chilling**, can release seeds from dormancy. Many seeds require a period of cold treatment (0–10°C) while in a fully hydrated (imbibed) state to germinate. Chilling seeds in moist conditions to stimulate germination is referred to as **stratification**; this not only breaks the dormancy of seeds but also has the added benefit of synchronizing germination to ensure that plants will mature simultaneously. A subset of GA biosynthetic genes is upregulated in response to low temperature in Arabidopsis, resulting in an increase in the level of bioactive GAs in imbibed Arabidopsis seeds (Yamauchi et al., 2004).

Some seeds may require **after-ripening**: a period of dry storage at room temperature before they can germinate. The duration of the after-ripening requirement may range from as short as a few weeks to as long as a few years. In Arabidopsis seeds, the sensitivity to GAs and light increases and ABA content decreases during after-ripening (Ali-Rachedi et al., 2004). The changes in ABA content or sensitivity are likely related to a progressive decrease of dormancy (hence an increasing potential of germination) of after-ripened seeds.

5.4.3   Stress responses

ABA rapidly accumulates in plants in response to environmental stresses such as drought, low temperature, or high salinity, and plays important roles in plant adaptation to these stresses (reviewed in Yamaguchi-Shinozaki and Shinozaki, 2006; Qin et al., 2011). Therefore ABA is considered a "stress hormone." ABA levels increase during stress but decrease when the stress is relieved (Zeevaart, 1980). Under drought conditions, ABA concentrations can increase up to 50-fold, which is the most dramatic change in concentration reported for any plant hormone in response to an environmental signal. Drought, low temperature, and high salinity all impose cellular osmotic and oxidative stresses, but they differ in other aspects, and consequently, plant responses to these stresses also differ.

At the whole-plant level, slightly elevated ABA levels, characteristic of mild water stress, promote root growth but inhibit shoot growth, leading to an increased **root:shoot ratio**. The root:shoot ratio appears to be governed by a balance between water uptake by the roots and photosynthesis by the shoots. Under water stress, an increased root:shoot ratio allows the roots to grow at the expense of the growth of the leaves. However, under extended drought stress, the growth of both roots and shoots is inhibited and many lateral roots are initiated but their growth stops until the stress is relieved. This phenomenon is known as **drought rhizogenesis** (Vartanian et al., 1994). These repressive effects of stress and the redistribution of nutrients depend at least partly on ABA signaling because ABA-deficient mutants and some ABA signaling mutants do not exhibit these effects.

The water flow across cell membranes to regulate growth and transpiration is largely controlled by aquaporins present in the tonoplast and plasma membranes. Arabidopsis has 13 plasma membrane-localized aquaporins (known as plasma membrane intrinsic proteins, PIPs) and 10 tonoplast-localized aquaporins (known as tonoplast intrinsic proteins). The expression of most aquaporin genes is downregulated during drought and salt stresses to limit water loss and potentially create a hydraulic signal to aerial parts to induce stomatal closure. In addition, ABA-induced dephosphorylation of aquaporins also leads to reduced water flux (Kline et al., 2010).

In addition to modulating water flow, ABA induces the accumulation of protectants in cells, such as chaperones, LEA proteins, antifreeze proteins, sugars, and proline. This stress hormone also activates detoxifying mechanisms that confer stress tolerance by regulating redox balance or modifying ion transport to reestablish homeostasis.

5.4.4 Stomatal movement

Stomata, the openings formed by pairs of specialized epidermal guard cells, regulate gas exchange in plants. Stomatal apertures affect photosynthesis, water use efficiency, and hence crop yields. Several environmental factors induce opening of stomata, including blue light, high humidity, and low CO_2. Other factors induce closure of stomata, including lack of light, low humidity, elevated CO_2, and drought. During drought, plants accumulate ABA, which induces rapid closure of the stomata to prevent water loss by transpiration. ABA affects stomata by promoting closure and inhibiting opening of stomata. Although both effects lead to closed stomata, they are not simple reversals of the same process since they involve different ion channels controlled by different signaling mechanisms (reviewed in Schroeder et al., 2001; Kim et al., 2010).

The driving force for stomatal movement is turgor pressure. When guard cells perceive increased ABA levels, an efflux of anions and K^+ ions is triggered, which decreases their turgor and volume. ABA triggers an increase of cytosolic Ca^{2+}, which activates slow-activating (S-type) and rapid-transient (R-type) anion channels (Schroeder et al., 2001). The S-type anion channels generate a slow and sustained anion efflux, whereas R-type anion channels are transiently activated within 50 ms. Thus the two types of anion channels provide distinct mechanisms for anion efflux. Anion efflux via anion channels results in membrane depolarization, which subsequently causes an efflux of K^+ from guard cells through outward K^+ channels in the plasma membrane (Kim et al., 2010).

Stomatal opening requires the activation of H^+-ATPase in the plasma membrane of the guard cells. Blue light induces the phosphorylation and activation of the guard cell plasma membrane H^+-ATPase, and the membrane hyperpolarization caused by H^+-ATPase induces the uptake of K^+ through inward K^+ channels (reviewed in Shimazaki et al., 2007). The accumulation of positively charged K^+ ions in guard cells must be compensated for by anions, mainly in the form of the organic acid malate^{2-}, which is formed in the chloroplasts of guard cells. Guard cells also use Cl^- and NO_3^- as counterions for K^+. The influx of K^+, Cl^-, and NO_3^- and the production of malate increase turgor and volume in the guard cells, which induces stomatal opening. ABA inhibits stomatal opening by the downregulation of inward K^+ channels and H^+-ATPase (Kim et al., 2010).

5.5   ABA signal transduction

5.5.1   ABA receptors

Microinjection studies and treatments with impermeant ABA analogs performed in the 1990s suggested that ABA may be perceived at both intracellular and extracellular sites. However, since ABA can be quickly transported into cells through plasma membrane–localized ABA transporters (see earlier discussion), the existence of plasma membrane–localized ABA receptors remains a matter of debate. Several proteins were suggested to be ABA receptors before 2009, but their identities remain controversial and they will not be discussed further in this chapter. In 2009, two groups independently reported a family of 14 *Arabidopsis thaliana* START proteins, called PYR/PYL/RCARs, as ABA receptors (reviewed in Hubbard et al., 2010; Raghavendra et al., 2010).

Two separate screens simultaneously identified the previously uncharacterized PYR/PYL/RCAR proteins as soluble ABA receptors. In one approach, a chemical-genetic screen for mutants resistant to pyrabactin, a synthetic growth inhibitor that functions as a selective ABA agonist, led to the isolation of *pyrabactin resistance 1* (*pyr1*) mutants (Park et al., 2009). In the other approach, a yeast two-hybrid screen using ABI2 as the bait identified RCAR1/PYL9 (Ma et al., 2009). The ABI1 and ABI2 proteins are phosphoprotein phosphatases of clade A phosphoprotein phosphatase 2Cs (PP2Cs) (see later discussion), which function as negative regulators of ABA signaling and act upstream of all known rapid ABA signaling responses. The identities of these ABA receptors are unequivocal and have been repeatedly confirmed by independent research groups. The 14 PYR/PYL/RCAR proteins are highly conserved at the amino acid sequence level, which may explain why they had not been identified in many previous attempts to screen for ABA-insensitive mutants.

PYR/PYL/RCARs are members of the soluble START [steroidogenic acute regulatory (StAR)-related lipid transfer] domain superfamily proteins, which contain a central hydrophobic ligand-binding pocket (Iyer et al., 2001). The initial evidence for ABA binding of PYR1 was obtained by heteronuclear single quantum coherence nuclear magnetic resonance experiments (Park et al., 2009) and ABA binding of RCAR1/PYL9 was confirmed by isothermal titration calorimetry assays (Ma et al., 2009). The crystal structures of ABA-bound PYR/PYL/RCARs were then quickly reported for PYR1 (Nishimura et al., 2009; Santiago et al., 2009), PYL1 (Miyazono et al., 2009), and PYL2 (Melcher et al., 2009; Yin et al., 2009), and for coreceptor complexes PYL1-ABA-ABI1 (Miyazono et al., 2009; Yin et al., 2009) and PYL2–ABA–HAB1 (Melcher et al., 2009). The ligand-binding site of PYR/PYL/RCAR proteins lies within a large internal cavity and the binding of ABA by PYR/PYL/RCARs induces conformational changes of the receptor proteins, which seal ABA inside the receptor and reshape the protein surface, thus providing a novel site for interaction with PP2C proteins. A conserved tryptophan residue of the PP2Cs inserts into the PYR/PYL/RCAR proteins to interact with the ABA molecule; therefore, PP2Cs further stabilize the ABA binding of PYR/PYL/RCARs, with an approximately 10-fold increase in affinity in the presence of

PP2Cs (Ma et al., 2009; Santiago et al., 2009). The PYR/PYL/RCARs also occlude access to the Mg^{2+}-containing active site of the PP2C, inhibiting its phosphatase activity in the presence of ABA (Melcher et al., 2009; Miyazono et al., 2009; Yin et al., 2009).

Structural studies also indicated that some PYR/PYL/RCAR proteins exist as dimers and convert into monomers in response to ABA; this probably has an important signaling function, as the final PYR/PYL/RCAR–PP2C complex is a heterodimer (Nishimura et al., 2009; Santiago et al., 2009; Yin et al., 2009). In addition, although the PYR/PYL/RCAR–PP2C interaction is ABA-dependent for PYR1 and PYL1–4, yeast two-hybrid analyses showed that PYL5–12 interact constitutively with PP2C (Park et al., 2009). A subsequent structural and biochemical study further confirmed these observations and revealed that the ABA-independent inhibition of PP2Cs by PYLs requires: (1) monomeric PYLs, which appear to be necessary for constitutive association with and inhibition of PP2Cs, and (2) bulky and hydrophobic residues guarding the ligand-binding pocket of PYLs (Hao et al., 2011). In addition to ABA, PA, the oxidative catabolite of ABA, can also be perceived by a subset of ABA receptors, and thus activate these ABA receptors (Weng et al., 2016) (Fig. 5.2).

5.5.2   ABA coreceptors

The first ABA-responsive loci identified in Arabidopsis, designated *ABA INSENSITIVE 1-3* (*ABI1 - ABI3*), were first reported by Koornneef et al. (1984). The *ABI1* gene, which encodes a PP2C, was cloned a decade later (Leung et al., 1994; Meyer et al., 1994). The *ABI2* gene was cloned thereafter and surprisingly, ABI2 was shown to be a close homolog of ABI1. Also the initial alleles (*abi1-1* and *abi2-1*) reported by Koornneef et al. (1984) harbor identical substitutions at an equivalent position in their catalytic domains (G180D for *abi1-1* and G168D for *abi2-1*) (Leung et al., 1997; Rodriguez et al., 1998). Later work showed that ABI1 acts as a negative regulator of ABA signaling (Gosti et al., 1999) and both *abi1-1* and *abi2-1* are dominant mutations that block the interactions between ABI1 or ABI2 and the PYR/PYL/RCAR receptors (Ma et al., 2009; Miyazono et al., 2009; Park et al., 2009).

The ABI1 and ABI2 proteins belong to the clade A PP2Cs and Arabidopsis has a total of nine phosphoprotein phosphatases in this clade (Schweighofer et al., 2004). In addition to ABI1 and ABI2, four other members (HAB1, HAB2, AHG1, and AHG3/PP2CA) also have established functions as negative regulators of ABA signaling (Leonhardt et al., 2004; Saez et al., 2004; Kuhn et al., 2006; Yoshida et al., 2006; Nishimura et al., 2007). The remaining three members, HAI1, HAI2, and HAI3, appear to have functionally differentiated from other clade A PP2C members since the *hai1 hai2 hai3* triple mutants were ABA insensitive in germination but moderately hypersensitive in postgermination ABA responses (Bhaskara et al., 2012). However, a more recent report showed that HAI2 also negatively regulates the ABA response in germination (Kim et al., 2013). An outstanding feature of clade A *PP2C* genes is that they are all rapidly induced by exogenous ABA treatments (Rodriguez et al., 1998; Saez et al., 2004; Kuhn et al., 2006; Fujita et al., 2009), which may be part of an ABA desensitization mechanism that functions under abiotic stress to adjust ABA signaling to strongly increased levels of ABA (Raghavendra et al., 2010).

As mentioned earlier, the ABA-binding affinities of some PYR/PYL/RCARs are enhanced when they interact with clade A PP2Cs; therefore, PP2Cs are also considered ABA coreceptors. Formation of the PYR/PYL/RCAR-ABA-PP2C complex blocks substrate entry into PP2C, thus relieving PP2C-mediated inhibition of the downstream SnRK2 kinases (see later discussion; Melcher et al., 2009; Miyazono et al., 2009; Yin et al., 2009) (Fig. 5.2). Some PYR/PYL/RCARs can also interact with PP2Cs in an ABA-independent manner, but exogenous ABA strongly increases their inhibition of PP2C activity (Hao et al., 2011). The ABA-independent interaction of these PYR/PYL/RCARs with PP2Cs might be explained by: (1) these PYR/PYL/RCARs potentially being involved in processes other than ABA responses, or (2) their maintaining a basal level of ABA signaling since ABA is essential for plant growth and survival (Hao et al., 2011).

5.5.3   Protein kinases

The identification and characterization of clade A PP2Cs as ABA coreceptors indicated the importance of protein phosphorylation/dephosphorylation in ABA signaling. Consistent with this concept, several classes of protein kinases were identified and characterized as important regulators of ABA signaling, including SnRK2s, SnRK3s/CIPKs, CDPKs/CPKs, MAPKs, and receptor-like kinases.

5.5.3.1   SNF1-related protein kinase 2s

The *A. thaliana* genome encodes 38 protein kinases that are related to sucrose nonfermenting 1 (SNF1) from yeast. These were named SNF1-related protein kinases (SnRKs) (Hrabak et al., 2003). Based on sequence similarity and domain structures, Arabidopsis SnRKs can be further divided into SnRK1 (3 members), SnRK2 (10 members), and SnRK3 families (25 members) (Hrabak et al., 2003). In the SnRK2 family (SnRK2.1 to SnRK2.10), SnRK2.2, SnRK2.3, and SnRK2.6 [also known as OPEN STOMATA 1 (OST1)] function as central, positive regulators of ABA signaling (Fig. 5.2). The first report regarding the involvement of this family of protein kinases in the ABA signaling pathway described PKABA1, which participates in ABA suppression of gene expression in barley aleurone layers (Gomez-Cadenas et al., 1999). Subsequently, a guard-cell-specific ABA-activated serine–threonine protein kinase (AAPK) from *Vicia faba* was shown to be involved in the regulation of ABA-induced stomatal closure (Li et al., 2000). The putative AAPK ortholog in Arabidopsis, SnRK2.6/OST1, was shown to modulate ABA regulation of stomatal aperture (Mustilli et al., 2002; Yoshida et al., 2002). The two other kinases most closely related to SnRK2.6/OST1, namely SnRK2.2 and SnRK2.3, were later shown to play important roles in mediating ABA-regulated seed germination, root growth, and gene expression (Fujii et al., 2007). An Arabidopsis triple mutant defective in all three of these SnRK2

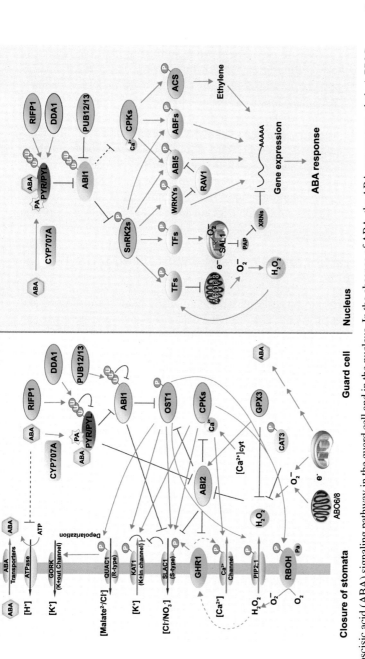

Figure 5.2 The abscisic acid (ABA) signaling pathway in the guard cell and in the nucleus. In the absence of ABA, the ABA coreceptors, clade A PP2Cs, are active and repress SnRK2 activity and downstream events, whereas in the presence of ABA, ABA-bound PYR/PYL/RCARs interact with PP2Cs and inhibit their activity, thus allowing the activation of SnRK2s, CPKs, and GHR1 and the phosphorylation of target proteins. Green arrow, positive regulation; red bar, negative regulation. *ABA*, abscisic acid; *ABF*, ABRE-binding factor; *ABO*, ABA OVERLY-SENSITIVE; *ACS*, 1–aminocyclopropane-1–carboxylate synthase; *ABI*, ABA INSENSITIVE; *CAT*, catalase; *CPK*, calcium-dependent protein kinase; *CYP*, cytochrome P450; *GHR*, GUARD CELL HYDROGEN PEROXIDE-RESISTANT; *GORK*, guard cell outward rectifying K+ channel; *GPX*, glutathione peroxidase; *KAT*, K+ CHANNEL IN ARABIDOPSIS THALIANA; *OST*, OPEN STOMATA; *PA*, phaseic acid; *Pa*, phosphatidic acid; *PAP*, phosphonucleotide 3′-phosphoad-enosine 5′-phosphate; *PIP*, plasma membrane intrinsic protein; *PUB*, PLANT U-BOX E3 LIGASE; *pyr*, *pyrabactin resistance*; *QUAC*, quickly activating anion channel; *RAV*, related to ABI3/VP1; *RBOH*, RESPIRATORY BURST OXIDASE HOMOLOG; *RIFP*, RCAR3 INTERACTING F-BOX PROTEIN; *SLAC*, SLOW ANION CHANNEL-ASSOCIATED; *SnRK*, SNF1-related protein kinase; *XRNs*, 5′ to 3′ exoribonucleases; *TF*, transcription factor; *WRKY*, each letter represents an amino acid.

proteins (known as subclass III) exhibited severe ABA-insensitive phenotypes that indicated strong defects in ABA signaling (Fujii and Zhu, 2009; Fujita et al., 2009; Nakashima et al., 2009a). The seeds of the *snrk2.2 snrk2.3 snrk2.6* triple mutant could even germinate on 300 μM ABA, indicating that the mutant is totally insensitive to ABA (Fujita et al., 2009). In addition, the *snrk2.2 snrk2.3 snrk2.6* triple mutant showed strong defects in plant growth and seed production, indicating that ABA plays important roles in regulating plant growth and reproduction (Fujii and Zhu, 2009). The activities of SnRK2.2, SnRK2.3, and SnRK2.6/OST1 were strongly activated by ABA, whereas those of SnRK2.7 and SnRK2.8 were weakly activated by ABA (Boudsocq et al., 2004; Furihata et al., 2006). Interestingly, all of the SnRK2s, except SnRK2.9, can be activated by osmotic stress and an Arabidopsis mutant impaired in all 10 SnRK2 kinases exhibited defects in gene regulation and ABA accumulation under osmotic stress, indicating that SnRK2s are critical for osmotic stress responses in plants (Fujii et al., 2011).

5.5.3.2 CBL-interacting protein kinase/SNF1-related protein kinase 3s

Calcium serves as a critical, versatile messenger in many adaptive responses and developmental processes in plants, and cellular calcium signals are detected and transmitted by calcium-binding proteins that function as sensor molecules (Weinl and Kudla, 2009). Calcineurin B-like proteins (CBLs) are a unique group of calcium sensors that form a complex network with their target kinases, CBL-interacting protein kinases (CIPKs), also classified as SnRK3s (Hrabak et al., 2003; Weinl and Kudla, 2009). The *A. thaliana* genome encodes 10 CBL and 26 CIPK proteins. Each CBL protein consists of four EF-hand motifs and each CIPK harbors a kinase domain at the N-terminal end and a regulatory domain at the C-terminal end (Weinl and Kudla, 2009). CBLs bind to Ca^{2+} ions and interact with the C-terminal regulatory domain of CIPKs, which leads to the activation of CIPKs. Thus Ca^{2+} sensing and kinase activity form two flexible, combinable modules, which allow the formation of a complex and dynamic Ca^{2+}-decoding signaling network (Weinl and Kudla, 2009).

Some CBL-CIPK complexes are implicated in ABA signaling. For example, the complex between SCaBP5 (CBL1) and PKS3 (CIPK15) functions as a negative regulator of ABA-regulated seed germination, stomatal aperture, and gene expression (Guo et al., 2002). Thus SCaBP5 (CBL1) and PKS3 (CIPK15) may constitute a calcium-sensing system involved in feedback inhibition of ABA responses. In addition, PKS3 (CIPK15) physically interacts with the ABA coreceptors ABI1 and ABI2. The dominant mutations *abi1-1* and *abi2-1* suppress the ABA-hypersensitive phenotypes of *pks3 (cipk15)* or *scabp5 (cbl1)* in seed germination and seedling growth, suggesting that the complex of SCaBP5 (CBL1) with PKS3 (CIPK15) acts in the same pathway as ABI1 and ABI2 in ABA signaling (Guo et al., 2002). CIPK3 acts as a negative regulator of ABA signaling during seed germination, but ABA-induced stomatal closure was not affected in a loss-of-function *cipk3* mutant (Kim et al., 2003). In addition, CIPK3 regulates the expression of cold- and salt-induced genes, but not drought-induced genes, indicating that CIPK3 regulates selective pathways in response to abiotic stresses (Kim et al., 2003). CIPK1 and CIPK23 are also negative regulators of ABA signaling, but CIPK23 appears to play a dominant role in ABA regulation of stomatal movement, but not of germination, whereas CIPK1 regulates ABA responses in seed germination (D'Angelo et al., 2006; Cheong et al., 2007). One report showed that CIPK26 interacts with several ABA signaling components, ABI1, ABI2, and ABI5, and its stability may be regulated by the E3 ubiquitin ligase KEEP ON GOING (KEG) (Lyzenga et al., 2013). However, in contrast with the negative roles of CIPKs reported in most studies, the *cipk26* mutants did not show any ABA-related phenotypes, whereas the CIPK26-overexpression lines were hypersensitive to ABA during seed germination, implying that CIPK26 may play a positive role in ABA signaling (Lyzenga et al., 2013).

Two independent reverse genetic analyses showed that CBL1 regulates salt, drought, and low temperature stress responses in Arabidopsis, implying that CBL1 acts as a central integrator of responses of the plant to abiotic stresses (Albrecht et al., 2003; Cheong et al., 2003). However, the *cbl1* mutant did not show any obvious ABA-related phenotypes in these reports, whereas the loss of function of CBL9, a closely related homolog of CBL1, renders plants hypersensitive to ABA during seed germination and seedling growth, suggesting that CBL9 plays a negative role in ABA signaling (Pandey et al., 2004).

5.5.3.3 Calcium-dependent protein kinases (CDPKs)

CDPKs contain a kinase catalytic domain and an autoinhibitory junction domain, with a calmodulin-like calcium-binding regulatory domain at their C-termini (Hrabak et al., 2003). Therefore CDPKs combine both kinase and calcium sensor domains in one protein and can be directly activated by calcium (Weinl and Kudla, 2009). The *A. thaliana* genome contains 34 genes predicted to encode CDPKs (Cheng et al., 2002a) and CDPK genes and proteins are indicated by CPK, followed by a number.

The first evidence for CDPK involvement in ABA signaling was provided by a report that the constitutively active forms of CPK10 (CDPK1) and CPK30 (CDPK1a) could activate an ABA-inducible promoter in maize leaf protoplasts (Sheen, 1996). Later work showed that CPK32 interacts with ABA-responsive element-binding factor 4 (ABF4) and can phosphorylate ABF4 at Serine 110, which is highly conserved among ABF family members (Choi et al., 2005). Overexpression of CPK32 resulted in ABA-hypersensitive inhibition of seed germination, indicating that CPK32 positively regulates ABA signaling, possibly by modulating ABF4 activity (Choi et al., 2005). Analyses of the mutants impaired in CPK3 and CPK6 functions showed that these two CDPKs play positive roles in ABA-mediated regulation of stomatal aperture; however, the mutants did not show ABA-related phenotypes in seed germination and seedling growth (Mori et al., 2006). By contrast, the *cpk4* and *cpk11* mutants exhibited pleiotropic ABA-insensitive phenotypes in seed germination, seedling growth, and stomatal movement, and the *cpk4 cpk11* double mutants showed stronger phenotypes. In addition, CPK4 and CPK11 can phosphorylate ABF1 and ABF4 in vitro, suggesting that these two kinases may regulate ABA signaling through these transcription factors. Therefore CPK4 and CPK11 play positive roles in the CDPK/calcium-mediated ABA signaling pathway (Zhu et al., 2007).

In contrast to the positive roles of CDPKs identified in most reports, one study using *CPK12-RNAi* lines suggested that CPK12 may function as a negative regulator of ABA signaling (Zhao et al., 2011).

5.5.3.4 Mitogen-activated protein kinases

Evidence obtained in the last decade strongly supports the involvement of mitogen-activated protein kinases (MAPKs) in ABA signaling. An MAPK module typically includes combinations of at least three protein kinases: an MAP3K (MAPKKK), an MAP2K (MAPKK, MKK, or MEK), and an MAPK (MPK) (reviewed in de Zelicourt et al., 2016). The *A. thaliana* genome encodes at least 80 MAP3Ks, 10 MAP2Ks, and 20 MAPKs. The ABA-mediated activation of MAPKs has been reported in several plant species, including barley, maize, pea, and rice (reviewed in Danquah et al., 2014). In Arabidopsis seedlings, H_2O_2 and ABA can activate MPK3 and overexpression of *MPK3* increases ABA sensitivity in postgermination growth (Lu et al., 2002). ABA can also increase MPK1 and MPK2 activity (Ortiz-Masia et al., 2007), and the ABA-dependent MKK1-mediated activation of MPK6 regulates *CATALASE1* (*CAT1*) expression and H_2O_2 production (Xing et al., 2008). MPK9 and MPK12, which are preferentially expressed in guard cells, positively regulate ROS-mediated ABA signaling in guard cells (Jammes et al., 2009). Phosphor-proteomic analyses revealed a significant fraction of ABA-induced phosphorylation sites as potential MAPK targets (Umezawa et al., 2013; Wang et al., 2013). Moreover, an entire ABA-activated MAPK module, composed of MAP3K17/18-MKK3-MPK1/2/7/14, was identified and the expression of *MAP3K17* and *MAP3K18* is dramatically induced by ABA (Danquah et al., 2015).

5.5.3.5 Receptor-like kinases

Other types of protein kinases, such as the receptor-like kinases, have also been implicated in ABA signaling. The *RECEPTOR-LIKE KINASE1* (*RPK1*) gene, encoding a leucine-rich repeat (LRR) receptor kinase in the plasma membrane, is upregulated by ABA. The loss-of-function mutants of *RPK1* displayed decreased sensitivity to ABA during germination, seedling growth, and stomatal closure, indicating that RPK1 is a positive regulator of ABA signaling (Osakabe et al., 2005). A mutant deficient in GUARD CELL HYDROGEN PEROXIDE-RESISTANT1 (GHR1) was isolated by its fast water-loss phenotype (Hua et al., 2012). The *ghr1* mutant was impaired in ABA and H_2O_2 induction of stomatal closure, but did not exhibit any apparent differences in seed dormancy or in ABA inhibition of seed germination and seedling growth (Hua et al., 2012). Further analyses indicated that GHR1 physically interacts with and phosphorylates SLOW ANION CHANNEL-ASSOCIATED1 (SLAC1), a key anion channel in stomatal signaling, and this activation was inhibited by the ABA coreceptor ABI2, but not ABI1 (Hua et al., 2012) (Fig. 5.2). The homologs of GHR1 are conserved in monocots and dicots, indicating that a wide variety of crop plants might have a similar molecular mechanism for regulating ABA- and H_2O_2-mediated guard cell movement under drought stress (Hua et al., 2012).

5.5.4   Transcription factors

5.5.4.1   ABI transcription factors

The *abi3* mutation was first reported by Koornneef et al. (1984) along with two dominant mutants insensitive to ABA, *abi1-1* and *abi2-1* (see previous discussion). The *ABI3* gene was cloned (Giraudat et al., 1992) and the encoded protein displayed high sequence similarity to maize VP1 (McCarty et al., 1991), both of which belong to the B3-domain transcription factor family. The *abi3* loss-of-function mutants had severe defects in seed development, exhibited a viviparous phenotype under humid conditions, and produced mature seeds that remained green after desiccation, suggesting that the maturation program is defective and that seed germination occurs prematurely (Nambara et al., 1992, 1995). Of the 87 B3-domain family transcription factors encoded in the *A. thaliana* genome, ABI3 groups in a distinct clade with two other members, FUS3 and LEC2 (Suzuki and McCarty, 2008; Romanel et al., 2009). Loss of function of either *FUS3* or *LEC2* blocked seed maturation, similar to what was observed in the *abi3* mutants. In addition, the *lec2* and *fus3* mutations caused a partial transformation of embryo cotyledons to true leaves (Meinke et al., 1994). However, unlike *abi3*, the *fus3* and *lec2* single mutants were still sensitive to ABA, but the *fus3* mutation combined with the *abi1-1*, *abi2-1*, or *abi3* mutations exhibited increased vivipary and decreased desiccation tolerance (Keith et al., 1994). In addition, FUS3 negatively regulates active GA accumulation through repression of *GA3ox2* and positively regulates ABA accumulation (Curaba et al., 2004; Gazzarrini et al., 2004). Moreover, ABA promotes the stability of ABI3 and FUS3 (Gazzarrini et al., 2004; Zhang et al., 2005).

The *abi4* and *abi5* mutations were isolated in a genetic screen for ABA-insensitive mutants (Finkelstein, 1994). The *ABI4* gene encodes a transcription factor of the APETELLA2/ETHYLENE RESPONSE FACTOR (AP2/ERF) family (Finkelstein et al., 1998) and is a unique, nonredundant gene in Arabidopsis. Searches for *ABI4* homologs in sequenced plant genomes revealed that most plants possess a single gene encoding ABI4 (Wind et al., 2013). The ABI4 protein functions as a positive regulator in the ABA signaling pathway, especially during seed development and germination (Finkelstein, 1994; Finkelstein et al., 1998; Soderman et al., 2000). In addition, ABI4 is a versatile factor that functions in diverse signaling pathways, including lipid biosynthesis and breakdown (Penfield et al., 2006; Yang et al., 2011b), sugar signaling (Arenas-Huertero et al., 2000; Laby et al., 2000), salt responses (Quesada et al., 2000), the mitochondrial and chloroplast-nucleus retrograde signaling pathways (Koussevitzky et al., 2007; Giraud et al., 2009; Sun et al., 2011), and light signaling (Xu et al., 2016). Reports showed that ABI4 also mediates the cross talk of different phytohormone pathways such as ABA and cytokinin inhibition of lateral root development (Shkolnik-Inbar and Bar-Zvi, 2010), ABA- and jasmonate-mediated plant defense responses (Kerchev et al., 2011), and ABA and GA control of seed dormancy (Shu et al., 2013, 2016).

The *ABI5* gene encodes a transcription factor of the bZIP family (Finkelstein and Lynch, 2000) that is grouped in a clade with ABFs/AREBs/DPBFs (see later discussion).

The ABI5 protein can integrate signals of several pathways, such as ABA (Finkelstein, 1994; Finkelstein and Lynch, 2000; Lopez-Molina et al., 2001), sugar (Arenas-Huertero et al., 2000; Laby et al., 2000; Brocard et al., 2002; Arroyo et al., 2003), and light (Chen et al., 2008; Tang et al., 2013; Xu et al., 2014). Multiple transcription factors such as HY5 (Chen et al., 2008), FHY3 and FAR1 (Tang et al., 2013), ABI4 (Bossi et al., 2009), RAV1 (Feng et al., 2014), WRKY18, WRKY40, and WRKY60 (Shang et al., 2010; Liu et al., 2012), and ABI5 itself (Xu et al., 2014) can directly bind to the *ABI5* promoter and regulate its expression. In addition, *ABI5* also appeared as a target of PIF1 in a chromatin immunoprecipitation-microarray study (Oh et al., 2009). Therefore multiple signals converge on the *ABI5* promoter and fine-tune its expression by coordinated actions of the transcription factors of the respective pathways. The stability of ABI5 is enhanced by ABA, whereas AFP (ABI FIVE BINDING PROTEIN) negatively regulates ABI5 stability (Lopez-Molina et al., 2001, 2003).

The *ABI3*, *ABI4*, and *ABI5* genes are tightly coexpressed while modulating ABA-regulated processes (Wind et al., 2013) and physiological studies indicated that *ABI3*, *ABI4*, and *ABI5* have similar qualitative effects on seed development and ABA sensitivity, but the null mutations of *ABI3* have more severe phenotypes than those of *ABI4* or *ABI5* (Finkelstein, 2013). The relationships among ABI3, ABI4, and ABI5 have been explored extensively, especially during seed germination, but their relationships are complicated. ABI5 interacts with both ABI3 (Nakamura et al., 2001; Lim et al., 2013) and ABI4 (Finkelstein, 2013); however, ABI5 seems to act downstream of ABI3 and ABI4 (Lopez-Molina et al., 2002; Bossi et al., 2009). Biochemical approaches revealed the important *cis*-elements to be the RY motif (CATGCA) for ABI3 (Suzuki et al., 1997; Ezcurra et al., 2000; Monke et al., 2004), Coupling Element1 (CE1)-like sequence (CACCG) or CCAC motif for ABI4 (Niu et al., 2002; Koussevitzky et al., 2007), and ABA-responsive element (ABRE) (ACGT) for ABI5 (Carles et al., 2002). The genes producing the most abundant transcripts in mature seeds have potential binding sites for at least two ABI transcription factors, and ABREs are overrepresented in all highly expressed genes at this stage (Nakabayashi et al., 2005). Consistent with this observation, transcriptome comparisons also revealed that ABI3, ABI4, and ABI5 regulate overlapping sets of genes (Suzuki et al., 2003; Nakabayashi et al., 2005; Reeves et al., 2011). Specifically, ABI3 and ABI5 interact to activate the expression of *SOMNUS* to inhibit seed germination at high temperatures (Lim et al., 2013).

5.5.4.2   ABRE-binding factor/ABA-responsive element binding protein/Dc3 promoter-binding factors

The bZIP transcription factor family has a total of 13 bZIP proteins in the ABI5 clade, members of which share at least three conserved motifs in addition to the bZIP domain (Jakoby et al., 2002). However, only ABI5 was identified by forward genetic screens whereas five of them, known as ABRE-binding factors (ABFs or AREBs), were isolated by using ABREs as bait in yeast one-hybrid screening (Choi et al., 2000; Uno et al., 2000). In addition, five Arabidopsis bZIPs in this clade were also named Dc3 promoter-binding factors (DPBFs) (Kim et al., 2002), but DPBF1 is identical to ABI5, and DPBF3 and DPBF5 are identical to other known ABF/AREB proteins.

Most of the related bZIPs also mediate ABA- or stress-regulated gene expression. Overexpression using the 35S promoter of *AREB1 (ABF2)*, *AREB2 (ABF4)*, or *ABF3* resulted in ABA hypersensitivity and other ABA-associated phenotypes and the transgenic plants exhibited enhanced drought tolerance (Kang et al., 2002; Kim et al., 2004; Fujita et al., 2005; Oh et al., 2005). The *areb1 areb2 abf3* triple mutant and *areb1 areb2 abf3 abf1* quadruple mutant displayed impaired expression of ABA- and osmotic stress-responsive genes, resulting in increased sensitivity to drought and decreased sensitivity to ABA (Yoshida et al., 2010, 2015). Thus these AREBs/ABFs are likely the master transcription factors in ABA-regulated drought tolerance.

ABA-dependent phosphorylation of AREB/ABF transcription factors also functions as a crucial mechanism in transducing ABA signals. Multiple conserved RXXS/T sites in AREBs/ABFs are phosphorylated by SnRK2 protein kinases in an ABA-dependent manner, as was shown by in-gel kinase assays (Uno et al., 2000; Furihata et al., 2006; Fujii et al., 2007) (Fig. 5.2). Substitution of the Ser or Thr residues to Asp in conserved RXXS/T sites resulted in high transactivation activity, and transgenic plants overexpressing the phosphorylated active form of AREB1 (ABF2) expressed many ABA-inducible genes without ABA treatment, indicating that ABA-activated phosphorylation of AREBs/ABFs by SnRK2 enzymes regulates the transactivation activity of these transcription factors (Furihata et al., 2006).

5.5.4.3   C-repeat-binding factor/dehydration-responsive element-binding proteins

The C-repeat-binding factor/dehydration-responsive element-binding protein 1 (CBF/ DREB1) proteins belong to a small group of the AP2/ERF transcription factor family (reviewed in Qin et al., 2011). Another subgroup of the AP2/ERF transcription factor family includes the DREB2 proteins, which were isolated together with DREB1 in a yeast one-hybrid screen (Liu et al., 1998). These CBF/DREB proteins recognize the dehydration-responsive element (DRE)/C-repeat (CRT) *cis*-acting element (core sequence: CCGAC) located in the promoters of many low temperature- and drought-inducible genes (Stockinger et al., 1997; Liu et al., 1998). Most of these *CBF/ DREB* genes are transcriptionally induced by stress, and then transactivate the expression of downstream genes that confer stress tolerance to plants (Qin et al., 2011). However, currently little is known regarding how these CBF/DREB proteins are regulated by ABA signaling. The *CBF* genes are induced by ABA and DREB proteins interact with ABF proteins to coordinately regulate the expression of ABA-responsive genes (Haake et al., 2002; Knight et al., 2004; Lee et al., 2010b). Therefore, although CBFs/DREBs were originally thought to mediate ABA-independent gene expression (Yamaguchi-Shinozaki and Shinozaki, 2006), it is evident that they are also involved in regulating ABA-responsive gene expression.

5.5.4.4   Other types of transcription factors

Additional transcription factors involved in ABA- or stress-induced gene expression have been identified, such as MYB2 and MYC2 (Abe et al., 1997, 2003), several

WRKY family proteins (Rushton et al., 2012; Huang et al., 2016), NAC family factors RD26 and VNI2 (Fujita et al., 2004; Yang et al., 2011a), and B3 domain factor RAV1 (Feng et al., 2014). It is widely accepted that the transcription factors, including those mentioned previously and those not mentioned or not yet identified, constitute a transcriptional network that rapidly changes the expression of a large number of ABA-responsive genes.

5.5.5 Ion channels

As described in Section 5.4.4, ABA promotes stomatal closure and prevents stomatal opening by inducing depolarization of the plasma membrane through (1) inhibiting the plasma membrane H^+-ATPase, (2) inhibiting the inward K^+ channels, and (3) activating the slow (S-type) anion channels such as SLOW ANION CHANNEL-ASSOCIATED 1 (SLAC1). Two dominant alleles of Arabidopsis *OPEN STOMATA 2 (OST2)*, which cause the plant to completely lose the stomatal response to ABA, were identified (Merlot et al., 2007). The *OST2* gene encodes the major plasma membrane H^+-ATPase, ARABIDOPSIS H^+-ATPASE 1 (AHA1), and both mutant alleles produce constitutively activated H^+-ATPases, persistent stomatal opening, and thus, ABA insensitivity (Merlot et al., 2007). The inward-rectifying K^+ channels KAT1 (K^+ CHANNEL IN ARABIDOPSIS THALIANA) and KAT2 were shown to facilitate K^+ influx in guard cells (Kwak et al., 2001; Lebaudy et al., 2008), and KAT1 was inhibited by OST1 or its homolog AAPK in *V. faba* (Mori et al., 2000; Sato et al., 2009), or by CPK13 (Ronzier et al., 2014). The SLAC1 protein was identified from independent genetic screens for ozone-sensitive mutants (Vahisalu et al., 2008) and CO_2-insensitive stomatal closure mutants (Negi et al., 2008). The *SLAC1* gene belongs to a small family of five genes in Arabidopsis that encode proteins with 10 predicted transmembrane domains with similarity to bacterial and fungal dicarboxylate/malate transporters (Negi et al., 2008; Vahisalu et al., 2008). ABA activation of S-type anion channels is impaired in the guard cells of *slac1* mutants, thus providing genetic evidence that *SLAC1* encodes a major anion-transporting component of S-type anion channels in guard cells (Vahisalu et al., 2008). Numerous kinases including GHR1, OST1, and several CDPKs, such as CPK3, CPK5, CPK6, CPK11, CPK21, and CPK23, can phosphorylate the extended cytosolic N-terminal region of SLAC1 and activate its channel activity (Mori et al., 2006; Geiger et al., 2009, 2010; Lee et al., 2009; Vahisalu et al., 2010; Kollist et al., 2014; Brandt et al., 2015). SLAC1 and its Arabidopsis homolog SLAH3 interact with KAT1 and directly inhibit KAT1 activity (Zhang et al., 2016). The ALMT12 protein also known as QUICKLY ACTIVATING ANION CHANNEL1 (QUAC1) is an R-type anion channel (Meyer et al., 2010) that interacts with and is activated by OST1, likely through protein phosphorylation (Imes et al., 2013) (Fig. 5.2).

5.5.6 The role of ROS in ABA signaling

In Arabidopsis, NADPH oxidases, named respiratory burst oxidase homologues (RBOHs), are encoded by a small family of 10 genes and are crucial components in ABA signaling (Kwak et al., 2003). The plasma membrane–localized RBOHs produce extracellular superoxide that can be easily converted into H_2O_2, which can be transported into the cytoplasm by aquaporin PIP2;1 to trigger stomatal closure (Grondin et al., 2015). The OST1 kinase is able to phosphorylate PIP2;1 at Ser121 and enhance its water transport activity. Apoplastic H_2O_2 signaling is also likely transduced into cells through GHR1 (Hua et al., 2012). OST1 and CIPK26 can phosphorylate and regulate AtrbohF NADPH oxidase (Sirichandra et al., 2009; Drerup et al., 2013). AtrbohD and AtrbohF can be activated by the phospholipase Dα1 product phosphatidic acid in ABA signaling (Zhang et al., 2009).

In addition to RBOHs, ABA also promotes the production of ROS in mitochondria and chloroplasts. The *ABA OVERLY-SENSITIVE 5* (*ABO5*) and *ABO8* genes encode pentatricopeptide repeat (PPR) proteins that are required for *cis*-splicing of mitochondrial *nad2* intron 3 and *nad4* intron 3, respectively, in the formation of complex I (an NADH oxidase) of the electron transport chain (Liu et al., 2010; Yang et al., 2014). The *ABO6* gene encodes a DExH-box RNA helicase that regulates the splicing of several genes of complex I (He et al., 2012). Mutations in these three genes increase the ABA-promoted production of ROS, suggesting that ABA signaling mediates the production of ROS in mitochondria (Miller et al., 2010). In chloroplasts, SAL1/FRY1 is an oxidative stress sensor. The activity of SAL1 phosphoadenosine phosphatase can be redox regulated by intramolecular disulfide bond formation and dimerization, which causes the accumulation of the phosphonucleotide 3′-phosphoadenosine 5′-phosphate (PAP) in chloroplasts. PAP is transferred from chloroplasts to nuclei to inhibit 5′ to 3′ exoribonucleases (XRNs) that are involved in mediating mRNA metabolism, resulting in changes in gene expression (Estavillo et al., 2011; Chan et al., 2016).

Sensing of ROS is also achieved by glutathione peroxidase 3 (GPX3), which interacts with ABI1 and ABI2, thus modulating PP2C activities (Miao et al., 2006). H_2O_2 inhibits the activities of ABI1, ABI2, and HAB1, which releases PP2C inhibition of the downstream target kinases (Meinhard and Grill, 2001; Meinhard et al., 2002; Sridharamurthy et al., 2014). ABA also induces the production of nitric oxide (NO) in guard cells, which S-nitrosylates cysteine 137 of OST1 and impairs OST1 activity (Wang et al., 2015). PYR1 can be nitrated at the tyrosine residue by NO. Tyrosine nitration reduces the ABA-induced activity of PYR1 and promotes proteasomal-mediated degradation of PYR1 (Castillo et al., 2015).

5.5.7 ABA signaling pathway and evolution of the core components

Rapid progress over the last decade has led to the construction of a simplified model of ABA signaling in which a central ABA signaling module is made up of three protein classes, ABA receptors, coreceptors, and SnRK2 kinases, and is responsible for the earliest events of ABA signaling (Fig. 5.2). In the absence of ABA, the PP2Cs are active and repress SnRK2 activity and modify downstream events. In the presence of ABA, ABA-bound PYR/PYL/RCAR proteins interact with PP2C enzymes and inhibit their activity, thus allowing activation of SnRK2 and phosphorylation of target proteins. Several SnRK2 targets have been identified to date, both at the plasma membrane and in the nucleus, such as the transcription factors ABI5 and ABFs/AREBs, the ion channels SLAC1 and KAT1, and the ROS-generating enzyme RESPIRATORY BURST OXIDASE HOMOLOG F (RbohF) (Sirichandra et al., 2009). Phosphorylation of these target proteins by SnRK2s transduces ABA signals and allows for ABA control of gene expression, ion channels, and production of secondary messengers such as ROS (Fig. 5.2). In support of this model, an in vitro reconstitution study demonstrated that the PYR/PYL/RCARs, clade-A PP2Cs, SnRK2s, and ABFs are the core components to complete ABA regulation of gene expression (Fujii et al., 2009).

With the sequencing and genome analyses of the green alga *Chlamydomonas reinhardtii*, the moss (bryophyte) *Physcomitrella patens*, and the fern (lycophyte) *Selaginella moellendorffii*, the core components of the ABA signaling pathway have been found in all classifications of land plants, and the development of an ABA signaling system appears to be highly correlated with the evolution from aquatic to terrestrial plants (Umezawa et al., 2010). Therefore it is believed that the achievement of drought tolerance is one of the critical steps in the evolution of land plants, in which establishment of the core ABA signaling pathway is critical (Umezawa et al., 2010).

5.6   ABA control of nuclear gene expression

The expression of numerous genes is regulated by ABA during seed maturation and acclimation to various stresses such as drought, high salinity, and low temperatures. Transcriptome profiling analyses revealed genome-wide gene expression changes regulated by ABA. About 10% of the genes display ABA-regulated expression in Arabidopsis seedlings, with approximately equal numbers of ABA-induced and ABA-repressed genes (Nemhauser et al., 2006). ABA regulates two to six times more genes, compared with most of the other plant hormones (Nemhauser et al., 2006). It should be noted that the set of ABA-regulated genes depends on cell type and developmental stage, but it is generally believed that the ABA-regulated genes contribute to the tolerance of dehydration conditions, which can occur as part of development as in seed or pollen maturation, or in response to environmental stresses such as drought, high salinity, or low temperatures (Cutler et al., 2010).

Comparisons of the transcriptomes of Arabidopsis and rice exposed to ABA, drought, high salinity, and other abiotic stresses have revealed that 5%–10% of the transcriptome is changed by these treatments. More than half of the drought-inducible genes are also induced by ABA and high salinity, whereas only approximately 10% of the drought-inducible genes are also induced by low temperatures (Shinozaki et al., 2003; Nakashima et al., 2009b). Therefore it is evident that there is significant cross talk among the ABA-mediated responses to drought and high salinity. However, many genes that are induced by drought and low temperature stresses do not respond to exogenous application of ABA in Arabidopsis, suggesting the existence of ABA-dependent and ABA-independent signal transduction pathways in response to drought and cold stresses (Yamaguchi-Shinozaki and Shinozaki, 2006).

The genome of the moss *P. patens* encodes all of the components of the core ABA signaling pathway identified in angiosperms (Rensing et al., 2008) and ABA regulates similar classes of genes in Arabidopsis and *P. patens* (Cuming et al., 2007), suggesting that the regulation of ABA signaling may be evolutionarily conserved (Komatsu et al., 2009; Khandelwal et al., 2010). In addition, transcriptome analyses using Arabidopsis genome tiling arrays showed that approximately 8000 transcriptionally active regions (TARs) are present in the "intergenic" regions, and 5%–10% of these TARs are regulated by ABA (Matsui et al., 2008; Zeller et al., 2009). The biological roles of these ABA-inducible "intergenic" TARs in mediating ABA signaling remain largely unknown.

5.7   Ubiquitin–proteasome system in ABA signaling

Tight control of the turnover of key regulatory components in the ABA signaling path-way is critical for plant growth and development, as well as for survival of stress conditions (reviewed in Yu et al., 2016). Ubiquitin-26S proteasome system (UPS)-mediated proteolysis is a prominent mechanism for removing proteins from the cell. The covalent attachment of ubiquitin to its target proteins is accomplished through sequential reactions catalyzed by the ubiquitin-activating enzyme (E1), the ubiquitin conjugation enzyme (E2), and ubiquitin ligase (E3). Ubiquitin is activated by E1 in an ATP-dependent manner and E2 accepts the activated ubiquitin on a cysteine and forms a thioester bond with ubiquitin. The E3 ligase mediates the transfer of ubiquitin from E2 to the substrate by forming an isopeptide bond between the glycine of ubiquitin and the lysine of the target protein (Vierstra, 2009). Therefore E3 ligases are respon-sible for the specificity of ubiquitination by recruiting the appropriate target proteins.

The Arabidopsis E3 ubiquitin ligase SENESCENCE-ASSOCIATED E3 UBIQUITIN LIGASE1/PLANT U-BOX E3 LIGASE 44 (SAUL1/AtPUB44) directly targets ABSCISIC ALDEHYDE OXIDASE 3 (AAO3), an enzyme that converts abscisic aldehyde to ABA, for ubiquitin-dependent degradation via the 26S protea-some (Raab et al., 2009). The *saul1* mutants accumulated more AAO3 protein and ABA, and thus exhibited premature senescence. Thus SAUL1 prevents premature senescence by targeting AAO3 for degradation (Raab et al., 2009).

Accumulating evidence shows that the PYR/PYL/RCARs receptors are regulated by 26S proteasome-mediated degradation. RING FINGER OF SEED LONGEVITY1 (RSL1) is a single subunit RING-type E3 ligase that was shown to mediate ubiquitina-tion of PYL4 and PYR1 in vitro and promote the degradation of these receptors in vivo (Bueso et al., 2014). Overexpression using the 35S promoter of *RSL1* reduces ABA sensitivity and *rsl1* RNAi transgenic lines that silenced at least three members of the *RSL1*-like gene family show enhanced sensitivity to ABA. The RSL1 protein interacts with two ABA receptors, PYR1 and PYL4, on the plasma membrane. Furthermore, following brefeldin A treatment, PYL4 localized to the microsomal fraction when coexpressed with RSL1. These results suggest that RSL1 acts as a negative regula-tor in the ABA signaling pathway by mediating the ubiquitination of ABA recep-tors and modulating their half-life, protein interactions, or trafficking (Bueso et al., 2014). DET1-, DDB1-ASSOCIATED1 (DDA1), part of the COP10-DET1-DDB1 (CDD) complex, binds to PYL8, PYL4, and PYL9 and promotes their proteasomal degradation. Accordingly, DDA1 negatively regulates ABA-mediated developmental responses such as inhibition of seed germination, seedling establishment, and root growth (Irigoyen et al., 2014). The F-box E3 ligase RCAR3 INTERACTING F-BOX PROTEIN 1 (RIFP1) interacts with the ABA receptor PYL8/RCAR3 in the nucleus and facilitates its proteasomal degradation. The *rifp1* mutant plants showed increased ABA-mediated inhibition of seed germination, whereas the *RIFP1* overexpressing plants displayed the opposite phenotypes (Li et al., 2016). Together, these different types of E3 ligases act as negative regulators of ABA signaling by targeting different members of the PYR/PYL/RCARs for 26S proteasome-mediated degradation.

In terms of ABA coreceptors, work showed that the U-box E3 ligases PUB12 and PUB13 interact with ABI1, but the ubiquitination occurs only when ABI1 is interacting with the ABA receptors (Kong et al., 2015). The *pub12 pub13* double mutant, which accumulates more ABI protein than the wild type, exhibits reduced sensitivity to ABA. Introduction of the *abi1-3* loss-of-function mutation into the *pub12 pub13* double mutant recovers its ABA-insensitive phenotypes. Interestingly, PUB12 and PUB13 only interact with ABI1 but not with other clade A PP2Cs (Kong et al., 2015). Thus PUB12 and PUB13 play important roles in ABA signaling by modulating the key negative regulator ABI1.

Several transcription factors, both positive and negative players in ABA signaling, have been shown to be regulated by UPS-mediated proteolysis in the nucleus. For example, the ABI5 transcription factor is targeted by several E3 ubiquitin ligases including KEG (KEEP ON GOING, a RING-type E3 ligase) and three substrate receptors for CULLIN4-based E3 ligases including DWA1 (DWD HYPERSENSITIVE TO ABA1) and DWA2, and ABD1 (ABA-HYPERSENSITIVE DCAF1). The ABI5 protein accumulates in seedlings treated with 26S proteasome inhibitors and in mutant plants lacking RPN10 (a subunit of the 26S proteasome), suggesting that ABI5 turnover is dependent on the 26S proteasome pathway (Lopez-Molina et al., 2001; Smalle et al., 2003). KEG directly interacts with and targets ABI5 for ubiquitination in vitro, indicating that ABI5 is a substrate of KEG. The *keg* mutants accumulate more ABI5 and are hypersensitive to ABA compared with the wild type, and loss of *ABI5* partially rescues the ABA-hypersensitive phenotype of the *keg* mutants (Stone et al., 2006; Liu and Stone, 2010). Moreover, ABA promotes ABI5 accumulation by inducing the ubiquitination and proteasomal degradation of KEG (Liu and Stone, 2010). Interestingly, it was suggested that KEG mainly targets ABI5 for degradation in the cytoplasm in the absence of ABA, whereas in response to ABA, ABI5 accumulates in the nucleus (Liu and Stone, 2013) and its stability may be controlled by DWA1, DWA2, and ABD1. The DWA1, DWA2, and ABD1 proteins function as substrate receptors in the CUL4-DDB1 E3 ligases and they all interact with ABI5 and regulate ABI5 protein stability (Lee et al., 2010a; Seo et al., 2014). Therefore DWA1, DWA2, and ABD1 act as negative regulators of ABA signaling, and their mutants are hypersensitive to ABA (Lee et al., 2010a; Seo et al., 2014).

The stability of ABI3 is directly regulated through polyubiquitination by ABI3-INTERACTING PROTEIN2 (AIP2), a RING-type E3 ligase (Zhang et al., 2005). The *aip2* mutant shows higher ABI3 protein levels and is hypersensitive to ABA, similar to the phenotype of *ABI3*-overexpressing plants, whereas the *AIP2*-overexpressing plants exhibit the opposite phenotypes. Therefore AIP2 plays a negative role in ABA signaling by modulating the stability of ABI3 (Zhang et al., 2005).

The class I homeobox-leucine zipper (HD-ZIP) transcription factor ATHB6 plays a negative role in the ABA signaling pathway (Himmelbach et al., 2002). The ATHB6 protein interacts with Arabidopsis MATH (Meprin and TRAF homology)-BTB (bric-a-brac/tramtrack/broad) proteins (BPMs), which act as substrate adaptors in CUL3-based E3 ligase complexes (Lechner et al., 2011). Genetic and biochemical data showed that BPMs directly interact with and target ATHB6 for proteasomal degradation. Consistent with this model, the ATHB6 level increased and the ubiquitinated form of ATHB6 decreased in an artificial microRNA *amiR-bpm* line, whereas a faster turnover of ATHB6 was observed in transgenic plants overexpressing *BPM6* (Lechner et al., 2011). The *amiR-bpm* line and *ATHB6* overexpression lines showed larger stomatal apertures than the wild type in darkness or after ABA treatment (Lechner et al., 2011). Together, CUL3-based E3 ligase complexes, utilizing BPMs as substrate adaptors, target ATHB6 for proteasomal degradation and thus play an important role in ABA control of stomatal aperture.

5.8   Summary points

- In plants, ABA biosynthesis begins in the plastids with the conversion of C5 isopentenyl diphosphate (IPP) to xanthoxin (a C15 intermediate), which moves to the cytoplasm where it is converted into ABA via oxidative steps involving the intermediate abscisic aldehyde.
- ABA catabolism involves hydroxylation and conjugation reactions and functions as a major mechanism for regulating ABA levels in vivo.
- During drought stress, ABA accumulates first in shoot vascular tissues and only later appears in roots and guard cells. The identification of AtABCG25 as an ABA efflux transporter and AtABCG40 and AIT1/NRT1.2 as ABA influx transporters emphasizes the importance of ABA transport for ABA signaling throughout the plant.
- In response to environmental stresses, such as drought, low temperature, or high salinity, ABA levels increase dramatically and play important roles in the adaptation to these abiotic stresses.
- The molecular identification of PYR/PYL/RCAR proteins as ABA receptors led to the current model of the ABA signaling pathway including a central ABA signaling module made up of ABA receptors, coreceptors, and SnRK2s, acting upon transcription factors and downstream regulators such as ion channels and ROS-generating enzymes.
- Several classes of transcription factors are involved in ABA signaling. ABA regulates about 10% of the genes in Arabidopsis seedlings through the action of these transcription factors.
- Several types of E3 ubiquitin ligases have been identified in Arabidopsis that are involved in ABA-regulated proteolysis of ABA biosynthetic enzymes and of the key components of ABA signaling, including ABA receptors, coreceptors, and transcription factors.

5.9   Future perspectives

- To elucidate how ABA biosynthesis is triggered during plant responses to drought stress.
- To investigate whether plants have additional ABA receptors and to determine the other functions of the PYR/PYL/RCAR ABA receptors.
- To understand how ABA signaling factors interact with other hormone signaling pathways in plant growth and development, and in response to pathogens.
- To identify which metabolites are crucial for drought tolerance in ABA signaling.
- To dissect how ROS and Ca^{2+} signaling are transduced in ABA signaling.

- To study why different plants have different levels of drought tolerance even though the key components of the ABA signaling pathway are highly conserved among higher plants.
- To improve water use efficiency and resistance to drought stress in crops based on the understanding of ABA signaling.

Abbreviations

AAO	Abscisic aldehyde oxidase
ABA	Abscisic acid
ABA-GE	Glucosyl ester
ABCG25	ATP-BINDING CASSETTE G25
ABD	ABA-HYPERSENSITIVE DCAF
ABF	ABRE-binding factor
ABI	ABA INSENSITIVE
ABO	ABA OVERLY-SENSITIVE
ABRE	ABA-responsive element
AHA	ARABIDOPSIS H$^+$-ATPASE
AIP	ABI3-INTERACTING PROTEIN
AIT1	ABA-IMPORTING TRANSPORTER 1
BCH	β-Carotene hydroxylase
BR	Brassinosteroid
CAT	Catalase
CBF	C-repeat-binding factor
CBL	Calcineurin B-like protein
CDD	COP10-DET1-DDB1
CDPK	Calcium-dependent protein kinase
CE1	Coupling element 1
CIPK	CBL-interacting protein kinase
CRT	C-repeat
CYP450	Cytochrome P450
DDA	DET1-DDB1-ASSOCIATED
DMAPP	Dimethylallyl diphosphate
DPBF	Dc3 promoter-binding factor
DRE	Dehydration-responsive element
DREB	Dehydration-responsive element-binding protein
DWA	DWD HYPERSENSITIVE TO ABA
E1	Ubiquitin-activating enzyme
E2	Ubiquitin conjugation enzyme
E3	Ubiquitin ligase
FR	Far-red
FUS	FUSCA
GA	Gibberellin
GGPP	Geranylgeranyl diphosphate
GHR	GUARD CELL HYDROGEN PEROXIDE-RESISTANT
GPX	Glutathione peroxidase
HD-ZIP	Homeobox-leucine zipper

IPP	Isopentenyl diphosphate
KAT	K+ CHANNEL IN ARABIDOPSIS THALIANA
KEG	KEEP ON GOING
LEA	LATE EMBRYOGENESIS ABUNDANT
LEC	LEAFY COTYLEDON
LRR	Leucine-rich repeat
MAPK	Mitogen-activated protein kinase
MoCo	Molybdenum cofactor
NCED	9-*cis*-Epoxycarotenoid dioxygenase
NF-Y	Nuclear factor Y
NO	Nitric oxide
NRT	Nitrate transporter
OST	OPEN STOMATA
PA	Phaseic acid
Pa	Phosphatidic acid
PAP	5′-phosphate
PIF	PHYTOCHROME-INTERACTING FACTOR
PIP	PLASMA MEMBRANE INTRINSIC PROTEIN
PUB	PLANT U-BOX E3 LIGASE
pyr	*pyrabactin resistance*
QUAC	QUICKLY ACTIVATING ANION CHANNEL
R	Red
RBOH	RESPIRATORY BURST OXIDASE HOMOLOG
RIFP	RCAR3-INTERACTING F-BOX PROTEIN
ROS	Reactive oxygen species
RPK	RECEPTOR-LIKE PROTEIN KINASE
RSL	RING FINGER OF SEED LONGEVITY
SAUL	SENESCENCE-ASSOCIATED E3 UBIQUITIN LIGASE
SDR	Short-chain dehydrogenase/reductase
SLAC	SLOW ANION CHANNEL-ASSOCIATED
SNF	Sucrose nonfermenting
SnRK	SNF1-related protein kinase
TAR	Transcriptionally active region
TIP	TONOPLAST INTRINSIC PROTEIN
UPS	Ubiquitin-26S proteasome system
VDE	Violaxanthin de-epoxidase
vp	*viviparous*
XRNs	5′ to 3′ exoribonucleases
ZEP	Zeaxanthin epoxidase

Acknowledgments

We thank Dr. Yiting Shi and Ms. Jingjing Meng for their help in preparing the figures. This work was supported by grants from the National Natural Science Foundation of China (91317301, 31121002 to Z.G., and 31371221 to J.L.), and the Recruitment Program of Global Youth Experts of China to J.L.

References

Abe, H., Yamaguchi-Shinozaki, K., Urao, T., et al., 1997. Role of *Arabidopsis* MYC and MYB homologs in drought- and abscisic acid regulated gene expression. Plant Cell 9, 1859–1868.

Abe, H., Urao, T., Ito, T., et al., 2003. *Arabidopsis* AtMYC2 (bHLH) and AtMYB2 (MYB) function as transcriptional activators in abscisic acid signaling. Plant Cell 15, 63–78.

Albrecht, V., Weinl, S., Blazevic, D., et al., 2003. The calcium sensor CBL1 integrates plant responses to abiotic stresses. Plant J. 36, 457–470.

Ali-Rachedi, S., Bouinot, D., Wagner, M.H., et al., 2004. Changes in endogenous abscisic acid levels during dormancy release and maintenance of mature seeds: studies with the Cape Verde Islands ecotype, the dormant model of *Arabidopsis thaliana*. Planta 219, 479–488.

Arenas-Huertero, F., Arroyo, A., Zhou, L., et al., 2000. Analysis of *Arabidopsis* glucose insensitive mutants, *gin5* and *gin6*, reveals a central role of the plant hormone ABA in the regulation of plant vegetative development by sugar. Genes Dev. 14, 2085–2096.

Arroyo, A., Bossi, F., Finkelstein, R.R., et al., 2003. Three genes that affect sugar sensing (*abscisic acid insensitive 4, abscisic acid insensitive 5, and constitutive triple response 1*) are differentially regulated by glucose in *Arabidopsis*. Plant Physiol. 133, 231–242.

Barrero, J.M., Piqueras, P., Gonzalez-Guzman, M., et al., 2005. A mutational analysis of the *ABA1* gene of *Arabidopsis thaliana* highlights the involvement of ABA in vegetative development. J. Exp. Bot. 56, 2071–2083.

Bauer, H., Ache, P., Lautner, S., et al., 2013. The stomatal response to reduced relative humidity requires guard cell-autonomous ABA synthesis. Curr. Biol. 23, 53–57.

Bentsink, L., Koornneef, M., 2008. Seed dormancy and germination. Arabidopsis Book 6, e0119.

Bhaskara, G.B., Nguyen, T.T., Verslues, P.E., 2012. Unique drought resistance functions of the highly ABA-induced clade A protein phosphatase 2Cs. Plant Physiol. 160, 379–395.

Bittner, F., Oreb, M., Mendel, R.R., 2001. ABA3 is a molybdenum cofactor sulfurase required for activation of aldehyde oxidase and xanthine dehydrogenase in *Arabidopsis thaliana*. J. Biol. Chem. 276, 40381–40384.

Borthwick, H.A., Hendricks, S.B., Parker, M.W., et al., 1952. A reversible photoreaction controlling seed germination. Proc. Natl. Acad. Sci. U.S.A. 38, 662–666.

Bossi, F., Cordoba, E., Dupre, P., et al., 2009. The *Arabidopsis* ABA-INSENSITIVE (ABI) 4 factor acts as a central transcription activator of the expression of its own gene, and for the induction of *ABI5* and *SBE2.2* genes during sugar signaling. Plant J. 59, 359–374.

Boudsocq, M., Barbier-Brygoo, H., Lauriere, C., 2004. Identification of nine sucrose nonfermenting 1-related protein kinases 2 activated by hyperosmotic and saline stresses in *Arabidopsis thaliana*. J. Biol. Chem. 279, 41758–41766.

Boursiac, Y., Leran, S., Corratge-Faillie, C., et al., 2013. ABA transport and transporters. Trends Plant Sci. 18, 325–333.

Brandt, B., Munemasa, S., Wang, C., et al., 2015. Calcium specificity signaling mechanisms in abscisic acid signal transduction in *Arabidopsis* guard cells. eLife 4.

Brocard, I.M., Lynch, T.J., Finkelstein, R.R., 2002. Regulation and role of the *Arabidopsis Abscisic Acid-Insensitive 5* gene in abscisic acid, sugar, and stress response. Plant Physiol. 129, 1533–1543.

Bueso, E., Rodriguez, L., Lorenzo-Orts, L., et al., 2014. The single-subunit RING-type E3 ubiquitin ligase RSL1 targets PYL4 and PYR1 ABA receptors in plasma membrane to modulate abscisic acid signaling. Plant J. 80, 1057–1071.

Carles, C., Bies-Etheve, N., Aspart, L., et al., 2002. Regulation of *Arabidopsis thaliana Em* genes: role of ABI5. Plant J. 30, 373–383.

Castillo, M.C., Lozano-Juste, J., Gonzalez-Guzman, M., et al., 2015. Inactivation of PYR/PYL/ RCAR ABA receptors by tyrosine nitration may enable rapid inhibition of ABA signaling by nitric oxide in plants. Sci. Signal 8, ra89.

Chan, K.X., Mabbitt, P.D., Phua, S.Y., et al., 2016. Sensing and signaling of oxidative stress in chloroplasts by inactivation of the SAL1 phosphoadenosine phosphatase. Proc. Natl. Acad. Sci. U.S.A. 113, E4567–E4576.

Chen, H., Zhang, J., Neff, M.M., et al., 2008. Integration of light and abscisic acid signaling during seed germination and early seedling development. Proc. Natl. Acad. Sci. U.S.A. 105, 4495–4500.

Cheng, S.H., Willmann, M.R., Chen, H.C., et al., 2002a. Calcium signaling through protein kinases. The *Arabidopsis* calcium-dependent protein kinase gene family. Plant Physiol. 129, 469–485.

Cheng, W.H., Endo, A., Zhou, L., et al., 2002b. A unique short-chain dehydrogenase/reductase in *Arabidopsis* glucose signaling and abscisic acid biosynthesis and functions. Plant Cell 14, 2723–2743.

Cheong, Y.H., Kim, K.N., Pandey, G.K., et al., 2003. CBL1, a calcium sensor that differentially regulates salt, drought, and cold responses in *Arabidopsis*. Plant Cell 15, 1833–1845.

Cheong, Y.H., Pandey, G.K., Grant, J.J., et al., 2007. Two calcineurin B-like calcium sensors, interacting with protein kinase CIPK23, regulate leaf transpiration and root potassium uptake in *Arabidopsis*. Plant J. 52, 223–239.

Choi, H., Hong, J., Ha, J., et al., 2000. ABFs, a family of ABA-responsive element binding factors. J. Biol. Chem. 275, 1723–1730.

Choi, H.I., Park, H.J., Park, J.H., et al., 2005. *Arabidopsis* calcium-dependent protein kinase AtCPK32 interacts with ABF4, a transcriptional regulator of abscisic acid-responsive gene expression, and modulates its activity. Plant Physiol. 139, 1750–1761.

Christmann, A., Hoffmann, T., Teplova, I., et al., 2005. Generation of active pools of abscisic acid revealed by in vivo imaging of water-stressed *Arabidopsis*. Plant Physiol. 137, 209–219.

Christmann, A., Weiler, E.W., Steudle, E., et al., 2007. A hydraulic signal in root-to-shoot signalling of water shortage. Plant J. 52, 167–174.

Cracker, L.E., Abeles, F.B., 1969. Abscission: role of abscisic acid. Plant Physiol. 44, 1144–1149.

Cuming, A.C., Cho, S.H., Kamisugi, Y., et al., 2007. Microarray analysis of transcriptional responses to abscisic acid and osmotic, salt, and drought stress in the moss, *Physcomitrella patens*. New Phytol. 176, 275–287.

Curaba, J., Moritz, T., Blervaque, R., et al., 2004. AtGA3ox2, a key gene responsible for bioactive gibberellin biosynthesis, is regulated during embryogenesis by LEAFY COTYLEDON2 and FUSCA3 in *Arabidopsis*. Plant Physiol. 136, 3660–3669.

Cutler, A.J., Krochko, J.E., 1999. Formation and breakdown of ABA. Trends Plant Sci. 4, 472–478.

Cutler, S.R., Rodriguez, P.L., Finkelstein, R.R., et al., 2010. Abscisic acid: emergence of a core signaling network. Ann. Rev. Plant Biol. 61, 651–679.

D'Angelo, C., Weinl, S., Batistic, O., et al., 2006. Alternative complex formation of the Ca-regulated protein kinase CIPK1 controls abscisic acid-dependent and independent stress responses in *Arabidopsis*. Plant J. 48, 857–872.

Danquah, A., de Zelicourt, A., Colcombet, J., et al., 2014. The role of ABA and MAPK signaling pathways in plant abiotic stress responses. Biotechnol. Adv. 32, 40–52.

Danquah, A., de Zelicourt, A., Boudsocq, M., et al., 2015. Identification and characterization of an ABA-activated MAP kinase cascade in *Arabidopsis thaliana*. Plant J. 82, 232–244.

de Zelicourt, A., Colcombet, J., Hirt, H., 2016. The role of MAPK modules and ABA during abiotic stress signaling. Trends Plant Sci. 21, 677–685.

Drerup, M.M., Schlucking, K., Hashimoto, K., et al., 2013. The calcineurin B-like calcium sensors CBL1 and CBL9 together with their interacting protein kinase CIPK26 regulate the *Arabidopsis* NADPH oxidase RBOHF. Mol. Plant 6, 559–569.

Estavillo, G.M., Crisp, P.A., Pornsiriwong, W., et al., 2011. Evidence for a SAL1-PAP chloroplast retrograde pathway that functions in drought and high light signaling in *Arabidopsis*. Plant Cell 23, 3992–4012.

Ezcurra, I., Wycliffe, P., Nehlin, L., et al., 2000. Transactivation of the *Brassica napus* napin promoter by ABI3 requires interaction of the conserved B2 and B3 domains of ABI3 with different *cis*-elements: B2 mediates activation through an ABRE, whereas B3 interacts with an RY/G-box. Plant J. 24, 57–66.

Feng, C.Z., Chen, Y., Wang, C., et al., 2014. *Arabidopsis* RAV1 transcription factor, phosphorylated by SnRK2 kinases, regulates the expression of *ABI3, ABI4,* and *ABI5* during seed germination and early seedling development. Plant J. 80, 654–668.

Finkelstein, R.R., Lynch, T.J., 2000. The *Arabidopsis* abscisic acid response gene *ABI5* encodes a basic leucine zipper transcription factor. Plant Cell 12, 599–609.

Finkelstein, R.R., Wang, M.L., Lynch, T.J., et al., 1998. The *Arabidopsis* abscisic acid response locus *ABI4* encodes an APETALA 2 domain protein. Plant Cell 10, 1043–1054.

Finkelstein, R.R., 1994. Mutations at two new *Arabidopsis* ABA response loci are similar to the *abi3* mutations. Plant J. 5, 765–771.

Finkelstein, R.R., 2013. Abscisic acid synthesis and response. Arabidopsis Book 11, e0166.

Frey, A., Godin, B., Bonnet, M., et al., 2004. Maternal synthesis of abscisic acid controls seed development and yield in *Nicotiana plumbaginifolia*. Planta 218, 958–964.

Frey, A., Effroy, D., Lefebvre, V., et al., 2012. Epoxycarotenoid cleavage by NCED5 fine-tunes ABA accumulation and affects seed dormancy and drought tolerance with other NCED family members. Plant J. 70, 501–512.

Fujii, H., Zhu, J.K., 2009. *Arabidopsis* mutant deficient in 3 abscisic acid-activated protein kinases reveals critical roles in growth, reproduction, and stress. Proc. Natl. Acad. Sci. U.S.A. 106, 8380–8385.

Fujii, H., Verslues, P.E., Zhu, J.K., 2007. Identification of two protein kinases required for abscisic acid regulation of seed germination, root growth, and gene expression in *Arabidopsis*. Plant Cell 19, 485–494.

Fujii, H., Chinnusamy, V., Rodrigues, A., et al., 2009. *In vitro* reconstitution of an abscisic acid signalling pathway. Nature 462, 660–664.

Fujii, H., Verslues, P.E., Zhu, J.K., 2011. *Arabidopsis* decuple mutant reveals the importance of SnRK2 kinases in osmotic stress responses *in vivo*. Proc. Natl. Acad. Sci. U.S.A. 108, 1717–1722.

Fujita, M., Fujita, Y., Maruyama, K., et al., 2004. A dehydration-induced NAC protein, RD26, is involved in a novel ABA-dependent stress-signaling pathway. Plant J. 39, 863–876.

Fujita, Y., Fujita, M., Satoh, R., et al., 2005. AREB1 is a transcription activator of novel ABRE-dependent ABA signaling that enhances drought stress tolerance in *Arabidopsis*. Plant Cell 17, 3470–3488.

Fujita, Y., Nakashima, K., Yoshida, T., et al., 2009. Three SnRK2 protein kinases are the main positive regulators of abscisic acid signaling in response to water stress in *Arabidopsis*. Plant Cell Physiol. 50, 2123–2132.

Furihata, T., Maruyama, K., Fujita, Y., et al., 2006. Abscisic acid-dependent multisite phosphorylation regulates the activity of a transcription activator AREB1. Proc. Natl. Acad. Sci. U.S.A. 103, 1988–1993.

Gazzarrini, S., Tsuchiya, Y., Lumba, S., et al., 2004. The transcription factor FUSCA3 controls developmental timing in *Arabidopsis* through the hormones gibberellin and abscisic acid. Dev. Cell 7, 373–385.

Geiger, D., Scherzer, S., Mumm, P., et al., 2009. Activity of guard cell anion channel SLAC1 is controlled by drought-stress signaling kinase-phosphatase pair. Proc. Natl. Acad. Sci. U.S.A. 106, 21425–21430.

Geiger, D., Scherzer, S., Mumm, P., et al., 2010. Guard cell anion channel SLAC1 is regulated by CDPK protein kinases with distinct Ca^{2+} affinities. Proc. Natl. Acad. Sci. U.S.A. 107, 8023–8028.

Giraud, E., Van Aken, O., Ho, L.H., et al., 2009. The transcription factor ABI4 is a regulator of mitochondrial retrograde expression of *ALTERNATIVE OXIDASE1a*. Plant Physiol. 150, 1286–1296.

Giraudat, J., Hauge, B.M., Valon, C., et al., 1992. Isolation of the *Arabidopsis ABI3* gene by positional cloning. Plant Cell 4, 1251–1261.

Gomez-Cadenas, A., Verhey, S.D., Holappa, L.D., et al., 1999. An abscisic acid-induced protein kinase, PKABA1, mediates abscisic acid-suppressed gene expression in barley aleurone layers. Proc. Natl. Acad. Sci. U.S.A. 96, 1767–1772.

Gonzalez-Guzman, M., Apostolova, N., Belles, J.M., et al., 2002. The short-chain alcohol dehydrogenase ABA2 catalyzes the conversion of xanthoxin to abscisic aldehyde. Plant Cell 14, 1833–1846.

Gosti, F., Beaudoin, N., Serizet, C., et al., 1999. ABI1 protein phosphatase 2C is a negative regulator of abscisic acid signaling. Plant Cell 11, 1897–1910.

Grondin, A., Rodrigues, O., Verdoucq, L., et al., 2015. Aquaporins contribute to ABA-triggered stomatal closure through OST1-mediated phosphorylation. Plant Cell 27, 1945–1954.

Guo, Y., Xiong, L., Song, C.P., et al., 2002. A calcium sensor and its interacting protein kinase are global regulators of abscisic acid signaling in *Arabidopsis*. Dev. Cell 3, 233–244.

Haake, V., Cook, D., Riechmann, J.L., et al., 2002. Transcription factor CBF4 is a regulator of drought adaptation in *Arabidopsis*. Plant Physiol. 130, 639–648.

Hao, Q., Yin, P., Li, W., et al., 2011. The molecular basis of ABA-independent inhibition of PP2Cs by a subclass of PYL proteins. Mol. Cell 42, 662–672.

He, J., Duan, Y., Hua, D., et al., 2012. DEXH box RNA helicase-mediated mitochondrial reactive oxygen species production in *Arabidopsis* mediates crosstalk between abscisic acid and auxin signaling. Plant Cell 24, 1815–1833.

Himmelbach, A., Hoffmann, T., Leube, M., et al., 2002. Homeodomain protein ATHB6 is a target of the protein phosphatase ABI1 and regulates hormone responses in *Arabidopsis*. EMBO J. 21, 3029–3038.

Hrabak, E.M., Chan, C.W., Gribskov, M., et al., 2003. The *Arabidopsis* CDPK-SnRK superfamily of protein kinases. Plant Physiol. 132, 666–680.

Hua, D., Wang, C., He, J., et al., 2012. A plasma membrane receptor kinase, GHR1, mediates abscisic acid- and hydrogen peroxide-regulated stomatal movement in *Arabidopsis*. Plant Cell 24, 2546–2561.

Huang, Y., Feng, C.Z., Ye, Q., et al., 2016. *Arabidopsis* WRKY6 transcription factor acts as a positive regulator of abscisic acid signaling during seed germination and early seedling development. PLoS Genet. 12, e1005833.

Hubbard, K.E., Nishimura, N., Hitomi, K., et al., 2010. Early abscisic acid signal transduction mechanisms: newly discovered components and newly emerging questions. Genes Dev. 24, 1695–1708.

Imes, D., Mumm, P., Bohm, J., et al., 2013. Open stomata 1 (OST1) kinase controls R-type anion channel QUAC1 in *Arabidopsis* guard cells. Plant J. 74, 372–382.

Irigoyen, M.L., Iniesto, E., Rodriguez, L., et al., 2014. Targeted degradation of abscisic acid receptors is mediated by the ubiquitin ligase substrate adaptor DDA1 in *Arabidopsis*. Plant Cell 26, 712–728.

Iyer, L.M., Koonin, E.V., Aravind, L., 2001. Adaptations of the helix-grip fold for ligand binding and catalysis in the START domain superfamily. Proteins 43, 134–144.

Jakoby, M., Weisshaar, B., Droge-Laser, W., et al., 2002. bZIP transcription factors in *Arabidopsis*. Trends Plant Sci. 7, 106–111.

Jammes, F., Song, C., Shin, D., et al., 2009. MAP kinases MPK9 and MPK12 are preferentially expressed in guard cells and positively regulate ROS-mediated ABA signaling. Proc. Natl. Acad. Sci. U.S.A. 106, 20520–20525.

Kang, J.Y., Choi, H.I., Im, M.Y., et al., 2002. *Arabidopsis* basic leucine zipper proteins that mediate stress-responsive abscisic acid signaling. Plant Cell 14, 343–357.

Kang, J., Hwang, J.U., Lee, M., et al., 2010. PDR-type ABC transporter mediates cellular uptake of the phytohormone abscisic acid. Proc. Natl. Acad. Sci. U.S.A. 107, 2355–2360.

Kanno, Y., Hanada, A., Chiba, Y., et al., 2012. Identification of an abscisic acid transporter by functional screening using the receptor complex as a sensor. Proc. Natl. Acad. Sci. U.S.A. 109, 9653–9658.

Karssen, C.M., Brinkhorst-van der Swan, D.L., Breekland, A.E., et al., 1983. Induction of dormancy during seed development by endogenous abscisic acid: studies on abscisic acid deficient genotypes of *Arabidopsis thaliana* (L.) Heynh. Planta 157, 158–165.

Keith, K., Kraml, M., Dengler, N.G., et al., 1994. fusca3: A heterochronic mutation affecting late embryo development in *Arabidopsis*. Plant Cell 6, 589–600.

Kerchev, P.I., Pellny, T.K., Vivancos, P.D., et al., 2011. The transcription factor ABI4 is required for the ascorbic acid-dependent regulation of growth and regulation of jasmonate-dependent defense signaling pathways in *Arabidopsis*. Plant Cell 23, 3319–3334.

Khandelwal, A., Cho, S.H., Marella, H., et al., 2010. Role of ABA and ABI3 in desiccation tolerance. Science 327, 546.

Kim, S.Y., Ma, J., Perret, P., et al., 2002. *Arabidopsis* ABI5 subfamily members have distinct DNA-binding and transcriptional activities. Plant Physiol. 130, 688–697.

Kim, K.N., Cheong, Y.H., Grant, J.J., et al., 2003. CIPK3, a calcium sensor-associated protein kinase that regulates abscisic acid and cold signal transduction in *Arabidopsis*. Plant Cell 15, 411–423.

Kim, S., Kang, J.Y., Cho, D.I., et al., 2004. ABF2, an ABRE-binding bZIP factor, is an essential component of glucose signaling and its overexpression affects multiple stress tolerance. Plant J. 40, 75–87.

Kim, T.H., Bohmer, M., Hu, H., et al., 2010. Guard cell signal transduction network: advances in understanding abscisic acid, CO_2, and Ca^{2+} signaling. Ann. Rev. Plant Biol. 61, 561–591.

Kim, W., Lee, Y., Park, J., et al., 2013. HONSU, a protein phosphatase 2C, regulates seed dormancy by inhibiting ABA signaling in *Arabidopsis*. Plant Cell Physiol. 54, 555–572.

Kirby, J., Keasling, J.D., 2009. Biosynthesis of plant isoprenoids: perspectives for microbial engineering. Ann. Rev. Plant Biol. 60, 335–355.

Kline, K.G., Barrett-Wilt, G.A., Sussman, M.R., 2010. *In planta* changes in protein phosphorylation induced by the plant hormone abscisic acid. Proc. Natl. Acad. Sci. U.S.A. 107, 15986–15991.

Knight, H., Zarka, D.G., Okamoto, H., et al., 2004. Abscisic acid induces CBF gene transcription and subsequent induction of cold-regulated genes via the CRT promoter element. Plant Physiol. 135, 1710–1717.

Kollist, H., Nuhkat, M., Roelfsema, M.R., 2014. Closing gaps: linking elements that control stomatal movement. New Phytol. 203, 44–62.

Komatsu, K., Nishikawa, Y., Ohtsuka, T., et al., 2009. Functional analyses of the ABI1-related protein phosphatase type 2C reveal evolutionarily conserved regulation of abscisic acid signaling between *Arabidopsis* and the moss *Physcomitrella patens*. Plant Mol. Biol. 70, 327–340.

Kong, L., Cheng, J., Zhu, Y., et al., 2015. Degradation of the ABA co-receptor ABI1 by PUB12/13 U-box E3 ligases. Nat. Commun. 6, 8630.

Koornneef, M., Jorna, M.L., Brinkhorst-van der Swan, D.L., et al., 1982. The isolation of abscisic acid (ABA) deficient mutants by selection of induced revertants in non-germinating gibberellin sensitive lines of *Arabidopsis thaliana* (L.) heynh. Theor. Appl. Genet. 61, 385–393.

Koornneef, M., Reuling, G., Karssen, C.M., 1984. The isolation and characterization of abscisic acid-insensitive mutants of *Arabidopsis thaliana*. Physiol. Plant 61, 377–383.

Koussevitzky, S., Nott, A., Mockler, T.C., et al., 2007. Signals from chloroplasts converge to regulate nuclear gene expression. Science 316, 715–719.

Kuhn, J.M., Boisson-Dernier, A., Dizon, M.B., et al., 2006. The protein phosphatase AtPP2CA negatively regulates abscisic acid signal transduction in *Arabidopsis*, and effects of *abh1* on *AtPP2CA* mRNA. Plant Physiol. 140, 127–139.

Kuromori, T., Miyaji, T., Yabuuchi, H., et al., 2010. ABC transporter AtABCG25 is involved in abscisic acid transport and responses. Proc. Natl. Acad. Sci. U.S.A. 107, 2361–2366.

Kushiro, T., Okamoto, M., Nakabayashi, K., et al., 2004. The *Arabidopsis* cytochrome P450 CYP707A encodes ABA 8'-hydroxylases: key enzymes in ABA catabolism. EMBO J. 23, 1647–1656.

Kwak, J.M., Murata, Y., Baizabal-Aguirre, V.M., et al., 2001. Dominant negative guard cell K+ channel mutants reduce inward-rectifying K+ currents and light-induced stomatal opening in *Arabidopsis*. Plant Physiol. 127, 473–485.

Kwak, J.M., Mori, I.C., Pei, Z.M., et al., 2003. NADPH oxidase *AtrbohD* and *AtrbohF* genes function in ROS-dependent ABA signaling in *Arabidopsis*. EMBO J. 22, 2623–2633.

Laby, R.J., Kincaid, M.S., Kim, D., et al., 2000. The *Arabidopsis* sugar-insensitive mutants *sis4* and *sis5* are defective in abscisic acid synthesis and response. Plant J. 23, 587–596.

Lebaudy, A., Vavasseur, A., Hosy, E., et al., 2008. Plant adaptation to fluctuating environment and biomass production are strongly dependent on guard cell potassium channels. Proc. Natl. Acad. Sci. U.S.A. 105, 5271–5276.

Lechner, E., Leonhardt, N., Eisler, H., et al., 2011. MATH/BTB CRL3 receptors target the homeodomain-leucine zipper ATHB6 to modulate abscisic acid signaling. Dev. Cell 21, 1116–1128.

Lee, K.H., Piao, H.L., Kim, H.Y., et al., 2006. Activation of glucosidase via stress-induced polymerization rapidly increases active pools of abscisic acid. Cell 126, 1109–1120.

Lee, S.C., Lan, W., Buchanan, B.B., et al., 2009. A protein kinase-phosphatase pair interacts with an ion channel to regulate ABA signaling in plant guard cells. Proc. Natl. Acad. Sci. U.S.A. 106, 21419–21424.

Lee, J.H., Yoon, H.J., Terzaghi, W., et al., 2010a. DWA1 and DWA2, two *Arabidopsis* DWD protein components of CUL4-based E3 ligases, act together as negative regulators in ABA signal transduction. Plant Cell 22, 1716–1732.

Lee, S.J., Kang, J.Y., Park, H.J., et al., 2010b. DREB2C interacts with ABF2, a bZIP protein regulating abscisic acid-responsive gene expression, and its overexpression affects abscisic acid sensitivity. Plant Physiol. 153, 716–727.

Lefebvre, V., North, H., Frey, A., et al., 2006. Functional analysis of *Arabidopsis NCED6* and *NCED9* genes indicates that ABA synthesized in the endosperm is involved in the induction of seed dormancy. Plant J. 45, 309–319.

Leonhardt, N., Kwak, J.M., Robert, N., et al., 2004. Microarray expression analyses of *Arabidopsis* guard cells and isolation of a recessive abscisic acid hypersensitive protein phosphatase 2C mutant. Plant Cell 16, 596–615.

Leung, J., Bouvier-Durand, M., Morris, P.C., et al., 1994. *Arabidopsis* ABA response gene *ABI1*: features of a calcium-modulated protein phosphatase. Science 264, 1448–1452.

Leung, J., Merlot, S., Giraudat, J., 1997. The *Arabidopsis ABSCISIC ACID-INSENSITIVE2 (ABI2)* and *ABI1* genes encode homologous protein phosphatases 2C involved in abscisic acid signal transduction. Plant Cell 9, 759–771.

Li, J., Wang, X.Q., Watson, M.B., et al., 2000. Regulation of abscisic acid-induced stomatal closure and anion channels by guard cell AAPK kinase. Science 287, 300–303.

Li, Y., Zhang, L., Li, D., et al., 2016. The *Arabidopsis* F-box E3 ligase RIFP1 plays a negative role in abscisic acid signalling by facilitating ABA receptor RCAR3 degradation. Plant Cell Environ. 39, 571–582.

Lim, S., Park, J., Lee, N., et al., 2013. ABA-insensitive3, ABA-insensitive5, and DELLAs interact to activate the expression of *SOMNUS* and other high-temperature-inducible genes in imbibed seeds in *Arabidopsis*. Plant Cell 25, 4863–4878.

Liu, H., Stone, S.L., 2010. Abscisic acid increases *Arabidopsis* ABI5 transcription factor levels by promoting KEG E3 ligase self-ubiquitination and proteasomal degradation. Plant Cell 22, 2630–2641.

Liu, H., Stone, S.L., 2013. Cytoplasmic degradation of the *Arabidopsis* transcription factor abscisic acid insensitive 5 is mediated by the RING-type E3 ligase KEEP ON GOING. J. Biol. Chem. 288, 20267–20279.

Liu, Q., Kasuga, M., Sakuma, Y., et al., 1998. Two transcription factors, DREB1 and DREB2, with an EREBP/AP2 DNA binding domain separate two cellular signal transduction pathways in drought- and low-temperature-responsive gene expression, respectively, in *Arabidopsis*. Plant Cell 10, 1391–1406.

Liu, Y., He, J., Chen, Z., et al., 2010. *ABA overly-sensitive 5 (ABO5)*, encoding a pentatricopeptide repeat protein required for cis-splicing of mitochondrial nad2 intron 3, is involved in the abscisic acid response in *Arabidopsis*. Plant J. 63, 749–765.

Liu, Z.Q., Yan, L., Wu, Z., et al., 2012. Cooperation of three WRKY-domain transcription factors WRKY18, WRKY40, and WRKY60 in repressing two ABA-responsive genes *ABI4* and *ABI5* in *Arabidopsis*. J. Exp. Bot. 63, 6371–6392.

Lopez-Molina, L., Mongrand, S., Chua, N.H., 2001. A postgermination developmental arrest checkpoint is mediated by abscisic acid and requires the ABI5 transcription factor in *Arabidopsis*. Proc. Natl. Acad. Sci. U.S.A. 98, 4782–4787.

Lopez-Molina, L., Mongrand, S., McLachlin, D.T., et al., 2002. ABI5 acts downstream of ABI3 to execute an ABA-dependent growth arrest during germination. Plant J. 32, 317–328.

Lopez-Molina, L., Mongrand, S., Kinoshita, N., et al., 2003. AFP is a novel negative regulator of ABA signaling that promotes ABI5 protein degradation. Genes Dev. 17, 410–418.

Lu, C., Han, M.H., Guevara-Garcia, A., et al., 2002. Mitogen-activated protein kinase signaling in postgermination arrest of development by abscisic acid. Proc. Natl. Acad. Sci. U.S.A. 99, 15812–15817.

Luo, X., Chen, Z., Gao, J., et al., 2014. Abscisic acid inhibits root growth in *Arabidopsis* through ethylene biosynthesis. Plant J. 79, 44–55.

Lyzenga, W.J., Liu, H., Schofield, A., et al., 2013. *Arabidopsis* CIPK26 interacts with KEG, components of the ABA signalling network and is degraded by the ubiquitin-proteasome system. J. Exp. Bot. 64, 2779–2791.

Ma, Y., Szostkiewicz, I., Korte, A., et al., 2009. Regulators of PP2C phosphatase activity function as abscisic acid sensors. Science 324, 1064–1068.

Ma, B., Yin, C.C., He, S.J., et al., 2014. Ethylene-induced inhibition of root growth requires abscisic acid function in rice (*Oryza sativa* L.) seedlings. PLoS Genet. 10, e1004701.

Matsui, A., Ishida, J., Morosawa, T., et al., 2008. *Arabidopsis* transcriptome analysis under drought, cold, high-salinity and ABA treatment conditions using a tiling array. Plant Cell Physiol. 49, 1135–1149.

McAdam, S.A., Brodribb, T.J., Ross, J.J., 2016. Shoot-derived abscisic acid promotes root growth. Plant Cell Environ. 39, 652–659.

McCarty, D.R., Hattori, T., Carson, C.B., et al., 1991. The *Viviparous-1* developmental gene of maize encodes a novel transcriptional activator. Cell 66, 895–905.

Meinhard, M., Grill, E., 2001. Hydrogen peroxide is a regulator of ABI1, a protein phosphatase 2C from *Arabidopsis*. FEBS Lett. 508, 443–446.

Meinhard, M., Rodriguez, P.L., Grill, E., 2002. The sensitivity of ABI2 to hydrogen peroxide links the abscisic acid-response regulator to redox signalling. Planta 214, 775–782.

Meinke, D.W., Franzmann, L.H., Nickle, T.C., et al., 1994. Leafy cotyledon mutants of *Arabidopsis*. Plant Cell 6, 1049–1064.

Melcher, K., Ng, L.M., Zhou, X.E., et al., 2009. A gate-latch-lock mechanism for hormone signalling by abscisic acid receptors. Nature 462, 602–608.

Merlot, S., Leonhardt, N., Fenzi, F., et al., 2007. Constitutive activation of a plasma membrane H(+)-ATPase prevents abscisic acid-mediated stomatal closure. EMBO J. 26, 3216–3226.

Meyer, K., Leube, M.P., Grill, E., 1994. A protein phosphatase 2C involved in ABA signal transduction in *Arabidopsis thaliana*. Science 264, 1452–1455.

Meyer, S., Mumm, P., Imes, D., et al., 2010. AtALMT12 represents an R-type anion channel required for stomatal movement in *Arabidopsis* guard cells. Plant J. 63, 1054–1062.

Miao, Y., Lv, D., Wang, P., et al., 2006. An *Arabidopsis* glutathione peroxidase functions as both a redox transducer and a scavenger in abscisic acid and drought stress responses. Plant Cell 18, 2749–2766.

Miller, G., Suzuki, N., Ciftci-Yilmaz, S., et al., 2010. Reactive oxygen species homeostasis and signalling during drought and salinity stresses. Plant Cell Environ. 33, 453–467.

Miyazono, K., Miyakawa, T., Sawano, Y., et al., 2009. Structural basis of abscisic acid signalling. Nature 462, 609–614.

Monke, G., Altschmied, L., Tewes, A., et al., 2004. Seed-specific transcription factors ABI3 and FUS3: molecular interaction with DNA. Planta 219, 158–166.

Mori, I.C., Uozumi, N., Muto, S., 2000. Phosphorylation of the inward-rectifying potassium channel KAT1 by ABR kinase in Vicia guard cells. Plant Cell Physiol. 41, 850–856.

Mori, I.C., Murata, Y., Yang, Y., et al., 2006. CDPKs CPK6 and CPK3 function in ABA regulation of guard cell S-type anion- and Ca(2+)-permeable channels and stomatal closure. PLoS Biol. 4, e327.

Mustilli, A.C., Merlot, S., Vavasseur, A., et al., 2002. *Arabidopsis* OST1 protein kinase mediates the regulation of stomatal aperture by abscisic acid and acts upstream of reactive oxygen species production. Plant Cell 14, 3089–3099.

Nakabayashi, K., Okamoto, M., Koshiba, T., et al., 2005. Genome-wide profiling of stored mRNA in *Arabidopsis thaliana* seed germination: epigenetic and genetic regulation of transcription in seed. Plant J. 41, 697–709.

Nakamura, S., Lynch, T.J., Finkelstein, R.R., 2001. Physical interactions between ABA response loci of *Arabidopsis*. Plant J. 26, 627–635.

Nakashima, K., Fujita, Y., Kanamori, N., et al., 2009a. Three *Arabidopsis* SnRK2 protein kinases, SRK2D/SnRK2.2, SRK2E/SnRK2.6/OST1 and SRK2I/SnRK2.3, involved in ABA signaling are essential for the control of seed development and dormancy. Plant Cell Physiol. 50, 1345–1363.

Nakashima, K., Ito, Y., Yamaguchi-Shinozaki, K., 2009b. Transcriptional regulatory networks in response to abiotic stresses in *Arabidopsis* and grasses. Plant Physiol. 149, 88–95.

Nambara, E., Marion-Poll, A., 2005. Abscisic acid biosynthesis and catabolism. Ann. Rev. Plant Biol. 56, 165–185.

Nambara, E., Naito, S., McCourt, P., 1992. A mutant of *Arabidopsis* which is defective in seed development and storage protein accumulation is a new *abi3* allele. Plant J. 2, 435 441.

Nambara, E., Keith, K., McCourt, P., et al., 1995. A regulatory role for the *ABI3* gene in the establishment of embryo maturation in *Arabidopsis thaliana*. Development 121, 629–636.

Negi, J., Matsuda, O., Nagasawa, T., et al., 2008. CO_2 regulator SLAC1 and its homologues are essential for anion homeostasis in plant cells. Nature 452, 483–486.

Nemhauser, J.L., Hong, F., Chory, J., 2006. Different plant hormones regulate similar processes through largely nonoverlapping transcriptional responses. Cell 126, 467–475.

Nishimura, N., Yoshida, T., Kitahata, N., et al., 2007. *ABA-hypersensitive germination1* encodes a protein phosphatase 2C, an essential component of abscisic acid signaling in *Arabidopsis* seed. Plant J. 50, 935–949.

Nishimura, N., Hitomi, K., Arvai, A.S., et al., 2009. Structural mechanism of abscisic acid binding and signaling by dimeric PYR1. Science 326, 1373–1379.

Niu, X., Helentjaris, T., Bate, N.J., 2002. Maize ABI4 binds coupling element1 in abscisic acid and sugar response genes. Plant Cell 14, 2565–2575.

North, H.M., De Almeida, A., Boutin, J.P., et al., 2007. The *Arabidopsis* ABA-deficient mutant *aba4* demonstrates that the major route for stress-induced ABA accumulation is via neoxanthin isomers. Plant J. 50, 810–824.

Ogawa, M., Hanada, A., Yamauchi, Y., et al., 2003. Gibberellin biosynthesis and response during *Arabidopsis* seed germination. Plant Cell 15, 1591–1604.

Oh, E., Kim, J., Park, E., et al., 2004. PIL5, a phytochrome-interacting basic helix-loop-helix protein, is a key negative regulator of seed germination in *Arabidopsis thaliana*. Plant Cell 16, 3045–3058.

Oh, S.J., Song, S.I., Kim, Y.S., et al., 2005. *Arabidopsis* CBF3/DREB1A and ABF3 in transgenic rice increased tolerance to abiotic stress without stunting growth. Plant Physiol. 138, 341–351.

Oh, E., Yamaguchi, S., Hu, J., et al., 2007. PIL5, a phytochrome-interacting bHLH protein, regulates gibberellin responsiveness by binding directly to the *GAI* and *RGA* promoters in *Arabidopsis* seeds. Plant Cell 19, 1192–1208.

Oh, E., Kang, H., Yamaguchi, S., et al., 2009. Genome-wide analysis of genes targeted by PHYTOCHROME INTERACTING FACTOR 3-LIKE5 during seed germination in *Arabidopsis*. Plant Cell 21, 403–419.

Okamoto, M., Kuwahara, A., Seo, M., et al., 2006. *CYP707A1* and *CYP707A2*, which encode abscisic acid 8'-hydroxylases, are indispensable for proper control of seed dormancy and germination in *Arabidopsis*. Plant Physiol. 141, 97–107.

Ortiz-Masia, D., Perez-Amador, M.A., Carbonell, J., et al., 2007. Diverse stress signals activate the C1 subgroup MAP kinases of *Arabidopsis*. FEBS Lett. 581, 1834–1840.

Osakabe, Y., Maruyama, K., Seki, M., et al., 2005. Leucine-rich repeat receptor-like kinase1 is a key membrane-bound regulator of abscisic acid early signaling in *Arabidopsis*. Plant Cell 17, 1105–1119.

Pandey, G.K., Cheong, Y.H., Kim, K.N., et al., 2004. The calcium sensor calcineurin B-like 9 modulates abscisic acid sensitivity and biosynthesis in *Arabidopsis*. Plant Cell 16, 1912–1924.

Park, S.Y., Fung, P., Nishimura, N., et al., 2009. Abscisic acid inhibits type 2C protein phosphatases via the PYR/PYL family of START proteins. Science 324, 1068–1071.

Penfield, S., Li, Y., Gilday, A.D., et al., 2006. *Arabidopsis* ABA INSENSITIVE4 regulates lipid mobilization in the embryo and reveals repression of seed germination by the endosperm. Plant Cell 18, 1887–1899.

Qin, F., Shinozaki, K., Yamaguchi-Shinozaki, K., 2011. Achievements and challenges in understanding plant abiotic stress responses and tolerance. Plant Cell Physiol. 52, 1569–1582.

Quesada, V., Ponce, M.R., Micol, J.L., 2000. Genetic analysis of salt-tolerant mutants in *Arabidopsis thaliana*. Genetics 154, 421–436.

Raab, S., Drechsel, G., Zarepour, M., et al., 2009. Identification of a novel E3 ubiquitin ligase that is required for suppression of premature senescence in *Arabidopsis*. Plant J. 59, 39–51.

Raghavendra, A.S., Gonugunta, V.K., Christmann, A., et al., 2010. ABA perception and signalling. Trends Plant Sci. 15, 395–401.

Rajjou, L., Duval, M., Gallardo, K., et al., 2012. Seed germination and vigor. Ann. Rev. Plant Biol. 63, 507–533.

Raz, V., Bergervoet, J.H., Koornneef, M., 2001. Sequential steps for developmental arrest in *Arabidopsis* seeds. Development 128, 243–252.

Reeves, W.M., Lynch, T.J., Mobin, R., et al., 2011. Direct targets of the transcription factors ABA-Insensitive(ABI)4 and ABI5 reveal synergistic action by ABI4 and several bZIP ABA response factors. Plant Mol. Biol. 75, 347–363.

Rensing, S.A., Lang, D., Zimmer, A.D., et al., 2008. The *Physcomitrella* genome reveals evolutionary insights into the conquest of land by plants. Science 319, 64–69.

Rock, C.D., Zeevaart, J.A., 1991. The *aba* mutant of *Arabidopsis thaliana* is impaired in epoxy-carotenoid biosynthesis. Proc. Natl. Acad. Sci. U.S.A. 88, 7496–7499.

Rodriguez, P.L., Benning, G., Grill, E., 1998. ABI2, a second protein phosphatase 2C involved in abscisic acid signal transduction in *Arabidopsis*. FEBS Lett. 421, 185–190.

Romanel, E.A., Schrago, C.G., Counago, R.M., et al., 2009. Evolution of the B3 DNA binding superfamily: new insights into REM family gene diversification. PLoS One 4, e5791.

Ronzier, E., Corratge-Faillie, C., Sanchez, F., et al., 2014. CPK13, a noncanonical Ca2+-dependent protein kinase, specifically inhibits KAT2 and KAT1 shaker K+ channels and reduces stomatal opening. Plant Physiol. 166, 314–326.

Ruiz-Sola, M.A., Rodriguez-Concepcion, M., 2012. Carotenoid biosynthesis in *Arabidopsis*: a colorful pathway. Arabidopsis Book 10, e0158.

Rushton, D.L., Tripathi, P., Rabara, R.C., et al., 2012. WRKY transcription factors: key components in abscisic acid signalling. Plant Biotechnol. J. 10, 2–11.

Saez, A., Apostolova, N., Gonzalez-Guzman, M., et al., 2004. Gain-of-function and loss-of-function phenotypes of the protein phosphatase 2C HAB1 reveal its role as a negative regulator of abscisic acid signalling. Plant J. 37, 354–369.

Saito, S., Hirai, N., Matsumoto, C., et al., 2004. *Arabidopsis CYP707As* encode (+)-abscisic acid 8′-hydroxylase, a key enzyme in the oxidative catabolism of abscisic acid. Plant Physiol. 134, 1439–1449.

Santiago, J., Dupeux, F., Round, A., et al., 2009. The abscisic acid receptor PYR1 in complex with abscisic acid. Nature 462, 665–668.

Sato, A., Sato, Y., Fukao, Y., et al., 2009. Threonine at position 306 of the KAT1 potassium channel is essential for channel activity and is a target site for ABA-activated SnRK2/OST1/SnRK2.6 protein kinase. Biochem. J. 424, 439–448.

Schroeder, J.I., Allen, G.J., Hugouvieux, V., et al., 2001. Guard cell signal transduction. Ann. Rev. Plant Biol. 52, 627–658.

Schweighofer, A., Hirt, H., Meskiene, I., 2004. Plant PP2C phosphatases: emerging functions in stress signaling. Trends Plant Sci. 9, 236–243.

Seo, M., Hanada, A., Kuwahara, A., et al., 2006. Regulation of hormone metabolism in *Arabidopsis* seeds: phytochrome regulation of abscisic acid metabolism and abscisic acid regulation of gibberellin metabolism. Plant J. 48, 354–366.

Seo, K.I., Lee, J.H., Nezames, C.D., et al., 2014. ABD1 is an *Arabidopsis* DCAF substrate receptor for CUL4-DDB1-based E3 ligases that acts as a negative regulator of abscisic acid signaling. Plant Cell 26, 695–711.

Shang, Y., Yan, L., Liu, Z.Q., et al., 2010. The Mg-chelatase H subunit of *Arabidopsis* antagonizes a group of WRKY transcription repressors to relieve ABA-responsive genes of inhibition. Plant Cell 22, 1909–1935.

Sheen, J., 1996. Ca^{2+}-dependent protein kinases and stress signal transduction in plants. Science 274, 1900–1902.

Shimazaki, K., Doi, M., Assmann, S.M., et al., 2007. Light regulation of stomatal movement. Ann. Rev. Plant Biol. 58, 219–247.

Shinozaki, K., Yamaguchi-Shinozaki, K., Seki, M., 2003. Regulatory network of gene expression in the drought and cold stress responses. Curr. Opin. Plant Biol. 6, 410–417.

Shkolnik-Inbar, D., Bar-Zvi, D., 2010. ABI4 mediates abscisic acid and cytokinin inhibition of lateral root formation by reducing polar auxin transport in *Arabidopsis*. Plant Cell 22, 3560–3573.

Shu, K., Zhang, H., Wang, S., et al., 2013. ABI4 regulates primary seed dormancy by regulating the biogenesis of abscisic acid and gibberellins in *Arabidopsis*. PLoS Genet. 9, e1003577.

Shu, K., Chen, Q., Wu, Y., et al., 2016. ABI4 mediates antagonistic effects of abscisic acid and gibberellins at transcript and protein levels. Plant J. 85, 348–361.

Sirichandra, C., Gu, D., Hu, H.C., et al., 2009. Phosphorylation of the *Arabidopsis* AtrbohF NADPH oxidase by OST1 protein kinase. FEBS Lett. 583, 2982–2986.

Smalle, J., Kurepa, J., Yang, P., et al., 2003. The pleiotropic role of the 26S proteasome subunit RPN10 in *Arabidopsis* growth and development supports a substrate-specific function in abscisic acid signaling. Plant Cell 15, 965–980.

Soderman, E.M., Brocard, I.M., Lynch, T.J., et al., 2000. Regulation and function of the *Arabidopsis ABA-insensitive4* gene in seed and abscisic acid response signaling networks. Plant Physiol. 124, 1752–1765.

Sridharamurthy, M., Kovach, A., Zhao, Y., et al., 2014. H_2O_2 inhibits ABA-signaling protein phosphatase HAB1. PLoS One 9, e113643.

Steber, C.M., Cooney, S.E., McCourt, P., 1998. Isolation of the GA-response mutant *sly1* as a suppressor of ABI1-1 in *Arabidopsis thaliana*. Genetics 149, 509–521.

Stockinger, E.J., Gilmour, S.J., Thomashow, M.F., 1997. *Arabidopsis thaliana CBF1* encodes an AP2 domain-containing transcriptional activator that binds to the C-repeat/DRE, a cis-acting DNA regulatory element that stimulates transcription in response to low temperature and water deficit. Proc. Natl. Acad. Sci. U.S.A. 94, 1035–1040.

Stone, S.L., Williams, L.A., Farmer, L.M., et al., 2006. KEEP ON GOING, a RING E3 ligase essential for *Arabidopsis* growth and development, is involved in abscisic acid signaling. Plant Cell 18, 3415–3428.

Sun, X., Feng, P., Xu, X., et al., 2011. A chloroplast envelope-bound PHD transcription factor mediates chloroplast signals to the nucleus. Nat. Commun. 2, 477.

Suzuki, M., McCarty, D.R., 2008. Functional symmetry of the B3 network controlling seed development. Curr. Opin. Plant Biol. 11, 548–553.

Suzuki, M., Kao, C.Y., McCarty, D.R., 1997. The conserved B3 domain of VIVIPAROUS1 has a cooperative DNA binding activity. Plant Cell 9, 799–807.

Suzuki, M., Ketterling, M.G., Li, Q.B., et al., 2003. Viviparous1 alters global gene expression patterns through regulation of abscisic acid signaling. Plant Physiol. 132, 1664–1677.

Tan, B.C., Joseph, L.M., Deng, W.T., et al., 2003. Molecular characterization of the *Arabidopsis* 9-cis epoxycarotenoid dioxygenase gene family. Plant J. 35, 44–56.

Tang, W., Ji, Q., Huang, Y., et al., 2013. FAR-RED ELONGATED HYPOCOTYL3 and FAR-RED IMPAIRED RESPONSE1 transcription factors integrate light and abscisic acid signaling in *Arabidopsis*. Plant Physiol. 163, 857–866.

Umezawa, T., Nakashima, K., Miyakawa, T., et al., 2010. Molecular basis of the core regulatory network in ABA responses: sensing, signaling and transport. Plant Cell Physiol. 51, 1821–1839.

Umezawa, T., Sugiyama, N., Takahashi, F., et al., 2013. Genetics and phosphoproteomics reveal a protein phosphorylation network in the abscisic acid signaling pathway in *Arabidopsis thaliana*. Sci. Signal 6, rs8.

Uno, Y., Furihata, T., Abe, H., et al., 2000. *Arabidopsis* basic leucine zipper transcription factors involved in an abscisic acid-dependent signal transduction pathway under drought and high-salinity conditions. Proc. Natl. Acad. Sci. U.S.A. 97, 11632–11637.

Vahisalu, T., Kollist, H., Wang, Y.F., et al., 2008. SLAC1 is required for plant guard cell S-type anion channel function in stomatal signalling. Nature 452, 487–491.

Vahisalu, T., Puzorjova, I., Brosche, M., et al., 2010. Ozone-triggered rapid stomatal response involves the production of reactive oxygen species, and is controlled by SLAC1 and OST1. Plant J. 62, 442–453.

Vartanian, N., Marcotte, L., Giraudat, J., 1994. Drought rhizogenesis in *Arabidopsis thaliana* (differential responses of hormonal mutants). Plant Physiol. 104, 761–767.

Vierstra, R.D., 2009. The ubiquitin-26S proteasome system at the nexus of plant biology. Nat. Rev. Mol. Cell Biol. 10, 385–397.

Wang, P., Xue, L., Batelli, G., et al., 2013. Quantitative phosphoproteomics identifies SnRK2 protein kinase substrates and reveals the effectors of abscisic acid action. Proc. Natl. Acad. Sci. U.S.A. 110, 11205–11210.

Wang, P., Du, Y., Hou, Y.J., et al., 2015. Nitric oxide negatively regulates abscisic acid signaling in guard cells by S-nitrosylation of OST1. Proc. Natl. Acad. Sci. U.S.A. 112, 613–618.

Weinl, S., Kudla, J., 2009. The CBL-CIPK Ca(2+)-decoding signaling network: function and perspectives. New Phytol. 184, 517–528.

Weng, J.K., Ye, M., Li, B., et al., 2016. Co-evolution of hormone metabolism and signaling networks expands plant adaptive plasticity. Cell 166, 881–893.

Wind, J.J., Peviani, A., Snel, B., et al., 2013. ABI4: versatile activator and repressor. Trends Plant Sci. 18, 125–132.

Xing, Y., Jia, W., Zhang, J., 2008. AtMKK1 mediates ABA-induced CAT1 expression and H_2O_2 production via AtMPK6-coupled signaling in *Arabidopsis*. Plant J. 54, 440–451.

Xiong, L., Zhu, J.K., 2003. Regulation of abscisic acid biosynthesis. Plant Physiol. 133, 29–36.

Xiong, L., Ishitani, M., Lee, H., et al., 2001. The *Arabidopsis LOS5/ABA3* locus encodes a molybdenum cofactor sulfurase and modulates cold stress- and osmotic stress-responsive gene expression. Plant Cell 13, 2063–2083.

Xu, Z.Y., Lee, K.H., Dong, T., et al., 2012. A vacuolar beta-glucosidase homolog that possesses glucose-conjugated abscisic acid hydrolyzing activity plays an important role in osmotic stress responses in *Arabidopsis*. Plant Cell 24, 2184–2199.

Xu, D., Li, J., Gangappa, S.N., et al., 2014. Convergence of light and ABA signaling on the *ABI5* promoter. PLoS Genet. 10, e1004197.

Xu, X., Chi, W., Sun, X., et al., 2016. Convergence of light and chloroplast signals for de-etiolation through ABI4-HY5 and COP1. Nat. Plants 2, 16066.

Yamaguchi, S., Smith, M.W., Brown, R.G., et al., 1998. Phytochrome regulation and differential expression of gibberellin 3beta-hydroxylase genes in germinating *Arabidopsis* seeds. Plant Cell 10, 2115–2126.

Yamaguchi-Shinozaki, K., Shinozaki, K., 2006. Transcriptional regulatory networks in cellular responses and tolerance to dehydration and cold stresses. Ann. Rev. Plant Biol. 57, 781–803.

Yamauchi, Y., Ogawa, M., Kuwahara, A., et al., 2004. Activation of gibberellin biosynthesis and response pathways by low temperature during imbibition of *Arabidopsis thaliana* seeds. Plant Cell 16, 367–378.

Yamauchi, Y., Takeda-Kamiya, N., Hanada, A., et al., 2007. Contribution of gibberellin deactivation by AtGA2ox2 to the suppression of germination of dark-imbibed *Arabidopsis thaliana* seeds. Plant Cell Physiol. 48, 555–561.

Yang, S.D., Seo, P.J., Yoon, H.K., et al., 2011a. The Arabidopsis NAC transcription factor VNI2 integrates abscisic acid signals into leaf senescence via the *COR/RD* genes. Plant Cell 23, 2155–2168.

Yang, Y., Yu, X., Song, L., et al., 2011b. ABI4 activates *DGAT1* expression in *Arabidopsis* seedlings during nitrogen deficiency. Plant Physiol. 156, 873–883.

Yang, L., Zhang, J., He, J., et al., 2014. ABA-mediated ROS in mitochondria regulate root meristem activity by controlling PLETHORA expression in *Arabidopsis*. PLoS Genet. 10, e1004791.

Yin, P., Fan, H., Hao, Q., et al., 2009. Structural insights into the mechanism of abscisic acid signaling by PYL proteins. Nat. Struct. Mol. Biol. 16, 1230–1236.

Yoshida, R., Hobo, T., Ichimura, K., et al., 2002. ABA-activated SnRK2 protein kinase is required for dehydration stress signaling in *Arabidopsis*. Plant Cell Physiol. 43, 1473–1483.

Yoshida, T., Nishimura, N., Kitahata, N., et al., 2006. *ABA-hypersensitive germination3* encodes a protein phosphatase 2C (AtPP2CA) that strongly regulates abscisic acid signaling during germination among *Arabidopsis* protein phosphatase 2Cs. Plant Physiol. 140, 115–126.

Yoshida, T., Fujita, Y., Sayama, H., et al., 2010. AREB1, AREB2, and ABF3 are master transcription factors that cooperatively regulate ABRE-dependent ABA signaling involved in drought stress tolerance and require ABA for full activation. Plant J. 61, 672–685.

Yoshida, T., Fujita, Y., Maruyama, K., et al., 2015. Four *Arabidopsis* AREB/ABF transcription factors function predominantly in gene expression downstream of SnRK2 kinases in abscisic acid signalling in response to osmotic stress. Plant Cell Environ. 38, 35–49.

Yu, F., Wu, Y., Xie, Q., 2016. Ubiquitin-proteasome system in ABA signaling: from perception to action. Mol. Plant 9, 21–33.

Zeevaart, J.A., 1980. Changes in the levels of abscisic acid and its metabolites in excised leaf blades of *Xanthium strumarium* during and after water stress. Plant Physiol. 66, 672–678.

Zeller, G., Henz, S.R., Widmer, C.K., et al., 2009. Stress-induced changes in the *Arabidopsis thaliana* transcriptome analyzed using whole-genome tiling arrays. Plant J. 58, 1068–1082.

Zhang, X., Garreton, V., Chua, N.H., 2005. The AIP2 E3 ligase acts as a novel negative regulator of ABA signaling by promoting ABI3 degradation. Genes Dev. 19, 1532–1543.

Zhang, Y., Zhu, H., Zhang, Q., et al., 2009. Phospholipase dalpha1 and phosphatidic acid regulate NADPH oxidase activity and production of reactive oxygen species in ABA-mediated stomatal closure in *Arabidopsis*. Plant Cell 21, 2357–2377.

Zhang, A., Ren, H.M., Tan, Y.Q., et al., 2016. S-type anion channels SLAC1 and SLAH3 function as essential negative regulators of inward K+ channels and stomatal opening in *Arabidopsis*. Plant Cell 28, 949–965.

Zhao, R., Sun, H.L., Mei, C., et al., 2011. The *Arabidopsis* Ca(2+) -dependent protein kinase CPK12 negatively regulates abscisic acid signaling in seed germination and post-germination growth. New Phytol. 192, 61–73.

Zhu, S.Y., Yu, X.C., Wang, X.J., et al., 2007. Two calcium-dependent protein kinases, CPK4 and CPK11, regulate abscisic acid signal transduction in *Arabidopsis*. Plant Cell 19, 3019–3036.

Ethylene

6

Dongdong Hao[1,2], Xiangzhong Sun[1,2], Biao Ma[3], Jin-Song Zhang[3], Hongwei Guo[1,2]
[1]South University of Science and Technology of China, Shenzhen, China; [2]Peking University, Beijing, China; [3]Institute of Genetics and Developmental Biology, Chinese Academy of Sciences, Beijing, China

Summary

Ethylene is the first-identified plant hormone known to regulate numerous processes in plant growth, development, and response to biotic and abiotic stresses. Ethylene is best known for its effect on fruit ripening and organ abscission, and thus has great commercial importance in agriculture. As a gaseous hormone, ethylene can freely diffuse across membranes and is thought to be synthesized at or near its site of action, which is different from other plant hormones. By use of the typical "triple response" phenotype, significant progresses have been made on ethylene metabolism, perception, and signaling in the past two decades. To summarize the major aspects of ethylene biology in plants, this chapter is mainly divided into three parts. Part one deals with ethylene biosynthesis and its regulation covering the well-known "Yang cycle" and the transcriptional and posttranslational regulation of related enzymes. Part two deals with ethylene perception and signaling in the model plant Arabidopsis, including the key components and main breakthroughs of ethylene signal transduction. Part three deals with the ethylene perception and signaling and the cross talk of ethylene with other hormones in rice. Finally, intriguing questions related to ethylene action are proposed, which need to be further addressed in the future.

6.1 Ethylene biology

Ethylene gas is a plant hormone with diverse functions, ranging from being an environmental signal with profound effects on plant adaptation, to an endogenous regulator controlling growth, development, and reproduction, such as germination, cell elongation, fruit ripening, and senescence. After decades of research, the action mode underlying this simple gas has gradually been unveiled, wherein key components involved in the ethylene biosynthesis and signaling pathways were identified and characterized. These advances not only promote our understanding of ethylene biosynthesis and signal transduction, but also increase our knowledge about the signaling cross talk between ethylene and other plant regulators in response to environmental and internal cues (Fig. 6.1).

Background on the historical breakthrough in ethylene biology and on both ethylene biosynthesis and signaling is provided in reviews (Bakshi et al., 2015; Chang, 2016; Ju and Chang, 2015; McManus, 2012; Xu and Zhang, 2014). This chapter places emphasis on the advances made between 2011 and 2016, particularly on the findings made in *Arabidopsis thaliana* and *Oryza sativa* (rice).

Hormone Metabolism and Signaling in Plants. http://dx.doi.org/10.1016/B978-0-12-811562-6.00006-2

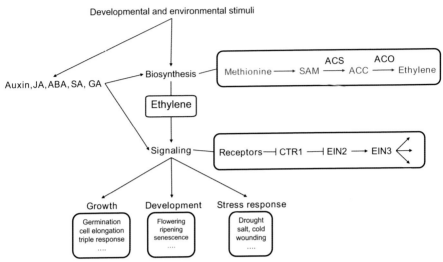

Figure 6.1 An overview of ethylene biology. Ethylene is a vital plant hormone involved in plant growth and development, as well as stress responses. Ethylene functions mainly through regulation of its biosynthesis and signaling, and interactions with other plant hormones in response to developmental and environmental stimuli. Simplified ethylene biosynthesis and signaling pathways are also shown in the figure.

6.2 Metabolism

6.2.1 Ethylene biosynthesis in Arabidopsis

Ethylene biosynthesis is under tight and complex regulation in response to environmental and internal stimuli (Abeles et al., 1992; Fluhr et al., 1996; Kende, 1993; Yang and Hoffman, 1984). Ethylene production is not tissue-specific, as almost all the plant body has the capacity to produce ethylene. Normally, ethylene production by the plant is maintained at low levels, but during certain developmental stages or stress events ethylene production is rapidly induced (Kende, 1993; Wang et al., 2002; Xu and Zhang, 2015; Yang and Hoffman, 1984). What is more, plants do not utilize other means, such as regulation of transport, metabolic inactivation, or degradation, to modulate ethylene levels, which makes the regulation of ethylene biosynthesis particularly important for its effects (Buerstenbinder and Sauter, 2012).

Ethylene biosynthesis in higher plants is well characterized, that is, it is comprised of a relatively simple metabolic pathway. Ethylene is synthesized from S-adenosylmethionine (SAM), an activated form of the amino acid methionine (Met), via 1-aminocyclopropane-1-carboxylate (ACC) as an intermediate. Two key reactions of this pathway are catalyzed by ACC synthase (ACS), which converts SAM to ACC, and ACC oxidase (ACO), which oxidases ACC to ethylene, CO_2, and cyanide (Yang and Hoffmann, 1984). Both ACS and ACO are encoded by small gene families, which are differentially regulated at both transcriptional and posttranscriptional levels in response to various signals, ensuring fine-tuning of ethylene synthesis (Xu and Zhang, 2014).

6.2.2   ACC synthase

It is generally considered that ACS is the rate-limiting enzyme of ethylene biosynthesis pathway during vegetative growth, and the abundance of different type of ACS enzymes is under tight regulation throughout plant development and in response to stresses (Lin et al., 2009b). Members of the *ACS* multigene family encode isoforms with high protein sequence conservation in the N-terminal catalytic domains, but relative divergence in their short C-termini. Based on the presence or absence of putative calcium-dependent protein kinase (CDPK) or mitogen-activating protein kinase (MAPK) phosphorylation sites in their C-terminal motifs, ACS proteins are classified into three groups (Fig. 6.2a). Type I ACS proteins have an extended C-terminal domain containing three MAPK and one CDPK phosphorylation sites. Type II ACS proteins have an intermediate length C-terminal domain containing a single potential CDPK phosphorylation site, which is within a domain known as TOE (Target of ETO1, the Ethylene Overproducer 1 protein). Type III ACS proteins have a minimal C-terminal extension which has no known phosphorylation sites.

6.2.2.1   Transcriptional regulation

ACS genes are transcriptionally and posttranscriptionally regulated in response to both developmental and environmental stimuli. Studies on *ACS* expression found that all *ACS* genes show tissue- and/or stage-specific expression patterns (Tsuchisaka et al., 2009), although how these *ACS* genes are differentially regulated is largely unclear. *ACS* genes are reported to be induced by auxin, implying that the transcriptional regulation of *ACS* genes can play a role in cross talk with other plant signals (Paponov et al., 2008). Transcriptional regulation of *ACS* genes is also a key mechanism regulating ethylene biosynthesis in response to biotic and abiotic stresses (Han et al., 2010; Li et al., 2012a; Peng et al., 2005). Wounding, hypoxia, herbivore and pathogen attack, and many other environmental stresses can induce ethylene production through upregulating *ACS* gene transcription, and different *ACS* genes show unique and overlapping expression patterns in response to different stresses (Han et al., 2010; Li et al., 2012a; Peng et al., 2005; Tsuchisaka and Theologis, 2004). To date, only a very limited number of transcription factors that regulate *ACS* gene expression in Arabidopsis have been reported (Li et al., 2012a), so the detailed mechanism of transcriptional regulation of *ACS* genes still needs to be further explored.

6.2.2.2   Posttranslational regulation

In addition to transcriptional regulation of the *ACS* genes, many studies have highlighted the role of posttranslational regulation of ACS proteins in the regulation of ethylene biosynthesis (Argueso et al., 2007; Chae and Kieber, 2005; Lyzenga and Stone, 2012; McClellan and Chang, 2008; Xu and Zhang, 2014). The regulation of ACS protein activity and stability through protein phosphorylation, dephosphorylation, and degradation, alter ethylene production rapidly, thus allowing plants to respond to environmental and internal stimuli quickly (Fig. 6.2).

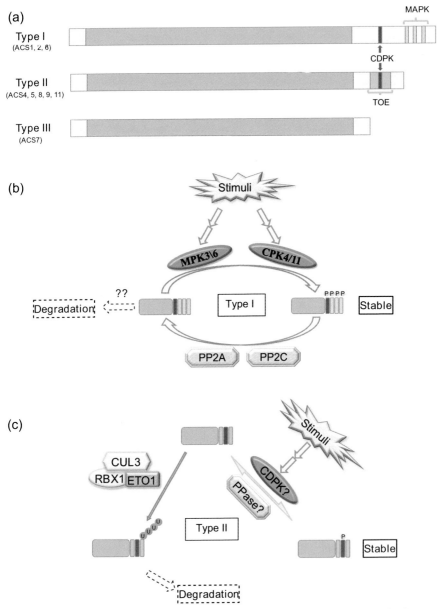

Figure 6.2 Proposed model for the regulation of ACS proteins in ethylene biosynthesis.
(a) Classification of ACS proteins into three subgroups based on the presence or absence
of MAPK and CDPK phosphorylation sites in the C termini. (b) Type I ACS proteins are
phosphorylated by MPK3/6 or CPK4/11 in response to environmental and internal stimuli,
which leads to stabilization of the proteins. Phosphatases involved in the dephosphorylation of ACS proteins have also been identified. Dephosphorylated proteins are unstable and
degraded rapidly. (c) Type II ACS proteins have a single potential CDPK phosphorylation
site. Its phosphorylation level is also under tight regulation, but the detailed mechanism is
still unclear. ETO1 binds to the TOE domain of type II ACS, targeting them for ubiquitination and degradation by the 26S proteasome.

The characterization of Arabidopsis ethylene-overproducing mutants (*eto1, eto2, eto3*) has revealed mechanisms underlying the posttranscriptional regulation of ACS proteins (Chae and Kieber, 2005). The BTB (broad-complex/tramtrack/bric-a-brac) domain containing E3 ligase component ETO1, which associates with the CRL (Cullin-RING ligase) complex subunit CUL3/E3 ligase, directly targets type II ACS proteins for ubiquitin-dependent degradation by the 26S proteasome (Wang et al., 2004). An additional E3 ligase component, a RING-type E3 ligase, XBAT32 (XB3 ortholog 2 in *Arabidopsis thaliana*), has also been identified and shown to regulate the stability of type II and type III ACS proteins (Lyzenga et al., 2012; Prasad et al., 2010). It was reported that MG132, a specific inhibitor of the 26S proteasome, can enhance the stability of type I ACS protein, implying the existence of E3 ligases targeting type I ACS proteins (Joo et al., 2008). Thus, the ubiquitin-proteasome pathway plays a vital role in regulating the turnover of all three types of ACS proteins.

Several studies using general kinase and phosphatase inhibitors provide evidence that the phosphorylation level of ACS proteins is a key determinant of its stability (Felix et al., 1991; Hansen et al., 2009; Spanu et al., 1994). Direct evidence that phosphorylation and dephosphorylation play crucial roles in ACS turnover has been obtained from studies on type I ACS proteins (Agnieszka et al., 2014; Han et al., 2010; Li et al., 2012a; Luo et al., 2014; Skottke et al., 2011). Arabidopsis MAPKs MPK3/MPK6 are able to phosphorylate ACS2 and ACS6, and two abscisic acid (ABA)-activated CDPKs CPK4/CPK11 also phosphorylate ACS6, which enhances ACS protein stability and promotes ethylene production under stresses (Han et al., 2010; Li et al., 2012a; Luo et al., 2014; Skottke et al., 2011). Dephosphorylation is also found to play roles in ACS regulation. Phosphoprotein phosphatases PP2C and PP2A reduce the phosphorylation level of ACS2/ACS6 and promote their degradation (Agnieszka et al., 2014; Skottke et al., 2011).

Type III ACS has no putative phosphorylation sites in its C-terminal end, but it was reported to be phosphorylated in vitro by a CDPK in its catalytic domain in the control of ethylene induction during root gravitropism (Huang et al., 2013). Although type II ACS proteins have one conserved CDPK phosphorylation site in their C termini, there is no report of CDPKs or phosphatases that target and influence the stability of type II ACS proteins. A study on 14-3-3 proteins and ACS provides further support for the involvement of phosphorylation in the control of ACS protein stability (Yoon and Kieber, 2013). The 14-3-3 proteins are a family of conserved regulatory proteins, which normally bind to the phosphorylated residues of their interaction proteins (Fu et al., 2000). Interestingly, Yoon and Kieber (2013) found that 14-3-3 interacts with both ETO1/EOLs (ETO1-LIKE) and ACS proteins, which consequentially increases the stability of ACS proteins in a yet unknown mechanism.

6.2.3 ACC oxidase

Similar to *ACS* genes, ACO proteins are also encoded by a small gene family that shows differential regulation in response to various developmental and environmental cues. Although ACS is generally considered as the rate-limiting enzyme in ethylene biosynthesis pathway, growing evidence reveals that ACO acts as a control point in ethylene biosynthesis under certain developmental and stress conditions in plants (Linkies et al., 2009; Qin et al., 2007; Rauf et al., 2013). The expression of *ACO* genes is induced rapidly and dramatically in a number of physiological processes, including fruit ripening, senescence, and wound responses (Barry et al., 1996; Blume and Grierson, 1997). In the apical hook of Arabidopsis etiolated seedlings, ethylene-inducible *ACO2* is specifically located in cells undergoing elongation (Raz and Ecker, 1999). During cotton (*Gossypium hirsutum*) fiber elongation, ethylene production is determined by the transcriptional induction of *ACO*, but not *ACS* (Qin et al., 2007). In Arabidopsis, the *ACO2* gene is involved in endosperm cap weakening and rupture during seed germination, and plays an important role in the ethylene-ABA antagonism in this process (Linkies et al., 2009). Another *ACO* gene, *ACO5*, is identified as a downstream target of the NAC (NAM, ATAF1/2, and CUC2 domain) transcription factor SHYG (SPEEDY HYPONASTIC GROWTH), and is required for SHYG-induced hyponastic growth upon flooding (Rauf et al., 2013). These findings demonstrate that *ACO* genes also play vital roles in regulation of ethylene biosynthesis in certain developmental or stressful situations. Further studies on the regulation of ACOs and their regulatory mechanisms are necessary for comprehensive understanding of regulation of ethylene biosynthesis.

6.3   Ethylene perception and signaling in Arabidopsis

6.3.1   Ethylene responses in Arabidopsis

Ethylene regulates a wide variety of plant growth and development processes, such as seed germination, flower development, sex determination, root elongation, root hair development, leaf senescence, and fruit ripening, as well as the responses to biotic and abiotic stresses, including drought, cold, high salinity, flooding, and pathogen attack (Chang, 2016). For instance, ethylene promotes the germination of dormant seeds and nondormant seeds under unfavorable situations (Corbineau et al., 2014). During seed germination, ethylene biosynthesis is upregulated mainly by the increased activity of ACO (Linkies et al., 2009). Ethylene is a major hormone that regulates plant response to salt. Under high salinity, the transcription factor EIN3 (ETHYLENE INSENSITIVE 3) is stabilized. Then the activated EIN3 scavenge excess reactive oxygen species (ROS) and increase salt tolerance (Peng et al., 2014). Ethylene is best known for its effect on the ripening of climacteric fruits and organ abscission, and thus has enormous importance in agriculture and horticulture. Costly methods including the use of chemical inhibitors to perturb ethylene biosynthesis, perception, and signaling are employed to prevent the spoilage of fruits, vegetables, and flowers during their transport and prolong their storage period (Chang, 2016). In Arabidopsis, dark-grown seedlings treated with ethylene exhibit a classical morphological response called the triple response, which consists of swelling of the hypocotyl, exaggeration of the apical hook, and inhibition of hypocotyl and root elongation (Fig. 6.3a). Based on the triple response phenotype, numerous ethylene-related mutants have been identified (Figs. 6.3b and c) in the past three decades (Bleecker et al., 1988; Hua and Meyerowitz, 1998; Kieber et al., 1993; Roman et al., 1995). A largely linear ethylene signaling pathway from endoplasmic reticulum (ER) to the nucleus has been established with the help of genetic analysis and biochemical approaches (Guo and Ecker, 2004).

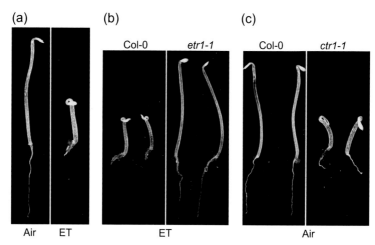

(a)   (b)   (c)

Air   ET   ET   Air

Figure 6.3 The Arabidopsis triple response and ethylene-related mutants. (a) The phenotype of Col-0 etiolated seedlings grown in the air without or with 10 ppm ethylene for 3 days. (b) The phenotype of etiolated Arabidopsis seedlings grown in the air supplemented with 10 ppm ethylene for 3 days. The seedling with long hypocotyl indicates the ethylene-insensitive mutant *etr1-1*. (c) The phenotype of etiolated Arabidopsis seedlings grown in the air for 3 days. The seedling with short hypocotyl and exaggerated hook represents the constitutive triple response mutant *ctr1-1*.

6.3.2 Ethylene perception

6.3.2.1 The first identified ethylene receptor ETR1

Ethylene is gaseous hormone, which can freely diffuse into the cells and is soluble in membranes. In Arabidopsis, ethylene is perceived by ER-localized receptor complexes, including ETR1, ETR2, ERS1, ERS2, and EIN4. Among all the plant hormone receptors, ETR1 was the first receptor identified in Arabidopsis through a genetic approach (Fig. 6.3b) (Bleecker et al., 1988; Chang et al., 1993; Guzmán and Ecker, 1990). Three lines of evidence have established ETR1 as an ethylene receptor. Firstly, specific mutations in the transmembrane domains of *ETR1* lead to ethylene insensitivity in almost all aspects of ethylene responses (Bleecker et al., 1988). Secondly, ETR1 acts upstream of the rest of the ethylene signaling components (Guzmán and Ecker, 1990; Kieber et al., 1993). Thirdly, specific ethylene-binding activity was detected in yeast cells expressing the wild-type ETR1 protein, but not a mutant ETR1 (Rodríguez et al., 1999; Schaller et al., 1995). Further studies using aqueous two-phase partitioning, sucrose density-gradient centrifugation, and immunoelectron microscopy have demonstrated that ETR1 is predominantly localized to the ER membrane (Chen et al., 2002). In addition to ETR1, four ETR1-related proteins with high ethylene-binding affinity have been identified, including ETR2, ERS1, ERS2, and EIN4 (Hua et al., 1995; Hua and Meyerowitz, 1998; Lacey and Binder, 2014; O'Malley et al., 2005; Sakai et al., 1998).

6.3.2.2 Structural and biochemical characteristics of ethylene receptors

Ethylene receptors show sequence similarity to bacterial two-component system regulators, which consist of two conserved proteins: a histidine protein kinase (HK) and a response regulator protein (RR). Phosphotransfer from HK to RR results in the activation of RR and the generation of signaling output (West and Stock, 2001). Of the five ethylene receptors in Arabidopsis, each protein contains an N-terminal transmembrane ethylene-binding domain, a GAF (cGMP-specific phosphodiesterases, adenylyl cyclases, and FhIA) domain that may facilitate heteromeric receptor interactions, and a C-terminal histidine kinase-like domain (Chang et al., 1993; Chang and Meyerowitz, 1995). ETR1, ETR2, and EIN4 contain an additional receiver domain that is commonly found in RR proteins (Chang and Shockey, 1999; Wang et al., 2002). In general, the ethylene receptors can be divided into two subfamilies mainly based on the feature of the histidine kinase-like domains. Subfamily I receptors, including ETR1 and ERS1, contain a three-transmembrane ethylene-binding domain and a conserved histidine kinase domain, which show both histidine kinase and serine/threonine kinase activity in vitro (Moussatche and Klee, 2004; Wang et al., 2006). Subfamily II receptors including ETR2, ERS2, and EIN4, contain a four-transmembrane ethylene-binding domain and a degenerate histidine kinase domain (Hua and Meyerowitz, 1998), which instead show serine/threonine kinase activity in vitro. So far, the biological importance of the histidine kinase-like domains in the ethylene receptors is still unclear. It seems likely that the histidine kinase activity has only a minor role in ethylene signaling because the kinase-dead version of etr1-1 or the truncated version of etr1-1 lacking the histidine kinase domain also confers ethylene insensitivity (Gamble et al., 2002). Nevertheless, the histidine kinase activity was shown to participate in the control of growth recovery rate after ethylene removal (Binder et al., 2004). Similar to the subfamily I receptors, the serine/threonine kinase activity of the subfamily II receptors is not required for ethylene signaling either, but may have a regulatory role in other responses independent of ethylene signal transduction (Chen et al., 2009b).

6.3.2.3 Distinct but overlapping functions of ethylene receptors in ethylene signaling

Initially, all genetically identified receptor mutants, including etr1-1, etr2-1, and ein4-1, were dominant and insensitive to ethylene (Chang et al., 1993; Roman et al., 1995; Sakai et al., 1998). The absence of ethylene-related phenotype in single recessive loss-of-function receptor mutants, except for the slightly ethylene-hypersensitive phenotype of the etr1 null mutant, suggests the existence of functional redundancy among the ethylene receptors (Hua and Meyerowitz, 1998). Different combinations of double, triple, and quadruple loss-of-function receptor mutants show different extents of the constitutive triple response phenotype, indicating that the receptors negatively and collectively regulate ethylene signaling (Hua and Meyerowitz, 1998). Despite their overlapping function, the ethylene receptors also exhibit functional divergence in mediating ethylene responses. For example, the constitutive triple response of *etr1 ers1* double mutant is more severe than that of *etr2 ers2 ein4* triple mutant (Qu et al., 2007). The *etr1 etr2 ein4 ers2* quadruple mutant has an extremely strong constitutive ethylene-response phenotype, implying that ERS1 alone cannot effectively suppress the ethylene response (Hua and Meyerowitz, 1998). In contrast, ETR1 is the only receptor in *ers1 etr2 ein4 ers2* background, and this quadruple mutant shows a moderate constitutive ethylene response, indicating that ETR1 alone may be sufficient to suppress the ethylene response to a greater extent than other receptors (Liu and Wen, 2012).

6.3.2.4 Regulation of ETR1

Biochemical studies also revealed the uniqueness of ETR1 in ethylene perception, as the activity of ETR1 was found to be solely affected by RTE1 (REVERSION-TO-ETHYLENE SENSITIVITY 1), which was identified in a genetic screen for suppressors of *etr1-2* (Resnick et al., 2006). While the *rte1* mutant suppresses the ethylene-insensitive phenotype of several dominant missense alleles of *etr1*, it has no effect on any of the other four ethylene receptor mutants (Resnick et al., 2006). RTE1 is primarily located in the Golgi apparatus and partially in the ER, and physically interacts with ETR1 (Dong et al., 2010, 2008). The precise molecular mechanism by which RTE1 affects the function of ETR1 is still unclear, but it was speculated to play a role in modulating ETR1 folding (Resnick et al., 2008; Zhou et al., 2007). Recently, a group of ER-localized small hemoproteins, called Cb5 (cytochrome b_5), were identified that physically interact with RTE1 in plants (Chang et al., 2014). Mutation in *Cb5* partially suppresses the ethylene insensitivity of dominant *etr1* alleles. Genetic analysis found that Cb5 may work upstream of RTE1 to promote the ETR1-mediated repression of ethylene response (Chang et al., 2014). Cytochrome *b5* serves as an electron transfer protein in various oxidation/reduction reactions (Hirayama et al., 1999). Disturbances in redox regulation upon sensing abiotic and biotic signals can lead to ER stress responses (Vembar and Brodsky, 2008). Given that ethylene is a well-known stress hormone, it is thus likely that the redox status of the ER may affect ethylene perception especially under stress conditions. This may be achieved by redox-related modification of ethylene receptors, at least for ETR1, through a mechanism involving RTE1 and Cb5 (Chang et al., 2014; Dong et al., 2010, 2008; Resnick et al., 2008, 2006; Zhou et al., 2007).

6.3.3 A negative regulatory kinase CTR1

The triple response of Arabidopsis has been used to identify mutants involved in ethylene signaling, which can be broadly divided into two groups: ethylene-insensitive mutants and constitutive triple response mutants (Fig. 6.3). Loss-of-function *ctr1* (*constitutive triple response 1*) mutants display a constitutive triple response even in the presence of inhibitors of ethylene biosynthesis or perception (Fig. 6.3c), indicating that CTR1 is a negative regulator of ethylene signal transduction (Kieber et al., 1993). The ethylene-insensitive phenotype of *etr1* is rescued by *ctr1* mutation, suggesting that CTR1 acts downstream of ethylene receptors (Kieber et al., 1993). CTR1 consists of an N-terminal regulatory domain and a C-terminal kinase domain, which shows similarity to the Raf family of serine/threonine protein kinases (Kieber et al., 1993). It was shown that CTR1 possesses intrinsic Ser/Thr protein kinase activity (Huang et al., 2003). Missense alleles of *ctr1* with disrupted kinase activity exhibit a similar phenotype to *ctr1* null alleles, highlighting the importance of the CTR1 kinase activity in ethylene signaling (Huang et al., 2003; Kieber et al., 1993). The N-terminal domain of CTR1, which has no effect on the kinase activity, physically interacts with the histidine kinase domain and the receiver domain of the ethylene receptors, and the interaction is also vital for ethylene signal transduction (Gao et al., 2003). Once it is disrupted, seedlings will show constitutive triple response, as exemplified by the phenotype of *ctr1-8*, whose mutation blocks the interaction of CTR1 with the receptors (Huang et al., 2003).

While CTR1 itself has no predicted transmembrane domain, it was revealed that CTR1 is associated with the ER membrane fraction in Arabidopsis (Gao et al., 2003). Further investigation found that the ER-localization of CTR1 is dependent on the interaction with ethylene receptors, suggesting the existence of a negative signaling complex including the receptors and CTR1 involved in the initial step of ethylene signal transduction (Gao et al., 2003). The receptor complex is active and the ethylene response is repressed in the absence of ethylene. Upon ethylene binding, the function of the receptor complex is inhibited and consequently the downstream ethylene signaling is derepressed (Gao et al., 2003). The exact molecular mechanism by which this receptor complex is inactivated upon ethylene binding remains elusive.

Meanwhile, there is some evidence for an alternative ethylene signaling pathway that bypasses CTR1 but depends on the ethylene receptors. Although displaying a constitutive ethylene-response phenotype, the *ctr1* null mutants are still mildly responsive to ethylene application (Guo and Ecker, 2003). Mutants lacking multiple receptors show a stronger constitutive triple response phenotype than the *ctr1* null mutant (Hua and Meyerowitz, 1998; Liu et al., 2010). Furthermore, expression of *ETR1(1-349)* or *etr1-1(1-349)*, which lack both the histidine kinase and receiver domains, suppressed the phenotype of *ctr1* mutant (Qiu et al., 2012; Xie et al., 2012). Currently the molecular basis of this CTR1-independent pathway, as well as its biological significance in ethylene-regulated processes, remains to be determined.

Based on sequence comparisons, CTR1 has been thought to function as a Raf-like kinase. However, the substrates of CTR1 kinase remained mysterious and controversial since its discovery in 1993 (Kiber et al., 1993; Guo and Ecker, 2004). In 2012, it was reported that CTR1 physically interacts with and phosphorylates EIN2 on a series of serine/threonine residues (Ju et al., 2012), which provides a definitive answer to this long-standing question. This finding represents a major advance in the understanding of ethylene signaling, and will be described in detail in the following section.

6.3.4   *The key positive regulator EIN2*

EIN2 (*ETHYLENE INSENSITIVE 2*) is a single-copy gene in the Arabidopsis genome whose loss-of-function mutations confer complete insensitivity to ethylene, demonstrating its extreme importance in transducing the ethylene signal (Alonso et al.,

1999). Genetic analyses show that EIN2 acts downstream of CTR1 and upstream of EIN3/EIL1 (Roman et al., 1995; Chao et al., 1997). *EIN2* encodes a polypeptide of 1294 amino acids that consist of a predicted 12-fold transmembrane region in its N terminus, which shows sequence similarity with Nramp (natural resistance associated macrophage protein) family of metal transporter proteins, and a structurally unknown hydrophilic C terminus (Alonso et al., 1999). However, no metal-transporting capacity of EIN2 has been reported to date. Overexpression of the C terminus of EIN2 (CEND, amino acids 459-1294) in *ein2* displays constitutive ethylene response in the light, indicating that CEND is sufficient to activate ethylene signaling in this condition (Alonso et al., 1999). Interestingly, *CEND* overexpression cannot restore the insensitive phenotype of *ein2* under dark conditions, suggesting that the N terminus is required for EIN2 function in darkness (Alonso et al., 1999).

In 2009, Bisson et al. reported that EIN2 is localized on the ER membrane when transiently expressed in tobacco leaves. The ER-localization of EIN2 was further confirmed using sucrose density-gradient centrifugation and transgenic plants expressing fluorescence-tagged *EIN2* (Qiao et al., 2012). Moreover, it was shown that the C terminus of EIN2 interacts with the kinase domain of all five ethylene receptors at the ER membrane. Mutational analysis demonstrates that the phosphorylation status of the receptors affects the formation of the EIN2-receptor complex (Bisson et al., 2009; Bisson and Groth, 2011). Given that the kinase activity of the receptors is dispensable for the typical ethylene response, the significance of this receptor-EIN2 interaction is puzzling.

The EIN2 protein undergoes proteasome-mediated turnover, and its accumulation is upregulated by ethylene (Qiao et al., 2009). Two F-box proteins ETP1 (EIN2 TARGETING PROTEIN 1) and ETP2 were identified that interact with EIN2-CEND. Reducing the expression level of both *ETP1* and *ETP2* leads to the accumulation of EIN2 protein, which confers a constitutive triple response phenotype. Conversely, overexpression of *ETP1* or *ETP2* results in a decrease in EIN2 accumulation and ethylene sensitivity (Qiao et al., 2009). Furthermore, the protein levels of ETP1 and ETP2 are downregulated and the interactions between ETP1/ETP2 and EIN2 are impaired upon ethylene treatment (Qiao et al., 2009). Since the kinetics of EIN2 protein stabilization is relatively slow, it is arguable about whether ETP1/2-mediated EIN2 protein degradation is a primary or feedback mechanism in ethylene signal transduction.

6.3.4.1 EIN2 is cleaved and transported into the nucleus

Though a linear ethylene signaling pathway was established, it remained unclear how the ethylene signal is transmitted from the ER-localized receptors, CTR1 and EIN2 to the nucleus. In 2012, research progress by three independent groups greatly advanced our understanding of the mechanisms underlying ethylene signal transduction. It was reported that EIN2 undergoes an ethylene-induced cleavage and translocation from the ER membrane to the nucleus, which is controlled by CTR1-mediated phosphorylation of the C terminus of EIN2 (Fig. 6.4) (Ju et al., 2012; Qiao et al., 2012; Wen et al., 2012).

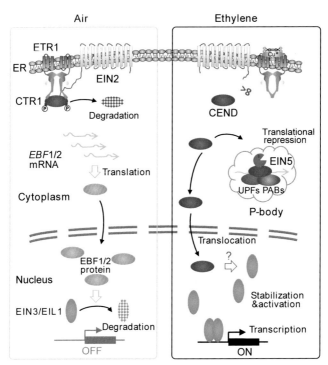

Figure 6.4 Two modes of EIN2 action in ethylene signal transduction. Two modes of EIN2 action in mediating ethylene signaling. In the absence of ethylene, EIN2 is phosphorylated by CTR1 and retained in the ER membrane. The transcription factors EIN3/EIL1 are degraded by the SCF complex containing EBF1/EBF2, and thus the ethylene responses are inhibited. Upon sensing ethylene, EIN2 is dephosphorylated, resulting in its cleavage and release from the ER membrane. On one hand, the cytoplasm-localized EIN2 CEND associates with the 3′UTR of *EBF1/EBF2* mRNAs and targets them to P-body through interacting with multiple P-body factors, resulting in the translational repression of *EBF1* and *EBF2* mRNA. On the other hand, EIN2 C-terminal fragment can also be transported into the nucleus to stabilize EIN3/EIL1 through a yet unknown mechanism. Together, EIN3/EIL1 proteins are accumulated in the nucleus followed by the activation of downstream gene expression and ethylene responses.

A conserved nuclear localization signal (NLS) motif was predicted in the C terminus of EIN2 (Alonso et al., 1999; Bisson and Groth, 2011). It was found that the NLS is required for the function of EIN2, as expression of an NLS-mutated *EIN2* was unable to complement the *ein2-5* mutant phenotype (Qiao et al., 2012). Although EIN2 is primarily localized on the ER membrane in the absence of ethylene, the nuclear localization of EIN2 was detected within 10 min of ethylene exposure, and a greater amount of nuclear EIN2 was observed with longer ethylene treatment. However, the nuclear localization was absent when the NLS was mutated, demonstrating that the ER-nucleus translocation of EIN2 is vital for ethylene signaling (Qiao et al., 2012; Wen et al., 2012). Given that EIN2 contains an N-terminal transmembrane domain, the researchers hypothesized that EIN2 may undergo proteolysis and its C-terminal cleavage fragment translocates to the nucleus upon ethylene sensing. A cleaved form of the native EIN2 protein called EIN2-C', was observed to gradually accumulate in the nucleus upon ethylene treatment (Qiao et al., 2012). Interestingly, Wen et al. (2012) detected at least five C-terminal fragments of EIN2 using an Arabidopsis suspension cell line stably expressing EIN2-GFP. By use of mass spectrometry, the cleavage site of EIN2 was identified between S645 and F646 (Qiao et al., 2012), while none of the precise cleavage sites were determined for the cleaved fragments observed by Wen et al. (2012).

The next apparent question is how ethylene induces the cleavage and translocation of EIN2. An earlier proteomic study identified multiple phosphorylation sites in the C terminus of EIN2 (Chen et al., 2011). Considering that CTR1 is a Raf-like protein kinase that is active in the absence of ethylene, Ju et al. (2012) proposed that CTR1 might be a good candidate kinase responsible for the phosphorylation of EIN2 in such conditions. This hypothesis turned out to be true, as CTR1 was able to interact with EIN2 through its kinase domain. An in vitro kinase assay followed by mass spectrum analysis identified S645 and S924 as the phosphorylation sites of EIN2 by CTR1 (Ju et al., 2012). Study by Qiao et al. (2012) also found the phosphorylation of S645, and identified a cleavage immediately after this site, but the phosphorylation of S924 was not observed in this study. Consistent with these results, the phosphorylation of S645 was significantly inhibited in *ctr1* mutants, suggesting that CTR1 is required for S645 phosphorylation of EIN2 (Ju et al., 2012; Qiao et al., 2012).

The importance of EIN2 phosphorylation is further supported by the phenotypic analysis of phosphorylation mutants. It was found that expression of *EIN2^{S645A}* rescues the insensitive phenotype of *ein2* and shows a constitutive triple response, whereas *EIN2^{S645E}* cannot (Qiao et al., 2012). Taken together, Qiao et al. (2012) suggested that the dephosphorylation of S645 is a critical step regulating EIN2 cleavage and ethylene response. However, Ju et al. (2012) suggested a more important role for S924 than S645 in regulating EIN2 function. Plants expressing *EIN2^{S924A}* show a stronger triple response than that of *EIN2^{S645A}*, and plants expressing *EIN2* with both substitutions (*EIN2^{AA}*) display the strongest phenotype, suggesting an additive effect of the phosphorylation of S645 and S924 (Ju et al., 2012).

6.3.4.2   EIN2 mediates translational repression in cytoplasm

The stabilization of EIN3/EIL1 by ethylene, which is mediated through inhibition of EBF1 (EIN3-BINDING PROTEIN 1)/EBF2-directed proteasomal degradation of EIN3/ EIL1 in an EIN2-dependent manner, is a primary mechanism for ethylene signaling (Guo and Ecker, 2003; Zhao and Guo, 2011). However, how ethylene and EIN2 inhibit the function of EBF1/EBF2 remained unknown until a novel mechanism of ethylene signaling was uncovered by two groups. Both groups reported that ethylene induces EIN2 to target *EBF1/EBF2* mRNAs to cytoplasmic P-bodies to repress their translation (Li et al., 2015a; Merchante et al., 2015).

The research of Li et al. (2015a) started from previous observations that the 3'UTR (Un-Translated Region) fragments of *EBF1* and *EBF2* transcripts are accumulated in *ein5,* an ethylene-hyposensitive mutant deficient in the *XRN4* (*EXORIBONUCLEASE 4*) gene (Olmedo et al., 2006; Souret et al., 2004). To investigate whether the accumulation of these 3'UTR fragments might confer the ethylene hyposensitivity of *ein5*, *EBF1* 3'UTR (*1U*) region was overexpressed in wild-type plants (Li et al., 2015a). They found that overexpression of *1U* displays significantly reduced ethylene sensitivity and impaired EIN3 protein accumulation. They further demonstrated that overexpression of *1U* promotes the translation of endogenous *EBF1/EBF2* mRNAs, which might arise from the competition or titration of translational repressors presumably binding to the 3'UTR regions of endogenous *EBF1/EBF2* transcripts (Li et al., 2015a). Using the ribosome footprinting technology, Merchante et al. (2015) observed similar results that the translation of *EBF1/EBF2* mRNAs is repressed by ethylene.

The inhibitory role of *EBF1/EBF2* 3'UTR in translation was further characterized by fusing a GFP coding region with the 3'UTR sequences. Both groups found that the GFP fluorescence and protein abundance in seedlings expressing *GFP-3'UTR* were significantly decreased upon treatment with ACC, indicating that *EBF1/2* 3'UTR sequences are sufficient to confer ethylene-induced translational repression (Li et al., 2015a; Merchante et al., 2015). Genetic analyses showed that the upstream signaling components including the ethylene receptors and EIN2 are required for this translational repression, whereas EIN3/EIL1 proteins are not (Li et al., 2015a; Merchante et al., 2015). Further investigation found that poly U motifs in *EBF1* and *EBF2* 3'UTRs, as well as amino acids 654-1272 of EIN2, including the NLS, are required for ethylene-induced and EIN2-mediated translational repression (Li et al., 2015a).

Next, both groups reported that EIN2 associates with *EBF1/EBF2* 3'UTR sequences, which is enhanced by ethylene treatment. In previous studies (Qiao et al., 2012; Wen et al., 2012), EIN2 was observed to form speckles in the cytoplasm in addition to nuclear accumulation after ethylene application. Both groups found that 3'UTR-containing mRNAs also form granules in the cytoplasm and co-localize with EIN2 and other processing-body (P-body) factors, implying that at least some of the speckles or granules in the cytoplasm are P-bodies. Consistent with these observations, mutants of these P-body factors displayed reduced ethylene sensitivity and EIN3 accumulation. These observations strongly suggest that P-bodies are involved in *EBF1/EBF2* 3'UTR-mediated translational repression by EIN2. Taken together, a new mode of EIN2 action was proposed (Li et al., 2015a; Merchante et al., 2015). After ethylene perception, EIN2 is cleaved and released from the ER membrane. The cytoplasm-localized EIN2 CEND associates with the 3'UTR of *EBF1/EBF2* mRNAs and targets *EBF1/EBF2* mRNAs to P-bodies through interaction with multiple P-body factors, resulting in the translational repression of *EBF1* and *EBF2* mRNAs (Fig. 6.4).

As mentioned before, the NLS of EIN2 is necessary for its ER-nucleus translocation and ethylene response. The NLS is also essential for its function in translational repression (Li et al., 2015a). Recently, it was reported that EIN2 lacking the NLS domain shows reduced interaction with the ethylene receptors (Bisson and Groth, 2015). It would be very interesting to determine how such a short motif is involved in seemingly distinctive subcellular signaling events. One possibility is that the NLS is vital for the normal conformation of EIN2 CEND. In addition to *EBF1* and *EBF2* mRNAs, many other mRNAs have been identified that are regulated by ethylene or EIN2-mediated translational repression (Merchante et al., 2015). Studies on these genes will help elucidate the action of EIN2 in other ethylene-dependent or ethylene-independent processes.

6.3.5 The master transcription factors EIN3 and EIL1 in the nucleus

Downstream of EIN2, ethylene signaling is mediated by two plant-specific transcription factors, EIN3 and EIL1 (EIN3-LIKE 1). The *ein3* mutant was identified in a genetic screen for ethylene-insensitive mutants (Roman et al., 1995) and was subsequently cloned and characterized by Chao et al. (1997). The *EIN3* gene encodes a nuclear-localized transcription factor, though the precise NLS is yet to be identified. Based on the structure and interaction analyses, at least three domains can be defined in the EIN3 protein: a DNA binding domain including amino acids 80-359, a dimerization domain comprising amino acids 113-257, and a C-terminal regulatory domain required for interaction with other proteins, such as EBF1/EBF2 (Chao et al., 1997; Li et al., 2012b; Solano et al., 1998; Yamasaki et al., 2005). There are five EIN3 homologs in Arabidopsis, EIL1-EIL5, among which EIN3 and EIL1 share the highest sequence similarity (Chao et al., 1997; An et al., 2010). Although the *eil1* mutant shows weak ethylene insensitivity, the *ein3 eil1* double mutant has almost complete ethylene insensitivity similar to that of *ein2* in many aspects of ethylene responses (Alonso et al., 2003). Moreover, overexpression of *EIL1* restores the partial ethylene-insensitive phenotype of *ein3* and leads to a constitutive triple response similar to overexpression of *EIN3* (Chao et al., 1997). These observations suggest that EIN3 and EIL1 are two major transcription factors in ethylene signaling with functional redundancy.

EIN3/EIL1 proteins are necessary and sufficient to activate the ethylene response. The ultimate step of ethylene signaling relayed from the ER membrane is the stabilization of EIN3/EIL1 proteins in the nucleus, followed by the induction of multiple downstream genes (Zhao and Guo, 2011). In the absence of ethylene, EIN3/EIL1 proteins are rapidly degraded by an SCF complex containing F-box proteins EBF1/EBF2 through a 26S proteasome-mediated protein degradation pathway (An et al., 2010; Guo and Ecker, 2003; Potuschak et al., 2003). Although differential phosphorylation of EIN3 by two MAP kinases MPK3/6 has been suggested as a determinant for EIN3 protein stability, it is widely believed that the MAPK pathway plays a role in ethylene biosynthesis under stress conditions and may not target EIN3 directly (An et al., 2010; Xu et al., 2008; Yoo et al., 2008).

6.3.5.1 The regulation of EIN3 activity

The regulation of EIN3 activity has emerged as an important mechanism controlling ethylene response and cross talk between ethylene and other signals. The regulation of EIN3/EIL1 activity was firstly reported by Zhu et al. (2011). They found that EIN3/EIL1 proteins are positive regulators in many aspects of jasmonic acid (JA) responses, because the *ein3 eil1* double mutant is defective in JA-induced gene expression, root hair development and inhibition of root growth. The transcription repressors JAZ (JA-Zim domain) proteins physically interact with EIN3/EIL1 and inhibit their transcriptional activity. JAZ protein failed to interfere with the DNA-binding activity of EIN3/EIL1 in vitro, suggesting that other components may be required for JAZ action. HDA6 (HISTONE DEACETYLASE 6) was then found to interact with both JAZ and EIN3, and the interaction between HDA6 and EIN3 was weakened by JA treatment. HDA6 negatively regulates JA responses in an EIN3/EIL1-dependent manner. Moreover, the histone acetylation level in the promoter region of *ERF1* (*ETHYLENE RESPONSE FACTOR 1*) is significantly increased upon JA treatment, which is more pronounced in the *hda6* mutant. Taken together, it is proposed that JAZ proteins recruit HDA6 to deacetylate histones and disrupt the chromatin binding of EIN3/EIL1 in the absence of JA. Upon JA treatment, JAZs-HDA6 corepressors are removed from EIN3/EIL1 through degradation of JAZ proteins, resulting in the derepression of EIN3/EIL1 activity (Zhu et al., 2011). Later, An et al. (2012) reported that cross talk between gibberellic acid (GA) and ethylene in regulating hook development also converges on the transcription activity of EIN3/EIL1. Similar to the aforementioned mechanism, DELLA proteins, which are key repressors of GA responses, interact with EIN3/EIL1 and repress their activity in a yet unknown mechanism. GA promotes hook curvature through degradation of DELLA proteins, thus releasing their inhibition on EIN3/EIL1 activity, which is followed by direct induction of *HLS1* (*HOOKLESS1*) expression (An et al., 2012).

Many aspects of plant growth and development are coordinately regulated by JA and ethylene. However, it has also been shown that JA and ethylene antagonistically regulate apical hook development. It has been found that the transcription factor MYC2, which is activated by JA, directly interacts with EIN3 and inhibits its DNA binding activity (Song et al., 2014; Zhang et al., 2014). Whereas MYC2 interacts with EIN3 to attenuate the transcriptional activity of EIN3 and repress ethylene-induced hook curvature, Song et al. (2014) found that EIN3 reciprocally acts to repress MYC2 to inhibit JA-regulated plant defense against generalist herbivores, further adding the complexity on hormone interplays.

The aforementioned JAZ, DELLA, and MYC2 proteins are all repressors of EIN3 transcriptional activity. On the other hand, FIT (FER-LIKE FE DEFICIENCY-INDUCED TRANSCRIPTION FACTOR), a key regulator of Fe acquisition in roots, is positively regulated by EIN3/EIL1 through direct binding (Lingam et al., 2011). Association with EIN3/EIL1 might prevent the proteasomal degradation of FIT, and thus positively regulate the response to Fe deficiency (Lingam et al., 2011). Whether FIT also has some effect on EIN3/EIL1 stability needs to be further addressed. In summary, the regulation of EIN3 stability and activity is subjected to more and more research in the cross talk between ethylene and other signals (Fig. 6.5).

6.3.5.2   The F-box proteins EBF1 and EBF2

The stability of EIN3/EIL1 proteins is tightly regulated by SCF complexes containing F-box protein EBF1 or EBF2 through the 26S proteasome-mediated protein degradation pathway (Guo and Ecker, 2003; Potuschak et al., 2003). It has been reported

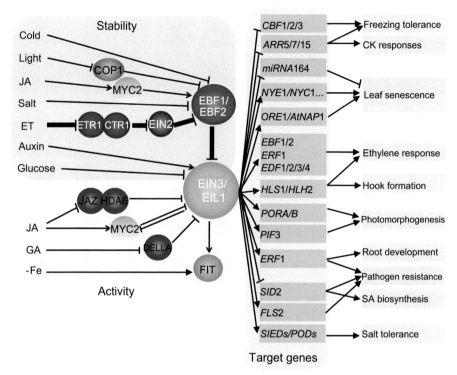

Figure 6.5 A summary on the EIN3/EIL1-regulated gene expression in different ethylene responses of Arabidopsis. In addition to ethylene, EIN3/EIL1 are also subjected to regulation by different types of signals and stimuli. Cold, light, and salt induce EIN3/EIL1 protein accumulation through destabilization of EBF1/EBF2 proteins. COP1 directly targets EBF1/EBF2 for degradation and ubiquitination, thus promoting the accumulation of EIN3/EIL1. MYC2 directly induces the expression of *EBF1* upon JA treatment, resulting in the increase of EBF1 protein followed by the degradation of EIN3/EIL1. Auxin promotes the accumulation of EIN3/EIL1 proteins and glucose accelerates the degradation of EIN3/EIL1 through unknown mechanisms. In addition to the protein stability, the transcriptional activity of EIN3/EIL1 is also affected by multiple signals. In the absence of JA, JAZ proteins recruit HDA6 to deacetylate histones and disrupt the chromatin binding of EIN3/EIL1, the effect of which is released upon JA treatment through degradation of JAZ proteins. Moreover, MYC2, which is activated by JA, directly interacts with EIN3 and inhibits its DNA binding activity. EIN3 reciprocally acts to repress the transcriptional activity of MYC2. DELLA proteins also directly repress the activity of EIN3/EIL1. The inhibition is released by GA application through degradation of DELLA proteins. EIN3/EIL1 positively regulates the stability of transcription factor FIT in response to Fe deficiency.

that EBF1 and EBF2 have distinct but overlapping roles in regulating EIN3 abundance (Binder et al., 2007; Gagne et al., 2004; Guo and Ecker, 2003). EBF1 plays a major role in the absence of ethylene and during the initial phase of signaling, whereas EBF2 exerts its effect primarily in the later stages of the response and the resumption of seedling growth following ethylene removal (Binder et al., 2007). The expression levels of *EBF1* and *EBF2* are upregulated by ethylene, while the induction of *EBF2* is greater than that of *EBF1* (Gagne et al., 2004; Konishi and Yanagisawa, 2008). It has been demonstrated that EIN3 directly binds to the promoter of *EBF2* gene, thus forming a negative feedback loop to fine-tune the abundance of EIN3/EIL1 protein and ethylene response (Konishi and Yanagisawa, 2008).

Given that the stability of EIN3/EIL1 is key to ethylene responses, the significance of the regulation on EBF1/EBF2 is apparent. The protein levels of EBF1/EBF2 are downregulated by ethylene but upregulated by MG132, a potent 26S proteasome inhibitor (An et al., 2010). These observations suggest that ethylene promotes the accumulation of EIN3/EIL1 at least partly by inducing EBF1/EBF2 proteasomal degradation. As described above, *EBF1/2* mRNAs are also under translational repression mediated by EIN2 and other P-body factors including EIN5 (Li et al., 2015a; Merchante et al., 2015). Interestingly, in *ein5* mutants, the levels of the full-length *EBF1* and *EBF2* transcripts are also over-accumulated, although the decay rate of these mRNAs is not apparently altered (Olmedo et al., 2006; Potuschak et al., 2006). It is thus intriguing to know which effect of EIN5 (P-body component in translation repression and transcripts abundance control) is the authentic cause for its role in ethylene signaling.

In addition to ethylene, several stress or hormone signals were shown to affect the level of EIN3 protein, some of which act by modulating the stability of EBF1/2 proteins. Cold temperature induces EIN3 protein accumulation through destabilizing EBF1 protein in an EIN2-dependent manner (Shi et al., 2012), while high salinity stabilizes EIN3 protein by promoting EBF1/EBF2 proteasomal degradation in both EIN2-dependent and EIN2-independent processes (Peng et al., 2014). It was reported that COP1, a RING finger ubiquitin E3 ligase, which negatively regulates seedling photomorphogenesis, directly targets EBF1/EBF2 for ubiquitination and degradation, thus promoting the accumulation of EIN3/EIL1 in a light-mediated process (Shi et al., 2016; Zhong et al., 2009). Overall, the abundance of EBF1/EBF2 protein is one of the key nodes in the cross talk between ethylene and other signals (Fig. 6.5).

6.3.6   *Transcriptional network regulated by EIN3/EIL1*

As described above, EIN3 and EIL1 are two master transcription factors that cooperatively and differentially regulate the vast majority of ethylene-responsive gene expression (An et al., 2010; Chang et al., 2013). The first gene that was identified as a direct target gene of EIN3 is *ERF1* (Solano et al., 1998). Together with other *ERF* genes that are also targeted by EIN3/EIL1, this family of transcription factors bind to the GCC-box elements and further regulate the expression of a myriad defense genes (Solano et al., 1998). EIN3 also directly activates the expression of *FLS2* (*FLAGELLIN SENSITIVE 2*), which encodes a leucine-rich repeat receptor kinase that participates in sensing the bacterial flagellin in plant immunity (Boutrot et al., 2010). In addition to enhancing pathogen-related defense, ethylene is also a major hormone in the tolerance or resistance of plants to abiotic stresses. For instance, ethylene application can increase plant tolerance to high salinity. It was found that EIN3 directly activates the transcription of a number of *SIED* (*SALT-INDUCED*

AND EIN3/EIL1-DEPENDENT) genes, including *SIED1, ZAT12 (ZINC FINGER OF ARABIDOPSIS THALIANA12), SZF2 (SALT-INDUCIBLE ZINC FINGER 2), AZF2 (ARABIDOPSIS ZINC FINGER PROTEIN 2)*, as well as the expression of numerous peroxidase (*POD*) genes, which collectively leads to a higher capacity to scavenge ROS under salt stress (Peng et al., 2014).

Studies have revealed that ethylene plays a critical role in plant photomorphogenesis and chlorophyll metabolism. In early seedling development, ethylene was shown to facilitate cotyledon greening of etiolated seedlings upon light irradiation. This is due to the EIN3-mediated induction of *PORA/B (PROTOCHLOROPHYLLIDE OXIDOREDUCTASE A and B)* that encodes the key enzymes in the chlorophyll biosynthesis pathway (Zhong et al., 2009). Meanwhile, *PIF3 (PHYTOCHROME-INTERACTING FACTOR 3)* encoding a basic helix-loop-helix (bHLH) transcription factor in light signaling, is also directly induced by EIN3, providing a nice explanation for how ethylene promotes hypocotyl elongation of light-grown seedlings (Zhong et al., 2012). As an "aging" signal, ethylene promotes chlorophyll degradation and leaf senescence at a later developmental stage. Several chlorophyll catabolic genes (*CCGs*), including *NYE1 (NON-YELLOWING 1), NYC1 (NON-YELLOW COLORING 1)*, and *PAO (PHEOPHORBIDE A OXYGENASE)*, are direct targets of EIN3 (Qiu et al., 2015). Additionally, two senescence-associated NAC transcription factors, ORE1 (ORESARA1) and AtNAP (NAC-LIKE, ACTIVATED BY AP3/PI), are directly induced by EIN3 at the transcription level (Kim et al., 2014).

While EIN3 and EIL1 are generally considered as transcriptional activators, there are several reports that implicate them as transcriptional repressors. The expression of *SID2 (SALICYLIC ACID INDUCTION DEFICIENT 2)*, which encodes isochorismate synthase required for the biosynthesis of SA (salicylic acid), is directly repressed by EIN3/EIL1, leading to a decrease in SA levels and downregulation of PAMP(pathogen-associated molecular pattern)-triggered immunity by ethylene (Chen et al., 2009a). EIN3 was also found to inhibit the transcription of *miRNA164* through directly binding to its promoter, and the decrease of *miRNA164* abundance under ethylene application results in acceleration of leaf senescence (Li et al., 2013). In addition, ethylene negatively regulates plant cold tolerance at least partially through directly binding of EIN3 to the promoters of cold-regulated *CBF (C-REPEAT BINDING FACTOR)* genes and type-A *ARR (ARABIDOPSIS RESPONSE REGULATOR)* genes and repressing their transcription (Shi et al., 2012).

A genome-wide study using ChIP-sequence and RNA-sequence has identified over a 1000 genes that can be directly targeted by EIN3 during a time-course of ethylene treatment (Chang et al., 2013). It was shown that EIN3 regulates the expression of ethylene-response genes in a sequence of four waves, each of which is represented by a distinct set of EIN3 targets. Interestingly, the majority of the known ethylene signaling components were identified as direct targets of EIN3, suggesting a widely existing feedback regulation circuitry in ethylene responses. They also found that EIN3 binding is involved in the constitution of a transcriptional network by integrating most of the hormone and stress signals that coordinate plant growth and defense (Fig. 6.5) (Chang et al., 2013).

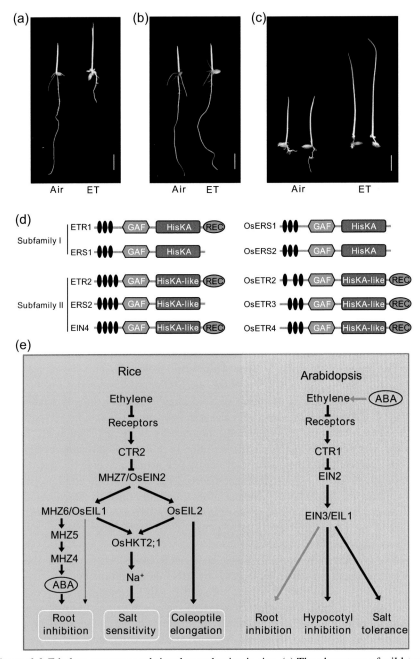

Figure 6.6 Ethylene response and signal transduction in rice. (a) The phenotype of wild-type rice (Nipponbare) seedlings grown in the dark without (air) or with 10 ppm ethylene (ET) for 3 days. Bar = 10 mm. (b) The phenotype of rice mutant *mhz7/Osein2* seedlings grown in the dark without (air) or with 10 ppm ethylene (ET) for 3 days. Bar = 10 mm. (c) The phenotype of *MHZ7/OsEIN2*-overexpressing rice seedlings grown in the dark without (air) or with 10 ppm

6.4 Ethylene perception and signaling in rice

6.4.1 Ethylene responses in rice

Rice is an important crop worldwide and lives in water-saturated environment in most of the life cycle. Ethylene plays essential roles in the processes of rice adaptation to the hypoxic condition, including shoot/coleoptile elongation, adventitious root emergence, and aerenchyma formation. The hormone also functions in rice growth and developmental processes, including seed germination, coleoptile elongation, leaf senescence, flowering, grain-filling, and yield trait formation, as well as abiotic and biotic stresses (Fukao and Bailey-Serres, 2008; Fukao et al., 2011; Ismail et al., 2009; Jackson, 2003; Ku et al., 1970; Ma et al., 2014; Rzewuski and Sauter, 2008; Satler and Kende, 1985; Yang et al., 2015a). Ethylene signaling components and mechanisms have been investigated based on homolog analysis and mutant screening in rice. Taking advantage of a double response of etiolated rice seedlings, that is, coleoptile elongation and root inhibition (Figs. 6.6a–c), which is different from the triple response of Arabidopsis etiolated seedlings, a set of rice ethylene-response mutants *mao huzi* (*mhz*, Chinese name with an English meaning of "cat whiskers") have been identified and the genes have been cloned and their functions dissected. These studies reveal conserved and diverged features of rice ethylene signaling. Given that the major components of ethylene signaling have been identified and the pathway is primarily established in Arabidopsis, we will summarize the advances of ethylene signaling in rice by comparing it with Arabidopsis in the next section.

6.4.2 Ethylene receptors and interacting proteins

In the rice genome, there are five ethylene receptor genes, namely *OsERS1*, *OsERS2*, *OsETR2*, *OsETR3*, and *OsETR4*. The OsERS1 and OsERS2 proteins belong to subfamily I, which have a conserved histidine kinase domain, whereas the other three are

ethylene (ET) for 3 days. Bar = 10 mm. (d) The ethylene receptor family of Arabidopsis and rice. The N-terminal transmembrane domains are indicated by *black bars*, followed by a GAF domain, a histidine kinase (HisKA) or histidine kinase-like (HisKA-like) domain, and a receiver (REC) domain if present. The left five receptors are from Arabidopsis and the right five are from rice. For both Arabidopsis and rice, the top two receptors belong to subfamily I whereas the bottom three belong to subfamily II. (e) Comparison of ethylene signaling pathway between rice and Arabidopsis. Ethylene signaling in rice and Arabidopsis shows both conserved and diverged aspects. For conserved features, both Arabidopsis and rice have five ethylene receptors or homologs, functional CTR1/or CTR1-like proteins, functional EIN2 or MHZ7/OsEIN2, functional EIN3 and EIL1 or MHZ6/OsEIL1 and OsEIL2. For diverged features, Arabidopsis etiolated seedling has triple response phenotype whereas rice etiolated seedling has double response phenotype upon ethylene treatment. Rice MHZ6/OsEIL1 and OsEIL2 have organ specificity in regulation of ethylene response whereas Arabidopsis EIN3 and EIL1 do not. Ethylene signaling facilitates salt tolerance in Arabidopsis but causes salt sensitivity in rice. In Arabidopsis, ABA requires ethylene pathway for root inhibition, whereas in rice, ethylene requires ABA pathway for root inhibition.

subfamily II members and contain a diverged kinase domain plus a receiver domain. This is the same number of receptor genes as in Arabidopsis where ETR1 and ERS1 belong to subfamily I with a conserved histidine kinase domain, while ERS2, ETR2, and EIN4 are subfamily II members with a diverged kinase domain (Ju and Chang, 2015, Fig. 6.6d). It should be noted that rice and many other monocotyledonous plants do not have ETR1-type ethylene receptor, which plays a strong role in Arabidopsis, suggesting that different plants may have distinct features. Although all the ethylene receptors show structural similarity to bacterial histidine kinases, only Arabidopsis ETR1 and tobacco NtETR1 have histidine kinase activity, whereas subfamily II members, for example, tobacco NTHK1, Arabidopsis ERS2, ETR2, and EIN4, and rice OsETR2 have Ser/Thr kinase activity (Chen et al., 2009b; Gamble et al., 1998; Moussatche and Klee, 2004; Wuriyanghan et al., 2009; Xie et al., 2003). Arabidopsis ERS1 and tobacco subfamily II member NTHK2 have His and Ser/Thr kinase activity in the presence of different ions (Moussatche and Klee, 2004; Zhang et al., 2004).

Roles of the ethylene receptor kinase activity are unclear and have been analyzed in detail in only a few cases. The kinase domain of rice OsETR2 can phosphorylate its receiver domain; however, the significance of this phosphorylation is not known (Wuriyanghan et al., 2009). Overexpression of the tobacco subfamily II receptor gene *NTHK1* in transgenic tobacco and Arabidopsis plants leads to large seedlings and rosettes, reduced ethylene response, and salt sensitivity (Cao et al., 2007; 2006; Zhou et al., 2006). When the kinase activity of NTHK1 was disrupted, the transgenic plants did not show the reduced ethylene response or the salt-sensitive response, indicating that the NTHK1 kinase activity is required for both. However, this kinase activity is only partially required for the control of rosette size (Chen et al., 2009b). Moreover, NTHK1 plays a greater role than NtETR1 in the regulation of these responses (Chen et al., 2009b). In tomato, the ethylene receptor LeETR4 is multiply phosphorylated during fruit ripening, and ethylene treatment results in accumulation of LeETR4 with reduced phosphorylation (Kamiyoshihara et al., 2012). Ethylene binding in vitro abolishes the phosphorylation of the purified recombinant ETR1 of Arabidopsis (Voet-Van-Vormizeele and Groth, 2008). It is reported that the histidine kinase activity of ETR1 plays a modulating role but is not required for the regulation of ethylene responses in Arabidopsis (Hall et al., 2012). Overall, the kinase activity of ethylene receptors seems to be dispensable for ethylene signaling, but rather regulates particular processes in different plant species.

Among the five ethylene receptors in rice, only OsETR2 has been fully characterized (Wuriyanghan et al., 2009). Overexpression of *OsETR2* in transgenic rice reduced ethylene sensitivity and delayed transition to flowering, but promoted starch accumulation in stems. In RNAi plants or loss-of-function mutant *Osetr2*, the opposite phenotypes were observed. The delay of flowering in *OsETR2*-overexpressing plants is associated with *OsGI* (*OsGIGANTEA*) and *RCN1* (*TERMINAL FLOWER 1/CENTRORADIALIS -like*) expression. The thousand-seed weight was enhanced in OsETR2-RNAi plants or the *Osetr2* mutant (Wuriyanghan et al., 2009). In the *Osetr3* mutant, early flowering and reduced starch accumulation in stems were observed. For the *Osers2*, *Osetr2*, and *Osetr3* mutants in Zonghua (ZH, a rice cultivar) background all showed mild enhanced ethylene response in coleoptile elongation (Wuriyanghan et al., 2009). For the *Osers1*, *Osers2*, and *Osetr2* mutants in Dongjin (DJ) background, all showed constitutive and enhanced ethylene responses in root growth inhibition (Ma et al., 2014; Yin et al., 2015). The expression of *OsETR4* is generally very low in rice (Wuriyanghan et al., 2009); however, this gene is abundantly expressed at 6 days after anthesis during spikelet development, indicating a possible role in grain-filling (Sekhar et al., 2015). Most of the single loss-of-function mutants of rice ethylene receptor genes have a phenotypic change, whereas in Arabidopsis, only double or multiple receptor loss-of-function mutants exhibited an apparent phenotype change (Hua and Meyerowitz, 1998), suggesting a species-specific feature.

The membrane protein RTE1 interacts with ETR1 for regulation of receptor signaling in Arabidopsis (Resnick et al., 2006; Dong et al., 2010). Tomato RTE1 homologs GR (GREEN RIPE) and SlGRL1 (Solanum lycopersicum GREEN-RIPE LIKE1) repress distinct and overlapping ethylene responses (Barry and Giovannoni, 2006). In rice, three RTE1 homologs were identified and only OsRTH1 seems to have functions (Zhang et al., 2012). Overexpression of the *OsRTH1* rescues the short hypocotyl phenotype of the Arabidopsis *etr1-2 rte1-2* double mutant. Transgenic rice overexpressing the *OsRTE1* showed inhibition of ethylene-related phenotypic changes, suggesting that rice OsRTE1 has conserved functions in ethylene signaling (Zhang et al., 2012).

Ethylene receptors have other associated proteins. Interacting proteins have been identified including the tobacco subfamily II ethylene receptor NTHK1-interacting proteins TCTP (translationally controlled tumor protein) (Tao et al., 2015a), NEIP2 (NTHK1 ETHYLENE-RECEPTOR-INTERACTING PROTEIN 2) (Cao et al., 2015), and Arabidopsis subfamily II receptor ETR2/EIN4-interacting proteins SAUR (Small auxin-up RNA) 76/77/78 (Li et al., 2015b). Analysis of the functions of these proteins reveals desensitizing mechanisms of ethylene responses. TCTP is conserved throughout the plant and animal systems. In animals, TCTP interacts with many proteins and acts as a mitotic regulator during cell cycle progression. In tobacco, TCTP interacts with subfamily II receptors NTHK1 and NTHK2, but not subfamily I member NtETR1. TCTP associates with the kinase domain of NTHK1. Ethylene induces TCTP protein accumulation, and overexpression of TCTP stabilized NTHK1. Mutation or deletion of the NTHK1 kinase domain, which is responsible for the interaction with TCTP, disrupted NTHK1 stability, indicating that interaction of the two proteins is required for the stability of the NTHK1 receptor. Overexpression of *NTHK1* or *TCTP* results in similar phenotypes including large seedlings and leaves and reduced ethylene responses. Genetically, TCTP acts downstream of NTHK1 to promote plant growth but reduces the ethylene response. From these analyses, it was proposed that ethylene can induce expression of TCTP to associate with and stabilize the NTHK1 receptor to desensitize the ethylene response, and TCTP further leads to plant growth recovery through promotion of cell proliferation (Tao et al., 2015a).

Tobacco ankyrin repeat protein NEIP2 also interacts with ethylene receptor NTHK1 but not with NTHK2 or NtETR1, and the ankyrin domain of NEIP2 is associated with the kinase domain of NTHK1 (Cao et al., 2015). NTHK1 phosphorylates NEIP2 in vitro. Ethylene and salt stress induce NEIP2 protein accumulation, and NTHK1 expression further promotes NEIP2 levels in response to ethylene or salt treatments. It is interesting to note that ethylene and salt stress also induced NEIP2 phosphorylation. The function of NEIP2 phosphorylation requires further study. Overexpression of the NEIP2 promotes seedling and leaf growth but reduces ethylene responses, similar to the NTHK1 function. However, NEIP2 overexpression enhances stress tolerance, in contrast to the role of NTHK1 which confers salt sensitivity. It is suggested that ethylene- or stress-induced NEIP2 associates with the ethylene receptor NTHK1 to reduce the ethylene response, promote plant growth, and enhance stress tolerance (Cao et al., 2015).

Three SAUR proteins SAUR76, SAUR77, and SAUR78 have been found to interact with Arabidopsis subfamily II ethylene receptors ETR2 and EIN4 (Li et al., 2015b). Ethylene and auxin induce expression of the three *SAUR* genes. Overexpression of each gene promotes cell size and rosette size, and reduces ethylene sensitivity. The gene overexpression also suppresses the small seedling phenotype of the *etr2-3 ein4-4* double loss-of-function mutant. Loss-of-function mutant *saur76* suppresses the ethylene insensitivity of gain-of-function mutant *etr2-2*. The results imply that ethylene- and auxin-induced SAUR76/77/78 proteins may act in association with or downstream of subfamily II receptors to enhance receptor signaling. It should be mentioned that SAUR78 was also identified during a screen for EIN2-interacting proteins (Lei et al., 2011), suggesting that this small class of SAUR proteins may function in signaling between subfamily II ethylene receptors ETR2/EIN4 and EIN2 (Li et al., 2015b). Another MA3 domain containing protein ECIP1 (EIN2 C-TERMINUS INTERACTING PROTEIN 1) was also found to interact with both EIN2 and subfamily II ethylene receptors ETR2 and EIN4 (Lei et al., 2011). Genetic analysis also supports the proposal that ECIP1 acts between subfamily II ethylene receptor ETR2/EIN4 and the downstream membrane protein EIN2 (Lei et al., 2011). Therefore, it is proposed that SAUR76/77/78 proteins and ECIP1 may function as "molecular glue" between subfamily II receptors ETR2/EIN4 and the central signaling protein EIN2 in ethylene signal transduction. While these proteins function downstream of subfamily II receptors, Lin et al. (2008, 2009b) found that tomato SlTPR1 (tetratricopeptide repeat) interacts only with subfamily I receptors NR (NEVER RIPE) and LeETR1 to promote ethylene responses, and Arabidopsis AtTPR1 interacts only with subfamily I member AtERS1 to enhance ethylene responses (Lin et al., 2009a, 2008). Therefore, it is possible that different plant species employ different proteins to interact with different receptor subfamilies for regulation of ethylene responses in a complex regulatory manner.

6.4.3   CTR1-like proteins in rice

In rice, three CTR-like proteins, namely OsCTR1, OsCTR2, and OsCTR3 were identified (Wang et al., 2013). The OsCTR1 and OsCTR2 proteins are more closely related to Arabidopsis CTR1 than is OsCTR3. Expression of *OsCTR2* partially rescued the Arabidopsis *ctr1-1* phenotype, and OsCTR2 interacts with Arabidopsis subfamily I ethylene receptors ETR1 and ERS1 in a yeast two-hybrid assay. The *Osctr2* loss-of-function rice seedlings showed short roots and early leaf senescence phenotypes. In addition, the *Osctr2* mutant had more tillers, reduced plant height, delayed flowering, and reduced seed weight compared to the control plants (Wang et al., 2013). It is not clear whether OsCTR1 and/or OsCTR3 have any role in ethylene responses.

6.4.4   EIN2-like proteins in rice

EIN2 is the central component in Arabidopsis ethylene signaling pathway (Alonso et al., 1999). In rice, there are four EIN2-like homologs (Yang et al., 2015a). However, only OsEIN2 (Os07g06130) appears to have functions in ethylene responses. Inhibition of *OsEIN2* expression by an antisense approach moderately reduced shoot elongation of seedlings and expression of ethylene-responsive genes *SC129* and *SC255* (Jun et al., 2004). Based on the "double response" phenotype of etiolated rice seedlings, a set of ethylene-response mutants (*mhz*) have been identified (Figs. 6.6a and b) (Ma et al., 2013). All the mutants showed ethylene insensitivity or reduced sensitivity in root growth inhibition. For ethylene-promoted coleoptile elongation, different mutants showed ethylene insensitivity, wild-type-like response, or enhanced ethylene response. It is interesting to note that while dark-grown rice seedlings showed root inhibition but coleoptile promotion in response to ethylene treatment, the seedlings of other monocot plants, for example, *Brachypodium distachyon*, maize, wheat, and sorghum, showed root inhibition and coleoptile inhibition with the ethylene treatment (Yang et al., 2015a). This analysis suggests that rice may have specific growth features. Through map-based cloning, *MHZ7/OsEIN2* (Os07g06130) has been identified as a component of ethylene signaling in rice (Ma et al., 2013). All *mhz7* allelic mutants are insensitive to ethylene in both root and coleoptile growth. This is the first report showing ethylene insensitivity in rice seedlings. In ethylene, etiolated *MHZ7*-overexpressing seedlings showed enhanced coleoptile elongation, increased mesocotyl growth, and extremely twisted short roots, representing a distinct feature of a hypersensitive ethylene-response phenotype (Fig. 6.6c). A number of ethylene-responsive marker genes were also identified and these genes showed no ethylene induction in mutants, but enhanced expression in *MHZ7*-overexpressing plants (Ma et al., 2013). Additionally, *MHZ7* overexpression increased grain size. Dark-induced leaf senescence was delayed in the *mhz7* mutants. The *mhz7* mutants will be valuable for the study of many other functions of ethylene signaling, including responses to abiotic stress, disease, nutrient, and heavy metals in rice.

In Arabidopsis, two F-box proteins ETP1 and ETP2 have been found to interact with the EIN2 C-terminal end and to trigger the proteasomal degradation of EIN2 in the absence of ethylene (Qiao et al., 2009). Putative homologs/paralogs have been identified in rice and other plants; however, these proteins shared very low (<30%) identity with ETP1/2 (Yang et al., 2015a). Possible functions of these proteins in the regulation of ethylene signaling require further investigation.

6.4.5   EIN3-like proteins in rice

In rice, six EIN3-like genes have been identified and named OsEIL1 to OsEIL6. Among these, OsEIL1 and OsEIL2 show the highest identity with EIN3, the master transcription factor of the ethylene signaling pathway in Arabidopsis (Chao et al., 1997). *OsEIL1* can partially rescue the Arabidopsis *ein3-1* mutant phenotype and OsEIL1 overexpression in transgenic Arabidopsis plants causes some ethylene-response phenotypes (Mao et al., 2006). Recently, through analysis of the rice ethylene-response mutant *mhz6*, it was found by map-based cloning that *MHZ6* cncodes OsEIL1 (Yang et al., 2015b). The *mhz6* mutant shows ethylene insensitivity mainly in root growth inhibition, whcreas silencing the *OsEIL2* leads to ethylene insensitivity mainly in coleoptile elongation in etiolated seedlings. Other *OsEIL* genes appear not to affect the ethylene response based on transgenic analysis in rice. The results indicate that MHZ6/OsEIL1 and OsEIL2 regulate ethylene responses of roots and coleoptiles respectively (Fig. 6.6e). This organ-specific functional divergence is different from that in Arabidopsis, where both EIN3 and EIL1 regulate similar ethylene responses in seedlings. Inhibition of both *MHZ6/OsEIL1* and *OsEIL2* expression leads to complete ethylene insensitivity of roots and coleoptiles (Yang et al., 2015b). This phenomenon is similar to that of the Arabidopsis *ein3 eil1* double mutant, which also shows complete ethylene insensitivity in the triple response.

Previous studies on the ethylene receptor gene *NTHK1* in tobacco and analyses of various Arabidopsis ethylene-response mutants, e.g., *etr1-1*, *ein2*, *ein3 eil1* and *eto1*, demonstrate that ethylene and its signaling play a positive role in salt stress tolerance (Cao et al., 2007, 2006; Chen et al., 2009b; Jiang et al., 2013; Lei et al., 2011; Peng et al., 2014; Tao et al., 2015b). An ankyrin domain protein NEIP2, which interacted with NTHK1, is also required for salt tolerance in tobacco (Cao et al., 2015). In contrast to the roles of ethylene signaling in tobacco and Arabidopsis, this pathway appears to play a negative role in salt stress tolerance in rice. The *mhz7/Osein2*, *mhz6/Oseil1* and *Oseil2* mutants are all tolerant to salt stress compared to wild-type plants, whereas overexpression of *MHZ7/OsEIN2*, *MHZ6/OsEIL1* or *OsEIL2* confers hypersensitivity to salt stress in transgenic rice (Yang et al., 2015b). Consistent with this observation, when rice plants are treated with 1-MCP (1-methylcyclopropene, an inhibitor of ethylene action) to block ethylene perception, the plants are more tolerant to salt stress than those in air. However, when the plants are exposed to ethylene, they are more sensitive to salt stress (Yang et al., 2015b). These analyses indicate that ethylene and its signaling cause susceptible response to salt stress in rice.

Further studies reveal the possible mechanism by which ethylene signaling regulates salt stress responses in rice (Yang et al., 2015b). The Na^+ contents are increased in the *MHZ6/OsEIL1*- and *OsEIL2*-overexpressing plants but are reduced in the *mhz6/Oseil1* mutant and *OsEIL2*-RNAi plants under salt stress. Ethylene induces expression of *OsHKT2;1*, a Na^+ transporter gene, and this induction requires *MHZ7/OsEIN2*, *MHZ6/OsEIL1* or *OsEIL2* functions. In the *MHZ6/OsEIL1*- and *OsEIL2*-overexpressing plants, *OsHKT2;1* gene expression is apparently increased. Both MHZ6/OsEIL1 and OsEIL2 can bind to specific elements in the promoter region of *OsHKT2;1* and activate its expression (Yang et al., 2015b). Therefore, it is most likely that rice adopts ethylene signaling to activate *OsHKT2;1* expression and promote Na^+ uptake. This feature should benefit plants for survival in water-saturated environments by maintaining a balanced ion homeostasis. However, under salt stress, high *OsHKT2;1* expression in transgenic plants with enhanced ethylene signaling would lead to uptake of excess Na^+ and hence cause hypersensitivity to salt (Fig. 6.6e). It is interesting to note that salt stress represses expression of *OsHKT2;1*, suggesting that rice plants may also mitigate excessive Na$^+$ uptake through inhibition of *OsHKT2;1* expression in high salinity conditions (Yang et al., 2015b).

Ethylene signaling affects seed traits in rice. Higher grain weight is observed in *OsETR2*-RNAi rice and in the *Osetr2* loss-of-function mutant (Wuriyanghan et al., 2009). *MHZ7/OsEIN2*-overexpressing plants also produce larger grains (Ma et al., 2013). Consistent with these observations, *MHZ6/OsEIL1* overexpression promotes seed size and seed weight, supporting a potential role of ethylene signaling in improvement of crop agronomic traits (Yang et al., 2015b).

In Arabidopsis, stability of EIN3 and EIL1 is regulated by EBF1 and EBF2 through the 26S proteasome-mediated degradation pathway (An et al., 2010; Gagne et al., 2004; Guo and Ecker, 2003; Potuschak et al., 2003). In rice, EBF-like proteins have been identified from genome sequences (Yang et al., 2015a); however, whether these proteins have any functions and whether they are regulated with similar mechanisms to those in Arabidopsis requires further investigation in relation to ethylene signaling.

6.4.6 Components downstream of EIN3/EIL1-like proteins in rice

In Arabidopsis, *ERF1* and other *ERF* genes represent the first identified and the most widely studied class of EIN3 target genes (Solano et al., 1998). In rice, ERF-type transcription factor genes including *OsERF1*, *OsEBP89*, *SUB1A* (*SUBMERGENCE1 A*), *SNORKEL1/2*, and *OsDERF1* (*DROUGHT-RESPONSIVE ERF1*) have been studied for their roles in ethylene responses and tolerance to stress including flooding (Hattori et al., 2009; Hu et al., 2008; Wan et al., 2011; Xu et al., 2006; Yang et al., 2002). SUB1A plays important roles in rice flooding tolerance by repressing shoot elongation and reducing carbohydrate consumption under flash flood conditions (Xu et al., 2006). SUB1A also improves drought resistance at vegetative stages and delays dark-induced leaf senescence (Fukao et al., 2011). SNORKEL1 and SNORKEL2 are involved in the "escape mechanism" in rice flooding tolerance. Under deep-water conditions, ethylene is accumulated in the plant and this accumulation induces expression of *SNORKEL1* and *SNORKEL2*, leading to internode elongation to escape from water submergence (Hattori et al., 2009). However, the *SUB1A* and *SNORKEL1/2* genes are only present in a few specific rice cultivars and whether these genes can be regulated by rice EIN3-like proteins MHZ6/OsEIL1 and/or OsEIL2 are not known. It should be noted that in *OsEIL1*-overexpressing plants, *OsEBP89* expression was enhanced (Mao et al., 2006), suggesting that *OsEBP89* may be the downstream gene of OsEIL1. In rice, there are more than 100 ERF-type transcription factor genes, and it is likely that only a portion of them is involved in ethylene responses and most of the members may participate in stress response or other processes.

In addition to ERF-type transcription factor genes, many other genes may also be regulated by MHZ6/OsEIL1. Through RNA-sequence analysis of rice ethylene-response to mutants *mhz7/Osein2*, *mhz6/Oseil1*, and/or *OsEIL2*-RNAi plants, 1046 ethylene-responsive genes were identified in rice roots (Yang et al., 2015b). Among these, MHZ7/OsEIN2 and MHZ6/OsEIL1 regulate 757 and 464 genes respectively, and 87% of the MHZ6/OsEIL1-regulated genes are also regulated by MHZ7/OsEIN2.

In shoots, there are 5112 ethylene-responsive genes. MHZ7/OsEIN2 and OsEIL2 regulate 4815 and 4515 genes respectively, and the 95% of the OsEIL2-regulated genes are also affected by MHZ7/OsEIN2. These results suggest that MHZ6/OsEIL1 and OsEIL2 may act downstream of MHZ7/OsEIN2 to regulate specific subsets of genes for the control of ethylene responses in roots and shoots/coleoptiles, respectively. Further analysis of these downstream genes may identify the biological processes and biochemical pathways involved.

6.4.7   Interactions of ethylene signaling with other hormones in rice

While screening for rice ethylene-response mutants, two mutants *mhz4* and *mhz5* showed similar phenotypes including reduced ethylene response in roots but enhanced ethylene response in coleoptiles. *MHZ4* encodes a chloroplast-localized protein homologous to Arabidopsis ABA4, which is functional in a branch of the ABA biosynthesis pathway (Ma et al., 2014). *MHZ5* encodes a carotenoid isomerase in the carotenoid biosynthesis pathway (Yin et al., 2015). One of the products of the carotenoid biosynthesis pathway is β-carotene, which is the precursor of ABA. Disruption of *MHZ5* or *MHZ4* in mutants drastically reduces ABA levels but promotes ethylene production, and ABA largely rescues the ethylene response of the two mutants (Ma et al., 2014; Yin et al., 2015). Ethylene induces expression of *MHZ5* and *MHZ4*, production of an ABA biosynthesis precursor neoxanthin, and accumulation of ABA in roots. Genetically, the *MHZ5*- and *MHZ4*-mediated ABA pathway acts downstream of the ethylene signaling to inhibit root growth of etiolated rice seedlings (Fig. 6.6e). However, in the aerial parts, the *MHZ5*- and *MHZ4*-mediated ABA pathway acts upstream and inhibits ethylene signaling to control coleoptile/shoot growth. When the *MHZ5* and *MHZ4* genes are mutated in *mhz5* and *mhz4* mutants, *EIN2* expression is apparently enhanced in coleoptiles/shoots, causing a hypersensitive ethylene response in coleoptile elongation (Ma et al., 2014; Yin et al., 2015). The interaction of ethylene and ABA pathways in rice is different from that in Arabidopsis. In rice, ethylene acts upstream of the ABA pathway to inhibit root growth, whereas in Arabidopsis, ethylene acts downstream of the ABA pathway to inhibit root growth (Beaudoin and Giraudat, 2000; Ghassemian et al., 2000; Luo et al., 2014), indicating specialization in hormone interactions. Further analysis of additional *mhz* mutants may reveal novel interactions of ethylene with other plant hormones in rice.

6.4.8   Possible novel components in rice ethylene signaling pathway

From analysis of rice ethylene-response mutants, MHZ7/OsEIN2, MHZ6/OsEIL1, MHZ5/carotenoid isomerase, and MHZ4/ABA4 have been identified. These components are either homologous to the EIN2 and EIN3 in the ethylene signaling pathway of Arabidopsis or are involved in ethylene interactions with the ABA pathway.

Mutants that are not allelic to the above dissected mutants have been screened (Ma et al., 2013). Identification of novel mutants in rice may reflect that key genes are present in only one copy in rice, whereas in Arabidopsis, the homologous genes commonly occur in two or more copies. Furthermore, individual genes may play more important roles in rice than in Arabidopsis. Further novel genes different from those operating in Arabidopsis, are expected to be found in rice through mutant analysis. Further investigation toward discovering such genes should give a full picture on the mechanisms of ethylene signaling in rice, Arabidopsis and other plants, and manipulation of the genes may improve agronomic traits in crops.

6.5 Summary points

- Both Arabidopsis and rice have those key factors of ethylene perception and signaling, including five ethylene receptors, CTR1-like, EIN2-like, EIN3/EILs-like transcription factors, ERF proteins, and RTE1-like proteins.
- Arabidopsis etiolated seedlings display triple response phenotype, whereas rice etiolated seedlings exhibit double response phenotype upon ethylene treatment.
- Arabidopsis has ETR1-type receptor in subfamily I members, whereas rice does not have this but has ERS-type.
- Rice MHZ6/OsEIL1 and OsEIL2 have organ specificity in regulation of ethylene response whereas Arabidopsis EIN3 and EIL1 are functional overlapping.
- Ethylene signaling facilitates salt tolerance in Arabidopsis but enhances salt sensitivity in rice.
- In Arabidopsis, ABA requires ethylene pathway for root inhibition, whereas in rice, ethylene requires ABA pathway for root inhibition.
- Rice is expected to have novel components based on mutant analysis.

6.6 Future perspectives

- What is the perception mechanism of ethylene receptors and how do those receptor-interacting proteins differentially modulate ethylene receptors in Arabidopsis?
- How do the ethylene receptors activate CTR1 in the air and inactivate it upon ethylene binding in Arabidopsis?
- What are the phosphatase(s) and protease(s) that are responsible for EIN2 dephosphorylation and cleavage in Arabidopsis?
- How does EIN2 C terminus stabilize EIN3/EIL1 after its nuclear translocation and how do EIN2 and P-body factors cooperatively inhibit the translation of *EBF1/EBF2* mRNAs in Arabidopsis?
- What's the function of EIN2 N terminal in ethylene signaling transduction?
- What's the mechanism of the different regulatory roles of MHZ6/OsEIL1 and OsEIL2 in ethylene response in rice?
- What are the branching points in the canonical linear ethylene signaling pathway in different aspects of plant growth and development?
- What are the responses of the *mhz7/Osein2* and *mhz6/Oseil1* mutants, and other *mhz* mutants to different abiotic stresses?

Abbreviations

1-MCP	1-Methylcyclopropene
ABA	Abscisic acid
ACC 1	Aminocyclopropane-1-carboxylate
ACO	ACC oxidase
ACS	ACC synthase
ARR	*ARABIDOPSIS RESPONSE REGULATOR*
AtNAP	NAC-LIKE, ACTIVATED BY AP3/PIAZF2 *ARABIDOPSIS ZINC FINGER PROTEIN 2*
bHLH	Basic helix-loop-helix
BTB	Broad-complex/tramtrack/bric-a-brac;
Cb5	Cytochrome b_5
CBF	*C-REPEAT BINDING FACTOR*
CCGs	Chlorophyll catabolic genes
CDPK	Calcium-dependent protein kinase
CRL	Cullin-RING ligase
Ctr1	*Constitutive triple response 1*
DERF1	*DROUGHT-RESPONSIVE ERF1*
DJ	Dongjin
EBF1	EIN3-BINDING PROTEIN 1
ECIP1	EIN2 C-TERMINUS INTERACTING PROTEIN 1
EIL1	EIN3-LIKE 1
EIN2	ETHYLENE INSENSITIVE 2
EIN3	ETHYLENE INSENSITIVE 3
EIN4	ETHYLENE INSENSITIVE 4
EOLs	ETO1-LIKE
ERF1	ETHYLENE RESPONSE FACTOR 1
ERS1	ETHYLENE RESPONSE SENSOR 1
ET	Ethylene
ETO1	Ethylene Overproducer 1
ETP1	EIN2 TARGETING PROTEIN 1
ETR1	ETHYLENE RESISTANT 1
FIT	FER-LIKE FE DEFICIENCY-INDUCED TRANSCRIPTION FACTOR
FLS2	*FLAGELLIN SENSITIVE 2*
GAF	cGMP-specific phosphodiesterases, adenylyl cyclases, and FhIA
GR	GREEN-RIPE
HDA6	HISTONE DEACETYLASE 6
HisKA	Histidine kinase
HK	Histidine protein kinase
HLS1	*HOOKLESS1*
JA	Jasmonic acid
JAZ	JA-Zim domain
MAPK	Mitogen-activating protein kinase
Met	Methionine
Mhz	*Mao huzi*

NAC TF	NAM, ATAF1, ATAF2, and CUC2 domain transcription factors
NEIP2	NTHK1 ETHYLENE-RECEPTOR-INTERACTING PROTEIN 2
NLS	Nuclear localization signal
NR	NEVER RIPE
Nramp	Natural resistance associated macrophage protein
NYC1	*NON-YELLOW COLORING 1*
NYE1	*NON-YELLOWING 1*
ORE1	ORESARA1
OsGI	*OsGIGANTEA*
P-body	Processing-body
PAMPs	Pathogen-associated molecular patterns
PAO	*PHEOPHORBIDE A OXYGENASE*
PIF3	*PHYTOCHROME-INTERACTING FACTOR 3*
POD	Peroxidase
PORA/B	*PROTOCHLOROPHYLLIDE OXIDOREDUCTASE A and B*
PP2C	Phosphoprotein phosphatases 2C
RCN1	*TERMINAL FLOWER 1/CENTRORADIALIS –like*
ROS	Reactive oxygen species
RR	Response regulator protein
RTE1	REVERSION-TO-ETHYLENE SENSITIVITY 1
SA	Salicylic acid
SAM	S-adenosylmethionine
SAUR	Small auxin-up RNA
SHYG	SPEEDY HYPONASTIC GROWTH
SID2	*SALICYLIC ACID INDUCTION DEFICIENT 2*
SIED	*SALT-INDUCED AND EIN3/EIL1-DEPENDENT*
SlGRL1	Solanum lycopersicum GREEN-RIPE LIKE1
SlTPR1	Solanum lycopersicum tetratricopeptide repeat
SUB1A	*SUBMERGENCE1 A*
SZF2	*SALT-INDUCIBLE ZINC FINGER 2*
TCTP	Translationally controlled tumor protein
TOE	Target of ETO1
UTR	Un-translated region
XBAT32	XB3 ortholog 2
XRN4	EXORIBONUCLEASE 4
ZAT12	*ZINC FINGER OF ARABIDOPSIS THALIANA12*

Acknowledgments

We thank Dr. Yichuan Wang (Department of Biology, South University of Science and Technology of China) and Dr. Wenyang Li (School of life sciences, Peking University) for kind help in preparing this review. This work is supported by National Natural Science Foundation of China (91217305 and 91017010) to H.G. and the National Natural Science Foundation of China (31530004) and the 973 project (2015CB755702) to J. Z.

References

Abeles, F.B., Morgan, P.W., Saltveit, M.E., 1992. Ethylene in Plant Biology, second ed. .

Agnieszka, L., Agata, C., Anna, K.-M., et al., 2014. *Arabidopsis* protein phosphatase 2C ABI1 interacts with type I ACC synthases and is involved in the regulation of ozone-induced ethylene biosynthesis. Mol. Plant 7, 960–976.

Alonso, J.M., Hirayama, T., Roman, G., et al., 1999. EIN2, a bifunctional transducer of ethylene and stress responses in *Arabidopsis*. Science 284, 2148–2152.

Alonso, J.M., Stepanova, A.N., Solano, R., et al., 2003. Five components of the ethylene-response pathway identified in a screen for weak ethylene-insensitive mutants in *Arabidopsis*. Proc. Natl. Acad. Sci. U.S.A. 100, 2992–2997.

An, F., Zhang, X., Zhu, Z., et al., 2012. Coordinated regulation of apical hook development by gibberellins and ethylene in etiolated *Arabidopsis* seedlings. Cell Res. 22, 915–927.

An, F., Zhao, Q., Ji, Y., et al., 2010. Ethylene-Induced stabilization of ETHYLENE INSENSITIVE3 and EIN3-LIKE1 is mediated by proteasomal degradation of EIN3 binding F-box 1 and 2 that requires EIN2 in *Arabidopsis*. The Plant Cell 22, 2384–2401.

Argueso, C.T., Hansen, M., Kieber, J.J., 2007. Regulation of ethylene biosynthesis. J. Plant Growth Regul. 26, 92–105.

Bakshi, A., Shemansky, J.M., Chang, C., et al., 2015. History of research on the plant hormone ethylene. J. Plant Growth Regul. 34, 809–827.

Barry, C.S., Blume, B., Bouzayen, M., et al., 1996. Differential expression of the 1-aminocyclo-propane-1-carboxylate oxidase gene family of tomato. Plant J. 9, 525–535.

Barry, C.S., Giovannoni, J.J., 2006. Ripening in the tomato Green-ripe mutant is inhibited by ectopic expression of a protein that disrupts ethylene signaling. Proc. Natl. Acad. Sci. U.S.A. 103, 7923–7928.

Beaudoin, N., Giraudat, J., 2000. Interactions between abscisic acid and ethylene signaling cascades. The Plant Cell 12, 1103–1115.

Binder, B.M., O'Malley, R.C., Wang, W., et al., 2004. *Arabidopsis* seedling growth response and recovery to ethylene. A kinetic analysis. Plant Physiol. 136, 2913–2920.

Binder, B.M., Walker, J.M., Gagne, J.M., et al., 2007. The *Arabidopsis* EIN3 binding F-box proteins EBF1 and EBF2 have distinct but overlapping roles in ethylene signaling. The Plant Cell 19, 509–523.

Bisson, M.M.A., Bleckmann, A., Allekotte, S., et al., 2009. EIN2, the central regulator of ethylene signalling, is localized at the ER membrane where it interacts with the ethylene receptor ETR1. Biochem. J. 424, 1–6.

Bisson, M.M.A., Groth, G., 2011. New paradigm in ethylene signaling: EIN2, the central regulator of the signaling pathway, interacts directly with the upstream receptors. Plant Signal. Behav. 6, 164–166.

Bisson, M.M.A., Groth, G., 2015. Targeting plant ethylene responses by controlling essential protein–protein interactions in the ethylene pathway. Mol. Plant 8, 1165–1174.

Bleecker, A.B., Estelle, M.A., Somerville, C., et al., 1988. Insensitivity to ethylene conferred by a dominant mutation in *Arabidopsis thaliana*. Science 241, 1086–1089.

Blume, B., Grierson, D., 1997. Expression of ACC oxidase promoter-GUS fusions in tomato and *Nicotiana plumbaginifolia* regulated by developmental and environmental stimuli. Plant J. 12, 731–746.

Boutrot, F., Segonzac, C., Chang, K.N., et al., 2010. Direct transcriptional control of the *Arabidopsis* immune receptor FLS2 by the ethylene-dependent transcription factors EIN3 and EIL1. Proc. Natl. Acad. Sci. U.S.A. 107, 14502–14507.

Buerstenbinder, K., Sauter, M., 2012. Early events in the ethylene biosynthetic pathway-regulation of the pools of methionine and S-Adenosylmethionine. Annu. Plant Rev. 44, 19–115.

Cao, W.H., Liu, J., He, X.J., et al., 2007. Modulation of ethylene responses affects plant salt-stress responses. Plant Physiol. 143, 707–719.

Cao, W.H., Liu, J., Zhou, Q.Y., et al., 2006. Expression of tobacco ethylene receptor NTHK1 alters plant responses to salt stress. Plant Cell Environ. 29, 1210–1219.

Cao, Y.R., Chen, H.W., Li, Z.G., et al., 2015. Tobacco ankyrin protein NEIP2 interacts with ethylene receptor NTHK1 and regulates plant growth and stress responses. Plant Cell Physiol. 56.

Chae, H.S., Kieber, J.J., 2005. Eto Brute ? Role of ACS turnover in regulating ethylene biosynthesis. Trends Plant Sci. 10, 291–296.

Chang, C., 2016. Q&A: How do plants respond to ethylene and what is its importance? BMC Biol. 14, 7.

Chang, C., Kwok, S., Bleecker, A., et al., 1993. Arabidopsis ethylene-response gene ETR1: similarity of product to two-component regulators. Science 262, 539–544.

Chang, C., Meyerowitz, E.M., 1995. The ethylene hormone response in Arabidopsis: a eukaryotic two-component signaling system. Proc. Natl. Acad. Sci. U.S.A. 92, 4129–4133.

Chang, C., Shockey, J.A., 1999. The ethylene-response pathway: signal perception to gene regulation. Curr. Opin. Plant Biol. 2, 352–358.

Chang, J., Clay, J.M., Chang, C., 2014. Association of cytochrome b(5) with ETR1 ethylene receptor signaling through RTE1 in Arabidopsis. Plant J. 77, 558–567.

Chang, K.N., Zhong, S., Weirauch, M.T., et al., 2013. Temporal transcriptional response to ethylene gas drives growth hormone cross-regulation in Arabidopsis. ELife 2, e00675.

Chao, Q., Rothenberg, M., Solano, R., et al., 1997. Activation of the ethylene gas response pathway in Arabidopsis by the nuclear protein ETHYLENE-INSENSITIVE3 and related proteins. Cell 89, 1133–1144.

Chen, H., Xue, L., Chintamanani, S., et al., 2009a. ETHYLENE INSENSITIVE3 and ETHYLENE INSENSITIVE3-LIKE1 repress SALICYLIC ACID INDUCTION DEFICIENT2 expression to negatively regulate plant innate immunity in Arabidopsis. The Plant Cell 21, 2527–2540.

Chen, R., Binder, B.M., Garrett, W.M., et al., 2011. Proteomic responses in Arabidopsis thaliana seedlings treated with ethylene. Mol. BioSyst. 7, 2637–2650.

Chen, T., Liu, J., Lei, G., et al., 2009b. Effects of tobacco ethylene receptor mutations on receptor kinase activity, plant growth and stress responses. Plant Cell Physiol. 50, 1636–1650.

Chen, Y.F., Randlett, M.D., Findell, J.L., et al., 2002. Localization of the ethylene receptor ETR1 to the endoplasmic reticulum of Arabidopsis. J. Biol. Chem. 277, 19861–19866.

Corbineau, F., Xia, Q., Bailly, C., et al., 2014. Ethylene, a key factor in the regulation of seed dormancy. Front. Plant Sci. 5, 539.

Dong, C.H., Jang, M., Scharein, B., et al., 2010. Molecular association of the Arabidopsis ETR1 ethylene receptor and a regulator of ethylene signaling, RTE1. J. Biol. Chem. 285, 40706–40713.

Dong, C.H., Rivarola, M., Resnick, J.S., et al., 2008. Subcellular co-localization of Arabidopsis RTE1 and ETR1 supports a regulatory role for RTE1 in ETR1 ethylene signaling. Plant J. 53, 275–286.

Felix, G., Grosskopf, D.G., Regenass, M., et al., 1991. Elicitor-induced ethylene biosynthesis in tomato cells: characterization and use as a bioassay for elicitor action. Plant Physiol. 97, 19–25.

Fluhr, R., Mattoo, A.K., Dilley, D.D.R., 1996. Ethylene — biosynthesis and perception. Crit. Rev. Plant Sci. 15, 479–523.

Fu, H., Subramanian, R.R., Masters, S.C., 2000. 14-3-3 proteins: structure, function, and regulation. Annu. Rev. Pharmacol. 40, 617–647.

Fukao, T., Bailey-Serres, J., 2008. Ethylene—a key regulator of submergence responses in rice. Plant Sci. 175, 43–51.

Fukao, T., Yeung, E., Baileyserres, J., 2011. The submergence tolerance regulator SUB1A mediates crosstalk between submergence and drought tolerance in rice. The Plant Cell 23, 412–427.

Gagne, J.M., Smalle, J., Gingerich, D.J., et al., 2004. *Arabidopsis* EIN3-binding F-box 1 and 2 form ubiquitin-protein ligases that repress ethylene action and promote growth by directing EIN3 degradation. Proc. Natl. Acad. Sci. U.S.A. 101, 6803–6808.

Gamble, R.L., Coonfield, M.L., Schaller, G.E., 1998. Histidine kinase activity of the ETR1 ethylene receptor from *Arabidopsis*. Proc. Natl. Acad. Sci. U.S.A. 95, 7825–7829.

Gamble, R.L., Qu, X., Schaller, G.E., 2002. Mutational analysis of the ethylene receptor ETR1. Role of the histidine kinase domain in dominant ethylene insensitivity. Plant Physiol. 128, 1428–1438.

Gao, Z., Chen, Y.F., Randlett, M.D., et al., 2003. Localization of the Raf-like kinase CTR1 to the endoplasmic reticulum of *Arabidopsis* through participation in ethylene receptor signaling complexes. J. Biol. Chem. 278, 34725–34732.

Ghassemian, M., Nambara, E., Cutler, S., et al., 2000. Regulation of abscisic acid signaling by the ethylene response pathway in *Arabidopsis*. The Plant Cell 12, 1117–1126.

Guo, H., Ecker, J.R., 2003. Plant responses to ethylene gas are mediated by SCF[EBF1/EBF2]-dependent proteolysis of EIN3 transcription factor. Cell 115, 667–677.

Guo, H., Ecker, J.R., 2004. The ethylene signaling pathway: new insights. Curr. Opin. Plant Biol. 7, 40–49.

Guzmán, P., Ecker, J.R., 1990. Exploiting the triple response of *Arabidopsis* to identify ethylene-related mutants. The Plant Cell 2, 513–523.

Hall, B.P., Shakeel, S.N., Amir, M., et al., 2012. Histidine kinase activity of the ethylene receptor ETR1 facilitates the ethylene response in *Arabidopsis*. Plant Physiol. 159, 682–695.

Han, L., Li, G.J., Yang, K.Y., et al., 2010. Mitogen-activated protein kinase 3 and 6 regulate *Botrytis cinerea* -induced ethylene production in *Arabidopsis*. Plant J. 64, 114–127.

Hansen, M., Chae, H.S., Kieber, J.J., 2009. Regulation of ACS protein stability by cytokinin and brassinosteroid. Plant J. 57, 606–614.

Hattori, Y., Nagai, K., Furukawa, S., et al., 2009. The ethylene response factors SNORKEL1 and SNORKEL2 allow rice to adapt to deep water. Nature 460, 1026–1030.

Hirayama, T., Kieber, J.J., Hirayama, N., et al., 1999. RESPONSIVE-TO-ANTAGONIST1, a Menkes/Wilson disease–related copper transporter, is required for ethylene signaling in *Arabidopsis*. Cell 97, 383–393.

Hu, Y., Zhao, L., Kang, C., et al., 2008. Overexpression of *OsERF1*, a novel rice *ERF* gene, up-regulates ethylene-responsive genes expression besides affects growth and development in *Arabidopsis*. J. Plant Physiol. 165, 1717–1725.

Hua, J., Chang, C., Sun, Q., et al., 1995. Ethylene insensitivity conferred by *Arabidopsis ERS* gene. Science 269, 1712–1714.

Hua, J., Meyerowitz, E.M., 1998. Ethylene responses are negatively regulated by a receptor gene family in *Arabidopsis thaliana*. Cell 94, 261–271.

Huang, S.J., Chang, C.L., Wang, P.H., et al., 2013. A type III ACC synthase, ACS7, is involved in root gravitropism in *Arabidopsis thaliana*. J. Exp. Bot. 64, 4343–4360.

Huang, Y., Li, H., Hutchison, C.E., et al., 2003. Biochemical and functional analysis of CTR1, a protein kinase that negatively regulates ethylene signaling in *Arabidopsis*. Plant J. 33, 221–233.

Ismail, A.M., Ella, E.S., Vergara, G.V., et al., 2009. Mechanisms associated with tolerance to flooding during germination and early seedling growth in rice (*Oryza sativa*). Ann. Bot. 103, 197–209.

Jackson, M.B., 2003. Ethylene and responses of plants to soil waterlogging and submergence. Annu. Rev. Plant Biol. 36, 145–174.

Jiang, C., Belfield, E.J., Cao, Y., et al., 2013. An *Arabidopsis* soil-salinity-tolerance mutation confers ethylene-mediated enhancement of sodium/potassium homeostasis. The Plant Cell 25, 3535–3552.

Joo, S., Liu, Y., Lueth, A., et al., 2008. MAPK phosphorylation-induced stabilization of ACS6 protein is mediated by the non-catalytic C-terminal domain, which also contains the cis-determinant for rapid degradation by the 26S proteasome pathway. Plant J. 54, 129–140.

Ju, C., Chang, C., 2015. Mechanistic insights in ethylene perception and signal transduction. Plant Physiol. 169, 85–95.

Ju, C., Yoon, G.M., Shemansky, J.M., et al., 2012. CTR1 phosphorylates the central regulator EIN2 to control ethylene hormone signaling from the ER membrane to the nucleus in *Arabidopsis*. Proc. Natl. Acad. Sci. U.S.A. 109, 19486–19491.

Jun, S.H., Han, M.J., Lee, S., et al., 2004. OsEIN2 is a positive component in ethylene signaling in rice. Plant Cell Physiol. 45, 281–289.

Kamiyoshihara, Y., Tieman, D.M., Huber, D.J., et al., 2012. Ligand-induced alterations in the phosphorylation state of ethylene receptors in tomato fruit. Plant Physiol. 160, 488–497.

Kende, H., 1993. Ethylene biosynthesis. Annu. Rev. Plant Biol. 44, 283–307.

Kieber, J.J., Rothenberg, M., Roman, G., et al., 1993. CTR1, a negative regulator of the ethylene response pathway in *Arabidopsis*, encodes a member of the raf family of protein kinases. Cell 72, 427–441.

Kim, H.J., Hong, S.H., Kim, Y.W., et al., 2014. Gene regulatory cascade of senescence-associated NAC transcription factors activated by ETHYLENE-INSENSITIVE2-mediated leaf senescence signalling in *Arabidopsis*. J. Exp. Bot. 65, 4023–4036.

Konishi, M., Yanagisawa, S., 2008. Ethylene signaling in *Arabidopsis* involves feedback regulation via the elaborate control of *EBF2* expression by EIN3. Plant J. 55, 821–831.

Ku, H.S., Suge, H., Rappaport, L., et al., 1970. Stimulation of rice coleoptile growth by ethylene. Planta 90, 333–339.

Lacey, R.F., Binder, B.M., 2014. How plants sense ethylene gas — the ethylene receptors. J. Inorg. Biochem. 133, 58–62.

Lei, G., Shen, M., Li, Z.G., et al., 2011. EIN2 regulates salt stress response and interacts with a MA3 domain-containing protein ECIP1 in *Arabidopsis*. Plant Cell Environ. 34, 1678–1692.

Li, G., Meng, X., Wang, R., et al., 2012a. Dual-level regulation of ACC synthase activity by MPK3/MPK6 cascade and its downstream WRKY transcription factor during ethylene induction in *Arabidopsis*. PLoS Genet. 8, 3202–3212.

Li, J., Li, Z., Tang, L., et al., 2012b. A conserved phosphorylation site regulates the transcriptional function of ETHYLENE-INSENSITIVE3-like1 in tomato. J. Exp. Bot. 63, 427–439.

Li, W., Ma, M., Feng, Y., et al., 2015a. EIN2-directed translational regulation of ethylene signaling in *Arabidopsis*. Cell 163, 670–683.

Li, Z.G., Chen, H.-W., Li, Q.-T., et al., 2015b. Three SAUR proteins SAUR76, SAUR77 and SAUR78 promote plant growth in *Arabidopsis*. Sci. Rep. 5, 12477.

Li, Z., Peng, J., Wen, X., et al., 2013. *ETHYLENE-INSENSITIVE3* is a senescence-associated gene that accelerates age-dependent leaf senescence by directly repressing *miR164* transcription in *Arabidopsis*. The Plant Cell 25, 3311–3328.

Lin, Z., Ho, C.W., Grierson, D., 2009a. AtTRP1 encodes a novel TPR protein that interacts with the ethylene receptor ERS1 and modulates development in *Arabidopsis*. J. Exp. Bot. 60, 3697–3714.

Lin, Z., Lucy, A., Rachel, H., et al., 2008. LeCTR2, a CTR1-like protein kinase from tomato, plays a role in ethylene signalling, development and defence. Plant J. 54, 1083–1093.

Lin, Z., Zhong, S., Grierson, D., 2009b. Recent advances in ethylene research. J. Exp. Bot. 60, 3311–3336.

Lingam, S., Mohrbacher, J., Brumbarova, T., et al., 2011. Interaction between the bHLH transcription factor FIT and ETHYLENE INSENSITIVE3/ETHYLENE INSENSITIVE3-LIKE1 reveals molecular linkage between the regulation of iron acquisition and ethylene signaling in *Arabidopsis*. The Plant Cell 23, 1815–1829.

Linkies, A., Müller, K., Morris, K., et al., 2009. Ethylene interacts with abscisic acid to regulate endosperm rupture during germination: a comparative approach using *Lepidium sativum* and *Arabidopsis thaliana*. The Plant Cell 21, 3803–3822.

Liu, Q., Wen, C.K., 2012. *Arabidopsis ETR1* and *ERS1* differentially repress the ethylene response in combination with other ethylene receptor genes. Plant Physiol. 158, 1193–1207.

Liu, Q., Xu, C., Wen, C.K., 2010. Genetic and transformation studies reveal negative regulation of ERS1 ethylene receptor signaling in *Arabidopsis*. BMC Plant Biol. 10, 60.

Luo, X., Chen, Z., Gao, J., et al., 2014. Abscisic acid inhibits root growth in *Arabidopsis* through ethylene biosynthesis. Plant J. 79, 44–55.

Lyzenga, W.J., Booth, J.K., Stone, S.L., 2012. The *Arabidopsis* RING-type E3 ligase XBAT32 mediates the proteasomal degradation of the ethylene biosynthetic enzyme, 1-aminocyclopropane-1-carboxylate synthase 7. Plant J. 71, 23–34.

Lyzenga, W.J., Stone, S.L., 2012. Regulation of ethylene biosynthesis through protein degradation. Plant Signal. Behav. 7, 1438–1442.

Ma, B., He, S.J., Duan, K.X., et al., 2013. Identification of rice ethylene-response mutants and characterization of MHZ7/OsEIN2 in distinct ethylene response and yield trait regulation. Mol. Plant 6, 1830–1848.

Ma, B., Yin, C.C., He, S.J., et al., 2014. Ethylene-induced inhibition of root growth requires abscisic acid function in rice (*Oryza sativa* L.) seedlings. PLoS Genet. 10, e1004701.

Mao, C., Wang, S., Jia, Q., et al., 2006. OsEIL1, a rice homolog of the *Arabidopsis* EIN3 regulates the ethylene response as a positive component. Plant Mol. Biol. 61, 141–152.

McClellan, C.A., Chang, C., 2008. The role of protein turnover in ethylene biosynthesis and response. Plant Sci. 175, 24–31.

McManus, M.T., 2012. The plant hormone ethylene. Annu. Plant Rev. 44.

Merchante, C., Brumos, J., Yun, J., et al., 2015. Gene-specific translation regulation mediated by the hormone-signaling molecule EIN2. Cell 163, 684–697.

Moussatche, P., Klee, H.J., 2004. Autophosphorylation activity of the *Arabidopsis* ethylene receptor multigene family. J. Biol. Chem. 279, 48734–48741.

Olmedo, G., Guo, H., Gregory, B.D., et al., 2006. *ETHYLENE-INSENSITIVE5* encodes a $5'\rightarrow3'$ exoribonuclease required for regulation of the EIN3-targeting F-box proteins EBF1/2. Proc. Natl. Acad. Sci. U.S.A. 103, 13286–13293.

O'Malley, R.C., Rodriguez, F.I., Esch, J.J., et al., 2005. Ethylene-binding activity, gene expression levels, and receptor system output for ethylene receptor family members from *Arabidopsis* and tomato. Plant J. 41, 651–659.

Paponov, I.A., Paponov, M., Teale, W., et al., 2008. Comprehensive transcriptome analysis of auxin responses in *Arabidopsis*. Mol. Plant 1, 321–337.

Peng, H.P., Lin, T.Y., Wang, N.N., et al., 2005. Differential expression of genes encoding 1-aminocyclopropane-1-carboxylate synthase in *Arabidopsis* during hypoxia. Plant Mol. Biol. 58, 15–25.

Peng, J., Li, Z., Wen, X., et al., 2014. Salt-induced stabilization of EIN3/EIL1 confers salinity tolerance by deterring ROS accumulation in *Arabidopsis*. PLoS Genet. 10, e1004664.

Potuschak, T., Lechner, E., Parmentier, Y., et al., 2003. EIN3-dependent regulation of plant ethylene hormone signaling by two *Arabidopsis* F box proteins: EBF1 and EBF2. Cell 115, 679–689.

Potuschak, T., Vansiri, A., Binder, B.M., et al., 2006. The exoribonuclease XRN4 is a component of the ethylene response pathway in *Arabidopsis*. The Plant Cell 18, 3047–3057.

Prasad, M.E., Schofield, A., Lyzenga, W., et al., 2010. *Arabidopsis* RING E3 ligase XBAT32 regulates lateral root production through its role in ethylene biosynthesis. Plant Physiol. 153, 1587–1596.

Qiao, H., Chang, K.N., Yazaki, J., et al., 2009. Interplay between ethylene, ETP1/ETP2 F-box proteins, and degradation of EIN2 triggers ethylene responses in *Arabidopsis*. Genes Dev. 23, 512–521.

Qiao, H., Shen, Z., Huang, S.S., et al., 2012. Processing and subcellular trafficking of ER-tethered EIN2 control response to ethylene gas. Science 338, 390–393.

Qin, Y.M., Hu, C.Y., Pang, Y., et al., 2007. Saturated very-long-chain fatty acids promote cotton fiber and *Arabidopsis* cell elongation by activating ethylene biosynthesis. The Plant Cell 19, 3692–3704.

Qiu, K., Li, Z., Yang, Z., et al., 2015. EIN3 and ORE1 accelerate degreening during ethylene-mediated leaf senescence by directly activating chlorophyll catabolic genes in *Arabidopsis*. PLoS Genet. 11, e1005399.

Qiu, L., Xie, F., Yu, J., et al., 2012. *Arabidopsis* RTE1 is essential to ethylene receptor ETR1 amino-terminal signaling independent of CTR1. Plant Physiol. 159, 1263–1276.

Qu, X., Hall, B.P., Gao, Z., et al., 2007. A strong constitutive ethylene-response phenotype conferred on *Arabidopsis* plants containing null mutations in the ethylene receptors ETR1 and ERS1. BMC Plant Biol. 7, 1–15.

Rauf, M., Arif, M., Fisahn, J., et al., 2013. NAC transcription factor SPEEDY HYPONASTIC GROWTH regulates flooding-induced leaf movement in *Arabidopsis*. The Plant Cell 25, 4941–4955.

Raz, V., Ecker, J.R., 1999. Regulation of differential growth in the apical hook of *Arabidopsis*. Development 126, 3661–3668.

Resnick, J.S., Rivarola, M., Chang, C., 2008. Involvement of RTE1 in conformational changes promoting ETR1 ethylene receptor signaling in *Arabidopsis*. Plant J. 56, 423–431.

Resnick, J.S., Wen, C.K., Shockey, J.A., et al., 2006. REVERSION-TO-ETHYLENE SENSITIVITY1, a conserved gene that regulates ethylene receptor function in *Arabidopsis*. Proc. Natl. Acad. Sci. U.S.A. 103, 7917–7922.

Rodríguez, F.I., Esch, J.J., Hall, A.E., et al., 1999. A copper cofactor for the ethylene receptor ETR1 from *Arabidopsis*. Science 283, 996–998.

Roman, G., Lubarsky, B., Kieber, J.J., et al., 1995. Genetic analysis of ethylene signal transduction in *Arabidopsis Thaliana*: five novel mutant loci integrated into a stress response pathway. Genetics 139, 1393–1409.

Rzewuski, G., Sauter, M., 2008. Ethylene biosynthesis and signaling in rice. Plant Sci. 175, 32–42.

Sakai, H., Hua, J., Chen, Q.G., et al., 1998. *ETR2* is an *ETR1*-like gene involved in ethylene signaling in *Arabidopsis*. Proc. Natl. Acad. Sci. U.S.A. 95, 5812–5817.

Satler, S.O., Kende, H., 1985. Ethylene and the growth of rice seedlings. Plant Physiol. 79, 194–198.

Schaller, G.E., Ladd, A.N., Lanahan, M.B., et al., 1995. The ethylene response mediator ETR1 from *Arabidopsis* forms a disulfide-linked dimer. J. Biol. Chem. 270, 12526–12530.

Sekhar, S., Panda, B.B., Mohapatra, T., et al., 2015. Spikelet-specific variation in ethylene production and constitutive expression of ethylene receptors and signal transducers during grain filling of compact- and lax-panicle rice (*Oryza sativa*) cultivars. J. Plant Physiol. 179, 21–34.

Shi, H., Liu, R., Xue, C., et al., 2016. Seedlings transduce the depth and mechanical pressure of covering soil using COP1 and ethylene to regulate EBF1/EBF2 for soil emergence. Curr. Biol. 26, 139–149.

Shi, Y., Tian, S., Hou, L., et al., 2012. Ethylene signaling negatively regulates freezing tolerance by repressing expression of *CBF* and *Type-A ARR* genes in *Arabidopsis*. The Plant Cell 24, 2578–2595.

Skottke, K.R., Yoon, G.M., Kieber, J.J., et al., 2011. Protein phosphatase 2A controls ethylene biosynthesis by differentially regulating the turnover of ACC synthase isoforms. PLoS Genet. 7, e1001370.

Solano, R., Stepanova, A., Chao, Q., et al., 1998. Nuclear events in ethylene signaling: a transcriptional cascade mediated by ETHYLENE-INSENSITIVE3 and ETHYLENE-RESPONSE-FACTOR1. Genes Dev. 12, 3703–3714.

Song, S., Huang, H., Gao, H., et al., 2014. Interaction between MYC2 and ETHYLENE INSENSITIVE3 modulates antagonism between jasmonate and ethylene signaling in Arabidopsis. The Plant Cell 26, 263–279.

Souret, F.F., Kastenmayer, J.P., Green, P.J., 2004. AtXRN4 degrades mRNA in Arabidopsis and its substrates include selected miRNA targets. Mol. Cell 15, 173–183.

Spanu, P., Grosskopf, D.G., Felix, G., et al., 1994. The apparent turnover of 1-aminocyclopropane-1-carboxylate synthase in tomato cells is regulated by protein phosphorylation and dephosphorylation. Plant Physiol. 106, 529–535.

Tao, J.J., Cao, Y.R., Chen, H., et al., 2015a. Tobacco TCTP interacts with ethylene receptor NTHK1 and enhances plant growth through promotion of cell proliferation. Plant Physiol. 169, 229–237.

Tao, J.J., Chen, H.W., Ma, B., et al., 2015b. The role of ethylene in plants under salinity stress. Front. Plant Sci. 6.

Tsuchisaka, A., Theologis, A., 2004. Unique and overlapping expression patterns among the Arabidopsis 1-amino-cyclopropane-1-carboxylate synthase gene family members. Plant Physiol. 136, 2982–3000.

Tsuchisaka, A., Yu, G., Jin, H., et al., 2009. A combinatorial interplay among the 1-aminocyclopropane-1-carboxylate isoforms regulates ethylene biosynthesis in Arabidopsis thaliana. Genetics 183, 979–1003.

Vembar, S.S., Brodsky, J.L., 2008. One step at a time: endoplasmic reticulum-associated degradation. Nat. Rev. Mol. Cell Biol. 9, 944–957.

Voet-Van-Vormizeele, J., Groth, G., 2008. Ethylene controls autophosphorylation of the histidine kinase domain in ethylene receptor ETR1. Mol. Plant 1, 380–387.

Wan, L., Zhang, J., Zhang, H., et al., 2011. Transcriptional activation of OsDERF1 in OsERF3 and OsAP2-39 negatively modulates ethylene synthesis and drought tolerance in rice. PloS One 6, e25216.

Wang, K.L., Li, H., Ecker, J.R., 2002. Ethylene biosynthesis and signaling networks. The Plant Cell 14 (Supp. l), S131–S151.

Wang, K.L., Yoshida, H., Lurin, C., et al., 2004. Regulation of ethylene gas biosynthesis by the Arabidopsis ETO1 protein. Nature 428, 945–950.

Wang, Q., Zhang, W., Yin, Z., et al., 2013. Rice CONSTITUTIVE TRIPLE-RESPONSE2 is involved in the ethylene-receptor signalling and regulation of various aspects of rice growth and development. J. Exp. Bot. 64, 4863–4875.

Wang, W., Esch, J.J., Shiu, S.-H., et al., 2006. Identification of important regions for ethylene binding and signaling in the transmembrane domain of the ETR1 ethylene receptor of Arabidopsis. The Plant Cell 18, 3429–3442.

Wen, X., Zhang, C., Ji, Y., et al., 2012. Activation of ethylene signaling is mediated by nuclear translocation of the cleaved EIN2 carboxyl terminus. Cell Res. 22, 1613–1616.

West, A.H., Stock, A.M., 2001. Histidine kinases and response regulator proteins in two-component signaling systems. Trends Biochem. Sci. 26, 369–376.

Wuriyanghan, H., Zhang, B., Cao, W.H., et al., 2009. The ethylene receptor ETR2 delays floral transition and affects starch accumulation in rice. The Plant Cell 21, 1473–1494.

Xie, C., Zhang, J.S., Zhou, H.L., et al., 2003. Serine/threonine kinase activity in the putative histidine kinase-like ethylene receptor NTHK1 from tobacco. Plant J. 33, 385–393.

Xie, F., Qiu, L., Wen, C.K., 2012. Possible modulation of *Arabidopsis* ETR1 N-terminal signaling by CTR1. Plant Signal. Behav. 7, 1243–1245.

Xu, J., Li, Y., Wang, Y., et al., 2008. Activation of MAPK kinase 9 induces ethylene and camalexin biosynthesis and enhances sensitivity to salt stress in *Arabidopsis*. J. Biol. Chem. 283, 26996–27006.

Xu, J., Zhang, S., 2014. Regulation of ethylene biosynthesis and signaling by protein kinases and phosphatases. Mol. Plant 7, 939–942.

Xu, J., Zhang, S., 2015. Ethylene biosynthesis and regulation in plants. Ethylene in Plants 1–25.

Xu, K., Xu, X., Fukao, T., et al., 2006. *Sub1A* is an ethylene-response-factor-like gene that confers submergence tolerance to rice. Nature 442, 705–708.

Yamasaki, K., Kigawa, T., Inoue, M., et al., 2005. Solution structure of the major DNA-binding domain of *Arabidopsis thaliana* ethylene-insensitive3-like3. J. Mol. Biol. 348, 253–264.

Yang, C., Lu, X., Ma, B., et al., 2015a. Ethylene signaling in rice and *Arabidopsis*: conserved and diverged aspects. Mol. Plant 8, 495–505.

Yang, C., Ma, B., He, S.J., et al., 2015b. MHZ6/OsEIL1 and OsEIL2 regulate ethylene response of roots and coleoptiles and negatively affect salt tolerance in rice. Plant Physiol. 169.

Yang, H.J., Shen, H., Chen, L., et al., 2002. The *OsEBP-89* gene of rice encodes a putative EREBP transcription factor and is temporally expressed in developing endosperm and intercalary meristem. Plant Mol. Biol. 50, 379–391.

Yang, S.A., Hoffman, N.E., 1984. Ethylene biosynthesis and its regulation in higher plants. Annu. Rev. Plant Physiol. Mol. Biol. 35, 155–189.

Yin, C.C., Ma, B., Collinge, D.P., et al., 2015. Ethylene responses in rice roots and coleoptiles are differentially regulated by a carotenoid isomerase-mediated abscisic acid pathway. The Plant Cell 27, 1061–1081.

Yoo, S.D., Cho, Y.-H., Tena, G., et al., 2008. Dual control of nuclear EIN3 by bifurcate MAPK cascades in C_2H_4 signalling. Nature 451, 789–795.

Yoon, G.M., Kieber, J.J., 2013. 14-3-3 regulates 1-aminocyclopropane-1-carboxylate synthase protein turnover in *Arabidopsis*. The Plant Cell 25, 1016–1028.

Zhang, W., Zhou, X., Wen, C.K., 2012. Modulation of ethylene responses by *OsRTH1* overexpression reveals the biological significance of ethylene in rice seedling growth and development. J. Exp. Bot. 63, 4151–4164.

Zhang, X., Zhu, Z., An, F., et al., 2014. Jasmonate-activated MYC2 represses ETHYLENE INSENSITIVE3 activity to antagonize ethylene-promoted apical hook formation in *Arabidopsis*. The Plant Cell 26, 1105–1117.

Zhang, Z., Zhou, H., Chen, T., et al., 2004. Evidence for serine/threonine and histidine kinase activity in the tobacco ethylene receptor protein NTHK2. Plant Physiol. 136, 2971–2981.

Zhao, Q., Guo, H.W., 2011. Paradigms and paradox in the ethylene signaling pathway and interaction network. Mol. Plant 4, 626–634.

Zhong, S., Shi, H., Xue, C., et al., 2012. A molecular framework of light-controlled phytohormone action in *Arabidopsis*. Curr. Biol. 22, 1530–1535.

Zhong, S., Zhao, M., Shi, T., et al., 2009. EIN3/EIL1 cooperate with PIF1 to prevent photo-oxidation and to promote greening of *Arabidopsis* seedlings. Proc. Natl. Acad. Sci. U.S.A. 106, 21431–21436.

Zhou, H.L., Cao, W.H., Cao, Y.R., et al., 2006. Roles of ethylene receptor NTHK1 domains in plant growth, stress response and protein phosphorylation. FEBS Lett. 580, 1239–1250.

Zhou, X., Liu, Q., Xie, F., et al., 2007. RTE1 is a golgi-associated and ETR1-dependent negative regulator of ethylene responses. Plant Physiol. 145, 75–86.

Zhu, Z., An, F., Feng, Y., et al., 2011. Derepression of ethylene-stabilized transcription factors (EIN3/EIL1) mediates jasmonate and ethylene signaling synergy in *Arabidopsis*. Proc. Natl. Acad. Sci. U.S.A. 108, 12539–12544.

Jasmonates

7

Qingzhe Zhai[1], Chun Yan[2], Lin Li[1], Daoxin Xie[2], Chuanyou Li[1]
[1]Institute of Genetics and Developmental Biology, Chinese Academy of Sciences, Beijing, China; [2]Tsinghua University, Beijing, China

Summary

As a major immunity hormone, the jasmonate family of oxylipins promote plant defense to mechanical wounding, chewing insects, and necrotrophic pathogens. In addition, jasmonates generally repress vegetative growth while promoting reproductive development. Molecular genetic studies, mainly conducted in the model systems of Arabidopsis (*Arabidopsis thaliana*) and tomato (*Solanum lycopersicum*), have significantly improved our understanding of the molecular mechanisms underlying jasmonate biosynthesis and signaling. Genes encoding nearly all enzymes involved in the octadecanoid pathway for jasmonate biosynthesis have been identified. Recent discoveries have illustrated a core jasmonate signaling module consisting of the F-box protein CORONATINE INSENSITIVE 1 (COI1) that forms a Skp-Cullin-F-box (SCF) E3 ubiquitin ligase complex, a group of jasmonate-ZIM domain (JAZ) proteins that function as transcriptional repressors, and various transcription factors that differentially regulate diverse aspects of JA responses. The COI1 protein serves as the jasmonate receptor to perceive the bioactive hormone jasmonoyl-isoleucine (JA-Ile) and recruit JAZ repressors for degradation, which leads to activation of JAZ-repressed transcription factors that regulate their corresponding jasmonate-responsive genes. Accumulating evidence shows that jasmonates also interact with other hormonal signals to regulate diverse plant defense responses and various developmental processes.

7.1 Introduction

7.1.1 Discovery of jasmonates

Jasmonates (JAs) are lipid-derived cyclopentanones that include jasmonic acid (JA) and its various derivatives. Compared with the so-called classical five plant hormones comprising auxin, gibberellin (GA), ethylene (ET), cytokinin, and abscisic acid (Kende and Zeevaart, 1997), JAs were among the "new" members of the plant hormone family. In 1962, the methyl ester of JA (MeJA) was first isolated as a major fragrance from the ethereal oil of *Jasminum grandiflorum* flowers (Demole et al., 1962), and the free acid was isolated subsequently from the culture filtrates of the fungus *Botryodiplodia theohromae* (Aldridge et al., 1971). Later, JA was isolated from higher plants including *Cucurbita pepo* (Fukui et al., 1977) and *Vicia faba* (Dathe et al., 1981). In 1980, JA and several structurally related compounds were chemically synthesized and their biological activity was tested on the growth of rice seedlings (Yamane et al., 1980).

Hormone Metabolism and Signaling in Plants. http://dx.doi.org/10.1016/B978-0-12-811562-6.00007-4

7.1.2   Physiological function of jasmonates

Since the discovery of JAs, diverse physiological functions of JAs have been described. Early observations of JA functions included promotion of senescence (Ueda and Kato, 1980) and inhibition of seedling growth (Dathe et al., 1981). A decade later, the group of Clarence Ryan found that the volatile MeJA acts as a potent inducer of defensive proteinase inhibitors (PI) in tomato leaves (Farmer and Ryan, 1990). Soon after the effect of MeJA in inhibiting root growth of Arabidopsis (*Arabidopsis thaliana*) was described and the first JA-response mutant named *jasmonic acid resistant 1* (*jar1*) was identified (Staswick et al., 1992). The group of Meinhart Zenk found that JAs are potent inducers of secondary metabolites in plant cell cultures (Gundlach et al., 1992). These pioneer discoveries aroused intense interests in the study of the biological functions of JAs.

Mutants of JA responses in Arabidopsis and tomato provided powerful tools to establish the chemistry, biosynthesis, and signaling pathway of JAs, which will be discussed in detail later. Mutant analysis in combination with advanced molecular, biochemical, and genomic approaches have also significantly advanced our understanding of the diverse functions of JAs. As one of the major plant immunity hormones, JAs positively regulate plant defense in response to mechanical wounding, chewing insects, and necrotrophic pathogens, whereas they negatively regulate plant defense to biotrophic pathogens. Grafting experiments with tomato mutants defective in JA synthesis or signaling support a scenario in which JAs may act as long-distance mobile signals in regulating systemic plant immunity (Li et al., 2002). As growth regulators, JAs generally promote reproductive development while repressing vegetative growth. In many cases, JA regulation of plant growth is achieved through collaboration with growth hormones. For example, JA-mediated inhibition of primary root growth is achieved through interaction with auxin to modulate root meristem activity (Chen et al., 2011). Emerging evidence indicated that, underling every aspect of JA function, matching pairs of transcription factors interact with a subset of JAZ repressors to orchestrate the expression of a spectrum of JA-responsive genes (Hu et al., 2013b; Jiang et al., 2014; Qi et al., 2011; Song et al., 2011; Zhai et al., 2015). Although JAs do not occur in the animal kingdom, intensive studies have revealed that JAs exhibit direct and selective inhibitory effects on several human tumor cells without affecting normal cells, raising the exciting possibility to develop JA-based anticancer drugs (Cohen and Flescher, 2009; Wasternack, 2015).

7.2   Biosynthesis of JA

7.2.1   The scheme of JA biosynthesis

JA is derived from α-linolenic acid (18:3) via the octadecanoid pathway. Following the first proposal of the chemistry of the JA biosynthetic pathway (Vick and Zimmerman, 1983, 1984), subsequent studies in plants confirmed the chemistry and added important details about the enzymology, regulation, and subcellular location of the pathway reactions (Browse, 2009; Wasternack, 2007; Wasternack and Hause, 2013). The pathway is initiated in the chloroplast by lipoxygenase (LOX)-catalyzed oxygenation of α-linolenic acid (Fig. 7.1). The resulting 13(S)-hydroperoxy-octadecatrienoic acid (13-HOPT) is substrate for allene oxide synthase (AOS), which generates an unstable

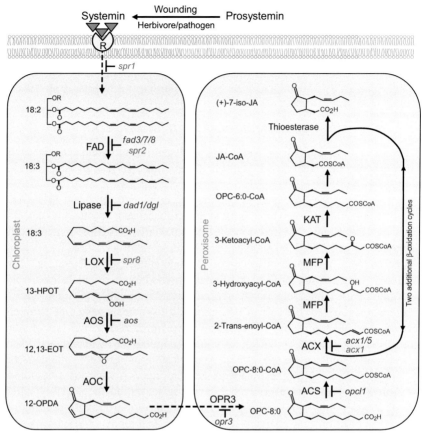

Figure 7.1 Octadecanoid pathway for jasmonic acid (JA) biosynthesis in Arabidopsis and tomato. JA is synthesized from 18:3 generated from galactolipids; the sequential activities of several enzymes involved in the initial steps of JA biosynthesis including lipoxygenase (LOX), allene oxide synthase (AOS), allene oxide cyclase (AOC) lead to the production of 12-oxo-phytodienoic acid (12-OPDA). 12-OPDA is then transported to the peroxisome and reduced to 3-oxo-2-(2′(Z)-pentenyl)-cyclopentane-1-octanoic acid (OPC-8:0); OPC-8:0 undergoes three rounds of β-oxidation to yield JA, see text for details. In tomato, insect attack, pathogen infection, or mechanical wounding triggers the cleavage of systemin from its precursor protein prosystemin. Interaction of systemin with its proposed receptor (R) in the plasma membrane activates the octadecanoid pathway for JA biosynthesis with a so-far unknown mechanism. Mutations that block JA biosynthesis in Arabidopsis (*blue*) and tomato (*red*) are indicated. *12,13-EOT*, 12,13(*S*)-epoxy-octadecatrienoic acid; *13-HOPT*, 13(*S*)-hydroperoxy-octadecatrienoic acid; *ACS*, acyl-CoA synthetase; *ACX*, acyl-CoA oxidase; *FAD*, fatty acid desaturases; *KAT*, 3-ketoacyl-CoA thiolase; *MFP*, multifunctional protein; *OPR3*, OPDA reductase3.

allene oxide, 12,13(*S*)-epoxy-octadecatrienoic acid (12,13-EOT). In the presence of allene oxide cyclase (AOC), 12,13-EOT is preferentially cyclized to the 9*S*, 13*S* isomer of 12-oxo-phytodienoic acid (12-OPDA), which is then exported to the peroxisome. In the peroxisome, the cyclopentenone ring of OPDA is reduced by OPDA reductase3 (OPR3) to yield 3-oxo-2-(2′(Z)-pentenyl)-cyclopentane-1-octanoic acid

(OPC-8:0). Three cycles of β-oxidation remove six carbons from the carboxyl side chain, completing the biosynthesis of JA (Schaller and Stintzi, 2009).

In a parallel hexadecanoid pathway, the same enzymes in the chloroplast act on hexadecatrienoic acid (16:3) to form *dinor*-oxophytodienoic acid (*dn*-OPDA). Assuming that *dn*-OPDA is a substrate for OPR3, this C_{16} compound could be further metabolized to JA by β-oxidation enzymes in the peroxisome (Weber et al., 1997).

7.2.2   Production of linolenic acid from linoleic acid

The JA precursor α-linolenic acid (18:3) is produced from linoleic acid (18:2), by a desaturation reaction catalyzed by fatty acid desaturase (FAD). In the Arabidopsis genome, there are three ω3 fatty acid desaturases (FAD3, FAD7, and FAD8) that convert dienoic fatty acids (16:2 and 18:2) to trienoic fatty acids (16:3 and 18:3). A *fad3 fad7 fad8* triple mutant is defective in JA synthesis because it produces no detectable 16:3 or 18:3 (McConn and Browse, 1996). Significantly, this triple mutant exhibits a characteristic male sterile phenotype, which can be readily restored by exogenous 18:3 and JA (McConn and Browse, 1996; McConn et al., 1997). Therefore, characterization of the *fad3 fad7 fad8* triple mutant provided the first conclusive evidence that JA plays an essential role in male reproductive development of Arabidopsis.

Further insights into the role of FADs in JA biosynthesis has come from characterization of the *suppressor of prosystemin-mediated responses2* (*spr2*) mutant of tomato (Li et al., 2003). The *spr2* mutation impairs wound-induced JA biosynthesis and defense gene expression. Map-based cloning studies demonstrated that *Spr2* encodes a chloroplast ω3 fatty acid desaturase, which is a homolog of the Arabidopsis *FAD7* gene product. Interestingly, despite the severely reduced levels of trienoic fatty acids, *spr2* plants exhibited normal growth, development, and reproduction. Further characterization of the Arabidopsis *fad3 fad7 fad8* triple mutant and the tomato *spr2* mutant should provide additional insight into the function of JAs in different plant species.

7.2.3   Release of linolenic acid from galactolipids

It is generally believed that the step to release the 18:3 precursor from membrane lipids plays a critical role to initiate JA biosynthesis. An important advance in this step came from the characterization of the Arabidopsis *delayed in anther dehiscence1* (*dad1*) mutant, which exhibits a male sterile phenotype. The *DAD1* gene encodes a chloroplastic phospholipase A_1 (PLA_1) that liberates 18:3 from the *sn1* position of chloroplast membrane glycerolipid (Ishiguro et al., 2001). The *dad1* mutant is defective in anther dehiscence and pollen development, but the levels of JA accumulation after wounding are largely normal, suggesting that DAD1-dependent JA biosynthesis is mainly involved in reproductive development (Ishiguro et al., 2001). Different from DAD1, it was shown that *DONGLE* (*DGL*), which encodes a homolog of DAD1, is highly expressed in leaves but not in flowers, siliques, and roots (Hyun et al., 2008). Expression of *DGL* was transiently induced by wounding and a *DGL*-overexpressing mutant exhibited elevated basal JA accumulation and increased expression of JA-responsive genes, suggesting that DGL-dependent JA biosynthesis is mainly involved in early wound response (Hyun et al., 2008). Together, these studies exemplify a notion that JA biosynthesis involved in wound response or development is differentially regulated.

7.2.4   Oxygenation of α-linolenic acid by 13-LOX

After release from galactolipids, free 18:3 is converted to 13-HPOT in reactions catalyzed by 13-LOX enzymes. All four 13-LOXs (LOX2, LOX3, LOX4, and LOX6) encoded in the Arabidopsis genome are capable of oxygenating 18:3 in vitro (Bannenberg et al., 2009). Among them, LOX2 is mainly responsible for wound-induced JA production (Bell et al., 1995; Glauser et al., 2009; Schommer et al., 2008). LOX3 and LOX4 are mainly involved in the development of fertile flowers (Caldelari et al., 2011). Genetic and physiological studies have provided evidence that LOX6 is involved in wound-induced and drought-induced JA production (Chauvin et al., 2013; Grebner et al., 2013).

In tomato, characterization of the *spr8* mutant, which is also a suppressor of prosystemin-mediated responses, revealed that *Spr8* encodes tomato lipoxygenase D (TomLoxD), a 13-LOX that catalyzes the hydroperoxidation of linolenic acid in JA biosynthesis. Interestingly, the general plant growth and flower development of *spr8* plants are largely normal, indicating that *TomLoxD* is specifically involved in the wound response (Yan et al., 2013b). The TomLoxD protein shows high sequence similarity to the maize TASSELSEED1 (TS1) protein, which plays an essential role for sex determination (Acosta et al., 2009; Yan et al., 2013b). These studies support the view that different lipoxygenase isoforms have distinct physiological functions.

Significantly, overexpression of *TomLoxD* leads to enhanced expression of wound-responsive genes and increased resistance to insects and necrotrophic pathogens, but shows minor effects on general plant growth, indicating that there was no "fitness cost" associated with overexpression of *TomLoxD*. This is important because the continuous mounting of defenses usually results in severe impacts on plant growth (Bostock, 2005). The finding that overexpression of *TomLoxD* does not lead to a significant "fitness cost" suggests that the *TomLoxD* gene has application potential for crop protection (Yan et al., 2013b).

7.2.5   Dehydration of 13-HPOT by AOS

Conversion of 13-HPOT to the unstable 12,13-EOT is catalyzed by the plastid-localized 13-AOS, which is an atypical cytochrome P450 enzyme (CYP74A) usually using oxygenated fatty acid hydroperoxide substrates as oxygen donor (Howe and Schilmiller, 2002; Werck-Reichhart et al., 2002). In the Arabidopsis genome there is a single *AOS* gene, whereas in the tomato genome there are two (Howe et al., 2000; Laudert et al., 1996).

The *AOS* genes from both Arabidopsis and tomato are induced by wounding, suggesting that they are involved in wound-induced JA biosynthesis (Howe et al., 2000; Park et al., 2002). Direct evidence that AOS is important for JA biosynthesis came from the characterization of the Arabidopsis *aos* mutant, which disrupts the expression of the *AOS* gene (Park et al., 2002). The *aos* mutant is defective in basal and wound-induced JA accumulation and, as expected, is compromised in wound-induced defense gene expression. In addition, the *aos* mutant exhibits a severe male sterile phenotype that did not recover during development but was rescued by exogenous MeJA spraying (Park et al., 2002). These results demonstrated that AOS-dependent JA biosynthesis is important for the wound response as well as for reproductive development of Arabidopsis.

7.2.6   Synthesis of OPDA by AOC

Products of AOS are allylic epoxides, which are unstable in water. Hydrolysis of allylic epoxides leads to the formation of α- and γ-ketols. AOC is involved in the cyclization of allylic epoxides to produce OPDA, which is usually in a racemic mixture of *cis* (+) and *cis* (−) enantiomers (Brash et al., 1988). AOC exclusively catalyzes the stereospecific cyclization of allene oxide into the *cis* (+) enantiomer OPDA, indicating that AOC plays a critical role in the establishment of the enantiomeric structure of the JA precursor (Schaller and Stintzi, 2009). AOC is encoded by four genes in Arabidopsis (Stenzel et al., 2003b) and by a single gene in tomato (Ziegler et al., 2000). By using the corresponding *promoter::glucuronidase* (*GUS*) fusion lines, it was shown that the four Arabidopsis *AOC* genes are highly and redundantly expressed in the aerial tissues including leaves and flowers (Stenzel et al., 2012). These observations, together with the fact that none of the single or double mutants of *AOC1*, *ACO3*, and *AOC4* displayed a JA-related phenotype, raised the possibility that the four *AOC* genes act redundantly in JA biosynthesis. Interestingly, this study found that *AOC3* and *AOC4* were also highly expressed in the meristematic region of the root, implying that JA synthesis may actively occur in root tissues (Stenzel et al., 2012).

7.2.7   Completion of JA synthesis in peroxisomes

The final product of the plastid-localized part of the JA biosynthesis pathway is OPDA. A peroxisomal OPDA reductase (OPR) catalyzes the reduction of the cyclopentenone ring of OPDA and *dn*OPDA to produce OPC-8:0 and OPC-6:0, respectively (Schaller and Stintzi, 2009). Among the six OPRs of Arabidopsis, only OPR3 accepts the (9S, 13S)-enantiomer of OPDA as a substrate, suggesting that OPR3 is involved in JA production. Indeed, the *opr3* mutant of Arabidopsis, which disrupts *OPR3* expression, displayed compromised defense responses and male sterility resulting from defective JA production, indicating that OPR3 plays an essential role in JA biosynthesis (Sanders et al., 2000; Stintzi and Browse, 2000).

Activation of OPC-8:0 is achieved by OPCL1, which is a carboxyl-CoA ligase (Koo et al., 2006). In response to wounding, the *opcl1* null mutant of Arabidopsis shows elevated accumulation of OPC-8:0 and reduced accumulation of JA, indicating that OPCL1 is indeed involved in JA biosynthesis. However, depletion of OPCL1 only leads to about a 50% reduction of wound-induced JA accumulation, suggesting that additional CoA-ligase isozymes are involved in the activation of OPC-8:0 during JA biosynthesis (Kienow et al., 2008; Koo et al., 2006).

The shortening of the octanoic acid side chains of OPC-8:0 involves three rounds of β-oxidation to yield (+)-7-*iso*-JA. These final steps of JA synthesis are catalyzed by acyl-CoA oxidase (ACX), the multifunctional protein (MFP), and 3-ketoacyl-CoA thiolase (KAT), which are core enzymes of the β-oxidation cycle. Experimental evidence for the contribution of these enzymes in JA biosynthesis came from characterization of mutants in tomato and Arabidopsis (Li et al., 2005; Schilmiller et al., 2007). Biochemical data revealed that Arabidopsis ACX1A is able to catalyze the first step in β-oxidation of OPC-8:0-CoA (Li et al., 2005). Consistent with this observation, the *acx1* tomato mutant is deficient in wound-induced JA biosynthesis and is impaired in defense response to wounding and insect attack (Li et al., 2005). Similarly, depletion of ALTERED INFLORESCENCE MERSITEM 1 (AMI1), one of the two peroxiso-mal MFP proteins of Arabidopsis, also leads to impaired wound-induced JA accumulation and defense gene expression (Delker et al., 2007).

7.3   Derivatives and metabolites of JA

7.3.1   Different types of JA derivatives

In the cytosol, JA is subjected to various modifications to form a series of metabolites: conjugation with isoleucine to give JA-Ile (Staswick and Tiryaki, 2004; Suza and Staswick, 2008); methylation to MeJA; formation of JA glucose ester (Swiatek et al., 2004); decarboxylation to *cis*-jasmone (Koch et al., 1997), and hydroxylation to 12-OH-JA (Gidda et al., 2003). These derivatives differ in their biological functions (Fig. 7.2).

7.3.2   Jasmonoyl-isoleucine

In Arabidopsis, JA-Ile is synthesized by the enzyme JAR1, a member of the Gretchen Hagen 3 (GH3) subfamily of conjugating enzymes that belongs to the acyl adenylate-forming firefly luciferase superfamily (Staswick et al., 2002; Suza and Staswick, 2008). The *jar1* mutant is the first JA-insensitive mutant identified by the root growth inhibition assay (Staswick et al., 1992). In the *jar1* mutant, JA-Ile levels are reduced (Suza and Staswick, 2008) and JA-inducible defense gene expression is decreased (Staswick et al., 1992). Subsequent work revealed that JA-Ile, but not free JA or other JA metabolites, is sufficient to promote the interaction between JAZ proteins with the F-box protein COI1 (Katsir et al., 2008; Thines et al., 2007). Later on, crystallization and stereoisomer purification approaches proved that COI1-JAZ acts as the hormone coreceptor and (+)-7-*iso*-JA-L-Ile is its genuine ligand (Fonseca et al., 2009; Sheard et al., 2010). This genetic and biochemical evidence indicates that JA-Ile is the receptor-active form of the hormone.

7.3.3   Methyl ester of jasmonic acid

MeJA, initially found as an odorant of *J. grandiflorum* flowers (Demole et al., 1962), is another important JA derivative shown to participate in the transport of the hormone (Wasternack and Strnad, 2015). In Arabidopsis, an S-adenosyl-L-methionine:jasmonic acid carboxyl methyltransferase (JMT) catalyzes the methylation of JA to form MeJA (Seo et al., 2001). Similar to other JA biosynthetic genes, the expression of *JMT* is under the regulation of developmental cues as well as external stimuli including mechanical wounding and exogenous JA application (Seo et al., 2001). Overexpression of *JMT* leads to elevated MeJA levels, increased expression of JA-responsive genes and enhanced disease resistance (Seo et al., 2001). These results support the view that JMT-dependent methylation of JA is an important regulatory step of JA-signaled plant immunity.

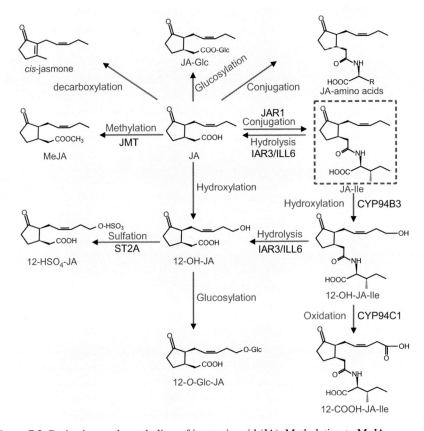

Figure 7.2 Derivatives and metabolites of jasmonic acid (JA). Methylation to MeJA, glucosylation to JA-Glc, decarboxylation to *cis*-jasmone, hydroxylation to 12-OH-JA, and conjugation with amino acids, preferentially isoleucine to give JA-Ile, are indicated. JA-Ile is the active form of the hormone, which is outlined with *dotted lines* (---). Two biochemical pathways convert JA-Ile to inactive compounds including oxidation and hydrolysis. The known enzymes involved are: *CYP94B3*, JA-Ile-12-hydroxylase; *CYP94C1*, 12-OH-JA-Ile carboxylase; *IAR3/ILL6*, amidohydrolases; *JAR1*, jasmonoyl-isoleucine synthetase; *JMT*, JA methyl transferase; *ST2A*, 12-OH-JA sulfotransferase.

7.3.4 Coronatine, a structural and functional mimic of JA-Ile

Coronatine is a non-host-specific phytotoxin produced by several pathovars of *Pseudomonas syringae* (Bender et al., 1999; Geng et al., 2014). It is formed by the amide linkage of two moieties, the polyketide coronafacic acid and the ethylcyclopropyl amino acid coronamic acid (Ichihara et al., 1977; Mitchell, 1985; Parry et al., 1994).

Coronatine enhances bacterial virulence by facilitating bacterial invasion through stomata, promoting bacterial multiplication, and inducing disease symptoms. A wealth of evidence indicates that the virulence property of coronatine is largely attributed to its ability to mimic the function of JA-Ile (Geng et al., 2014; Xin and He, 2013; Uppalapati et al., 2005; Staswick and Tiryaki, 2004; Zhao et al., 2003; Tamogami and Kodama, 2000; Palmer and Bender, 1995; Weiler et al., 1994; Koda et al., 1992). Three-dimensional structural and pharmacological studies revealed that both JA-Ile and coronatine are perceived by the same COI1-dependent receptor complex (Fonseca et al., 2009; Katsir et al., 2008; Sheard et al., 2010). Notably, coronatine is even more active than JA-Ile in promoting the COI1-JAZ interaction and subsequent JAZ degradation (Sheard et al., 2010).

7.4 Regulation of JA biosynthesis

7.4.1 Jasmonate burst

It is well recognized that JA levels vary widely in different organs, different developmental stages, and different environmental conditions (Creelman and Mullet, 1997; Wasternack, 2007). Both wounding and insect attack trigger a burst of JA biosynthesis (Blechert et al., 1995; Glauser et al., 2008, 2009; Koo et al., 2009; McCloud and Baldwin, 1997; Strassner et al., 2002). Work in different laboratories demonstrated that mechanical wounding of Arabidopsis leaves results in a very rapid (<5 min) increase of JA-Ile accumulation in systemic unwounded leaves (Glauser et al., 2008, 2009; Koo et al., 2009). These observations support a hypothesis that a rapidly transmitted wound signal triggers the synthesis of JA-Ile in unwounded leaves, which activates the expression of wound-responsive genes through the JA signaling pathway.

7.4.2 Amplification of JA production by systemin

It has been shown that both the 18-amino-acid peptide signal systemin and JA are essential regulatory signals for wound-induced systemic defense responses in tomato (Farmer et al., 1992; Ryan, 2000; Ryan and Pearce, 1998). It was proposed that systemin and JA act in the same signaling pathway to regulate systemic defense responses (Farmer et al., 1992; Ryan, 2000; Ryan and Pearce, 1998). Grafting experiments using tomato mutants defective in JA biosynthesis or signaling indicated that JA, rather than systemin, is the long-distance mobile signal for systemic defense response (Li et al., 2002). This hypothesis challenged an existing model and raised an interesting question about how systemin interacts with the JA pathway to activate systemic defense responses. Characterization of the tomato *spr1* mutant, which is insensitive to systemin, shows that systemin mainly acts as a local mediator for JA accumulation (Lee and Howe, 2003). Different from other tomato wound response mutants, *spr1* is mainly defective in the systemic defense response but its local defense response remains largely normal. Furthermore, *spr1* is defective in JA accumulation in response to systemin treatment (Lee and Howe, 2003). Reciprocal grafting experiments between *spr1* and wild-type plants demonstrated that *spr1* is defective in the production, but not the perception, of the long-distance wound signal, suggesting that Spr1 is involved in the integration of systemin perception with the activation of the JA pathway (Lee and Howe, 2003). These observations suggest that systemin acts at the wounded parts to amplify the accumulation of JA to a threshold level, which in turn functions as a systemic signal to activate defense responses in distal unwounded parts (Schilmiller and Howe, 2005; Lee and Howe, 2003; Stenzel et al., 2003a).

7.4.3   Positive feedback regulation of JA biosynthetic genes

An important feature of the JA synthesis pathway is that most of the JA biosynthetic genes are rapidly induced by JA treatment, mechanical wounding, or other stimuli, indicating that JA synthesis is positively regulated by the JA signaling pathway. In Arabidopsis, the basic-helix-loop-helix (bHLH) transcription factor MYC2, which specifically binds to the JA-responsive G-box (CACGTG) or G-box-like motif (Dombrecht et al., 2007; Lorenzo et al., 2004), acts as a master transcription factor of JA signaling and regulates a wide range of JA responses (Dombrecht et al., 2007; Kazan and Manners, 2013; Zhai et al., 2013). It was reported that MYC2 directly binds the promoter of *LOX2* and regulates its expression (Dombrecht et al., 2007; Hou et al., 2010). In tomato, the wound-induced expression of *TomLoxD* is also directly regulated by SlMYC2, the homolog of MYC2 in Arabidopsis (Yan et al., 2013b). Additional regulation of JA biosynthesis including substrate availability and tissue specificity has been described in several excellent reviews (Acosta and Farmer, 2010; Browse, 2009; Wasternack, 2007; Wasternack and Hause, 2013).

7.5   Jasmonate signaling

7.5.1   A core JA signaling module

Decades of studies in the model system of Arabidopsis have illustrated a core JA signaling module consisting of the F-box protein COI1 that forms a functional Skp-Cullin-F-box (SCF) E3 ubiquitin ligase complex, a group of JAZ proteins that function as transcriptional repressors, and a battery of transcription factors that differentially regulate diverse aspects of JA responses. In response to developmental signals or environmental cues, JA-Ile is rapidly synthesized and perceived by the F-box protein COI1, which further recruits JAZ repressors for degradation through the 26S proteasome. Degradation of JAZ proteins releases transcription factors to modulate expression of their corresponding target genes, thereby leading to regulation of JA-regulated defense and growth (Fig. 7.3).

7.5.2   The COI1-based SCF complex plays a central role in JA signaling

An important advance in our understanding of the JA signaling came from the analysis of the Arabidopsis *coronatine insensitive 1* (*coi1*) mutant that is insensitive to JA (Feys et al., 1994). The *coi1-1* mutant displays defects in all aspects of the JA response, such as pest and pathogen resistance, wound responses, root inhibition, fertility, and secondary metabolite biosynthesis (Mewis et al., 2005; Song et al., 2011; Thomma et al., 1998; Vijayan et al., 1998; Zimmerli et al., 2001). Identification of COI1 as an F-box protein that is closely related to the F-box protein TRANSPORT INHIBITOR RESPONSE 1 (TIR1), which is involved in auxin signaling, suggests that COI1 might act as part of an SCF E3 ubiquitin ligase to mediate JA signaling (Xie et al., 1998). Indeed, COI1 was

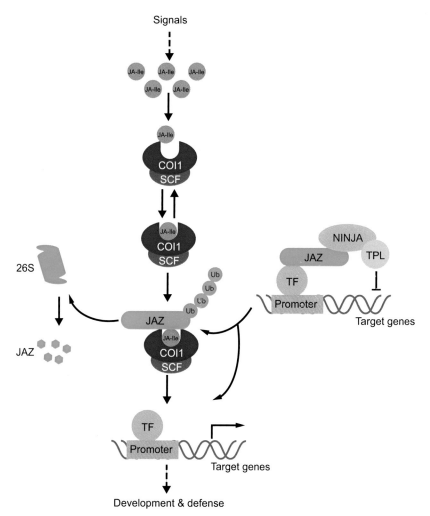

Figure 7.3 JA-Ile perception and signal transduction. In response to developmental signals or environmental cues, JA-Ile is rapidly biosynthesized and perceived by the F-box protein CORONATINE INSENSITIVE 1 (COI1), which further recruits different combinations of jasmonate-ZIM domain (JAZ) proteins for ubiquitination and degradation through the 26S proteasome. Degradation of JAZ proteins causes derepression of the NOVEL INTERACTOR OF JAZ (NINJA)–TOPLESS (TPL) inhibition on expression of the responsive genes targeted by transcription factors, and releases transcription factors to regulate the corresponding plant developmental processes and defense responses. *SCF*, Skp-Cullin-F-box; *TF*, transcription factor.

shown to be associated with components of the SCF complex, including ASK1, ASK2, Cullin1, and Rbx1 (Devoto et al., 2002; Ren et al., 2005; Xu et al., 2002). COI1 provides specificity to the SCF^{COI1} complex by binding target proteins that modulate the transcriptional response to JA.

The COI1 protein is stabilized by the integrity of the SCF^{COI1} complex and the disassociated COI1 is degraded via the 26S proteasome pathway in an SCF^{COI1}-independent manner (Yan et al., 2013a). The dynamic balance between the SCF^{COI1}-mediated stabilization and 26S proteasome-mediated degradation of COI1 maintains this protein at a suitable level, which is essential for exerting its biological function (Yan et al., 2013a). COI1 homologs have been reported in many plant species, including *Oryza sativa*, *Solanum lycopersicum*, *Nicotiana attenuata*, and major functions of COI1 in the plant kingdom are largely conserved (Lee et al., 2013, 2015; Li et al., 2004, 2006; Liao et al., 2015; Paschold et al., 2007; Wang et al., 2005; Ye et al., 2012).

7.5.3 JAZ proteins act as repressors of JA signaling

Based on the fact that auxin promotes SCF^{TIR1} to bind AUXIN/INDOLE-3-ACETIC ACID (AUX/IAA) transcriptional repressors for degradation (Dharmasiri et al., 2005; Kepinski and Leyser, 2005), it was hypothesized that SCF^{COI1} targets the repressor proteins of JA signaling for degradation and, therefore, activates JA-mediated transcriptional reprogramming. Three independent research groups identified the JAZ proteins as targets of COI1 (Chini et al., 2007; Thines et al., 2007; Yan et al., 2007). JAZ proteins serve as repressors of JA signaling, and are recruited by SCF^{COI1} in the presence of JA-Ile or coronatine for ubiquitin-dependent degradation (Chini et al., 2007; Thines et al., 2007; Yan et al., 2007) (Fig. 7.3). In Arabidopsis, 13 JAZ repressors are reported to date and they belong to the plant-specific TIFY family that is defined by the presence of the TIF[F/Y]XG motif within a larger (~28 amino acids) conserved region known as the ZIM domain (Chini et al., 2007; Chung and Howe, 2009; Pauwels and Goossens, 2011; Shyu et al., 2012; Thines et al., 2007; Thireault et al., 2015; Vanholme et al., 2007; Yan et al., 2007). In addition to containing a central ZIM domain, JAZ proteins also contain a relatively conserved N-terminal (NT) domain and a highly conserved C-terminal JA-associated (Jas) domain (Pauwels and Goossens, 2011). Within the Jas domain, the minimal amino acid sequence sufficient for coronatine or JA-Ile binding is called JAZ degron. The JAZ degron forms a bipartite structure consisting of a loop and an amphiphatic α-helix. Crystal structure studies indicated that the loop of the JAZ degron acts as a lid of the JA-Ile binding pocket and that the α-helix of the JAZ degron binds COI1 (Sheard et al., 2010).

The different domains present in JAZ proteins provide the specificity for protein–protein interactions that determine the differential formation of complexes in the absence or presence of the active hormone (Gimenez-Ibanez et al., 2015). The ZIM domain is required for formation of homo- and heterodimers among JAZ proteins and interaction with the NOVEL INTERACTOR OF JAZ (NINJA), which contains an ERF-associated amphiphilic repression motif (EAR) and recruits TOPLESS (TPL) to form the JAZ-NINJA-TPL repressor complex (Acosta et al., 2013; Pauwels et al., 2010). The Jas domain mediates interaction with COI1 and transcription factors (Chini et al., 2007; Melotto et al., 2008; Sheard et al., 2010). Five of the 13 JAZ proteins of the Arabidopsis genome, JAZ5, JAZ6, JAZ7, JAZ8, and JAZ13, contain an additional EAR motif that directly interacts with TPL even in the absence of the NINJA adaptor (Shyu et al., 2012; Thireault et al., 2015).

The lack of obvious JA-hypersensitive phenotypes among most *jaz* mutants reflects the functional redundancy between JAZ proteins (Chini et al., 2007; Chung et al., 2010; Chung and Howe, 2009; Thines et al., 2007; Yan et al., 2007). However, several studies have shown that Arabidopsis JAZ repressors have finely separated functions in JA signaling. First, JAZ8 contains a degenerate Jas degron that is unable to associate with COI1 in the presence of JA-Ile, but interacts with most transcription factors that regulate JA responses (Shyu et al., 2012). Second, alternative splicing of the *JAZ10* gene generates several stable JAZ10ΔJas isomers that are resistant to JA-induced degradation (Chung and Howe, 2009). It was proposed that alternative splicing of *JAZ* genes provides plants with a mechanism to tune the JA-signaling output to proper levels (Chung et al., 2010). Third, the transcription factors TARGET OF EAT1 (TOE1) and TOE2 and their interacting JAZ proteins (JAZ1, JAZ3, JAZ4, and JAZ9) specifically regulate JA-mediated delay of flowering, but not other aspects of JA responses (Zhai et al., 2015). These results support a view that the functional diversity of JAZ proteins underpins the complexity and specificity of JA responses.

7.5.4   COI1 serves as a receptor of JA-Ile

A key step in our understanding of JA signaling concerns the identification of the receptor. A ligand-binding assay with ^{3}H-coronatine demonstrates that coronatine binds directly to the COI1-JAZ complex. The ability of JA-Ile to compete with coronatine for specific binding indicates that JA-Ile and coronatine are recognized by the same receptor. These studies indicate that COI1 is a critical component of a receptor for JA-Ile and coronatine (Katsir et al., 2008).

To determine which protein serves as the receptor of JA-Ile and coronatine, high-quality structural model of COI1 was built and molecular docking simulation was performed (Yan et al., 2009). Molecular docking revealed that the COI1 protein has a binding pocket to retain the biologically active JA molecules (JA-Ile and coronatine), but not MeJA or OPDA. Biochemical assays show that COI1 is able to specifically bind the JA-immobilized sepharose, and directly interacts with the biotin-tagged photoaffinity probe for coronatine. Surface plasmon resonance assays demonstrate that, upon JA-Ile or coronatine binding, COI1 interacts with the JAZ repressors. These results demonstrate that COI1 serves as the JA receptor, which initially binds the JA active molecule to form a platform suitable for interaction with the JAZ repressor proteins (Yan et al., 2009).

Crystal structure studies further revealed detailed interactions within the COI1-coronatine-JAZ complex (Sheard et al., 2010). The leucine-rich repeats (LRRs) of COI1 form a horseshoe-shaped solenoid domain, housing the JA-Ile binding pocket, whereas the JAZ1 degron interacts with the surface of the coronatine-bound COI1 pocket. In addition, the inositol pentakisphosphate (InsP$_5$) was shown to be essential for COI1 function (Sheard et al., 2010).

Taken together, COI1 perceives JA-Ile to form a platform, which further recruits different combinations of JAZ proteins for ubiquitination and degradation, which otherwise bind and repress various transcription factors. Targeting of such JAZ proteins regulates the corresponding plant developmental processes and defense responses (Fig. 7.3).

7.5.5   MYC2 acts as a key transcription factor of JA signaling

The primary function of JAZ proteins is to bind and repress the activity of MYC2, a key regulator of JA signaling. MYC2 encodes a bHLH-type transcription factor and binds to the G-box (CACGTG) and G-box-related hexamers (Abe et al., 1997; Boter et al., 2004; Dombrecht et al., 2007; Fernández-Calvo et al., 2011; Saijo et al., 1997). In Arabidopsis, MYC2 differentially regulates two branches of JA-responsive genes (Boter et al., 2004; Lorenzo et al., 2004). One branch of JA-responsive genes is involved in plant responses to wounding. The other branch is associated with plant responses to pathogen infection (Boter et al., 2004; Lorenzo et al., 2004). In response to JA, MYC2 positively regulates a group of intermediate transcription factors, which in turn regulate the expression of downstream wound-responsive genes (Zheng et al., 2012; Zhai et al., 2013). MYC2 also negatively regulates a group of intermediate transcription factors, which in turn regulate the expression of downstream pathogen-responsive genes (Dombrecht et al., 2007; Zhai et al., 2013). It is worth to note that the positive regulation of wound response by MYC2 occurs relatively early (peak expression occurs within 6 h after JA application), whereas the negative regulation of pathogen response by MYC2 occurs relatively late (peak expression occurs 48 h after JA application) (Zhai et al., 2013). MYC2 also directly regulates the expression of a group of immediate JA-responsive genes (peak expression occurs within 1 h of JA application) including JAZs and JA biosynthetic genes (Chini et al., 2007; Grunewald et al., 2009). Furthermore, phosphorylation-coupled proteolysis of MYC2 stimulates the transcriptional activity of this transcription factor (Zhai et al., 2013).

MYC2 plays a critical role in JA-mediated inhibition of primary root growth. It was shown that, underlying the long-standing observation that JA inhibits primary root growth, JA modulates root meristem activity by repressing the expression of the stem cell transcription factors PLETHORA1 (PLT1) and PLT2, which act in the auxin pathway to regulate the maintenance of root stem cell niche. During this process, MYC2 represses *PLT1* and *PLT2* expression through promoter binding (Chen et al., 2011). These studies illustrate a molecular framework for JA-regulated root growth inhibition through interaction with the growth hormone auxin.

Studies showed that the Mediator (MED) transcriptional coactivator complex plays an essential role in JA-triggered activation of MYC2 (Çevik et al., 2012; Chen et al., 2012). The *bestatin-resistant 6* (*ber6*) mutant, which is a loss-of-function mutant of MED25 in Arabidopsis, exhibits defective JA responses in both root growth inhibition and defense gene expression (Chen et al., 2012; Zheng et al., 2006). MED25, which is a subunit of the Arabidopsis Mediator complex (Bäckström et al., 2007), interacts with the transcriptional activation domain (TAD) of MYC2 and therefore serves as a coactivator of MYC2. In addition, MED25 is required for the recruitment of RNA polymerase II (Pol II) to the core promoter of MYC2 targets. These data support a hypothesis that, during hormone-triggered transcription of JA-responsive genes, MED25 directly bridges the communication of the gene-specific transcription factor MYC2 and the Pol II general transcription machinery for preinitiation complex assembly (Chen et al., 2012).

7.5.6   Two subgroups of bHLH transcription factors differentially regulate JA signaling

As a key transcription factor in JA signaling, MYC2 differentially regulates diverse aspects of JA responses through interacting with most of the JAZ proteins. However, MYC2 was not the only transcription factor targeted by JAZ repressors, because not all JA responses were affected in the *myc2* loss-of-function mutant (Gimenez-Ibanez et al., 2015). Indeed, studies have identified a growing number of transcription factor-JAZ interaction pairs that convert the JA signals into specific context-dependent responses (Fig. 7.4).

Studies showed that the bHLH subgroup IIIe factors (MYC2, MYC3, MYC4, and MYC5) act as activators, whereas the bHLH subgroup IIId factors (bHLH3/JAM3, bHLH13/JAM2, bHLH14, and bHLH17/JAM1) act as repressors in JA-mediated gene expression (Cheng et al., 2011; Fernández-Calvo et al., 2011; Lorenzo et al., 2004; Nakata et al., 2013; Niu et al., 2011; Sasaki-Sekimoto et al., 2013; Song et al., 2013). The IIIe bHLH factors MYC2, MYC3, and MYC4 are direct targets of JAZ repressors and function redundantly to regulate JA-inhibitory root growth (Cheng et al., 2011; Fernández-Calvo et al., 2011; Niu et al., 2011), defense against insect attack (Fernández-Calvo et al., 2011), resistance to pathogens (Fernández-Calvo et al., 2011),

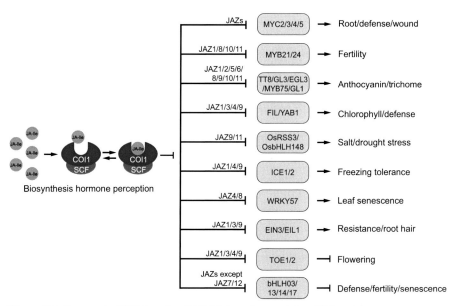

Figure 7.4 Jasmonate-ZIM domains (JAZs) interacting proteins. JAZ proteins serve as repressors of jasmonic acid (JA) signaling, which bind and repress a series of transcription factors essential for the corresponding JA responses. Upon perception of the JA signal, CORONATINE INSENSITIVE 1 (COI1) recruits JAZs for ubiquitin-mediated degradation and subsequently releases transcription factors to modulate expression of their corresponding target genes, thereby leading to regulation of JA-regulated plant development and defense. *SCF*, Skp-Cullin-F-box.

hook formation (Song et al., 2014; Zhang et al., 2014), leaf senescence (Qi et al., 2015b), and JA-induced glucosinolate biosynthesis (Schweizer et al., 2013).

MYC5 is reported as a novel target of JAZ repressors to function redundantly with MYC2, MYC3, and MYC4 in the regulation of stamen development and seed production (Figueroa and Browse, 2015; Qi et al., 2015a). In contrast to MYCs, bHLH3/JAM3, bHLH13/JAM2, bHLH14, and bHLH17/JAM1 negatively regulate JA-mediated root growth, resistance to bacterial pathogens, anthocyanin accumulation, chlorophyll loss, and leaf senescence (Nakata et al., 2013; Qi et al., 2015b; Sasaki-Sekimoto et al., 2013; Song et al., 2013). The antagonistic function between activators and repressors determines the outputs of growth, secondary metabolism, defense, and senescence, and dynamically modulates progression for plant survival according to the ever changing environment.

7.5.7 MYB21/MYB24/MYB57 regulates JA-mediated male fertility

Stamen development in Arabidopsis requires JA (Feys et al., 1994; Sanders et al., 2000; Stintzi and Browse, 2000; Xie et al., 1998). The R2R3 MYB transcription factors, MYB21 and MYB24, were found to redundantly regulate stamen development. A subset of JAZ proteins (JAZ1, JAZ8, and JAZ11) bind the N-terminal R2R3 domain of MYB21 and MYB24 to repress the transcriptional activity of MYB21 and MYB24 (Song et al., 2011). In addition, GA also regulates JA biosynthesis to control the expression of MYB21, MYB24, and MYB57 for coordination of normal stamen development (Cheng et al., 2009). The MYB transcription factors MYB21 and MYB24 also interact with IIIe subgroup bHLH factors (MYC2, MYC3, MYC4, and MYC5) to form an MYC-MYB transcription complex and cooperatively regulate stamen development (Qi et al., 2015a).

7.5.8 WD40/bHLH/MYB complexes regulate JA-dependent anthocyanin and trichome development

Complexes comprising WD-repeat, bHLH, and MYB proteins are key transcription complexes in the regulation of plant trichome initiation and anthocyanin accumulation. The specific transcription factors involved are WD-repeat protein Transparent Testa Glabra1 (TTG1), bHLH transcription factors TRANSPARENT TESTA8 (TT8), GLABRA3 (GL3) or ENHANCER of GL3 (EGL3), and R2R3-MYB transcription factors MYB75 or GLABRA1 (GL1). A subset of JAZ proteins (JAZ1, JAZ8, and JAZ11) interact with TT8, GL3, EGL3, MYB75, and GL1 to inhibit assembly of the WD-repeat/bHLH/MYB complexes and attenuate their transcriptional activity, leading to repression of trichome initiation and anthocyanin accumulation (Qi et al., 2011). In response to developmental signals or environmental cues, plants biosynthesize JA-Ile to induce degradation of JAZ proteins, which activates the WD-repeat/bHLH/MYB complexes and mediates downstream gene expression essential for trichome initiation and anthocyanin accumulation (Boter et al., 2015; Qi et al., 2011).

7.5.9   TOE1 and TOE2 regulate JA-mediated flowering time

Yeast two-hybrid screening to identify JAZ1-interacting proteins led to the identification of the transcription factors TOE1 and TOE2, which negatively regulate the expression of *FLOWERING LOCUS T* (*FT*) through interacting with a subset of JAZ proteins (JAZ1, JAZ3, JAZ4, and JAZ9) (Zhai et al., 2015). Interestingly, TOE1 and TOE2 specifically regulate JA-dependent control of flowering, but show negligible effect on other aspects of JA responses, suggesting that the specificity of the JA responses is determined by matching pairs of transcription factors and their interacting JAZ repressors (Zhai et al., 2015).

7.5.10   Other transcription factors regulate diverse JA responses

Studies have identified a growing number of pairs of transcription factors interacting with JAZ proteins that regulate different aspects of JA-dependent transcriptional responses (Fig. 7.4). For example, the bHLH-type transcription factors INDUCER OF CBF EXPRESSION1 (ICE1) and ICE2 interact with JAZ4 and JAZ9 to regulate JA-dependent freezing tolerance (Hu et al., 2013b). The WRKY family transcription factor WRKY57 interacts with JAZ4 and JAZ8 to regulate JA-induced leaf senescence (Jiang et al., 2014). The YABBY (YAB) family transcription factor FILAMENTOUS FLOWER (FIL), also known as YAB1, interacts with JAZ3 to regulate JA-mediated responses such as anthocyanin accumulation, chlorophyll loss, and susceptibility to *P. syringae* DC3000 infection (Boter et al., 2015). The rice nuclear factor, RICE SALT SENSITIVE3 (RSS3), interacts not only with a subset of JAZ proteins (JAZ9 and JAZ11), but also with non-R/B-like bHLH transcription factors and forms an RSS3-JAZ-bHLH ternary complex to regulate JA-mediated salt stress (Toda et al., 2013). It is reasonable to speculate that future studies will identify more interacting pairs of transcription factors and JAZ proteins that regulate different aspects of JA-dependent transcriptional responses, which will deepen our understanding of the molecular mechanisms of JA actions.

In response to wounding, the Arabidopsis GLUTAMATE RECEPTOR-LIKE (GLR) membrane proteins GLR3.3 and GLR3.6 mediate the transmission of wound-induced surface potential charge to distal leaves, and ultimately activate the JA pathway in distal unwounded leaves (Mousavi et al., 2013).

Most recently, JAV1, a unique VQ motif-containing protein, was found to specifically control plant defense without obviously affecting plant development. Plants in which JAV1 is suppressed by RNAi exhibit enhanced defense against pathogen infection and insect attack (Hu et al., 2013a).

7.6   Cross talk between JA and other phytohormones

7.6.1   Cross talk between JA and ethylene

JA and ET synergistically regulate plant resistance to necrotrophic pathogens. In the absence of JA signals, JAZ proteins interact with and repress EIN3 and

EIL1 (two homologous transcription factors in the ET pathway) to inhibit JA/ET-mediated plant resistance against *Botrytis cinerea* (Zhu et al., 2011). Further studies revealed that JAZ proteins recruit HDA6 as a corepressor to inhibit the function of EIN3 and EIL1 (Zhu et al., 2011). In response to JA and ET signals, JA-induced EIN3 and EIL1 activation (Zhu et al., 2011) and ET-induced EIN3 and EIL1 stabilization (Gagne et al., 2004; Guo and Ecker, 2003; Potuschak et al., 2003) mediate JA and ET signaling synergy in the regulation of root hair development and resistance against *B. cinerea* infection.

In addition to acting synergistically, JA and ET also act antagonistically in regulating apical hook formation, wound response, and defense against insect attack. JA-activated MYC2 physically interacts with and suppresses EIN3 and EIL1 to inhibit ET-enhanced hook curvature (Song et al., 2014; Zhang et al., 2014) and plant resistance to *B. cinerea* (Song et al., 2014). Conversely, ET-stabilized EIN3 and EIL1 interact with and attenuate MYC2 to repress plant defense against the generalist insect *Spodoptera exigua* (Song et al., 2014).

7.6.2   Cross talk between JA and gibberellin

JA and GA antagonize each other in regulating seedling growth and resistance to pathogens via interaction between JAZs and DELLAs. JAZ proteins interact with and repress DELLAs to derepress the bHLH factor PHYTOCHROME INTERACTING FACTOR 3 (PIF3) in the GA pathway, while JA signals induce JAZ degradation to release DELLAs that repress PIF3 and inhibit GA-enhanced hypocotyl elongation (Wild et al., 2012; Yang et al., 2012). Conversely, DELLAs interact with and repress JAZs to release MYC2 that positively regulates JA-mediated root growth inhibition, while GA-induced degradation of DELLAs frees JAZs to bind MYC2 and suppresses MYC2-dependent JA-signaling outputs (Hong et al., 2012; Hou et al., 2010).

JA and GA also synergistically regulate trichome initiation, stamen development, and sesquiterpene biosynthesis. Both DELLA and JAZ proteins interact with and repress the WD-repeat/bHLH/MYB complex to mediate synergism between GA and JA signaling in regulating trichome development. GA and JA induce degradation of DELLA and JAZ proteins, respectively, to coordinately activate the WD-repeat/bHLH/MYB complex to induce trichome initiation in a synergistic and mutually dependent manner (Qi et al., 2014).

7.6.3   Cross talk between JA and other signals

JAs play important roles in regulating plant defense against herbivores, and against necrotrophic pathogens such as *B. cinerea*, whereas salicylic acid (SA) primarily regulates plant defense against biotrophic and hemibiotrophic pathogens such as *P. syringae* (Gimenez-Ibanez and Solano, 2013; Glazebrook, 2005). JA-mediated and SA-mediated defense pathways generally antagonize each other and thus elevated resistance against necrotrophs is often correlated with increased susceptibility to biotrophs, and vice versa (Gimenez-Ibanez et al., 2016; Grant and Lamb, 2006).

Microbial pathogens have evolved strategies to manipulate plant hormone signaling pathways to cause hormonal imbalance for their own benefit. Production of phytotoxin coronatine, a structural and functional mimic of JA-Ile, by several pathovars of *P. syringae* is a well-characterized example of JA hormonal manipulation by pathogens (Chini et al., 2007; Gimenez-Ibanez and Solano, 2013; Glazebrook, 2005; Haider et al., 2000; Sheard et al., 2010; Thines et al., 2007; Weiler et al., 1994).

7.7 Summary points

1. Following the elucidation of the octadecanoid pathway for JA biosynthesis from α-linolenic acid in the plastid, conventional biochemical analyses in combination with the powerful molecular genetics approaches have confirmed and defined the molecular details of JA synthesis and its conversion to the active hormone JA-Ile.
2. Discoveries in Arabidopsis have illustrated a core JA signaling module consisting of the SCF^{COI1} E3 ubiquitin ligase complex, a group of JAZ transcriptional repressors, and the master transcription factor MYC2 that differentially regulates diverse aspects of JA responses.
3. The F-box protein COI1 serves as a JA receptor, which recruits the JAZ repressors for degradation and thereby leads to activation of JAZ-repressed transcription factors that regulate their corresponding JA-responsive genes.
4. During hormone-triggered activation of JA signaling, the MED25 subunit of the Mediator coactivator complex bridges the communication of the master transcription factor MYC2 with the RNA polymerase Pol II general transcription machinery through physical interaction.
5. Accumulating evidence shows that underlying a specific aspect of JA actions, interacting pairs of transcription factors and JAZ proteins orchestrate the expression of a specific spectrum of JA-responsive genes, suggesting that the JA signals are converted into specific context-dependent responses by matching pairs of transcription factors and JAZ proteins.
6. JA cross talks with other signals to regulate diverse plant defense responses and various developmental processes.

7.8 Future issues

1. To identify the JA transporters and elucidate the molecular mechanisms controlling JA movement.
2. To determine how JAs control resource allocation between growth- and defense-related processes.
3. To investigate the contribution of chromatin-based epigenetic mechanisms in conveying and integrating responses to JAs.
4. To elucidate the mechanism of JA burst and understand how plants rapidly accumulate JA upon wounding, insect attack, and pathogen infection.
5. To define synergistic or antagonistic interactions between JA signaling and other hormonal signaling pathways to appreciate diverse JA functions.
6. To elucidate the mechanism of action of JAs on cancer cells as well as on healthy cells.

Abbreviations

12,13-EOT	12,13(S)-epoxy-octadecatrienoic acid
12-COOH-JA-Ile	Dicarboxy-JA-Ile
12-OH-JA-Ile	12-Hydroxy-JA-Ile
12-OPDA	12-Oxo-phytodienoic acid
13-HOPT	13(S)-hydroperoxy-octadecatrienoic acid
ABA	Abscisic acid
ABC	ATP-binding cassette
ACX	Acyl-CoA oxidase
AOC	Allene oxide cyclase
AOS	Allene oxide synthase
ber6	*bestatin-resistant 6*
COI1	CORONATINE INSENSITIVE 1
dad1	*Delayed in anther dehiscence1*
DGL	*DONGLE*
dn-OPDA	Dinor-oxophytodienoic acid
EAR	ETHYLENE RESPONSE FACTOR (ERF)-associated amphiphilic repression motif
EGL3	ENHANCER of GL3
ET	Ethylene
FAD	Fatty acid desaturases
FIL	FILAMENTOUS FLOWER
FT	*FLOWERING LOCUS T*
GA	Gibberellin
GL1	GLABRA1
GL3	GLABRA3
GUS	Glucuronidase
HLS1	HOOKLESS1
ICE1	INDUCER OF CBF EXPRESSION1
InsP$_5$	Inositol pentakisphosphate
JA	Jasmonic acid
jai1	*jasmonate-insensitive 1*
JA-Ile	Jasmonoyl-isoleucine
jar1	*jasmonic acid resistant 1*
JAs	Jasmonates
JAZ	Jasmonate-ZIM domain
jin1	*Methyl jasmonate-insensitive 1*
JMT	Methyltransferase
KAT	3-Ketoacyl-CoA thiolase
LOX	Lipoxygenase
LRRs	Leucine-rich repeats
MeJA	Methyl ester of JA
MFP	Multifunctional protein
NINJA	NOVEL INTERACTOR OF JAZ
OPC	3-Oxo-2-(2′(Z)-pentenyl)-cyclopentane-1-octanoic
OPC-8:0	3-Oxo-2-(2′(Z)-pentenyl)-cyclopentane-1-octanoic acid

OPR	OPDA reductase
OPR3	OPDA reductase3
PI	*Proteinase inhibitors*
PIC	Preinitiation complex
PIF3	PHYTOCHROME INTERACTING FACTOR 3
PLA_1	Phospholipase A_1
PLT1	PLETHORA1
Pol II	RNA polymerase II
RSS3	RICE SALT SENSITIVE3
SA	Salicylic acid
SCF	Skp-Cullin-F-box
spr2	*Suppressor of prosystemin-mediated responses2*
TAD	Transcriptional activation domain
TOE1	TARGET OF EARLY ACTIVATION TAGGED 1
TomLoxD	Tomato lipoxygenase D
TPL	TOPLESS
TS1	TASSELSEED1
TT8	TRANSPARENT TESTA8
TTG1	Transparent Testa Glabra1
VSPs	Vegetative storage proteins
YAB	YABBY

Acknowledgments

This work is supported by grants from the National Natural Science Foundation of China to C.L. (grant nos. 90717007, 91117013, 91317309) and D.X. (grant nos. 31230008 and 31421001).

References

Abe, H., YamaguchiShinozaki, K., Urao, T., et al., 1997. Role of *Arabidopsis* MYC and MYB homologs in drought- and abscisic acid-regulated gene expression. Plant Cell 9, 1859–1868.

Acosta, I.F., Farmer, E.E., 2010. Jasmonates. Arabidopsis Book 8, e0129.

Acosta, I.F., Gasperini, D., Chetelat, A., et al., 2013. Role of NINJA in root jasmonate signaling. Proc. Natl. Acad. Sci. U.S.A. 110, 15473–15478.

Acosta, I.F., Laparra, H., Romero, S.P., et al., 2009. *tasselseed1* is a lipoxygenase affecting jasmonic acid signaling in sex determination of maize. Science 323, 262–265.

Aldridge, D.C., Galt, S., Giles, D., et al., 1971. Metabolites of *Lasiodiplodia theobromae*. J. Chem. Soc. C 1623–1627.

Bäckström, S., Elfving, N., Nilsson, R., et al., 2007. Purification of a plant mediator from *Arabidopsis thaliana* identifies PFT1 as the Med25 subunit. Mol. Cell 26, 717–729.

Bannenberg, G., Martinez, M., Hamberg, M., et al., 2009. Diversity of the enzymatic activity in the lipoxygenase gene family of *Arabidopsis thaliana*. Lipids 44, 85–95.

Bell, E., Creelman, R.A., Mullet, J.E., 1995. A chloroplast lipoxygenase is required for wound-induced jasmonic acid accumulation in *Arabidopsis*. Proc. Natl. Acad. Sci. U.S.A. 92, 8675–8679.

Bender, C.L., Alarcon-Chaidez, F., Gross, D.C., 1999. *Pseudomonas syringae* phytotoxins: mode of action, regulation, and biosynthesis by peptide and polyketide synthetases. Microbiol. Mol. Biol. Rev. 63, 266–292.

Blechert, S., Brodschelm, W., Holder, S., et al., 1995. The octadecanoic pathway: signal molecules for the regulation of secondary pathways. Proc. Natl. Acad. Sci. U.S.A. 92, 4099–4105.

Bostock, R.M., 2005. Signal crosstalk and induced resistance: straddling the line between cost and benefit. Annu. Rev. Phytopathol. 43, 545–580.

Boter, M., Golz, J.F., Gimenez-Ibanez, S., et al., 2015. FILAMENTOUS FLOWER is a direct target of JAZ3 and modulates responses to jasmonate. Plant Cell 27, 3160–3174.

Boter, M., Ruiz-Rivero, O., Abdeen, A., et al., 2004. Conserved MYC transcription factors play a key role in jasmonate signaling both in tomato and *Arabidopsis*. Gene Dev. 18, 1577–1591.

Brash, A.R., Baertschi, S.W., Ingram, C.D., et al., 1988. Isolation and characterization of natural allene oxides: unstable intermediates in the metabolism of lipid hydroperoxides. Proc. Natl. Acad. Sci. U.S.A. 85, 3382–3386.

Browse, J., 2009. Jasmonate passes muster: a receptor and targets for the defense hormone. Annu. Rev. Plant Biol. 60, 183–205.

Caldelari, D., Wang, G., Farmer, E.E., et al., 2011. *Arabidopsis lox3 lox4* double mutants are male sterile and defective in global proliferative arrest. Plant Mol. Biol. 75, 25–33.

Çevik, V., Kidd, B.N., Zhang, P.J., et al., 2012. MEDIATOR25 acts as an integrative hub for the regulation of jasmonate-responsive gene expression in *Arabidopsis*. Plant Physiol. 160, 541–555.

Chauvin, A., Caldelari, D., Wolfender, J.L., et al., 2013. Four 13-lipoxygenases contribute to rapid jasmonate synthesis in wounded *Arabidopsis thaliana* leaves: a role for lipoxygenase 6 in responses to long-distance wound signals. New Phytol. 197, 566–575.

Chen, Q., Sun, J.Q., Zhai, Q.Z., et al., 2011. The basic helix-loop-helix transcription factor MYC2 directly represses *PLETHORA* expression during jasmonate-mediated modulation of the root stem cell niche in *Arabidopsis*. Plant Cell 23, 3335–3352.

Chen, R., Jiang, H.L., Li, L., et al., 2012. The *Arabidopsis* mediator subunit MED25 differentially regulates jasmonate and abscisic acid signaling through interacting with the MYC2 and ABI5 transcription factors. Plant Cell 24, 2898–2916.

Cheng, H., Song, S.S., Xiao, L.T., et al., 2009. Gibberellin acts through jasmonate to control the expression of MYB21, MYB24, and MYB57 to promote stamen filament growth in *Arabidopsis*. PLoS Genet. 5.

Cheng, Z.W., Sun, L., Qi, T.C., et al., 2011. The bHLH transcription factor MYC3 interacts with the jasmonate ZIM-domain proteins to mediate jasmonate response in *Arabidopsis*. Mol. Plant 4, 279–288.

Chini, A., Fonseca, S., Fernández, G., et al., 2007. The JAZ family of repressors is the missing link in jasmonate signalling. Nature 448, 666–671.

Chung, H.S., Cooke, T.F., Depew, C.L., et al., 2010. Alternative splicing expands the repertoire of dominant JAZ repressors of jasmonate signaling. Plant J. 63, 613–622.

Chung, H.S., Howe, G.A., 2009. A critical role for the TIFY motif in repression of jasmonate signaling by a stabilized splice variant of the JASMONATE ZIM-domain protein JAZ10 in *Arabidopsis*. Plant Cell 21, 131–145.

Cohen, S., Flescher, E., 2009. Methyl jasmonate: a plant stress hormone as an anti-cancer drug. Phytochemistry 70, 1600–1609.

Creelman, R.A., Mullet, J.E., 1997. Biosynthesis and action of jasmonates in plants. Annu. Rev. Plant Physiol. Plant Mol. Biol. 48, 355–381.

Dathe, W., Ronsch, H., Preiss, A., et al., 1981. Endogenous plant hormones of the broad bean, Vicia-faba L (−)-jasmonic acid, a plant-growth inhibitor in pericarp. Planta 153, 530–535.

Delker, C., Zolman, B.K., Miersch, O., et al., 2007. Jasmonate biosynthesis in Arabidopsis thaliana requires peroxisomal beta-oxidation enzymes – additional proof by properties of pex6 and aim1. Phytochemistry 68, 1642–1650.

Demole, E., Lederer, E., Mercier, D., 1962. Isolement et détermination de la structure du jasmonate de méthyle, constituant odorant caractéristique de lessence de jasmin. Helv. Chim. Acta 45, 675–685.

Devoto, A., Nieto-Rostro, M., Xie, D.X., et al., 2002. COI1 links jasmonate signalling and fertility to the SCF ubiquitin-ligase complex in Arabidopsis. Plant J. 32, 457–466.

Dharmasiri, N., Dharmasiri, S., Estelle, M., 2005. The F-box protein TIR1 is an auxin receptor. Nature 435, 441–445.

Dombrecht, B., Xue, G.P., Sprague, S.J., et al., 2007. MYC2 differentially modulates diverse jasmonate-dependent functions in Arabidopsis. Plant Cell 19, 2225–2245.

Farmer, E.E., Johnson, R.R., Ryan, C.A., 1992. Regulation of expression of proteinase inhibitor genes by methyl jasmonate and jasmonic acid. Plant Physiol. 98, 995–1002.

Farmer, E.E., Ryan, C.A., 1990. Interplant communication: airborne methyl jasmonate induces synthesis of proteinase inhibitors in plant leaves. Proc. Natl. Acad. Sci. U.S.A. 87, 7713–7716.

Fernández-Calvo, P., Chini, A., Fernandez-Barbero, G., et al., 2011. The Arabidopsis bHLH transcription factors MYC3 and MYC4 are targets of JAZ repressors and act additively with MYC2 in the activation of jasmonate responses. Plant Cell 23, 701–715.

Feys, B.F., Benedetti, C.E., Penfold, C.N., et al., 1994. Arabidopsis mutants selected for resistance to the phytotoxin coronatine are male-sterile, insensitive to methyl jasmonate, and resistant to a bacterial pathogen. Plant Cell 6, 751–759.

Figueroa, P., Browse, J., 2015. Male sterility in Arabidopsis induced by overexpression of a MYC5-SRDX chimeric repressor. Plant J. 81, 849–860.

Fonseca, S., Chini, A., Hamberg, M., et al., 2009. (+)-7-iso-jasmonoyl-L-isoleucine is the endogenous bioactive jasmonate. Nat. Chem. Biol. 5, 344–350.

Fukui, H., Koshimizu, K., Usuda, S., et al., 1977. Isolation of plant-growth regulators from seeds of Cucurbita-pepo L. Agric. Biol. Chem. Tokyo 41, 175–180.

Gagne, J.M., Smalle, J., Gingerich, D.J., et al., 2004. Arabidopsis EIN3-binding F-box 1 and 2 form ubiquitin-protein ligases that repress ethylene action and promote growth by directing EIN3 degradation. Proc. Natl. Acad. Sci. U.S.A. 101, 6803–6808.

Geng, X., Jin, L., Shimada, M., et al., 2014. The phytotoxin coronatine is a multifunctional component of the virulence armament of Pseudomonas syringae. Planta 240, 1149–1165.

Gidda, S.K., Miersch, O., Levitin, A., et al., 2003. Biochemical and molecular characterization of a hydroxyjasmonate sulfotransferase from Arabidopsis thaliana. J. Biol. Chem. 278, 17895–17900.

Gimenez-Ibanez, S., Boter, M., Solano, R., 2015. Novel players fine-tune plant trade-offs. Essays Biochem. 58, 83–100.

Gimenez-Ibanez, S., Chini, A., Solano, R., 2016. How microbes twist jasmonate signaling around their little fingers. Plants (Basel) 5.

Gimenez-Ibanez, S., Solano, R., 2013. Nuclear jasmonate and salicylate signaling and crosstalk in defense against pathogens. Front. Plant Sci. 4, 72.

Glauser, G., Dubugnon, L., Mousavi, S.A., et al., 2009. Velocity estimates for signal propagation leading to systemic jasmonic acid accumulation in wounded *Arabidopsis*. J. Biol. Chem. 284, 34506–34513.

Glauser, G., Grata, E., Dubugnon, L., et al., 2008. Spatial and temporal dynamics of jasmonate synthesis and accumulation in *Arabidopsis* in response to wounding. J. Biol. Chem. 283, 16400–16407.

Glazebrook, J., 2005. Contrasting mechanisms of defense against biotrophic and necrotrophic pathogens. Annu. Rev. Phytopathol. 43, 205–227.

Guo, H., Ecker, J.R., 2003. Plant responses to ethylene gas are mediated by SCF[EBF1/EBF2]-dependent proteolysis of EIN3 transcription factor. Cell 115, 667–677.

Grant, M., Lamb, C., 2006. Systemic immunity. Curr. Opin. Plant Biol. 9, 414–420.

Grebner, W., Stingl, N.E., Oenel, A., et al., 2013. Lipoxygenase6-dependent oxylipin synthesis in roots is required for abiotic and biotic stress resistance of *Arabidopsis*. Plant Physiol. 161, 2159–2170.

Grunewald, W., Vanholme, B., Pauwels, L., et al., 2009. Expression of the *Arabidopsis* jasmonate signalling repressor JAZ1/TIFY10A is stimulated by auxin. EMBO Rep. 10, 923–928.

Gundlach, H., Muller, M.J., Kutchan, T.M., et al., 1992. Jasmonic acid is a signal transducer in elicitor-induced plant cell cultures. Proc. Natl. Acad. Sci. U.S.A. 89, 2389–2393.

Haider, G., von Schrader, T., Fusslein, M., et al., 2000. Structure–activity relationships of synthetic analogs of jasmonic acid and coronatine on induction of benzo[c]phenanthridine alkaloid accumulation in *Eschscholzia californica* cell cultures. Biol. Chem. 381, 741–748.

Hong, G.J., Xue, X.Y., Mao, Y.B., et al., 2012. *Arabidopsis* MYC2 interacts with DELLA proteins in regulating sesquiterpene synthase gene expression. Plant Cell 24, 2635–2648.

Hou, X., Lee, L.Y., Xia, K., et al., 2010. DELLAs modulate jasmonate signaling via competitive binding to JAZs. Dev. Cell 19, 884–894.

Howe, G.A., Lee, G.I., Itoh, A., et al., 2000. Cytochrome P450-dependent metabolism of oxylipins in tomato. Cloning and expression of allene oxide synthase and fatty acid hydroperoxide lyase. Plant Physiol. 123, 711–724.

Howe, G.A., Schilmiller, A.L., 2002. Oxylipin metabolism in response to stress. Curr. Opin. Plant Biol. 5, 230–236.

Hu, P., Zhou, W., Cheng, Z., et al., 2013a. JAV1 controls jasmonate-regulated plant defense. Mol. Cell 50, 504–515.

Hu, Y.R., Jiang, L.Q., Wang, F., et al., 2013b. Jasmonate regulates the inducer of CBF expression-C-repeat binding factor/DRE binding factor1 cascade and freezing tolerance in *Arabidopsis*. Plant Cell 25, 2907–2924.

Hyun, Y., Choi, S., Hwang, H.J., et al., 2008. Cooperation and functional diversification of two closely related galactolipase genes for jasmonate biosynthesis. Dev. Cell 14, 183–192.

Ichihara, A., Shiraishi, K., Sato, H., et al., 1977. Structure of coronatine. J. Am. Chem. Soc. 99, 636–637.

Ishiguro, S., Kawai-Oda, A., Ueda, J., et al., 2001. The DEFECTIVE IN ANTHER DEHISCENCE gene encodes a novel phospholipase A1 catalyzing the initial step of jasmonic acid biosynthesis, which synchronizes pollen maturation, anther dehiscence, and flower opening in *Arabidopsis*. Plant Cell 13, 2191–2209.

Jiang, Y.J., Liang, G., Yang, S.Z., et al., 2014. *Arabidopsis* WRKY57 functions as a node of convergence for jasmonic acid- and auxin-mediated signaling in jasmonic acid-induced leaf senescence. Plant Cell 26, 230–245.

Katsir, L., Schilmiller, A.L., Staswick, P.E., et al., 2008. COI1 is a critical component of a receptor for jasmonate and the bacterial virulence factor coronatine. Proc. Natl. Acad. Sci. U.S.A. 105, 7100–7105.

Kazan, K., Manners, J.M., 2013. MYC2: the master in action. Mol. Plant 6, 686–703.

Kende, H., Zeevaart, J., 1997. The five "classical" plant hormones. Plant Cell 9, 1197–1210.

Kepinski, S., Leyser, O., 2005. The *Arabidopsis* F-box protein TIR1 is an auxin receptor. Nature 435, 446–451.

Kienow, L., Schneider, K., Bartsch, M., et al., 2008. Jasmonates meet fatty acids: functional analysis of a new acyl-coenzyme A synthetase family from *Arabidopsis thaliana*. J. Exp. Bot. 59, 403–419.

Koch, T., Bandemer, K., Boland, W., 1997. Biosynthesis of *cis*-jasmone: a pathway for the inactivation and the disposal of the plant stress hormone jasmonic acid to the gas phase? Helv. Chim. Acta 80, 838–850.

Koda, Y., Kikuta, Y., Kitahara, T., et al., 1992. Comparisons of various biological-activities of stereoisomers of methyl jasmonate. Phytochemistry 31, 1111–1114.

Koo, A.J., Chung, H.S., Kobayashi, Y., et al., 2006. Identification of a peroxisomal acyl-activating enzyme involved in the biosynthesis of jasmonic acid in *Arabidopsis*. J. Biol. Chem. 281, 33511–33520.

Koo, A.J., Gao, X., Jones, A.D., et al., 2009. A rapid wound signal activates the systemic synthesis of bioactive jasmonates in *Arabidopsis*. Plant J. 59, 974–986.

Laudert, D., Pfannschmidt, U., Lottspeich, F., et al., 1996. Cloning, molecular and functional characterization of *Arabidopsis thaliana* allene oxide synthase (CYP74), the first enzyme of the octadecanoid pathway to jasmonates. Plant Mol. Biol. 31, 323–335.

Lee, G.I., Howe, G.A., 2003. The tomato mutant *spr1* is defective in systemin perception and the production of a systemic wound signal for defense gene expression. Plant J. 33, 567–576.

Lee, H.Y., Seo, J.S., Cho, J.H., et al., 2013. *Oryza sativa COI* homologues restore jasmonate signal transduction in *Arabidopsis coi1-1* mutants. PLoS One 8, e52802.

Lee, S., Sakuraba, Y., Lee, T., et al., 2015. Mutation of *Oryza sativa CORONATINE INSENSITIVE 1b* (*OsCOI1b*) delays leaf senescence. J. Integr. Plant Biol. 57, 562–576.

Li, C., Liu, G., Xu, C., et al., 2003. The tomato *suppressor of prosystemin-mediated responses2* gene encodes a fatty acid desaturase required for the biosynthesis of jasmonic acid and the production of a systemic wound signal for defense gene expression. Plant Cell 15, 1646–1661.

Li, C., Schilmiller, A.L., Liu, G., et al., 2005. Role of beta-oxidation in jasmonate biosynthesis and systemic wound signaling in tomato. Plant Cell 17, 971–986.

Li, C., Zhao, J., Jiang, H., et al., 2006. The wound response mutant *suppressor of prosystemin-mediated responses6* (*spr6*) is a weak allele of the tomato homolog of *CORONATINE-INSENSITIVE1* (*COI1*). Plant Cell Physiol. 47, 653–663.

Li, L., Li, C., Lee, G.I., et al., 2002. Distinct roles for jasmonate synthesis and action in the systemic wound response of tomato. Proc. Natl. Acad. Sci. U.S.A. 99, 6416–6421.

Li, L., Zhao, Y., McCaig, B.C., et al., 2004. The tomato homolog of CORONATINE-INSENSITIVE1 is required for the maternal control of seed maturation, jasmonate-signaled defense responses, and glandular trichome development. Plant Cell 16, 126–143.

Liao, Y., Wei, J., Xu, Y., et al., 2015. Cloning, expression and characterization of *COI1* gene (*AsCOI1*) from *Aquilaria sinensis* (Lour.) Gilg. Acta Pharm. Sin. B 5, 473–481.

Lorenzo, O., Chico, J.M., Sanchez-Serrano, J.J., et al., 2004. *JASMONATE-INSENSITIVE1* encodes a MYC transcription factor essential to discriminate between different jasmonate-regulated defense responses in *Arabidopsis*. Plant Cell 16, 1938–1950.

McCloud, E.S., Baldwin, I.T., 1997. Herbivory and caterpillar regurgitants amplify the wound-induced increases in jasmonic acid but not nicotine in *Nicotiana sylvestris*. Planta 203, 430–435.

McConn, M., Browse, J., 1996. The critical requirement for linolenic acid is pollen development, not photosynthesis, in an *Arabidopsis* mutant. Plant Cell 8, 403–416.

McConn, M., Creelman, R.A., Bell, E., et al., 1997. Jasmonate is essential for insect defense in *Arabidopsis*. Proc. Natl. Acad. Sci. U.S.A. 94, 5473–5477.

Melotto, M., Mecey, C., Niu, Y., et al., 2008. A critical role of two positively charged amino acids in the Jas motif of *Arabidopsis* JAZ proteins in mediating coronatine- and jasmonoyl isoleucine-dependent interactions with the COI1 F-box protein. Plant J. 55, 979–988.

Mewis, I., Appel, H.M., Hom, A., et al., 2005. Major signaling pathways modulate *Arabidopsis* glucosinolate accumulation and response to both phloem-feeding and chewing insects. Plant Physiol. 138, 1149–1162.

Mitchell, R.E., 1985. Coronatine biosynthesis – incorporation of l-[U-14C]isoleucine and l-[U-14C] threonine into the 1-amido-1-carboxy-2-ethylcyclopropyl moiety. Phytochemistry 24, 247–249.

Mousavi, S.A., Chauvin, A., Pascaud, F., et al., 2013. *GLUTAMATE RECEPTOR-LIKE* genes mediate leaf-to-leaf wound signalling. Nature 500, 422–426.

Nakata, M., Mitsuda, N., Herde, M., et al., 2013. A bHLH-type transcription factor, ABA-INDUCIBLE BHLH-TYPE TRANSCRIPTION FACTOR/JA-ASSOCIATED MYC2-LIKE1, acts as a repressor to negatively regulate jasmonate signaling in *Arabidopsis*. Plant Cell 25, 1641–1656.

Niu, Y.J., Figueroa, P., Browse, J., 2011. Characterization of JAZ-interacting bHLH transcription factors that regulate jasmonate responses in *Arabidopsis*. J. Exp. Bot. 62, 2143–2154.

Palmer, D.A., Bender, C.L., 1995. Ultrastructure of tomato leaf tissue treated with the pseudomonad phytotoxin coronatine and comparison with methyl jasmonate. Mol. Plant-Microbe Interact. 8, 683–692.

Park, J.H., Halitschke, R., Kim, H.B., et al., 2002. A knock-out mutation in allene oxide synthase results in male sterility and defective wound signal transduction in *Arabidopsis* due to a block in jasmonic acid biosynthesis. Plant J. 31, 1–12.

Parry, R.J., Mhaskar, S.V., Lin, M.T., et al., 1994. Investigations of the biosynthesis of the phytotoxin coronatine. Can. J. Chem. 72, 86–99.

Paschold, A., Halitschke, R., Baldwin, I.T., 2007. Co(i)-ordinating defenses: NaCOI1 mediates herbivore- induced resistance in *Nicotiana attenuata* and reveals the role of herbivore movement in avoiding defenses. Plant J. 51, 79–91.

Pauwels, L., Barbero, G.F., Geerinck, J., et al., 2010. NINJA connects the co-repressor TOPLESS to jasmonate signalling. Nature 464, 788–791.

Pauwels, L., Goossens, A., 2011. The JAZ proteins: a crucial interface in the jasmonate signaling cascade. Plant Cell 23, 3089–3100.

Potuschak, T., Lechner, E., Parmentier, Y., et al., 2003. EIN3-dependent regulation of plant ethylene hormone signaling by two *Arabidopsis* F-box proteins: EBF1 and EBF2. Cell 115, 679–689.

Qi, T.C., Huang, H., Song, S.S., et al., 2015a. Regulation of jasmonate-mediated stamen development and seed production by a bHLH-MYB complex in *Arabidopsis*. Plant Cell 27, 1620–1633.

Qi, T.C., Huang, H., Wu, D., et al., 2014. *Arabidopsis* DELLA and JAZ proteins bind the WD-repeat/bHLH/MYB complex to modulate gibberellin and jasmonate signaling synergy. Plant Cell 26, 1118–1133.

Qi, T.C., Song, S.S., Ren, Q.C., et al., 2011. The jasmonate-ZIM-domain proteins interact with the WD-repeat/bHLH/MYB complexes to regulate jasmonate-mediated anthocyanin accumulation and trichome initiation in *Arabidopsis thaliana*. Plant Cell 23, 1795–1814.

Qi, T.C., Wang, J.J., Huang, H., et al., 2015b. Regulation of jasmonate-induced leaf senescence by antagonism between bHLH subgroup IIIe and IIId factors in *Arabidopsis*. Plant Cell 27, 1634–1649.

Ren, C., Pan, J., Peng, W., et al., 2005. Point mutations in *Arabidopsis Cullin1* reveal its essential role in jasmonate response. Plant J. 42, 514–524.

Ryan, C.A., 2000. The systemin signaling pathway: differential activation of plant defensive genes. Biochim. Biophys. Acta 1477, 112–121.

Ryan, C.A., Pearce, G., 1998. Systemin: a polypeptide signal for plant defensive genes. Annu. Rev. Cell Dev. Biol. 14, 1–17.

Saijo, Y., Uchiyama, B., Abe, T., et al., 1997. Contiguous four-guanosine sequence in c-myc antisense phosphorothioate oligonucleotides inhibits cell growth on human lung cancer cells: possible involvement of cell adhesion inhibition. Jpn. J. Cancer Res. 88, 26–33.

Sanders, P.M., Lee, P.Y., Biesgen, C., et al., 2000. The *Arabidopsis DELAYED DEHISCENCE1* gene encodes an enzyme in the jasmonic acid synthesis pathway. Plant Cell 12, 1041–1061.

Sasaki-Sekimoto, Y., Jikumaru, Y., Obayashi, T., et al., 2013. Basic helix-loop-helix transcription factors JASMONATE-ASSOCIATED MYC2-LIKE1 (JAM1), JAM2, and JAM3 are negative regulators of jasmonate responses in *Arabidopsis*. Plant Physiol. 163, 291–304.

Schaller, A., Stintzi, A., 2009. Enzymes in jasmonate biosynthesis – structure, function, regulation. Phytochemistry 70, 1532–1538.

Schilmiller, A.L., Howe, G.A., 2005. Systemic signaling in the wound response. Curr. Opin. Plant Biol. 8, 369–377.

Schilmiller, A.L., Koo, A.J., Howe, G.A., 2007. Functional diversification of acyl-coenzyme A oxidases in jasmonic acid biosynthesis and action. Plant Physiol. 143, 812–824.

Schommer, C., Palatnik, J.F., Aggarwal, P., et al., 2008. Control of jasmonate biosynthesis and senescence by miR319 targets. PLoS Biol. 6, e230.

Schweizer, F., Fernández-Calvo, P., Zander, M., et al., 2013. *Arabidopsis* basic helix-loop-helix transcription factors MYC2, MYC3, and MYC4 regulate glucosinolate biosynthesis, insect performance, and feeding behavior. Plant Cell 25, 3117–3132.

Seo, H.S., Song, J.T., Cheong, J.J., et al., 2001. Jasmonic acid carboxyl methyltransferase: a key enzyme for jasmonate-regulated plant responses. Proc. Natl. Acad. Sci. U.S.A. 98, 4788–4793.

Sheard, L.B., Tan, X., Mao, H., et al., 2010. Jasmonate perception by inositol-phosphate-potentiated COI1-JAZ co-receptor. Nature 468, 400–405.

Shyu, C., Figueroa, P., DePew, C.L., et al., 2012. JAZ8 lacks a canonical degron and has an EAR motif that mediates transcriptional repression of jasmonate responses in *Arabidopsis*. Plant Cell 24, 536–550.

Song, S.S., Huang, H., Gao, H., et al., 2014. Interaction between MYC2 and INSENSITIVE3 modulates antagonism between jasmonate and ethylene signaling in *Arabidopsis*. Plant Cell 26, 263–279.

Song, S.S., Qi, T.C., Fan, M., et al., 2013. The bHLH subgroup IIId factors negatively regulate jasmonate-mediated plant defense and development. PLoS Genet. 9, e1003653.

Song, S.S., Qi, T.C., Huang, H., et al., 2011. The jasmonate-ZIM domain proteins interact with the R2R3-MYB transcription factors MYB21 and MYB24 to affect jasmonate-regulated stamen development in *Arabidopsis*. Plant Cell 23, 1000–1013.

Staswick, P.E., Su, W., Howell, S.H., 1992. Methyl jasmonate inhibition of root growth and induction of a leaf protein are decreased in an *Arabidopsis thaliana* mutant. Proc. Natl. Acad. Sci. U.S.A. 89, 6837–6840.

Staswick, P.E., Tiryaki, I., 2004. The oxylipin signal jasmonic acid is activated by an enzyme that conjugates it to isoleucine in *Arabidopsis*. Plant Cell 16, 2117–2127.

Staswick, P.E., Tiryaki, I., Rowe, M.L., 2002. Jasmonate response locus JAR1 and several related *Arabidopsis* genes encode enzymes of the firefly luciferase superfamily that show activity on jasmonic, salicylic, and indole-3-acetic acids in an assay for adenylation. Plant Cell 14, 1405–1415.

Stenzel, I., Hause, B., Maucher, H., et al., 2003a. Allene oxide cyclase dependence of the wound response and vascular bundle-specific generation of jasmonates in tomato – amplification in wound signalling. Plant J. 33, 577–589.

Stenzel, I., Hause, B., Miersch, O., et al., 2003b. Jasmonate biosynthesis and the allene oxide cyclase family of *Arabidopsis thaliana*. Plant Mol. Biol. 51, 895–911.

Stenzel, I., Otto, M., Delker, C., et al., 2012. ALLENE OXIDE CYCLASE (AOC) gene family members of *Arabidopsis thaliana*: tissue- and organ-specific promoter activities and in vivo heteromerization. J. Exp. Bot. 63, 6125–6138.

Stintzi, A., Browse, J., 2000. The *Arabidopsis* male-sterile mutant, *opr3*, lacks the 12-oxophytodienoic acid reductase required for jasmonate synthesis. Proc. Natl. Acad. Sci. U.S.A. 97, 10625–10630.

Strassner, J., Schaller, F., Frick, U.B., et al., 2002. Characterization and cDNA-microarray expression analysis of 12-oxophytodienoate reductases reveals differential roles for octadecanoid biosynthesis in the local versus the systemic wound response. Plant J. 32, 585–601.

Suza, W.P., Staswick, P.E., 2008. The role of JAR1 in Jasmonoyl-L-isoleucine production during *Arabidopsis* wound response. Planta 227, 1221–1232.

Swiatek, A., Van Dongen, W., Esmans, E.L., et al., 2004. Metabolic fate of jasmonates in tobacco bright yellow-2 cells. Plant Physiol. 135, 161–172.

Tamogami, S., Kodama, O., 2000. Coronatine elicits phytoalexin production in rice leaves (*Oryza sativa* L.) in the same manner as jasmonic acid. Phytochemistry 54, 689–694.

Thines, B., Katsir, L., Melotto, M., et al., 2007. JAZ repressor proteins are targets of the SCF[COI1] complex during jasmonate signalling. Nature 448, 661–665.

Thireault, C., Shyu, C., Yoshida, Y., et al., 2015. Repression of jasmonate signaling by a non-TIFY JAZ protein in *Arabidopsis*. Plant J. 82, 669–679.

Thomma, B.P.H.J., Eggermont, K., Penninckx, I.A.M.A., et al., 1998. Separate jasmonate-dependent and salicylate-dependent defense-response pathways in *Arabidopsis* are essential for resistance to distinct microbial pathogens. Proc. Natl. Acad. Sci. U.S.A. 95, 15107–15111.

Toda, Y., Tanaka, M., Ogawa, D., et al., 2013. RICE SALT SENSITIVE3 forms a ternary complex with JAZ and class-C bHLH factors and regulates jasmonate-induced gene expression and root cell elongation. Plant Cell 25, 1709–1725.

Ueda, J., Kato, J., 1980. Isolation and identification of a senescence-promoting substance from wormwood (*Artemisia absinthium* L.). Plant Physiol. 66, 246–249.

Uppalapati, S.R., Ayoubi, P., Weng, H., et al., 2005. The phytotoxin coronatine and methyl jasmonate impact multiple phytohormone pathways in tomato. Plant J. 42, 201–217.

Vanholme, B., Grunewald, W., Bateman, A., et al., 2007. The tify family previously known as ZIM. Trends Plant Sci. 12, 239–244.

Vick, B.A., Zimmerman, D.C., 1983. The biosynthesis of jasmonic acid: a physiological role for plant lipoxygenase. Biochem. Biophys. Res. Commun. 111, 470–477.

Vick, B.A., Zimmerman, D.C., 1984. Biosynthesis of jasmonic acid by several plant species. Plant Physiol. 75, 458–461.

Vijayan, P., Shockey, J., Levesque, C.A., et al., 1998. A role for jasmonate in pathogen defense of *Arabidopsis*. Proc. Natl. Acad. Sci. U.S.A. 95, 7209–7214.

Wang, Z., Dai, L., Jiang, Z., et al., 2005. GmCOI1, a soybean F-box protein gene, shows ability to mediate jasmonate-regulated plant defense and fertility in *Arabidopsis*. Mol. Plant-Microbe Interact. 18, 1285–1295.

Wasternack, C., 2007. Jasmonates: an update on biosynthesis, signal transduction and action in plant stress response, growth and development. Ann. Bot. 100, 681–697.

Wasternack, C., 2015. How jasmonates earned their laurels: past and present. J. Plant Growth Regul. 34, 761–794.

Wasternack, C., Hause, B., 2013. Jasmonates: biosynthesis, perception, signal transduction and action in plant stress response, growth and development. An update to the 2007 review in Annals of Botany. Ann. Bot. 111, 1021–1058.

Wasternack, C., Strnad, M., 2015. Jasmonate signaling in plant stress responses and development – active and inactive compounds. New Biotechnol.

Weber, H., Vick, B.A., Farmer, E.E., 1997. Dinor-oxo-phytodienoic acid: a new hexadecanoid signal in the jasmonate family. Proc. Natl. Acad. Sci. U.S.A. 94, 10473–10478.

Weiler, E.W., Kutchan, T.M., Gorba, T., et al., 1994. The *Pseudomonas* phytotoxin coronatine mimics octadecanoid signalling molecules of higher plants. FEBS Lett. 345, 9–13.

Werck-Reichhart, D., Bak, S., Paquette, S., 2002. Cytochromes p450. Arabidopsis Book 1, e0028.

Wild, M., Daviere, J.M., Cheminant, S., et al., 2012. The *Arabidopsis* DELLA RGA-LIKE3 is a direct target of MYC2 and modulates jasmonate signaling responses. Plant Cell 24, 3307–3319.

Xie, D.X., Feys, B.F., James, S., et al., 1998. COI1: an *Arabidopsis* gene required for jasmonate-regulated defense and fertility. Science 280, 1091–1094.

Xin, X.F., He, S.Y., 2013. *Pseudomonas syringae* pv. *tomato* DC3000: a model pathogen for probing disease susceptibility and hormone signaling in plants. Ann. Rev. Phytopathol. 51, 473–498.

Xu, L.H., Liu, F.Q., Lechner, E., et al., 2002. The SCF^{COI1} ubiquitin-ligase complexes are required for jasmonate response in *Arabidopsis*. Plant Cell 14, 1919–1935.

Yamane, H., Sugawara, J., Suzuki, Y., et al., 1980. Syntheses of jasmonic acid related-compounds and their structure-activity-relationships on the growth of rice seedings. Agric. Biol. Chem. Tokyo 44, 2857–2864.

Yan, J.B., Li, H., Li, S., et al., 2013a. The *Arabidopsis* F-box protein CORONATINE INSENSITIVE1 is stabilized by SCF^{COI1} and degraded via the 26S proteasome pathway. Plant Cell 25, 486–498.

Yan, J.B., Zhang, C., Gu, M., et al., 2009. The *Arabidopsis* CORONATINE INSENSITIVE1 protein is a jasmonate receptor. Plant Cell 21, 2220–2236.

Yan, L.H., Zhai, Q.Z., Wei, J.N., et al., 2013b. Role of tomato lipoxygenase D in wound-induced jasmonate biosynthesis and plant immunity to insect herbivores. PLoS Genet. 9, e1003964.

Yan, Y.X., Stolz, S., Chetelat, A., et al., 2007. A downstream mediator in the growth repression limb of the jasmonate pathway. Plant Cell 19, 2470–2483.

Yang, D.L., Yao, J., Mei, C.S., et al., 2012. Plant hormone jasmonate prioritizes defense over growth by interfering with gibberellin signaling cascade. Proc. Natl. Acad. Sci. U.S.A. 109, E1192–E1200.

Ye, M., Luo, S.M., Xie, J.F., et al., 2012. Silencing COI1 in rice increases susceptibility to chewing insects and impairs inducible defense. PLoS One 7, e36214.

Zhai, Q.Z., Yan, L.H., Tan, D., et al., 2013. Phosphorylation-coupled proteolysis of the transcription factor MYC2 is important for jasmonate-signaled plant immunity. PLoS Genet. 9.

Zhai, Q.Z., Zhang, X., Wu, F.M., et al., 2015. Transcriptional mechanism of jasmonate receptor COI1-mediated delay of flowering time in *Arabidopsis*. Plant Cell 27, 2814–2828.

Zhang, X., Zhu, Z.Q., An, F.Y., et al., 2014. Jasmonate-activated MYC2 represses ETHYLENE INSENSITIVE3 activity to antagonize ethylene-promoted apical hook formation in *Arabidopsis*. Plant Cell 26, 1105–1117.

Zhao, Y., Thilmony, R., Bender, C.L., et al., 2003. Virulence systems of *Pseudomonas syringae* pv. *tomato* promote bacterial speck disease in tomato by targeting the jasmonate signaling pathway. Plant J. 36, 485–499.

Zheng, W.G., Zhai, Q.Z., Sun, J.Q., et al., 2006. Bestatin, an inhibitor of aminopeptidases, provides a chemical genetics approach to dissect jasmonate signaling in *Arabidopsis*. Plant Physiol. 141, 1400–1413.

Zheng, X.Y., Spivey, N.W., Zeng, W.Q., et al., 2012. Coronatine promotes *Pseudomonas syringae* virulence in plants by activating a signaling cascade that inhibits salicylic acid accumulation. Cell Host Microbe 11, 587–596.

Zhu, Z., An, F., Feng, Y., et al., 2011. Derepression of ethylene-stabilized transcription factors (EIN3/EIL1) mediates jasmonate and ethylene signaling synergy in *Arabidopsis*. Proc. Natl. Acad. Sci. U.S.A. 108, 12539–12544.

Ziegler, J., Stenzel, I., Hause, B., et al., 2000. Molecular cloning of allene oxide cyclase. The enzyme establishing the stereochemistry of octadecanoids and jasmonates. J. Biol. Chem. 275, 19132–19138.

Zimmerli, L., Metraux, J.P., Mauch-Mani, B., 2001. Beta-aminobutyric acid-induced protection of *Arabidopsis* against the necrotrophic fungus *Botrytis cinerea*. Plant Physiol. 126, 517–523.

Salicylic acid

8

Jing Bo Jin[1], Bin Cai[1], Jian-Min Zhou[2]
[1]Institute of Botany, Chinese Academy of Sciences, Beijing, China; [2]Institute of Genetics and Developmental Biology, Chinese Academy of Sciences, Beijing, China

Summary

The phytohormone salicylic acid (SA) plays important roles in the regulation of responses to biotic and abiotic stress, as well as developmental processes, but the most extensive research has been focused on characterization of SA signaling in plant–microbial pathogen interactions. In response to pathogen infection, plants activate SA biosynthesis, which is essential for defense responses, such as PAMP (pathogen-associated molecular pattern)-triggered immunity (also known as basal resistance), effector-triggered immunity, and establishment of systemic acquired resistance. Studies revealed that Nonexpresser of Pathogenesis-Related protein 1 (NPR1), 3, and 4 are SA receptors, which are responsible for SA-mediated defense gene expression as well as regulation of cell fate in local infected tissues and systemic noninfected tissues. Defense gene expression is activated by SA mainly through NPR1, the activity of which is tightly regulated by cytoplasm-to-nucleus translocation and post-translational modifications, such as phosphorylation, ubiquitination, and sumoylation. This chapter mainly summarizes advances concerning SA biosynthesis and signaling pathways.

8.1   Discovery and roles of salicylic acid

Salicylic acid (SA) is a simple phenolic compound synthesized in a wide range of prokaryotic and eukaryotic organisms, including plants. Leaf and bark of willow tree (*Salix* sp.) contain large amounts of SA, which was widely used as a medication for pain relief in the ancient world. In 1828, German scientist Johann A. Buchner purified salicyl alcohol glucoside (an SA derivative called salicin) from willow bark. Ten years later, an Italian chemist Raffaele Piria working in Paris converted salicin into an acidic aromatic compound that he named salicylic acid. In 1859 Hermann Kolbe et al. chemically synthesized SA, but the bitter taste and side effects limited the long-term use of SA as a medication. In 1897 Felix Hoffmann working in the Bayer pharmaceutical company synthesized acetyl salicylic acid (originally produced by a French chemist, Charles Frederic Gerhardt, in 1853), which became known as Aspirin, to reduce the side effects (Weissmann, 1991). Now Aspirin is widely used to treat pain, fever, inflammation, heart attacks, strokes, and blood clot formation.

In 1979 a role of SA in the defense response of plants was discovered. Treatment of tobacco plants with Aspirin resulted in enhanced resistance to tobacco mosaic virus (White, 1979). Subsequent studies revealed that pathogen infection increases the level of SA and promotes transcription of genes encoding pathogenesis-related proteins in plants, which confers disease resistance. In the pathogen-infected local tissues, plants

Hormone Metabolism and Signaling in Plants. http://dx.doi.org/10.1016/B978-0-12-811562-6.00008-6

recognize the conserved pathogen-associated molecular patterns (PAMPs), such as bacterial flagellin or fungal chitin, through plasma membrane-localized pattern recognition receptors (PRRs), which triggers PAMP-triggered immunity (PTI) (Monaghan and Zipfel, 2012). Successful phytopathogens have evolved to secrete effector proteins into the plant to interfere with PTI and promote pathogenesis (Dou and Zhou, 2012). This forced plants to evolve NOD-Like Receptors (NLRs) to recognize these effectors and activate effector-triggered immunity (ETI), which is usually associated with the hypersensitive cell death response (HR) (Rajamuthiah and Mylonakis, 2014). Activation of defense responses in local infected tissues also induces the generation of mobile signals that lead to establishment of long-lasting systemic acquired resistance (SAR) in systemic tissues to protect plants from secondary infection (Fu and Dong, 2013; Mishina and Zeier, 2007). Blocking SA accumulation by expressing the bacterial salicylate hydroxylase (NahG, converts SA to catechol) in plants causes compromised SA-dependent immune responses, including local defense response and establishment of SAR (Delaney et al., 1994; Gaffney et al., 1993; Wildermuth et al., 2001). Studies have identified Nonexpresser of Pathogenesis-Related protein (NPR) 1, 3, and 4 as SA receptors and characterized their roles in SA-dependent defense responses (Fu et al., 2012; Wu et al., 2012).

In addition to defense responses, SA is implicated in the regulation of a variety of biological processes, such as seed germination, seedling development, nodulation in legumes, plant vegetative growth, senescence-associated gene expression, flowering time, fruit yield, respiration, as well as response to ultraviolet (UV)-B radiation, ozone, metals, drought, temperature, and salinity stresses (Khan et al., 2015; Vlot et al., 2009).

8.2 Biosynthesis of SA

To date, two distinct SA biosynthetic pathways have been identified in plants, including the phenylalanine ammonia lyase (PAL)-mediated phenylalanine pathway and the isochorismate synthase (ICS)-mediated isochorismate pathway (Chen et al., 2009) (Fig. 8.1). Both pathways require the shikimic acid pathway-derived chorismate as a precursor. In the phenylalanine pathway, chorismate-derived phenylalanine is converted to cinnamic acid by PAL. Subsequently, cinnamic acid is hydroxylated to form *ortho*-courmarate or oxidized to benzoic acid. Finally, *ortho*-courmarate undergoes oxidation or benzoic acid undergoes hydroxylation to produce SA (Metraux, 2002). PAL plays an important role in both basal and pathogen-induced SA biosynthesis (Huang et al., 2010). Although benzoic acid-2-hydroxylase (BA2H) has been suggested to catalyze the conversion of benzoic acid to SA (León et al., 1995), most of the enzymes involved in the conversion of cinnamic acid to SA have not been identified in plants.

In some bacteria such as *Pseudomonas aeruginosa*, ICS catalyzes the conversion of chorismate to isochorismate, which is further metabolized by isochorismate pyruvate lyase (IPL) to form SA (Serino et al., 1995). Genetic evidence indicates that the function of ICS in SA biosynthesis is conserved in plants (Wildermuth et al., 2001). The Arabidopsis (*Arabidopsis thaliana*) genome contains two genes encoding ICS, *ICS1* (also known as *SALICYLIC ACID INDUCTION DEFICIENT 2 or SID2*) and *ICS2*.

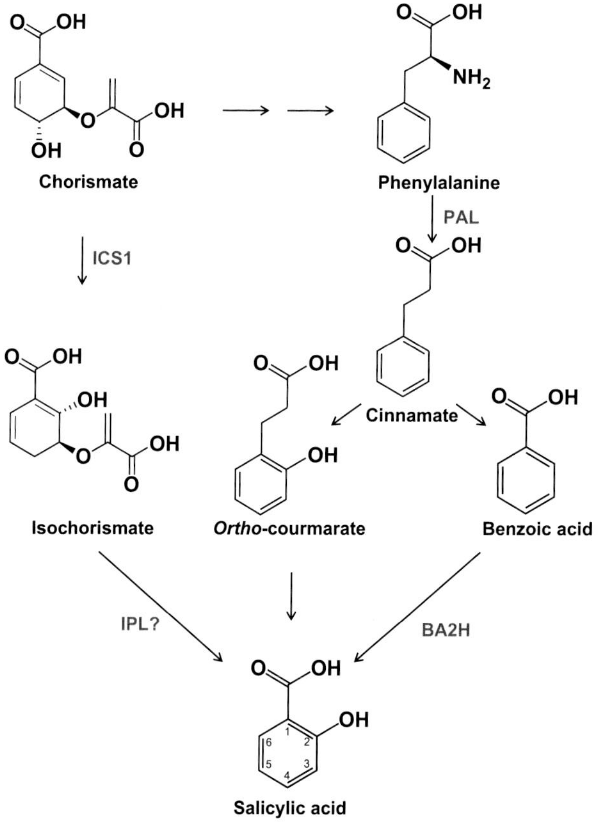

Figure 8.1 Salicylic acid biosynthetic pathways.

The ICS1 enzyme plays a major role in isochorismate-dependent SA biosynthesis under biotic and abiotic stress conditions, whereas ICS2 is mainly involved in basal SA biosynthesis (Garcion et al., 2008; Wildermuth et al., 2001). Upon pathogen infection, two plant-specific transcription factors, SAR Deficient1 (SARD1) and Cam-Binding Protein 60-like g (CBP60g), are recruited to the *ICS1* promoter to induce *ICS1* expression and SA biosynthesis (Wang et al., 2011; Zhang et al., 2010b). To date, the putative IPL enzyme to convert isochorismate to SA in plants has not been identified. Plant genomes do not contain genes that encode recognizable IPL enzymes, raising the possibility that plants may contain other proteins with IPL-like enzyme activity or they convert isochorismate to SA through an IPL-independent pathway. Both ICS1 and ICS2 are localized in plastids, suggesting that this is the site of SA production (Strawn et al., 2007). The plastid-synthesized SA is transported to the cytosol by Enhanced Disease Susceptibility 5 (EDS5), a Multidrug and Toxin Extrusion (MATE) transporter family protein (Serrano et al., 2013).

It has been suggested that the ICS pathway plays a major role in SA biosynthesis, since the *ics1 ics2* double mutant plants accumulate about 20% of wild-type levels of SA under nonstressed conditions and 5%–10% of wild-type levels of SA upon pathogen infection or UV exposure (Garcion et al., 2008; Wildermuth et al., 2001). However, the *pal1 pal2 pal3 pal4* quadruple mutant plants accumulate 25% of wild-type basal levels of SA and 50% of wild-type levels of pathogen-induced SA (Huang et al., 2010). Thus, both phenylalanine and isochorismate pathways appear to be required for basal and pathogen-induced SA biosynthesis, but the individual contributions to SA biosynthesis in different physiological processes remain to be characterized.

8.3   NPR1-dependent SA signaling

NPR1 is a transcriptional coregulator, acts downstream of SA, and plays an essential role in SA-dependent defense signaling pathways (Cao et al., 1997). Mutations in the *NPR1* gene causes impaired SA-induced transcriptional reprogramming and establishment of SAR, and enhanced HR (Cao et al., 1994, 1997; Rate and Greenberg, 2001). Under noninfection conditions, to prevent constitutive activation of defense signaling, the majority of NPR1 is sequestered in the cytoplasm as a redox-sensitive oligomeric complex. Upon pathogen challenge, accumulation of SA causes the cellular environment to become reduced, resulting in the NPR1 oligomeric complex to become converted to monomers catalyzed by thioredoxins and then translocated into the nucleus through the NPR1 bipartite nuclear localization signal (Mou et al., 2003; Tada et al., 2008). Interestingly, SA treatment not only induces NPR1 monomer release but also promotes S-nitrosylation-mediated oligomerization of NPR1 to prevent depletion of the NPR1 protein pool. Blocking oligomerization or monomer accumulation of NPR1 caused impaired defense responses, indicating that regulation of NPR1 protein assembly is required for proper immune responses (Tada et al., 2008). Within the nucleus, NPR1 monomers physically interact with basic leucine zipper transcription factors, such as TGACG Sequence-specific Binding Proteins (TGAs), to activate expression of disease resistance genes (Després et al., 2003, 2000; Fan and Dong, 2002; Rochon et al., 2006).

Activity of NPR1 is also tightly regulated by other posttranslational modifications, such as phosphorylation, ubiquitination, and sumoylation (Saleh et al., 2015; Spoel et al., 2009). Sumoylation is a rapid and reversible posttranslational covalent modification of proteins with SUMO (small ubiquitin-related modifier), required for differentiation, development, and hormonal and environmental responses in metazoans (Miura et al., 2007). Two SUMO1 and SUMO2 paralogs function redundantly to suppress SA accumulation in uninfected cells, whereas SUMO3 paralog functions downstream of SA to activate plant defense response (van den Burg et al., 2010). In uninfected cells, the NPR1 monomer is constantly degraded in the nucleus by Cullin3-based ubiquitin ligase-mediated proteasome degradation, which prevents inappropriate activation of defense genes expression and SAR. In addition, NPR1 is phosphorylated at Ser55 and Ser59, which prevents SUMO3 modification of NPR1. Nonsumoylated NPR1 interacts with WRKY70 (Wang et al., 2006), a transcriptional repressor, to down-regulate expression of defense genes. Upon pathogen infection, accumulation of SA promotes dephosphorylation of Ser55 and Ser59 and SUMO3 modification of NPR1, which causes dissociation of NPR1 from WRKY70 possibly resulting in derepression of defense genes. Moreover, SUMO3 conjugation of NPR1 induces phosphorylation

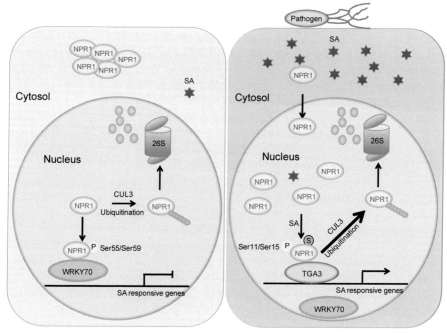

Figure 8.2 Activation of NPR1 by SA.

of NPR1 at Ser11 and Ser15 and promotes physical interaction between NPR1 and a transcription factor TGA3 to activate expression of defense genes. Conjugation with SUMO3 and phosphorylation of NPR1 at Ser11 and Ser15 facilitate Cullin3-based ubiquitin ligase-dependent degradation of NPR1. Interestingly, the SA-induced rapid degradation of NPR1 is required for full activation of defense gene expression and establishment of SAR (Saleh et al., 2015; Spoel et al., 2009). It has been proposed that SA-induced rapid turnover of the activated NPR1 is required for exchange of "exhausted" NPR1 for "fresh" NPR1 on the target gene promoter to accelerate transcription cycle (Spoel et al., 2009) (Fig. 8.2).

8.4   Perception of SA by NPR proteins

The past 20 years of research has seen great effort put into identifying SA receptor(s). Since *npr1* mutants are defective in SA-mediated defense response, NPR1 is a good candidate as an SA receptor. In addition to *NPR1*, the Arabidopsis genome contains five *NPR1* paralogs, *NPR2*, *NPR3*, *NPR4*, *BLADE-ON-PETIOLE1* (*BOP1*), and *BOP2* (Zhang et al., 2006). The *npr3 npr4* double mutant accumulates a higher level of NPR1 and exhibits enhanced basal disease resistance, but compromised SAR and ETI (Fu et al., 2012; Zhang et al., 2006). On the other hand, BOP1 and BOP2 function redundantly to regulate developmental processes, such as leaf and floral patterning in Arabidopsis (Norberg et al., 2005). Breakthrough studies suggest that NPR1, and its paralogs NPR3 and NPR4, are SA receptor(s) (Fu et al., 2012; Wu et al., 2012). In one study (Fu et al., 2012) it was shown that NPR3 and NPR4 proteins bind to SA with different affinities, i.e., NPR3 had lower SA binding affinity (K_d=981 nM) than that of NPR4 (K_d=46 nM). Both NPR3 and NPR4 contain BTB ("bric à brac, tramtrack, broad-complex") and ankyrin repeat domains. These two domains are often found in adaptors of Cullin 3 E3 ligases, which mediate ubiquitination and proteasomal-dependent protein degradation (Pintard et al., 2004). Indeed, NPR3 and NPR4 interact with both Cullin 3 and NPR1, and function as Cullin 3 adaptors to facilitate NPR1 degradation (Fu et al., 2012). Interestingly, a higher level of SA appears to promote interaction between NPR1 and NPR3, but disrupt interaction between NPR1 and NPR4. It was proposed that in the infected local tissues, SA is accumulated to a higher level and binds to NPR3, facilitating NPR3-NPR1 interaction and subsequent Cullin 3-mediated degradation of NPR1. The turnover of NPR1 facilitated by NPR3 results in programmed cell death in local tissues and establishment of SAR in systemic tissues (Fu et al., 2012; Rate and Greenberg, 2001; Spoel et al., 2009). In the systemic tissues, the lower level of induced SA binds preferentially to NPR4 rather than to NPR3, and disrupts the NPR4-NPR1 interaction, resulting in the accumulation of NPR1. Consequently, NPR1 suppresses HR in the systemic tissues and facilitates expression of defense genes (Cao et al., 1997; Fu et al., 2012; Rate and Greenberg, 2001) (Fig. 8.3).

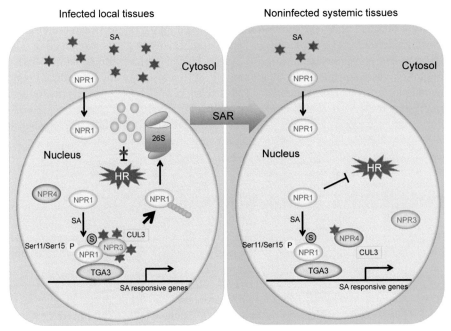

Figure 8.3 NPR1 suppresses hypersensitive cell death response in the systemic tissues and facilitates expression of defense genes.

In a traditional ligand-binding assay, ^{3}H-SA does not bind to NPR1 (Fu et al., 2012). In an equilibrium dialysis assay, however, SA binds to NPR1 through two cysteine residues (Cys521 and Cys529) with a K_d of about 140 nM, and this binding is dependent on the transition metal copper. These two Cys residues are located in the C-terminal transactivation domain of NPR1 and are critical for SA-induced defense gene expression (Rochon et al., 2006; Wu et al., 2012). It was thus proposed that in the absence of SA, the N-terminal BTB domain of NPR1 interacts with the C-terminal transactivation domain to inhibit activity of the transactivation domain. When present, SA binds to NPR1 and disrupts the BTB–transactivation domain interaction, resulting in derepression of the transactivation domain and expression of defense genes (Wu et al., 2012). Further studies are needed to resolve the two competing models (Boatwright and Pajerowska-Mukhtar, 2013; Yan and Dong, 2014).

8.5 PAMP- and effector-triggered immunity

PTI is the first layer of active defense following perception by the plant cell surface-localized PRRs of the molecular patterns coded by either the pathogen or the host plant but released specifically during pathogen infection (Monaghan and Zipfel, 2012). PAMPs are slowly evolving conserved molecules associated with entire groups of pathogens. Known PAMPs include bacterial flagellin and elongation factor Tu (EF-Tu), and fungal chitin and xylanase. In addition to PAMPs, pathogen infection also leads to the release of plant molecules such as small proteins called "Peps" and pectin fragments of oligogalacturonic acids. These molecules are collectively called damage-associated molecular patterns (DAMPs). PAMPs and DAMPs are recognized by two classes of plasma membrane-localized PRRs, receptor-like kinases (RLKs), or receptor-like proteins (RLPs). RLKs generally contain a variable extracellular receiver domain such as a leucine-rich repeat (LRR) or lysine motif (LysM) domain, a single-span transmembrane domain, and an intracellular serine/threonine kinase domain. RLPs lack the intracellular kinase domain but contain an extracellular domain, a single-span transmembrane domain, and a short cytosolic domain (Trdá et al., 2015). Bacterial Flagellin and EF-Tu are recognized by LRR-RLKs Flagellin-Sensing 2 (FLS2) and the *Brassicaceae*-specific EF-Tu Receptor (EFR), respectively (Monaghan and Zipfel, 2012). Chitin is sensed by a LysM-RLK known as LYK5 and Chitin Elicitor Receptor Kinase 1 (CERK1) (Cao et al., 2014), whereas xylanase is sensed by the tomato (*Lycopersicon esculentum*) LRR-RLPs *L. esculentum* Ethylene-inducing xylanase1 (LeEix1) and LeEix2 (Monaghan and Zipfel, 2012).

The best characterized RLK is the flagellin receptor FLS2, which is conserved in Arabidopsis, tomato, tobacco (*Nicotiana benthamiana*), and rice (*Oryza sativa*) (Boller and Felix, 2009). In the prestimulation state, FLS2 constitutively interacts with cytoplasmic kinase Botrytis-Induced Kinase 1 (BIK1) (Lu et al., 2010; Zhang et al., 2010a). The Arabidopsis FLS2 perceives a 22-amino acid epitope (flg22) conserved in the N terminus of flagellin. The binding of flg22 to the LRR domain of FLS2 recruits the coreceptor protein Brassinosteroid insensitive 1-Associated Kinase 1 (BAK1),

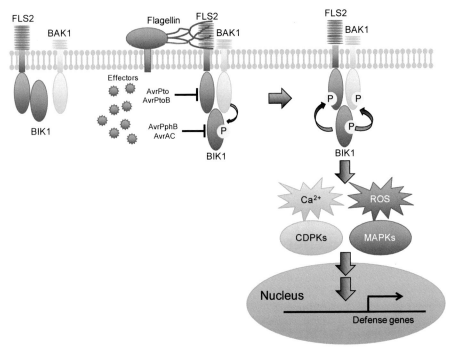

Figure 8.4 Activated receptor complex transduces the flagellin-induced signal via bursts of calcium and reactive oxygen species, mitogen-activated protein kinases, and calcium-dependent protein kinases, resulting in transcriptional reprogramming.

forming an active receptor complex (Sun et al., 2013) and resulting in phosphorylation of all three kinases. This activated receptor complex transduces the flagellin-induced signal via bursts of calcium and reactive oxygen species, mitogen-activated protein kinases, and calcium-dependent protein kinases, resulting in transcriptional reprogramming (Boller and Felix, 2009; Segonzac and Zipfel, 2011) (Fig. 8.4).

Phytopathogens are known to secrete a large number of effector proteins into the plant apoplast or cytoplasm to promote pathogenesis (Dou and Zhou, 2012). For example, gram-negative bacterial pathogens secrete effectors directly into the host cells through the bacterial type III secretion system (Desveaux et al., 2006). Many of these effectors target and inhibit PTI signaling components to bring about increased virulence (Feng and Zhou, 2012). For instance, *Pseudomonas syringae* effectors AvrPto and AvrPtoB target the FLS2-BAK1 receptor complex to block PTI signaling. In detail, AvrPtoB targets FLS2 for ubiquitination and degradation, and AvrPto and AvrPtoB interact with BAK1 and interfere with ligand-dependent FLS2-BAK1 interaction upon pathogen infection (Göhre et al., 2008; Shan et al., 2008; Xiang et al., 2011, 2008). *Xanthomonas campestris* pathovar *campestris* effector AvrAC uridylylates (add uridine 5′-monophosphate to a substrate) and blocks phosphorylation of BIK1 to prevent activation of its kinase activity (Feng et al., 2012). BIK1 is also targeted by *P. syringae* effector AvrPphB, a cysteine protease, and undergoes degradation (Zhang et al., 2010a) (Fig. 8.4).

To combat pathogen effectors, plants have evolved NLRs to recognize these effectors in the cytoplasm and activate disease resistance that is often associated with HR. Activation of ETI results in the induction of SA biosynthesis and secretion of antimicrobial proteins and hydrolytic enzymes to confer pathogen resistance at the infection site (Rajamuthiah and Mylonakis, 2014). Most of the NLR proteins contain an N-terminal TIR (toll interleukin 1 receptor, resistance protein) or CC (coiled-coil) domain, an NB (nucleotide-binding) domain, and an LRR domain. NLR proteins recognize pathogen effectors directly or indirectly (Dodds and Rathjen, 2010). For example, NLR proteins encoded by the *L* locus of flax specify rust disease resistance by directly interacting with fungal AvrL proteins in the cytosol (Dodds et al., 2006). The majority of NLRs studied to date, however, indirectly recognize pathogen effectors by monitoring effector-induced modifications of NLR-associated accessory proteins. For example, the NLR proteins RPM1 and RPS2 monitor posttranslational modification of the RPM1-Interacting Protein 4 (RIN4) induced by *P. syringae* effectors AvrB, AvrRpm1, and AvrRpt2 (Belkhadir et al., 2004; Boyes et al., 1998; Day et al., 2005; Mackey et al., 2003, 2002). Bacterial pathogen effectors AvrRpm1 and AvrB induce phosphorylation of RIN4 by enhancing expression of the receptor-like cytoplasmic kinase RPM1-Induced Protein Kinase (RIPK), which facilitates phosphorylation of RIN4 at a conserved threonine residue T166. The T166 phosphorylation of RIN4 is sensed by RPM1 and subsequently activates RPM1-dependent ETI (Chung et al., 2011; Liu et al., 2011a; Mackey et al., 2002). On the other hand, RPS2 senses cleavage of RIN4 by the pathogen effector AvrRpt2 to activate RPS2-dependent ETI (Mackey et al., 2003; Day et al., 2005).

ETI often leads to induced biosynthesis of SA through two pathways. A number of NLRs containing a CC domain in the N terminus induce SA synthesis through Non-race specific Disease Resistance 1 (NDR1), a glycophosphatidyl-inositol-anchored plasma membrane protein previously known as positive regulator of PTI (Century et al., 1995, 1997; Shapiro and Zhang, 2001). NLR proteins containing a TIR domain in the N terminus, however, often induce SA synthesis through Enhanced Disease Susceptibility1 (EDS1), a lipase-like protein (Falk et al., 1999; Wiermer et al., 2005). Similar to RIN4, EDS1 also functions as an accessory protein of TIR-NLRs, such as RPS4 and RPS6, to sense pathogen effectors AvrRps4 and HopA1, respectively (Bhattacharjee et al., 2011; Heidrich et al., 2011). EDS1 interacts with three TIR-NB-LRR proteins, RPS4, RPS6, and Suppressor of *npr1-1* constitutive 1 (SNC1), as well as Suppressor of *rps4*-RLD1 (SRFR1) in microsomal compartment. Bacterial effectors AvrRps4 and HopA1 interact with EDS1 and disrupt the resting state of EDS1-R proteins-SRFR1 complex. The activated EDS1 and RPS4 as well as AvrRps4 shuttle between cytosol and nucleus and activate ETI.

8.6 Systemic acquired resistance

The systemic induction of long-lasting and broad-spectrum disease resistance is known as SAR, but is not associated with Programmed Cell Death. Activation of PTI and ETI in local infected tissues induces the generation of mobile signals that lead to establishment of SAR (Fu and Dong, 2013; Mishina and Zeier, 2007). Although local and systemic induction of SA biosynthesis is required for establishment of SAR, SA itself is not the mobile signal (Gaffney et al., 1993). Potential SAR mobile signals include methyl salicylic acid (MeSA), Glycerol-3-phosphate (G3P), Azelaic acid (AzA), and dehydroabietinal (DA) (Fu and Dong, 2013).

The SA-derivative MeSA does not have the same biological effects as SA, but it plays an important role in the accumulation of SA and establishment of SAR in systemic tissues upon pathogen infection (Park et al., 2007). Pathogen-induced SA is converted to MeSA by SA methyltransferase in local infected tissues. MeSA is then transported to distal pathogen-free tissues, where Salicylic Acid-Binding Protein 2 (SABP2) converts MeSA back to SA through its MeSA esterase activity (Forouhar et al., 2005; Park et al., 2007, 2009). Interestingly, MeSA-triggered SAR requires a short period of light exposure after the primary infection in Arabidopsis. Under a long period of light exposure after the primary infection, MeSA has little effect on the establishment of SAR (Liu et al., 2011b).

The Arabidopsis *SUPPRESSOR OF FATTY ACID DESATURASE DEFICIENCY1* (*SFD1*, also known as *GLY1*, which was originally identified from genetic screening for the mutants with reduced levels of hexadecatrienoic acid) encodes a dihydroxyacetone phosphate reductase, which synthesizes G3P (Nandi et al., 2004). Mutant plants defective in G3P biosynthesis, such as *gly1*, exhibit compromised SAR (Chanda et al., 2011). The level of G3P is rapidly increased in local and distal tissues upon pathogen infection. Exogenous application of G3P alone could not induce SA biosynthesis and SAR, but coinfiltration of G3P together with petiole exudates from mock-inoculated leaves could induce SAR, indicating that G3P requires helpers to establish SAR. Consistent with this notion, it has been reported that putative lipid-transfer proteins Defective in Induced Resistance 1 (DIR1) and Azelaic Acid Induced 1 (AZI1) are required for G3P-induced SAR (Champigny et al., 2013; Yu et al., 2013).

The nine-carbon dicarboxylic acid AzA is found in petiole exudates of pathogen-infected plants (Jung et al., 2009). Although application of AzA alone could not induce SA accumulation and signaling, AzA pretreated plants accumulate higher levels of SA and exhibit enhanced pathogen resistance upon subsequent pathogen infection. Upon infection by SAR-inducing pathogens, the level of AzA is increased in the local tissues and AzA is transported to distal tissues, where it facilitates systemic defense priming (Jung et al., 2009). Like G3P, AzA also requires DIR1 and AZI1 to induce systemic defense priming. However, one study demonstrated that AzA induces G3P biosynthesis and promotes SAR possibly through the activity of G3P (Yu et al., 2013).

The C20 diterpenoid DA acts synergistically with AzA to induce SAR, and requires the presence of DIR1 (Chaturvedi et al., 2012). Local application of DA induces SA accumulation and *PR1* gene expression in distal tissues. Unlike MeSA, G3P, and AzA, SAR-inducing pathogen infection does not induce DA accumulation. Instead, pathogen infection shifts the DA from SAR-noninducible low-molecular-weight fraction (<30 kDa) to SAR-inducible high-molecular-weight fraction (>100 kDa), suggesting that the pathogen infection causes a change from a biologically inactive form of DA to an active signaling form of DA.

The H3K4 demethylase Flowering Locus D (FLD) plays a central role in SAR induction by DA, AzA, and pathogens (He et al., 2003; Shah et al., 2014; Singh et al.,

2013, 2014). Pathogen infection induces SA accumulation and generation of SAR signal in local infected tissues of the *fld* mutant. The SAR signal is transported to systemic tissues of the *fld*; however, the *fld* plant is defective in SA accumulation and priming of defense-related genes, such as *PR1*, *WRKY6*, and *WRKY29*, in systemic tissues (He et al., 2003; Shah et al., 2014; Singh et al., 2013, 2014). These results indicate that FLD is required for induction of SA accumulation and priming of defense genes in systemic tissues, but is not required for induction of SA accumulation in local tissues (Singh et al., 2013, 2014).

8.7   Summary points

Since the discovery of a role for SA in plant immune response, tremendous progress has been made in the understanding of its biosynthesis pathways, key signaling components, as well as underlying molecular and biochemical mechanisms.

1. Two distinct SA biosynthetic pathways, PAL-mediated phenylalanine and ICS-mediated isochorismate pathways, as well as key enzymes PALs, BA2H, and ICS1/2, have been identified in plants.
2. Molecular mechanisms of NPR1-dependent SA signaling have been extensively characterized. NPR1 is a key SA signaling component; its activity and protein stability are dynamically regulated by posttranslational modifications, such as S-nitrosylation, phosphorylation, ubiquitination, and sumoylation.
3. Identification of NPR3, NPR4, and/or NPR1 as likely SA receptors strongly expanded our knowledge about activation of distinct defense responses in infected local and noninfected systemic tissues.
4. The role of SA has been characterized in PTI, ETI, and SAR.

8.8   Future perspectives

1. Further studies are required to identify remaining SA biosynthesis-related enzymes, and clarify redundancy and potential interplay between phenylalanine and isochorismate pathways.
2. SA not only activates NPR1, but also induces rapid NPR1 degradation, which is required for full-scale defense gene expression and establishment of SAR. The SA-induced NPR1 turnover is thought to be required for continuous delivering of "fresh" active NPR1 to target gene promoters, but the hypothesis remains to be proved by an elegant experimental design.
3. NPR1-SA interaction was detected in an equilibrium dialysis assay, but not in a conventional ligand-binding assay. In the future, more sensitive technologies are required to reconcile these differences.
4. SA activates defense gene expression in part through an NPR1-independent pathway, raising the possibility that other SA receptor(s) may exist.

Collectively, although considerable amount of signaling components as well as mobile signals implicated in defense response are identified, our understanding of this pathway is still incomplete.

Abbreviations

ASA	Acetyl salicylic acid
AzA	Azelaic acid
AZI1	Azelaic acid induced 1
BA2H	Benzoic acid-2-hydroxylase
BAK1	Brassinosteroid insensitive 1-associated kinase 1
BIK1	Botrytis-induced kinase 1
BOP	Blade-on-petiole
CBP60g	Cam-binding protein 60-like g
CC	Coiled-coil
CDPK	Calcium-dependent protein kinase
CERK	Chitin elicitor receptor kinase
DA	Dehydroabietinal
DAMP	Damage-associated molecular pattern
DIR1	Defective in induced resistance 1
EDS	Enhanced disease susceptibility
EF-Tu	Elongation factor Tu
EFR	EF-Tu receptor
ETI	Effector-triggered immunity
FLD	Flowering locus D
FLS2	Flagellin-sensing 2
G3P	Glycerol-3-phosphate
HR	Hypersensitive cell death response
ICS	Isochorismate synthase
IPL	Isochorismate pyruvate lyase
LeEix1	*Lycopersicon esculentum* ethylene-inducing xylanase1
LRR	Leucine-rich repeat
LysM	Lysine motif
MAPK	Mitogen-activated protein kinases
MATE	Multidrug and toxin extrusion
MeSA	Methyl salicylic acid
NB	Nucleotide-binding
NDR1	Non-race specific disease resistance 1
NLR	NOD-like receptor
NPR1	Nonexpresser of pathogenesis-related protein 1
PAL	Phenylalanine ammonia lyase
PAMP	Pathogen-associated molecular pattern
PCD	Programmed cell death
PR	Pathogen-related protein
PRR	Pattern recognition receptor
PTI	PAMP-triggered immunity
RIN4	RPM1-interacting protein 4
RIPK	RPM1-induced protein kinase
RLK	Receptor-like kinase
RLP	Receptor-like protein

ROS	Reactive oxygen species
SA	Salicylic acid
SABP2	Salicylic acid-binding protein 2
SAR	Systemic acquired resistance
SARD1	SAR deficient1
SFD1	Suppressor of fatty acid desaturase deficiency1
SID2	Salicylic acid induction deficient 2
SNC1	Suppressor of *npr1-1*, constitutive 1
SRFR1	Suppressor of *rps4*-RLD1
SUMO	Small ubiquitin-related modifier
TGA	TGACG sequence-specific binding protein
TIR	Toll interleukin 1 receptor, resistance protein
TMV	Tobacco mosaic virus
TRX	Thioredoxin

References

Belkhadir, Y., Nimchuk, Z., Hubert, D.A., et al., 2004. Arabidopsis RIN4 negatively regulates disease resistance mediated by RPS2 and RPM1 downstream or independent of the NDR1 signal modulator and is not required for the virulence functions of bacterial type III effectors AvrRpt2 or AvrRpm1. Plant Cell 16, 2822–2835.

Bhattacharjee, S., Halane, M.K., Kim, S.H., et al., 2011. Pathogen effectors target Arabidopsis EDS1 and alter its interactions with immune regulators. Science 334, 1405–1408.

Boatwright, J.L., Pajerowska-Mukhtar, K., 2013. Salicylic acid: an old hormone up to new tricks. Mol. Plant Pathol. 14, 623–634.

Boller, T., Felix, G., 2009. A renaissance of elicitors: perception of microbe-associated molecular patterns and danger signals by pattern-recognition receptors. Annu. Rev. Plant Biol. 60, 379–406.

Boyes, D.C., Nam, J., Dangl, J.L., 1998. The *Arabidopsis thaliana RPM1* disease resistance gene product is a peripheral plasma membrane protein that is degraded coincident with the hypersensitive response. Proc. Natl. Acad. Sci. U.S.A. 95, 15849–15854.

Cao, H., Bowling, S.A., Gordon, A.S., et al., 1994. Characterization of an Arabidopsis mutant that is nonresponsive to inducers of systemic acquired resistance. Plant Cell 6, 1583–1592.

Cao, H., Glazebrook, J., Clarke, J.D., et al., 1997. The Arabidopsis *NPR1* gene that controls systemic acquired resistance encodes a novel protein containing ankyrin repeats. Cell 88, 57–63.

Cao, Y.R., Liang, Y., Tanaka, K., et al., 2014. The kinase LYK5 is a major chitin receptor in Arabidopsis and forms a chitin-induced complex with related kinase CERK1. eLIFE 3, e03766.

Century, K.S., Holub, E.B., Staskawicz, B.J., 1995. NDR1, a locus of *Arabidopsis thaliana* that is required for disease resistance to both a bacterial and a fungal pathogen. Proc. Natl. Acad. Sci. U.S.A. 92, 6597–6601.

Century, K.S., Shapiro, A.D., Repetti, P.P., et al., 1997. NDR1, a pathogen-induced component required for Arabidopsis disease resistance. Science 278, 1963–1965.

Champigny, M.J., Isaacs, M., Carella, P., et al., 2013. Long distance movement of DIR1 and investigation of the role of DIR1-like during systemic acquired resistance in Arabidopsis. Front. Plant Sci. 4, 230.

Chanda, B., Xia, Y., Mandal, M.K., et al., 2011. Glycerol-3-phosphate is a critical mobile inducer of systemic immunity in plants. Nat. Genet. 43, 421–427.

Chaturvedi, R., Venables, B., Petros, R.A., et al., 2012. An abietane diterpenoid is a potent activator of systemic acquired resistance. Plant J. 71, 161–172.

Chen, Z., Zheng, Z., Huang, J., et al., 2009. Biosynthesis of salicylic acid in plants. Plant Signal. Behav. 4, 493–496.

Chung, E.H., da Cunha, L., Wu, A.J., et al., 2011. Specific threonine phosphorylation of a host target by two unrelated type III effectors activates a host innate immune receptor in plants. Cell Host & Microbe 9, 125–136.

Day, B., Dahlbeck, D., Huang, J., et al., 2005. Molecular basis for the RIN4 negative regulation of RPS2 disease resistance. Plant Cell 17, 1292–1305.

Delaney, T.P., Uknes, S., Vernooij, B., et al., 1994. A central role of salicylic acid in plant disease resistance. Science 266, 1247–1250.

Després, C., Chubak, C., Rochon, A., et al., 2003. The Arabidopsis NPR1 disease resistance protein is a novel cofactor that confers redox regulation of DNA binding activity to the basic domain/leucine zipper transcription factor TGA1. Plant Cell 15, 2181–2191.

Després, C., DeLong, C., Glaze, S., et al., 2000. The Arabidopsis NPR1/NIM1 protein enhances the DNA binding activity of a subgroup of the TGA family of bZIP transcription factors. Plant Cell 12, 279–290.

Desveaux, D., Singer, A.U., Dangl, J.L., 2006. Type III effector proteins: doppelgangers of bacterial virulence. Curr. Opin. Plant Biol. 9, 376–382.

Dodds, P.N., Lawrence, G.J., Catanzariti, A.M., et al., 2006. Direct protein interaction underlies gene-for-gene specificity and coevolution of the flax resistance genes and flax rust avirulence genes. Proc. Natl. Acad. Sci. U.S.A. 103, 8888–8893.

Dodds, P.N., Rathjen, J.P., 2010. Plant immunity: towards an integrated view of plant-pathogen interactions. Nat. Rev. Genet. 11, 539–548.

Dou, D., Zhou, J.M., 2012. Phytopathogen effectors subverting host immunity: different foes, similar battleground. Cell Host & Microbe 12, 484–495.

Falk, A., Feys, B.J., Frost, L.N., et al., 1999. EDS1, an essential component of R gene-mediated disease resistance in Arabidopsis has homology to eukaryotic lipases. Proc. Natl. Acad. Sci. U.S.A. 96, 3292–3297.

Fan, W.H., Dong, X.N., 2002. In vivo interaction between NPR1 and transcription factor TGA2 leads to salicylic acid-mediated gene activation in Arabidopsis. Plant Cell 14, 1377–1389.

Feng, F., Yang, F., Rong, W., et al., 2012. A Xanthomonas uridine 5'-monophosphate transferase inhibits plant immune kinases. Nature 485, 114–118.

Feng, F., Zhou, J.M., 2012. Plant-bacterial pathogen interactions mediated by type III effectors. Curr. Opin. Plant Biol. 15, 469–476.

Forouhar, F., Yang, Y., Kumar, D., et al., 2005. Structural and biochemical studies identify tobacco SABP2 as a methyl salicylate esterase and implicate it in plant innate immunity. Proc. Natl. Acad. Sci. U.S.A. 102, 1773–1778.

Fu, Z.Q., Dong, X.N., 2013. Systemic acquired resistance: turning local infection into global defense. Annu. Rev. Plant Biol. 64, 839–863.

Fu, Z.Q., Yan, S.P., Saleh, A., et al., 2012. NPR3 and NPR4 are receptors for the immune signal salicylic acid in plants. Nature 486, 228–232.

Göhre, V., Spallek, T., Haweker, H., et al., 2008. Plant pattern-recognition receptor FLS2 is directed for degradation by the bacterial ubiquitin ligase AvrPtoB. Curr. Biol. 18, 1824–1832.

Gaffney, T., Friedrich, L., Vernooij, B., et al., 1993. Requirement of salicylic acid for the induction of systemic acquired resistance. Science 261, 754–756.

Garcion, C., Lohmann, A., Lamodiere, E., et al., 2008. Characterization and biological function of the *ISOCHORISMATE SYNTHASE2* gene of Arabidopsis. Plant Physiol. 147, 1279–1287.

He, Y.H., Michaels, S.D., Amasino, R.M., 2003. Regulation of flowering time by histone acetylation in Arabidopsis. Science 302, 1751–1754.

Heidrich, K., Wirthmueller, L., Tasset, C., et al., 2011. Arabidopsis EDS1 connects pathogen effector recognition to cell compartment-specific immune responses. Science 334, 1401–1404.

Huang, J.L., Gu, M., Lai, Z.B., et al., 2010. Functional analysis of the Arabidopsis *PAL* gene family in plant growth, development, and response to environmental stress. Plant Physiol. 153, 1526–1538.

Jung, H.W., Tschaplinski, T.J., Wang, L., et al., 2009. Priming in systemic plant immunity. Science 324, 89–91.

Khan, M.I.R., Fatma, M., Per, T.S., et al., 2015. Salicylic acid-induced abiotic stress tolerance and underlying mechanisms in plants. Front. Plant Sci. 6, 462.

León, J., Shulaev, V., Yalpani, N., et al., 1995. Benzoic acid 2-hydroxylase, a soluble oxygenase from tobacco, catalyzes salicylic acid biosynthesis. Proc. Natl. Acad. Sci. U.S.A. 92, 10413–10417.

Liu, J., Elmore, J.M., Lin, Z.J., et al., 2011a. A receptor-like cytoplasmic kinase phosphorylates the host target RIN4, leading to the activation of a plant innate immune receptor. Cell Host & Microbe 9, 137–146.

Liu, P.P., von Dahl, C.C., Klessig, D.F., 2011b. The extent to which methyl salicylate is required for signaling systemic acquired resistance is dependent on exposure to light after infection. Plant Physiol. 157, 2216–2226.

Lu, D.P., Wu, S.J., Gao, X.Q., et al., 2010. A receptor-like cytoplasmic kinase, BIK1, associates with a flagellin receptor complex to initiate plant innate immunity. Proc. Natl. Acad. Sci. U.S.A. 107, 496–501.

Mackey, D., Belkhadir, Y., Alonso, J.M., et al., 2003. Arabidopsis RIN4 is a target of the type III virulence effector AvrRpt2 and modulates RPS2-mediated resistance. Cell 112, 379–389.

Mackey, D., Holt 3rd, B.F., Wiig, A., et al., 2002. RIN4 interacts with *Pseudomonas syringae* type III effector molecules and is required for RPM1-mediated resistance in Arabidopsis. Cell 108, 743–754.

Metraux, J.P., 2002. Recent breakthroughs in the study of salicylic acid biosynthesis. Trends Plant Sci. 7, 332–334.

Mishina, T.E., Zeier, J., 2007. Pathogen-associated molecular pattern recognition rather than development of tissue necrosis contributes to bacterial induction of systemic acquired resistance in Arabidopsis. Plant J. 50, 500–513.

Miura, K., Jin, J.B., Hasegawa, P.M., 2007. Sumoylation, a post-translational regulatory-process in plants. Curr. Opin. Plant Biol. 10, 495–502.

Monaghan, J., Zipfel, C., 2012. Plant pattern recognition receptor complexes at the plasma membrane. Curr. Opin. Plant Biol. 15, 349–357.

Mou, Z., Fan, W.H., Dong, X.N., 2003. Inducers of plant systemic acquired resistance regulate NPR1 function through redox changes. Cell 113, 935–944.

Nandi, A., Welti, R., Shah, J., 2004. The *Arabidopsis thaliana* dihydroxyacetone phosphate reductase gene *SUPPRESSOR OF FATTY ACID DESATURASE DEFICIENCY1* is required for glycerolipid metabolism and for the activation of systemic acquired resistance. Plant Cell 16, 465–477.

Norberg, M., Holmlund, M., Nilsson, O., 2005. The BLADE ON PETIOLE genes act redundantly to control the growth and development of lateral organs. Development 132, 2203–2213.

Park, S.W., Kaimoyo, E., Kumar, D., et al., 2007. Methyl salicylate is a critical mobile signal for plant systemic acquired resistance. Science 318, 113–116.

Park, S.W., Liu, P.P., Forouhar, F., et al., 2009. Use of a synthetic salicylic acid analog to investigate the roles of methyl salicylate and its esterases in plant disease resistance. J. Biol. Chem. 284, 7307–7317.

Pintard, L., Willems, A., Peter, M., 2004. Cullin-based ubiquitin ligases: Cul3-BTB complexes join the family. Embo J. 23, 1681–1687.

Rajamuthiah, R., Mylonakis, E., 2014. Effector triggered immunity. Virulence 5, 697–702.

Rate, D.N., Greenberg, J.T., 2001. The Arabidopsis aberrant growth and *death2* mutant shows resistance to *Pseudomonas syringae* and reveals a role for NPR1 in suppressing hypersensitive cell death. Plant J. 27, 203–211.

Rochon, A., Boyle, P., Wignes, T., et al., 2006. The coactivator function of Arabidopsis NPR1 requires the core of its BTB/POZ domain and the oxidation of C-terminal cysteines. Plant Cell 18, 3670–3685.

Saleh, A., Withers, J., Mohan, R., et al., 2015. Posttranslational modifications of the master transcriptional regulator NPR1 enable dynamic but tight control of plant immune responses. Cell Host & Microbe 18, 169–182.

Segonzac, C., Zipfel, C., 2011. Activation of plant pattern-recognition receptors by bacteria. Curr. Opin. Microbiol. 14, 54–61.

Serino, L., Reimmann, C., Baur, H., et al., 1995. Structural genes for salicylate biosynthesis from chorismate in *Pseudomonas aeruginosa*. Mol. Gen. Genet. 249, 217–228.

Serrano, M., Wang, B.J., Aryal, B., et al., 2013. Export of salicylic acid from the chloroplast requires the multidrug and toxin extrusion-like transporter EDS5. Plant Physiol. 162, 1815–1821.

Shah, J., Chaturvedi, R., Chowdhury, Z., et al., 2014. Signaling by small metabolites in systemic acquired resistance. Plant J. 79, 645–658.

Shan, L.B., He, P., Li, J.M., et al., 2008. Bacterial effectors target the common signaling partner BAK1 to disrupt multiple MAMP receptor-signaling complexes and impede plant immunity. Cell Host & Microbe 4, 17–27.

Shapiro, A.D., Zhang, C., 2001. The role of NDR1 in avirulence gene-directed signaling and control of programmed cell death in Arabidopsis. Plant Physiol. 127, 1089–1101.

Singh, V., Roy, S., Giri, M.K., et al., 2013. *Arabidopsis thaliana* FLOWERING LOCUS D is required for systemic acquired resistance. Mol. Plant Microbe. Interact. 26, 1079–1088.

Singh, V., Roy, S., Singh, D., et al., 2014. Arabidopsis FLOWERING LOCUS D influences systemic-acquired-resistance-induced expression and histone modifications of *WRKY* genes. J. Biosci. 39, 119–126.

Spoel, S.H., Mou, Z.L., Tada, Y., et al., 2009. Proteasome-mediated turnover of the transcription coactivator NPR1 plays dual roles in regulating plant immunity. Cell 137, 860–872.

Strawn, M.A., Marr, S.K., Inoue, K., et al., 2007. Arabidopsis isochorismate synthase functional in pathogen-induced salicylate biosynthesis exhibits properties consistent with a role in diverse stress responses. J. Biol. Chem. 282, 5919–5933.

Sun, Y.D., Li, L., Macho, A.P., et al., 2013. Structural basis for flg22-induced activation of the Arabidopsis FLS2-BAK1 immune complex. Science 342, 624–628.

Tada, Y., Spoel, S.H., Pajerowska-Mukhtar, K., et al., 2008. Plant immunity requires conformational changes of NPR1 via S-nitrosylation and thioredoxins. Science 321, 952–956.

Trdá, L., Boutrot, F., Claverie, J., et al., 2015. Perception of pathogenic or beneficial bacteria and their evasion of host immunity: pattern recognition receptors in the frontline. Front. Plant Sci. 6, 219.

van den Burg, H.A., Kini, R.K., Schuurink, R.C., et al., 2010. Arabidopsis small ubiquitin-like modifier paralogs have distinct functions in development and defense. Plant Cell 22, 1998–2016.

Vlot, A.C., Dempsey, D.A., Klessig, D.F., 2009. Salicylic acid, a multifaceted hormone to combat disease. Annu. Rev. Phytopathol. 47, 177–206.

Wang, D., Amornsiripanitch, N., Dong, X.N., 2006. A genomic approach to identify regulatory nodes in the transcriptional network of systemic acquired resistance in plants. PLoS Pathog. 2, 1042–1050.

Wang, L., Tsuda, K., Truman, W., et al., 2011. CBP60g and SARD1 play partially redundant critical roles in salicylic acid signaling. Plant J. 67, 1029–1041.

Weissmann, G., 1991. Aspirin. Sci. Am. 264, 84–90.

White, R.F., 1979. Acetylsalicylic-acid (aspirin) induces resistance to tobacco mosaic virus in tobacco. Virology 99, 410–412.

Wiermer, M., Feys, B.J., Parker, J.E., 2005. Plant immunity: the EDS1 regulatory node. Curr. Opin. Plant Biol. 8, 383–389.

Wildermuth, M.C., Dewdney, J., Wu, G., et al., 2001. Isochorismate synthase is required to synthesize salicylic acid for plant defence. Nature 414, 562–565.

Wu, Y., Zhang, D., Chu, J.Y., et al., 2012. The Arabidopsis NPR1 protein is a receptor for the plant defense hormone salicylic acid. Cell Rep. 1, 639–647.

Xiang, T., Zong, N., Zhang, J., et al., 2011. BAK1 is not a target of the *Pseudomonas syringae* effector AvrPto. Mol. Plant Microbe Interact. 24, 100–107.

Xiang, T., Zong, N., Zou, Y., et al., 2008. *Pseudomonas syringae* effector AvrPto blocks innate immunity by targeting receptor kinases. Curr. Biol. 18, 74–80.

Yan, S., Dong, X., 2014. Perception of the plant immune signal salicylic acid. Curr. Opin. Plant Biol. 20, 64–68.

Yu, K.S., Soares, J.M., Mandal, M.K., et al., 2013. A feedback regulatory loop between G3P and lipid transfer proteins DIR1 and AZI1 mediates azelaic-acid-induced systemic immunity. Cell Rep. 3, 1266–1278.

Zhang, J., Li, W., Xiang, T.T., et al., 2010a. Receptor-like cytoplasmic kinases integrate signaling from multiple plant immune receptors and are targeted by a *Pseudomonas syringae* effector. Cell Host & Microbe 7, 290–301.

Zhang, Y.L., Cheng, Y.T., Qu, N., et al., 2006. Negative regulation of defense responses in Arabidopsis by two NPR1 paralogs. Plant J. 48, 647–656.

Zhang, Y.X., Xu, S.H., Ding, P.T., et al., 2010b. Control of salicylic acid synthesis and systemic acquired resistance by two members of a plant-specific family of transcription factors. Proc. Natl. Acad. Sci. U.S.A. 107, 18220–18225.

Brassinosteroids

Haijiao Wang[1], Zhuoyun Wei[2], Jia Li[2], Xuelu Wang[1]
[1]Huazhong Agricultural University, Wuhan, China; [2]Lanzhou University, Lanzhou, China

9

Summary

Brassinosteroids (BRs) are a class of growth-promoting steroidal phytohormones. BRs control almost all aspects of plant growth and development, and also play significant role in plant adaptation to biotic and abiotic stresses. Their biosynthetic and signaling pathways have been well characterized by forward and reverse genetics. The entire synthetic pathway includes the general cycloartenol-to-campesterol sterol biosynthesis pathway and a specific campesterol-to-brassinolide (BL) biosynthetic pathway in Arabidopsis. Campesterol converts to BL with a campestanol-dependent or a campestanol-independent pathway. BRs are perceived by a plasma membrane localized receptor and co-receptor complex including BRI1 and BAK1. The activated BRI1/BAK1 complex inactivates BIN2, which is one of the GSK3-like protein kinases, and negatively regulates BR signaling, to promote the activity of two critical transcription factors, BES1 and BZR1, and BR-responsive gene expression. BR-regulated plant developmental processes include cell elongation, root hair initiation, stomatal development, cell division, and reproductive development. BRs also interact with many other hormonal and environmental cues to regulate plant growth and development. In this chapter, we focus on the BR biosynthetic pathway, the BR signaling pathway, BR-regulated plant growth and development, and the cross talk between BRs and other signaling pathways.

9.1   The history of brassinosteroids

Steroid hormones have been studied in animals earlier than in plants. In animals, steroid hormones include the sex hormones (such as estradiol, testosterone, and progesterone) and the adrenal cortex hormones (glucocorticoids and mineralocorticoids), mediating a wide range of pivotal developmental and physiological functions. The receptors of steroid hormones are a class of nuclear receptors. In plants, brassinosteroids (BRs) are also a group of steroid hormones which play important roles in plant growth and development. The identification of plant steroid hormones started with the investigation of novel plant growth-promoting substances in pollen from many plant species. Researchers found that the organic solvent extract of rape (*Brassica napus* L.) pollen has the greatest growth-promoting activity in a bean second-internode bioassay. The bioactive substances were named brassins (Mitchell et al., 1970). Bioassay also indicated that the biological function of brassins is different from other growth-promoting hormones such as gibberellic acid, and brassins are likely a family of new plant hormones. In 1979, from 227 kg of honeybee-collected rape pollen, 4 mg active growth-promoting substance was purified and crystallized (Grove et al., 1979). X-ray analysis revealed that

Hormone Metabolism and Signaling in Plants. http://dx.doi.org/10.1016/B978-0-12-811562-6.00009-8

it is a polyhydroxylated steroid similar to animal steroid hormones, with the systematic name (22R, 23R, 24S)-2α-3α, 22, 23-tetrahydroxy-24-methyl-6, 7-s-5α-cholestano-6, 7-lactone and common name brassinolide (BL). BL and its derivatives are designated as BRs. Now, more than 70 naturally occurring BRs have been identified in the plant kingdom. Due to the high synthetic cost of BL, 24-epibrassinolide (24-epiBL) is often used in bioassay experiments in the laboratory.

Although low-level BRs have obvious physiological function, for example, 10 ng BL can produce an approximately 200% increase in elongation of the bean second-internode, not until the mid-1990s were BRs widely accepted as plant hormones following the identification of several biosynthetic and signaling mutants in the model plants Arabidopsis and rice. From 1979 to 1996, scientists mainly engaged in the identification of intermediates in the BR biosynthesis pathway. In 1996, the phenotype and crude mapping of the *brassinosteroid-insensitive1* (*bri1*) mutation which is insensitive to BL treatment was reported (Clouse et al., 1996). In 1997, the *BRI1* gene, encoding the receptor of BRs, was cloned (Li and Chory, 1997). Later, researchers tried to define the signaling pathway of BRs. Now, a relatively intact BR signaling pathway from receptor to downstream transcription factors has been elucidated.

9.2   The biosynthesis and catabolism of brassinosteroids

9.2.1   Sterol synthesis

The BL biosynthesis pathway can be divided into two parts, the general cycloartenol-to-campesterol sterol synthesis pathway and the specific biosynthesis of BL from campesterol (Fig. 9.1). Three molecules of acetyl-CoA are joined together to form mevalonate, which is then converted to isopentenyl pyrophosphate, geranyl pyrophosphate, and farnesyl pyrophosphate. Two molecules of farnesyl pyrophosphate are condensed to form squalene, which is then converted via squalene-2,3-oxide to cycloartenol. SQUALENE EPOXIDASE 1 (SQE1) and CYCLOARTENOL SYNTHASE 1 (CAS1) catalyze the conversion of squalene to squalene-2,3-oxide and squalene-2,3-oxide to cycloartenol in Arabidopsis (Babiychuk et al., 2008; Pose et al., 2009). STEROL METHYLTRANSFERASE 1 (SMT1) catalyzes the conversion of cycloartenol to 24-methylene cycloartenol by adding a single methyl at the C-24 position of the side chain (Diener et al., 2000). 24-methylene cycloartenol is then converted to 4α-methyl-5α-ergosta-8,14,24(28)-trien-3β-ol through C-4 demethylation, isomerization, and C-14 demethylation. Conversion of 4α-methyl-5α-ergosta-8,14,24(28)-trien-3β-ol to 4α-methylfecosterol is accomplished by FACKEL (FK), a sterol C-14 reductase in Arabidopsis (Jang et al., 2000; Schrick et al., 2000). *HYDRA* (*HYD1*) encodes a Δ^8-Δ^7sterol isomerase, which is responsible for converting 4α-methylfecosterol to 24-methylenelophenol (Schrick et al., 2002; Souter et al., 2002). From 24-methylenelophenol, the sterol biosynthesis pathway is divided into two parallel branches, one leading to campesterol via episterol, 5-dehydroepisterol, and 24-methylenecholesterol and the other leading to sitosterol via citrostadienol, avenasterol, 5-dehydroavenasterol, and isofucosterol. Two C-24 methyltransferases,

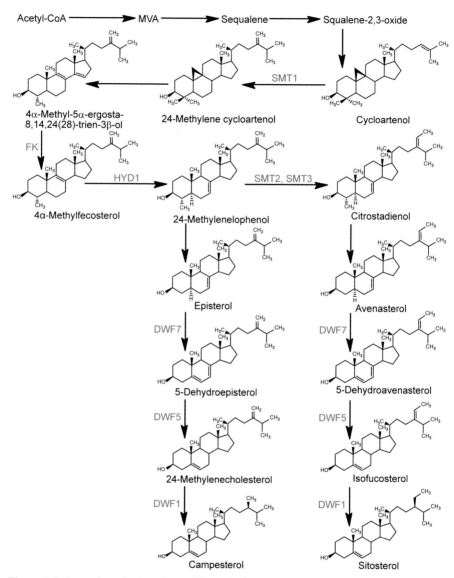

Figure 9.1 A sterol synthesis pathway. Biosynthetic routes bypassing acetyl-CoA, MVA, cycloartenol to campasterol and sitosterol are shown. Enzymes involved in the related steps are indicated by red words.

SMT2 and SMT3, redundantly catalyze the reaction from 24-methylenelophenol to citrostadienol (Carland et al., 2010, 2002). Genetic and biochemical analyses of the *dwarf7* (*dwf7*) mutant identified DWF7 as a Δ^7C-5-desaturase of Arabidopsis, catalyzing the reaction from episterol to 5-dehydroepisterol and from avenasterol to 5-dehydroavenasterol (Choe et al., 1999b). Conversion of 5-dehydroepisterol to

24-methylenecholesterol, and of 5-dehydroavenasterol to isofucosterol are accomplished by Δ^7-sterol reductase DWF5 in Arabidopsis (Choe et al., 2000). Reduction of 24-methylenecholesterol to campesterol, and of isofucosterol to sitosterol is catalyzed by DWF1 in parallel reactions (Choe et al., 1999a; Klahre et al., 1998).

9.2.2   Specific brassinosteroid biosynthesis pathway

9.2.2.1   Campestanol-dependent pathway of BL biosynthesis

The specific BR biosynthetic pathway has been elucidated mainly by feeding deuterated and tritiated putative precursors to cell suspension cultures of *Catharanthus roseus* followed by GC-MS analysis of the labeled products. Campesterol is the first compound specifically entering into the BR biosynthetic pathway and can be converted to BL via a campestanol (CN)-dependent and a CN-independent pathway.

In the CN-dependent BL biosynthesis pathway (Fig. 9.2), the first four reactions lead to conversion of campesterol to CN via 4-en-3β-ol, (24*R*)-ergostan-4-en-3-one and (24*R*)-5α-ergostan-3-one. One of these steps is catalyzed by DE-ETIOLATED 2 (DET2). The *det2-1* mutants grown in total darkness show a de-etiolated morphology (Chory et al., 1991). The *DET2* gene encodes a protein that shares about 40% sequence identity with mammalian 5α-steroid reductases (Li et al., 1996). Subsequent biochemical analysis found that the DET2 protein expressed in human embryonic kidney 293 cells reduced several animal steroid substrates and showed similar kinetic properties to the mammalian 5α-steroid reductases (Li et al., 1997). Moreover, the human steroid reductases can rescue the *det2* mutant phenotypes, indicating that DET2 is an ortholog of the mammalian 5α-steroid reductases (Li et al., 1997). Feeding experiments revealed that DET2 catalyzes the conversion of (24*R*)-ergostan-4-en-3-one to (24*R*)-5α-ergostan-3-one (Fujioka et al., 1997; Noguchi et al., 1999b).

Two parallel pathways, the early and the late C-6 oxidation pathways, convert CN to castasterone in 6-oxo and 6-deoxo forms, respectively. In the early C-6 oxidation pathway, CN is first oxidized to 6-oxocampestanol via 6α-hydroxy-campestanol, which is then converted into cathasterone, teasterone, 3-dehydro-teasterone, typhasterol, and castasterone (Fujioka and Sakurai, 1997). The biosynthetic route for the late C-6 oxidation pathway is as follows: CN→6-deoxocathasterone→6-deoxoteasterone→3-dehydro-6-deoxoteasterone→6-deoxotyphasterol→6-deoxocastasterone (Noguchi et al., 2000). Castasterone is ultimately converted to BL, the final and the most bioactive product of the BR biosynthesis pathway. Both the early and the late C-6 oxidation pathways have been demonstrated in numerous plant species, such as *Catharanthus roseus*, Arabidopsis, tomato, and rice, suggesting that these two pathways are common BR biosynthetic routes in plants (Choi et al., 1997; Fujioka et al., 1996; Fujioka and Sakurai, 1997; Kim et al., 2004; Noguchi et al., 2000).

The CN-to-6-deoxocathasterone conversion in the late C-6oxidation pathway and the 6-oxocampestanol-to-cathasterone conversion in early C-6 oxidation pathway are both catalyzed by DWARF4 (DWF4), a cytochrome P450 enzyme that acts as a 22α-hydroxylase (Choe et al., 1998). The next step in both branches involves a C-23 hydroxylation catalyzed by ROTUNDIFOLIA3 (ROT3, also known as CYP90C1)

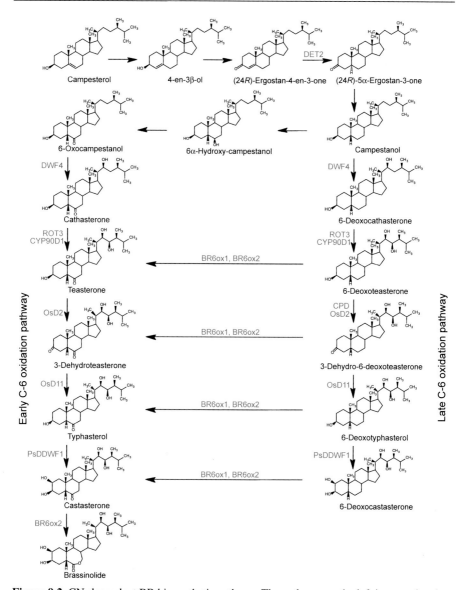

Figure 9.2 CN-dependent BR biosynthetic pathway. The pathway on the left is termed early C-6 oxidation and that on the right, late C-6 oxidation. C-6 oxidation steps connect the early and the late C-6 oxidation branches. Enzymes involved in the related steps are indicated by red words.

and its homolog CYP90D1 (Ohnishi et al., 2006), which is different from a previous report suggesting that CONSTITUTIVE PHOTOMORPHOGENIC DWARF(CPD) is responsible for the C-23 hydroxylation of BR intermediates (Szekeres et al., 1996). Recently, Ohnishi et al. (2012) found that CPD participates in the oxidation of

6-deoxoteasterone to 3-dehydro-6-deoxoteasterone in the late C-6 oxidation branch. The rice *D2* gene encodes a cytochrome P450, categorized as CYP90D, which has been shown to convert 6-deoxoteasterone to3-dehydro-6-deoxoteasterone and teasterone to 3-dehydroteasterone (Hong et al., 2003). However, a biochemical assay revealed that D2 and its homolog D3 catalyze C-23 hydroxylation of brassinosteroids in vitro (Sakamoto et al., 2012). The *D11* gene encodes another cytochrome P450 enzyme in rice, identified as CYP724B1, which has been shown to catalyze the C-3 reduction steps from 3-dehydro-teasterone to typhasterol and from 3-dehydro-6-deoxoteasterone to 6-deoxotyphasterol (Tanabe et al., 2005). DARK-INDUCED DWF-LIKE PROTEIN 1 (DDWF1) catalyzes C-2 hydroxylation steps converting typhasterol to castasterone and 6-deoxotyphasterol to 6-deoxocastasterone in pea (Kang et al., 2001). The enzymes responsible for the C-3 reduction and C-2 hydroxylation steps in Arabidopsis have not yet been reported.

Steps connecting the early and the late C-6 oxidation pathways have been characterized in different plant species. The tomato DWARF cytochrome P450 catalyzes the C-6 oxidation of 6-deoxocastasterone to castasterone (Bishop et al., 1999). A rice BR6ox enzyme (BRD1) with a high sequence similarity to the tomato DWARF regulates multiple C-6 oxidation steps in the rice BR biosynthetic pathway (Hong et al., 2002). Subsequently, the Arabidopsis AtBR6ox (CYP85A1) with 68% identity to DWARF was characterized by homology searches (Shimada et al., 2001). Analysis of the ability of yeast cells transformed with *DWARF* and *AtBR6ox* to metabolize 6-deoxo-BRs found that both enzymes catalyze multiple steps in BR biosynthesis: 6-deoxoteasterone to teasterone, 3-dehydro-6-deoxoteasterone to 3-dehydroteasterone, 6-deoxotyphasterol to typhasterol, and 6-deoxocastasterone to castasterone (Shimada et al., 2001). CYP85A2, a homolog of CYP85A1 in Arabidopsis, also mediates the bridge reactions that connect the late and early C-6 oxidation pathways by converting 6-deoxoBR to 6-oxoBRs (Kwon et al., 2005). In addition, CYP85A2 differs from CYP85A1 in that it can convert castasterone to BL via a Baeyer-Villiger oxidation (Kim et al., 2005; Nomura et al., 2005).

9.2.2.2   *Campestanol-independent pathway of BL biosynthesis (Fig. 9.3)*

An early C-22 oxidation branch was identified, in which campesterol is converted to (22*S*)-22-hydroxy-campesterol catalyzed by DWF4 (Fujioka et al., 2002; Fujita et al., 2006). Measurement of C-22 hydroxylated BR and analysis of DWF4 enzyme catalytic efficiency demonstrated that the primary substrate of DWF4 is campesterol rather than CN, suggesting that C-22 hydroxylation of campesterol before 5α-reduction is the main route of BR biosynthesis in Arabidopsis (Fujioka et al., 2002; Fujita et al., 2006). C-3 oxidation of (22*S*)-22-hydroxy-campesterol to (22*S*, 24*R*)-22-hydroxy-ergost-4-en-3-one, and of (22*R*,23*R*)-22,23-dihydroxycampesterol to (22*R*,23*R*)-22,23-dihydroxy-campest-4-en-3-one, by CPD was identified by analyzing the enzyme activity of CPD expressed in a baculovirus-insect cell system (Ohnishi et al., 2012). DET2 catalyzes 5α-reduction of (22*S*,24*R*)-22-hydroxy-ergost-4-en-3-one and (22*R*,23*R*)-22,23-dihydroxy-campest-4-en-3-one to (22*S*,24*R*)-22-hydroxy-5α-ergostan-3-one and 3-dehydro-6-deoxoteasterone, respectively (Fujioka et al., 2002; Noguchi et al., 1999b). The C-22-hydroxylated form of the 4-en-3-one was found to be

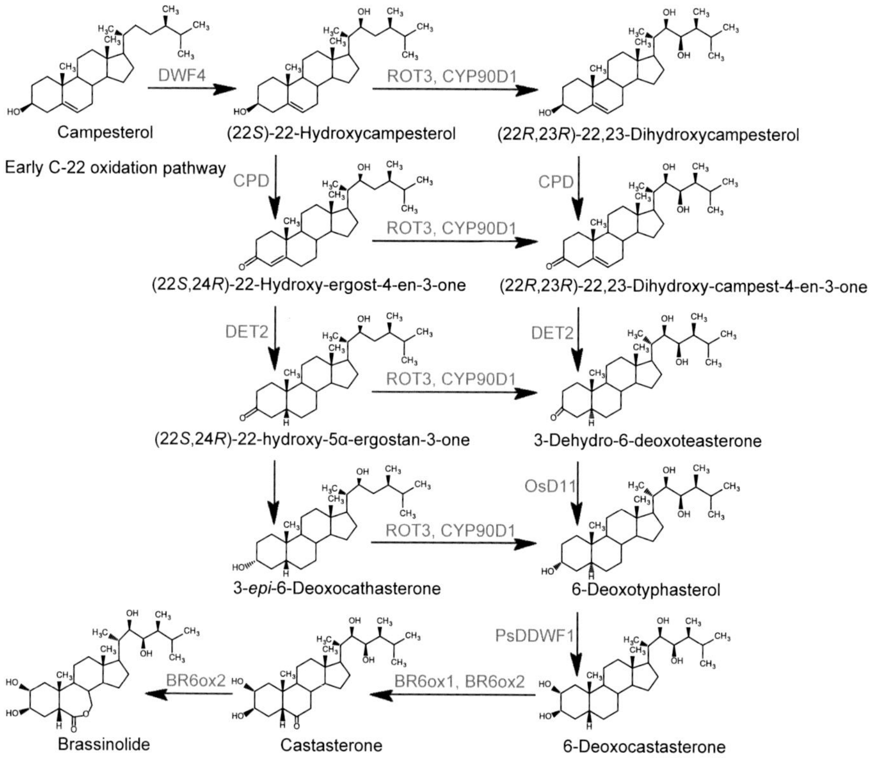

Figure 9.3 CN-independent BR biosynthetic pathway. The early C-22 oxidation pathway merges into the late C-6 oxidation pathway, forming a shortcut BR biosynthesis pathway. Biochemical experiments in vitro coupled with measurement of endogenous BRs suggest that this is a prominent pathway in Arabidopsis. Enzymes involved in the related steps are indicated by red words.

a primary substrate of DET2 in Arabidopsis (Fujioka et al., 2002). Both CYP90C1 and CYP90D1 convert 3-*epi*-6-deoxocathasterone, (22S,24R)-22-hydroxy-5a-ergostan-3-one, (22S,24R)-22-hydroxyergost-4-en-3-one, and (22S)-22-hydroxy-campesterol to 23-hydroxylated products (Ohnishi et al., 2006). The pathway then merges into the late C-6 oxidation pathway to produce biologically active BRs. Thus, a shortcut BR biosynthetic pathway, also called the CN-independent pathway, was established. Biochemical experiments coupled with measurement of endogenous BRs suggest that this is a prominent pathway in Arabidopsis.

9.2.3 Regulation of brassinosteroid biosynthesis

9.2.3.1 Feedback inhibition of BR biosynthesis

BR biosynthesis can be regulated by feedback inhibition by the pathway end product. For example, it was found that five BR-specific biosynthesis genes (*DET2*, *DWF4*, *CPD*, *BR6ox1*, and *ROT3*) and two sterol biosynthesis genes (*FK* and *DWF5*) were upregulated in BR-depleted wild-type Arabidopsis plants treated with BRZ, a BR biosynthesis inhibitor (Tanaka et al., 2005). Furthermore, four BR-specific biosynthesis genes (*DWF4, CPD, BR6ox1,* and *ROT3*) and one sterol-specific biosynthesis gene (*DWF7*) were downregulated in wild-type plants that were fed with BL (Tanaka et al., 2005). These results suggested that feedback regulation occurs at multiple BR biosynthesis steps. In addition, the BL-induced *CPD* downregulation is controlled by the BRI1-mediated signaling pathway, because such regulation was not observed in *bri1* mutants (Li et al., 2001).

9.2.3.2   Positive regulation of BR biosynthesis

Genetic and biochemical analyses also identified a number of factors positively mediating BR biosynthesis. For instance, TEOSINTE BRANCHED1/CYCLOIDEA/PROLIFERATING CELL FACTOR1 (TCP1), a basic helix-loop-helix (bHLH)-containing transcription factor, upregulates the expression of *DWF4* by interacting with GGNCCC motifs in its promoter region (Gao et al., 2015; Guo et al., 2010). Similarly, CESTA, another bHLH transcription factor, elevates the expression of *CPD* by directly binding to G-box motifs in its promoter region (Poppenberger et al., 2011). It was found that auxin signaling also induces the expression of *DWF4*, possibly through reducing BZR1 binding to the *DWF4* promoter (Chung et al., 2011). In rice, RELATED TO ABI3/VP1, ABA INSENSITIVE 3/VIVAPARIOUS 1 (RAVL1), a B3-domain-containing transcription factor, activates the expression of *D2, D11,* and *BRD1* through E-box sequences (Je et al., 2010).

9.2.3.3   BR biosynthesis inhibitor

Chemical inhibitors of BR biosynthesis have also been found. They play significant roles in BR-related research. For instance, brassinazole (BRZ) is a triazole derivative that can inhibit BR biosynthesis by blocking cytochrome P450 enzymes (Asami and Yoshida, 1999). Plants treated with BRZ show dwarfism with downward-curling dark green leaves, and these morphological changes can be reversed by BL treatment (Asami and Yoshida, 1999). BRZ2001, a modified form of BRZ, can induce similar morphological changes to those seen in BRZ-treated plants (Sekimata et al., 2001). Propiconazole (Pcz) is also a triazole compound, and has an effect similar to BRZ with low cost (Hartwig et al., 2012). Another triazole-type BR biosynthesis inhibitor is YCZ-18, which binds to CYP90D1, and inhibits BR-induced cell elongation. This effect can be rescued by BL but not CS, indicating that YCZ-18 functions differently from BRZ (Oh et al., 2015).

9.2.4   Brassinosteroid catabolism

Feeding experiments showed that exogenously applied excessive amounts of BRs are rapidly metabolized (Fujioka and Yokota, 2003). In most of the cases, conjugates of plant hormones represent inactive forms. Conjugation with glucose at 23-OH of BL and castasterone was found to be catalyzed by an Arabidopsis UDP glycosyltransferase termed UGT73C5 (Poppenberger et al., 2005). In addition, conjugation with glucose at 2α-OH, 3β-OH, 25-OH, 26-OH; conjugation with fatty acids at 3β-OH; and conjugation with 6-*O*-β glucosylglucose or 4-*O*-β glucosylglucose at 3β-OH were reported (Fujioka and Yokota, 2003). However, the enzymes involved in these reactions are still unknown.

The *BRASSICA NAPUS* SULFOTRANSFERASE 3 (BNST3) protein from *B. napus* and AtST4a and AtST1 from Arabidopsis catalyze the O-sulfonation of BRs (Marsolais et al., 2007; Rouleau et al., 1999). In Arabidopsis, *PHYB ACTIVATION-TAGGED SUPPRESSOR 1 (BAS1)* encodes a cytochrome P450 enzyme capable of hydroxyling bioactive CS or BL into bioinactive 26-OH-CS or 26-OH-BL (Neff et al., 1999). Besides C-26, hydroxylation of C-12, C-20, and C-25 has also been observed. The *BEN1*gene encodes a dihydroflavonol 4-reductase (DFR)-like protein that likely catalyzes the conversion of TY, CS, and BL into the biologically inactive 6-OH-TY, 6-OH-CS, and 6-OH-BL, respectively (Yuan et al., 2007). In Arabidopsis DWARF AND ROUND LEAF-1 (DRL1), a putative CoA-dependent acyltransferase is involved in BR metabolism likely by promoting esterification of certain BRs (Zhu et al., 2013). Also observed were β-epimerization of 2α-OH and 3α-OH, demethylation of C-26 CH$_3$ and C-28 CH$_3$, side chain cleavage of C-20/22, and oxidation of 23-OH. However, the enzymes responsible for these predicted reactions have not yet been identified.

9.3 The signaling pathway of brassinosteroids

9.3.1 Discovery and characterization of the brassinosteroid receptor

9.3.1.1 The discovery of BRI1

To investigate the BR receptor and components of the BR signaling pathway, forward genetic analysis has been used to screen for brassinosteroid-insensitive (*bri*) mutants from EMS-mutagenized M2 seedlings in Arabidopsis. One mutant, termed *bri1*, was reported first by screening the mutants that are unable to respond to exogenously added BL in root elongation analyses (Clouse et al., 1996). The *bri1* mutant shows severe BR-defective phenotypes, including dark green and curled leaves, short petioles, reduced apical dominance, delayed flowering time, and senescence. The *bri1* phenotype was caused by a recessive mutation in a single gene on chromosome IV, but the gene was not initially identified. A subsequent independent screen for BL insensitive (brassinosteroid insensitive, or *bin*) mutants which have *det2*-like phenotypes, was conducted with EMS-mutagenized seedlings (Li and Chory, 1997). The sensitivity to exogenously applied BL was examined, and 18 *bin* mutants were isolated. Interestingly, it was found that all of the 18 *bin* mutants and the original *bri1* mutant (Clouse et al., (1996) are caused by mutation of the same gene, namely *BRI1* (Li and Chory, 1997). Later, more *bri* mutants were isolated by mutant screening, and strikingly, all of these mutants are different *BRI1* alleles (Table 9.1; Friedrichsen et al., 2000; Jiang et al., 2013; Noguchi et al., 1999a).

9.3.1.2 Characterization of BRI1

The *BRI1*gene encodes a leucine-rich repeat (LRR) receptor-like kinase localized on the plasma membrane (Li and Chory, 1997; Friedrichsen et al., 2000). It has an extracellular domain, a transmembrane domain, and a cytoplasmic kinase domain (Fig. 9.4a). The extracellular domain is composed of an N-terminal signal peptide and 25 LRR motifs in which a 70-amino-acid island domain (ID) is located between the

Table 9.1 **Various mutated alleles of BRI1**

Alleles	Background	Mutation sites	Phenotype
bri1-1	Col-0	Ala-909-Thr	Strong
bri1-2/cbb2	Col-0	Ac/Ds Transposon insertion	Strong
bri1-3	Ws-2	4 bp deletion in the kinase domain	Strong
bri1-4	Ws-2	10 bp deletion in the LRR	Strong
bri1-5	Ws-2	Cys-69-Tyr	Weak
bri1-R1/bri1-5R1	Ws-2/*bri1-5*	Gly-87-Glu	*bri1-5* intragenic suppressor
bri1-6/bri1-119	En2	Gly-644-Asp	Weak
bri1-7	Ws-2	Gly-613-Ser	Weak
bri1-8	Ws-2	Arg-983-Gln	Intermediate
bri1-9	Ws-2/Col-0	Ser-662-Phe	Weak
bri1-101	Col-0	Glu-1078-Lys	Strong
bri1-102	Col-0	Thr-750-Ile	Strong
bri1-103,104	Col-0	Ala-1031-Thr	Strong
bri1-105,106, 107	Col-0	Gln-1059-Stop	Strong
bri1-108,109, 110,111,112	Col-0/Ws-2	Arg-983-Gln	Intermediate
bri1-113	Col-0	Gly-611-Glu	Strong
bri1-114, bri1-116	Col-0	Gln-583-Stop	Strong
bri1-115	Col-0	Gly-1048-Asp	Strong
bri1-117,118	Col-0	Asp-1139-Asn	Strong
bri1-120/cp3	Ler	Ser-399-Phe	Weak
bri1-201	*ld-3*/Ws	Gly-611-Arg	Strong, late flowering *ld-3* enhancer
bri1-202	*ld-3*/Ws	Arg-952-Trp	Strong, late flowering *ld-3* enhancer
bri1-301	Col-0	Gly-989-Ile	Weak
bri1-302^D/sud1	*det2-1*	Gly-643-Glu	*det2-1* gain-of-function suppressor
bri1-701	Col-0	T-DNA insertion (SALK_003371)	Strong
salade	Col-0	Transposon insertion, genome deletion (*BRI1* and *WRKY13*)	Strong

21st and 22nd LRR. Binding assays using a biotin-tagged photoaffinity castasterone showed that BL binds directly to the extracellular domain of BRI1 including the ID and the 22nd LRR (Kinoshita et al., 2005). Further, the crystal structure of the extracellular domain of BRI1, confirmed that BRI1 is the receptor of BRs (Fig. 9.4b). In addition, the cytoplasmic domain is divided into three parts: the juxtamembrane domain (JM), the kinase domain (KD), and the C-terminal tail. The cytoplasmic domain of BRI1 is responsible for substrate activation mainly by phosphorylation.

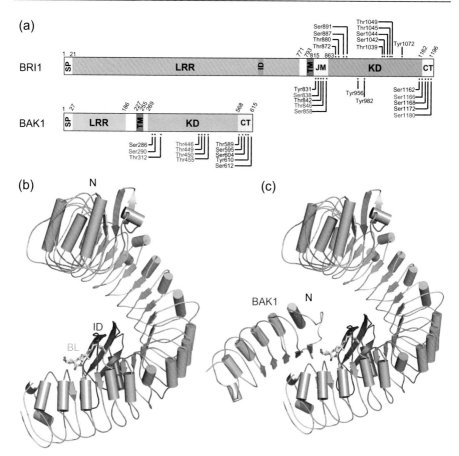

Figure 9.4 The domain and crystal structure of BRI1 and BAK1. (a) The domain structure of BRI1 and BAK1. The number indicates the location of amino acid. * stands for phosphorylation. Red (gray in print version) colored words stand for transphosphorylation sites. (b) The overall structure of brassinolide-bound BRI1 (LRR). Brassinolide colored in yellow and ID domain colored in purple. (c) The overall structure of BRI1LRR-BL-BAK1LRR.
(b and c) Courtesy of Jijie Chai.

9.3.1.3 Phosphorylation of BRI1

The BRI1 protein is a dual-specificity receptor kinase with serine and threonine phosphorylation and tyrosine phosphorylation activities. Identified serine/threonine/tyrosine phosphorylation residues in the cytoplasmic domain are essential for BR signal transduction (Oh et al., 2000, 2009; Wang et al., 2005a,b).

9.3.1.4 The homologs of BRI1

In Arabidopsis, *BRI1* has three homologous *BRI1*-like genes, named *BRL1*, *BRL2*, and *BRL3*. Overexpression of *BRL1* and *BRL3* can partially rescue the phenotype of *bri1-301*, but *BRL2* cannot, indicating that *BRL1* and *BRL3* also are BR receptors

(Cano-Delgado et al., 2004). BRL1 and BRL3 are localized on the plasma membrane and can bind to BL. Crystal structure of the extracellular domain indicates that BRL1 has a higher binding affinity for BL than does BRI1 (She et al., 2013). BRL1 and BRL3 are mainly expressed in the vascular cells to regulate vascular differentiation, which complement the function of BRI1 (Cano-Delgado et al., 2004; Zhou et al., 2004).

9.3.2   The brassinosteroid co-receptor BAK1

9.3.2.1   The discovery of BAK1

After the identification of BR receptor BRI1, researchers tried to identify additional regulatory components in BR signaling. Forward and reverse genetic screening was used to identify the downstream components of BR signaling. Two research groups independently discovered the same co-receptor *bri1*-associated receptor kinase 1 (BAK1) through different approaches. One group identified BAK1 by screening for the genetic suppressors of *bri1-5* using an activation tagging approach (Li et al., 2002); Another group found BAK1 by isolating the interacting proteins of BRI1 using the KD of BRI1 as bait in a yeast two-hybrid screen (Nam and Li, 2002).

9.3.2.2   BAK1 is a co-receptor of BRI1

BAK1 is also a member of the LRR receptor-like kinase II (LRR RLKII) family containing five LRRs in its extracellular domain (Fig. 9.4a). BAK1 localizes on the plasma membrane, and directly interacts with BRI1 through their cytoplasmic domains. The biosynthesis of BRs and the kinase activity of BRI1 are essential for the function of BAK1. The interaction of BRI1 and BAK1 can lead to activation of the kinases of both, and to *trans*-phosphorylation of each other (Fig. 9.4a; Karlova et al., 2009; Li et al., 2002; Nam and Li, 2002; Wang et al., 2014a, 2008). In Arabidopsis BAK1 is also named SOMATIC EMBRYOGENESIS RECEPTOR-like KINASE3 (SERK3). The SERK family contains five members, named AtSERK1, AtSERK2, AtSERK3 (BAK1), AtSERK4 (BKK1), and AtSERK5. Besides of BAK1, AtSERK1, AtSERK2, and AtSERK4 also interact with BRI1 and activate BR signal transduction (Gou et al., 2012; He et al., 2007; Karlova et al., 2006; van Esse et al., 2013). In addition to the indispensable role of SERKs in the BR signaling pathway, SERKs also function in other pathways which are independent of the BR signaling pathway (Chinchilla et al., 2007; Du et al., 2012; He et al., 2007; Heese et al., 2007; Kemmerling et al., 2007; Meng et al., 2015; Roux et al., 2011).

9.3.3   Regulation of the brassinosteroid receptor complex

9.3.3.1   The structure of BRI1 and BAK1

Researchers found that part of BRI1 can form homodimers in vivo (Hink et al., 2008; Russinova et al., 2004; Wang et al., 2005b). However, structural studies indicated that the LRR domain of BRI1 exists as a monomer in vitro. The 25 LRRs packed in tandem assemble into a highly curved solenoid structure. BL binds to a hydrophobicity-dominating surface pocket which is composed of a70-amino-acid ID (Fig. 9.4b; Hothorn et al., 2011; She et al., 2011). As a co-receptor, BAK1 is essential to the activation of early BR signaling. Mutation analysis showed that the extracellular LRR domain of BAK1 provides a platform for BRI1/BAK1 heterodimer formation (Jaillais et al., 2011a). The BRI1/BAK1 heterodimers were also observed in cowpea protoplasts and Arabidopsis root epidermal cells (Bucherl et al., 2013; Russinova et al., 2004). Structural studies indicated that BL can bind with the LRR domain of BAK1 and SERK1 and function as a "molecular glue" to promote the association of BRI1 with BAK1 (Fig. 9.4c) or BRI1 with SERK1 (Santiago et al., 2013; Sun et al., 2013). However, structural studies also found that the cytoplasmic domain of BRI1 and SERKs can form homodimers when they are expressed alone, and the C-terminal region of BRI1 provides a substrate binding platform (Bojar et al., 2014).

9.3.3.2   Inactivation of BRI1

BR binding with the LRR domains of BRI1/BAK1 leads to the phosphorylation of both BRI1 and BAK1, resulting in the activation of the BR signaling cascade. How the BRI1/BAK1 receptor complex is kept silent at low BR level is an important aspect of the regulation of BR signal transduction. Firstly, it was found that the kinase activity of BRI1 is self-regulated. The C-terminal tail (residues 1156–1196) can inhibit the kinase activity of BRI1, and BR binding with BRI1 leads to phosphorylation of the C terminus, releasing its inhibitory effect on BRI1 activity (Wang et al., 2005b).

Secondly, the activity of BRI1 is inhibited by a negative regulator, BRI1 kinase inhibitor 1(BKI1). BKI1 was identified as a BRI1 KD interacting protein by yeast two-hybrid assay (Wang and Chory, 2006). BKI1 is localized to the plasma membrane by its N-terminal lysine/arginine-rich membrane targeting motif. The unphosphorylated BKI1 can interact with BRI1 and prevent the interaction of BAK1 with BRI1, to inhibit the activation of BR signaling (Jiang et al., 2015a; Wang and Chory, 2006). When BRs bind to BRI1, BKI1 is phosphorylated and then dissociates from the plasma membrane to release its inhibition of BRI1 (Fig. 9.5; Jaillais et al., 2011b; Wang et al., 2011). The phosphorylating sites and interaction motif of BKI1 are independent from each other (Wang et al., 2011). A 20-amino-acid BRI1-interacting motif (BIM) region (residues 306–325) at the C terminus of BKI1 is enough to bind to the KD of BRI1

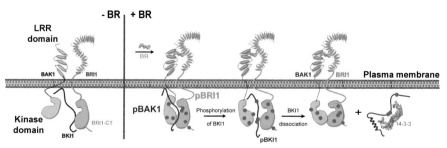

Figure 9.5 The current model for BRI1 receptor complex inhibition and activation. Orange dots indicate phosphorylation.

(Jaillais et al., 2011b; Wang et al., 2014a). Structural analysis of BIM and BRI1-KD showed that BIM folds into a four-turn α-helix and binds to αH and αG in the C-lobe of BRI1-KD (Wang et al., 2014a). The residues S270 and S274 sites are phosphorylated by BRI1, and the S270/S274 phosphorylated BKI1 can dissociate from the plasma membrane with the help of 14-3-3 proteins (Wang et al., 2011).The phosphorylated BKI1 also can compete for 14-3-3 proteins with BRI1 EMS SUPPRESSOR1 (BES1) and BRASSINAZOLE-RESISTANT 1 (BZR1) proteins, relieving14-3-3 inhibition of the phosphorylated BES1 or BZR1 proteins to rapidly enhance BR signaling.

In addition, the endocytosis of BRI1 can attenuate BR signaling. The endocytosed BRI1 is localized to the early endosomal compartment, mediated by the microdomain at the plasma membrane (Geldner et al., 2007; Irani et al., 2012; Wang et al., 2015), ADAPTOR PROTEIN COMPLEX-2 (AP-2; Di Rubbo et al., 2013), and ubiquitination (Martins et al., 2015). Impaired BRI1 endocytosis can enhance BR signaling (Di Rubbo et al., 2013; Russinova et al., 2004), but the endocytosis of BRI1 is BR-independent. However, whether the regulation of BRI1 endocytosis is critical to keep BRI1 quiet needs further study.

9.3.4 GSK3: A negative regulator in brassinosteroid signal transduction

Glycogen synthase kinase 3 (GSK3) family proteins are key downstream negative regulators of the BR signaling pathway. GSK3 proteins, also known as SHAGGY protein kinases in *Drosophila*, have 10 members in Arabidopsis, which are termed shaggy kinases (AtSK or ASK). Based on sequence alignments, they are classified into four subgroups (group I: AtSK11/ASKα, AtSK12/ASKγ, AtSK13/ASKε; group II: AtSK21/ASKη/BIN2/UCU1, AtSK22/ASKι/BIL1/AtGSK1, AtSK23/ASKζ/BIL2; group III: AtSK31/ASKθ, AtSK32/ASKβ; group IV: AtSK41/ASKκ/AtK-1, AtSK42/ ASKδ). Current knowledge indicates that all members except AtSK32, AtSK41, and AtSK42, play a role in BR signaling (Saidi et al., 2012; Youn and Kim, 2014).

9.3.4.1 The function of BIN2

BIN2 is the first reported GSK3 protein involved in BR signaling. Screening for dwarf and semidwarf mutants from the EMS mutagenesis population which is used for *bri1* mutant screening found two BR-insensitive 2 (*bin2*) mutants (Li et al., 2001). Cloning of genes from the *bin2* mutants found mutations in a TREE domain (E-K in *bin2-1*, and T-I in *bin2-2*), which lead to gain-of-function mutations (Li and Nam, 2002). Two other laboratories independently also found *bin2* mutants, named *dwarf12*, and *ucu1* (Choe et al., 2002; Perez-Perez et al., 2002). Interestingly, *dwarf12* and *ucu1* mutants were also caused by mutations in the TREE domain, indicating that this domain is critical for the function of BIN2. BIN2 interacts with and phosphorylates downstream transcription factors BES1 and BZR1 to inhibit BR signaling (He et al., 2002; Wang et al., 2002; Yin et al., 2002). A 12-amino-acid motif in the C terminus of BZR1 is the interacting region between BIN2 and BZR1, and this motif is enough for the phosphorylation of BZR1 by BIN2 (Peng et al., 2010). The phosphorylated BES1 and

BZR1 proteins have low DNA binding activity, leading to the inhibition of BR signaling (Vert and Chory, 2006).

9.3.4.2 Regulation of BIN2

There are several ways in which the activity of BIN2 is regulated. First, BR binding to the BRI1/BAK1 complex leads to phosphorylation and activation of the plasma membrane localized BR signaling kinase (BSK) and constitutive differential growth (CDG) proteins, members of receptor-like cytoplasmic kinase subfamily receptor-like cytoplasmic kinase (RLCK)-XII, and subsequent activation of phosphatase BRI1 suppressor 1 (BSU1) to dephosphorylate BIN2 (Kim et al., 2011, 2009). Second, BR early signaling can cause rapid BIN2 degradation by the 26S proteasome by an unknown mechanism (Peng et al., 2008). Third, BRs can induce the plasma membrane localization of BIN2 by OCTOPUS (OPS) to prevent the inhibitory effect of BIN2 on BES1 and BZR1 in the nucleus (Anne et al., 2015). However, OPS is mainly expressed in phloem, and the regulation of BIN2 by OPS may just occur in the phloem to promote its differentiation. Furthermore, HSP90 interacts with BIN2 to keep BIN2 in the nucleus. BRs lead to trafficking of the BIN2-HSP90 complex into the cytoplasm to promote BR signaling output (Samakovli et al., 2014).

9.3.5 Downstream transcription factors BES1/BZR1 regulate brassinosteroid-responsive gene expression

9.3.5.1 The discovery of BES1 and BZR1

BES1 and BZR1 are the major downstream transcription factors in the BR signaling pathway. They were identified by two independent genetic screens, and they share 88% protein homology. The *bzr1-1D* mutant was identified by screening for brassinazole-insensitive mutants from a EMS mutation population (Wang et al., 2002). The *bes1-D* mutant, also named *bzr2*, was identified from suppressors of *bri1-119* (Yin et al., 2002). Both *bzr1-1D* and *bes1-D* are dominant mutants, caused by a mutation in PEST domain which mediates protein degradation. The mutation of P233L in *bes1-D* and P234L in *bzr1-1D* can stabilize BES1 and BZR1 proteins, leading to their accumulation to enhance BR signaling.

9.3.5.2 The function of BES1 and BZR1

BES1 and BZR1 can directly bind to many primary BR-responsive genes at E-box (CANNTG) and BRRE (BR response element, CGTGT/CG) motifs, respectively, to promote or repress gene expression (He et al., 2005; Vert and Chory, 2006). Chromatin-immunoprecipitation microarray (ChIP-chip) analysis shows that BZR1 binds to more than 950 BR-regulated target genes (Sun et al., 2010) and BES1 binds to more than 1600 putative target genes (Yu et al., 2011). BES1/BZR1 1-4 (BEH1-4) are homologs of BES1 and BZR1, and all function redundantly in BR signaling (Yin et al., 2005).

9.3.5.3   The regulation of BES1 and BZR1

The activity of BES1 and BZR1 is regulated by several different ways. First, BES1 and BZR1 are phosphorylated by BIN2, and the phosphorylated BES1 can be degraded by the 26S proteosome. BL treatments can rapidly induce accumulation of the dephosphorylated BES1 and BZR1 to promote BR signal outputs (Wang et al., 2002; Yin et al., 2002; Zhao et al., 2002). The phosphorylation status of BES1 has been widely used as a key biochemical marker to evaluate BR signaling output. Second, the 14-3-3 proteins interact with phosphorylated BES1 and BZR1 to maintain their phosphorylation and retain them in the cytoplasm, thus reducing their function in the nucleus (Gampala et al., 2007; Ryu et al., 2007). Third, BRZ-SENSITIVE-SHORT HYPOCOTYL1 (BSS1), a BTB-POZ domain protein, interacts with the phosphorylated BES1 and BZR1 proteins to keep the BSS1/BZR1 and BSS1/BES1 complexes located in the cytoplasm to inhibit BR signaling. BL induces the disassembly of the complex and nuclear localization of BES1 and BZR1 (Shimada et al., 2015). Fourth, the B' regulatory subunits of phosphoprotein phosphatase PP2A interact with the PEST domain of BZR1 to dephosphorylate BZR1 to promote BR signaling (Tang et al., 2011). In addition, plants overexpressing BES1 showed no obvious phenotypes as compared with the wild-type plants. A report found that the *BES1* gene encodes another form of BES1, 22 amino acids longer than BES1 (BES1-L). Overexpression of *BES1-L* showed an obvious increased BR signaling phenotype (Jiang et al., 2015b). BES1-L can induce the nuclear localization of BES1-S (the canonical short BES1) to enhance BR signaling.

9.3.5.4   BES1 interacting proteins

BES1 also interacts with other proteins to coordinately regulate BR-responsive gene expression to modulate plant growth and development. The BES1-INTERACTING MYC-LIKE 1(BIM1) protein and AtMYB30 are well-known positive transcription factors, which interact with BES1 and co-bind to BR target gene promoters to activate BR signaling (Li et al., 2009; Yin et al., 2005). The expression of *AtMYB30* is also directly regulated by BES1. Furthermore, MYELOBLASTOSIS FAMILY TRANSCRIPTION FACTOR-LIKE 2 (MYBL2) and HOMEODOMAIN LEUCINE ZIPPER PROTEIN1 (HAT1) are negative transcription factors. They interact with BES1 to downregulate BR-repressed gene expression. BIN2 phosphorylates MYBL2 and HAT1 to stabilize them (Ye et al., 2012; Zhang et al., 2014).

    In addition, the transcription regulators, including *ARABIDOPSIS THALIANA* INTERACT-WITH-SPT6 (AtIWS1), SET DOMAIN GROUP 8 (SDG8), EARLY FLOWERING 6 (ELF6), and RELATIVE OF EARLY FLOWERING 6 (REF6), can also interact with BES1 and may be involved in BR signaling. AtIWS1, a conserved protein involved in RNA polymerase II post-recruitment and the transcriptional elongation processes, interacts with BES1 to enhance the transcriptional activation of the BR-induced genes (Li et al., 2010). SDG8 is a histone lysine methyl transferase, which targets H3K36 di- and tri-methylation modifications to activate gene expression. SDG8 interacts with BES1to co-regulate BR-responsive gene expression (Wang et al., 2014b). SDG8 also interacts with AtIWS1, indicating that BES1 may recruit these transcription regulators implicated in chromatin modifications to further regulate BR-responsive gene expression. ELF6 and its homolog REF6, Jumonji N/C domain-containing proteins, can inhibit H3K9me3 modification. BES1 interacts with ELF6 and REF6 to positively coordinate BR-responses and other developmental processes (Yu et al., 2008).

9.3.5.5   BR signaling pathway in other plants

Compared to Arabidopsis, the BR signaling pathway is not well characterized in other plant species. The BR signaling pathway has been studied in rice, tomato, wheat, pea, and barley. An expanding number of components of BR signaling in rice have been identified. They include OsBRI1, OsBAK1, OsBSK3, OsBZR1, Os14-3-3 proteins, and GSK2. In addition, some rice-specific components have been found, for example, DWARF AND LOW-TILLERING (OsDLT) and LEAF AND TILLER ANGLE INCREASED CONTROLLER (OsLIC). In general, the BR signaling pathway is highly conserved in higher plants, including dicot and monocot species.

9.3.5.6   The current model of BR signaling

Based on this progress, we describe a model for the BR signaling pathway (Fig. 9.6). When BR level is low, BR signal transduction is repressed at multiple levels. In the upstream steps, BRI1 activity is not only *cis*-inhibited by its C-terminal tail and by phosphorylation, but is also *trans*-inhibited by the negative regulator BKI1 to prevent the interaction of BRI1 with BAK1. In the downstream steps, the GSK3 family proteins phosphorylate transcription factors BES1 and BZR1 to reduce their DNA binding activity. Furthermore, 14-3-3 and BSS1 proteins interact with the phosphorylated BES1 and BZR1 proteins to maintain them in the cytoplasm. When BR level is increased, BRs bind to BRI1 and BAK1, the C-terminal autoinhibition of BRI1 is released by phosphorylation, and BKI1 is phosphorylated by BRI1 to dissociate it from the plasma membrane. BRI1 and BAK1 transphosphorylate each other to form an active receptor complex, and the activated BRI1 phosphorylates BSKs and CDGs to promote the phosphorylation of the phosphatase BSU1 family. The phosphorylated BSU1 is activated leading to dephosphorylation and degradation of BIN2. Furthermore, PP2A dephosphorylates BES1 and BZR1, which then bind to the promoter regions of many BR-responsive genes together with other transcription factors to regulate plant growth and development. In addition, the phosphorylated BKI1 mediates a rapid BR signaling pathway by interacting with 14-3-3 proteins, releasing them from BES1 proteins.

9.4   Roles of brassinosteroids in physiology and development

9.4.1   The diversity of responses to brassinosteroid

The dwarf stature of the BR-deficient mutants and the specific activity of BRs to regulate plant development demonstrate that BRs are essential for normal plant growth and development. In light, the BR mutants show a number of altered

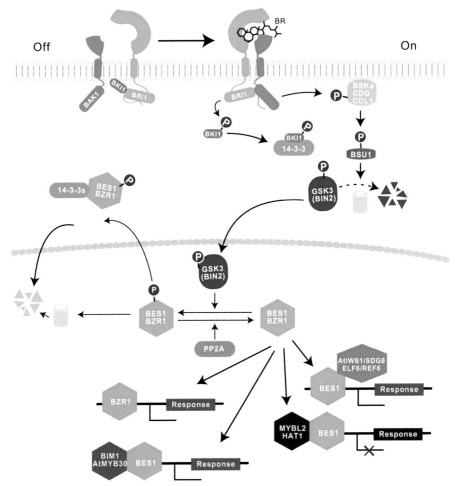

Figure 9.6 The current model of BR signaling pathway in Arabidopsis.

developmental phenotypes, including dwarfed inflorescence, dark green and
rounded leaves, prolonged life-span, reduced fertility, altered vascular development, and delayed flowering and senescence (Chory et al., 1991; Li and Chory,
1997). In the dark, BR mutants exhibit some characteristics of light-grown plants,
such as shortened hypocotyls, de-etiolated and open cotyledons (Chory et al.,
1991; Li and Chory, 1997). Analysis of BR biosynthetic and signaling mutants
coupled with the effects of the exogenously applied BRs on plant growth and
development have uncovered the physiological roles of BRs in many aspects of
plant developmental and environmental responses, including cell elongation, cell
division, root growth and development, vascular differentiation, reproductive
growth and development, senescence, and responses to biotic and abiotic stresses
(Clouse and Sasse, 1998).

9.4.2   Cell elongation and cell division

Electron microscopy observation of wild-type Arabidopsis and BR mutants provides direct evidence that cell length and cell number are greatly reduced in BR mutants (Nakaya et al., 2002). Studies found that BIN2 can phosphorylate PIF4, as well as BZR1, leading to their degradation by the 26S proteasome (Bernardo-Garcia et al., 2014; He et al., 2002). BR signal transduction leads to inactivation of BIN2, releasing BZR1 and PIF4 which interact with each other to coordinately regulate the expression of numerous genes involved in cell-wall loosening to promote cell elongation (Oh et al., 2012). In addition, a rapid BL-regulated signaling pathway within the plasma membrane that links BRI1 with P-ATPase for the regulation of cell wall expansion has been discovered (Caesar et al., 2011). BRs can substitute for cytokinins in the culture of Arabidopsis callus and suspension cells to induce transcription of the *CycD3* gene, encoding a D-type plant cyclin through which cytokinin activates cell division (Hu et al., 2000). Interestingly, in rice, BRs inhibit abaxial sclerenchyma cell division in the leaf lamina joints through OsGSK2 and OsBZR1 (Sun et al., 2015). OsGSK2 can interact and phosphorylate OsCYC U4;1 to promote cell proliferation, leading to leaf erectness. In addition, OsBZR1 can also bind to the promoter region of *OsCYC U4;1* to inhibit its expression.

9.4.3   Root growth and development

Both the loss-of-function and the gain-of-function BR mutants display a significantly shortened root phenotype, suggesting that an appropriate level of BR signaling is important for optimal root growth and development (Clouse et al., 1996; Gonzalez-Garcia et al., 2011; Li et al., 2002). Physiological assays found that a very low concentration of BL promotes root growth, whereas a high concentration of BL inhibits root growth by regulating root meristem size (Gonzalez-Garcia et al., 2011). The behavior of the cell division markers *pCYCB1;1*, *pICK2/KRP2* (interactor of Cdc2 kinase 1/Kip-related protein 2), and *KNOLLE* (a cytokinesis-specific syntaxin), together with the recapitulation of the *bri1-116* meristem defects by overexpression of *CYCD3;1*, revealed that BRs play a regulatory role in the control of cell cycle progression and of differentiation in the Arabidopsis root meristem (Gonzalez-Garcia et al., 2011). A study found that BRs affect root cell elongation depending on the cells responding to BR (Fridman et al., 2014). Ectopic expression of *BRI1* in hair cells promotes the elongation of all different cell types within the root elongation zone, whereas expression of *BRI1* in nonhair cells inhibits root cell elongation, suggesting that the spatial distribution rather than an absolute level of BRI1 determines root cell elongation (Fridman et al., 2014). The effects of BRs on root hair formation, lateral root development, root response to gravity, nodulation, and mycorrhiza formation in different plant species have also been reported (Bao et al., 2004; Bitterlich et al., 2014; Cheng et al., 2014; Ferguson et al., 2005; Kim et al., 2000, 2007; Kuppusamy et al., 2009; Li et al., 2005; Terakado et al., 2005).

9.4.4   Vascular development

Physiological studies using cell cultures of *Zinnia elegans* showed that BRs play an important role in promoting xylem differentiation (Fukuda, 1997). In Arabidopsis, the number of vascular bundles was reduced in BR-deficient mutants compared to those of wild-type, and the spacing between the vascular bundles in BR mutants is irregular (Choe et al., 1999b; Szekeres et al., 1996). Histological analysis of the vasculature in *brl1* showed an increase of the number of phloem cells but a decrease of xylem cells, indicating that BRL1-mediated signaling promotes xylem differentiation and represses phloem differentiation in the inflorescence stem in Arabidopsis (Cano-Delgado et al., 2004).

9.4.5   Reproductive development

Mutants impaired in BR biosynthesis or signaling, such as *dwf4*, *cpd*, and *bri1-5*, all showed a delayed flowering time (Choe et al., 1998; Noguchi et al., 1999a; Szekeres et al., 1996), suggesting an important role for BRs in regulating flowering. It has been found that BR signaling promotes flowering by downregulating the expression of *FLC* (Flowering Locus C), a repressor of flowering (Domagalska et al., 2007). REF6, a transcription regulator repressing *FLC* expression, has been shown to interact with BES1 (Yu et al., 2008). It is possible that the BES1-REF6 dimer mediates the BR signaling to regulate *FLC* expression and hence flowering time.

BRs are essential for male fertility since pollen is a rich source of endogenous BRs. Defective male fertility is a common feature of BR mutants. Reduced length of stamen filaments in BR mutants results in the deposition of pollen on the ovary wall rather than on the stigmatic surface (Azpiroz et al., 1998). More importantly, BR mutants show reduced pollen number, viability, and efficiency of release (Ye et al., 2010). It was found that BES1 could directly bind to the promoter regions of genes encoding transcription factors essential for anther and pollen development, such as *SPOROCYTELESS/NOZZLE* (*SPL/NZZ*), defective in *Tapetal Development and Function 1* (*TDF1*), *ABORTED MICROSPORES* (*AMS*), *MALE STERILITY 1* (*MS1*), and *MS2*.

BRs also play a critical role in ovule initiation and development. In BR-deficient and BR-insensitive mutants, the siliques are shorter and thinner, whereas the siliques are longer and thicker in the BR signaling-enhanced mutant *bzr1-1D*. Detailed observation showed that ovule and seed numbers are significantly reduced in BR mutants. It has been found that BR signaling regulates expression of genes involved in ovule and seed development, such as *HUELLENLOS* (*HLL*), *AINTEGUMENTA* (*ANT*), and *APETALA2* (*AP2*) (Huang et al., 2013).

9.4.6   Stomata development

Stomata are important organs for gas exchange between the plant and the atmosphere. In BR-deficient mutants, clustered stomata can be observed, but stomata are always distributed with at least one pavement cell between them. BL treatment can obviously reduce stomatal density. BR negatively regulates stomatal development through BIN2 at several different levels. BIN2 can interact with and phosphorylate mitogen-activated protein kinase kinase kinase (MAPKKK), YDA, to inhibit the phosphorylation of the YDA substrate, MAPK kinase 4 (MKK4) (Kim et al., 2012). BIN2 also interacts with and phosphorylates the transcription factor SPEECHLESS (SPCH) to inactivate SPCH by an unknown mechanism (Gudesblat et al., 2012). Furthermore, BIN2 can interact with and phosphorylate MKK4 and MKK5 to reduce their activity toward their substrate MPK6 (Khan et al., 2013).

9.5 Cross talk of brassinosteroids and other signals

9.5.1 Complexity of brassinosteroid cross talk

BRs interact with multiple phytohormones and environmental signals, such as auxin, abscisic acid (ABA), ethylene (ET), cytokinin (CK), jasmonic acid (JA), gibberellin (GA), salicylic acid (SA), and light (Nemhauser et al., 2006; Yang et al., 2011; Zhang et al., 2009b). BRs and other hormones or environmental cues coordinately regulate plant growth, development, and responses to biotic and abiotic stresses by co-regulating the expression of hundreds of genes or through direct interaction of the primary signaling components of their signaling pathways.

9.5.2 Brassinosteroids and abscisic acid

The mechanisms of the cross talks between BR and ABA signaling pathways have been intensively investigated. BRs and ABA antagonistically regulate plant development, for example, seed germination and responses to abiotic stresses. It was reported that the BR biosynthesis mutant *det2-1* and BR signaling mutants *bri1* and *bin2-1* are more sensitive to ABA, indicating that BRs inhibit ABA signaling. BRs likely inhibit ABA signaling at different levels (Fig. 9.7). BAK1 interacts with and phosphorylates SnRK2.6 (OST1) to enhance ABA signaling, which is inhibited by additional BL (Shang et al., 2016). GSK3 proteins (especially BIN2, BIL1 and BIL2) can interact with and phosphorylate SnRK2 or ABI5 proteins to promote ABA signaling (Cai et al., 2014; Hu and Yu, 2014). Finally, BR signaling can activate BES1 to form a transcriptional repressor complex with TOPLESS-HISTONE DEACETYLASE 19 (TPL-HDA19) and inhibit ABA signaling by deacetylating *ABI3* chromatin (Ryu et al., 2014). ABA may also inhibit BR signaling at the protein level via interactions between the receptor BRII complexes and the GSK3 kinase BIN2 (Yang et al., 2014; Zhang et al., 2009a). Under abiotic stress, the antagonistic interactions between BR and ABA in plant development may help the plant to respond to a hostile environment.

9.5.3 Brassinosteroids and auxin

BRs and auxin both are growth-promoting hormones, and act synergistically to regulate physiological and developmental processes (Nemhauser et al., 2004). BRs increase seedling sensitivity to auxin via the interaction of BIN2 with AUXIN-RESPONSE FACTOR 2 (ARF2), which is phosphorylated by BIN2 (Vert et al., 2008). BRs may regulate auxin polar transport by positively regulating the actin cytoskeleton (Lanza et al., 2012). Auxin promotes the expression of *DWF4* to stimulate BR biosynthesis, independently of the BR primary signaling pathway (Chung et al., 2011). In rice, some ARF proteins regulate

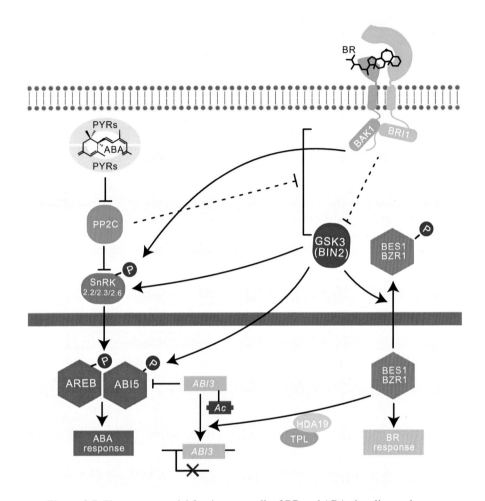

Figure 9.7 The current model for the cross talk of BR and ABA signaling pathways.

the expression of *OsBRI1* by binding to its promoter at an auxin-responsive element (Sakamoto and Fujioka, 2013). However it was reported that BRs and auxin maintain the spatiotemporal balance between stem cell maintenance and differentiation in the root meristems by an antagonistic action, which may be controlled by the transcription factor BZR1 (Chaiwanon and Wang, 2015). Notably, under low-blue light which induces shade avoidance, BRs and auxin work in a nonredundant and non-synergistic manner to promote cell elongation in Arabidopsis (Keuskamp et al., 2011).

9.5.4 Brassinosteroids and gibberellins

BRs and GAs act synergistically to promote hypocotyl elongation and skotomorphogenic developmental programs. A defect in either of these hormones leads to reduced plant growth and dwarfism. BRs and GAs also interact at multiple levels.

BRs act downstream of GAs in the etiolated Arabidopsis seedlings. The negative regulator DELLA proteins in the GA signaling pathway interact with transcription factor BZR1 to inhibit the binding of BZR1 to its promoters (Gallego-Bartolome et al., 2012). BRs can also increase the accumulation of DELLAs to increase the expression of the BR-induced genes *GID1a, GID1b, bHLH137*, and *XERICO* by a mechanism which is unclear (Stewart Lilley et al., 2013). Studies in rice showed that BRs regulate GA biosynthesis by BZR1 directly binding to the promoters of GA biosynthetic genes (Tong et al., 2014). In addition, studies in Arabidopsis showed that BRs regulate GA biosynthesis by BES1 binding to the non-E-box motifs of the promoters of GA biosynthetic genes (Unterholzner et al., 2015).

9.5.5   Brassinosteroids and light

Light and BRs antagonistically regulate the developmental switch from etiolation in the dark to photomorphogenesis in the light at multiple levels (Oh et al., 2014; Song et al., 2009). BRs repress *GATA-type transcription factor* (*GATA2*) transcription through the transcription factor BZR1, whereas light causes accumulation of GATA2 proteins (Luo et al., 2010). BRs inhibit light-regulated hypocotyl elongation by BIN2 interacting with and phosphorylating PIF4 to mark PIF4 for proteasome-mediated degradation (Bernardo-Garcia et al., 2014). Although BZR1 and PIF4 directly interact in vitro and in vivo, and co-regulate nearly 2000 target genes, their interaction is independent in the promotion of cell elongation in response to BRs (Oh et al., 2012). In contrast, light signaling acts via HY5 to inhibit BR signaling on regulating cotyledon development. HY5 specifically interacts with the dephosphorylated form of BZR1 and attenuates the transcriptional activity of BZR1toward its target genes, to promote cotyledon opening (Li and He, 2016).

9.5.6   Brassinosteroids and other hormones or signals

The interactions of BRs with ET, JA, SA, sugar, and temperature have also been reported, but the underlying mechanisms are not well studied. BRs and ET interact to regulate different aspects of plant growth and development, including gravitropic reorientation, cell expansion, hyponastic growth, and stomatal closure. BRs and ET antagonize each other to regulate gravitropic reorientation and cell expansion, but they act synergistically to regulate hyponastic growth and stomatal closure. FERONIA kinase may be a bridge between BRs and ET acting on cell expansion (Deslauriers and Larsen, 2010), but, there is no indication that FERONIA is a component of the ET or BR signaling pathways. BRs can induce ET biosynthesis to close stomata and induce hyponastic growth, but the mechanism is unknown (Shi et al., 2015).

The interaction of BRs and JA plays a crucial role in plant development, and in biotic and abiotic responses. BRs attenuate the inhibition of root growth by JA, by controlling unknown components downstream CORONATINE INSENSITIVE 1 (COI1) (Huang et al., 2010; Ren et al., 2009). BRs can reduce JA-induced anthocyanin biosynthesis by regulating anthocyanin biosynthetic gene expression (Peng et al., 2011). BR and JA directly affect trichome density and allelochemical content but with opposite effects, such that JA promotes while BRs inhibit these processes (Campos et al., 2009). The antagonistic interaction between BR and JA in rice roots has also been reported. However, the mechanism of interaction still needs to be further studied. BRs and SA coordinately regulate responses to stresses. BR promotes seedling survival under salt stress in a NPR1-dependent manner (Divi et al., 2010). Although it is still unknown whether BRs interact with SL signaling, it was reported that BR transcription factor BES1 interacts with MAX2 to regulate plant branching (Wang et al., 2013). However, it is unknown whether the MAX2-mediated degradation of BES1 has an effect on BR signaling. In addition, sugar signaling promotes the accumulation of the phosphorylated BZR1 through TARGET OF RAPAMYCIN (TOR) to balance the BR-induced plant growth and sugar availability (Zhang et al., 2016).

9.6   Summary points

- The BR biosynthetic pathway contains a general cycloartenol-to-campesterol sterol biosynthesis pathway and a specific campesterol-to-brassinolide (BL) biosynthesis pathway. The campesterol converts to BL via campestanol-dependent and CN-independent pathways.
- BRs are perceived by a plasma membrane-located receptor-like kinase, BRI1. BRI1 and co-receptor BAK1 are activated by *trans*-phosphorylation, then phosphorylate their substrates, BSKs and CDG1 proteins.
- BKI1 and BIN2 are important negative regulators in the BR signaling pathway. BKI1 inhibits the interaction of BRI1 with BAK1, and so inhibits their activation. BIN2 phosphorylates key transcription factors BES1 and BZR1 to inhibit BR signaling outputs. BR binding to BRI1 can induce BKI1 phosphorylation and its dissociation from the plasma membrane to the cytosol. The activated BRI1 complex can inhibit BIN2 kinase activity through the activity of phosphatase BSU1, relieving its inhibitory effect on BES1 and BZR1 to activate BR-induced gene expression.
- BRs control almost all aspects of plant growth and development, including cell elongation, cell division, root growth and development, vascular differentiation, reproductive growth and development, senescence, and responses to biotic and abiotic stresses.
- BRs interact with other hormonal or environmental signals at different levels to coordinate plant growth and development and responses to stresses.

9.7   Future perspectives

- As a key negative regulator in the BR signaling, BIN2 plays a hub role in cross talk with many other signaling pathways. The regulation of BIN2 remains to be investigated.
- The underlying molecular and genetic mechanisms in the cross talk between BR signaling and many other hormonal or environmental signaling pathways needs to be further studied in the future.
- The cell-specific or tissue-specific functions and impacts of BR biosynthesis and signaling remain to be investigated.
- The biosynthetic and signaling pathways of BRs have been extensively studied, but the application of these components in agricultural science is rare. In the future, we should find ways to apply BR-related components in crop science and agriculture.

Abbreviations

BR	Brassinosteroid
BL	Brassinolide
24-epiBL	24-Epibrassinolide
SQE1	SQUALENE EPOXIDASE 1
CAS1	CYCLOARTENOL SYNTHASE 1
SMT1	STEROL METHYLTRANSFERASE 1
FK	FACKEL
HYD1	HYDRA
DWF	DWARF
DET2	DE-ETIOLATED 2
ROT3	ROTUNDIFOLIA3
CPD	Constitutive photomorphogenic dwarf
DDWF1	DARK-INDUCED DWF-LIKE PROTEIN 1
TCP1	TEOSINTE BRANCHED1/CYCLOIDEA/PROLIFERATING CELL FACTOR1
RAVL1	RELATED TO ABI3/VP1, ABA INSENSITIVE 3/VIVAPARIOUS 1
BRZ	Brassinazole
Pcz	Propiconazole
BNST3	*BRASSICA NAPUS* SULFOTRANSFERASE 3
BAS1	PHYB ACTIVATION-TAGGED SUPPRESSOR 1
DRL1	DWARF AND ROUND LEAF-1
bri	Brassinosteroid-insensitive
bin1	Brassinosteroid-insensitive 1
LRR	Leucine-rich repeat
ID	Island domain
JM	Juxtamembrane
KD	Kinase domain
BAK1	*bri1*-associated receptor kinase 1
SERK	SOMATIC EMBRYOGENESIS RECEPTOR-like KINASE
BKI1	BRI1 kinase inhibitor 1
BIM	BRI1 interacting motif
AP-2	ADAPTOR PROTEIN COMPLEX-2
GSK3	Glycogen synthase kinase 3
AtSK or ASK	Shaggy kinase
bin2	BR-insensitive 2
OPS	OCTOPUS
bzr1-1D	Brassinazole-resistant 1-1D
bes1-D	bri1-EMS-suppressor 1
ChIP–chip	Chromatin-immunoprecipitation microarray
BIM1	BES1-interacting Myc-like 1
MYBL2	Myeloblastosis family transcription factor-like 2
HAT1	Homeodomain leucine zipper protein 1
AtIWS1	*Arabidopsis thaliana* Interact-With-Spt6
SDG8	SET Domain Group 8
ELF6	Early flowering 6

REF6	Relative of early flowering 6
OsDLT	DWARF AND LOW-TILLERING
OsLIC	Leaf and tiller angle increased controller
ABA	Abscisic acid
ET	Ethylene
CK	Cytokinin
JA	Jasmonic acid
GA	Gibberellin
SA	Salicylic acid
ARF2	Auxin-response factor 2
COI1	CORONATINE INSENSITIVE 1
TOR	Target of rapamycin

Acknowledgments

This work is supported by grant 91535104(to X.W.), 31271300 (to X.W.), 31470380 (to J.L.), 31530005 (to J.L.), and 31401032 (to H.W.) of the National Natural Science Foundation of China, fundamental research funds for the central universities lzujbky-2016-bt05 (to J.L. and Z.W.) of Lanzhou University, initiative grants 2662015PY020 (to X.W.), 2014RC002 (to X. W.), and 2662014PY068 (to H.W.) of Huazhong Agricultural University, and general financial grant 2016M602889 (to Z.W.) of China Postdoctoral Science Foundation. We thank J. Jiang and C. Liu for help in drawing Figures.

References

Anne, P., Azzopardi, M., Gissot, L., et al., 2015. OCTOPUS negatively regulates BIN2 to control phloem differentiation in *Arabidopsis thaliana*. Curr. Biol. 25, 2584–2590.

Asami, T., Yoshida, S., 1999. Brassinosteroid biosynthesis inhibitors. Trends Plant Sci. 4, 348–353.

Azpiroz, R., Wu, Y., LoCascio, J.C., et al., 1998. An *Arabidopsis* brassinosteroid-dependent mutant is blocked in cell elongation. Plant Cell 10, 219–230.

Babiychuk, E., Bouvier-Nave, P., Compagnon, V., et al., 2008. Allelic mutant series reveal distinct functions for *Arabidopsis* cycloartenol synthase 1 in cell viability and plastid biogenesis. Proc. Natl. Acad. Sci. U.S.A. 105, 3163–3168.

Bao, F., Shen, J., Brady, S.R., et al., 2004. Brassinosteroids interact with auxin to promote lateral root development in *Arabidopsis*. Plant Physiol. 134, 1624–1631.

Bernardo-Garcia, S., de Lucas, M., Martinez, C., et al., 2014. BR-dependent phosphorylation modulates PIF4 transcriptional activity and shapes diurnal hypocotyl growth. Genes Dev. 28, 1681–1694.

Bishop, G.J., Nomura, T., Yokota, T., et al., 1999. The tomato DWARF enzyme catalyses C-6 oxidation in brassinosteroid biosynthesis. Proc. Natl. Acad. Sci. U.S.A. 96, 1761–1766.

Bitterlich, M., Krugel, U., Boldt-Burisch, K., et al., 2014. The sucrose transporter SlSUT2 from tomato interacts with brassinosteroid functioning and affects arbuscular mycorrhiza formation. Plant J. 78, 877–889.

Bojar, D., Martinez, J., Santiago, J., et al., 2014. Crystal structures of the phosphorylated BRI1 kinase domain and implications for brassinosteroid signal initiation. Plant J. 78, 31–43.

Bucherl, C.A., van Esse, G.W., Kruis, A., et al., 2013. Visualization of BRI1 and BAK1(SERK3) membrane receptor heterooligomers during brassinosteroid signaling. Plant Physiol. 162, 1911–1925.

Caesar, K., Elgass, K., Chen, Z., et al., 2011. A fast brassinolide-regulated response pathway in the plasma membrane of *Arabidopsis thaliana*. Plant J. 66, 528–540.

Cai, Z., Liu, J., Wang, H., Yang, C., et al., 2014. GSK3-like kinases positively modulate abscisic acid signaling through phosphorylating subgroup III SnRK2s in *Arabidopsis*. Proc. Natl. Acad. Sci. U.S.A. 111, 9651–9656.

Campos, M.L., de Almeida, M., Rossi, M.L., et al., 2009. Brassinosteroids interact negatively with jasmonates in the formation of anti-herbivory traits in tomato. J. Exp. Bot. 60, 4347–4361.

Cano-Delgado, A., Yin, Y., Yu, C., et al., 2004. BRL1 and BRL3 are novel brassinosteroid receptors that function in vascular differentiation in *Arabidopsis*. Development 131, 5341–5351.

Carland, F., Fujioka, S., Nelson, T., 2010. The sterol methyltransferases SMT1, SMT2, and SMT3 influence *Arabidopsis* development through nonbrassinosteroid products. Plant Physiol. 153, 741–756.

Carland, F.M., Fujioka, S., Takatsuto, S., et al., 2002. The identification of CVP1 reveals a role for sterols in vascular patterning. Plant Cell 14, 2045–2058.

Chaiwanon, J., Wang, Z.Y., 2015. Spatiotemporal brassinosteroid signaling and antagonism with auxin pattern stem cell dynamics in *Arabidopsis* roots. Curr. Biol. 25, 1031–1042.

Cheng, Y., Zhu, W., Chen, Y., et al., 2014. Brassinosteroids control root epidermal cell fate via direct regulation of a MYB-bHLH-WD40 complex by GSK3-like kinases. Elife (Cambridge) e02525.

Chinchilla, D., Zipfel, C., Robatzek, S., et al., 2007. A flagellin-induced complex of the receptor FLS2 and BAK1 initiates plant defence. Nature 448, 497–500.

Choe, S., Dilkes, B.P., Fujioka, S., et al., 1998. The DWF4 gene of *Arabidopsis* encodes a cytochrome P450 that mediates multiple 22alpha-hydroxylation steps in brassinosteroid biosynthesis. Plant Cell 10, 231–243.

Choe, S., Dilkes, B.P., Gregory, B.D., et al., 1999a. The *Arabidopsis* dwarf1 mutant is defective in the conversion of 24-methylenecholesterol to campesterol in brassinosteroid biosynthesis. Plant Physiol. 119, 897–907.

Choe, S., Noguchi, T., Fujioka, S., et al., 1999b. The *Arabidopsis* dwf7/ste1 mutant is defective in the delta7 sterol C-5 desaturation step leading to brassinosteroid biosynthesis. Plant Cell 11, 207–221.

Choe, S., Schmitz, R.J., Fujioka, S., et al., 2002. *Arabidopsis* brassinosteroid-insensitive dwarf12 mutants are semidominant and defective in a glycogen synthase kinase 3beta-like kinase. Plant Physiol. 130, 1506–1515.

Choe, S., Tanaka, A., Noguchi, T., et al., 2000. Lesions in the sterol delta reductase gene of *Arabidopsis* cause dwarfism due to a block in brassinosteroid biosynthesis. Plant J. 21, 431–443.

Choi, Y.H., Fujioka, S., Nomura, T., et al., 1997. An alternative brassinolide biosynthetic pathway via late C-6 oxidation. Phytochemistry 44, 609–613.

Chory, J., Nagpal, P., Peto, C.A., 1991. Phenotypic and genetic analysis of det2, a new mutant that affects light-regulated seedling development in *Arabidopsis*. Plant Cell 3, 445–459.

Chung, Y., Maharjan, P.M., Lee, O., et al., 2011. Auxin stimulates DWARF4 expression and brassinosteroid biosynthesis in *Arabidopsis*. Plant J. 66, 564–578.

Clouse, S.D., Langford, M., McMorris, T.C., 1996. A brassinosteroid-insensitive mutant in *Arabidopsis thaliana* exhibits multiple defects in growth and development. Plant Physiol. 111, 671–678.

Clouse, S.D., Sasse, J.M., 1998. Brassinosteroids: essential regulators of plant growth and development. Annu. Rev. Plant Phys. 49, 427–451.

Deslauriers, S.D., Larsen, P.B., 2010. FERONIA is a key modulator of brassinosteroid and eth-ylene responsiveness in *Arabidopsis* hypocotyls. Mol. Plant 3, 626–640.

Di Rubbo, S., Irani, N.G., Kim, S.Y., et al., 2013. The clathrin adaptor complex AP-2 mediates endocytosis of brassinosteroid insensitive1 in *Arabidopsis*. Plant Cell 25, 2986–2997.

Diener, A.C., Li, H., Zhou, W., et al., 2000. Sterol methyltransferase 1 controls the level of cholesterol in plants. Plant Cell 12, 853–870.

Divi, U.K., Rahman, T., Krishna, P., 2010. Brassinosteroid-mediated stress tolerance in *Arabidopsis* shows interactions with abscisic acid, ethylene and salicylic acid pathways. BMC Plant Biol. 10, 151.

Domagalska, M.A., Schomburg, F.M., Amasino, R.M., Vierstra, R.D., Nagy, F., Davis, S.J., 2007. Attenuation of brassinosteroid signaling enhances FLC expression and delays flow-ering. Development 134, 2841–2850.

Du, J., Yin, H., Zhang, S., et al., 2012. Somatic embryogenesis receptor kinases control root development mainly via brassinosteroid-independent actions in *Arabidopsis thaliana*. J. Integr. Plant Biol. 54, 388–399.

Ferguson, B.J., Ross, J.J., Reid, J.B., 2005. Nodulation phenotypes of gibberellin and brassino-steroid mutants of pea. Plant Physiol. 138, 2396–2405.

Fridman, Y., Elkouby, L., Holland, N., et al., 2014. Root growth is modulated by differential hormonal sensitivity in neighboring cells. Genes Dev. 28, 912–920.

Friedrichsen, D.M., Joazeiro, C.A., Li, J., et al., 2000. Brassinosteroid-insensitive-1 is a ubiquitously expressed leucine-rich repeat receptor serine/threonine kinase. Plant Physiol. 123, 1247–1256.

Fujioka, S., Choi, Y.H., Takatsuto, S., et al., 1996. Identification of castasterone, 6-deoxo-castasterone, typhasterol and 6-deoxotyphasterol from the shoots of *Arabidopsis thaliana*. Plant Cell Physiol. 37, 1201–1203.

Fujioka, S., Li, J., Choi, Y.H., et al., 1997. The *Arabidopsis* deetiolated2 mutant is blocked early in brassinosteroid biosynthesis. Plant Cell 9, 1951–1962.

Fujioka, S., Sakurai, A., 1997. Biosynthesis and metabolism of brassinosteroids. Physiol. Plant. 100, 710–715.

Fujioka, S., Takatsuto, S., Yoshida, S., 2002. An early C-22 oxidation branch in the brassinos-teroid biosynthetic pathway. Plant Physiol. 130, 930–939.

Fujioka, S., Yokota, T., 2003. Biosynthesis and metabolism of brassinosteroids. Annu. Rev. Plant Biol. 54, 137–164.

Fujita, S., Ohnishi, T., Watanabe, B., et al., 2006. *Arabidopsis* CYP90B1 catalyses the early C-22 hydroxylation of C27, C28 and C29 sterols. Plant J. 45, 765–774.

Fukuda, H., 1997. Tracheary element differentiation. Plant Cell 9, 1147–1156.

Gallego-Bartolome, J., Minguet, E.G., Grau-Enguix, F., et al., 2012. Molecular mechanism for the interaction between gibberellin and brassinosteroid signaling pathways in *Arabidopsis*. Proc. Natl. Acad. Sci. U.S.A. 109, 13446–13451.

Gampala, S.S., Kim, T.W., He, J.X., et al., 2007. An essential role for 14-3-3 proteins in brassi-nosteroid signal transduction in *Arabidopsis*. Dev. Cell 13, 177–189.

Gao, Y., Zhang, D., Li, J., 2015. TCP1 modulates DWF4 expression via directly interacting with the GGNCCC motifs in the promoter region of DWF4 in *Arabidopsis thaliana*. J. Genet. Genomics 42, 383–392.

Geldner, N., Hyman, D.L., Wang, X., et al., 2007. Endosomal signaling of plant steroid receptor kinase BRI1. Genes Dev. 21, 1598–1602.

Gonzalez-Garcia, M.P., Vilarrasa-Blasi, J., Zhiponova, M., et al., 2011. Brassinosteroids control meristem size by promoting cell cycle progression in *Arabidopsis* roots. Development 138, 849–859.

Gou, X., Yin, H., He, K., et al., 2012. Genetic evidence for an indispensable role of somatic embryogenesis receptor kinases in brassinosteroid signaling. PLoS Genet. 8, e1002452.

Grove, M.D., Spencer, G.F., Rohwedder, W.K., et al., 1979. Brassinolide, a plant growth-promoting steroid isolated from *Brassica napus* pollen. Nature 281, 216–217.

Gudesblat, G.E., Schneider-Pizon, J., Betti, C., et al., 2012. SPEECHLESS integrates brassinosteroid and stomata signalling pathways. Nat. Cell Biol. 14, 548–554.

Guo, Z., Fujioka, S., Blancaflor, E.B., 2010. TCP1 modulates brassinosteroid biosynthesis by regulating the expression of the key biosynthetic gene DWARF4 in *Arabidopsis thaliana*. Plant Cell 22, 1161–1173.

Hartwig, T., Corvalan, C., Best, N.B., et al., 2012. Propiconazole is a specific and accessible brassinosteroid (BR) biosynthesis inhibitor for *Arabidopsis* and maize. PLoS One 7, e36625.

He, J.X., Gendron, J.M., Sun, Y., et al., 2005. BZR1 is a transcriptional repressor with dual roles in brassinosteroid homeostasis and growth responses. Science 307, 1634–1638.

He, J.X., Gendron, J.M., Yang, Y., et al., 2002. The GSK3-like kinase BIN2 phosphorylates and destabilizes BZR1, a positive regulator of the brassinosteroid signaling pathway in *Arabidopsis*. Proc. Natl. Acad. Sci. U.S.A. 99, 10185–10190.

He, K., Gou, X., Yuan, T., et al., 2007. BAK1 and BKK1 regulate brassinosteroid-dependent growth and brassinosteroid-independent cell-death pathways. Curr. Biol. 17, 1109–1115.

Heese, A., Hann, D.R., Gimenez-Ibanez, S., et al., 2007. The receptor-like kinase SERK3/BAK1 is a central regulator of innate immunity in plants. Proc. Natl. Acad. Sci. U.S.A. 104, 12217–12222.

Hink, M.A., Shah, K., Russinova, E., et al., 2008. Fluorescence fluctuation analysis of *Arabidopsis thaliana* somatic embryogenesis receptor-like kinase and brassinosteroid insensitive 1 receptor oligomerization. Biophys. J. 94, 1052–1062.

Hong, Z., Ueguchi-Tanaka, M., Shimizu-Sato, S., et al., 2002. Loss-of-function of a rice brassinosteroid biosynthetic enzyme, C-6 oxidase, prevents the organized arrangement and polar elongation of cells in the leaves and stem. Plant J. 32, 495–508.

Hong, Z., Ueguchi-Tanaka, M., Umemura, K., et al., 2003. A rice brassinosteroid-deficient mutant, ebisu dwarf (d2), is caused by a loss of function of a new member of cytochrome P450. Plant Cell 15, 2900–2910.

Hothorn, M., Belkhadir, Y., Dreux, M., et al., 2011. Structural basis of steroid hormone perception by the receptor kinase BRI1. Nature 474, 467–471.

Hu, Y., Bao, F., Li, J., 2000. Promotive effect of brassinosteroids on cell division involves a distinct CycD3-induction pathway in *Arabidopsis*. Plant J. 24, 693–701.

Hu, Y., Yu, D., 2014. BRASSINOSTEROID INSENSITIVE2 interacts with ABSCISIC ACID INSENSITIVE5 to mediate the antagonism of brassinosteroids to abscisic acid during seed germination in *Arabidopsis*. Plant Cell 26, 4394–4408.

Huang, H.Y., Jiang, W.B., Hu, Y.W., et al., 2013. BR signal influences *Arabidopsis* ovule and seed number through regulating related genes expression by BZR1. Mol. Plant 6, 456–469.

Huang, Y., Han, C., Peng, W., et al., 2010. Brassinosteroid negatively regulates jasmonate inhibition of root growth in *Arabidopsis*. Plant Signal. Behav. 5, 140–142.

Irani, N.G., Di Rubbo, S., Mylle, E., et al., 2012. Fluorescent castasterone reveals BRI1 signaling from the plasma membrane. Nat. Chem. Biol. 8, 583–589.

Jaillais, Y., Belkhadir, Y., Balsemao-Pires, E., et al., 2011a. From the cover: extracellular leucine-rich repeats as a platform for receptor/coreceptor complex formation. Proc. Natl. Acad. Sci. U.S.A. 108, 8503–8507.

Jaillais, Y., Hothorn, M., Belkhadir, Y., et al., 2011b. Tyrosine phosphorylation controls brassinosteroid receptor activation by triggering membrane release of its kinase inhibitor. Genes
Dev. 25, 232–237.

Jang, J.C., Fujioka, S., Tasaka, M., et al., 2000. A critical role of sterols in embryonic patterning
and meristem programming revealed by the fackel mutants of Arabidopsis thaliana. Genes
Dev. 14, 1485–1497.

Je, B.I., Piao, H.L., Park, S.J., et al., 2010. RAV-Like1 maintains brassinosteroid homeostasis via the coordinated activation of BRI1 and biosynthetic genes in rice. Plant Cell 22,
1777–1791.

Jiang, J., Wang, T., Wu, Z., et al., 2015a. The intrinsically disordered protein BKI1 is essential
for inhibiting BRI1 signaling in plants. Mol. Plant 8, 1675–1678.

Jiang, J., Zhang, C., Wang, X., 2013. Ligand perception, activation, and early signaling of plant
steroid receptor brassinosteroid insensitive 1. J. Integr. Plant Biol. 55, 1198–1211.

Jiang, J., Zhang, C., Wang, X., 2015b. A recently evolved isoform of the transcription factor
BES1 promotes brassinosteroid signaling and development in Arabidopsis thaliana. Plant
Cell 27, 361–374.

Kang, J.G., Yun, J., Kim, D.H., et al., 2001. Light and brassinosteroid signals are integrated via
a dark-induced small G protein in etiolated seedling growth. Cell 105, 625–636.

Karlova, R., Boeren, S., Russinova, E., et al., 2006. The Arabidopsis SOMATIC EMBRYOG-
ENESIS RECEPTOR-LIKE KINASE1 protein complex includes BRASSINOSTEROID-
INSENSITIVE1. Plant Cell 18, 626–638.

Karlova, R., Boeren, S., van Dongen, W., et al., 2009. Identification of in vitro phosphorylation
sites in the Arabidopsis thaliana somatic embryogenesis receptor-like kinases. Proteomics
9, 368–379.

Kemmerling, B., Schwedt, A., Rodriguez, P., et al., 2007. The BRI1-associated kinase 1, BAK1,
has a brassinolide-independent role in plant cell-death control. Curr. Biol. 17, 1116–1122.

Keuskamp, D.H., Sasidharan, R., Vos, I., et al., 2011. Blue-light-mediated shade avoidance
requires combined auxin and brassinosteroid action in Arabidopsis seedlings. Plant J. 67,
208–217.

Khan, M., Rozhon, W., Bigeard, J., et al., 2013. Brassinosteroid-regulated GSK3/Shaggy-like
kinases phosphorylate mitogen-activated protein (MAP) kinase kinases, which control stomata development in Arabidopsis thaliana. J. Biol. Chem. 288, 7519–7527.

Kim, S.K., Chang, S.C., Lee, E.J., et al., 2000. Involvement of brassinosteroids in the gravitropic response of primary root of maize. Plant Physiol. 123, 997–1004.

Kim, T.W., Chang, S.C., Lee, J.S., et al., 2004. Cytochrome P450-catalyzed brassinosteroid
pathway activation through synthesis of castasterone and brassinolide in Phaseolus vulgaris. Phytochemistry 65, 679–689.

Kim, T.W., Guan, S., Burlingame, A.L., et al., 2011. The CDG1 kinase mediates brassinosteroid signal transduction from BRI1 receptor kinase to BSU1 phosphatase and GSK3-like
kinase BIN2. Mol. Cell 43, 561–571.

Kim, T.W., Guan, S., Sun, Y., et al., 2009. Brassinosteroid signal transduction from cell-surface
receptor kinases to nuclear transcription factors. Nat. Cell Biol. 11, 1254–1260.

Kim, T.W., Hwang, J.Y., Kim, Y.S., et al., 2005. Arabidopsis CYP85A2, a cytochrome P450,
mediates the Baeyer-Villiger oxidation of castasterone to brassinolide in brassinosteroid
biosynthesis. Plant Cell 17, 2397–2412.

Kim, T.W., Lee, S.M., Joo, S.H., et al., 2007. Elongation and gravitropic responses of
Arabidopsis roots are regulated by brassinolide and IAA. Plant Cell Environ. 30, 679–689.

Kim, T.W., Michniewicz, M., Bergmann, D.C., et al., 2012. Brassinosteroid regulates stomatal development by GSK3-mediated inhibition of a MAPK pathway. Nature 482,
419–422.

Kinoshita, T., Cano-Delgado, A., Seto, H., et al., 2005. Binding of brassinosteroids to the extra-cellular domain of plant receptor kinase BRI1. Nature 433, 167–171.

Klahre, U., Noguchi, T., Fujioka, S., et al., 1998. The *Arabidopsis* DIMINUTO/DWARF1 gene encodes a protein involved in steroid synthesis. Plant Cell 10, 1677–1690.

Kuppusamy, K.T., Chen, A.Y., Nemhauser, J.L., 2009. Steroids are required for epidermal cell fate establishment in *Arabidopsis* roots. Proc. Natl. Acad. Sci. U.S.A. 106, 8073–8076.

Kwon, M., Fujioka, S., Jeon, J.H., et al., 2005. A double mutant for the CYP85A1 and CYP85A2 genes of *Arabidopsis* exhibits a brassinosteroid dwarf phenotype. J. Plant Biol. 48, 237–244.

Lanza, M., Garcia-Ponce, B., Castrillo, G., et al., 2012. Role of actin cytoskeleton in brassi-nosteroid signaling and in its integration with the auxin response in plants. Dev. Cell 22, 1275–1285.

Li, J., Biswas, M.G., Chao, A., et al., 1997. Conservation of function between mammalian and plant steroid 5alpha-reductases. Proc. Natl. Acad. Sci. U.S.A. 94, 3554–3559.

Li, J., Chory, J., 1997. A putative leucine-rich repeat receptor kinase involved in brassinosteroid signal transduction. Cell 90, 929–938.

Li, J., Nagpal, P., Vitart, V., et al., 1996. A role for brassinosteroids in light-dependent develop-ment of *Arabidopsis*. Science 272, 398–401.

Li, J., Nam, K.H., 2002. Regulation of brassinosteroid signaling by a GSK3/SHAGGY-like kinase. Science 295, 1299–1301.

Li, J., Nam, K.H., Vafeados, D., et al., 2001. BIN2, a new brassinosteroid-insensitive locus in *Arabidopsis*. Plant Physiol. 127, 14–22.

Li, J., Wen, J., Lease, K.A., et al., 2002. BAK1, an *Arabidopsis* LRR receptor-like protein kinase, interacts with BRI1 and modulates brassinosteroid signaling. Cell 110, 213–222.

Li, L., Xu, J., Xu, Z.H., et al., 2005. Brassinosteroids stimulate plant tropisms through mod-ulation of polar auxin transport in *Brassica* and *Arabidopsis*. Plant Cell 17, 2738–2753.

Li, L., Ye, H., Guo, H., et al., 2010. *Arabidopsis* IWS1 interacts with transcription factor BES1 and is involved in plant steroid hormone brassinosteroid regulated gene expression. Proc. Natl. Acad. Sci. U.S.A. 107, 3918–3923.

Li, L., Yu, X., Thompson, A., et al., 2009. *Arabidopsis* MYB30 is a direct target of BES1 and cooperates with BES1 to regulate brassinosteroid-induced gene expression. Plant J. 58, 275–286.

Li, Q.F., He, J.X., 2016. BZR1 interacts with HY5 to mediate brassinosteroid- and light-regu-lated cotyledon opening in *Arabidopsis* in darkness. Mol. Plant 9, 113–125.

Luo, X.M., Lin, W.H., Zhu, S., et al., 2010. Integration of light and brassinosteroid signaling pathways by a GATA transcription factor in *Arabidopsis*. Dev. Cell 19, 872–883.

Marsolais, F., Boyd, J., Paredes, Y., et al., 2007. Molecular and biochemical characterization of two brassinosteroid sulfotransferases from *Arabidopsis*, AtST4a (At2g14920) and AtST1 (At2g03760). Planta 225, 1233–1244.

Martins, S., Dohmann, E.M., Cayrel, A., et al., 2015. Internalization and vacuolar targeting of the brassinosteroid hormone receptor BRI1 are regulated by ubiquitination. Nat. Commun. 6, 6151.

Meng, X., Chen, X., Mang, H., et al., 2015. Differential function of *Arabidopsis* SERK family receptor-like kinases in stomatal patterning. Curr. Biol. 25, 2361–2372.

Mitchell, J.W., Mandava, N., Worley, J.F., et al., 1970. Brassins-a new family of plant hormones from rape pollen. Nature 225, 1065–1066.

Nakaya, M., Tsukaya, H., Murakami, N., et al., 2002. Brassinosteroids control the proliferation of leaf cells of *Arabidopsis thaliana*. Plant Cell Physiol. 43, 239–244.

Nam, K.H., Li, J., 2002. BRI1/BAK1, a receptor kinase pair mediating brassinosteroid signal-ing. Cell 110, 203–212.

Neff, M.M., Nguyen, S.M., Malancharuvil, E.J., et al., 1999. BAS1: A gene regulating brassinosteroid levels and light responsiveness in *Arabidopsis*. Proc. Natl. Acad. Sci. U.S.A. 96, 15316–15323.

Nemhauser, J.L., Hong, F., Chory, J., 2006. Different plant hormones regulate similar processes through largely nonoverlapping transcriptional responses. Cell 126, 467–475.

Nemhauser, J.L., Mockler, T.C., Chory, J., 2004. Interdependency of brassinosteroid and auxin signaling in *Arabidopsis*. PLoS Biol. 2, e258.

Noguchi, T., Fujioka, S., Choe, S., et al., 2000. Biosynthetic pathways of brassinolide in *Arabidopsis*. Plant Physiol. 124, 201–209.

Noguchi, T., Fujioka, S., Choe, S., et al., 1999a. Brassinosteroid-insensitive dwarf mutants of *Arabidopsis* accumulate brassinosteroids. Plant Physiol. 121, 743–752.

Noguchi, T., Fujioka, S., Takatsuto, S., et al., 1999b. *Arabidopsis* det2 is defective in the conversion of (24R)-24-methylcholest-4-En-3-one to (24R)-24-methyl-5alpha-cholestan-3-one in brassinosteroid biosynthesis. Plant Physiol. 120, 833–840.

Nomura, T., Kushiro, T., Yokota, T., et al., 2005. The last reaction producing brassinolide is catalyzed by cytochrome P-450s, CYP85A3 in tomato and CYP85A2 in *Arabidopsis*. J. Biol. Chem. 280, 17873–17879.

Oh, E., Zhu, J.Y., Bai, M.Y., et al., 2014. Cell elongation is regulated through a central circuit of interacting transcription factors in the *Arabidopsis* hypocotyl. Elife (Cambridge) e03031.

Oh, E., Zhu, J.Y., Wang, Z.Y., 2012. Interaction between BZR1 and PIF4 integrates brassinosteroid and environmental responses. Nat. Cell Biol. 14, 802–809.

Oh, K., Matsumoto, T., Yamagami, A., et al., 2015. YCZ-18 is a new brassinosteroid biosynthesis inhibitor. PLoS One 10, e0120812.

Oh, M.H., Ray, W.K., Huber, S.C., et al., 2000. Recombinant brassinosteroid insensitive 1 receptor-like kinase autophosphorylates on serine and threonine residues and phosphorylates a conserved peptide motif in vitro. Plant Physiol. 124, 751–766.

Oh, M.H., Wang, X., Kota, U., et al., 2009. Tyrosine phosphorylation of the BRI1 receptor kinase emerges as a component of brassinosteroid signaling in *Arabidopsis*. Proc. Natl. Acad. Sci. U.S.A. 106, 658–663.

Ohnishi, T., Godza, B., Watanabe, B., et al., 2012. CYP90A1/CPD, a brassinosteroid biosynthetic cytochrome P450 of *Arabidopsis*, catalyzes C-3 oxidation. J. Biol. Chem. 287, 31551–31560.

Ohnishi, T., Szatmari, A.M., Watanabe, B., et al., 2006. C-23 hydroxylation by *Arabidopsis* CYP90C1 and CYP90D1 reveals a novel shortcut in brassinosteroid biosynthesis. Plant Cell 18, 3275–3288.

Peng, P., Yan, Z., Zhu, Y., et al., 2008. Regulation of the *Arabidopsis* GSK3-like kinase BRASSINOSTEROID-INSENSITIVE 2 through proteasome-mediated protein degradation. Mol. Plant 1, 338–346.

Peng, P., Zhao, J., Zhu, Y., et al., 2010. A direct docking mechanism for a plant GSK3-like kinase to phosphorylate its substrates. J. Biol. Chem. 285, 24646–24653.

Peng, Z., Han, C., Yuan, L., et al., 2011. Brassinosteroid enhances jasmonate-induced anthocyanin accumulation in *Arabidopsis* seedlings. J. Integr. Plant Biol. 53, 632–640.

Perez-Perez, J.M., Ponce, M.R., Micol, J.L., 2002. The UCU1 *Arabidopsis* gene encodes a SHAGGY/GSK3-like kinase required for cell expansion along the proximodistal axis. Dev. Biol. 242, 161–173.

Poppenberger, B., Fujioka, S., Soeno, K., et al., 2005. The UGT73C5 of *Arabidopsis thaliana* glucosylates brassinosteroids. Proc. Natl. Acad. Sci. U.S.A. 102, 15253–15258.

Poppenberger, B., Rozhon, W., Khan, M., et al., 2011. CESTA, a positive regulator of brassinosteroid biosynthesis. EMBO J. 30, 1149–1161.

Pose, D., Castanedo, I., Borsani, O., et al., 2009. Identification of the *Arabidopsis* dry2/sqe1-5 mutant reveals a central role for sterols in drought tolerance and regulation of reactive oxygen species. Plant J. 59, 63–76.

Ren, C., Han, C., Peng, W., et al., 2009. A leaky mutation in DWARF4 reveals an antagonistic role of brassinosteroid in the inhibition of root growth by jasmonate in *Arabidopsis*. Plant Physiol. 151, 1412–1420.

Rouleau, M., Marsolais, F., Richard, M., et al., 1999. Inactivation of brassinosteroid biological activity by a salicylate-inducible steroid sulfotransferase from *Brassica napus*. J. Biol. Chem. 274, 20925–20930.

Roux, M., Schwessinger, B., Albrecht, C., et al., 2011. The *Arabidopsis* leucine-rich repeat receptor-like kinases BAK1/SERK3 and BKK1/SERK4 are required for innate immunity to hemibiotrophic and biotrophic pathogens. Plant Cell 23, 2440–2455.

Russinova, E., Borst, J.W., Kwaaitaal, M., et al., 2004. Heterodimerization and endocytosis of *Arabidopsis* brassinosteroid receptors BRI1 and AtSERK3 (BAK1). Plant Cell 16, 3216–3229.

Ryu, H., Cho, H., Bae, W., et al., 2014. Control of early seedling development by BES1/TPL/HDA19-mediated epigenetic regulation of ABI3. Nat. Commun. 5, 4138.

Ryu, H., Kim, K., Cho, H., et al., 2007. Nucleocytoplasmic shuttling of BZR1 mediated by phosphorylation is essential in *Arabidopsis* brassinosteroid signaling. Plant Cell 19, 2749–2762.

Saidi, Y., Hearn, T.J., Coates, J.C., 2012. Function and evolution of 'green' GSK3/Shaggy-like kinases. Trends Plant Sci. 17, 39–46.

Sakamoto, T., Fujioka, S., 2013. Auxins increase expression of the brassinosteroid receptor and brassinosteroid-responsive genes in *Arabidopsis*. Plant Signal. Behav. 8, e23509.

Sakamoto, T., Ohnishi, T., Fujioka, S., et al., 2012. Rice CYP90D2 and CYP90D3 catalyze C-23 hydroxylation of brassinosteroids in vitro. Plant Physiol. Biochem. 58, 220–226.

Samakovli, D., Margaritopoulou, T., Prassinos, C., et al., 2014. Brassinosteroid nuclear signaling recruits HSP90 activity. New Phytol. 203, 743–757.

Santiago, J., Henzler, C., Hothorn, M., 2013. Molecular mechanism for plant steroid receptor activation by somatic embryogenesis co-receptor kinases. Science 341, 889–892.

Schrick, K., Mayer, U., Horrichs, A., 2000. FACKEL is a sterol C-14 reductase required for organized cell division and expansion in *Arabidopsis* embryogenesis. Genes Dev. 14, 1471–1484.

Schrick, K., Mayer, U., Martin, G., et al., 2002. Interactions between sterol biosynthesis genes in embryonic development of *Arabidopsis*. Plant J. 31, 61–73.

Sekimata, K., Kimura, T., Kaneko, I., et al., 2001. A specific brassinosteroid biosynthesis inhibitor, Brz2001: evaluation of its effects on *Arabidopsis*, cress, tobacco, and rice. Planta 213, 716–721.

Shang, Y., Dai, C., Lee, M.M., et al., 2016. BRI1-Associated receptor kinase 1 regulates guard cell ABA signaling mediated by open stomata 1 in *Arabidopsis*. Mol. Plant 9, 447–460.

She, J., Han, Z., Kim, T.W., et al., 2011. Structural insight into brassinosteroid perception by BRI1. Nature 474, 472–476.

She, J., Han, Z., Zhou, B., et al., 2013. Structural basis for differential recognition of brassinolide by its receptors. Protein Cell 4, 475–482.

Shi, C., Qi, C., Ren, H., et al., 2015. Ethylene mediates brassinosteroid-induced stomatal closure via Galpha protein-activated hydrogen peroxide and nitric oxide production in *Arabidopsis*. Plant J. 82, 280–301.

Shimada, S., Komatsu, T., Yamagami, A., et al., 2015. Formation and dissociation of the BSS1 protein complex regulates plant development via brassinosteroid signaling. Plant Cell 27, 375–390.

Shimada, Y., Fujioka, S., Miyauchi, N., et al., 2001. Brassinosteroid-6-oxidases from *Arabidopsis* and tomato catalyze multiple C-6 oxidations in brassinosteroid biosynthesis. Plant Physiol. 126, 770–779.

Song, L., Zhou, X.Y., Li, L., et al., 2009. Genome-wide analysis revealed the complex regulatory network of brassinosteroid effects in photomorphogenesis. Mol. Plant 2, 755–772.

Souter, M., Topping, J., Pullen, M., et al., 2002. *hydra* Mutants of *Arabidopsis* are defective in sterol profiles and auxin and ethylene signaling. Plant Cell 14, 1017–1031.

Stewart Lilley, J.L., Gan, Y., Graham, I.A., et al., 2013. The effects of DELLAs on growth change with developmental stage and brassinosteroid levels. Plant J. 76, 165–173.

Sun, S., Chen, D., Li, X., et al., 2015. Brassinosteroid signaling regulates leaf erectness in *Oryza sativa* via the control of a specific U-type cyclin and cell proliferation. Dev. Cell 34, 220–228.

Sun, Y., Fan, X.Y., Cao, D.M., et al., 2010. Integration of brassinosteroid signal transduction with the transcription network for plant growth regulation in *Arabidopsis*. Dev. Cell 19, 765–777.

Sun, Y., Han, Z., Tang, J., et al., 2013. Structure reveals that BAK1 as a co-receptor recognizes the BRI1-bound brassinolide. Cell Res. 23, 1326–1329.

Szekeres, M., Nemeth, K., Koncz-Kalman, Z., et al., 1996. Brassinosteroids rescue the deficiency of CYP90, a cytochrome P450, controlling cell elongation and de-etiolation in *Arabidopsis*. Cell 85, 171–182.

Tanabe, S., Ashikari, M., Fujioka, S., et al., 2005. A novel cytochrome P450 is implicated in brassinosteroid biosynthesis via the characterization of a rice dwarf mutant, dwarf11, with reduced seed length. Plant Cell 17, 776–790.

Tanaka, K., Asami, T., Yoshida, S., et al., 2005. Brassinosteroid homeostasis in *Arabidopsis* is ensured by feedback expressions of multiple genes involved in its metabolism. Plant Physiol. 138, 1117–1125.

Tang, W., Yuan, M., Wang, R., et al., 2011. PP2A activates brassinosteroid-responsive gene expression and plant growth by dephosphorylating BZR1. Nat. Cell Biol. 13, 124–131.

Terakado, J., Fujihara, S., Goto, S., et al., 2005. Systemic effect of a brassinosteroid on root nodule formation in soybean as revealed by the application of brassinolide and brassinazole. Soil Sci. Plant Nutr. 51, 389–395.

Tong, H., Xiao, Y., Liu, D., et al., 2014. Brassinosteroid regulates cell elongation by modulating gibberellin metabolism in rice. Plant Cell 26, 4376–4393.

Unterholzner, S.J., Rozhon, W., Papacek, M., 2015. Brassinosteroids are master regulators of gibberellin biosynthesis in *Arabidopsis*. Plant Cell 27, 2261–2272.

van Esse, W., van Mourik, S., Albrecht, C., et al., 2013. A mathematical model for the coreceptors SOMATIC EMBRYOGENESIS RECEPTOR-LIKE KINASE1 and SOMATIC EMBRYOGENESIS RECEPTOR-LIKE KINASE3 in BRASSINOSTEROID INSENSITIVE1-mediated signaling. Plant Physiol. 163, 1472–1481.

Vert, G., Chory, J., 2006. Downstream nuclear events in brassinosteroid signalling. Nature 441, 96–100.

Vert, G., Walcher, C.L., Chory, J., et al., 2008. Integration of auxin and brassinosteroid pathways by auxin response factor 2. Proc. Natl. Acad. Sci. U.S.A. 105, 9829–9834.

Wang, H., Yang, C., Zhang, C., et al., 2011. Dual role of BKI1 and 14-3-3 s in brassinosteroid signaling to link receptor with transcription factors. Dev. Cell 21, 825–834.

Wang, J., Jiang, J., Wang, J., et al., 2014a. Structural insights into the negative regulation of BRI1 signaling by BRI1-interacting protein BKI1. Cell Res. 24, 1328–1341.

Wang, L., Li, H., Lv, X., et al., 2015. Spatiotemporal Dynamics of the BRI1 receptor and its regulation by membrane microdomains in Living *Arabidopsis* cells. Mol. Plant 8, 1334–1349.

Wang, X., Chen, J., Xie, Z., et al., 2014b. Histone lysine methyltransferase SDG8 is involved in brassinosteroid regulated gene expression in *Arabidopsis thaliana*. Mol. Plant 7, 1303–1315.

Wang, X., Chory, J., 2006. Brassinosteroids regulate dissociation of BKI1, a negative regulator of BRI1 signaling, from the plasma membrane. Science 313, 1118–1122.

Wang, X., Goshe, M.B., Soderblom, E.J., et al., 2005a. Identification and functional analysis of in vivo phosphorylation sites of the *Arabidopsis* BRASSINOSTEROID-INSENSITIVE1 receptor kinase. Plant Cell 17, 1685–1703.

Wang, X., Kota, U., He, K., et al., 2008. Sequential transphosphorylation of the BRI1/BAK1 receptor kinase complex impacts early events in brassinosteroid signaling. Dev. Cell 15, 220–235.

Wang, X., Li, X., Meisenhelder, J., et al., 2005b. Autoregulation and homodimerization are involved in the activation of the plant steroid receptor BRI1. Dev. Cell 8, 855–865.

Wang, Y., Sun, S., Zhu, W., et al., 2013. Strigolactone/MAX2-induced degradation of brassinosteroid transcriptional effector BES1 regulates shoot branching. Dev. Cell 27, 681–688.

Wang, Z.Y., Nakano, T., Gendron, J., et al., 2002. Nuclear-localized BZR1 mediates brassinosteroid-induced growth and feedback suppression of brassinosteroid biosynthesis. Dev. Cell 2, 505–513.

Yang, C., Liu, J., Dong, X., et al., 2014. Short-term and continuing stresses differentially interplay with multiple hormones to regulate plant survival and growth. Mol. Plant 7, 841–855.

Yang, C.J., Zhang, C., Lu, Y.N., et al., 2011. The mechanisms of brassinosteroids' action: from signal transduction to plant development. Mol. Plant 4, 588–600.

Ye, H., Li, L., Guo, H., et al., 2012. MYBL2 is a substrate of GSK3-like kinase BIN2 and acts as a corepressor of BES1 in brassinosteroid signaling pathway in *Arabidopsis*. Proc. Natl. Acad. Sci. U.S.A. 109, 20142–20147.

Ye, Q., Zhu, W., Li, L., et al., 2010. Brassinosteroids control male fertility by regulating the expression of key genes involved in *Arabidopsis* anther and pollen development. Proc. Natl. Acad. Sci. U.S.A. 107, 6100–6105.

Yin, Y., Vafeados, D., Tao, Y., et al., 2005. A new class of transcription factors mediates brassinosteroid-regulated gene expression in *Arabidopsis*. Cell 120, 249–259.

Yin, Y., Wang, Z.Y., Mora-Garcia, S., et al., 2002. BES1 accumulates in the nucleus in response to brassinosteroids to regulate gene expression and promote stem elongation. Cell 109, 181–191.

Youn, J.H., Kim, T.W., 2014. Functional insights of plant GSK3-like kinases: multi-taskers in diverse cellular signal transduction pathways. Mol. Plant 8, 552–565.

Yu, X., Li, L., Guo, M., et al., 2008. Modulation of brassinosteroid-regulated gene expression by Jumonji domain-containing proteins ELF6 and REF6 in *Arabidopsis*. Proc. Natl. Acad. Sci. U.S.A. 105, 7618–7623.

Yu, X., Li, L., Zola, J., et al., 2011. A brassinosteroid transcriptional network revealed by genome-wide identification of BESI target genes in *Arabidopsis thaliana*. Plant J. 65, 634–646.

Yuan, T., Fujioka, S., Takatsuto, S., et al., 2007. BEN1, a gene encoding a dihydroflavonol 4-reductase (DFR)-like protein, regulates the levels of brassinosteroids in *Arabidopsis thaliana*. Plant J. 51, 220–233.

Zhang, D., Ye, H., Guo, H., et al., 2014. Transcription factor HAT1 is phosphorylated by BIN2 kinase and mediates brassinosteroid repressed gene expression in *Arabidopsis*. Plant J. 77, 59–70.

Zhang, S., Cai, Z., Wang, X., 2009a. The primary signaling outputs of brassinosteroids are regulated by abscisic acid signaling. Proc. Natl. Acad. Sci. U.S.A. 106, 4543–4548.

Zhang, S., Wei, Y., Lu, Y., et al., 2009b. Mechanisms of brassinosteroids interacting with multiple hormones. Plant Signal. Behav. 4, 1117–1120.

Zhang, Z., Zhu, J.Y., Roh, J., et al., 2016. TOR signaling promotes accumulation of BZR1 to balance growth with carbon availability in *Arabidopsis*. Curr. Biol. 26, 1854–1860.

Zhao, J., Peng, P., Schmitz, R.J., et al., 2002. Two putative BIN2 substrates are nuclear components of brassinosteroid signaling. Plant Physiol. 130, 1221–1229.

Zhou, A., Wang, H., Walker, J.C., et al., 2004. BRL1, a leucine-rich repeat receptor-like protein kinase, is functionally redundant with BRI1 in regulating *Arabidopsis* brassinosteroid signaling. Plant J. 40, 399–409.

Zhu, W., Wang, H., Fujioka, S., et al., 2013. Homeostasis of brassinosteroids regulated by DRL1, a putative acyltransferase in *Arabidopsis*. Mol. Plant 6, 546–558.

Strigolactones

10

Bing Wang, Yonghong Wang, Jiayang Li
Institute of Genetics and Developmental Biology, Chinese Academy of Sciences, Beijing, China

Summary

Strigolactones (SLs) are carotenoid-derived signaling molecules and plant hormones that enable root-parasitic plants and symbiotic fungi to detect their host plants. They also regulate several developmental processes that adapt shoot and root architecture to the environment. Highly branching or tillering mutants in *Arabidopsis thaliana* petunia, pea, and rice have greatly facilitated the identification of SL biosynthetic enzymes and signaling components. The carotenoid precursor undergoes isomerization and cleavage to generate carlactone, which then undergoes oxidation and further modification to produce a range of strigolactone structures. Perception of SLs involves a novel mechanism in which a serine hydrolase-type enzyme receptor attacks the SLs and becomes covalently modified. This triggers the interaction of the receptor with other proteins including an F-box protein, leading to ubiquitination and destruction of target proteins such as transcriptional regulators. In this chapter, we introduce the discovery of SLs, summarize progress in understanding the SL biosynthesis, transport of SLs, and their signaling pathway. We also highlight their functions in plant development, symbiosis, and parasitism, as well as the cross talk with environmental signals and other phytohormones.

10.1 Discovery and functions of strigolactones

Strigolactones (SLs) are a group of carotenoid-derived plant metabolites that were originally characterized as rhizosphere signals that enable root-parasitic plants to detect and colonize their hosts (Cook et al., 1966). Subsequently, SLs were shown to serve as signals for arbuscular mycorrhizal (AM) fungi to form symbiotic associations with host roots (Akiyama et al., 2005). Besides the functions of SLs in the rhizosphere, studies of highly shoot branching and tillering mutants in both dicotyledonous and monocotyledonous plants demonstrated the presence of a root-to-shoot signaling molecule that suppresses shoot branching (Bennett and Leyser, 2014; Booker et al., 2004, 2005; Ferguson and Beveridge, 2009; Simons et al., 2007; Snowden et al., 2005; Sorefan et al., 2003; Xie et al., 2010). In 2008, SLs were characterized as new phytohormones that regulate above-ground plant architecture by inhibiting bud outgrowth, and function in underground communication with neighboring organisms (Gomez-Roldan et al., 2008; Umehara et al., 2008).

Genetic screens of *more axillary growth* (*max*) mutants in *Arabidopsis thaliana*, *high-tillering dwarf* (*htd* and *d*) mutants in rice (*Oryza sativa*), *ramosus* (*rms*) mutants in pea (*Pisum sativum*), and *decreased apical dominance* (*dad*) mutants in petunia (*Petunia hybrida*) have led to the isolation of key genes required for SL biosynthesis

Hormone Metabolism and Signaling in Plants. http://dx.doi.org/10.1016/B978-0-12-811562-6.00010-4

and signaling. Through grafting studies, exogenous application of synthetic SL analog GR24, and analysis of metabolism, *D10, D17* (also known as *HTD1*), and *D27* in rice; *MAX1, MAX3, MAX4,* and *LATERAL BRANCHING OXIDOREDUCTASE (LBO)* in Arabidopsis; *RMS1* and *RMS5* in pea; and *DAD1* and *DAD3* in petunia were identified to participate in the SL biosynthetic pathway (Alder et al., 2012; Arite et al., 2007; Booker et al., 2004, 2005; Brewer et al., 2016; Drummond et al., 2009; Lin et al., 2009; Simons et al., 2007; Snowden et al., 2005; Sorefan et al., 2003; Waters et al., 2012a; Zhang et al., 2014; Zou et al., 2006). Meanwhile *D3, D14,* and *D53* in rice; *MAX2, AtD14, D53-Like SMXLs (SUPPRESSOR OF MAX2-LIKEs)* in Arabidopsis; *RMS3* and *RMS4* in pea; and *DAD2* in petunia were shown to be involved in SL perception and/or signal transduction (Arite et al., 2009; Beveridge et al., 1996; Gao et al., 2009; Hamiaux et al., 2012; Ishikawa et al., 2005; Jiang et al., 2013; Liu et al., 2009; Soundappan et al., 2015; Stirnberg et al., 2002; Umehara et al., 2008; Wang et al., 2015; Waters et al., 2012b; Zhou et al., 2013).

In addition to the shoot branch-regulating function, SLs are now recognized as a new class of phytohormones involved in many aspects of plant development (Al-Babili and Bouwmeester, 2015). SLs are involved in the regulation of internode length (de Saint Germain et al., 2013; Snowden et al., 2005), leaf morphology (Scaffidi et al., 2013; Stirnberg et al., 2002), leaf senescence (Snowden et al., 2005; Yamada et al., 2014), shoot gravitropism (Sang et al., 2014), stem thickness (Agusti et al., 2011), seed germination, early seedling development (Toh et al., 2012; Tsuchiya et al., 2010), as well as colony growth in the moss *Physcomitrella patens* (Hoffmann et al., 2014; Proust et al., 2011). In the root system, SLs enhance the growth of primary roots, the elongation of root hairs, and the growth of rice crown roots (Arite et al., 2012; Kagiyama et al., 2013; Kapulnik et al., 2011), but inhibit adventitious root formation in Arabidopsis, tomato, and pea (Kohlen et al., 2012; Rasmussen et al., 2013; Urquhart et al., 2015). SLs also play important roles in adaptive responses to environmental factors such as phosphate, nitrogen, light, drought, and high salinity (Brewer et al., 2013; Rasmussen et al., 2013; Ruyter-Spira et al., 2013; Sun et al., 2016; Waldie et al., 2014; Zwanenburg et al., 2016).

10.2   Strigolactone biosynthesis

10.2.1   Structures and nomenclature

The most common naturally occurring SLs share a common tricyclic lactone structure composed of an ABC-ring and a D-ring butenolide group, which are connected with an enol-ether bridge (Fig. 10.1). The SL 5-deoxystrigol (5DS) provides a simple example as it contains no additional oxygen-containing groups on the A- and B-rings (Alder et al., 2012). It is isolated from root exudates or tissues of Fabaceae plants, rice, Arabidopsis, and the basal embryophyte *Marchantia* (Akiyama and Hayashi, 2006; Akiyama et al., 2005; Alder et al., 2012; Delaux et al., 2012; Yoneyama et al., 2008). Strigol and orobanchol stimulate seed germination of *Striga* and *Orobanche* spp. respectively, and are usually considered as references to designate other SLs according to two families with different stereochemistry of the B-C ring junction (Al-Babili and Bouwmeester, 2015; Boyer et al., 2012; Cook et al., 1966; Flematti et al., 2016;

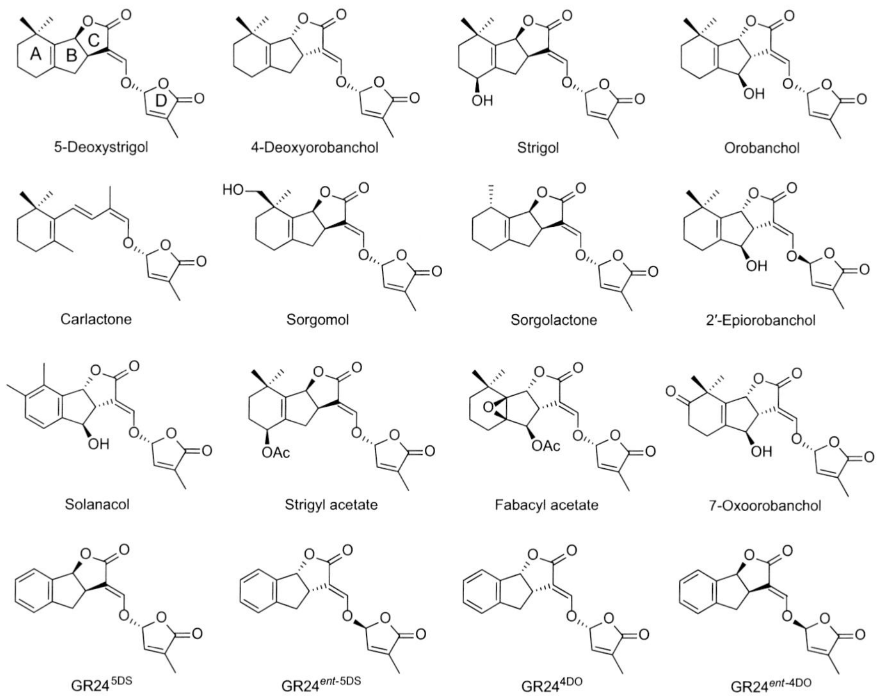

Figure 10.1 Structures of the main natural strigolactones and GR24 stereoisomers. In naturally occurring canonical SLs, a common tricyclic lactone structure (ABC-ring) is connected with a D-ring butenolide group through an enol-ether bridge. Carlactone is a SL precursor and lacks B and C rings. Chemical structures of natural SLs and the GR24 stereoisomers are given.

Yokota et al., 1998). Several strigol- and orobanchol-like compounds have been identified with hydroxyl or acetate groups at different positions of the A- or B-ring (Fig. 10.1). The synthetic SL analog GR24 is commonly used for the studies on the biosynthesis and signaling of SLs and karrikins (see below). It is usually prepared as a racemic mixture of two stereoisomers (*rac*-GR24); one has the configuration of 5DS and another is its enantiomer (*ent*-5DS) (Fig. 10.1). The purified GR24 stereoisomers with the configuration of 4-deoxyorobanchol (4DO) and *ent*-4DO were also used to trigger SL and karrikin signaling specifically (Scaffidi et al., 2014).

10.2.2 Biosynthetic pathways of strigolactones

10.2.2.1 The role of carotenoids

Root exudates of maize, sorghum, and cowpea plants that had been treated with the carotenoid biosynthesis inhibitor, fluridone, were impaired in stimulating seed germination, suggesting that SLs are derived from carotenoids (Matusova et al., 2005). This has now been confirmed following the characterization of SL biosynthetic mutants. Carotenoids exist extensively in heterotrophic microorganisms and photosynthetic organisms, and are involved in many important biological processes, such as photosynthesis, lipid peroxidation, attraction of pollinating insects, as well as biosynthesis of phytohormones SLs and abscisic acid (ABA) (Fraser and Bramley, 2004; Walter and Strack, 2011). In the plant carotenoid biosynthetic pathway, isopentenyl pyrophosphate (IPP) and its isomer, dimethylallyl pyrophosphate (DMAPP), are the predominant precursor compounds, with geranylgeranyl pyrophosphate (GGPP), 15-*cis*-phytoene, 9,15,9′-tri-*cis*-ζ-carotene, 9,9′-di-*cis*-ζ-carotene, pro-lycopene, all-*trans*-lycopene, α-carotene, β-carotene, and γ-carotene as intermediate compounds (Al-Babili and Bouwmeester, 2015). Biosynthesis of SLs from β-carotene starts with the reversible, 9-*cis*/all-*trans*-β-carotene isomerization catalyzed by D27, then undergoes successive cleavage reactions catalyzed by carotenoid cleavage dioxygenase 7 (CCD7) and CCD8, and oxidations by cytochrome P450 enzymes of the MAX1 family and LBO for further modification (Fig. 10.2).

10.2.2.2   Isomerization by D27

The first dedicated step of SL biosynthesis involves a unique all-*trans*-β-carotene isomerization at the C-9 position to form 9-*cis*-β-carotene (Alder et al., 2012) (Fig. 10.2). This reaction is catalyzed by an iron-containing enzyme, D27, the first all-*trans*-β-carotene/9-*cis*-β-carotene isomerase identified initially in rice and then in Arabidopsis (Alder et al., 2012; Lin et al., 2009; Waters et al., 2012a). D27 encodes a chloroplast-localized protein that is mainly expressed in vascular cells of shoots and roots. The tillering and dwarf phenotypes of *d27* are correlated with enhanced polar auxin transport (PAT) from shoot apex in the uppermost internodes. Furthermore, *ent-2′-epi*-5-deoxystrigol (now named 4DO), an identified SL in root exudates of rice seedlings, was undetectable in *d27*, and GR24 treatment could rescue the phenotypes of *d27* (Lin et al., 2009). The *Atd27* mutant also forms more branches, which can be rescued by exogenous application of GR24. Grafting experiments indicated that AtD27 operates on a nonmobile precursor upstream of MAX1 in the SL biosynthesis pathway (Waters et al., 2012a).

10.2.2.3   Cleavage by CCD7 and CCD8

After isomerization, 9-*cis*-β-carotene is converted into carlactone (CL) by sequential actions of CCD7 and CCD8 (Booker et al., 2005; Seto et al., 2014; Sorefan et al., 2003). The CCD7 enzyme performs a stereospecific cleavage of 9-*cis*-β-carotene at the C9′-C10′ double bond in the *trans*-moiety of the substrate, yielding the intermediate 9-*cis*-β-apo-10′-carotenal and β-ionone. Subsequently, CCD8 adds three oxygens and rearranges the backbone of 9-*cis*-β-apo-10′-carotenal, forming the characteristic D-ring and the enol-ether bridge linked to the A-ring, and thereby producing CL (Alder et al., 2012) (Fig. 10.2). The Arabidopsis *MAX3*, rice *D17/HTD1* and pea *RMS5* genes encode CCD7, while the Arabidopsis *MAX4*, rice *D10,* and pea *RMS1* genes encode CCD8 respectively (Alder et al., 2012; Arite et al., 2007; Booker et al., 2004; Drummond et al., 2009; Simons et al., 2007; Snowden et al., 2005; Sorefan et al., 2003; Zou et al., 2006).

The high-tillering phenotypes of rice SL biosynthesis mutants, such as *d10* and *d27*, are suppressed by CL, which also induces the germination of *Striga hermonthica* seeds, suggesting that CL is a biosynthetic precursor for SLs (Alder et al., 2012).

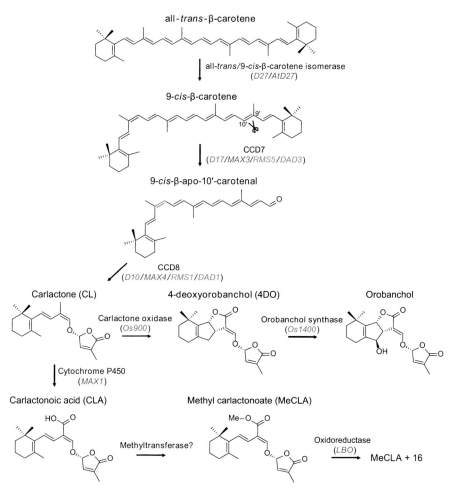

Figure 10.2 Plant strigolactone biosynthetic pathways. All-*trans*-β-carotene is converted by all-*trans*/9-*cis*-β-carotene isomerase, carotenoid cleavage dioxygenase 7 (CCD7) and CCD8 in sequence to carlactone (CL). In rice, CO converts (*Z*)-(*11R*)-CL into 4-deoxyorobanchol, and subsequently orobanchol synthase converts 4-deoxyorobanchol (4DO) to orobanchol. In Arabidopsis, CL is oxidized to CLA by the cytochrome P450 monooxygenase MAX1. An unknown methyltransferase is suggested to transform CLA to methyl carlactonoate (MeCLA), and the oxidoreductase LATERAL BRANCHING OXIDOREDUCTASE (LBO) subsequently converts MeCLA to an unknown strigolactone-like product (MeCLA + 16 Da). Red, purple, green, and brown colors represent genes of rice, Arabidopsis, pea, and petunia, respectively.

This is supported by suppression of branching in the Arabidopsis *max3* mutant by CL (Scaffidi et al., 2013). Recently, CL has been isolated from plant tissues and its absolute stereochemistry has been identified, indicating that CL functions as the precursor of tricyclic lactone-containing SLs in vivo and is a critical intermediate of the plant SL biosynthetic pathway (Seto et al., 2014).

10.2.2.4   Oxidation by cytochrome P450 enzymes

The cytochrome P450 MAX1 works downstream of D27, CCD7, and CCD8 and is required for biosynthesis of active SLs (Booker et al., 2005; Kohlen et al., 2011; Scaffidi et al., 2013). The Arabidopsis MAX1 is predicted to localize in cytosol and predominantly expressed in the cambial region and xylem-associated parenchyma (Booker et al., 2005). When *max1* rootstocks are grafted to *Atd27*, *max3* or *max4* scions, the highly branching phenotypes are rescued, whereas reciprocal grafts form plants with more branches, indicating that MAX1 works downstream of D27, MAX3, and MAX4 and participates in the biosynthesis of a mobile SL intermediate (Booker et al., 2005; Waters et al., 2012a). In Arabidopsis, MAX1 has been demonstrated to mediate the conversion of CL to carlactonoic acid (CLA), which is further converted to methyl carlactonoate (MeCLA) by an unknown enzyme (Abe et al., 2014). CL is accumulated in *max1* and cannot rescue the shoot branching phenotypes of *max1* (Scaffidi et al., 2013; Seto et al., 2014). In rice, a quantitative trait locus (QTL) containing two *MAX1* orthologs is present in a low-tillering, high SL-containing cultivar, but absent from a high-tillering, low SL-containing cultivar, suggesting that *MAX1* orthologs are responsible for SL-mediated shoot branching regulation in rice (Cardoso et al., 2014). Recently, two MAX1 orthologs in rice have been found to convert CL to orobanchol instead of CLA. The first is named carlactone oxidase (CO) and catalyzes the conversion of (*Z*)-(*11R*)-CL predominately into 4DO, and the second, orobanchol synthase (OS), converts 4DO to orobanchol (Zhang et al., 2014). Intriguingly, in different species CL is produced by conserved enzymes, while MAX1 orthologs from Arabidopsis and rice convert CL to different subsets of SLs, suggesting that the observed diversification in MAX1 may reflect the production of different SLs with specific functions in diverse species.

10.2.2.5   LBO and unknown enzymes

The existence of diverse natural SLs suggests that additional enzymes are required for SL biosynthesis and metabolism. Recently, LBO has been identified as an oxidoreductase-like enzyme that belongs to the 2-oxoglutarate and Fe(II)-dependent dioxygenase superfamily (Brewer et al., 2016). The expression levels of *LBO* show appreciable coexpression with *MAX3* and are significantly increased in *max1*, *max2*, *max3*, and *max4* mutants. The *lbo* mutants show moderate increase in cauline branch number and do not enhance the shoot branching phenotype of *max4*. Similar to other SL biosynthesis mutants, grafting the *lbo* scion to wild-type rootstocks reduces branching in shoots of *lbo*, whereas branching of wild-type shoots is not effected by grafting to the *lbo* rootstocks. Meanwhile, the *lbo* rootstocks rescue shoot branching phenotypes of *max1* scions, but the *max1* rootstocks cannot reduce shoot branching in *lbo* scions, suggesting that LBO converts a mobile product of MAX1 to a product involved in branching inhibition. Furthermore, CL represses the shoot branching phenotype of *max4* but has little effect on the *lbo max4* double mutant. The CL and MeCLA levels are dramatically increased in *lbo*, and LBO converts MeCLA to an unidentified SL-like compound, suggesting that LBO is involved in SL biosynthesis downstream of MAX1 and regulates shoot branching in Arabidopsis (Brewer et al., 2016). Unknown enzymes may function in further steps in SL biosynthesis following

MAX1 to yield a variety of identified SL structures and potentially undiscovered SL active forms (Ruyter-Spira et al., 2013; Seto and Yamaguchi, 2014; Xie et al., 2010).

10.2.2.6   Karrikins

Karrikins are isolated from burned plant materials and shown to activate seed germination of fire-following plant species that otherwise remain dormant in the soil for long periods of time. Karrikins and SLs have common lactone (butenolide) and enol-ether moieties, thus showing similarity in chemical structures. In addition, SLs also trigger the germination of seeds that remain dormant in the soil for long periods of time. However the seeds awakened by SLs are typically those of root-parasitic plants such as *Striga* (witchweed), *Orobanche* and *Phelipanche* spp., and *Aletra* spp. (Smith and Li, 2014). In Arabidopsis, karrikins control seed germination after high temperature treatment and regulate photomorphogenesis of young seedling, including inhibition of hypocotyl elongation, promotion of cotyledon expansion and greening, as well as improvement of seedling vigor (Nelson et al., 2011; Smith and Li, 2014). Comprehensive reviews on karrikins are given elsewhere (Flematti et al., 2015; Waters et al., 2014).

10.3   Strigolactone transport

10.3.1   Strigolactone exudation from root into the rhizosphere

After biosynthesis, a portion of SLs is exuded from the root and enters the rhizosphere, where SLs can induce symbiosis between the plant and AM fungi. The mechanism of SL transport and exudation was firstly investigated in petunia (*P. hybrid*). In petunia, the PLEIOTROPIC DRUG RESISTANCE 1 (PDR1) protein plays a key role in regulating the development of AM and axillary branches (Kretzschmar et al., 2012). PDR1 belongs to the ATP BINDING CASSETTE (ABC) family of transporters, which is reported to be involved in the transport of other phytohormones, such as ABA and auxin (Li et al., Chapter 5 of this book; Kang et al., 2010; Kuromori et al., 2010; Petrasek and Friml, 2009; Qu et al., Chapter 2 of this book). Compared to the wild-type, SL levels are similar in root extracts of the *Phpdr1* mutants, but severely reduced in the root exudate of *Phpdr1*, resulting in a reduced symbiotic interaction with the AM fungal species *Glomus intraradices*. Overexpression of *P. axillaris PDR1* (*PaPDR1*) in *A. thaliana* resulted in an increased tolerance to synthetic SL, suggesting that SL export from the roots was increased in the transgenic plants (Kretzschmar et al., 2012).

Consistent with a role in SL exudation into the soil, PaPDR1 is present at the outer lateral membrane in the hypodermal passage cells that form gates for the entry of mycorrhizal fungi. The *Papdr1* mutant displays impaired transport of SLs out of the root tip to the rhizosphere (Sasse et al., 2015). Furthermore, the *PhPDR1* transcript levels increase in response to phosphate starvation, colonization by the AM fungus, and treatment with GR24 or the auxin analog 1-naphthaleneacetic acid (NAA) (Kretzschmar et al., 2012). PDR1 is the first known component in SL transport, providing new opportunities for investigating and manipulating the SL-dependent processes (Kretzschmar et al., 2012; Sasse et al., 2015).

10.3.2 Long-distance transport of strigolactones from root to shoot

In grafting experiments, the inhibition of shoot branching in SL-deficient scions by wild-type rootstocks demonstrates that SLs and their precursors are transported from root to shoot (Beveridge, 2006; Kohlen et al., 2011; Sorefan et al., 2003; Turnbull et al., 2002). Because *max1* rootstocks rescue the branching phenotypes of *max3* and *max4* scions, the substrate of MAX1 was considered likely to also be transported upward from the roots (Booker et al., 2005), and this has been confirmed following the discovery and synthesis of CL (Alder et al., 2012; Scaffidi et al., 2013). However the mechanism underlying how active SLs are transported to the shoot is still unclear. Functional analyses of PhPDR1 and PaPDR1 have shed light on this issue. In root tips, the expression of *PaPDR1* strongly overlaps with the SL biosynthetic gene *DAD1* (*CCD8*), and GFP-PDR1 is localized at the apical membrane of root hypodermal cells. Moreover, the SL transport from root tip to the shoot is impaired in *Papdr1*, suggesting that the polar localization of PaPDR1 mediates directional shootward SL transport (Sasse et al., 2015). In *Nicotiana tabacum*, *NtPDR6*, a homolog of *PhPDR1*, is preferentially expressed in roots and is strongly induced by phosphate starvation, as well as by NAA. Transgenic plants with knock-down of *NtPDR6* displayed a significantly increased branching phenotype, suggesting that *NtPDR6* plays a key role in regulation of shoot branching (Xie et al., 2015).

Further questions remain to be investigated in this area. In petunia, the *Phpdr1* mutant displays increased bud outgrowth but only at basal nodes, while other *dad* mutants are highly branched at more aerial nodes (Kretzschmar et al., 2012; Napoli, 1996; Snowden and Napoli, 2003). It is not clear whether the shoot phenotypes of *Phpdr1* are due to a deficiency in transport of SLs from root to shoot or in the transport of SLs produced in the shoot. Furthermore, there is no ortholog of PDR1 in *A. thaliana* which cannot establish symbiosis with AM fungi, suggesting that unknown transporter proteins may be involved in SL transport from root to shoot. Thus, the role of active SL transport in the shoot and the implications for branching control need further study.

10.4 Strigolactone signaling in plants

10.4.1 Perception of strigolactones

Main components of SL perception and signaling have been identified by characterizing SL-insensitive mutants in rice and Arabidopsis. The rice *d14* mutant is a typical high-tillering and dwarf mutant that is insensitive to SL treatment and accumulates an excess amount of SLs in plants (Arite et al., 2009; Umehara et al., 2008). In Arabidopsis, petunia, and pea, mutations in *D14* orthologs confer promotion of bud outgrowth and these phenotypes cannot be rescued by exogenous SL treatment (Gao et al., 2009; Hamiaux et al., 2012; Hoffmann et al., 2014; Liu et al., 2009; de Saint Germain et al., 2016; Waters et al., 2012b). Similar to *D14*, mutations in Arabidopsis *MAX2*, rice *D3* or pea *RMS4* promote bud outgrowth and cause more branching phenotypes, which cannot be rescued by SL treatment or by grafting to wild-type roots

(Beveridge et al., 1996; Gomez-Roldan et al., 2008; Ishikawa et al., 2005; Johnson et al., 2006; Stirnberg et al., 2002; Umehara et al., 2008). Moreover, the *max2* mutant displays additional phenotypes such as reduced germination and elongated hypocotyl (Nelson et al., 2011; Scaffidi et al., 2013; Shen et al., 2007; Stirnberg et al., 2002; Waters et al., 2012b). The *MAX2, D3,* or *RMS4* gene encodes an F-box protein, which is the substrate-recognition subunit of the SCF (Skp-Cullin-F-box) ubiquitin E3 ligase for proteasome-mediated proteolysis (Ishikawa et al., 2005; Johnson et al., 2006; Stirnberg et al., 2002). Perception and signaling of SLs requires the hormone-dependent interaction between D14 and D3, DAD2 and PhMAX2, or AtD14 and MAX2 in rice, petunia, or Arabidopsis, respectively.

The crystal structures and biochemical analyses of D14, DAD2, and AtD14 have revealed that these α/β-fold hydrolases have a large internal cavity and that the canonical catalytic triad (Ser, His, Asp) is required for SL perception and signal transduction (Hamiaux et al., 2012; Kagiyama et al., 2013; Zhao et al., 2013, 2015). The highly conserved catalytic triad of DAD2 and D14 are critical for hydrolysis of SLs which can induce the interaction between DAD2 and PhMAX2 in petunia and the interaction between D14 and D3 in rice (Hamiaux et al., 2012; Jiang et al., 2013; Smith and Waters, 2012; Zhao et al., 2014, 2015; Zhou et al., 2013). Furthermore, the crystal structure analysis of D14 incubated with *rac*-GR24 revealed that *rac*-GR24 was hydrolyzed into an intermediate 2,4,4,-trihydroxy-3-methyl-3-butenal (TMB), which was covalently attached to the active site Ser97 residue. The final products of hydrolysis, tricyclic lactone (ABC-OH) and deduced hydroxymethyl butenolide (D-OH), lose their SL biological activities (Hamiaux et al., 2012; Jiang et al., 2013; Nakamura et al., 2013; Zhao et al., 2013).

In 2016, a crystal structure of the *rac*-GR24-induced AtD14-D3-ASK1 complex was successfully achieved. In-depth analysis of this complex structure revealed that the binding of SLs to the receptor AtD14 induces an open-to-closed state transition to trigger SL signaling. In this process, SLs are hydrolyzed into a covalently linked intermediate molecule (CLIM) to trigger a conformational change of AtD14 to facilitate its interaction with D3 (Yao et al., 2016). CLIM is a small D-ring-derived intermediate with a molecular weight corresponding to the $C_5H_5O_2$, sealed inside the D3-bound AtD14, and is covalently linked with His247 and possibly Ser97 residues of AtD14. The AtD14 protein in the *rac*-GR24-induced AtD14-D3 complex (closed state) has experienced a significant conformational change. Genetic analysis of a highly branched Arabidopsis mutant *d14-5*, which results from a Gly158Glu substitution in AtD14, showed that *d14-5* exhibited reduced sensitivity in *rac*-GR24-inhibitory hypocotyl elongation and axillary bud growth, suggesting that SL signaling is severely impaired in *d14-5*. The AtD14 (Gly158Glu) mutant maintains enzyme activity to hydrolyze SLs but fails to efficiently interact with D3 or MAX2, and greatly attenuates the ability to act as a receptor *in planta*, indicating that the hydrolytic activity of AtD14 does not provide the signaling function of AtD14 and therefore it is the AtD14 interaction with D3 or MAX2 that initiates SL signaling (Yao et al., 2016).

In another study, enzymatic analyses using the fluorescent SL analogs indicate that the initial hydrolysis of these compounds is relatively fast, but the turnover to release the D-ring seems to be extremely slow. Furthermore, several substrates that do not have biological activity are cleaved by the receptor, but these substrates could not form the covalent linkage with the receptor due to lack of a methyl group on the D-ring (de Saint Germain et al., 2016). Taken together, these studies have led to the conclusion that CLIM functions as an active SL-derived molecule that initiates SL signaling and represents a novel active hormone molecule generated during AtD14-mediated SL hydrolysis and is sealed covalently inside AtD14 (de Saint Germain et al., 2016; Snowden and Janssen, 2016; Wang and Smith, 2016; Yao et al., 2016). This mechanism is different from all known active phytohormones that are generated by biosynthesis enzymes and then reversibly bound by their receptors to initiate signal transduction (see Chapters 2–9 in this book; Hothorn et al., 2011a,b; Murase et al., 2008; Nishimura et al., 2009; Santiago et al., 2009; She et al., 2011; Sheard et al., 2010; Tan et al., 2007).

10.4.2 Targets of strigolactone signaling

Several approaches have been taken to screen for the target proteins of the D14-SCF^{D3} ubiquitination complex. A major breakthrough in this field has come from the research on a rice dominant mutant, *d53*, which shows typical high-tillering and dwarf phenotypes and is insensitive to exogenous application of *rac*-GR24. Treatment of wild-type plants with *rac*-GR24 causes ubiquitination and degradation of D53 through the 26S proteasome system in a D14- and D3-dependent manner, whereas the dominant form of D53 is resistant to SL-mediated degradation. Upon SL treatment of rice callus, D53 interacts with D14 in a manner dependent on SL concentration, and mutations in the Ser-His-Asp catalytic triad of D14 could abolish or significantly attenuate the SL-induced degradation of D53 and reduce the interaction between D53 and D14. D53 could also interact with D3 with or without the presence of D14 and *rac*-GR24, suggesting that D53 could be directly targeted by D3 and that hormone-dependent association of D14 with D53 could trigger D53 ubiquitination and degradation (Jiang et al., 2013; Zhou et al., 2013). The D53 protein was further shown to contain three ethylene responsive element binding factor-associated amphiphilic repression (EAR) motifs and could interact with transcriptional corepressors known as TOPLESS (TPL) and TPL-related proteins (TPRs), which could potentially repress the activities of their target transcription factors (TFs) (Jiang et al., 2013; Smith and Li, 2014; Xiong et al., 2014).

Through a large-scale screening in Arabidopsis seedlings for suppressors of *max2*, *SUPPRESSOR OF MAX2 1* (*SMAX1*) was identified and characterized as a gene downstream of *MAX2* in the karrikin or SL signaling pathway. The *smax1* mutation rescues the elongated hypocotyl, small cotyledon size, and reduced seed germination of *max2*, but does not rescue the shoot branching phenotypes of *max2*. The growth of *smax1* seeds and seedlings mimics the wild-type treated with karrikin or with the SL analog *rac*-GR24, but *smax1* seedlings are still responsive to karrikin or *rac*-GR24, suggesting that other members of a *SMXL* gene family in Arabidopsis may act downstream of MAX2 to control the diverse developmental responses to karrikins and SLs (Stanga et al., 2013).

A. thaliana contains seven genes closely related to *SMAX1* and are designated *SMXLs* (*SMXL2-SMXL8*). Three of these proteins (SMXL6, SMXL7, and SMXL8) share 36%–41% identity with rice D53 and so were also designated as D53-like SMXLs (Jiang et al., 2013; Stanga et al., 2013). The D53-like *SMXL* genes are expressed in leaves and axillary branches, and they are induced by *rac*-GR24, suggesting that they may have a role in shoot development. Recently, two independent studies indicated that D53-like SMXLs regulate shoot branching and leaf development through the SL signaling pathway. In *A. thaliana*, these SMXL proteins can form a complex with the TPR2 protein and function as transcription repressors, but are targeted for degradation by SL-dependent interaction of AtD14 with SMXLs and with MAX2 (Soundappan et al., 2015; Wang et al., 2015). The *smxl6 smxl7 smxl8* triple mutant suppresses SL-related phenotypes in *max2*, including the highly branched phenotypes, increased auxin transport, and PIN-FORMED1 (PIN1) accumulation, as well as increased lateral root density, but shows little effect on the karrikin-regulated phenotypes of *max2*, such as seed germination and hypocotyl elongation. Exogenous application of *rac*-GR24, but not KAR$_1$, causes ubiquitination and degradation of SMXL6, 7, and 8 in a process that requires AtD14 and MAX2. The D53-like SMXL proteins form complexes with MAX2 and TPR2, and also interact with AtD14 in a *rac*-GR24-responsive manner. Furthermore, D53-like SMXLs exhibit TPR2-dependent transcriptional repression activity and repress the expression of *BRANCHED1* (*BRC1*). Altogether, these findings reveal that in Arabidopsis, D53-like proteins SMXL6, 7, and 8 act with TPR2 to repress transcription and so allow the outgrowth of lateral buds, but that SL-induced degradation of D53-like proteins activates transcription to inhibit outgrowth (Mach, 2015; Soundappan et al., 2015; Wang et al., 2015).

10.4.3   Strigolactone responsive genes

A better understanding of the SL signaling pathway requires the identification of downstream SL responsive genes, which could be causally linked to the control of shoot branching and to other biological functions of SLs. Up to now, only a few SL early responsive genes have been identified. In Arabidopsis and pea, *BRC1* is rapidly upregulated after SL treatment and BRC1 negatively regulates the development of rosette and cauline branches (Aguilar-Martinez et al., 2007; Braun et al., 2012; Dun et al., 2012). The BRC1 protein belongs to the TEOSINTE BRANCHED 1 (TB1), CYCLOIDEA (CYC), and PROLIFERATING CELL FACTORS (PCF) 1 and 2 (TCP) family. Loss-of-function mutations in *BRC1* and its orthologs result in an increase in shoot branches, exemplified by *tb1* in maize, *fine culm 1* (*fc1*) in rice, and *brc1* in Arabidopsis and pea (Aguilar-Martinez et al., 2007; Braun et al., 2012; Doebley et al., 1997; Finlayson, 2007; Takeda et al., 2003). In pea, *BRC1* expression is dramatically reduced in the buds of SL mutants compared to the wild-type buds, and exogenous SL treatment of buds causes upregulation of *BRC1* within 6 h in a cycloheximide-independent manner, suggesting that *BRC1* is a primary responsive gene of SLs (Braun et al., 2012; Dun et al., 2012). In Arabidopsis, *BRC1* expression is dramatically increased in the non-elongated axillary buds of primary rosette branches and secondary cauline branches in *smxl6/7/8*, *max2 smxl6/7/8*, and *max3 smxl6/7/8* triple and quadruple mutants, while *BRC1* expression is significantly reduced in the non-elongated buds of *max2* and *max3*, suggesting that SLs probably regulate shoot branching through activating the expression of *BRC1* (Soundappan et al., 2015; Wang et al., 2015). Whether *TB1* in maize and *FC1* in rice function in a similar manner as the primary responsive downstream genes in pea and Arabidopsis needs further investigation.

Negative feedback on several SL biosynthesis and signaling genes has been observed in rice, pea, and Arabidopsis. In Arabidopsis, *MAX3* and *MAX4* are downregulated by *rac*-GR24 or GR24^{5DS} treatment within 2 h and this repression is dependent on *D14* and *MAX2*. Compared with the wild-type, *MAX3* and *MAX4* expression levels are higher in *d14*, *max2*, and SL biosynthesis mutants, but lower in *smxl6/7/8*, *max2 smxl6/7/8*, and *max3 smxl6/7/8* (Mashiguchi et al., 2009; Wang et al., 2015). Similar results have been observed in pea and rice (Arite et al., 2007; Johnson et al., 2006). Moreover, the SL biosynthesis genes are upregulated by auxin in Arabidopsis, rice, and pea, which provide a mechanism for synergistic repression of bud outgrowth by auxin and SLs (Arite et al., 2007; Foo et al., 2005; Sorefan et al., 2003; Zhang et al., 2010). Furthermore, activation of SL signaling by *rac*-GR24 treatment upregulates transcription of *D53* and *D53-like SMXLs* genes after 1 h or longer time period, while deficiency in SL biosynthesis and signaling as shown in *d* mutants results in downregulation of *D53* transcription, suggesting that *D53* expression is subject to a negative feedback control by SLs (Jiang et al., 2013; Stanga et al., 2013). Microarray analyses revealed that 31 genes (including *BRC1*) were induced by exogenous *rac*-GR24 and 33 genes (including *MAX3* and *MAX4*) were repressed by *rac*-GR24. Most of the SL-repressible genes belong to auxin-inducible genes, and two SL-inducible genes might be involved in light signaling (Mashiguchi et al., 2009). However genetic evidence of the candidate SL responsive genes is absent, and the reported candidate genes cannot explain diverse functions of SLs on plant developments. Further characterization of SL responsive genes is critical for investigating the SL signaling pathway.

10.4.4 *Model of strigolactone signaling*

Based on existing evidence to date, a working model has been proposed for the SL perception and signaling pathway (Fig. 10.3). In the presence of SLs, D14 binds SLs and then achieves a nucleophilic attack to form a D-ring-derived molecule (CLIM), which is covalently sealed in the catalytic active site pocket of D14. This triggers a conformational change of D14, leading to interaction with the D3/MAX2-based SCF complex and D53/D53-like SMXL proteins. D53 and D53-like SMXLs then undergo ubiquitination and degradation which relieves the transcriptional repression of key downstream genes such as *D53* and *BRC1*. In the absence or an extreme low level of SLs, D53 and D53-like SMXL proteins interact with TPL/TPR proteins and repress downstream target genes through repressing the activities of unknown TFs (Jiang et al., 2013; Soundappan et al., 2015; Smith and Li, 2014; Wang et al., 2015; Yao et al., 2016; Zhou et al., 2013).

In addition to the proposed role of D53-like proteins in the control of transcription, SMXL6, 7, and 8 have putative chaperoning functions since TPL proteins have been reported to participate in vesicle trafficking, suggesting that these proteins might regulate plant development through SL-induced posttranscriptional events (Soundappan et al., 2015; Waldie et al., 2014). Recently, genetic analyses using a series of SMXL7 protein variants with altered subcellular localization and SL-induced degradation indicate that different developmental processes show differential sensitivity to the loss of the EAR motif from SMXL7, raising the possibility that there may be several distinct mechanisms that play downstream of SMXL7 (Liang et al., 2016).

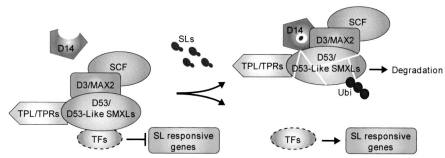

Figure 10.3 Model of the strigolactone signaling pathway. In the absence of SLs, D14 stays in an open state with a vacant active site pocket, and D53/D53-like SMXLs interact with TOPLESS/TPL-RELATED (TPL/TPR) proteins to repress the expression of SL responsive genes. In the presence of SLs, D14 docks and attacks SLs to form a D-ring-derived molecule (CLIM), which is covalently linked with the catalytic center of D14 and triggers a conformational change to form the D14-SCF$^{D3/MAX2}$ complex. Meanwhile, D53/D53-like SMXLs form a complex with D14 and D3/MAX2, and subsequently undergo ubiquitination and degradation which relieves the transcriptional repression of SL responsive genes. TFs are assumed to participate in this process, but it is unknown if the repression of SL responsive genes is direct or indirect. *TFs*, transcription factors (unidentified); *Ubi*, ubiquitin.

10.4.5   Similarities and differences between strigolactone and karrikin signaling pathways

The Arabidopsis protein KARRIKIN INSENSITIVE2 (KAI2), also known as HYPOSENSITIVE TO LIGHT (HTL) and DWARF14-LIKE (D14L), is required for response to karrikins, which are compounds containing a butenolide similar to SLs, but present in smoke of wildfires (Guo et al., 2013; Sun and Ni, 2011; Waters et al., 2012b). The activities of KAI2/HTL/D14L protein include control of seed germination and seedling photomorphogenesis. It has an essential catalytic triad, and its function requires MAX2 (the ortholog of D3) and SMAX1 (a D53-like protein). Genetic studies in Arabidopsis have identified KAI2 and MAX2 as two key players in karrikin signaling (Nelson et al., 2011; Waters et al., 2012b). Meanwhile, KAI2/HTL/D14L does not respond in vivo to canonical SLs produced by the known SL biosynthetic pathway (Al-Babili and Bouwmeester, 2015; Waters et al., 2014, 2015). Thus, SLs and karrikins share a structural similarity (the butenolide ring), bind to two homologous receptors (D14 and KAI2/HTL/D14L, respectively), and use MAX2 for signal transduction. Moreover, the specificity of SL and karrikin pathways is ensured by signaling through either D14 or KAI2/HTL/D14L. Differential responses to SLs and karrikins are mediated by members of the eight-gene family of SMXL proteins in *A. thaliana*. As described above SMXL6, SMXL7, and SMXL8 are targets of SL signaling, whereas SMAX1 and SMXL2 might mediate responses to karrikins (Stanga et al., 2016). The *smxl6/7/8* triple mutant and *smax1* mutants restore different aspects of the *max2* mutant phenotypes, indicating that different SMXL family members enable distinct MAX2-dependent responses to SLs and karrikins respectively and play diverse roles in plant development (Smith and Li, 2014; Soundappan et al., 2015; Wang et al., 2015).

From an evolutionary perspective, *KAI2/HTL/D14L* homologs are present in basal land plants such as *Physcomitrella patens*, *Marchantia polymorpha*, and *Selaginella moellendorffii*. An ancient KAI2-type protein from *S. moellendorffii* can function in Arabidopsis seedlings, but does not respond to SLs or karrikins (Waters et al., 2015). Nevertheless, the close parallels between D14 and KAI2/HTL/D14L signaling have led to speculation that KAI2/HTL/D14L proteins could respond to SLs under some circumstances or respond to specific types of SLs. Some algae and mosses can respond to SL analogs such as *rac*-GR24 and are reported to contain canonical SLs. Since they do not have recognizable D14-type proteins, it is likely that they use KAI2-type proteins for perception of SL-type signals (Delaux et al., 2012; Proust et al., 2011).

10.5   Strigolactones and parasitism

10.5.1   Strigolactones induce germination of parasitic plants

The first identified natural SL was strigol, isolated from root exudates as a germination stimulant of the root-parasitic plant witchweed *Striga lutea* (Cook et al., 1966). Since then, a range of SLs have been identified as germination stimulants of root-parasitic plants that belong to the broomrapes (*Orobanche* and *Phelipanche* spp.), witchweeds (*Striga* spp.) and *Alectra* spp. (Xie et al., 2009, 2010; Yokota et al., 1998). Root-parasitic plants exploit other flowering plants for water, assimilates, and nutrients and can be harmful to their host plants (Bouwmeester et al., 2003). Such root-parasitic plants cause significant losses to agricultural production all over the world. Species of the *Striga* genus are hemiparasites that parasitize important staple crops including maize, millet, sorghum, and rice (Westwood et al., 2010). In sub-Saharan Africa, nearly 300 million people are adversely affected by *Striga* infestation, which causes annual crop losses of about 10 billion dollars (Ejeta and Gressel, 2007). Furthermore, in semiarid areas of sub-Saharan Africa, *Alectra vogelii* causes severe yield losses of leguminous crops, especially cowpea (Joel et al., 2007). Broomrapes also constitute a serious threat to agriculture and are weeds in crops such as legumes, tomato, sunflower, and rapeseed growing under more temperate climatic conditions (Rubiales et al., 2009). Excellent reviews on the biology (Musselman, 1980), physiology and biochemistry (Stewart and Press, 1990), and biology and management (Joel et al., 2007) of root parasites have been published. Although immense progress has been made on the biology, genetics, biochemistry, physiology, and ecology of *Striga*, *Orobanche*, and *Phelipanche*, understanding of the mechanisms underlying their host recognition remain challenging.

10.5.2   Strigolactone perception by parasitic plants

Exudation of SLs from the roots of host plants can induce seed germination of parasitic plants at extremely low concentration. However the SL perception system in such seeds was unknown until recently. Although Arabidopsis and rice have single copies of *D14* and *KAI2/HTL/D14L* genes, in root-parasitic plants the *KAI2/HTL/D14L* gene has undergone extensive amplification and is present in up to 11 copies in *Striga* (Conn et al., 2015; Toh et al., 2015). This raises the possibility that these genes are required for SL perception. Because *Striga* itself is difficult to study and, as a noxious weed, is generally inaccessible to most researchers, a new bioassay system was developed to systematically discriminate the roles of *Striga HTL/KAI2* homologs (*ShHTLs* or *ShKAI2s*) in SL perception, using Arabidopsis as a surrogate (Toh et al., 2015). Seeds of wild-type Arabidopsis germinate poorly after exposing to high temperature, but this thermoinhibition is alleviated by addition of *rac*-GR24 and is dependent on a functional *HTL/KAI2* gene (Toh et al., 2012). Thus, the 11 *ShHTLs* were individually expressed in an Arabidopsis *HTL/KAI2* loss-of-function mutant *htl-3* and their effects on seed germination after exposing to high temperature were assayed. Surprisingly, the *ShHTL7* transgenic seeds were four to five orders of magnitude more responsive to *rac*-GR24 than other lines including *AtHTL* transgenic seeds. This picomolar level of sensitivity is in the realm of *Striga* SL sensitivity, suggesting that ShHTL7 functions as a highly sensitive SL receptor from *Striga* and that different ShHTL hydrolases are sufficient to confer the extreme sensitivity of parasitic plants to SLs. Further analysis of crystal structure revealed that the binding pocket of ShHTL7 is larger than that of AtHTL, which could explain its ability to recognize SLs and the increased range of SL sensitivity. In an SL bioassay system, the thermoinhibited *ShHTL7* seed responded to *rac*-GR24 concentrations as low as 2 pM. The seeds also germinated after being sprinkled on the roots of rice that secrete either low or high levels of SLs, with more germination on the latter (Toh et al., 2015). The Arabidopsis seeds expressing *ShHTL7* could function as a bioassay material to identify chemicals that are effective on the control of *Striga* infestations with low cost and to breed crops with altered SL levels (Toh et al., 2015).

Independent research (Conn et al., 2015) showed that the *KAI2/HTL* paralogs in parasites are divided into three phylogenetic clades which are different in evolutionary selection rates, ligand-binding pocket shape and ligand-specificity. These clades were described as conserved (c), divergent (d), and intermediate (i). Homology modeling based on an AtKAI2 structure and crystal structure of ShHTL5, a highly sensitive SL receptor from *S. hermonthica*, indicates that the ShKAI2d paralogs have larger ligand-binding pockets than ShKAI2c or ShKAI2i in parasites, as well as AtKAI2. The parasite-specific KAI2d clade has undergone the fastest evolution and contains the majority of KAI2 paralogs. Furthermore, parasite-specific *KAI2d* transgenes confer SL-specific germination on the Arabidopsis *kai2* seed. Thus, the KAI2 paralog D14 and the parasite-specific KAI2d apparently underwent convergent evolution of SL recognition, enabling developmental responses to SLs in angiosperms and host detection in parasites, respectively (Conn et al., 2015).

Other research (Tsuchiya et al., 2015) developed two fluorogenic agonists, Yoshimulactone Green (YLG) and Yoshimulactone Green Double (YLGW), which could be recognized by SL receptors, with subsequent hydrolysis leading to the generation of fluorescent products. Biochemical and physiological analyses indicate that YLG works as an in vitro and in vivo fluorogenic agonist for AtD14 in Arabidopsis. YLG also stimulates *Striga* germination and functions as a fluorogenic agonist in *Striga*, which cleaves the ligand as perceived. YLG and *rac*-GR24 stimulated germination in the Arabidopsis *ShHTL7* transgenic lines with the *htl-3* background but not in the parental *htl-3* mutant, indicating that *ShHTL7* is a functional SL receptor that can support germination in Arabidopsis. Moreover, live imaging using YLGW revealed that a dynamic wavelike propagation of SL hydrolysis occurs during *Striga* seed germination. This research potentially tracks signal perception by the SL receptor in intact *Striga* seeds, and indicates that *ShHTLs* probably function as the SL receptors mediating seed germination in *Striga* (Tsuchiya et al., 2015).

Taken together, these exciting studies indicate that the sensitivity of *Striga* to SLs from host plants is driven by receptor sensitivity, and they enable observation of the regulatory dynamics for SL signal transduction in *Striga*. The transgenic plants expressing *ShHTL7* and *KAI2d*, as well as the small-molecule fluorogenic reporter, also provide useful systems for screening chemical libraries for KAI2d agonists and for developing approaches to improve crops. These studies will greatly accelerate research to combat parasite infestation and will also aid investigations of how parasites rapidly evolve new host specificities and become virulent agricultural weeds (Conn et al., 2015; Jamil et al., 2011b; Toh et al., 2015; Tsuchiya et al., 2015).

10.6 Strigolactones and symbiosis

10.6.1 Strigolactones stimulate symbiosis between plants and AM fungi

Secretion of SLs from roots mediates symbiosis with AM fungi which facilitates the absorption of water and nutrients, especially phosphate and nitrate (Al-Babili and Bouwmeester, 2015). In more than 80% of land plants, AM fungi form mutualistic associations with roots of the host plants to obtain photoassimilates for growth (Parniske, 2008). The hyphae of the fungus form an extensive network and greatly extend the root system of host plants, facilitating the absorption of water and nutrients from soil (Ruyter-Spira et al., 2013).

The spores of AM fungi can germinate and grow in the absence of a host root, but cannot complete their life cycles. In the first stages of host recognition, the hyphae of AM fungi show extensive branching near host roots and then form the appressorium to penetrate the plant root. Meanwhile, host roots release signaling molecules, such as 5DS, to trigger hyphal branching. The natural SLs (5DS, sorgolactone, and strigol) and the synthetic SL analog *rac*-GR24, stimulate extensive hyphal branching in germinating spores of the AM fungus *Gigaspora margarita* at very low concentrations, which is critical for the root colonization process (Akiyama et al., 2005). Consistent with their role in activating AM fungus symbiosis, SL biosynthesis and exudation from roots are upregulated in response to nutrient deficiency and mutants defective in SL biosynthesis or exudation are impaired in their ability to form a symbiosis with AM fungi (Jamil et al., 2011a, 2012; Kohlen et al., 2011; Lopez-Raez et al., 2008; Umehara et al., 2008; Yoneyama et al., 2007).

10.6.2   Perception of symbiotic AM fungi in rice

The mechanism by which SLs are perceived by plants or fungi is a great challenge for this research area. Recent research has indicated that the α/β-fold hydrolase D14L in rice is required for establishing symbiosis with either of two AM fungi, *Rhizophagus irregularis* and *Gigaspora rosea* (Gutjahr et al., 2015). The D14L protein is homologous to Arabidopsis KAI2/HTL, which acts together with MAX2 in karrikin perception. The rice *d3* mutant is impaired in AM colonization, while the *d14* mutant is not perturbed in AM symbiosis, indicating that the karrikin receptor complex is important for the earliest stages of AM development. Furthermore, the germinated spore exudates (GSEs) of AM fungi activate pre-contact plant responses, which rely on a functional D14L. However there is no evidence for karrikin perception in rice and exogenous application of karrikins could not stimulate colonization of wild-type roots by *R. irregularis* (Gutjahr et al., 2015). It is likely that the natural substrate for D14L is an unknown karrikin-like molecule.

This study reveals an unexpected strategy for recognition of AM fungi and a previously unknown signaling link between symbiosis and plant development. On poor natural soils, plants rely on AM fungi for water and mineral nutrient supply and need to coordinate AM fungi development with their physiological and developmental needs. The karrikin receptor complex may represent a node in the cross talk between plant developments and AM signaling (Gutjahr et al., 2015). Further characterization of other components in this signaling pathway could facilitate crop breeding in infertile areas.

10.7   Cross talk between strigolactones and other signals

10.7.1   Strigolactones and phosphate

Under phosphate deficiency higher plants increase SL production, which changes root architecture and promotes fungal symbiosis to enhance phosphate uptake (Akiyama et al., 2005; Kapulnik and Koltai, 2016). Meanwhile to channel resources away from unessential shoot tissues into the main stem and root system, the shoot also changes its architecture and development, such as decreased shoot branching, enhanced leaf senescence, and increased root-to-shoot ratio. This hormonal response represents a critical adaptive change to whole plant architecture to optimize development under diverse environmental conditions (Brewer et al., 2013).

Generally, plants growing under favorable conditions in the laboratory display extremely low endogenous SL levels. However, when plants encounter phosphate deprivation, SL biosynthesis and release are promoted, which is revealed at the transcriptional level (Kohlen et al., 2011; Umehara et al., 2008). For example, SLs accumulate dramatically in red clover grown under low phosphate conditions (Yoneyama et al., 2007). Under phosphate deprivation, the SL levels in plant samples could increase as much as 100,000-fold, which greatly facilitates research on SL detection and metabolism (Yoneyama et al., 2011). Elevated levels of SLs could inhibit shoot branching (Kohlen et al., 2011; Umehara et al., 2010), promote lateral root formation (Ruyter-Spira et al., 2011), and improve root hair density (Mayzlish-Gati et al., 2012). In rice, phosphate limitation enhances the expression of *D10*, *D17*, and *D27* and stimulates SL production (Sun et al., 2014). In *Petunia hybrida*, low- and minus-phosphate treatment increases the abundance of *CCD7*, *CCD8*, and *PDR1* transcripts (Drummond et al., 2015). In *Medicago truncatula*, low phosphate and nitrogen also up-regulate the expression of *MtD27*, *MtCCD7*, and *MtCCD8* (Bonneau et al., 2013).

Consistent with these observations, SL biosynthetic or signaling mutants are impaired in response to low phosphate conditions (Kohlen et al., 2011; Mayzlish-Gati et al., 2012; Umehara et al., 2008). In Arabidopsis, mutants of SL signaling (*max2*) and biosynthesis (*max4*) that were subjected to phosphate starvation, show reduced root hair density and reduced expression levels of genes that are normally induced under phosphate starvation relative to wild-type plants. Application of *rac*-GR24 could compensate for the reduction in response to low phosphate in *max4*, but not in *max2* (Mayzlish-Gati et al., 2012). In wild-type rice, phosphate starvation leads to elongation of crown roots and seminal roots and a decrease in lateral root density. The rice dwarf mutants that are SL-deficient or SL-insensitive exhibit reduced response to phosphate deficiency including short crown root phenotypes. Exogenous application of *rac*-GR24 restores the crown root and seminal root length and lateral root density in the SL-deficient mutants but not in SL-insensitive mutants, suggesting that SLs are induced by nutrient-limiting conditions and lead to changes in rice root growth through the SL signaling pathway (Arite et al., 2012; Sun et al., 2014).

10.7.2   Strigolactones and nitrogen

Besides the roles of SLs in regulating phosphate acquisition, they are required for the regulation of shoot branching by nitrogen supply in Arabidopsis. Limited nitrate availability can have profound effect on the architecture and biomass allocation of both root and shoot systems, shifting the balance in favor of the root (de Jong et al., 2014). In shoots, nitrogen deficiency affects anthocyanin accumulation, plant growth and senescence of leaves (Vidal and Gutierrez, 2008). Treatment of wild-type seedlings with SL also induces anthocyanin accumulation, promotes leaf senescence, and reduces plant growth, suggesting that SL signaling may have overlapping functions with nitrogen signaling pathways in Arabidopsis (Ito et al., 2015; Yamada et al., 2014). In addition, SL signaling regulates nitrogen-deficiency-induced anthocyanin accumulation, growth retardation, and reduction of chlorophyll content (Ito et al., 2015, 2016). Compared to wild-type, nitrogen deficiency has a lower impact on the growth of *max1-1* (SL biosynthesis mutant) and *Atd14-1* (SL-insensitive mutant) plants. Furthermore, nitrogen deficiency enhances the expression of SL biosynthesis genes such as *MAX3* and *MAX4* (Ito et al., 2016).

Recently, auxin and SLs were reported to be important mediators in shoot branching suppression under low nitrate conditions (de Jong et al., 2014). Physiological analysis indicates that nitrate limitation delays axillary bud activation and attenuates the basipetal sequence of bud activation, thus reducing shoot branching. The response of branching to nitrogen limitation in wild-type plants was compared with that of the auxin response mutant *axr1* and the SL biosynthesis mutant *max1*. The single mutants respond to nitrogen limitation by reducing branch numbers but the *axr1 max1* double mutant displays no significant change in branch numbers on high- and low-nitrate conditions, suggesting that both auxin and SL signaling are involved in the shoot branch repression in response to low nitrate. In addition, nitrogen deprivation and SL treatment have similar effects on bud growth, while auxin transport in the main stem is unaffected by nitrogen availability. Under limiting nitrate supply, plants show a shift toward an increased proportion of biomass in the root, while in the *max1* mutant this shift is attenuated, suggesting that SL deficiency changes the relative allocation of biomass to the root and shoot (de Jong et al., 2014).

10.7.3 Strigolactones regulate drought and high salinity response

Plants, including agriculturally important crops, are continuously exposed to changing environments. Drought and high salinity adversely affect plant growth and productivity. Various phytohormones including ABA, brassinosteroid, and cytokinin cooperatively regulate plant stress responses (Choudhary et al., 2012; Feng et al., Chapter 3 of this book; Fujita et al., 2011; Ha et al., 2012; Li et al., Chapter 5 of this book; Wang et al., Chapter 9 of this book). Recently, the roles of SLs in adaptive responses to environmental stresses have been discovered and potentially provide opportunities for the improvement of stress tolerance in crops by manipulating SL biosynthetic and signaling pathways (Ha et al., 2014; Lopez-Raez, 2016).

As mentioned above, SLs secreted from roots mediate symbiosis with AM fungi, which could improve water absorption of the host plants and alleviate drought stress (Evelin et al., 2009). It has been estimated that about 20% of the total plant water uptake can be accomplished by hyphae of fungi (Ruth et al., 2011). The fungal hyphae are thinner than roots and are able to enter the water-filled pores in soil that are not accessible to the roots (Smith et al., 2010). Furthermore, AM symbiosis alleviates drought stress by altering the hormonal profiles and affecting plant physiology in the host plant (Ruiz-Lozano et al., 2016). In tomato and lettuce, root colonization by *Rhizophagus irregularis* is greater in plants subjected to drought stress treatments than those under well-watered conditions. Under drought stress conditions, the AM plants show an improved growth rate and efficiency of photosystem II from very early stages of plant colonization compared to non-AM plants. Meanwhile, ABA accumulation and upregulation of transcripts of ABA-biosynthesis and ABA-responsive genes upon drought treatment are greater in AM plants. The levels of SLs and the expression of corresponding marker genes are affected by both AM symbiosis and drought (Ruiz-Lozano et al., 2016).

In addition, in Arabidopsis that cannot establish symbiosis with AM fungi, SLs also act as positive regulators of plant response to drought and salt stresses (Ha et al., 2014). The SL-deficient and SL-response mutants in Arabidopsis exhibit hypersensitivity to drought and salt stresses. Exogenous SL treatment rescues the drought-sensitive phenotypes of the SL-deficient mutants but not of the SL-response mutants, and moderately enhances drought tolerance of wild-type plants. Compared with the wild-type, the SL-deficient and SL-response mutants exhibit increased rate of water loss in leaf during dehydration, slower ABA-induced stomatal closure, and decreased ABA responsiveness, suggesting that SLs and ABA may cooperatively regulate stomatal development and drought response. Comparative transcriptome analysis suggests that plants integrate multiple hormone response pathways for drought stress adaptation, but critical genes working downstream of SLs to regulate drought response need to be further identified and analyzed (Ha et al., 2014). Taken together, SLs alleviate drought and salt stresses under unfavorable environmental conditions by promoting the establishment of AM symbiosis and positively regulating drought and high salinity responses in shoot.

10.7.4 Strigolactones and light

Light is another environmental factor that affects SL levels or signaling and is particularly relevant to shading responses (Brewer et al., 2013). Under red light stimulation, PHYTOCHROME B (PHYB), the key photoperiod-determining protein, is activated and enters the nucleus to regulate gene expression (Martinez-Garcia et al., 2000). Under normal light conditions with plenty of red light, the *phyB* mutant shows impairment in response to a high red:far-red ratio, and displays a tall slender plant phenotype with reduced branching (Finlayson et al., 2010). The biosynthesis mutant *max4*; an SL-insensitive mutant, *max2*; and *brc1*, a mutant defective in downstream response could repress the *phyB* phenotypes under high red:far-red light, suggesting that SLs may act downstream of the PHYB-dependent response to high red:far-red. Under low red:far-red light conditions, shoot branching of the wild-type plants is reduced, but the *brc1* mutant plants still display higher numbers of branches (Finlayson et al., 2010). Thus, SLs may be involved in growth regulation under low and high red:far-red light conditions.

In addition, the *max2* mutant shows deficiency in both light response and seedling development, such as decreased expression of *ELONGATED HYPOCOTYL 5* (*HY5*), increased hypocotyl length, and attenuated seed germination (Nelson et al., 2011; Shen et al., 2007; Stirnberg et al., 2002; Tsuchiya et al., 2010). The F-box protein MAX2 functions as a key component in both SL and karrikin signaling (Nelson et al., 2011; Stirnberg et al., 2002). Thus, it is unclear whether the *max2*-related light response phenotypes are mediated by the SL signaling pathway or other mechanisms. Moreover, SLs are linked to chlorophyll production in leaves and function as positive regulators of light-harvesting genes in tomato (Mayzlish-Gati et al., 2010). Further characterization of the SL responsive genes that are involved in light response and investigation of the relationships among CONSTITUTIVE PHOTOMORPHOGENIC1 (COP1), HY5, and SL signaling components will help us understand the cross talk between SLs and light response.

10.7.5 Strigolactones and auxin

Auxin and SLs play important roles in the regulation of shoot branching and thus have profound effect on plant architecture. Auxin is mainly synthesized in the young expanding leaves at the shoot apex and is transported downward in the PAT stream to regulate bud outgrowth (Ljung et al., 2001; Stirnberg et al., 2010). Apically produced auxin cannot move upward and so a second upwardly mobile signal is needed that moves upward into the buds to repress their growth. It has been suggested that SLs and cytokinin act as long-distance second messengers of auxin to repress bud outgrowth, but play antagonistic roles with each other (Brewer et al., 2009; Dun et al., 2012). SLs directly inhibit bud outgrowth (Gomez-Roldan et al., 2008; Umehara et al., 2008), while cytokinin could directly promote bud outgrowth even in the presence of apical auxin (Sachs and Thimann, 1967; Wickson and Thimann, 1958). In garden pea, SLs and cytokinin display antagonistic effects on *BRC1,* which is specifically expressed in buds and represses bud outgrowth (Aguilar-Martinez et al., 2007; Braun et al., 2012; Dun et al., 2012; Finlayson, 2007).

Auxin promotes the transcription of SL biosynthetic genes *MAX3, MAX4,* and their homologs in pea, petunia, and rice, and enhances the production of SLs (Arite et al., 2007; Brewer et al., 2009; Foo et al., 2005; Hayward et al., 2009; Johnson et al., 2006; Liang et al., 2010; Sorefan et al., 2003). Removing the source of auxin by apical decapitation leads to a strong decrease in the expression of these SL biosynthesis genes and decreases the levels of SLs (Arite et al., 2007; Brewer et al., 2009; Foo et al., 2005; Hayward et al., 2009; Johnson et al., 2006). Meanwhile, auxin reduces cytokinin synthesis by regulating the expression of *ADENYLATE ISOPENTENYLTRANSFERASE* (*IPT*) family members, which have key roles in cytokinin biosynthesis (Feng et al., Chapter 3 of this book; Nordstrom et al., 2004; Tanaka et al., 2006). The effects on both SLs and cytokinins are mediated by the classical AUXIN RESISTANCE PROTEIN 1 (AXR1)-AUXIN SIGNALING F-BOX PROTEIN (AFB)-dependent auxin signaling pathway (Hayward et al., 2009; Nordstrom et al., 2004; Qu et al., Chapter 2 of this book). Thus, in the secondary messenger model SLs and cytokinin are considered as the second messengers that move directly into the bud to control its activity (Domagalska and Leyser, 2011).

Meanwhile, an auxin transport canalization-based model has been proposed to explain the SL inhibition of bud outgrowth (Domagalska and Leyser, 2011). In Arabidopsis, the auxin efflux transporter PIN1 is localized at the basal membrane of cells in the PAT stream and facilitates the basipetal (downwards) transport of auxin (Galweiler et al., 1998; Wisniewska et al., 2006). Auxin transport in the main stem prevents axillary buds from establishing the PAT channel to the main stem, thus inhibiting bud activation and outgrowth (Balla et al., 2011; Bennett et al., 2006; Li and Bangerth, 1999; Prusinkiewicz et al., 2009; Shinohara et al., 2013). In the auxin transport canalization model, buds act as auxin sources and the stem acts as an auxin sink. The auxin transport canalization hypothesis proposes that an initial auxin flux between a source and a sink positively regulates and polarizes auxin transport and helps direct the initial flux to a canal, thus establishing efficient auxin transport from the source to the sink (Sachs, 1981, 2000). To activate dormant buds, they need to set up an efficient auxin flow into the main stem (Li and Bangerth, 1999). Thus buds from different nodes compete for the common auxin transport pathway through the main stem to root. This allows auxin transport from shoot apex or the apical buds to regulate bud activation in an indirect way (Domagalska and Leyser, 2011). In rice and Arabidopsis SL biosynthesis and signaling mutants, the PAT is elevated and the PIN1 protein is accumulated on basal membranes of xylem parenchyma cells in a MAX2-dependent manner (Bennett et al., 2006; Crawford et al., 2010; Lazar and Goodman, 2006; Lin et al., 2009). Thus, SLs act systemically to dampen PAT and reduce PIN1 accumulation on cell membranes, which enhances the competition between buds for the common auxin sink in the stem (Domagalska and Leyser, 2011). However, there are also some conflicting data in pea, in which some experiments show an SL-mediated regulation on PAT but others show SL inhibition of branching is independent of PAT. In SL-deficient mutants, the auxin transport inhibitor N-1-naphthylphthalamic acid (NPA) has little effect on bud outgrowth in pea. SLs and NPA show additive effects on inhibition of bud outgrowth, and the *yucca1D* mutants with enhanced auxin biosynthesis suppress branching in wild-type plants but not in the SL-deficient mutant *max3* background, suggesting that SLs may inhibit bud outgrowth through the direct action model (Brewer et al., 2009, 2015).

10.8   Summary points

- SLs were originally discovered in root exudates and serve as rhizosphere signals that stimulate seed germination of parasitic plants and enable AM fungi to form associations with their hosts.
- The naturally occurring SLs comprise a common structure in which a tricyclic lactone ABC-ring is connected with a butenolide D-ring group through an enol-ether bridge.
- Highly branching or tillering mutants in Arabidopsis, petunia, pea, and rice led to the discovery of SL biosynthetic enzymes and signaling components.
- SLs are derived from a carotenoid precursor, which undergoes isomerization, successive cleavage, oxidation, and further modification to produce diverse SL structures.
- SLs are transported from root to shoot where they play systemic and critical roles in the regulation of shoot architecture; meanwhile SLs in roots influence the development of lateral roots and root hairs.
- Perception of SLs depends on a novel mechanism employing a serine hydrolase-type enzyme-receptor D14 which attacks SLs and becomes covalently modified, and subsequently triggers the interaction of D14 with the F-box protein D3 (in rice) or MAX2 (in Arabidopsis), leading to ubiquitination and degradation of transcriptional regulator D53 or D53-like SMXLs, respectively.
- SLs interact with various environmental signals and other phytohormones to control whole plant architecture and optimize the root and shoot development under changing environments.

10.9   Future perspectives

- The key components of SL biosynthesis have been identified from genetic studies using shoot branching mutants, in combination with biochemical analyses of important enzymes and metabolic intermediates. However, the active forms of SLs and their functions in plant development remain to be elucidated. Understanding the biochemical function of MAX1 and its orthologs in different species, as well as identification of new enzymes involved in SL biosynthesis, will provide critical information for the whole picture of the SL biosynthetic pathway.

- CLIM has been characterized as an active SL-derived intermediate molecule that is covalently linked to AtD14 and triggers a conformational change of AtD14. It is still unknown whether other structures of the intermediate molecules exist in vivo, and whether D14 recognizes different forms of SLs and transduces the signal for diverse roles of SLs.
- Due to the covalent linkage between D14 and CLIM, the mechanism of inactivation of the SL signal triggered by the D14-SCF$^{D3/MAX2}$-D53 complex becomes an open question. Recently, AtD14 has been shown to undergo 26S-proteasome-dependent degradation which may function as part of a negative feedback mechanism (Chevalier et al., 2014), but the molecular mechanism of D14 degradation is unknown.
- The sequence in which the SL receptor, the SCF complex, and the proteins targeted for degradation are assembled into a super complex is not clear. It is unknown whether D53 and D53-like proteins interact with TFs to regulate the expression of downstream target genes and diverse developmental processes.
- The genes responding rapidly to SL treatment need to be further explored. Moreover, it is still unknown how SLs regulate different aspects of plant development and responses to environmental signals. Elucidation of signaling mechanisms downstream of the D14-SCF$^{D3/MAX2}$-D53 complex are critical to understanding the perception and signal transduction of SLs in plant cells.

Abbreviations

4DO	4-Deoxyorobanchol
5DS	5-Deoxystrigol
ABA	Abscisic acid
ABC	ATP BINDING CASSETTE
AM	Arbuscular mycorrhizal
AXR1-AFB	AUXIN RESISTANCE PROTEIN 1-AUXIN SIGNALING F-BOX PROTEIN
BRC1	*BRANCHED1*
CCD7	Carotenoid cleavage dioxygenase 7
CCD8	Carotenoid cleavage dioxygenase 8
CL	Carlactone
CLA	Carlactonoic acid
CLIM	Covalently linked intermediate molecule
CO	Carlactone oxidase
COP1	CONSTITUTIVE PHOTOMORPHOGENIC1
CYC	CYCLOIDEA
d	*dwarf*
D14L	DWARF14 LIKE
dad	*decreased apical dominance*
DMAPP	Dimethylallyl pyrophosphate
EAR	Ethylene responsive element binding factor-associated amphiphilic repression
fc1	*fine culm 1*
GGPP	Geranylgeranyl pyrophosphate
GSEs	Germinated spore exudates
htd	*high-tillering dwarf*

HTL	HYPOSENSITIVE TO LIGHT
HY5	*ELONGATED HYPOCOTYL 5*
IPP	Isopentenyl pyrophosphate
IPT	*ADENYLATE ISOPENTENYLTRANSFERASE*
KAI2	KARRIKIN INSENSITIVE2
LBO	LATERAL BRANCHING OXIDOREDUCTASE
max	*more axillary growth*
MeCLA	Methyl carlactonoate
NAA	1-Naphthaleneacetic acid
NPA	N-1-naphthylphthalamic acid
OS	Orobanchol synthase
PAT	Polar auxin transport
PCF	PROLIFERATING CELL FACTORS
PDR1	PLEIOTROPIC DRUG RESISTANCE 1
PHYB	PHYTOCHROME B
PIN1	PIN-FORMED1
QTL	Quantitative trait locus
rms	*ramosus*
SCF	Skp-Cullin-F-box
SLs	Strigolactones
SMAX1	*SUPPRESSOR OF MAX2 1*
SMXL	*SMAX1-LIKE*
TB1	TEOSINTE BRANCHED 1
TFs	Transcription factors
TMB	2,4,4,-Trihydroxy-3-methyl-3-butenal
TPL	TOPLESS
TPRs	TPL-related proteins
Ubi	Ubiquitin
YLG	Yoshimulactone Green
YLGW	Yoshimulactone Green Double

Acknowledgments

We thank Dr. Lin Li for preparing figures. This work is supported by the National Natural Science Foundation of China (Grants 90917021, 91335204 and 91117014). No conflicts of interest declared.

References

Abe, S., Sado, A., Tanaka, K., et al., 2014. Carlactone is converted to carlactonoic acid by MAX1 in Arabidopsis and its methyl ester can directly interact with AtD14 in vitro. Proc. Natl. Acad. Sci. U.S.A. 111, 18084–18089.
Aguilar-Martinez, J.A., Poza-Carrion, C., Cubas, P., 2007. *Arabidopsis BRANCHED1* acts as an integrator of branching signals within axillary buds. Plant Cell 19, 458–472.

Agusti, J., Herold, S., Schwarz, M., et al., 2011. Strigolactone signaling is required for aux-in-dependent stimulation of secondary growth in plants. Proc. Natl. Acad. Sci. U.S.A. 108, 20242–20247.

Akiyama, K., Hayashi, H., 2006. Strigolactones: chemical signals for fungal symbionts and parasitic weeds in plant roots. Ann. Bot. 97, 925–931.

Akiyama, K., Matsuzaki, K., Hayashi, H., 2005. Plant sesquiterpenes induce hyphal branching in arbuscular mycorrhizal fungi. Nature 435, 824–827.

Al-Babili, S., Bouwmeester, H.J., 2015. Strigolactones, a novel carotenoid-derived plant hormone. Annu. Rev. Plant Biol. 66, 161–186.

Alder, A., Jamil, M., Marzorati, M., et al., 2012. The path from beta-carotene to carlactone, a strigolactone-like plant hormone. Science 335, 1348–1351.

Arite, T., Iwata, H., Ohshima, K., et al., 2007. *DWARF10*, an *RMS1/MAX4/DAD1* ortholog, controls lateral bud outgrowth in rice. Plant J. 51, 1019–1029.

Arite, T., Umehara, M., Ishikawa, S., et al., 2009. *d14*, a strigolactone-insensitive mutant of rice, shows an accelerated outgrowth of tillers. Plant Cell Physiol. 50, 1416–1424.

Arite, T., Kameoka, H., Kyozuka, J., 2012. Strigolactone positively controls crown root elongation in rice. J. Plant Growth Regul. 31, 165–172.

Balla, J., Kalousek, P., Reinohl, V., et al., 2011. Competitive canalization of PIN-dependent auxin flow from axillary buds controls pea bud outgrowth. Plant J. 65, 571–577.

Bennett, T., Leyser, O., 2014. Strigolactone signalling: standing on the shoulders of DWARFs. Curr. Opin. Plant Biol. 22, 7–13.

Bennett, T., Sieberer, T., Willett, B., et al., 2006. The Arabidopsis *MAX* pathway controls shoot branching by regulating auxin transport. Curr. Biol. 16, 553–563.

Beveridge, C.A., Ross, J.J., Murfet, I.C., 1996. Branching in pea, action of genes *Rms3* and *Rms4*. Plant Physiol. 110, 859–865.

Beveridge, C.A., 2006. Axillary bud outgrowth: sending a message. Curr. Opin. Plant Biol. 9, 35–40.

Bonneau, L., Huguet, S., Wipf, D., et al., 2013. Combined phosphate and nitrogen limitation generates a nutrient stress transcriptome favorable for arbuscular mycorrhizal symbiosis in *Medicago truncatula*. New Phytol. 199, 188–202.

Booker, J., Auldridge, M., Wills, S., et al., 2004. MAX3/CCD7 is a carotenoid cleavage dioxygenase required for the synthesis of a novel plant signaling molecule. Curr. Biol. 14, 1232–1238.

Booker, J., Sieberer, T., Wright, W., et al., 2005. *MAX1* encodes a cytochrome P450 family member that acts downstream of *MAX3/4* to produce a carotenoid-derived branch-inhibiting hormone. Dev. Cell 8, 443–449.

Bouwmeester, H.J., Matusova, R., Zhongkui, S., et al., 2003. Secondary metabolite signalling in host-parasitic plant interactions. Curr. Opin. Plant Biol. 6, 358–364.

Boyer, F.D., de Saint Germain, A., Pillot, J.P., et al., 2012. Structure-activity relationship studies of strigolactone-related molecules for branching inhibition in garden pea: molecule design for shoot branching. Plant Physiol. 159, 1524–1544.

Braun, N., de Saint Germain, A., Pillot, J.P., et al., 2012. The pea TCP transcription factor PsBRC1 acts downstream of strigolactones to control shoot branching. Plant Physiol. 158, 225–238.

Brewer, P.B., Dun, E.A., Ferguson, B.J., et al., 2009. Strigolactone acts downstream of auxin to regulate bud outgrowth in pea and Arabidopsis. Plant Physiol. 150, 482–493.

Brewer, P.B., Koltai, H., Beveridge, C.A., 2013. Diverse roles of strigolactones in plant development. Mol. Plant 6, 18–28.

Brewer, P.B., Dun, E.A., Gui, R., et al., 2015. Strigolactone inhibition of branching independent of polar auxin transport. Plant Physiol. 168, 1820–1829.

Brewer, P.B., Yoneyama, K., Filardo, F., et al., 2016. *LATERAL BRANCHING OXIDORE-DUCTASE* acts in the final stages of strigolactone biosynthesis in Arabidopsis. Proc. Natl. Acad. Sci. U.S.A. 113, 6301–6306.

Cardoso, C., Zhang, Y., Jamil, M., et al., 2014. Natural variation of rice strigolactone biosynthesis is associated with the deletion of two *MAX1* orthologs. Proc. Natl. Acad. Sci. U.S.A. 111, 2379–2384.

Chevalier, F., Nieminen, K., Sanchez-Ferrero, J.C., et al., 2014. Strigolactone promotes degradation of DWARF14, an α/β hydrolase essential for strigolactone signaling in Arabidopsis. Plant Cell 26, 1134–1150.

Choudhary, S.P., Yu, J.Q., Yamaguchi-Shinozaki, K., et al., 2012. Benefits of brassinosteroid crosstalk. Trends Plant Sci. 17, 594–605.

Conn, C.E., Bythell-Douglas, R., Neumann, D., et al., 2015. Convergent evolution of strigolactone perception enabled host detection in parasitic plants. Science 349, 540–543.

Cook, C.E., Whichard, L.P., Turner, B., et al., 1966. Germination of witchweed (*Striga lutea* Lour.): isolation and properties of a potent stimulant. Science 154, 1189–1190.

Crawford, S., Shinohara, N., Sieberer, T., et al., 2010. Strigolactones enhance competition between shoot branches by dampening auxin transport. Development 137, 2905–2913.

Delaux, P.M., Xie, X., Timme, R.E., et al., 2012. Origin of strigolactones in the green lineage. New Phytol. 195, 857–871.

Doebley, J., Stec, A., Hubbard, L., 1997. The evolution of apical dominance in maize. Nature 386, 485–488.

Domagalska, M.A., Leyser, O., 2011. Signal integration in the control of shoot branching. Nat. Rev. Mol. Cell Biol. 12, 211–221.

Drummond, R.S., Martinez-Sanchez, N.M., Janssen, B.J., et al., 2009. *Petunia hybrida CAROTENOID CLEAVAGE DIOXYGENASE7* is involved in the production of negative and positive branching signals in petunia. Plant Physiol. 151, 1867–1877.

Drummond, R.S., Janssen, B.J., Luo, Z., et al., 2015. Environmental control of branching in petunia. Plant Physiol. 168, 735–751.

Dun, E.A., de Saint Germain, A., Rameau, C., et al., 2012. Antagonistic action of strigolactone and cytokinin in bud outgrowth control. Plant Physiol. 158, 487–498.

Ejeta, G., Gressel, J. (Eds.), 2007. Intergrating New Technologies for *Striga* Control: Towards Ending the Witch-hunt. World Scientific Publishing Company, Singapore.

Evelin, H., Kapoor, R., Giri, B., 2009. Arbuscular mycorrhizal fungi in alleviation of salt stress: a review. Ann. Bot. 104, 1263–1280.

Ferguson, B.J., Beveridge, C.A., 2009. Roles for auxin, cytokinin, and strigolactone in regulating shoot branching. Plant Physiol. 149, 1929–1944.

Finlayson, S.A., Krishnareddy, S.R., Kebrom, T.H., et al., 2010. Phytochrome regulation of branching in Arabidopsis. Plant Physiol. 152, 1914–1927.

Finlayson, S.A., 2007. Arabidopsis Teosinte Branched1-like 1 regulates axillary bud outgrowth and is homologous to monocot Teosinte Branched1. Plant Cell Physiol. 48, 667–677.

Flematti, G.R., Dixon, K.W., Smith, S.M., 2015. What are karrikins and how were they 'discovered' by plants? BMC Biol. 13, 108.

Flematti, G.R., Scaffidi, A., Waters, M.T., et al., 2016. Stereospecificity in strigolactone biosynthesis and perception. Planta 243, 1361–1373.

Foo, E., Bullier, E., Goussot, M., et al., 2005. The branching gene *RAMOSUS1* mediates interactions among two novel signals and auxin in pea. Plant Cell 17, 464–474.

Fraser, P.D., Bramley, P.M., 2004. The biosynthesis and nutritional uses of carotenoids. Prog. Lipid Res. 43, 228–265.

Fujita, Y., Fujita, M., Shinozaki, K., et al., 2011. ABA-mediated transcriptional regulation in response to osmotic stress in plants. J. Plant Res. 124, 509–525.

Galweiler, L., Guan, C., Muller, A., et al., 1998. Regulation of polar auxin transport by AtPIN1 in Arabidopsis vascular tissue. Science 282, 2226–2230.

Gao, Z., Qian, Q., Liu, X., et al., 2009. *Dwarf 88*, a novel putative esterase gene affecting architecture of rice plant. Plant Mol. Biol. 71, 265–276.

Gomez-Roldan, V., Fermas, S., Brewer, P.B., et al., 2008. Strigolactone inhibition of shoot branching. Nature 455, 189–194.

Guo, Y., Zheng, Z., La Clair, J.J., et al., 2013. Smoke-derived karrikin perception by the α/β-hydrolase KAI2 from Arabidopsis. Proc. Natl. Acad. Sci. U.S.A. 110, 8284–8289.

Gutjahr, C., Gobbato, E., Choi, J., et al., 2015. Rice perception of symbiotic arbuscular mycorrhizal fungi requires the karrikin receptor complex. Science 350, 1521–1524.

Ha, S., Vankova, R., Yamaguchi-Shinozaki, K., et al., 2012. Cytokinins: metabolism and function in plant adaptation to environmental stresses. Trends Plant Sci. 17, 172–179.

Ha, C.V., Leyva-Gonzalez, M.A., Osakabe, Y., et al., 2014. Positive regulatory role of strigolactone in plant responses to drought and salt stress. Proc. Natl. Acad. Sci. U.S.A. 111, 851–856.

Hamiaux, C., Drummond, R.S., Janssen, B.J., et al., 2012. DAD2 is an α/β hydrolase likely to be involved in the perception of the plant branching hormone, strigolactone. Curr. Biol. 22, 2032–2036.

Hayward, A., Stirnberg, P., Beveridge, C., et al., 2009. Interactions between auxin and strigolactone in shoot branching control. Plant Physiol. 151, 400–412.

Hoffmann, B., Proust, H., Belcram, K., et al., 2014. Strigolactones inhibit caulonema elongation and cell division in the moss *Physcomitrella patens*. PLoS One 9, e99206.

Hothorn, M., Belkhadir, Y., Dreux, M., et al., 2011a. Structural basis of steroid hormone perception by the receptor kinase BRI1. Nature 474, 467–471.

Hothorn, M., Dabi, T., Chory, J., 2011b. Structural basis for cytokinin recognition by *Arabidopsis thaliana* histidine kinase 4. Nat. Chem. Biol. 7, 766–768.

Ishikawa, S., Maekawa, M., Arite, T., et al., 2005. Suppression of tiller bud activity in tillering dwarf mutants of rice. Plant Cell Physiol. 46, 79–86.

Ito, S., Nozoye, T., Sasaki, E., et al., 2015. Strigolactone regulates anthocyanin accumulation, acid phosphatases production and plant growth under low phosphate condition in Arabidopsis. PLoS One 10, e0119724.

Ito, S., Ito, K., Abeta, N., et al., 2016. Effects of strigolactone signaling on Arabidopsis growth under nitrogen deficient stress condition. Plant Signal. Behav. 11, e1126031.

Jamil, M., Charnikhova, T., Cardoso, C., et al., 2011a. Quantification of the relationship between strigolactones and *Striga hermonthica* infection in rice under varying levels of nitrogen and phosphorus. Weed Res. 51, 373–385.

Jamil, M., Rodenburg, J., Charnikhova, T., et al., 2011b. Pre-attachment *Striga hermonthica* resistance of New Rice for Africa (NERICA) cultivars based on low strigolactone production. New Phytol. 192, 964–975.

Jamil, M., Kanampiu, F.K., Karaya, H., et al., 2012. *Striga hermonthica* parasitism in maize in response to N and P fertilisers. Field Crop Res. 134, 1–10.

Jiang, L., Liu, X., Xiong, G., et al., 2013. DWARF 53 acts as a repressor of strigolactone signalling in rice. Nature 504, 401–405.

Joel, D.M., Hershenhorn, J., Eizenburg, H., et al., 2007. Biology and Management of Weedy Root Parasites. John Wiley & Sons, London.

Johnson, X., Brcich, T., Dun, E.A., et al., 2006. Branching genes are conserved across species. Genes controlling a novel signal in pea are coregulated by other long-distance signals. Plant Physiol. 142, 1014–1026.

de Jong, M., George, G., Ongaro, V., et al., 2014. Auxin and strigolactone signaling are required for modulation of Arabidopsis shoot branching by nitrogen supply. Plant Physiol. 166, 384–395.

Kagiyama, M., Hirano, Y., Mori, T., et al., 2013. Structures of D14 and D14L in the strigolactone and karrikin signaling pathways. Genes Cells 18, 147–160.

Kang, J., Hwang, J.U., Lee, M., et al., 2010. PDR-type ABC transporter mediates cellular uptake of the phytohormone abscisic acid. Proc. Natl. Acad. Sci. U.S.A. 107, 2355–2360.

Kapulnik, Y., Koltai, H., 2016. Fine-tuning by strigolactones of root response to low phosphate. J. Integr. Plant Biol. 58, 203–212.

Kapulnik, Y., Delaux, P.M., Resnick, N., et al., 2011. Strigolactones affect lateral root formation and root-hair elongation in Arabidopsis. Planta 233, 209–216.

Kohlen, W., Charnikhova, T., Liu, Q., et al., 2011. Strigolactones are transported through the xylem and play a key role in shoot architectural response to phosphate deficiency in nonarbuscular mycorrhizal host Arabidopsis. Plant Physiol. 155, 974–987.

Kohlen, W., Charnikhova, T., Lammers, M., et al., 2012. The tomato *CAROTENOID CLEAVAGE DIOXYGENASE8* (*SlCCD8*) regulates rhizosphere signaling, plant architecture and affects reproductive development through strigolactone biosynthesis. New Phytol. 196, 535–547.

Kretzschmar, T., Kohlen, W., Sasse, J., et al., 2012. A petunia ABC protein controls strigolactone-dependent symbiotic signalling and branching. Nature 483, 341–344.

Kuromori, T., Miyaji, T., Yabuuchi, H., et al., 2010. ABC transporter AtABCG25 is involved in abscisic acid transport and responses. Proc. Natl. Acad. Sci. U.S.A. 107, 2361–2366.

Lazar, G., Goodman, H.M., 2006. *MAX1*, a regulator of the flavonoid pathway, controls vegetative axillary bud outgrowth in Arabidopsis. Proc. Natl. Acad. Sci. U.S.A. 103, 472–476.

Li, C.J., Bangerth, F., 1999. Autoinhibition of indoleacetic acid transport in the shoots of two-branched pea (*Pisum sativum*) plants and its relationship to correlative dominance. Physiol. Plant 106, 415–420.

Liang, J., Zhao, L., Challis, R., et al., 2010. Strigolactone regulation of shoot branching in chrysanthemum (*Dendranthema grandiflorum*). J. Exp. Bot. 61, 3069–3078.

Liang, Y., Ward, S., Li, P., et al., 2016. SMAX1-LIKE7 signals from the nucleus to regulate shoot development in Arabidopsis via partially EAR motif-independent mechanisms. Plant Cell 28, 1581–1601.

Lin, H., Wang, R., Qian, Q., et al., 2009. DWARF27, an iron-containing protein required for the biosynthesis of strigolactones, regulates rice tiller bud outgrowth. Plant Cell 21, 1512–1525.

Liu, W., Wu, C., Fu, Y., et al., 2009. Identification and characterization of *HTD2*: a novel gene negatively regulating tiller bud outgrowth in rice. Planta 230, 649–658.

Ljung, K., Bhalerao, R.P., Sandberg, G., 2001. Sites and homeostatic control of auxin biosynthesis in Arabidopsis during vegetative growth. Plant J. 28, 465–474.

Lopez-Raez, J.A., Charnikhova, T., Gomez-Roldan, V., et al., 2008. Tomato strigolactones are derived from carotenoids and their biosynthesis is promoted by phosphate starvation. New Phytol. 178, 863–874.

Lopez-Raez, J.A., 2016. How drought and salinity affect arbuscular mycorrhizal symbiosis and strigolactone biosynthesis? Planta 243, 1375–1385.

Mach, J., 2015. Strigolactones regulate plant growth in Arabidopsis via degradation of the DWARF53-like proteins SMXL6, 7, and 8. Plant Cell 27, 3022–3023.

Martinez-Garcia, J.F., Huq, E., Quail, P.H., 2000. Direct targeting of light signals to a promoter element-bound transcription factor. Science 288, 859–863.

Mashiguchi, K., Sasaki, E., Shimada, Y., et al., 2009. Feedback-regulation of strigolactone biosynthetic genes and strigolactone-regulated genes in Arabidopsis. Biosci. Biotechnol. Biochem. 73, 2460–2465.

Matusova, R., Rani, K., Verstappen, F.W., et al., 2005. The strigolactone germination stimulants of the plant-parasitic *Striga* and *Orobanche* spp. are derived from the carotenoid pathway. Plant Physiol. 139, 920–934.

Mayzlish-Gati, E., LekKala, S.P., Resnick, N., et al., 2010. Strigolactones are positive regulators of light-harvesting genes in tomato. J. Exp. Bot. 61, 3129–3136.

Mayzlish-Gati, E., De-Cuyper, C., Goormachtig, S., et al., 2012. Strigolactones are involved in root response to low phosphate conditions in Arabidopsis. Plant Physiol. 160, 1329–1341.

Murase, K., Hirano, Y., Sun, T.P., et al., 2008. Gibberellin-induced DELLA recognition by the gibberellin receptor GID1. Nature 456, 459–463.

Musselman, L.J., 1980. The biology of *Striga, Orobanche*, and other root-parasitic weeds. Annu. Rev. Phytopathol. 18, 463–489.

Nakamura, H., Xue, Y.L., Miyakawa, T., et al., 2013. Molecular mechanism of strigolactone perception by DWARF14. Nat. Commun. 4, 2613.

Napoli, C., 1996. Highly branched phenotype of the petunia *dad1-1* mutant is reversed by grafting. Plant Physiol. 111, 27–37.

Nelson, D.C., Scaffidi, A., Dun, E.A., et al., 2011. F-box protein MAX2 has dual roles in karrikin and strigolactone signaling in *Arabidopsis thaliana*. Proc. Natl. Acad. Sci. U.S.A. 108, 8897–8902.

Nishimura, N., Hitomi, K., Arvai, A.S., et al., 2009. Structural mechanism of abscisic acid binding and signaling by dimeric PYR1. Science 326, 1373–1379.

Nordstrom, A., Tarkowski, P., Tarkowska, D., et al., 2004. Auxin regulation of cytokinin biosynthesis in *Arabidopsis thaliana*: a factor of potential importance for auxin-cytokinin-regulated development. Proc. Natl. Acad. Sci. U.S.A. 101, 8039–8044.

Parniske, M., 2008. Arbuscular mycorrhiza: the mother of plant root endosymbioses. Nat. Rev. Microbiol. 6, 763–775.

Petrasek, J., Friml, J., 2009. Auxin transport routes in plant development. Development 136, 2675–2688.

Proust, H., Hoffmann, B., Xie, X., et al., 2011. Strigolactones regulate protonema branching and act as a quorum sensing-like signal in the moss *Physcomitrella patens*. Development 138, 1531–1539.

Prusinkiewicz, P., Crawford, S., Smith, R.S., et al., 2009. Control of bud activation by an auxin transport switch. Proc. Natl. Acad. Sci. U.S.A. 106, 17431–17436.

Rasmussen, A., Depuydt, S., Goormachtig, S., et al., 2013. Strigolactones fine-tune the root system. Planta 238, 615–626.

Rubiales, D., Verkleij, J., Vurro, M., et al., 2009. Parasitic plant management in sustainable agriculture. Weed Res. 49, 1–5.

Ruiz-Lozano, J.M., Aroca, R., Zamarreno, A.M., et al., 2016. Arbuscular mycorrhizal symbiosis induces strigolactone biosynthesis under drought and improves drought tolerance in lettuce and tomato. Plant Cell Environ. 39, 441–452.

Ruth, B., Khalvati, M., Schmidhalter, U., 2011. Quantification of mycorrhizal water uptake via high-resolution on-line water content sensors. Plant Soil 342, 459–468.

Ruyter-Spira, C., Kohlen, W., Charnikhova, T., et al., 2011. Physiological effects of the synthetic strigolactone analog GR24 on root system architecture in Arabidopsis: another belowground role for strigolactones? Plant Physiol. 155, 721–734.

Ruyter-Spira, C., Al-Babili, S., van der Krol, S., et al., 2013. The biology of strigolactones. Trends Plant Sci. 18, 72–83.

Sachs, T., 1981. The control of the patterned differentiation of vascular tissues. Adv. Bot. Res. 9, 151–162.

Sachs, T., 2000. Integrating cellular and organismic aspects of vascular differentiation. Plant Cell Physiol. 41, 649–656.

Sachs, T., Thimann, V., 1967. Role of auxins and cytokinins in release of buds from dominance. Am. J. Bot. 54, 136–144.

de Saint Germain, A., Clave, G., Badet-Denisot, M.A., et al., 2016. An histidine covalent receptor and butenolide complex mediates strigolactone perception. Nat. Chem. Biol. 12, 787–794.

de Saint Germain, A., Ligerot, Y., Dun, E.A., et al., 2013. Strigolactones stimulate internode elongation independently of gibberellins. Plant Physiol. 163, 1012–1025.

Sang, D., Chen, D., Liu, G., et al., 2014. Strigolactones regulate rice tiller angle by attenuating shoot gravitropism through inhibiting auxin biosynthesis. Proc. Natl. Acad. Sci. U.S.A. 111, 11199–111204.

Santiago, J., Dupeux, F., Round, A., et al., 2009. The abscisic acid receptor PYR1 in complex with abscisic acid. Nature 462, 665–668.

Sasse, J., Simon, S., Gubeli, C., et al., 2015. Asymmetric localizations of the ABC transporter PaPDR1 trace paths of directional strigolactone transport. Curr. Biol. 25, 647–655.

Scaffidi, A., Waters, M.T., Ghisalberti, E.L., et al., 2013. Carlactone-independent seedling morphogenesis in Arabidopsis. Plant J. 76, 1–9.

Scaffidi, A., Waters, M.T., Sun, Y.K., et al., 2014. Strigolactone hormones and their stereoisomers signal through two related receptor proteins to induce different physiological responses in Arabidopsis. Plant Physiol. 165, 1221–1232.

Seto, Y., Yamaguchi, S., 2014. Strigolactone biosynthesis and perception. Curr. Opin. Plant Biol. 21, 1–6.

Seto, Y., Sado, A., Asami, K., et al., 2014. Carlactone is an endogenous biosynthetic precursor for strigolactones. Proc. Natl. Acad. Sci. U.S.A. 111, 1640–1645.

She, J., Han, Z., Kim, T.W., et al., 2011. Structural insight into brassinosteroid perception by BRI1. Nature 474, 472–476.

Sheard, L.B., Tan, X., Mao, H., et al., 2010. Jasmonate perception by inositol-phosphate-potentiated COI1–JAZ co-receptor. Nature 468, 400–405.

Shen, H., Luong, P., Huq, E., 2007. The F-box protein MAX2 functions as a positive regulator of photomorphogenesis in Arabidopsis. Plant Physiol. 145, 1471–1483.

Shinohara, N., Taylor, C., Leyser, O., 2013. Strigolactone can promote or inhibit shoot branching by triggering rapid depletion of the auxin efflux protein PIN1 from the plasma membrane. PLoS Biol. 11, e1001474.

Simons, J.L., Napoli, C.A., Janssen, B.J., et al., 2007. Analysis of the *DECREASED APICAL DOMINANCE* genes of petunia in the control of axillary branching. Plant Physiol. 143, 697–706.

Smith, S.M., Li, J., 2014. Signalling and responses to strigolactones and karrikins. Curr. Opin. Plant Biol. 21, 23–29.

Smith, S.M., Waters, M.T., 2012. Strigolactones: destruction-dependent perception? Curr. Biol. 22, R924–R927.

Smith, S.E., Facelli, E., Pope, S., et al., 2010. Plant performance in stressful environments: interpreting new and established knowledge of the roles of arbuscular mycorrhizas. Plant Soil 326, 3–20.

Snowden, K.C., Janssen, B.J., 2016. Structural biology: signal locked in. Nature 536, 402–404.

Snowden, K.C., Napoli, C.A., 2003. A quantitative study of lateral branching in petunia. Funct. Plant Biol. 30, 987–994.

Snowden, K.C., Simkin, A.J., Janssen, B.J., et al., 2005. The *decreased apical dominance1/Petunia hybrida CAROTENOID CLEAVAGE DIOXYGENASE8* gene affects branch production and plays a role in leaf senescence, root growth, and flower development. Plant Cell 17, 746–759.

Sorefan, K., Booker, J., Haurogne, K., et al., 2003. *MAX4* and *RMS1* are orthologous dioxygenase-like genes that regulate shoot branching in Arabidopsis and pea. Genes Dev. 17, 1469–1474.

Soundappan, I., Bennett, T., Morffy, N., et al., 2015. SMAX1-LIKE/D53 family members enable distinct MAX2-dependent responses to strigolactones and karrikins in Arabidopsis. Plant Cell 27, 3143–3159.

Stanga, J.P., Smith, S.M., Briggs, W.R., et al., 2013. *SUPPRESSOR OF MORE AXILLARY GROWTH2 1* controls seed germination and seedling development in Arabidopsis. Plant Physiol. 163, 318–330.

Stanga, J.P., Morffy, N., Nelson, D.C., 2016. Functional redundancy in the control of seedling growth by the karrikin signaling pathway. Planta 243, 1397–1406.

Stewart, G.R., Press, M.C., 1990. The physiology and biochemistry of parasitic angiosperms. Annu. Rev. Plant Physiol. Plant Mol. Biol. 41, 127–151.

Stirnberg, P., van De Sande, K., Leyser, H.M., 2002. *MAX1* and *MAX2* control shoot lateral branching in Arabidopsis. Development 129, 1131–1141.

Stirnberg, P., Ward, S., Leyser, O., 2010. Auxin and strigolactones in shoot branching: intimately connected? Biochem. Soc. Trans. 38, 717–722.

Sun, X.D., Ni, M., 2011. HYPOSENSITIVE TO LIGHT, an alpha/beta fold protein, acts downstream of ELONGATED HYPOCOTYL 5 to regulate seedling de-etiolation. Mol. Plant 4, 116–126.

Sun, H., Tao, J., Liu, S., et al., 2014. Strigolactones are involved in phosphate- and nitrate-deficiency-induced root development and auxin transport in rice. J. Exp. Bot. 65, 6735–6746.

Sun, H., Tao, J., Gu, P., et al., 2016. The role of strigolactones in root development. Plant Signal. Behav. 11, e1110662.

Takeda, T., Suwa, Y., Suzuki, M., et al., 2003. The *OsTB1* gene negatively regulates lateral branching in rice. Plant J. 33, 513–520.

Tan, X., Calderon-Villalobos, L.I., Sharon, M., et al., 2007. Mechanism of auxin perception by the TIR1 ubiquitin ligase. Nature 446, 640–645.

Tanaka, M., Takei, K., Kojima, M., et al., 2006. Auxin controls local cytokinin biosynthesis in the nodal stem in apical dominance. Plant J. 45, 1028–1036.

Toh, S., Kamiya, Y., Kawakami, N., et al., 2012. Thermoinhibition uncovers a role for strigolactones in Arabidopsis seed germination. Plant Cell Physiol. 53, 107–117.

Toh, S., Holbrook-Smith, D., Stogios, P.J., et al., 2015. Structure-function analysis identifies highly sensitive strigolactone receptors in *Striga*. Science 350, 203–207.

Tsuchiya, Y., Vidaurre, D., Toh, S., et al., 2010. A small-molecule screen identifies new functions for the plant hormone strigolactone. Nat. Chem. Biol. 6, 741–749.

Tsuchiya, Y., Yoshimura, M., Sato, Y., et al., 2015. Probing strigolactone receptors in *Striga hermonthica* with fluorescence. Science 349, 864–868.

Turnbull, C.G., Booker, J.P., Leyser, H.M., 2002. Micrografting techniques for testing long-distance signalling in Arabidopsis. Plant J. 32, 255–262.

Umehara, M., Hanada, A., Yoshida, S., et al., 2008. Inhibition of shoot branching by new terpenoid plant hormones. Nature 455, 195–200.

Umehara, M., Hanada, A., Magome, H., et al., 2010. Contribution of strigolactones to the inhibition of tiller bud outgrowth under phosphate deficiency in rice. Plant Cell Physiol. 51, 1118–1126.

Urquhart, S., Foo, E., Reid, J.B., 2015. The role of strigolactones in photomorphogenesis of pea is limited to adventitious rooting. Physiol. Plant 153, 392–402.

Vidal, E.A., Gutierrez, R.A., 2008. A systems view of nitrogen nutrient and metabolite responses in Arabidopsis. Curr. Opin. Plant Biol. 11, 521–529.

Waldie, T., McCulloch, H., Leyser, O., 2014. Strigolactones and the control of plant development: lessons from shoot branching. Plant J. 79, 607–622.

Walter, M.H., Strack, D., 2011. Carotenoids and their cleavage products: biosynthesis and functions. Nat. Prod. Rep. 28, 663–692.

Wang, L., Smith, S.M., 2016. Strigolactones redefine plant hormones. Sci. China Life Sci. 59, 1083–1085.

Wang, L., Wang, B., Jiang, L., et al., 2015. Strigolactone signaling in Arabidopsis regulates shoot development by targeting D53-Like SMXL repressor proteins for ubiquitination and degradation. Plant Cell 27, 3128–3142.

Waters, M.T., Brewer, P.B., Bussell, J.D., et al., 2012a. The Arabidopsis ortholog of rice DWARF27 acts upstream of MAX1 in the control of plant development by strigolactones. Plant Physiol. 159, 1073–1085.

Waters, M.T., Nelson, D.C., Scaffidi, A., et al., 2012b. Specialisation within the DWARF14 protein family confers distinct responses to karrikins and strigolactones in Arabidopsis. Development 139, 1285–1295.

Waters, M.T., Scaffidi, A., Sun, Y.M.K., et al., 2014. The karrikin response system of Arabidopsis. Plant J. 79, 623–631.

Waters, M.T., Scaffidi, A., Moulin, S.L., et al., 2015. A *Selaginella moellendorffii* ortholog of KARRIKIN INSENSITIVE2 functions in Arabidopsis development but cannot mediate responses to karrikins or strigolactones. Plant Cell 27, 1925–1944.

Westwood, J.H., Yoder, J.I., Timko, M.P., et al., 2010. The evolution of parasitism in plants. Trends Plant Sci. 15, 227–235.

Wickson, M., Thimann, K.V., 1958. The antagonism of auxin and kinetin in apical dominance. Physiol. Plant 11, 62–74.

Wisniewska, J., Xu, J., Seifertova, D., et al., 2006. Polar PIN localization directs auxin flow in plants. Science 312, 883.

Xie, X., Yoneyama, K., Harada, Y., et al., 2009. Fabacyl acetate, a germination stimulant for root parasitic plants from *Pisum sativum*. Phytochemistry 70, 211–215.

Xie, X., Yoneyama, K., Yoneyama, K., 2010. The strigolactone story. Annu. Rev. Phytopathol. 48, 93–117.

Xie, X., Wang, G., Yang, L., et al., 2015. Cloning and characterization of a novel *Nicotiana tabacum* ABC transporter involved in shoot branching. Physiol. Plant 153, 299–306.

Xiong, G., Wang, Y., Li, J., 2014. Action of strigolactones in plants. Enzymes 35, 57–84.

Yamada, Y., Furusawa, S., Nagasaka, S., et al., 2014. Strigolactone signaling regulates rice leaf senescence in response to a phosphate deficiency. Planta 240, 399–408.

Yao, R., Ming, Z., Yan, L., et al., 2016. DWARF14 is a non-canonical hormone receptor for strigolactone. Nature 536, 469–473.

Yokota, T., Sakai, H., Okuno, K., et al., 1998. Alectrol and orobanchol, germination stimulants for *Orobanche minor*, from its host red clover. Phytochemistry 49, 1967–1973.

Yoneyama, K., Yoneyama, K., Takeuchi, Y., et al., 2007. Phosphorus deficiency in red clover promotes exudation of orobanchol, the signal for mycorrhizal symbionts and germination stimulant for root parasites. Planta 225, 1031–1038.

Yoneyama, K., Xie, X., Sekimoto, H., et al., 2008. Strigolactones, host recognition signals for root parasitic plants and arbuscular mycorrhizal fungi, from Fabaceae plants. New Phytol. 179, 484–494.

Yoneyama, K., Xie, X.N., Kisugi, T., et al., 2011. Characterization of strigolactones exuded by Asteraceae plants. Plant Growth Regul. 65, 495–504.

Zhang, S., Li, G., Fang, J., et al., 2010. The interactions among *DWARF10*, auxin and cytokinin underlie lateral bud outgrowth in rice. J. Integr. Plant Biol. 52, 626–638.

Zhang, Y., van Dijk, A.D., Scaffidi, A., et al., 2014. Rice cytochrome P450 MAX1 homologs catalyze distinct steps in strigolactone biosynthesis. Nat. Chem. Biol. 10, 1028–1033.

Zhao, L.H., Zhou, X.E., Wu, Z.S., et al., 2013. Crystal structures of two phytohormone signal-transducing α/β hydrolases: karrikin-signaling KAI2 and strigolactone-signaling DWARF14. Cell Res. 23, 436–439.

Zhao, J., Wang, T., Wang, M., et al., 2014. DWARF3 participates in an SCF complex and associates with DWARF14 to suppress rice shoot branching. Plant Cell Physiol. 55, 1096–1109.

Zhao, L.H., Zhou, X.E., Yi, W., et al., 2015. Destabilization of strigolactone receptor DWARF14 by binding of ligand and E3-ligase signaling effector DWARF3. Cell Res. 25, 1219–1236.

Zhou, F., Lin, Q., Zhu, L., et al., 2013. D14-SCF[D3]-dependent degradation of D53 regulates strigolactone signalling. Nature 504, 406–410.

Zou, J., Zhang, S., Zhang, W., et al., 2006. The rice *HIGH-TILLERING DWARF1* encoding an ortholog of Arabidopsis MAX3 is required for negative regulation of the outgrowth of axillary buds. Plant J. 48, 687–698.

Zwanenburg, B., Pospisil, T., Cavar Zeljkovic, S., 2016. Strigolactones: new plant hormones in action. Planta 243, 1311–1326.

Peptide hormones

11

Xiu-Fen Song[1], Shi-Chao Ren[1], Chun-Ming Liu[1,2]
[1]Institute of Botany, Chinese Academy of Sciences, Beijing, China;
[2]Institute of Crop Science, Chinese Academy of Agricultural Sciences, Beijing, China

Summary

Intercellular communications among plant cells are regulated mainly by metabolite-based hormones such as auxin, cytokinin, gibberellins, brassinosteroids, abscisic acid, and ethylene. However, in recent years small peptides with a few to dozens of amino acids have been discovered as important cell-to-cell communication signals underlying many plant biological processes. These peptides act in a non-cell autonomous manner to coordinate defense and developmental processes including meristem maintenance, cell division, stomata development, reproduction, and nodulation. Although more than 1000 potential peptide hormones have been predicted in plant genomes, only dozens of them have been functionally characterized. From knowledge obtained so far, it seems evident that signals of these peptide hormones are exclusively perceived by leucine-rich repeat receptor kinases that represent the largest receptor family in plants. In this chapter, we provide a general overview on peptide coding genes, peptide processing and modification, and functions of known peptides. The challenge and perspective in peptide research are discussed.

11.1 Introduction

Peptide hormones have been recognized as important signaling molecules in animals for more than 100 years, for their critical roles in neural and endocrine systems (Edlund and Jessell, 1999; Takei and Hirose, 2002). The involvement of peptide hormones in plants, however, has been realized for less than 30 years. One of the reasons is the difficulty in identifying these peptides since they are usually small in size and low in abundance, typically present in plants in nano- to picomolar concentrations (Kondo et al., 2006; Ohyama et al., 2009; Ito et al., 2006). Peptide hormones known so far in plants play critical roles in both short-range and long-range signaling in developmental and defense processes (Marmiroli and Maestri, 2014; Grienenberger and Fletcher, 2015; De Coninck and De Smet, 2016).

11.2 The identification of peptide hormones

11.2.1 Biochemical identification

It has been difficult to identify peptide hormones since they are usually small, containing only few to tens of amino acids, and low abundance, at nano- to picomolar

Hormone Metabolism and Signaling in Plants. http://dx.doi.org/10.1016/B978-0-12-811562-6.00011-6

concentrations in plants (Kondo et al., 2006; Ohyama et al., 2009; Ito et al., 2006). Recently, with the technological advances in peptide isolation techniques and mass spectrometry, together with genetic and bioinformatic tools available, new strategies have been successfully developed in the identification and characterization of peptide hormones involved in a range of plant developmental and defense responses (Marmiroli and Maestri, 2014) (Table 11.1).

Isolation and identification of peptide hormones based on biochemistry methods also depend on a good biological assay system for detection of the peptide activity. Using such a biochemical approach, several peptide hormones have been characterized in plants (Table 11.1). Examples include systemin, rapid alkalinization factor (RALF), AtPep1, PSK, and tracheary element differentiation inhibitor factor (TDIF), which are described below.

11.2.1.1 Systemin and rapid alkalinization factor

When tomato leaves are attacked by insects, the plant not only shows local response to the attack but also produces a systematic signal that is transmitted to other leaves, generating systemic resistance. Based on this phenomenon, the signal molecule from wounded leaves was characterized (Pearce et al., 1991), and found to be a small polypeptide, named TomSystemin (TomSys). This is the first peptide hormone discovered in plants. TomSys induces the biosynthesis of two wound-inducible proteinase inhibitor proteins (PI) when applied to young tomato plants. TomSys is comprised of 18 amino acids (AVQSKPPSKRDPPKMQTD) and in vitro biochemically synthesized TomSys possesses a full PI-inducing activity (Pearce et al., 1991). Since the identification of TomSys, wound-induced peptides have also been found in other Solanaceae species including potato and sweet pepper (Constabel et al., 1998). Using the same biochemical strategy, TobSys has been isolated from tobacco, which also has been shown to comprise 18 amino acids but remarkably has no sequence similarity to TomSys. TobSys has two isoforms, TobSys I and TobSys II, with sequences similar to each other (Pearce et al., 2001). Later, three TomSys isoforms, with hydroxyproline and glycosylation modifications, were purified from tomato and named TomHypSys I, II, and III (Pearce and Ryan, 2003). Further studies showed that genes encoding preproproteins of these three TomHypSys peptides were induced by wounding and expressed in vascular parenchyma cells, with the preprotein localized in cell wall matrix. It was deduced that the TomHypSys preproproteins may be processed to mature peptides in the extracellular space (Narváez-Vásquez et al., 2005).

Several peptide hormones in plants were identified in cell suspension cultures in which the interaction between peptides and cell surface receptors causes alkalinization of the medium (Pearce et al., 2001). Based on the assay, a 49-amino acid peptide was identified in tobacco, which could induce the alkalinization of media but did not induce defense responses. This factor was named RALF and 23 RALF-like (RALFL) genes have been identified in the Arabidopsis genome (Pearce et al., 2001; Olsen et al., 2002; Lease and Walker, 2006).

11.2.1.2 Arabidopsis peptides

Since systemins were only found in a member of Solanaceae species, which activate antiherbivore defense responses, it was predicted that there should be other peptide hormones responsible for defenses in other taxa. Based on the ability to cause alkalinization of suspension culture media at sub-nanomolar concentrations, the peptide hormone AtPep1 was detected in Arabidopsis leaf extracts and purified using high-performance liquid chromatography (HPLC). The AtPep1 comprises 23 amino acid residues and chemically synthesized AtPep1 peptide was found to be active as native AtPep1 in the alkalinization assay (Huffaker et al., 2006). There are six AtPep family members in the Arabidopsis genome and they are differentially expressed in leaves in response to methyl jasmonic acid and ethylene (Huffaker and Ryan, 2007).

11.2.1.3 Phytosulfokine

Plant cell cultures have been known for decades to have a density effect, such that suspension cells cultured at a low density did not divide even if provided with additional hormones or nutrients. However, if condition media from a high-density culture is added to low-density cultured cells, the latter is able to divide by sensing some kinds of division-promoting factor released from the former. It was thus deduced that a "density factor" may be secreted by the suspension culture cells, which provides a bioassay for its isolation. The cell division-promoting factor was purified and named phytosulfokine (PSK), which comprises two related peptides containing four $(Y(SO_3H)-I-Y(SO_3H)-T)$ or five $(Y(SO_3H)-I-Y(SO_3H)-T-Q)$ amino acid residues with sulfated tyrosine modifications (named PSKα and PSKβ, respectively). PSKs are the first sulfate-modified peptide hormones discovered in plants, which triggers cell division at nanomolar concentration (Matsubayashi and Sakagami, 1996).

11.2.1.4 Tracheary element differentiation inhibitor factor

To identify signals responsible for cell–cell interactions, an in vitro *Zinnia elegans* tracheary element-inducible system was developed (Ito et al., 2006). Using this system, it was showed that tracheary element differentiation was inhibited by extracellular factors. Using HPLC analyses in association with a bioassay, this factor was isolated, characterized, and named TDIF. TDIF is a dodecapeptide with two hydroxyproline residues (HEVHypSGHypNPISN). Chemically synthesized TDIF inhibited tracheary element differentiation by 50% at 30 nM concentration (Ito et al., 2006). The sequence of TDIF was found to be the same as the CLE motifs in CLAVATA3/Endosperm Surrounding Region 41 (CLE41) and CLE44 peptides in Arabidopsis, and highly homologous to CLE42 and CLE46 (Ito et al., 2006). The CLE peptide family contains 32 members in Arabidopsis genome. Knock-out of CLE41, or even CLE41 and CLE42 simultaneously, did not give any visible phenotype, which is possibly due to functional redundancy among *CLE* genes. Overexpression of either the *CLE41* or the *CLE42* gene showed misaligned cell divisions and organizational defects in newly generated vascular tissue at the top of inflorescence stems (Etchells and Turner, 2010).

Table 11.1 **List of peptide hormones**

Category	Peptide	Sequence of peptides	Function
Defense response	Systemin	AVQSKPPSKRDPPKMQTD	Induce defense response
	TomHypSysI	RTOYKTOOOOTSSSOTHQ	
	TomHypSysII	GRHDYVASOOOOKPQDE	
	TomHypSysIII	GRHDSVLPOOSOKTD	
	TobHypSysI	RGANLPOOSOASSOOSKE	
	TobHypSysII	NRKPLSOOSOKPADGQRP	
	RALF	ATKKYISYGALQKNS VPCSRRGASYYNCKPGAQ ANPYSRGCSAITRCRS	Induce rapid alkalinization of suspension cell culture media; ABA and stress response
	AtPep1	ATKVKAKQRGKEKVSS GRPGQHN	Induce immune responses
	POLARIS	36 amino acid	Root growth and leaf vascular patterning
Cell division	PSK α	sYIsYTQ	Promote cellular proliferation
	PSK β	sYIsYT	
	4 KD peptide	ADCNGACSPFEVPPCRSRD CRCVPIGLFVGFCIHPTG	
	PSY1	DsYGDPSANPKHDPGV (L-Ara3-)OO-S	
Meristem maintenance	CLV3	RTVOSG(L-Ara3)ODPLHH or RTVOSG(L-Ara3) ODPLHHH	Meristem maintenance and differentiation
	FON4	RSVPAGPDPMHH(H)	
	CLE40	RQVPTGSDPLHH	Promote distal stem cell differentiation
	TDIF/ CLE41/44	HEVOSGONPISN	Repress vascular stem cell differentiation
	RGF1/CLEL8/ GLV11	DYSNPGHHPPRHN	Root meristem maintenance
	CEP1	DFROTNPGNSOGVGH	Root growth
Nodule formation	GrCLE1	RVTPGGPDPLHN	Nematode interaction with plant
	GmRIC1	RLAPEGPDPHHN	Soybean nodule formation
	GmRIC2	RLAPGGPDPQHN	
	GmNIC1	RLSPGGPDQKHH	
	MtCLE12	RLSPHGPNIHN	Medicago nodule formation
	MtCLE13	RLSPAGPDPQHN	
	ENOD40	MELCWLTTIHGS	Symbiotic interactions before nodule formation

Precursor ([a]aa)	Posttranslational modification	Receptors	Source	References
200 aa	Unknown	SR160	Tomato	Pearce et al. (1991)
146 aa	Hydroxylation, glycosylation			Pearce and Ryan (2003)
165 aa	Hydroxylation, glycosylation		Tobacco	Pearce et al. (2001)
115 aa	Unknown	FER	Tobacco, Arabidopsis	Pearce et al., (2001), Haruta et al. (2014), and Chen et al. (2016)
92 aa	Unknown	AtPEPR1, AtPEPR2	Arabidopsis	Huffaker et al. (2006), Yamaguchi et al. (2006), and Krol et al. (2010)
36 aa	Unknown	25 and 120 KD protein	Tomato, Tobacco, Arabidopsis	Topping and Lindsey (1997) and Casson et al. (2002)
89 aa	Tyrosine sulfation	PSKR	Carrot, Rice, *Asparagus officinalis*	Matsubayashi and Sakagami (1996) and Yang et al. (1999)
119 aa	Unknown	Unknown	Soybean	Watanabe et al. (1994) and Yamazaki et al. (2003)
75 aa	Tyrosine sulfation, glycosylation	PSY1R	Arabidopsis	Amano et al. (2007)
96 aa	Hydroxylation, arabinosylation	CLV1, CLV2, CRN/SOL2, RPK2, BAMs		Clark et al. (1995), Kondo et al. (2006), Fiers et al. (2005), Ohyama et al. (2009), Zhu et al. (2009)
122 aa		FON1	Rice	Chu et al. (2006)
80 aa	Unknown	ACR4	Arabidopsis	Hobe et al. (2003) and Stahl et al. (2009)
99 aa	Hydroxylation	TDR/PXY	*Zinnia elegans*	Ohyama et al. (2008)
116 aa		RGFR1, RGFR2, RGFR3	Arabidopsis	Matsuzaki et al. (2010), Meng et al. (2012), and Whitford et al. (2012)
91 aa	Arabinosylation and hydroxylation	Unknown		Ohyama et al. (2008)
	Unknown	CLV2, BAM1, BAM2	*Heterodera schachtii*	Guo et al. (2011)
95 aa		GmNARK	Soybean	Reid et al. (2011)
93 aa				
87 aa				
82 aa		SUNN (RLK)	Medicago	Mortier et al., Plant J. (2010)
85 aa				
	Unknown	Unknown	Medicago	Crespi et al. (1994), Kouchi and Hata (1993), Yang et al. (1993), and Rohrig et al. (2002)

Continued

Table 11.1 **Continued**

Category	Peptide	Sequence of peptides	Function
Cell fate determination and organogenesis	EPF1 EPF2 CHAL/EPFL6	C-terminal 52 amino acid C-terminal 68 amino acid C-terminal 105 amino acid	Repress stomata development
	STOMAGEN	IGSTAPTCTYNECRG CRYKCRAEQVPVEG NDPINSAYHYRCVCHR	Promote stomata development
	EPFL4 EPFL6/CHAL	C-terminal 58 amino acid C-terminal 105 amino acid	Inflorescence structure and plant height
	IDA	FGYLPKGVPIPPSAP SKRHN	Inhibit floral organ abscission
	CLE45	RRVRRGSDPIHN	Suppress protophloem differentiation
	EPFL2	C-terminal 52 amino acid	Regulate leaf margin morphogenesis
	GAD1	Arabidopsis EPF/EPFL-like	Regulate grain number, grain length, and awn development
Gamogenesis and embryogenesis	TPD1	Cysteine-rich protein	Promote tapetum formation in anther development
	CLE8	RRVPTGPNPLHH	Embryo and endosperm development
	CLE19	RVIPTGPNPLHN	development
	SCR/SP11	Variant S8 sequence: KRCTRGFRKLGKCT TLEEEKCKTLYPRGQCTC SDSKMNTHSCDCKSC	Prevent self-fertilization
	LURE	Cysteine-rich protein	Promote synergid attraction of pollen tube
	Chemocyanin	Cysteine-rich protein	Pollen tube attraction
	SCAs	Cysteine-rich protein	
	CLE45	RRVRRGSDPIHN	Promote pollen tube growth
	EC1	Cysteine-rich protein	Activate sperm during fertilization
	ESF1	Cysteine-rich protein	Early embryo patterning

[a]*aa*, amino acid.

Precursor ([a]aa)	Posttranslational modification	Receptors	Source	References
104 aa 120 aa 230 aa 101 aa	Disulfide bond on cysteine	TMM, ERF ERF TMM	Arabidopsis	Hara et al. (2007, 2009) and Hunt and Gray (2009) Hunt et al. (2010) and Sugano et al. (2010)
109 aa 230 aa		ER		Uchida et al. (2012)
230 aa	Hydroxylation	HAESA		Butenko et al. (2003), Jinn et al. (2000), and Stenvik et al. (2008)
124 aa	Unknown	BAM3		Depuydt et al. (2013)
128 aa		ER, ERL1, ERL2		Tameshige et al. (2016)
127 aa		Unknown	Rice	Jin et al. (2016)
176 aa	Unknown	EMS1	Arabidopsis	Yang et al. (2003) and Jia et al. (2008)
86 aa		Unknown	Arabidopsis	Fiume and Fletcher (2012)
74 aa				Xu et al. (2015)
47 KD protein	Disulfide bond formation among cysteines	SRK (RLK)	Brassica	Schopfer et al. (1999) and Takayama et al. (2000, 2001), and Mishima et al. (2003)
9–10 KD protein	Unknown	MDIS1, MIK1, MIK2, PRK6	*Torenia fournieri*, Arabidopsis	Okuda et al. (2009), Wang et al. (2016), and Takeuchi and Higashiyama (2016)
10 KD protein		Unknown	Lily	Kim et al. (2003)
9 KD protein				Park et al. (2000) and Park and Lord (2003)
124 aa		SKM1, SKM2	Arabidopsis	Endo et al. (2013)
158 aa		Unknown	Arabidopsis	Sprunck et al. (2012)
86 aa		Unknown	Arabidopsis	Costa et al. (2014)

Most known peptides have posttranslational modifications and these modifications may confer particular functions on peptides. A procedure for enrichment of sulfated peptides from complex mixtures was developed based on an ion-selective interaction of sulfated peptides with anion exchangers (Amano et al., 2005). Using this procedure, an 18-amino acid tyrosine-sulfated glycopeptide, named plant peptide containing sulfated tyrosine 1 (PSY1), was isolated and identified in Arabidopsis T87 suspension culture cells. Activity assays showed that PSY1 promotes cellular proliferation and expansion at nanomolar concentrations. The *PSY1* gene is widely expressed in various Arabidopsis tissues and is upregulated by wounding (Amano et al., 2007).

Since peptide hormones are present at low concentrations in plants and bioassays are usually interfered by the presence of secondary metabolites, limited numbers of peptides have been identified with bioassay guided methods. Other strategies such as genetic and bioinformatic tools have been applied to mine plant genomes for new peptide hormones.

11.2.2 Genetic identification

Genetic analysis, including forward- and reverse-genetics, is a common and powerful way for discovery and functional dissection of genes and their regulatory mechanisms in plants. Unsurprisingly, many peptide hormones have been identified using genetic approaches (Table 11.1). These include CLAVATA3 (CLV3), TAPETUM DETERMINANT1 (TPD1), and INFLORESCENCE DEFICIENT IN ABSCISSION (IDA) that are described in detail below.

11.2.2.1 CLV3 and CLE peptides

The *clv* mutants, named for their club-shaped siliques, of Arabidopsis were first identified by the well-known geneticist Maarten Koornneef for increased organ numbers in flowers (Koornneef et al., 1983). Phenotypic and genetic analyses of the *clv1* mutant were characterized in detail, and the potential role of the *CLV1* gene in regulating the homeostasis of shoot apical meristem (SAM) was thus proposed (Leyser and Furner, 1992). The *CLV1* gene that encodes a leucine-rich repeat family receptor kinase (LRR-RK) is expressed in the central domain of the SAM (Clark et al., 1997). The *clv3* mutant was obtained and characterized later (Clark et al., 1995). Compared with the 2-carpel silique in the wild-type, *clv3* mutants showed increased carpel number (ranging from 3 to 7 per silique), enlarged shoot apical and floral meristems, and increased flower number in inflorescences (Clark et al., 1995). Map-based cloning showed that *CLV3* encodes a small putative extracellular proprotein comprising 96 amino acids, with a predicted role of intercellular communications (Fletcher et al., 1999). The CLV3 protein shares a conserved 14-amino acid C terminal or near C-terminal CLE domain with the previously reported *ESR* gene in corn, and a family of small genes, later named *CLE* genes, were identified in Arabidopsis (Cock and McCormick, 2001). Overexpression of several of these *CLE* genes such as *CLE19* and *CLE40* showed a common short root phenotype, with premature termination of root meristems (Hobe et al., 2003; Fiers et al., 2005). Expression of *CLE40* under the control of the *CLV3* promoter can fully complement the defects of *clv3* mutants (Hobe et al., 2003). Since CLV3 and CLE40 are not conserved beyond the CLE motif, suggesting a critical role of the CLE motif in regulating SAM homeostasis. This proposition was further supported by a domain deletion analysis, showing that deletions of either the non-conserved region between the N-terminal secretion peptide and the CLE motif or the 15-amino acid C-terminal tail did not affect the function of CLV3 (Fiers et al., 2006).

The first evidence, illustrating that *CLE* genes encode peptide hormones, was provided by an in vitro peptide assay (Fiers et al., 2005). Treatments of Arabidopsis seedlings with chemically synthesized CLV3, CLE19, or CLE40 peptides containing the 14-amino acid CLE domain mimicked the overexpression phenotypes of these genes (Fiers et al., 2005). Cell biological studies showed that the peptide treatment led to cell layer identity confusion and premature stem cell differentiation in root meristems (Fiers et al., 2005). The *clv2* mutant was not sensitive to the peptide treatment, suggesting a role of CLV2 in perception of these peptides in roots (Fiers et al., 2005). It is thus deduced that the conserved 14-amino acid CLE motif is the functional domain of CLV3. Biochemical analysis using MALDI-TOF-TOF MS allowed identification of a 12-amino acid peptide with hydroxylation modifications at the fourth and seventh proline residues in transgenic Arabidopsis plants overexpressing the *CLV3* gene (Kondo et al., 2006). Later, another group showed that the mature CLV3 peptide may have 13 amino acid residues, with arabinosylation on the seventh proline residue (Ohyama et al., 2009). A homolog of CLV3 in rice, known as *FLORAL ORGAN NUMBER 4* (*FON4*), was identified, and the *fon4* mutant showed enlarged SAM and floral meristems, and an increased floral number (Suzaki et al., 2006; Chu et al., 2006), suggesting a conserved function of CLV3 in monocots and dicots.

Functions of several other CLE members have also been studied. The *CLE8* is expressed exclusively in young embryos and endosperms, and acts cell autonomously and non-cell autonomously in regulating embryo and endosperm development, respectively (Fiume and Fletcher, 2012). Similarly, *CLE19* is expressed specifically in cotyledon primordia during embryogenesis and plays a critical role in promoting the cellularization of endosperm and the establishment of the cotyledon (Xu et al., 2015). The antagonistic *CLE19* expressed in endosperms is able to inhibit cotyledon development even when expressed under an endosperm-specific promoter, demonstrating that CLE19 peptide acts as a mobile signal regulating cotyledon development (Xu et al., 2015).

CLE45 was found to regulate both phloem development and pollen tube growth (Depuydt et al., 2013; Endo et al., 2013). Expressions of *CLE1, CLE3, CLE4,* and *CLE7* were upregulated under nitrogen deficiency, and overexpression of these genes resulted in repression of lateral root emergence and the growth of these lateral roots (Araya et al., 2014).

Homologs of *CLE* genes have also been found in *Lotus japonicus*, *Medicago truncatula,* and *Glycine max* genomes for their roles in regulating nodulation. Twenty-five *CLE* genes have been identified in the genome of *Medicago truncatula*, and 39 in *L. japonica*. Overexpression of *LjCLE-RS1* and *LjCLE-RS2* inhibited nodulation in *L. japonica* (Okamoto et al., 2009). Similarly, two *CLE* genes, *MtCLE12* and *MtCLE13*, have been shown to regulate nodulation in *M. truncatula* (Mortier et al., 2010). In soybean, it has been indicated that ectopic expressions of *GmRIC1, GmRIC2,* and *GmNIC1* inhibited nodulation (Reid et al., 2011).

11.2.2.2   Tapetum determinant1

A severe male sterile mutant named *tapetum determinant1* (*tpd1*) was identified in Arabidopsis, showing no production of pollen grains. Tapetum precursor cells are formed in *tpd1* mutant, but fail to develop to a functional tapetum; instead, these precursor cells differentiate to microsporocytes. *TPD1* encodes a small extracellular preproprotein with 176 amino acid residues (Yang et al., 2003).

11.2.2.3   Inflorescence deficient in abcission

Abscission is an active process in plants, which enables plants to lose unwanted organs. A specialized cell layer called abscission zone forms between the abscission organs and the main plant body, and the abscission organs enter into a programmed cell death before abscission. An ethylene-sensitive mutant *inflorescence deficient in abscission* (*ida*) was identified in Arabidopsis in which floral organs remain attached to flowers even after the shedding of mature seeds. *IDA*, which encodes a 77-amino acid preproprotein with an N-terminal secretion signal, is specifically expressed in the abscission zone (Butenko et al., 2003). In the genome of Arabidopsis, five *IDA-like* genes named *IDL1* to *IDL5* have been identified, and a conserved signature sequence (pv/iPpSa/gPSk/rk/rHN), which is termed PIP peptide, has been found in C terminus of the IDA and these IDLs. Treatment of chemically synthesized PIP peptide is able to rescue the *ida* mutant phenotype, implying that IDA acts as a peptide hormone in regulating organ abscission (Stenvik et al., 2008).

Results obtained so far show that most peptide hormones are encoded by a family of functionally redundant genes (Marmiroli and Maestri, 2014; Grienenberger and Fletcher, 2015). Mutation of only a few of peptide hormone-coding genes showed visible phenotypes. Therefore, the traditional genetic approach is not efficient in elucidating roles of peptide hormones in plants.

11.2.3   Bioinformatics identification

With the rapid development of genomics and bioinformatics tools, putative peptide hormones have been identified rapidly over the past 20 years (Table 11.1). For example, two peptides named TfLURE1 and TfLURE2 have been identified through expressed sequence tag analyses in the stigma of *Torenia fournieri* (Okuda et al., 2009). In Arabidopsis, the AtLURE1 peptide is reported to be pollen tube attractant, guiding pollen tubes to the ovular micropyle (Higashiyama, 2010). These *LURE* genes are expressed specifically in synergid cells, and encode 9-KD extracellular cysteine-rich proteins (CRP) with six cysteine residues. Ectopic expression of the *AtLURE1* in synergid cells of *T. fournieri* was sufficient to guide *Arabidopsis thaliana* pollen tubes to the *T. fournieri* embryo sac (Higashiyama, 2010), suggesting functional conservation among AtLURE1 and TfLUREs. Disruption of disulphide bonds in the LURE peptides reduces the pollen attraction ability, indicating that the formation of disulphide bonds is important for the LURE stability and/or activity (Okuda et al., 2009; Higashiyama, 2010). With RNA sequence analysis in Arabidopsis, other genes encoding small peptides have been identified in the reproductive process. Most of those genes that are highly expressed in ovules are *CRPs*, in addition to some non-*CRPs* such as *CLEs*, *RGFs*, *PSKs,* and *IDAs* (Huang et al., 2015) and some glycine-rich and proline-rich peptides (Huang et al., 2015).

Whole genome searching in Arabidopsis has identified a family of 9 *CLE*-like genes that encode polypeptides named *ROOT GROWTH FACTOR1* (*RGF1*) to *RGF9*. Similar to the *tpst* mutant, the *rgf1 rgf2 rgf3* triple mutant of Arabidopsis showed reduced size of the root meristem zone, and treatment with chemically synthesized RGF1 peptide restored the short root and small root meristem phenotypes of the triple mutant, indicating that the RGF peptides are required for the maintenance of the root stem cell niche (Matsuzaki et al., 2010).

The importance of these *RGF* genes has also been uncovered by another independent group using a different strategy. By analysis of the CLE18 preprotein, an additional CLE motif with 13 amino acid residues was identified in the C terminus of CLE18. As the new motif is potentially responsible for the long-root phenotype triggered by overexpression of *CLE18* (Meng et al., 2012), the peptide derived from the C-terminal motif is designated as CLE-Like (CLEL) peptide. By homology search using this CLEL peptide as a query sequence, a family of 10 proteins, including CLE18, was identified. CLEL8 is the same as the RGF1 peptide described above. Studies showed that the CLEL8/RGF1 peptide without the sulfation modification is still active (Meng et al., 2012), which is inconsistent with the previous results obtained from RGF1 study that tyrosine sulfation is critical for the function of this peptide (Matsuzaki et al., 2010). Purification and identification of the endogenous RGF1 (also known as CLEL8) peptide in the wild-type and the *tpst* mutant plants may help to elucidate the discrepancy.

In an in silico study, a novel *C-terminally encoded peptide* (*CEP*) gene family with five members was identified in Arabidopsis. The 15-amino acid C-terminal regions of *CEP* proteins show significant sequence similarity among each other. Mature CEP1 peptide produced from the C-terminal motif of the *CEP1* may have its two proline residues hydroxylated. The *CEP1* is mainly expressed in lateral root primordia, and root growth was severely arrested when *CEP1* was overexpressed or when synthesized CEP1 peptide was applied to Arabidopsis seedlings in vitro (Ohyama et al., 2008).

To identify peptide hormones regulating stomatal development, 153 Arabidopsis genes that are predicted to encode small extracellular proteins (<150 amino acids) were overexpressed in Arabidopsis, and transgenic plants were examined for potential defects in guard cells. Detailed analyses of transgenic lines with decreased stomatal density led to the identification of a gene-designated *EPIDERMAL PATTERNING FACTOR1* (*EPF1*). Loss of function of *EPF1* showed a defect in the one-cell spacing rule of the stomatal pattern, exhibiting the formation of clustered guard cells. The *EPF1*, expressed in guard cells and their precursors, seemed to control stomatal patterning through regulating asymmetric cell division of these stomatal precursors (Hara et al., 2007). Eleven EPF homologs including EPF1, EPF2 and EPFL1 to EPFL9 were identified in the Arabidopsis genome (Hara et al., 2009). Similar to the function of *EPF1*, overexpression of *EPF2* also showed reduced stomatal numbers in transgenic plants. *EPF2* negatively regulates the ratio of guard cells and non-guard cells in the epidermis by limiting the number of meristemoid mother cells (MMCs) (Hara et al.,

2009). Furthermore, CHALLAH (CHAL), also known as EPFL6, is showed to inhibit stomatal development in a similar manner as the EPF1 and EPF2. Mutation of *CHAL/EPFL6* suppressed the phenotype of *too many mouths* (*tmm*) in a tissue-specific manner, leading to restored stomatal formation in stems but not in leaves (Abrash and Bergmann, 2010).

STOMAGEN, that is also called EPFL9, is another regulator of stomata development. The mature STOMAGEN is a 45-amino acid cysteine-rich peptide generated from its 102-amino acid preproprotein. The STOMAGEN peptide showed stomata inducing activity in a dosage-dependent manner (Sugano et al., 2010).

Recently, 180 putative peptide-encoding genes preferentially expressed in developing seeds have been identified based on microarray analyses performed during fertilization, early seed development, and seed maturation stages (Costa et al., 2014). Among them, *EMBRYO SURROUNDING FACTOR1* (*ESF1*) shares homology with the maize *MATERNALLY EXPRESSED GENE1* (*MEG1*). The *ESF1* encodes a CRP, which is highly expressed in central cells before fertilization and subsequently in embryo-surrounding endosperm cells (Costa et al., 2014). The ESF1 peptide acts in a non-cell autonomous manner in regulating early embryonic patterning and cell elongation in the suspensor (Costa et al., 2014).

A mass spectrometry (MS)-based analytical platform was developed to discover defense-related peptides by comparing globally endogenous peptides before and after stress induction. Pathogenesis-related protein 1b (PR-1b) was identified in tomato leaves after wounding. As a CRP family member, the *PR-1b* shows a similar expression response to the *systemin* gene (Chen et al., 2014). Antigen 5 is a pathogenesis-related 1 protein (CAP) superfamily member presented in both plants and animals (Gibbs et al., 2008), and is therefore designated as CAP-derived peptide 1 (CAPE1). CAPE1 is able to induce a significant antipathogen response and a minor antiherbivore response in tomato (Chen et al., 2014).

11.3   The cleavage and modifications of peptide hormones

11.3.1   The cleavage of peptide hormones

Peptide hormones in plants usually have few to dozens of amino acid residues. These peptides are often produced from a preproprotein with an N-terminal secretion peptide (Fig. 11.1). Mature peptides are then produced after cleavage and modifications during or after secretion. The cleavage removes nonfunctional peptide fragments from a preproprotein to generate small and functional peptides with fixed lengths. Peptide cleavage after secretion usually includes specific removals of peptide fragments by endopeptidases, and a removal of amino acid residues from the C-terminal end by carboxypeptidases (Breddam, 1986). Subtilase is a common type of endopeptidase in plants, with 56 members (called AtSBTs) in the Arabidopsis genome (Rautengarten et al., 2005). *AtSBT1.1*, as a subtilase, is predicted to be a secreted protein and functions in extracellular spaces. An early study has showed that expression of *AtSBT1.1* correlated with conditions for efficient shoot regeneration (Lall et al., 2004).

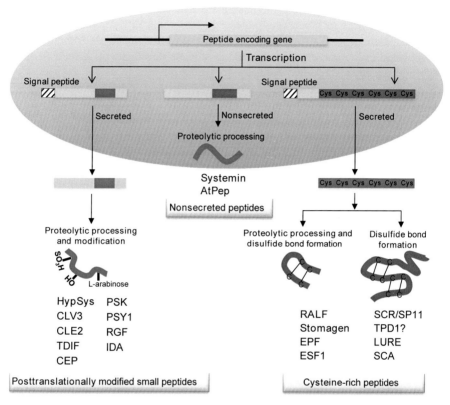

Figure 11.1 The cleavage and modification of peptide hormones in plants. Four types of posttranslational cleavage and modification are known for plant peptide hormones. Most preproteins for peptide hormones have an N-terminal secretory peptide sequences (in white with *slash lines*). Small peptides are released from their precursor proteins through internal cleavages and external removals by different proteases, and most of these peptides are then modified with different chemical groups (with *black mark*), such as hydroxylations on proline residues, sulfations on tyrosine residues and glycosylations on hydroxylated prolines; other cystine-rich peptides may form intramolecular disulfide bonds.

Since PSK is a peptide hormone identified in cell suspension culture media as a density factor that is able to promote callus propagation (Matsubayashi and Sakagami, 1996), the possibility of AtSBT1.1 in proteolytic processing of AtPSKs was investigated. Expression analyses showed that *AtSBT1.1* was upregulated after root explants were transferred to tissue culture media. A chemically synthesized fluorescence-labeled proAtPSK4 peptide was designed to characterize the AtSBT1.1 subtilase recognition site in vitro. In combination with alanine scanning experiment, the putative AtSBT1.1 cleavage site in proAtPSK4 proprotein was identified (Srivastava et al., 2008). Of course, the direct evidence of AtSBT1.1 as being the proAtPSK4 processing enzyme is not yet available.

The *stomata density and distribution1-1* (*sdd1-1*) mutant showed a two- to four-fold increase in stomatal density and the formation of clustered stomata as compared to the wild-type. The *SDD1* encodes a subtilase-like serine protease. It was thus speculated that SDD1 may act as a processing protease in processing EPF peptides that regulate the development of stomata cell lineages (Berger and Altmann, 2000). Although EPF1 is implicated in the one-cell spacing rule in stomata development, it remains to be determined whether SDD1 is responsible for the cleavage of EPF1.

Apparently, *AtSBTs* are functionally redundant as knock-out mutants of *AtSBTs* examined so far did not show any visible phenotypes (Rautengarten et al., 2005). Interestingly, though loss of function of *AtSBT5.4* did not exhibit apparent phenotypes, transgenic plants overexpressing *AtSBT5.4* showed *clv*-like phenotype, exhibiting fasciated inflorescence stems and compound inflorescences with increased number of floral buds. Genetic analyses suggest that AtSBT5.4 interacts with the CLV signaling pathway to regulate the number of stem cells in SAMs. With data obtained so far, it is not clear if AtSBT5.4 is involved in cleaving CLV3 preproprotein, or other peptides with a similar function (Liu et al., 2009). The possibility of elucidating peptide cleavage in vitro has also been pursued using peptideomic tools. When CLV3 fusion proproteins produced in *E. coli* or chemically synthesized CLV3 peptides with several additional N-terminal amino acid residues were added to culture of Arabidopsis seedlings, a predicted cleavage of CLV3 between Leu69 and Arg70 residues is detected using MALDI-TOF mass spectrometry (Xu et al., 2013). Furthermore, analysis of CLV3 peptides with different N-terminal extension sequences showed that a length of four amino acid residues upstream of the Arg70 is necessary and sufficient for effective cleavage, whereas shorter extensions hampered the cleavage. Furthermore, it has been demonstrated that both Leu69 and Arg70 are important for the CLV3 cleavage as substitutions of either of these two residues by alanine compromised the cleavage in vitro, and reduced significantly the CLV3 activity, as shown by in vivo complementation analyses (Ni et al., 2010; Xu et al., 2013).

Metacaspases are a class of cysteine-dependent proteases which are widely distributed in plants, fungi, and protozoa. An extracellular protein GRIM REAPER (GRI) in Arabidopsis is involved in reactive oxygen species (ROS)-mediated cell death (Wrzaczek et al., 2009). An in vitro leaf infiltration assay has shown that a 20-amino acid peptide GRIp[65–84], corresponding to the amino acid residues 65 to 84, induced ion leakage and cell death similar to the GST-GRIp[31–96] fusion protein produced in *E. coli*. Further analysis suggested that an Arabidopsis metacaspase named AtMC9 might be responsible for the cleavage of GRI since bacterially produced GRI fusion protein can be cleaved by a recombinant AtMC9 at the expected site (Wrzaczek et al., 2015).

Carboxypeptidases are another type of peptidases that are involved in peptide maturation by removing amino acid residues one by one from the C terminus of preproproteins. The suppressor of LLP1/CLE19 (SOL1) protein is a putative Zn^{2+} carboxypeptidase that is thought to be involved in CLE19 cleavage since the *sol1* mutant suppressed the *CLE19* overexpression phenotype (Casamitjana-Martínez et al., 2003). Genetic studies showed that *sol1* suppressed only the phenotype of *CLE19* overexpression, not that of *CLV3* overexpression, indicating that SOL1 may specifically catalyze the maturation of the CLE19 peptide (Casamitjana-Martínez et al., 2003). A study showed that SOL1 catalyzes the removal of a single C-terminal residue of arginine from the CLE19 preproprotein in vitro, and the removal is necessary for the optimal activity of the CLE19 peptide (Tamaki et al., 2013). The endosomal localization of the SOL1 protein suggests that the C-terminal processing of CLE19 by SOL1 may occur in endosomes in the secretory pathway (Tamaki et al., 2013).

CRPs are a large group of putative peptide hormones in plants (Marshall et al., 2011). These peptides are stabilized by the presence of intramolecular disulfide bonds (Marshall et al., 2011). Several of these CRPs such as the S-locus cysteine-rich protein (SCR)/S-locus protein 11 (SP11), TPD1, LURE, and EPFs play important roles in several reproductive processes (Mishima et al., 2003; Takayama et al., 2001; Yang et al., 2003; Okuda et al., 2009). Mature peptides detected for CRPs are usually larger in size than predicted (Silverstein et al., 2005). It is possible that, beside the removal of the N-terminal secretion peptide and the formation of disulfide bounds, other processing such as glycosylation modification may be involved in the maturation of these peptides.

Another kind of CRPs such as RALFs may require additional cleavage before maturation to remove some N-terminal sequences by endopeptidases for producing the mature peptide hormones (Pearce et al., 2001; Covey et al., 2010). Furthermore, a 45-amino acid STOMAGEN/EPFL9 mature peptide was identified in Arabidopsis from a 102-amino acid preproprotein. It is thus deduced that the maturation of EPFL9 peptide may involve the N-terminal cleavage of a fragment and the formation of internal disulfide bonds (Sugano et al., 2010).

11.3.2   The modifications of peptide hormones

Modifications of peptide hormones occur usually after cleavage and involve addition of groups through the enzymatic formation of bivalent bonds, to change their structures and potentially enhance their stabilities. So far, three types of modification, including sulfonation, hydroxylation, and arabinosylation, have been identified in mature peptides (Fig. 11.1).

Sulfonation is mediated by tyrosylprotein sulfotransferase (TPST). Arabidopsis TPST is a Golgi-localized 62 KD transmembrane protein. Loss of function of *AtTPST* results in a marked dwarf phenotype, accompanied by other phenotypes such as stunted roots, pale green leaves, reduced higher order of veins, early senescence, and a decreased number of flowers and siliques (Komori et al., 2009). Both PSK and PSY1 are tyrosine-sulfated peptide hormones, but neither PSK nor PSY1 can rescue the phenotype of *tpst*. By screening the Arabidopsis genome, a new tyrosine-sulfated peptide hormone RGF1 was identified, which contains a sulfonation modification at its tyrosine residue (Section 11.2.3). The *rgf1 rgf2 rgf3* triple mutant showed a reduced root meristem zone, which is similar to the *tpst* mutant, indicating that TPST may be responsible for the tyrosine sulfonation of PSK and PSY1 (Matsuzaki et al., 2010).

Hydroxylation modification is mediated by prolyl 4-hydroxylase, which catalyzes the oxidation of proline residues. Many peptide hormones have been found to be hydroxylated. The TomHypSys peptide has four hydroxyl modifications, and also forms two cysteine pairs within the peptide (Pearce and Ryan, 2003). Using transgenic calli overexpressing *CLV3*, a putative CLV3 mature peptide was identified and shown to have hydroxyl modifications on two proline residues (Hyp) at the fourth and seventh positions (Kondo et al., 2006). Similarly, TDIF/CLE41/CLE44 was found to have Hyp modifications at positions equivalent to those of the CLV3 peptide (Ito et al., 2006). TobHypSys, PSY1, CEP1, and RGF1 mature peptides are also found to have Hyp modifications (Pearce et al., 2001; Amano et al., 2007; Ohyama et al., 2008; Matsuzaki et al., 2010), suggesting that Hyp is a common type of modification for peptide hormones in plants.

Interestingly, biochemical analysis revealed that these Hyp residues in several peptide hormones such as PSY1, CLV3, and CLE2 are further modified with an O-linked L-arabinose (Amano et al., 2007; Ohyama et al., 2009). Nano-LC-MS/MS analyses performed in secreted peptides accumulated in culture media of *CLV3*-overexpression plants showed that, beside the hydroxylation of the fourth and seventh prolines, the seventh hydroxylated proline residue was further modified with three arabinoses, and the length of the mature CLV3 might have 13 instead of 12 amino acid residues (Ohyama et al., 2009). In vitro peptide assay showed that the 13-amino acid arabinosylated CLV3 peptide showed a higher activity than the non-glycosylated one. However, the importance of the arabinosylation is still under dispute since replacement of the seventh proline residue by alanine did not apparently affect the complementation efficiency in transgenic assays performed in vivo (Song et al., 2012).

The putative enzyme that catalyzes the transfer of the L-arabinose to the hydroxyl group of hydroxyl proline residues is Hyp O-arabinosyltransferase (HPAT). In Arabidopsis, three HPTAs have been identified, and loss of function of these HPTAs resulted in enhanced hypocotyl elongation, defects in cell wall thickening, early flowering, early senescence, and impaired pollen tube growth (Ogawa-Ohnishi et al., 2013), indicating the importance of the enzyme in plant development. In tomato, the *fasciated inflorescence* (*fin*) mutant that is defective in an arabinosyltransferase showed enlarged meristems and increased number of flowers and fruits (Xu et al., 2015). The phenotype of the strong *fin* allele can be partially rescued by application of the arabinosylated CLV3, suggesting that arabinosylation of CLV3 in tomato is important for its function in meristem maintenance (Xu et al., 2015).

11.4   The function of peptide hormones

Increasing number of peptide hormones have been identified and their downstream signaling pathways have been studied. In plants, peptide hormones usually diffuse and function within a distance of several cell layers, to interact with corresponding receptors and activate downstream signaling pathways (Fletcher et al., 1999). Usually, these peptide hormones are perceived by LRR-RKs localized in plasma membranes and function in many defense and developmental processes (Marmiroli and Maestri, 2014; Han et al., 2014; Grienenberger and Fletcher, 2015).

11.4.1   Peptides in regulating defense responses

Systemin is able to trigger defense responses in plants over a long distance to induce the production of a PI, suggesting that systemin may be transported through vascular tissues from wounded organs to non-wounded ones (Pearce et al., 1991). Studies on jasmonic acid (JA)-insensitive mutants showed that JA is involved in systemin signaling, facilitating the long-distance signaling (Li et al., 2002, 2003).

Alanine substitutions of individual amino acids in the entire 18-amino acid systemin revealed that two residues, namely the proline-13 and the threonine-17, are important for its function. The activity of the systemin was reduced to less than 0.2% when proline-13 was replaced by alanine, and the activity of systemin was lost completely when the threonine-17 was replaced by alanine. Whereas substitutions of other residues with alanine had very little effect on the activity of systemin. A synthetic tetrapeptide (Met-Gln-Thr-Asp) corresponding to the C terminal of systemin retained a low proteinase inhibitor-inducing activity, indicating that the C-terminal end of systemin is critical for its activity (Pearce et al., 1993). In contrast, systemin that contains only the N-terminal 14 amino acid residues showed an antagonistic activity on the intact systemin in a competitive manner (Meindl et al., 1998).

The putative receptor of systemin, SR160, was isolated from photoaffinity radiolabeled suspension culture cells of tomato (*Lycopersicon peruvianum*). The *SR160* encodes a 160-KD membrane-localized LRR family RK (Scheer and Ryan, 2002). Binding of systemin with the SR160 in the plasma membrane activates a signaling cascade. Alkalinization of culture cell media is the first response triggered by systemin due to the inhibition of H^+-ATPase in the plasma membrane, and therefore, a higher H^+ concentration outside of the cells (Felix and Boller, 1995; Schaller and Oecking, 1999; Moyen and Johannes, 1996). Binding of systemin with SR160 also triggered a mitogen-activated protein kinase (MPK) cascade and activated a downstream phospholipase that cleaves linolenic acid and activates the JA biosynthesis pathway (Narváez-Vásquez et al., 1999). Downregulation of the *MPK1* and *MPK2* expressions in tomato reduces systemin-mediated defense responses to hornworm (*Manduca sexta*), indicating that these MPKs are involved in systemin-mediated defense responses. The MPK cascade activation did not correlate with the inhibition of H^+-ATPase (Higgins et al., 2007), suggesting the presence of another signaling pathway located downstream of systemin. In addition, the MPK cascade signaling activated by systemin is also induced by oligosaccharide stimulating factor and UV-B, suggesting the presence of cross talk among downstream signaling pathways triggered by different signals (Stratmann and Ryan, 1997; Holley et al., 2003).

AtPep1 is isolated from Arabidopsis and regulates defense responses, which is a 23-amino acid peptide, produced from a 92-amino acid preproprotein. Expression of the *AtPep1* is induced by wounding, JA and ethylene (Huffaker et al., 2006). Biochemically synthesized AtPep1 peptide exhibits an activity similar to the one isolated from Arabidopsis, activating the expression of the defense-related *PDF1.2* gene and the production of H_2O_2 (Huffaker and Ryan, 2007). Deletion analysis showed that the C-terminal, but not the N-terminal end, of AtPep1 is important for its activity. Results from alanine scanning experiments indicated that serine-15 and glycine-17 are critical for the activity of AtPep1 (Pearce et al., 2008).

Using a photoaffinity method, a 170 KD membrane-associated LRR-RK, named Pep1 Receptor 1 (PEPR1), was identified as the putative AtPep1 receptor (Yamaguchi et al., 2006). Another LRR-RK protein, designated as PEPR2, with 76% similarity to PEPR1 at the amino acid level, was also identified as a putative AtPep receptor because it can sense Pep peptides and trigger defense responses. Both PEPR1 and PEPR2 are transcriptionally induced by wounding, JA, Pep peptides, and pathogen-associated molecular patterns. The PEPR1 is able to sense Pep1 to Pep6, while PEPR2 senses only Pep1 and Pep2. Mutations of either PEPR1 or PEPR2 inhibited seedling growth, elicited an oxidative burst, and induced ethylene biosynthesis after the AtPep1 treatment. However, a *pepr1 pepr2* double mutant was completely insensitive to AtPep1, indicating that PEPR1 and PEPR2 are redundant receptors for sensing the AtPep1. This *pepr1 pepr2* double mutant also failed to respond to either AtPep2 or AtPep3, implying that PEPR1 and PEPR2 are also responsible for perception of AtPep2 and AtPep3 (Krol et al., 2010). Structural analyses of the extracellular LRR domain of AtPEPR1 revealed that AtPep1 adopts a fully extended conformation and binds to the inner surface of the super helical LRR domain of the AtPEPR1 (Tang et al., 2015). Consistent with the previous results obtained from alanine scanning and amino acid deletion analysis, the C-terminal portion of AtPep1 binds to AtPEPR1. The C-terminal Aspartic acid-23 of the AtPep1 is required for the interaction with the LRR domain of AtPEPR1 since deletion of the Aspartic acid-23 compromises the interaction significantly (Tang et al., 2015).

The AtPEPR1 was predicted to have a guanylyl cyclase (GC) domain in the cytosolic kinase region, which was confirmed by its GC activities in conversion of GTP to cGMP and the cyclic nucleotide-gated channel (CNGC)-dependent elevation of cytosolic Ca^{2+}. Application of the AtPep3 peptide to Arabidopsis leaves resulted in an AtPEPR1-dependent cytosolic Ca^{2+} elevation (Qi et al., 2010). Expression of pathogen defense genes such as *PDF1.2*, *MPK3*, and *WRKY33* is mediated by the Ca^{2+} signaling pathway associated with AtPep peptides and their receptors.

11.4.2 Peptides in regulating shoot apical meristem maintenance

SAM is a collection of cells that has the capacity to continuously renew itself by cell division, and to generate new above-ground tissues and organs in leaves, stems, flowers, and fruits. Based on the developmental fates, cells in the SAM are divided into three layers: the epidermal cell layer (L1) that forms the epidermis of all above-ground organs such as shoots, leaves, and flowers; the subepidermal cell layer (L2) that produces mesophyll cells; the underlying cell layer (L3) that develops into the vascular and internal tissues (Carles and Fletcher, 2003). Based on relative locations, cells in the SAM are also divided into three zones, the peripheral zone (PZ), the central zone (CZ), and the rib zone (RZ). Multipotent stem cells are located in L1, L2, and part of the L3 layer of the CZ, with a relatively low rate of cell divisions, and function as the source cells for the PZ and RZ. The continuous division and differentiation of cells in the SAM maintains the number of stem cells in the SAM, revealing a tight balance between cell division and cell differentiation.

CLV3 peptide plays an important role in regulating the balance between cell division and cell differentiation in the SAM. Treatment of Arabidopsis seedlings with CLV3 peptide resulted in a consumption of the root apical meristem (Fiers et al., 2005). Alanine scanning of chemically synthesized CLV3 peptide in combination with in vitro peptide treatment showed that different amino acid residues in the peptide contribute differently to the CLV3 function (Kondo et al., 2008). Results from in vivo transgenic complementation analyses with alanine-substituted CLV3 constructs establish a precise contribution map of individual amino acid residues in the CLV3 peptide in SAM maintenance (Song et al., 2012). One unexpected result is that the replacement of proline-7 with alanine did not affect the complementation efficiency of the CLV3, indicating that the hydroxylation and arabinosylation modifications of CLV3 on proline-7 is not essential for the function of CLV3 in vivo (Song et al., 2012).

Perception of the CLV3 peptide in the SAM involves multiple receptor-related proteins. Both CLV1 and CLV2 are identified by map-based cloning of genes underlying the multi-carpel mutants of *clv* in Arabidopsis. *CLV1* encodes an LRR-RK protein with 21 LRRs in its extracellular domain (Clark et al., 1993, 1997), and *CLV2* encodes an LRR-receptor-like protein (LRR-RLP) that lacks an intracellular kinase domain (Kayes and Clark, 1998; Jeong et al., 1999). Expression analyses showed that *CLV1* is expressed in the CZ of the SAM, particularly in the stem cell-organizing center cells (OC) and in L3 stem cells, whereas *CLV2* is expressed in the whole SAM region (Fletcher et al., 1999). Using a competitive binding assay, it has been shown that CLV3 binds directly to the external LRR domain of CLV1 (Ogawa et al., 2008). Treatment of Arabidopsis seedlings with CLV3 peptide caused a reduced level of plasma membrane-localized CLV1 and internalization of the CLV1 to lytic vacuoles in deeper cell layers of the SAM (Nimchuk et al., 2011).

Screening for suppressors using CLV3 and CLE19 overexpression plants led to the identification of two genes, *RECEPTOR-LIKE PROTEIN KINASE2* (*RPK2*; also known as *TOADSTOOL2*, *TOAD2*) and *CORYNE* (*CRN*; or called *SUPPRESSOR OF LLP1 2*, *SOL2*) (Kinoshita et al., 2010; Casamitjana-Martínez et al., 2003). Plants with mutations in either *RPK2/TOAD2* or *CRN/SOL2* showed weak *clv*-like phenotypes (Kinoshita ct al., 2010; Müller et al., 2008). The CRN/SOL2 is a membrane-localized protein with an intracellular kinase domain but lacks an extracellular domain, which is therefore unlikely to be a receptor by itself. CRN/SOL2 forms a complex with CLV2, which is required to localize the complex to the plasma membrane and bind with the extracellular CLV3 peptide ligand (Bleckmann et al., 2010; Zhu et al., 2009). However, biochemical studies revealed that CRN/SOL2 might be a pseudokinase that is unable to execute autophosphorylation and signal transduction, suggesting that CRN/SOL2 may act as a scaffold protein, similar to animal pseudokinases (Nimchuk et al., 2011). The *RPK2/TOAD2* encodes an LRR-RLK, that is expressed constitutively in shoot and root meristems. In transient expression analyses performed in tobacco, RPK2/*TOAD2* forms homo-oligomers by itself, and is not associated with CLV1 or CLV2 (Kinoshita et al., 2010). A *clv1 clv2 rpk2* triple mutant displays a stronger phenotype than any single mutants, and shows a similar phenotype as *clv3*, implying that in the SAM these three receptor-like proteins may perceive the CLV3 peptide in independent pathways.

Additionally, another two genes encoding LRR-RLKs, *BAM1* and *BAM2*, with high homology to *CLV1*, participate in SAM regulation (DeYoung et al., 2006; Guo et al., 2010). Both of them are broadly expressed in the periphery of the meristem. The *bam1 bam2* double mutant showed an arrested SAM, which is opposite to *clv* mutants, indicating that BAM1-BAM2 may act antagonistically to CLVs in regulating the homeostasis of the SAM (DeYoung et al., 2006).

As mentioned just above, CLV3 binds directly with the extracellular domain of CLV1.

By photoaffinity labeling with photoactivatable arabinosylated CLV3, it has also been shown that CLV2 and RPK2 do not bind directly to the CLV3 peptide, but BAM1 did (Shinohara and Matsubayashi, 2015). It is plausible that, among multiple receptors expressed in the SAM, some may participate in one signaling pathway such as the SAM maintenance, and others may function in multiple pathways.

The *WUSCHEL* (*WUS*) encodes a plant-specific homeodomain transcription factor that plays a critical role in promoting stem cell identity in the SAM (Mayer et al., 1998; Fletcher et al., 1999; Schoof et al., 2000; Nimchuk et al., 2011). The *wus* mutant lacks SAM or display partially differentiated cells in the SAM. Overexpression of the *WUS* leads to enlarged meristems, suggesting that *WUS* is required for maintaining stem cell identity (Laux et al., 1996; Mayer et al., 1998). *WUS* is expressed in the stem cell OC region of the SAM (Laux et al., 1996). The *clv wus* double mutant displays a *wus*-like phenotype, with reduced stem cell population, suggesting that *WUS* acts downstream of the *CLV* pathway. The *WUS* expression domain is enlarged in the *clv3* mutant, which could be suppressed by overexpression of *CLV3*.

A negative feedback loop between WUS and CLV3 is thus established: when *CLV3* expression becomes lower as the number of stem cells decreases, *WUS* expression increases, promoting stem cells to divide; when stem cells become too abundant, *CLV3* expression is upregulated, leading to suppression of *WUS* expression and restriction of stem cell divisions (Brand et al., 2000; Schoof et al., 2000; Müller et al., 2008; Yadav et al., 2011). WUS protein produced in the stem cell OC is able to move to the nuclei of adjacent stem cells, where it may directly bind to the *CLV3* promoter and trigger *CLV3* expression, thus positively regulate the *CLV3* transcription (Kieffer et al., 2006; Yadav et al., 2011). If WUS mobility was restricted by fusion with a nuclear localization signal peptide, it fails to complement the *wus* phenotype, indicating that the intercellular migration is critical for the stem cell maintenance.

Furthermore, WUS may act together with another homeodomain transcription factor SHOOTMERISTEMLESS (STM) to regulate the expression level instead of the expression location of the *CLV3* (Brand et al., 2002). WUS can also form a heterodimer with the HAIRY MERISTEM (HAM) transcription factor, leading to increased activities of both WUS and HAM synergistically. The HAM and WUS share common targets in vivo, and their physical interaction is important for activating downstream genes to promote stem cell proliferation in the SAM (Zhou et al., 2015).

Taken together, results obtained so far indicate that the mature CLV3 peptide hormones are perceived by receptor complexes, which are composed of CLV1, CLV2, CRN/SOL2, RPK2/TOAD2, BAM1, and BAM2. These receptor components transduce the CLV3 signal to repress the expression of *WUS* transcription factor. WUS may move through several cell layers and bind directly to the *CLV3* promoter to enhance the expression of *CLV3*. Thus, the CLV3-WUS pathway forms a feedback regulation loop to regulate stem cell maintenance and cell differentiation in the SAM (Fig. 11.2).

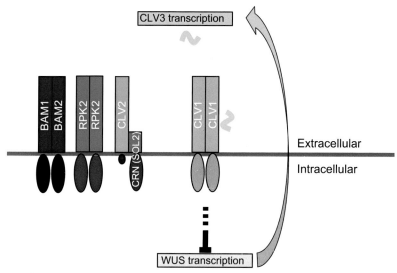

Figure 11.2 The role of the CLV3-WUS signaling pathway in plants. The CLV3 prepro-protein is translated, processed, and secreted to extracellular spaces. Receptor complexes perceiving the CLV3 peptide include CLV1/CLV1, CLV2/CRN (SOL2), RPK2/RPK2, and BAM1/BAM2. Most of these receptor proteins such as CLV1, RPK2, and BAM1/BAM2 are plasma membrane-localized LRR-RLKs that have an extracellular LRR domain, a single transmembrane domain, and an intracellular kinase domain. CLV3 signal is perceived by these receptor complexes and repress *WUS* expression. WUS then moves several cell layers to activate *CLV3* expression.

11.4.3   Peptides in regulating root apical meristem maintenance

Similar to the SAM, the root apical meristem (RAM) in plants also has a stem cell niche and the continuous renewal and differentiation of these stem cells maintain the long-term function of a root (Scheres et al., 2002). Among peptides identified in RAM development so far, CLE and RGF/GLV/CLEL peptides are the most prominent (Matsuzaki et al., 2010; Meng et al., 2012; Whitford et al., 2012).

11.4.3.1   CLE40 peptide

The regulation of stem cell maintenance in the RAM by CLE40 peptide is similar to that of the CLV3 function in the SAM. The *cle40* mutant exhibits slightly shorter roots with slightly irregular cell patterning in the RAM (Hobe et al., 2003). A mutant of a RLK ARABIDOPSIS HOMOLOG OF CRINKLY4 (ACR4) was identified with an additional layer of columella stem cells (CSCs), and was partially insensitive to CLE40 peptide treatment, suggesting a possible role of ACR4 in sensing the CLE40 peptide (De Smet et al., 2008; Stahl et al., 2009). In the RAM, *CLE40* is expressed in the stele and differentiating columella cells of the root cap (Stahl et al., 2009), whereas *ACR4* is expressed in CSCs and columella cells, but not in the Quiescent Center (QC) and proximal stem cells. It has been proposed that CLE40 is secreted from differentiated cells to regulate the proliferation of CSCs via ACR4 RLK. Recently, it has been reported that CLV1 is activated by the CLE40, together with ACR4 to restrict root stem cell. Both CLV1 and ACR4 overlap in their expression domains in the distal root meristem (Stahl et al., 2013). Thus, the CLE40-ACR4 module constitutes a signal transduction pathway, regulating the number of CSCs and the CSC differentiation in a non-cell autonomous manner.

Similar to the CLV3-WUS feedback regulation loop in SAM, a CLE40-WUSCHEL-RELATED HOMEOBOX 5 (WOX5) feedback loop has been proposed, which controls the stem cell homeostasis in the RAM. *WOX5* is specifically expressed in QC of the RAM. The *wox5* mutant showed terminal differentiation, with enlarged QC cells and premature differentiation of CSCs, as indicated by starch accumulation in these cells (Sarkar et al., 2007). Conversely, overexpression of *WOX5* repressed CSC differentiation, resulting in overaccumulation of columella initial cells (Sarkar et al., 2007). Therefore, it is plausible that WOX5 expressed in the QC acts in a non-cell autonomous manner to maintain CSCs in a non-differentiation state (Sarkar et al., 2007).

The local expression of *WOX5* within the QC is under the control of *CLE40* and *ACR4* (De Smet et al., 2008; Stahl et al., 2009). The intracellular domain of ACR4 is able to interact with and phosphorylate WOX5 in vitro (Meyer et al., 2015). The WOX5 protein moves from QC into the adjacent CSCs, where it directly represses the expression of the transcription factor *CYCLING DOF FACTOR 4* (*CDF4*) by recruiting corepressors TOPLESS/TOPLESS-RELATED (TPL/TPR) and HISTONE DEACETYLASE 19 (HDA19) (Pi et al., 2015), suggesting that CLE40 may regulate chromatin-mediated repression of differentiation programs in RAM.

In addition to the distal root meristem phenotype, the *cle40* mutant also exhibits a reduction in the primary root length and proximal root meristem size (Pallakies and Simon, 2014; Stahl et al., 2009), although the underlying mechanism has yet to be elucidated.

11.4.3.2   RGF peptides

In order to identify more peptides regulating the meristematic activity of roots and RAM stem cell niche maintenance, a search of the Arabidopsis genome for genes that are likely to encode sulfated peptides has been performed, leading to the identification of the RGF peptide family (Matsuzaki et al., 2010). The PLETHORA 1 (PLT1) and PLT2 proteins, important regulators of the RAM, are predicted to control root meristem activity downstream of the RGF peptide signaling pathway. By combination of a custom-made RLK expression library and photoaffinity labeling approach, three LRR-RLKs, namely RGFR1, RGFR2, and RGFR3, have been identified and shown to interact directly with RGF peptides to regulate RAM development in Arabidopsis (Shinohara et al., 2016). The *RGFR1*, *RGFR2*, and *RGFR3* are expressed in root tissues including the proximal meristem, the elongation zone, and the differentiation zone. The *rgfr1 rgfr2 rgfr3* triple mutant, with a short root phenotype, is insensitive to treatment of RGF peptide (Shinohara et al., 2016), suggesting that these LRR-RLKs mediate the perception of the RGF peptides. Further studies of the RGF-RGFRs signaling pathway may facilitate our understanding of the molecular regulation framework in the RAM.

11.4.3.3   C-terminally encoded peptides

Using in silico analyses, a family of 14 members of peptide-coding genes, each with a conserved C-terminally encoded peptide (CEP) motif, has been identified in the Arabidopsis genome (Ohyama et al., 2008). The mature CEP peptides, derived from the CEP motif of the precursor proteins, are 15 amino acid residues in length, with two hydroxylated proline residues (Ohyama et al., 2008). Overexpression of the *CEP1* or exogenous application of chemically synthesized CEP1 peptide resulted in reduced primary root growth (Ohyama et al., 2008). Interestingly, expressions of most of these *CEP* genes are induced by nitrogen starvation, but not by phosphate or potassium starvation (Tabata et al., 2014).

Two LRR-RLK proteins, CEP RECEPTOR1 (CEPR1) and CEPR2, have been identified as putative CEP receptors (Tabata et al., 2014). The CEP-CEPR signaling pathway induces the expression of nitrate transporter genes *NRT2.1*, *NRT3.1*, and *NRT1.1*. In split-root culture experiments, these CEP peptides were found to be secreted from nitrogen-starved roots, and to move to shoots to induce nitrogen transport, resulting in regulated systemic nitrogen homeostasis. This mode of CEPs action in roots may influence root architecture to restrict root elongation in poor soil conditions, and promote an efficient mineral uptake (Tabata et al., 2014).

11.4.4   Peptides in regulating vascular bundle development

In addition to stem cells in the SAM and RAM, another group of stem cells is present in procambium and cambium, which form the vascular tissues in plants, to allow long-distance transport of water, sugars, and mineral nutrients throughout the plant body (Lucas et al., 2013). These procambial and cambial cells divide continuously to produce two major vascular tissues, xylem and phloem. Studies have shown that several CLE peptides play important roles in vascular bundle development.

11.4.4.1   Tracheary element differentiation inhibitor factor peptides

Xylem is a complex tissue, consisting of tracheary elements (TEs), xylem fibers and xylem parenchyma cells. Using an in vitro xylem differentiation system established in *Z. elegans* mesophyll cells, an extracellular factor TDIF inhibiting the differentiation of TEs has been isolated from the culture medium (Ito et al., 2006).

TDIF is a 12-amino acid peptide, HEVHypSGHypNPISN, with hydroxylation modifications on two proline residues at the fourth and seventh positions. Genes that encode the TDIF in Arabidopsis are *CLE41* and *CLE44*, which are highly homologous to *CLE42* and *CLE46* (Ito et al., 2006). CLE41 and CLE44 preproteins are identical in their CLE motifs, and thus are expected to produce an identical TDIF peptide hormone. CLE41/CLE44 and CLE42 peptides showed a strong activity in inhibiting the xylem differentiation in vitro, whereas other CLE peptides did not, suggesting that CLE41/CLE44 and CLE42 peptides function specifically in suppressing xylem differentiation (Hirakawa et al., 2008). The *cle41-1* mutant in Arabidopsis showed enhanced xylem differentiation (Hirakawa et al., 2010). Overexpression of *CLE41* or *CLE44* in Arabidopsis partially inhibited the differentiation of TEs, leading to the formation of discontinuous xylem strands and enhanced vascular stem cell division in the hypocotyl (Hirakawa et al., 2008; Whitford et al., 2008). Therefore, CLE41/CLE44 functions not only as a positive signal to promote the vascular stem cell division rate, but also as a negative signal to repress xylem differentiation.

In order to identify the receptor for sensing the TDIF/CLE41/CLE44 peptides, a collection of LRR-RLK mutants were treated with TDIF (Hirakawa et al., 2008). Among them, one mutant, carrying a T-DNA insertion in a gene later named *PHLOEM INTERCALATED WITH XYLEM* (*PXY*) or *TDIF Receptor* (*TDR*), was shown to be insensitive to TDIF treatment. Photoaffinity labeling studies revealed that TDIF binds directly to PXY/TDR (Hirakawa et al., 2008). The *pxy/tdr* mutant exhibits discontinuous vascular strand formation in leaves (Hirakawa et al., 2008), which is consistent with the function of TDIF in suppressing TEs differentiation in vitro (Ito et al., 2006).

PXY/TDR is expressed preferentially in procambium and cambium cells (Fisher and Turner, 2007; Hirakawa et al., 2008), while *CLE41* and *CLE44* are expressed specifically in the phloem and its neighboring cells. The CLE41 and CLE44 peptides are secreted to the apoplasts surrounding phloem cells (Hirakawa et al., 2008). It is thus deduced that TDIF/CLE41/CLE44 peptides produced in phloem and its neighboring cells diffuse toward the vascular stem cells to regulate their cell fates in a non-cell autonomous manner (Hirakawa et al., 2008). An analysis showed that the structure of TDIF adopts a "Ω"-like conformation and binds directly to the inner surface of the LRR domain of PXY/TDR (Zhang et al., 2016).

WOX4 is a target gene of the TDIF signaling pathway (Hirakawa et al., 2010). After treatment with TDIF peptide, the expression of *WOX4* was rapidly induced in wild-type seedlings, but not in *pxy/tdr* mutants, indicating that TDIF regulates *WOX4* expression through the PXY/TDR receptor (Hirakawa et al., 2010). *WOX4* is expressed in the vascular tissues of the whole plant, which overlaps with the *PXY/TDR* expression domain. The *wox4* mutant showed reduced procambial proliferation and continuous xylem differentiation, indicating that *WOX4* regulates procambial cell divisions but not the procambium to xylem differentiation.

WOX14, that is homologous to *WOX4*, may act redundantly with *WOX4* in the regulation of vascular cell divisions since a *wox14* mutant showed an enhanced procambial cell division defect in the *wox4* background (Etchells et al., 2013). It is thus proposed that the CLE-WOX module is a common mechanism in regulating the homeostasis of three major meristems, SAM, RAM, and procambium/cambium, in plants.

Identification of components involving the TDIF/CLE41/CLE44-PXY/TDR signal pathway gives insight into understanding of TDIF function on procambial stem cell division and xylem differentiation. The Arabidopsis *HAM* family transcription regulators are identified as conserved interacting cofactors with WOX proteins (Zhou et al., 2015). The quadruple mutant of *HAM* genes shows a reduced number of procambial cells, which is similar to the *wox4* mutant (Zhou et al., 2015), suggesting that the WOX4-HAMs transcription factor complexes may function in procambial cell proliferation downstream of TDIF-PXY/TDR signaling pathway.

Based on yeast two-hybrid screening, GLYCOGEN SYNTHASE KINASE 3 (GSK3), BRASSINOSTEROID INSENSITIVE 2 (BIN2), BIN2-LIKE 1 (BIL1), BIL2, SHAGGY-RELATED KINASE 11 (ATSK11) and ATSK13 are identified to interact with PXY (Kondo et al., 2014), and these interactions were confirmed by fluorescence resonance energy transfer in plants. In tobacco transient assays, BIN2 interacts closely with PXY at the plasmamembrane, but is released from PXY with TDIF perception, revealing that the perception of TDIF by PXY results in promotion of GSK3 activity (Kondo et al., 2014). One target of GSK3s in TDIF-PXY signaling pathway is the BRI1-EMS-SUPPRESSOR 1 (BES1) transcription factor, which is negatively regulated by phosphorylation in xylem cell differentiation (Kondo et al., 2014, 2015). The dominant-negative *bes* mutant, *bes-1d*, was reported to have reduced number of procambial cells. Thus, TDIF/CLE41/CLE44-PXY/TDR may activate BIN2, which in turn suppresses BES1 activity to inhibit xylem cell differentiation. In the *pxy* mutant transcriptome, several *ETHYLENE RESPONSE FACTOR* (*ERF*) genes are upregulated (Etchells et al., 2012). Mutations in these *ERF* genes reduced the radial growth and vascular bundle size, suggesting a cross talk between PXY/TDR and ethylene signaling. In addition, EPIDERMAL PATTERNING FACTOR-LIKE peptides (EPFLs) and their receptors ERECTA (ER) and ER-LIKE (ERL) are also shown to act with TDIF-PXY/TDR signaling during vascular development (Etchells et al., 2013; Uchida and Tasaka, 2013). Endodermis-produced peptides, EPFL4 and EPFL6, may redundantly regulate the procambial development in inflorescence stems via ER and ERL1.

11.4.4.2 CLE45 peptide

The root of Arabidopsis provides an excellent model for studying phloem due to the case of following the phloem development in defined cell files (Bauby et al., 2007; Truernit et al., 2008). All phloem pole cell files are derived from a common pro-cambium-sieve element (PSE) stem cell located next to the QC. The procambium-SE stem cell divides to produce a procambium-SE precursor cell, which then gives rise to one outer protophloem cell layer and an inner metaphloem cell layer (Mahonen et al., 2000; Rodriguez-Villalon et al., 2014). The commitment of specific phloem cell fates from PSE stem cells is expected to rely on intrinsic and extrinsic position signals. CLE45 peptide is involved in regulating the transition from cell proliferation to phloem differentiation in roots.

The study of CLE45 in phloem development can be traced back to the research in BREVIS RADIX (BRX), which is a positive regulator of protophloem formation (Scacchi et al., 2010). By screening the suppressor of *brx*, a *bam3* mutant is identified as it suppresses the postembryonic root meristem defect (Depuydt et al., 2013). In order to identify the putative ligand for BAM3, chemically synthesized CLE peptides that are able to induce the short root phenotype were applied to *bam3* mutants. Interestingly, *bam3* is insensitive to CLE45 peptide (Depuydt et al., 2013), suggesting that CLE45 is the candidate ligand for BAM3.

In roots, *CLE45* and *BAM3* are specifically expressed along the developing protophloem (Rodriguez-Villalon et al., 2014). It is thus believed that CLE45 acts to inhibit protophloem specification via BAM3 by preventing the differentiation from SE precursor cells to their preceding developmental programs. The inhibition of protophloem differentiation by CLE45 is further confirmed by the abolished expression of the phloem marker gene *ALTERED PHLOEM DEVELOPMENT* (*APL*) in the developing phloem of the primary roots when treated with CLE45 peptide.

11.4.4.3   CLE26 peptide

In addition to CLE45, a study showed that *CLE26* is expressed along the developing protophloem in roots, which partially overlaps with the *CLE45* expression. Similar to CLE45, treatment of the wild-type Arabidopsis seedlings with CLE26 peptide suppressed the differentiation of newly formed root protophloem SEs and produced a short root phenotype, indicating that CLE26 may work together with CLE45 to regulate the differentiation of protophloem (Rodriguez-Villalon et al., 2015).

11.4.5   Peptides in regulating stomata development

11.4.5.1   EPF1/2 peptides

EPF1 is expressed in stomatal cells and their precursors, and controls stomatal patterning by regulating the asymmetric cell divisions of guard mother or precursor cells (Hara et al., 2007). *EPF2* is expressed earlier than *EPF1* during stomata development, mainly in meristemoids and their sister cells, guard mother cells, and MMCs (Hara et al., 2009). Mutants that lose both EPF1 and EPF2 functions exhibit clustered guard cell phenotype, and overexpression of either of them shows decreased number of guard cells (Hara et al., 2007, 2009). An EPF1-EPF2 fusion protein produced in *E. coli* has the activity of triggering severe inhibition of asymmetric cell divisions in stomatal lineages (Lee et al., 2012).

Genetic analyses showed that the function of EPF1 is dependent on the TOO MANY MOUTHS (TMM) receptor-like protein and the ERECTA family RLKs, ER, ER-Like 1 (ERL1), and ERL2 (Hara et al., 2007). Co-immunoprecipitation analyses showed that EPF1 and EPF2 expressed in *Nicotiana benthamiana* are associated with these ER family RLKs, while TMM is co-immunoprecipitated with EPF2, not with EPF1 (Lee et al., 2012). Using quartz crystal microbalance and surface plasmon resonance analyses, it has been demonstrated that the binding of EPF1 to both ER and ERL1 is rapid and shows similar kinetics, while EPF2 showed higher binding affinity to ER than to ERL1. EPF2, but not EPF1, binds to TMM (Lee et al., 2012). These results together suggest complex interactions between peptide hormones and their receptors.

11.4.5.2   Stomagen peptide

Different from EPF1 and EPF2 peptides, stomagen is a positive regulator of stomatal development. Stomagen is expressed in mesophyll cells of immature leaves instead of epidermal cells from which stomata are formed, indicating that stomagen acts in a non-cell autonomous and a cross-tissue manner to regulate stomata patterning. With the help of nuclear magnetic resonance (NMR), the structure of stomagen was resolved as being a loop and a scaffold containing three disulphide bonds. Domain swapping between EPF2 and stomagen revealed that the loop confers the functional specificity, and the scaffold is structurally required for their activities (Ohki et al., 2011). Overexpression of stomagen with amino acid residue substitutions to remove one or all three disulphide bonds lost its effect on stomatal density, indicating that these disulphide bonds are important for the activity of the stomagen (Ohki et al., 2011). Biochemically synthesized stomagen is antagonized by EPF2 in enhancing the stomatal density, while EPF2 is not antagonized by stomagen.

Genetic studies showed that TMM is epistatic to both EPF2 and stomagen, implying that the negative regulator of EPF1/2 and the positive regulator of stomagen may competitively bind to the TMM protein to regulate stomatal development (Sugano et al., 2010). Further studies to combine results of STOMAGEN, EPF1, and EPF2 with their receptors of ER family members and TMM showed that stomagen requires ER family RLKs to promote stomatal development, and interferes with the inhibition of stomatal development mediated by the EPF2-ER module (Lee et al., 2015). EPF2 treatment triggers a rapid phosphorylation of downstream signaling components of MPK3 and MPK6 in vivo, indicating that mitogen-activated protein kinases (MAPK) cascades participate in EPF2 signal transduction to inhibit stomatal development.

Studies using co-immunoprecipitation indicate that both ER and ERL1 RLKs form both homo- and hetero-dimers, and they also form heterodimers with TMM, but TMM does not form a homodimer by itself (Lee et al., 2012). Ectopic expression of the *Pseudomonas syringae* tomato (Pst) effector AvrPto in Arabidopsis lead to excessively clustered stomata in the cotyledon epidermis. BAK1/SERK3 is one of the physiological targets of AvrPto and AvrPtoB. Genetic evidence showed that these SERKs redundantly regulate stomatal patterning downstream of EPF peptides and upstream of MPKs, while EPFs trigger the heterodimerization of ER and SERK family RLKs. SERKs associate with TMM in a ligand-independent manner. SERK and ER family RLKs phosphorylate each other, which trigger the downstream YDA-MKK4/5-MPK3/6 cascade for stomata patterning (Meng et al., 2015) (Fig. 11.3).

11.4.6 Peptides in regulating reproductive processes

11.4.6.1 TPD1 peptide

TPD1 encodes a small protein that regulates tapetum cell differentiation (Yang et al., 2003). Since a *tpd1* mutant showed a similar phenotype to *excess microsporocytes1/extra sporogenous cells* (*ems1/exs,* Zhao et al., 2002), and EMS1/EXS is an LRR-RLK, it is speculated that TPD1 may be the peptide hormone received by the EMS1/EXS receptor to regulate cell differentiation in anthers. Yeast two-hybrid experiments verified that TPD1 interacts with the LRR domain of EMS1/EXS (Yang et al., 2003). The

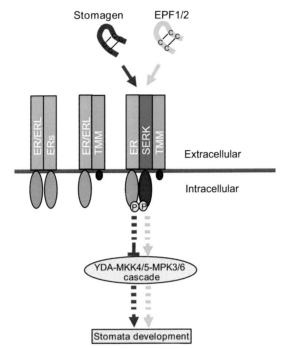

Figure 11.3 The model for stomagen and EPF1/2 peptide hormones signaling. Stomagen positively regulates stomata development, and EPF1/2 negatively regulates stomata development. Receptor complexes perceiving the stomagen signal are ERs/ERL, ER/ERL/TMM, and ER/SERK/TMM. ER and SERK can phosphorylate each other and promote stomata development through downregulating YDA-MKK4/5-MPK3/6 cascade. Functions of stomagen can be antagonized by EPF2 which suppresses stomata development by upregulation of the YDA-MKK4/5-MPK3/6 cascade.

interaction between TPD1 and EMS1/EXS triggers a self-phosphorylation of EMS1/EXS, which may be required for downstream signaling transduction (Jia et al., 2008).

11.4.6.2 CLE45 peptide

By screening the CLE family peptides, CLE45, CLE43, and CLV3 showed activities in promoting pollen tube growth. Among them, only *CLE45* is expressed in the stigma, and expands to the transmitting tract if the temperature is shifted from 22°C to 30°C, suggesting that CLE45 may facilitate pollen tube growth. *CLE45-RNAi* plants showed a significantly reduced seed number and seed size after higher temperature treatment. Among the LRR-RLK XI members, STERILITY-REGULATING KINASE MEMBER 1 (SKM1) and SKM2 were selected as candidate RLKs for CLE45 due to the expression of these genes in pollen tubes. The *skm1* mutant transformed with a "kinase-dead" version of SKM1 shows reduced seed production at 30°C, but not at 22°C, suggesting that the CLE45-SKM1/SKM2 signaling pathway may regulate pollen tube growth under higher temperature to ensure proper seed production (Endo et al., 2013).

11.4.6.3   LURE1 peptide

LURE1 is expressed in ovules and encodes a CRP that functions in pollen tube attraction (Okuda et al., 2009). Through screening of LRR-RLKs expressed specifically in pollen and pollen tubes, MDIS1, MDIS2, MIK1, and MIK2 were selected as candidate receptors for the LURE1 peptide. Further genetic and biochemical studies suggest that three plasma membrane-localized LRR-RLKs, MDIS1, MIK1, and MIK2, are receptors of LURE1 (Wang et al., 2016). LURE1 specifically binds to the extracellular domain of MDIS1, MIK1, and MIK2 and triggers dimerization of these three receptors, and activates the kinase activity of MIK1 (Wang et al., 2016).

At the same time, screening of T-DNA insertion lines of pollen-specific LRR-RLKs shows that PRK6 is also a key receptor for LURE1 (Takeuchi and Higashiyama, 2016). Further genetic analysis indicated that other PRK family members such as PRK1, PRK3, and PRK8 are also involved in perception of the LURE1 peptide, but with minor roles as compared with PRK6. PRK6 interacts with itself, PRK3 and two receptor-like cytoplasmic kinases, LIP1 and LIP2. Both LIP1 and LIP2 are involved in pollen tube growth and attraction, and in LURE1 signaling (Liu et al., 2013; Takeuchi and Higashiyama, 2016). PRK6 is able to interact with Rho of plant guanine-nucleotide exchange factors (GEF) (ROPGEFs) when examined in bimolecular fluorescence complementation assays, indicate that LURE1 may recruit the intracellular tip growth machinery such as ROPGEFs and ROP1 for directing pollen tube growth through asymmetrical re-localization of PRK6 on the plasma membrane of pollen tubes (Takeuchi and Higashiyama, 2016).

11.4.6.4   SCR/SP11 peptide

Self-incompatibility (SI) is a phenomenon in which pollen from the same species is recognized and rejected by the stigma. SI is controlled by genes at the S-locus. The S-LOCUS RECEPTOR KINASE (*SRK*) gene encodes a serine/threonine kinase localized in the plasma membrane of the stigma epidermis (Stein et al., 1996). A SCR protein, also called S-locus protein 11 (SP11), is a pollen-expressed peptide that functions as the male determinant of SI (Schopfer et al., 1999). Among SCR/SP11 variants, only a few amino acid residues are conserved. Structure determination and 3D modeling showed consistently that these diverged SCR/SP11 variants have a similar configuration (Chookajorn et al., 2004; Mishima et al., 2003). Domain swapping experiments indicated that four contiguous amino acid residues are enough to determine the specificity of each SCR/SP11 variant. SCR/SP11 is able to bind directly to its receptor SRK in a haplotype-specific manner (Takayama et al., 2001). Alanine scanning experiments in SCR/SP11 showed that most other residues are required for the interaction between SCR/SP11 and SRK (Chookajorn et al., 2004).

11.4.6.5   CLE8 and CLE19 peptides

One of the *CLE* family member in Arabidopsis, *CLE8,* is expressed in embryos and endosperms. Mutations of *CLE8* showed abnormal cell division in both embryos and suspensors. The CLE8 peptide acts non-cell autonomously in regulating early embryo development (Fiume and Fletcher, 2012). WOX8 is shown to be a downstream component of CLE8, and is the only *WOX* gene expressed in both endosperm and the basal lineage of the embryo (Haecker et al., 2004; Breuninger et al., 2008). Further genetic studies demonstrated that, in the *cle8* mutant, the expression of *WOX8* was detected only in the basal suspensor cells, and dramatically reduced or absent from other suspensor cells, indicating that WOX8 acts in the same signaling pathway with CLE8 (Fiume and Fletcher, 2012). In contrast to *CLE8*, *CLE19* is expressed only in embryos, starting in the cotyledon primordia in triangular-stage embryos, and in epidermal cells of the cotyledon in torpedo-stage embryos, and then at the edge of the cotyledon in cotyledonary embryos. Expression of an antagonistic *CLE19* construct under the control of *CLE19* regulatory elements in Arabidopsis lead to defective cotyledon development in embryos and delayed cellularization in endosperm, suggesting that CLE19 acts in a non-cell autonomous manner in regulating both embryo and endosperm development (Xu et al., 2015). This speculation is confirmed by the observation that expression of the antagonistic *CLE19* construct under the control of an endosperm-specific promoter also lead to defective cotyledon development (Xu et al., 2015). In addition, if CLE19 also functions through the WOX8 transcription factor still remains to be elucidated.

11.4.6.6 ESF1 peptide

The ESF1 peptide, a central cell-derived CRP, acts non-cell autonomously in regulating early seed development. NMR analysis indicated that the 68-amino acid ESF1 peptide consists of four loops and a scaffold supported by four disulfide bonds. Bioactivity assays showed that these disulfide bonds are necessary for the structural topology and the activity of ESF1. Since mutations of both YODA (YDA) and SHORT SUSPENSOR (SSP), a Pelle/IL-1R (an interleukin-1 receptor) associated kinase, exhibited abnormal suspensor phenotype, it is deduced that YDA and SSP may act together or downstream of ESF1 peptide. Genetic studies indicated that ESF1 functions as an additional component in the YDA/SSP-dependent signaling pathway in early embryo development (Costa et al., 2014).

11.4.7 Peptides in regulating cell divisions

11.4.7.1 PSK peptide

The PSK peptide regulates cell divisions by inducing cell dedifferentiation and reentry into the cell cycle. Using ligand-based affinity chromatography, a plasma membrane-localized protein was identified to interact with PSK peptide (Matsubayashi et al., 2002). The protein was defined as PSK receptor (named PSKR) through activity assay (Matsubayashi et al., 2002). Five *PSK* homologous genes in Arabidopsis genome showed cell division-promoting activities. Through sequence alignment, a homologous *AtPSKR1* gene was identified in the Arabidopsis genome, which encodes a 1008-amino acid LRR-RLK, with 60%

identity at the amino acid level to *Dacucus carota* PSKR1 (DcPSKR1) identified in carrot suspension culture. Cultured individual plant cells gradually lost their potential to form callus as the function of *AtPSKR1* was lost, indicating that AtPSKR1 activity is essential for the callus formation (Matsubayashi, 2006). With on-column photoaffinity labeling, the binding sites for interaction between DcPSKR and PSK were determined (Shinohara et al., 2007). A 15-amino acid peptide fragment (Glu^{503}-Lys^{517}) was identified, and deletion of this region abolished completely the binding capacity of DcPSKR to PSK (Hartmann et al., 2015). The phosphorylation sites of PSKR1 were determined with LC-ESI-MS/MS spectrometry and four conserved phosphorylation sites were identified in the activation domain of PSKRs (Hartmann et al., 2015).

The PSKR1 protein belongs to the same subfamily of LRR-RLK as BRI1 and SERKs, and may form a heterodimer with SERKs (Ladwig et al., 2015). Moreover, PSK promotes somatic embryogenesis, and DcSERK has been shown to be a marker of embryogenesis. Co-immunoprecipitation experiments showed that SERK1, SERK2, and SERK3/BAK1 form dimers individually with PSKR1/DcPSKR in the presence of PSK (Wang et al., 2015). The extracellular domain of PSKR1/DcPSKR, including an LRR domain and an island domain, is required for PSK perception (Matsubayashi et al., 2002; Amano et al., 2007). Structural analyses of the extracellular domain of PSKR1/DcPSKR and SERK1/SERK2, determined at the presence of PSK, showed that PSK interacts mainly with a β-strand from the island domain of PSKR1/DcPSKR. PSK is not directly involved in the PSKR1-SERK1 and DcPSKR-SERK2 interactions; instead, it stabilizes the island domain of PSKR1/DcPSKR for recruitment of a SERK (Wang et al., 2015).

11.4.7.2 RALF peptide

The RALF peptide was identified for its ability to increase the pH of the medium when the peptide was applied to cell suspensions (Pearce et al., 2001). Among the 23 *RALFL* genes in the Arabidopsis genome examined, RALF was shown to be highly expressed in roots. RALF treatment resulted in phosphorylation of the FERONIA (FER), a malectin family RLK, leading to increased activities of the plasmamembrane H^+-ATPase AHA2, the calcium-dependent protein kinase 9 (CPK9), and the PEN3/ABCG36 transporter, whereas the activity of a FER-related receptor-like kinase ERULUS was decreased. The *fer4* null mutant showed insensitivity to RALF treatment, while the *erulus* mutant showed no differences in sensitivity to RALF as compared to the wild-type, indicating that ERULUS may not be the receptor for RALF. Both RALF and FER are expressed in the root elongation zone, and regulate cell expansion. Transcriptome analysis indicates that expression of cell expansion-related genes in roots is downregulated after the RALF treatment, which is consistent with the functions of RALF and FER. Taken together, binding of RALF peptide to the FER receptor initiates a downstream phosphorylation signaling cascade that inhibits plasmamembrane H^+-ATPase AHA2 activity, leading to increased apoplasmic pH and reduced cell elongation (Haruta et al., 2014).

11.4.8 Peptides in regulating nodulation

Nodulation is essential for nitrogen fixation by rhizobial bacteria. Genetic analysis of mutants of *L. japonicus* with a supernodulation phenotype allowed identifying the *HYPERNODULATION ABERRANT ROOT FORMATION* (*HAR*) gene that is important for regulating the nodule number in roots. HAR is homologous to CLV1, and is found to regulate nodule development systemically (Nishimura et al., 2002). Since CLV1 is demonstrated to bind the CLV3 peptide in regulating SAM maintenance, HAR is speculated to bind a CLE peptide for regulating nodulation as well. Expression of 39 *LjCLE* genes from *L. japonicus* is analyzed after inoculation with *Mesorhizobium loti*, leading to the identification of three *LjCLE* genes *LjCLE-RS1*, *LjCLE-RS2*, and *LjCLE-RS3* with a significant upregulation of expression. Overexpression of *LjCLE-RS1* and *LjCLE-RS2* inhibits the nodulation systemically, and the nodulation suppression depends on the HAR1 receptor (Okamoto et al., 2009). With nano-LC-MS/MS analysis, LjCLE-RS2 was identified to be an arabinosylated glycopeptide with the hydroxylated proline at the seventh position that was modified further with three arabinose residues. The LjCLE-RS2 peptide synthesized in vitro binds directly to HAR1 at its arabinose chain in a sequence-dependent manner. LjCLE-RS2 produced in roots was found in xylem sap collected from shoots, implying that LjCLE-RSs may provide a long-distance mobile signal in the regulation of the initial step of nodulation (Okamoto et al., 2013).

Later, an LRR-RLK *KLAVIER* (*KLV*), which is highly homologous to the Arabidopsis RPK2 receptor kinase, was found to negatively regulate nodulation in *L. japonicus*. Double mutant analysis indicates that HAR and KLV act in the same signaling pathway. Biochemical analyses reveal a direct interaction between these two RLKs. Overexpression of *LjCLE-RS1* and *LjCLE-RS2* does not suppress the hypernodulation phenotype of the *klv* mutant, indicating that *KLV* is required for LjCLE-RS1 and LjCLE-RS2 signaling, and acts downstream of LjCLE-RS1 and LjCLE-RS2 (Miyazawa et al., 2010).

Three *CLE*-related genes, *GmRIC1*, *GmRIC2*, and *GmRIC3*, were found in soybean (*Glycine max*), with a conserved CLE motif of 12 amino acid residues. *GmRIC1*, *GmRIC2*, and *GmRIC3* regulate nodulation through the GmNARK RLK (Reid et al., 2011; Lim et al., 2011). Among 25 *CLE* genes identified in the *Medicago truncatula* genome, *MtCLE12* and *MtCLE13* regulate nodulation through an LRR-RLK called SUNN (Mortier et al., 2010). *WOX5* was expressed during nodule organogenesis in *M. truncatula*. Its expression level was increased in supernodulation mutants such as *har* and *klv*, indicating that WOX5 may be involved in the CLE peptide-mediated nodulation process (Osipova et al., 2012).

11.5 Summary points

- Since the first peptide hormone systemin was discovered in tomato in 1991, more peptide hormones have been identified in plants. They regulate many developmental and defense processes such as meristem maintenance, xylem and phloem differentiation, stomata patterning, pollination, embryo and endosperm development, cell division, nodulation, and systematic response.

- Preproproteins of peptide hormones are usually produced as a secreted extracellular protein, which is then processed by endopeptidases and carboxypeptidases to form small peptide containing several to dozens of amino acid residues. These peptides may be further modified by hydroxylation, glycosylation, and sulfonation to form mature peptide hormones.
- Peptide hormones usually bind with the extracellular domain of LRR-RLKs localized in plasma membrane to activate downstream signaling pathways.
- Among peptides hormones known so far, systemin, RALFs, and AtPeps regulate plant defense responses; CLV3 and CLE peptides regulate SAM and RAM maintenance, xylem and phloem differentiations, embryo–endosperm interactions, and nodulation; EPF1, EPF2 and Stomagen peptides regulate stomata development and patterning; CRPs such as LEUR1 regulate the pollen tube guidance during fertilization; and PSK regulates cell divisions.
- Diverse technologies such as bioinformatics analyses, in vitro functional assays, genetic analyses, MS-based biochemical analyses, antagonistic peptide technology have been developed to characterize the functions of peptide hormones. New technologies are still in need for identification of endogenous peptides in a high throughput manner.

11.6   Future issues

- Bioinformatics analyses predicts that more than 1000 peptide hormones are present in plant genomes, but less than 100 have been identified so far, and even fewer have been characterized for their functions. Many challenges have to be faced to characterize these peptide hormones: (1) genes encoding preproprotein of peptide hormones are small and their mature forms are even smaller, so it is difficult to knock them out using traditional genetic tools; (2) functional redundancy exists among peptide hormones as they usually occur in the plant genome as multiple members, and knock-out of individual ones produces no visible phenotype; (3) since peptide hormones are usually present in plants at nanomolar or even picomolar concentrations, it is very difficult to characterize them using traditional biochemical tools. Although several new technologies have been developed for characterizing peptide hormones (Song et al., 2013; Amano et al., 2007; Chen et al., 2014), it is necessary and important to develop new technologies that may facilitate the identification and functional characterization of peptide hormones.
- With the data available so far, it seems that peptide hormone signals are almost exclusively perceived by LRR-RLKs, with over 200 members in the Arabidopsis genome. It is a challenging task to identify RLKs and RLK complexes sensing a particular peptide hormone.
- Easy-to-use tools are needed to detect the interaction between peptide hormones and their receptors.
- As cleavage and modifications occur commonly for peptide hormones in plants, it is of great interest to know how these processes are regulated, which enzymes are involved, and if these cleavage activities and modifications contribute to their interactions with RLKs.
- Last but not least, it is important to know how RLKs relay their signals to downstream signaling pathways to execute their actions after the peptides are perceived. Knowledge obtained so far suggest that MAPK cascades are involved in peptide signal transduction. Detailed elucidation of these pathways may facilitate the functional dissection of peptide hormones in plants.

Abbreviations

ACR4	ARABIDOPSIS HOMOLOG OF CRINKLY4
CEP	C-terminally encoded peptide
CEPR1	CEP RECEPTOR 1
CLE	CLV3/endosperm surrounding region
CLV3	CLAVATA3
CRN	CORYNE
CRP	CYSTEINE RICH PEPTIDES
CSC	Columella stem cell
EMS1	EXCESS MICROSPOROCYTES1
EPF1	EPIDERMAL PATTERNING FACTOR 1
EPFLs	EPIDERMAL PATTERNING FACTOR-LIKE peptides
ESF1	EMBRYO SURROUNDING FACTOR 1
EXS	EXTRA SPROGENOUS CELLS
FON4	FLORAL ORGAN NUMBER 4
GLV	GOLVEN1
HAM	HAIRY MERISTEM
IDA	INFLORESCENCE DEFICIENT IN ABSCISSION
IDL	IDA-LIKE
KAPP	Kinase-associated protein phosphatase
LRR-RKs	Leucine-rich repeat receptor kinases
LRR-RLP	LRR-receptor protein
MALDI-TOF MS	Matrix-Assisted Laser Desorption/Ionization Time of Flight Mass Spectrometry
MMCs	Meristemoid mother cells
MPK	Mitogen-activated protein kinase
OC	Organizing center
PEPR1	PEP1 receptor
PLL1	POLTERGEIST LIKE 1
POL	POLTERGEIST
PR-1b	PATHOGENESIS-RELATED PROTEIN1b
PSK	PHYTOSULFOKINE
PSY1	PLANT PEPTIDE CONTAINING SULFATED TYROSINE 1
PXY	PHLOEM INTERCALATED WITH XYLEM
QC	Quiescent center
RALF	RAPID ALKALINIZATION FACTOR
RAM	Root Apical Meristem
RGF	ROOT GROWTH FACTOR
Rop	Rho GTPase-related protein
RPK2	RECEPTOR-LIKE PROTEIN KINASE 2
SAM	Shoot Apical Meristem
SBT	Subtilase
SKM1	STERILITY-REGULATING KINASE MEMBER1
SOL1	SUPPRESSOR OF LLP1
SOL2	SUPPRESSOR OF LLP1 2

STM	SHOOTMERISTEMLESS
TDIF	TRACHEARY ELEMENT DIFFERENTIATION INHIBITOR FACTOR
TDR	TDIF receptor
TEs	Tracheary elements
TMM	TOO MANY MOUTH
TOAD2	TOADSTOOL 2
TPD1	TAPETUM DETERMINANT1
TPST	TYROSYLPROTEIN SULFOTRANSFERASES
WOX	WUS-RELATED HOMEOBOX
WUS	WUSCHEL

Acknowledgments

We thank Ran Lu for help in preparing figures and arranging references. This work is supported by National Natural Science Foundation of China (grant no. 31370029) and the MOST projects (grant no. 2013CB967300 and 2014CB943401).

References

Abrash, E.B., Bergmann, D.C., 2010. Regional specification of stomatal production by the putative ligand CHALLAH. Development 137, 447–455.

Amano, Y., Shinohara, H., Sakagami, Y., et al., 2005. Ion-selective enrichment of tyrosine-sulfated peptides from complex protein digests. Anal. Biochem. 346, 124–131.

Amano, Y., Tsubouchi, H., Shinohara, H., et al., 2007. Tyrosine-sulfated glycopeptide involved in cellular proliferation and expansion in *Arabidopsis*. Proc. Natl. Acad. Sci. U.S.A. 104, 18333–18338.

Araya, T., Miyamoto, M., Wibowo, J., et al., 2014. CLE-CLAVATA1 peptide-receptor signaling module regulates the expansion of plant root systems in a nitrogen-dependent manner. Proc. Natl. Acad. Sci. U.S.A. 111, 2029–2034.

Bauby, H., Divol, F., Truernit, E., et al., 2007. Protophloem differentiation in early *Arabidopsis thaliana* development. Plant Cell Physiol. 48, 97–109.

Berger, D., Altmann, T., 2000. A subtilisin-like serine protease involved in the regulation of stomatal density and distribution in *Arabidopsis thaliana*. Genes Dev. 14, 1119–1131.

Bleckmann, A., Weidtkamp-Peters, S., Seidel, C.A., et al., 2010. Stem cell signaling in Arabidopsis requires CRN to localize CLV2 to the plasma membrane. Plant Physiol. 152, 166–176.

Brand, U., Fletcher, J.C., Hobe, M., et al., 2000. Dependence of stem cell fate in *Arabidopsis* on a feedback loop regulated by CLV3 activity. Science 289, 617–619.

Brand, U., Grünewald, M., Hobe, M., et al., 2002. Regulation of CLV3 expression by two homeobox genes in Arabidopsis. Plant Physiol. 129, 565–575.

Breddam, K., 1986. Serine carboxypeptidases. A review. Carlsberg Res. Commun. 51, 83–128.

Breuninger, H., Rikirsch, E., Hermann, M., et al., 2008. Differential expression of *WOX* genes mediates apical-basal axis formation in the *Arabidopsis* embryo. Dev. Cell 14, 867–876.

Butenko, M.A., Patterson, S.E., Grini, P.E., et al., 2003. Inflorescence of deficient in abscission controls floral organ abscission in Arabidopsis and identifies a novel family of putative ligands in plants. Plant Cell 15, 2296–2307.

Carles, C.C., Fletcher, J.C., 2003. Shoot apical meristem maintenance: the art of a dynamic balance. Trends Plant Sci. 8, 394–401.

Casson, S.A., Chilley, P.M., Topping, J.F., et al., 2002. The POLARIS gene of Arabidopsis encodes a predicted peptide required for correct root growth and leaf vascular patterning. Plant Cell 14, 1705–1721.

Casamitjana-Martínez, E., Hofhuis, H.F., Xu, J., et al., 2003. Root-specific CLE19 overexpression and the sol1/2 suppressors implicate a CLV-like pathway in the control of Arabidopsis root meristem maintenance. Curr. Biol. 13, 1435–1441.

Chen, Y.L., Lee, C.Y., Cheng, K.T., et al., 2014. Quantitative peptidomics study reveals that a wound-induced peptide from PR-1 regulates immune signaling in tomato. Plant Cell 26, 4135–4148.

Chen, J., Yu, F., Liu, Y., et al., 2016. FERONIA interacts with ABI2-type phosphatases to facilitate signaling cross-talk between abscisic acid and RALF peptide in Arabidopsis. Proc. Natl. Acad. Sci. U.S.A. 113, E5519–E5527.

Chookajorn, T., Kachroo, A., Ripoll, D.R., et al., 2004. Specificity determinants and diversification of the Brassica self-incompatibility pollen ligand. Proc. Natl. Acad. Sci. U.S.A. 101, 911–917.

Chu, H., Qian, Q., Liang, W., et al., 2006. The floral organ number4 gene encoding a putative ortholog of Arabidopsis CLAVATA3 regulates apical meristem size in rice. Plant Physiol. 142, 1039–1052.

Clark, S.E., Running, M.P., Meyerowitz, E.M., 1993. CLAVATA1, a regulator of meristem and flower development in Arabidopsis. Development 119, 397–418.

Clark, S.E., Running, M.P., Meyerowitz, E.M., 1995. CLAVATA3 is a specific regulator of shoot and floral meristem development affecting the same processes as CLAVATA1. Development 121, 2057–2067.

Clark, S.E., Williams, R.W., Meyerowitz, E.M., 1997. The CLAVATA1 gene encodes a putative receptor kinase that controls shoot and floral meristem size in Arabidopsis. Cell 89, 575–585.

Cock, J.M., McCormick, S., 2001. A large family of genes that share homology with CLAVATA3. Plant Physiol. 126, 939–942.

Constabel, C.P., Yip, L., Ryan, C.A., 1998. Prosystemin from potato, black nightshade, and bell pepper: primary structure and biological activity of predicted systemin polypeptides. Plant Mol. Biol. 36, 55–62.

Costa, L.M., Marshall, E., Tesfaye, M., et al., 2014. Central cell-derived peptides regulate early embryo patterning in flowering plants. Science 344, 168–172.

Covey, P.A., Subbaiah, C.C., Parsons, R.L., et al., 2010. A pollen-specific RALF from tomato that regulates pollen tube elongation. Plant Physiol. 153, 703–715.

Crespi, M.D., Jurkevitch, E., Poiret, M., et al., 1994. enod40, a gene expressed during nodule organogenesis, codes for a non-translatable RNA involved in plant growth. EMBO J 13, 5099–5112.

De Coninck, B., De Smet, I., 2016. Plant peptides – taking them to the next level. J. Exp. Bot. 67, 4791–4795.

De Smet, I., Vassileva, V., De Rybel, B., et al., 2008. Receptor-like kinase ACR4 restricts formative cell divisions in the Arabidopsis root. Science 322, 594–597.

Depuydt, S., Rodriguez-Villalon, A., Santuari, L., et al., 2013. Suppression of Arabidopsis protophloem differentiation and root meristem growth by CLE45 requires the receptor-like kinase BAM3. Proc. Natl. Acad. Sci. U.S.A. 110, 7074–7079.

DeYoung, B.J., Bickle, K.L., Schrage, K.J., et al., 2006. CLAVATA1-related BAM1, BAM2 and BAM3 receptor kinase-like proteins are required for meristem function in *Arabidopsis*. Plant J. 45, 1–16.

Edlund, T., Jessell, T.M., 1999. Progression from extrinsic to intrinsic signaling in cell fate specification: a view from the nervous system. Cell 22, 211–224.

Endo, S., Shinohara, H., Matsubayashi, Y., et al., 2013. A novel pollen–pistil interaction conferring high-temperature tolerance during reproduction via CLE45 signaling. Curr. Biol. 23, 1670–1676.

Etchells, J.P., Provost, C.M., Mishra, L., et al., 2013. WOX4 and WOX14 act downstream of the PXY receptor kinase to regulate plant vascular proliferation independently of any role in vascular organization. Development 140, 2224–2234.

Etchells, J.P., Provost, C.M., Turner, S.R., 2012. Plant vascular cell division is maintained by an interaction between PXY and ethylene signaling. PLoS Genet. 8, e1002997.

Etchells, J.P., Turner, S.R., 2010. The PXY-CLE41 receptor ligand pair defines a multifunctional pathway that controls the rate and orientation of vascular cell division. Development 137, 767–774.

Felix, G., Boller, T., 1995. Systemin induces rapid ion fluxes and ethylene biosynthesis in *Lycopersicon peruvianum* cells. Plant J. 7, 381–389.

Fiers, M., Golemiec, E., van der Schors, R., et al., 2006. The CLAVATA3/ESR motif of CLAVATA3 is functionally independent from the nonconserved flanking sequences. Plant Physiol. 141, 1284–1292.

Fiers, M., Golemiec, E., Xu, J., et al., 2005. The 14-amino acid CLV3, CLE19, and CLE40 peptides trigger consumption of the root meristem in *Arabidopsis* through a CLAVATA2-dependent pathway. Plant Cell 17, 2542–2553.

Fisher, K., Turner, S., 2007. PXY, a receptor-like kinase essential for maintaining polarity during plant vascular-tissue development. Curr. Biol. 17, 1061–1066.

Fletcher, J.C., Brand, U., Running, M.P., et al., 1999. Signaling of cell fate decisions by CLAVATA3 in *Arabidopsis* shoot meristems. Science 283, 1911–1914.

Fiume, E., Fletcher, J.C., 2012. Regulation of *Arabidopsis* embryo and endosperm development by the polypeptide signaling molecule CLE8. Plant Cell 24, 1000–1012.

Gibbs, G.M., Roelants, K., O'Bryan, M.K., 2008. The CAP superfamily: cysteine-rich secretory proteins, antigen 5, and pathogenesis-related 1 proteins—roles in reproduction, cancer, and immune defense. Endocr. Rev. 29, 865–897.

Grienenberger, E., Fletcher, J.C., 2015. Polypeptide signaling molecules in plant development. Curr. Opin. Plant Biol. 23, 8–14.

Guo, Y., Han, L., Hymes, M., et al., 2010. CLAVATA2 forms a distinct CLE-binding receptor complex regulating *Arabidopsis* stem cell specification. Plant J. 63, 889–900.

Guo, Y., Ni, J., Denver, R., et al., 2011. Mechanisms of molecular mimicry of plant CLE peptide ligands by the parasitic nematode Globodera rostochiensis. Plant Physiol. 157, 476–484.

Haecker, A., Gross-Hardt, R., Geiges, B., et al., 2004. Expression dynamics of *WOX* genes mark cell fate decisions during early embryonic patterning in *Arabidopsis thaliana*. Development 131, 657–668.

Han, Z., Sun, Y., Chai, J., 2014. Structural insight into the activation of plant receptor kinases. Curr. Opin. Plant Biol. 20, 55–63.

Hara, K., Kajita, R., Torii, K.U., et al., 2007. The secretory peptide gene *EPF1* enforces the stomatal one-cell-spacing rule. Genes Dev. 21, 1720–1725.

Hara, K., Yokoo, T., Kajita, R., et al., 2009. Epidermal cell density is autoregulated via a secretory peptide, EPIDERMAL PATTERNING FACTOR 2 in Arabidopsis leaves. Plant Cell Physiol. 50, 1019–1031.

Hartmann, J., Linke, D., Bönniger, C., et al., 2015. Conserved phosphorylation sites in the activation loop of the *Arabidopsis* phytosulfokine receptor PSKR1 differentially affect kinase and receptor activity. Biochem. J. 472, 379–391.

Haruta, M., Sabat, G., Stecker, K., et al., 2014. A peptide hormone and its receptor protein kinase regulate plant cell expansion. Science 343, 408–411.

Higashiyama, T., 2010. Peptide signaling in pollen–pistil interactions. Plant Cell Physiol. 51, 177–189.

Higgins, R., Lockwood, T., Holley, S., et al., 2007. Changes in extracellular pH are neither required nor sufficient for activation of mitogen-activated protein kinases (MAPKs) in response to systemin and fusicoccin in tomato. Planta 225, 1535–1546.

Hirakawa, Y., Kondo, Y., Fukuda, H., 2010. TDIF peptide signaling regulates vascular stem cell proliferation via the *WOX4* homeobox gene in *Arabidopsis*. Plant Cell 22, 2618–2629.

Hirakawa, Y., Shinohara, H., Kondo, Y., et al., 2008. Non-cell-autonomous control of vascular stem cell fate by a CLE peptide/receptor system. Proc. Natl. Acad. Sci. U.S.A. 105, 15208–15213.

Hobe, M., Muller, R., Grunewald, M., et al., 2003. Loss of CLE40, a protein functionally equivalent to the stem cell restricting signal CLV3, enhances root waving in *Arabidopsis*. Dev. Genes Evol. 213, 371–381.

Holley, S.R., Yalamanchili, R.D., Moura, D.S., et al., 2003. Convergence of signaling pathways induced by systemin, oligosaccharide elicitors, and ultraviolet-B radiation at the level of mitogen-activated protein kinases in *Lycopersicon peruvianum* suspension-cultured cells. Plant Physiol. 132, 1728–1738.

Huang, Q., Dresselhaus, T., Gu, H., et al., 2015. Active role of small peptides in *Arabidopsis* reproduction: expression evidence. J. Integr. Plant Biol. 57, 517–521.

Huffaker, A., Pearce, G., Ryan, C.A., 2006. An endogenous peptide signal in *Arabidopsis* activates components of the innate immune response. Proc. Natl. Acad. Sci. U.S.A. 103, 10098–10103.

Huffaker, A., Ryan, C.A., 2007. Endogenous peptide defense signals in *Arabidopsis* differentially amplify signaling for the innate immune response. Proc. Natl. Acad. Sci. U.S.A. 104, 10732–10736.

Hunt, L., Gray, J.E., 2009. The signaling peptide EPF2 controls asymmetric cell divisions during stomatal development. Curr Biol. 19, 864–869.

Hunt, L., Bailey, K.J., Gray, J.E., 2010. The signalling peptide EPFL9 is a positive regulator of stomatal development. New Phytol. 186, 609–614.

Ito, Y., Nakanomyo, I., Motose, H., et al., 2006. Dodeca-CLE peptides as suppressors of plant stem cell differentiation. Science 313, 842–845.

Jeong, S., Trotochaud, A.E., Clark, S.E., 1999. The Arabidopsis *CLAVATA2* gene encodes a receptor-like protein required for the stability of the CLAVATA1 receptor-like kinase. Plant Cell 11, 1925–1934.

Jia, G., Liu, X., Owen, H.A., et al., 2008. Signaling of cell fate determination by the TPD1 small protein and EMS1 receptor kinase. Proc. Natl. Acad. Sci. U.S.A. 105, 2220–2225.

Jin, J., Hua, L., Zhu, Z., et al., 2016. GAD1 Encodes a Secreted Peptide That Regulates Grain Number, Grain Length, and Awn Development in Rice Domestication. Plant Cell 28, 2453–2463.

Jinn, T.L., Stone, J.M., Walker, J.C., 2000. HAESA, an Arabidopsis leucine-rich repeat receptor kinase, controls floral organ abscission. Genes Dev. 14, 108–117.

Kayes, J.M., Clark, S.E., 1998. CLAVATA2, a regulator of meristem and organ development in *Arabidopsis*. Development 125, 3843–3851.

Kieffer, M., Stern, Y., Cook, H., et al., 2006. Analysis of the transcription factor WUSCHEL and its functional homologue in *Antirrhinum* reveal a potential mechanism for their roles in meristem maintenance. Plant Cell 18, 560–573.

Kim, S., Mollet, J.C., Dong, J., et al., 2003. From the cover: chemocyanin, a small basic protein from the lily stigma, induces pollen tube chemotropism. Proc. Natl. Acad. Sci. U.S.A. 100, 16125–16130.

Kinoshita, A., Betsuyaku, S., Osakabe, Y., et al., 2010. RPK2 is an essential receptor-like kinase that transmits the CLV3 signal in *Arabidopsis*. Development 137, 3911–3920.

Komori, R., Amano, Y., Ogawa-Ohnishi, M., et al., 2009. Identification of tyrosylprotein sulfotransferase in *Arabidopsis*. Proc. Natl. Acad. Sci. U.S.A. 106, 15067–15072.

Kondo, Y., Fujita, T., Sugiyama, M., et al., 2015. A novel system for xylem cell differentiation in *Arabidopsis thaliana*. Mol. Plant 8, 612–621.

Kondo, T., Nakamura, T., Yokomine, K., et al., 2008. Dual assay for MCLV3 activity reveals structure–activity relationship of CLE peptides. Biochem. Biophys. Res. Commun. 377, 312–316.

Kondo, T., Sawa, S., Kinoshita, A., et al., 2006. A plant peptide encoded by *CLV3* identified by in situ MALDI-TOF MS analysis. Science 313, 845–848.

Kondo, Y., Tamaki, T., Fukuda, H., 2014. Regulation of xylem cell fate. Front. Plant Sci. 5, 315.

Koornneef, M., Van Eden, J., Hanhart, C.J., et al., 1983. Linkage map of *Arabidopsis thaliana*. J. Hered. 74, 265–272.

Kouchi, H., Hata, S., 1993. Isolation and characterization of novel nodulin cDNAs representing genes expressed at early stages of soybean nodule development. Mol. Gen. Genet. 238, 106–119.

Krol, E., Mentzel, T., Chinchilla, D., et al., 2010. Perception of the *Arabidopsis* danger signal peptide 1 involves the pattern recognition receptor AtPEPR1 and its close homologue AtPEPR2. J. Biol. Chem. 285, 13471–13479.

Ladwig, F., Dahlke, R.I., Stührwohldt, N., et al., 2015. Phytosulfokine regulates growth in *Arabidopsis* through a response module at the plasma membrane that includes CYCLIC NUCLEOTIDE-GATED CHANNEL17, H+-ATPase, and BAK1. Plant Cell 27, 1718–1729.

Lall, S., Nettleton, D., DeCook, R., et al., 2004. Quantitative trait loci associated with adventitious shoot formation in tissue culture and the program of shoot development in Arabidopsis. Genetics 167, 1883–1892.

Laux, T., Mayer, K.F., Berger, J., et al., 1996. The *WUSCHEL* gene is required for shoot and floral meristem integrity in Arabidopsis. Development 122, 87–96.

Lease, K.A., Walker, J.C., 2006. The Arabidopsis unannotated secreted peptide database, a resource for plant peptidomics. Plant Physiol. 142, 831–838.

Lee, J.S., Hnilova, M., Maes, M., et al., 2015. Competitive binding of antagonistic peptides fine-tunes stomatal patterning. Nature 522, 439–443.

Lee, J.S., Kuroha, T., Hnilova, M., et al., 2012. Direct interaction of ligand-receptor pairs specifying stomatal patterning. Genes Dev. 26, 126–136.

Leyser, H.M.O., Furner, I.J., 1992. Characterisation of three shoot apical meristem mutants of *Arabidopsis thaliana*. Development 116, 397–403.

Li, C., Liu, G., Xu, C., et al., 2003. The tomato *suppressor of prosystemin-mediated responses2* gene encodes a fatty acid desaturase required for the biosynthesis of jasmonic acid and the production of a systemic wound signal for defense gene expression. Plant Cell 15, 1646–1661.

Li, L., Li, C., Lee, G.I., et al., 2002. Distinct roles for jasmonate synthesis and action in the systemic wound response of tomato. Proc. Natl. Acad. Sci. U.S.A. 99, 6416–6421.

Lim, C.W., Lee, Y.W., Hwang, C.H., 2011. Soybean nodule-enhanced CLE peptides in roots act as signals in GmNARK-mediated nodulation suppression. Plant Cell Physiol. 52, 1613–1627.

Liu, J., Zhong, S., Guo, X., et al., 2013. Membrane-bound RLCKs LIP1 and LIP2 are essential male factors controlling male-female attraction in *Arabidopsis*. Curr. Biol. 23, 993–998.

Liu, J.X., Srivastava, R., Howell, S., 2009. Overexpression of an Arabidopsis gene encoding a subtilase (AtSBT5.4) produces a clavata-like phenotype. Planta 230, 687–697.

Lucas, W.J., Groover, A., Lichtenberger, R., et al., 2013. The plant vascular system: evolution, development and functions. J. Integr. Plant Biol. 55, 294–388.

Mahonen, A.P., Bonke, M., Kauppinen, L., et al., 2000. A novel two-component hybrid molecule regulates vascular morphogenesis of the Arabidopsis root. Genes Dev. 14, 2938–2943.

Marmiroli, N., Maestri, E., 2014. Plant peptides in defense and signaling. Peptides 56, 30–44.

Marshall, E., Costa, L.M., Gutierrez-Marcos, J., et al., 2011. Cysteine-rich peptides (CRPs) mediate diverse aspects of cell-cell communication in plant reproduction and development. J. Exp. Bot. 62, 1677–1686.

Matsubayashi, Y., 2006. Disruption and overexpression of Arabidopsis phytosulfokine receptor gene affects cellular longevity and potential for growth. Plant Physiol. 142, 45–53.

Matsubayashi, Y., Sakagami, Y., 1996. Phytosulfokine, sulfated peptides that induce the proliferation of single mesophyll cells of *Asparagus officinalis* L. Proc. Natl. Acad. Sci. U.S.A. 93, 7623–7627.

Matsubayashi, Y., Ogawa, M., Morita, A., et al., 2002. An LRR receptor kinase involved in perception of a peptide plant hormone, phytosulfokine. Science 296, 1470–1472.

Matsuzaki, Y., Ogawa-Ohnishi, M., Mori, A., et al., 2010. Secreted peptide signals required for maintenance of root stem cell niche in *Arabidopsis*. Science 329, 1065–1067.

Mayer, K.F., Schoof, H., Haecker, A., et al., 1998. Role of *WUSCHEL* in regulating stem cell fate in the *Arabidopsis* shoot meristem. Cell 95, 805–815.

Meindl, T., Boller, T., Felix, G., 1998. The plant wound hormone systemin binds with the N-terminal part to its receptor but needs the C-terminal part to activate it. Plant Cell 10, 1561–1570.

Meng, L., Buchanan, B.B., Feldman, L.J., et al., 2012. CLE-like (CLEL) peptides control the pattern of root growth and lateral root development in *Arabidopsis*. Proc. Natl. Acad. Sci. U.S.A. 109, 1760–1765.

Meng, X., Chen, X., Mang, H., et al., 2015. Differential function of *Arabidopsis* SERK family receptor-like kinases in stomatal patterning. Curr. Biol. 25, 2361–2372.

Meyer, M.R., Shah, S., Zhang, J., et al., 2015. Evidence for intermolecular interactions between the intracellular domains of the Arabidopsis receptor-like kinase ACR4, its homologs and the Wox5 transcription factor. PLoS One 10, e0118861.

Mishima, M., Takayama, S., Sasaki, K., et al., 2003. Structure of the male determinant factor for *Brassica* self-incompatibility. J. Biol. Chem. 278, 36389–36395.

Miyazawa, H., Oka-Kira, E., Sato, N., et al., 2010. The receptor-like kinase KLAVIER mediates systemic regulation of nodulation and non-symbiotic shoot development in *Lotus japonicas*. Development 137, 4317–4325.

Mortier, V., Den Herder, G., Whitford, R., et al., 2010. CLE peptides control *Medicago truncatula* nodulation locally and systemically. Plant Physiol. 153, 222–237.

Moyen, C., Johannes, E., 1996. Systemin transiently depolarizes the tomato mesophyll cell membrane and antagonizes fusicoccin-induced extracellular acidification of mesophyll tissue. Plant Cell Environ. 19, 464–470.

Müller, R., Bleckmann, A., Simon, R., 2008. The receptor kinase CORYNE of *Arabidopsis* transmits the stem cell-limiting signal CLAVATA3 independently of CLAVATA1. Plant Cell 20, 934–946.

Narváez-Vásquez, J., Pearce, G., Ryan, C.A., 2005. The plant cell wall matrix harbors a precursor of defense signaling peptides. Proc. Natl. Acad. Sci. U.S.A. 102, 12974–12977.

Narváez-Vásquez, J., Florin-Christensen, J., Ryan, C.A., 1999. Positional specificity of a phospholipase A activity induced by wounding, systemin, and oligosaccharide elicitors in tomato leaves. Plant Cell 11, 2249–2260.

Ni, J., Guo, Y., Jin, H., et al., 2010. Characterization of a CLE processing activity. Plant Mol. Biol. 75, 67–75.

Nimchuk, Z.L., Tarr, P.T., Meyerowitz, E.M., 2011. An evolutionarily conserved pseudokinase mediates stem cell production in plants. Plant Cell 23, 851–854.

Nishimura, R., Hayashi, M., Wu, G.J., et al., 2002. HAR1 mediates systemic regulation of symbiotic organ development. Nature 420, 426–429.

Ogawa, M., Shinohara, H., Sakagami, Y., et al., 2008. *Arabidopsis* CLV3 peptide directly binds CLV1 ectodomain. Science 319, 294.

Ogawa-Ohnishi, M., Matsushita, W., Matsubayashi, Y., 2013. Identification of three hydroxyproline O-arabinosyltransferases in *Arabidopsis thaliana*. Nat. Chem. Biol. 9, 726–730.

Ohki, S., Takeuchi, M., Mori, M., 2011. The NMR structure of stomagen reveals the basis of stomatal density regulation by plant peptide hormones. Nat. Commun. http://dx.doi.org/10.1038/ncomms1520.

Ohyama, K., Ogawa, M., Matsubayashi, Y., 2008. Identification of a biologically active, small, secreted peptide in Arabidopsis by in silico gene screening, followed by LC-MS-based structure analysis. Plant J. 55, 152–160.

Ohyama, K., Shinohara, H., Ogawa-Ohnishi, M., et al., 2009. A glycopeptide regulating stem cell fate in *Arabidopsis thaliana*. Nat. Chem. Biol. 5, 578–580.

Okamoto, S., Ohnishi, E., Sato, S., et al., 2009. Nod factor/nitrate-induced *CLE* genes that drive HAR1-mediated systemic regulation of nodulation. Plant Cell Physiol. 50, 67–77.

Okamoto, S., Shinohara, H., Mori, T., et al., 2013. Root-derived CLE glycopeptides control nodulation by direct binding to HAR1 receptor kinase. Nat. Commun. 3191. http://dx.doi.org/10.1038/ncomms.

Okuda, S., Tsutsui, H., Shiina, K., et al., 2009. Defensin-like polypeptide LUREs are pollen tube attractants secreted from synergid cells. Nature 458, 357–361.

Olsen, A.N., Mundy, J., Skriver, K., 2002. Peptomics, identification of novel cationic Arabidopsis peptides with conserved sequence motifs. In Silico Biol. 2, 441–451.

Osipova, M.A., Mortier, V., Demchenko, K.N., et al., 2012. *Wuschel-related homeobox5* gene expression and interaction of CLE peptides with components of the systemic control add two pieces to the puzzle of autoregulation of nodulation. Plant Physiol. 158, 1329–1341.

Pallakies, H., Simon, R., 2014. The CLE40 and CRN/CLV2 signaling pathways antagonistically control root meristem growth in *Arabidopsis*. Mol. Plant 7, 1619–1636.

Park, S.Y., Jauh, G.Y., Mollet, J.C., et al., 2000. A lipid transfer-like protein is necessary for lily pollen tube adhesion to an in vitro stylar matrix. Plant Cell 12, 151–164.

Park, S.Y., Lord, E.M., 2003. Expression studies of SCA in lily and confirmation of its role in pollen tube adhesion. Plant Mol. Biol. 51, 183–189.

Pearce, G., Johnson, S., Ryan, C.A., 1993. Structure-activity of deleted and substituted systemin, an 18-amino acid polypeptide inducer of plant defensive genes. J. Biol. Chem. 268, 212–216.

Pearce, G., Moura, D.S., Stratmann, J., et al., 2001. Production of multiple plant hormones from a single polyprotein precursor. Nature 411, 817–820.

Pearce, G., Strydom, D., Johnson, S., et al., 1991. A polypeptide from tomato leaves induces wound-inducible inhibitor proteins. Science 253, 895–898.

Pearce, G., Ryan, C.A., 2003. Systemic signaling in tomato plants for defense against herbivores: isolation and characterization of three novel defense-signaling glycopeptide hormones coded in a single precursor gene. J. Biol. Chem. 278, 30044–30050.

Pearce, G., Yamaguchi, Y., Munske, G., et al., 2008. Structure–activity studies of AtPep1, a plant peptide signal involved in the innate immune response. Peptides 29, 2083–2089.

Pi, L., Aichinger, E., van der Graaff, E., et al., 2015. Organizer-derived WOX5 signal maintains root columella stem cells through chromatin-mediated repression of CDF4 expression. Dev. Cell 33, 576–588.

Qi, Z., Verma, R., Gehring, C., et al., 2010. Ca^{2+} signaling by plant *Arabidopsis thaliana* Pep peptides depends on AtPEPR1, a receptor with guanylyl cyclase activity, and cGMP-activated Ca^{2+} channels. Proc. Natl. Acad. Sci. U.S.A. 107, 21193–21198.

Rautengarten, C., Steinhauser, D., Büssis, D., et al., 2005. Inferring hypotheses on functional relationships of genes: analysis of the *Arabidopsis thaliana* subtilase gene family. PLoS Comput. Biol. 1, e40.

Reid, D.E., Ferguson, B.J., Gresshoff, P.M., 2011. Inoculation- and nitrate-induced CLE peptides of soybean control NARK-dependent nodule formation. Mol. Plant Microbe. Interact. 24, 606–618.

Rodriguez-Villalon, A., Gujas, B., Kang, Y.H., et al., 2014. Molecular genetic framework for protophloem formation. Proc. Natl. Acad. Sci. U.S.A. 111, 11551–11556.

Rodriguez-Villalon, A., Gujas, B., van Wijk, R., et al., 2015. Primary root protophloem differentiation requires balanced phosphatidylinositol-4,5-biphosphate levels and systemically affects root branching. Development 142, 1437–1446.

Rohrig, H., Schmidt, J., Miklashevichs, E., et al., 2002. Soybean ENOD40 encodes two peptides that bind to sucrose synthase. Proc. Natl. Acad. Sci. U.S.A. 99, 1915–1920.

Sarkar, A.K., Luijten, M., Miyashima, S., et al., 2007. Conserved factors regulate signaling in *Arabidopsis thaliana* shoot and root stem cell organizers. Nature 446, 811–814.

Scacchi, E., Salinas, P., Gujas, B., et al., 2010. Spatio-temporal sequence of cross-regulatory events in root meristem growth. Proc. Natl. Acad. Sci. U.S.A. 107, 22734–22739.

Schaller, A., Oecking, C., 1999. Modulation of plasma membrane H^+-ATPase activity differentially activates wound and pathogen defense responses in tomato plants. Plant Cell 11, 263–272.

Scheer, J.M., Ryan, C.A., 2002. The systemin receptor SR160 from *Lycopersicon peruvianum* is a member of the LRR receptor kinase family. Proc. Natl. Acad. Sci. U.S.A. 99, 9585–9590.

Scheres, B., Benfey, P., Dolan, L., 2002. Root development. Arabidopsis Book 1, e0101. http://dx.doi.org/10.1199/tab.0101.

Schoof, H., Lenhard, M., Haecker, A., et al., 2000. The stem cell population of *Arabidopsis* shoot meristems in maintained by a regulatory loop between the *CLAVATA* and *WUSCHEL* genes. Cell 100, 635–644.

Schopfer, C.R., Nasrallah, M.E., Nasrallah, J.B., 1999. The male determinant of self-incompatibility in Brassica. Science 286, 1697–1700.

Shinohara, H., Matsubayashi, Y., 2015. Reevaluation of the CLV3-receptor interaction in the shoot apical meristem: dissection of the CLV3 signaling pathway from a direct ligand-binding point of view. Plant J. 82, 328–336.

Shinohara, H., Mori, A., Yasue, N., et al., 2016. Identification of three LRR-RKs involved in perception of root meristem growth factor in Arabidopsis. Proc. Natl. Acad. Sci. U.S.A. 113, 3897–3902.

Shinohara, H., Ogawa, M., Sakagami, Y., et al., 2007. Identification of ligand binding site of phytosulfokine receptor by on-column photoaffinity labeling. J. Biol. Chem. 282, 124–131.

Silverstein, K.A., Graham, M.A., Paape, T.D., et al., 2005. Genome organization of more than 300 defensin-like genes in Arabidopsis. Plant Physiol. 138, 600–610.

Song, X.F., Yu, D.L., Xu, T.T., et al., 2012. Contributions of individual amino acid residues to the endogenous CLV3 function in shoot apical meristem maintenance in *Arabidopsis*. Mol. Plant 5, 515–523.

Song, X.F., Guo, P., Ren, S.C., et al., 2013. Antagonistic peptide technology for functional dissection of *CLV3/ESR* genes in Arabidopsis. Plant Physiol. 161, 1076–1085.

Sprunck, S., Rademacher, S., Vogler, F., et al., 2012. Egg cell-secreted EC1 triggers sperm cell activation during double fertilization. Science 338, 1093–1097.

Srivastava, R., Liu, J.X., Howell, S.H., 2008. Proteolytic processing of a precursor protein for a growth-promoting peptide by a subtilisin serine protease in *Arabidopsis*. Plant J. 56, 219–227.

Stahl, Y., Grabowski, S., Bleckmann, A., et al., 2013. Moderation of *Arabidopsis* root stemness by CLAVATA1 and ARABIDOPSIS CRINKLY4 receptor kinase complexes. Curr. Biol. 23, 362–371.

Stahl, Y., Wink, R.H., Ingram, G.C., et al., 2009. A signaling module controlling the stem cell niche in *Arabidopsis* root meristems. Curr. Biol. 19, 909–914.

Stein, J.C., Dixit, R., Nasrallah, M.E., et al., 1996. SRK, the stigma-specific S locus receptor kinase of *Brassica*, is targeted to the plasma membrane in transgenic tobacco. Plant Cell 8, 429–445.

Stenvik, G.E., Tandstad, N.M., Guo, Y., et al., 2008. The EPIP peptide of inflorescence deficient in abscission is sufficient to induce abscission in *Arabidopsis* through the receptor-like kinases HAESA and HAESA-LIKE2. Plant Cell 20, 1805–1817.

Stratmann, J.W., Ryan, C.A., 1997. Myelin basic protein kinase activity in tomato leaves is induced systemically by wounding and increases in response to systemin and oligosaccharide elicitors. Proc. Natl. Acad. Sci. U.S.A. 94, 11085–11089.

Sugano, S.S., Shimada, T., Imai, Y., et al., 2010. Stomagen positively regulates stomatal density in *Arabidopsis*. Nature 463, 241–244.

Suzaki, T., Toriba, T., Fujimoto, M., et al., 2006. Conservation and diversification of meristem maintenance mechanism in *Oryza sativa*: function of the *FLORAL ORGAN NUMBER2* gene. Plant Cell Physiol. 47, 1591–1602.

Tabata, R., Sumida, K., Yoshii, T., et al., 2014. Perception of root-derived peptides by shoot LRR-RKs mediates systemic N-demand signaling. Science 346, 343–346.

Takayama, S., Shiba, H., Iwano, M., et al., 2000. The pollen determinant of self-incompatibility in Brassica campestris. Proc. Natl. Acad. Sci. U.S.A. 97, 1920–1925.

Takayama, S., Shimosato, H., Shiba, H., et al., 2001. Direct ligand-receptor complex interaction controls *Brassica* self-incompatibility. Nature 413, 534–538.

Takei, Y., Hirose, S., 2002. The natriuretic peptide system in eels: a key endocrine system for euryhalinity? Am. J. Physiol. Regul. Integr. Comp. Physiol. 282, R940–R951.

Takeuchi, H., Higashiyama, T., 2016. Tip-localized receptors control pollen tube growth and LURE sensing in *Arabidopsis*. Nature 531, 245–248.

Tamaki, T., Betsuyaku, S., Fujiwara, M., et al., 2013. SUPPRESSOR OF LLP11-mediated C-terminal processing is critical for CLE19 peptide activity. Plant J. http://dx.doi.org/10.1111/tpj.12349.

Tang, J., Han, Z., Sun, Y., et al., 2015. Structural basis for recognition of an endogenous peptide by the plant receptor kinase PEPR1. Cell Res. 25, 110–120.

Topping, J.F., Lindsey, K., 1997. Promoter trap markers differentiate structural and positional components of polar development in Arabidopsis. Plant Cell 9, 1713–1725.

Truernit, E., Bauby, H., Dubreucq, B., et al., 2008. High-resolution whole-mount imaging of three-dimensional tissue organization and gene expression enables the study of phloem development and structure in *Arabidopsis*. Plant Cell 20, 1494–1503.

Uchida, N., Tasaka, M., 2013. Regulation of plant vascular stem cells by endodermis-derived EPFL-family peptide hormones and phloem-expressed ERECTA-family receptor kinases. J. Exp. Bot. 64, 5335–5343.

Uchida, N., Lee, J.S., Horst, R.J., et al., 2012. Regulation of inflorescence architecture by inter-tissue layer ligand-receptor communication between endodermis and phloem. Proc. Natl. Acad. Sci. U.S.A. 109, 6337–6342.

Wang, J., Li, H., Han, Z., et al., 2015. Allosteric receptor activation by the plant peptide hormone phytosulfokine. Nature 525, 265–268.

Wang, T., Liang, L., Xue, Y., et al., 2016. A receptor heteromer mediates the male perception of female attractants in plants. Nature 531, 241–244.

Watanabe, Y., Barbashov, S.F., Komatsu, S., et al., 1994. A peptide that stimulates phosphorylation of the plant insulin-binding protein. Isolation, primary structure and cDNA cloning. Eur. J. Biochem. 224, 167–172.

Whitford, R., Fernandez, A., De Groodt, R., et al., 2008. Plant CLE peptides from two distinct functional classes synergistically induce division of vascular cells. Proc. Natl. Acad. Sci. U.S.A. 105, 18625–18630.

Whitford, R., Fernandez, A., Tejos, R., et al., 2012. GOLVEN secretory peptides regulate auxin carrier turnover during plant gravitropic responses. Dev. Cell 22, 678–685.

Wrzaczek, M., Brosché, M., Kollist, H., et al., 2009. *Arabidopsis* GRI is involved in the regulation of cell death induced by extracellular ROS. Proc. Natl. Acad. Sci. U.S.A. http://dx.doi.org/10.1073/pnas.0808980106.

Wrzaczek, M., Vainonen, J.P., Stael, S., et al., 2015. GRIM REAPER peptide binds to receptor kinase PRK5 to trigger cell death in *Arabidopsis*. EMBO J. 34, 55–66.

Xu, C., Liberatore, K.L., MacAlister, C.A., et al., 2015. A cascade of arabinosyltransferases controls shoot meristem size in tomato. Nat. Genet. 47, 784–792.

Xu, T.T., Song, X.F., Ren, S.C., et al., 2013. The sequence flanking the N-terminus of the CLV3 peptide is critical for its cleavage and activity in stem cell regulation in *Arabidopsis*. BMC Plant Biol. http://dx.doi.org/10.1186/1471-2229-13-225.

Xu, T.T., Ren, S.C., Song, X.F., et al., 2015. CLE19 expressed in theembryo regulates both cotyledon establishment and endosperm development in Arabidopsis. J. Exp. Bot. 66, 5217–5227.

Yadav, R.K., Perales, M., Gruel, J., et al., 2011. WUSCHEL protein movement mediates stem cell homeostasis in the *Arabidopsis* shoot apex. Genes Dev. 25, 2025–2030.

Yamaguchi, Y., Pearce, G., Ryan, C.A., 2006. The cell surface leucine-rich repeat receptor for AtPep1, an endogenous peptide elicitor in Arabidopsis, is functional in transgenic tobacco cells. Proc. Natl. Acad. Sci. U.S.A. 103, 10104–10109.

Yamazaki, T., Takaoka, M., Katoh, E., et al., 2003. A possible physiological function and the tertiary structure of a 4-kDa peptide in legumes. Eur. J. Biochem. 270, 1269–1276.

Yang, H., Matsubayashi, Y., Nakamura, K., et al., 1999. Oryza sativa PSK gene encodes a precursor of phytosulfokine-alpha, a sulfated peptide growth factor found in plants. Proc. Natl. Acad. Sci. U.S.A. 96, 13560–13565.

Yang, W.C., Katinakis, P., Hendriks, P., et al., 1993. Characterization of GmENOD40 a gene showing novel patterns of cell-specific expression during soybean nodule development. Plant J 3, 573–585.

Yang, S.L., Xie, L.F., Mao, H.Z., et al., 2003. *TAPETUM DETERMINANT1* is required for cell specialization in the *Arabidopsis* anther. Plant Cell 15, 2792–2804.

Zhang, H., Lin, X., Han, Z., et al., 2016. Crystal structure of PXY-TDIF complex reveals a conserved recognition mechanism among CLE peptide-receptor pairs. Cell Res. 26, 543–555.

Zhao, D.Z., Wang, G.F., Speal, B., et al., 2002. The *EXCESS MICROSPOROCYTES1* gene encodes a putative leucine-rich repeat receptor protein kinase that controls somatic and reproductive cell fates in the *Arabidopsis* anther. Genes Dev. 16, 2021–2031.

Zhou, Y., Liu, X., Engstrom, E.M., et al., 2015. Control of plant stem cell function by conserved interacting transcriptional regulators. Nature 517, 377–380.

Zhu, Y., Wang, Y., Li, R., et al., 2009. Analysis of interactions among the CLAVATA3 receptors reveals a direct interaction between CLAVATA2 and CORYNE in Arabidopsis. Plant J. 61, 223–233.

Plant hormones and stem cells

12

Zhi Juan Cheng[1], Baoshuan Shang[2], Xian Sheng Zhang[1], Yuxin Hu[2]
[1]Shandong Agricultural University, Taian, China; [2]Institute of Botany, Chinese Academy of Sciences, Beijing, China

Summary

The remarkable developmental plasticity of plants greatly relies on the regulation of stem cell niches in two distal meristems, namely the shoot apical meristem (SAM) and the root apical meristem (RAM). Plants also have a remarkable capability to regenerate new stem cell niches in vitro and in vivo. These distinct features confer plants with an ability to survive and propagate successfully under ever-changing environmental conditions. Studies in the model flowering plant Arabidopsis have revealed that plant hormones, including auxin, cytokinin, and peptides, play key roles in the maintenance of stem cell niches in apical meristems and the de novo regeneration of a new SAM and RAM, which sheds a new light on the signals and molecular mechanisms underlying maintenance of plant stem cell niches and regeneration capacity.

12.1 Stem cells and hormonal regulation of stem cell activity

One of the key features in high plants is their continuous postembryonic organogenesis with remarkable developmental plasticity to cope with ever-changing environmental stimuli. This plasticity relies on the formation, maintenance, and differentiation of stem cells within plant bodies (Aichinger et al., 2012; Dinneny and Benfey, 2008; Scheres, 2007). Plants use stem cells to maintain their growth and development, and in some long-lived trees, these cells are active for thousands of years. The stem cells in plants are located in the shoot and root tips, namely shoot apical meristem (SAM) and root apical meristem (RAM). The stem cells in the SAM are responsible for the continuous formation of aerial organs, including leaf, stem and flower, while the stem cells in the RAM form the root system. Moreover, there are also other types of cells which are stem cell-like, such as those comprising the cambium that are responsible for generating cells for radial growth and the continuous increase in girth along the longitudinal axis of plants (Aichinger et al., 2012; De Rybel et al., 2016).

Another important feature in plants is that their somatic cells have long been considered to retain great potential to be induced into pluripotent stem cells and to form stem cell niches de novo (Gaillochet and Lohmann, 2015). Indeed, plants are not only able to form the axillary buds and lateral roots that contain stem cell niches resembling that of SAM and RAM (Domagalska and Leyser, 2011; Petricka et al., 2012; Tian et al., 2014; Wang and Li, 2008), but also can regenerate new stem cell niches by de novo organogenesis to replace lost or damaged tissues (Reinhardt et al., 2003; Sena

Hormone Metabolism and Signaling in Plants. http://dx.doi.org/10.1016/B978-0-12-811562-6.00012-8

et al., 2009; Xu et al., 2006). Moreover, detached organs or tissues can form pluripotent cell masses termed calli, and accomplish the regeneration of new plant bodies under in vitro culture conditions (Birnbaum and Sánchez Alvarado, 2008; Duclercq et al., 2011). Hence, the control of acquisition and maintenance of stem cell activity lies at the core of diverse developmental programs (Fisher and Sozzani, 2016).

Plant hormones play a key role to integrate environmental inputs and mediate plant developmental plasticity. Hormones, especially auxin and cytokinin, contribute to stem cell positioning and the balance of stem cell maintenance, proliferation, and differentiation (Galinha et al., 2009). Several other hormones, such as peptides, have a substantial input in controlling meristem activities. Moreover, auxin and cytokinin are principal players that mediate somatic pluripotency and de novo regeneration of new stem cell niches. This chapter will introduce the key factors and molecular networks involved in regulation of plant stem cells in SAM and RAM, mainly focusing on plant hormone signaling and their interactions in the maintenance of stem cell niches. We will also present the emerging insight into how plant hormones direct the acquisition of somatic cell pluripotency and de novo regeneration of new stem cell niches in vitro, attempting to outline the signals and molecular mechanisms underlying the plant regeneration program.

12.2   Hormones and stem cell niche maintenance

12.2.1   Hormonal regulation of shoot stem cell niche maintenance

12.2.1.1   Patterning and maintenance of the shoot apical meristem

The plant SAM is a group of elaborately organized cells positioned at the shoot tip (Sablowski, 2007). According to the function and cytological features, the SAM is divided into different zones (Murray et al., 2012). The central zone (CZ), residing at the summit of the SAM, harbors pluripotent stem cells (Tucker and Laux, 2007). The organizing center (OC) locates underneath the CZ and is required to maintain the stem cell fate (Clark, 2001). The stem cells in the CZ continuously divide and provide initials for the other two multipotent zones: the peripheral zone (PZ) which generates lateral organs at the flanks of the meristem, and the rib zone (RZ) which provides stem tissues (Fig. 12.1) (Xie et al., 2009). Therefore, an elaborate system coordinating stem cell proliferation in the CZ and differentiation in other functional sub-domains is required for the establishment and maintenance of the SAM (Gaillochet et al., 2015).

Over the past decades, a number of key regulators have been identified in Arabidopsis. A negative feedback loop between *WUSCHEL* (*WUS*) and *CLAVATA3* (*CLV3*) is the core pathway regulating the maintenance of shoot stem cell niche (Brand et al., 2000; Schoof et al., 2000). Expression of *WUS* defines the OC while *CLV3* marks the position of stem cells (Fig. 12.1) (Aichinger et al., 2012; Laux, 2003). The expression of *WUS* is detectable in 16-cell stage of embryos, preceding that of *CLV3* (Mayer et al., 1998). Mutations in *WUS* lead to differentiation of stem cells and loss

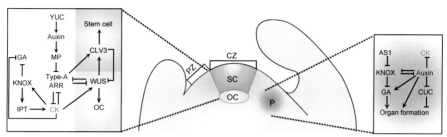

Figure 12.1 Hormone signaling in shoot-stem cell niche. Transcription factors KNOX and WUS facilitate high activity of cytokinin (CK; green), and are required for the maintenance of shoot stem cells (SC) in the meristem, but not for organ initiation. The *KNOX* genes have opposite regulatory effects on gibberellin (GA) and CK levels in the stem cell niche. CK downregulates GA response in the central zone (CZ) of SAM. Owing to the response of the type-A ARRs, auxin and CK are able to act synergistically in the CZ of the SAM. CK signaling positively regulates *WUS*, which is specifically expressed in the organizing center (OC). WUS represses several type-A *ARR* genes. High activities of GA and auxin (pink) might initiate and maintain populations of organ founder cells at primordium initiation sites (P) in the SAM periphery. Auxin antagonizes CK response but induces GA biosynthesis and activity for organ initiation in the peripheral zone (PZ) of the meristem.

of SAMs, indicating that *WUS* is required for the establishment and maintenance of stem cells (Aichinger et al., 2012; Mayer et al., 1998). Once produced in the OC, WUS protein moves to the CZ through the plasmodesmata to promote *CLV3* gene expression and activate stem cell activity therein (Daum et al., 2014; Yadav et al., 2011). CLV3 protein is secreted from the stem cells, and it in turn represses *WUS* transcription via the downstream signaling pathway including CLV1, CLV2, and CORYNE (Fig. 12.1). The WUS-CLV3 pathway has been found to be conserved in different species, such as maize, rice, and tomato (Brand et al., 2000; Fletcher et al., 1999; Gaillochet and Lohmann, 2015; Jeong et al., 1999; Müller and Sheen, 2008; Schoof et al., 2000). In parallel with WUS, another homeodomain transcription factor SHOOT MERISTEMLESS (STM) also functions in stem cell maintenance (Lenhard et al., 2002). The *STM* gene is expressed in the undifferentiated regions throughout the SAM and downregulates organ formation genes (Byrne et al., 2002). A severe mutant of *STM* lacks the SAM and is unable to initiate shoot meristem formation during postembryonic development (Long et al., 1996). Hormonal regulation of stem cell niche maintenance largely depends on modulating the expression and function of these key regulators.

12.2.1.2   Cytokinin and the maintenance of shoot stem cells

Cytokinin regulates the maintenance of stem cell niche through both inhibiting cell differentiation and creating the proper environment for stem cell activity (Gaillochet et al., 2015). Local biosynthesis and specific perception of cytokinin are important for exerting the above functions (Soyars et al., 2016). STM functions at least in part by promoting cytokinin biosynthesis. Ectopic expression of *STM* by using 35S promoter increases the transcription of *ISOPENTENYLTRANSFERASE7* (*IPT7*) and subsequent cytokinin accumulation (Jasinski et al., 2005; Yanai et al., 2005). Conversely, expressing an *IPT* 7 gene under control of the *STM* promoter, as well as exogenous application of cytokinin, can rescue the phenotype of a *stm* mutant (Yanai et al., 2005). The *LONELYGUY* (*LOG*) gene encoding a cytokinin activating enzyme acts downstream of *IPTs* and produces the biologically active cytokinin (Kurakawa et al., 2007). In rice, *log* mutants exhibit reduced size of the SAM, small panicles, and premature termination of floral meristems (Kurakawa et al., 2007; Matsuoka et al., 1993). The *LOG* gene is expressed in the CZ of both Arabidopsis and rice (Tokunaga et al., 2012). Cytokinin produced by LOG in the outermost cell layers might diffuse to the OC, where its perception provides positional information for *WUS* expression. The cytokinin degrading enzyme CYTOKININ OXIDASE 3 (CKX3) is expressed in the OC. A *ckx3 ckx5* double mutant shows increased *WUS* expression and enlarged SAM (Bartrina et al., 2011), whereas overexpressing *CKX* genes by using 35S promoter leads to reduced size of the SAM (Werner et al., 2003), demonstrating the importance of local cytokinin biosynthesis for shoot meristem maintenance.

In the SAM of Arabidopsis, *WUS* expression is positively correlated with cytokinin signaling. *WUS* expression closely overlaps with cytokinin signaling response and drops off sharply in the regions with lower cytokinin response signals (Gordon et al., 2009). The expression of *ARABIDOPSIS HISTIDINE KINASE 4* (*AHK4*), encoding a cytokinin receptor, correlates well with that of *WUS* in individual SAM cells. Increased *WUS* expression induced by local cytokinin can be explained by enhanced *AHK4* transcription in the OC. The WUS protein directly represses the expression of genes encoding type-A Arabidopsis response regulators (ARRs), which are negative regulators of cytokinin signaling (Leibfried et al., 2005). It is thus suggested that *WUS* expression and cytokinin complete a positive feedback loop (Gordon et al., 2009). Cytokinin signaling induces *WUS* expression, which in turn suppresses the transcription of type-A *ARRs* and thus further enhances cytokinin signaling. In addition, cytokinin signaling is involved in the activation of another WUS-related homeodomain transcription factor, WUSCHEL-RELATED HOMEOBOX 9 (WOX9, also known as STIMPY), which is required for SAM growth partially by positively regulating *WUS* expression (Skylar et al., 2010; Wu et al., 2005), indicating that properly regulated cytokinin signaling is required for OC activity.

12.2.1.3   Auxin and the maintenance of shoot stem cells

Auxin has been long considered to regulate cell differentiation and organ initiation in the SAM. As the progenies of stem cells leave the CZ and enter the PZ, they are programmed to undergo cell differentiation to form primordia of lateral organs (Soyars et al., 2016). This process is controlled by auxin accumulation at the flanks of the PZ. Relatively high concentration of auxin represses the expression of *STM* and *CUP-SHAPED COTYLEDON* (*CUC*) genes, which terminates the meristematic identity and starts differentiation (Heisler et al., 2005).

Studies have started to reveal the function of auxin in regulating stem cell niche maintenance. Inhibition of auxin transport by N-1-naphthylphthalamic acid (NPA) increases the expression levels of *CLV3* and reduces that of *WUS*, and these effects are largely dependent on the functions of *ARR7* and *ARR15*, which encode A-type ARRs (Zhao et al., 2010). Inducible silencing of *ARR7* and *ARR15* results in dramatic reduction of *CLV3* mRNA, moderately increased *WUS* transcription, and enlarged SAM, indicating that auxin participates in stem cell niche regulation through modulating cytokinin signaling (Zhao et al., 2010). The distribution of AUXIN RESPONSE FACTOR 5 (ARF5), also known as MONPTEROS (MP), a much investigated auxin signaling component, exhibits a gradient from the PZ into the CZ, and directly represses the expression of *ARR7* and *ARR15* within the CZ. Expressing *ARF5/MP* driven by the *CLV3* promoter decreases the expression of *CLV3*, consistent with the effect of silencing *ARR7* and *ARR15* (Zhao et al., 2010). Hence, auxin acts on the core regulatory pathway of stem cell niche through synergy with cytokinin signaling (Fig. 12.1). However, the molecular mechanisms underlying the regulation of ARRs on the WUS-CLV3 pathway remain to be investigated.

12.2.1.4   Roles of other hormones in shoot stem cell maintenance

Besides the fundamental roles of auxin and cytokinin, gibberellic acid (GA), brassinosteroid (BR), and ethylene also function to drive the patterning and growth of the SAM. GA is spatially excluded from the stem cell niche (Veit, 2009). In the central region of the SAM, class I KNOTTED1-like homeobox (KNOX) transcription factors negatively regulate GA biosynthesis by directly repressing the gene encoding a GA 20-oxidase (Chen et al., 2004; Hay et al., 2002; Sakamoto et al., 2001). The *STM* gene in Arabidopsis stimulates the transcription of *AtGA2ox2*, encoding an enzyme for deactivating GAs, at the base of the SAM and leaf primordia (Jasinski et al., 2005). GA accumulation is thus restricted to the incipient leaf primordia, where it participates in the initiation and morphogenesis of the leaf (Fleet and Sun, 2005; Veit, 2009). Ethylene has been reported to play an essential role in limiting cell division in the quiescent center (QC) (Ortega-Martínez et al., 2007). Constitutive ethylene response caused by mutation of *CONSTITUTIVE TRIPLE RESPONSE 1* (*CTR1*) disrupts SAM structure and reduces cell number (Hamant et al., 2002). However, whether it exerts similar functions in the CZ remains unknown. BR has been found to be a key factor in specifying the boundary domain between the SAM and the lateral organ, owing to the regulation of the boundary-specific genes *LATERAL ORGAN BOUNDARIES* and *CUC* family members (Bell et al., 2012; Gendron and Wang, 2007). Mutation of *KNOX family class 1 homeobox gene of rice* (*OsH1*) results in BR overproduction, causing defects in boundary formation between the SAM and first leaf primordium (Tsuda et al., 2014).

12.2.2   Hormonal regulation of root stem cell niche maintenance

12.2.2.1   Auxin-regulated formation of root apical meristem

The primary RAM is established during embryogenesis, which is of vital importance for postembryonic development (Petricka et al., 2012). Initially, the zygote undergoes an asymmetric division to yield a smaller apical cell and a larger basal cell. The smaller apical cell then divides vertically to form the apical part of the embryo, while the larger basal cell divides horizontally to produce the suspensor. Subsequently, the uppermost suspensor cell named hypophysis divides asymmetrically at the globular stage, generating an upper lens-shaped cell that finally forms the QC and a lower basal cell that gives rise to the columella initials (Laux et al., 2004; ten Hove et al., 2015).

The WOX transcription factors are crucial to regulate the formation of the embryonic root, and their expression coincides with the onset of RAM (Haecker et al., 2004). More importantly, extensive research shows that plant hormones especially auxin are crucial for establishment of stem cell niche of the primary RAM (Perilli et al., 2012; Petricka et al., 2012; Smit and Weijers, 2015). Formation of the root-stem cell niche occurs along with establishment of the auxin gradient, which is generated by the polar localization of PIN-FORMED (PIN) proteins (Blilou et al., 2005; Möller and Weijers, 2009). Several factors involved in auxin signaling or response are demonstrated to be critical for stem cell niche formation in the RAM. The ARF5/MP protein, a member of the auxin-dependent ARF family whose transcriptional activity is mediated by the transcription repressor BODENLOS (BDL), also known as INDOLE-3-ACETIC ACID INDUCIBLE 12 (IAA12), plays an important role in driving hypophysis specification during RAM formation (Hardtke and Berleth, 1998; Hamann et al., 2002). The basic helix–loop–helix (bHLH) transcription factors TARGET OF MP 5 (TMO5) and TMO7, which are direct targets of MP, act downstream of MP to mediate RAM formation (Schlereth et al., 2010). A study further demonstrated that the NO TRANSMITTING TRACT (NTT) and its two closely related paralogs WIP DOMAIN PROTEIN 4 (WIP4) and WIP5 whose expression is dependent on the MP-mediated auxin signaling pathway, are also required for the formation of the RAM (Crawford et al., 2015). Consistent with this information, RAM formation is greatly impaired in the mutants defective in auxin biosynthesis, transport, or signaling, including the quadruple mutant of YUCCA (YUC) flavin monooxygenases *yuc1 yuc4 yuc10 yuc11* (Cheng et al., 2007), the *tryptophan aminotransferase 1* (*taa1*) mutant defective in indole-3-pyruvic acid (IPA) branch of auxin biosynthesis (Stepanova et al., 2008), the *indole synthase* (*ins*) mutant disrupted in Trp-independent IAA biosynthetic pathway (Wang et al., 2015), the mutants of PIN efflux transporters (Benková et al., 2003; Blilou et al., 2005), and the auxin receptor mutant *transport inhibitor response 1* (*tir1*) (Mockaitis and Estelle, 2008).

12.2.2.2 WOX5 and hormones in maintenance of stem cell identity

WOX5 is exclusively expressed in the QC and plays a key role in the maintenance of stem cell identity and meristem size in the RAM (Fig. 12.2) (Haecker et al., 2004). A loss-of-function mutant of *WOX5* exhibits terminal differentiation of distal stem cells, while inducible overexpression of *WOX5* blocks stem cell differentiation (Sarkar et al., 2007). It is also notable that other regulatory circuits including plant hormones regulate QC identity mainly via WOX5. Auxin regulates distal stem cells by WOX5 through the IAA17-ARF10 and IAA17-ARF16 module, in which ARF10 and ARF16 restrict *WOX5* expression (Ding and Friml, 2010). A study reveals that the plant homeodomain (PHD)-containing protein REPRESSOR OF WUSCHEL 1 (ROW1)

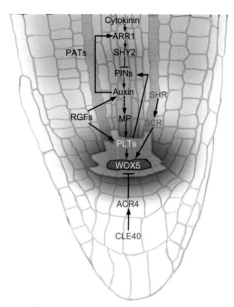

Figure 12.2 Hormone signaling in root-stem cell niche. An auxin maximum in root-stem cell niche (blue) directed by PINs facilitates a critical role in the maintenance of the root-stem cell niches. The auxin signaling is transduced by the auxin-dependent transcription factor AUXIN RESPONSE FACTOR 5 (ARF5)/MONPTEROS (MP) to promote the *PLETHORA* (*PLT*) expression in SC and thus maintain *WOX5* expression in the quiescent center (QC). The SCARECROW (SCR) and SHORT-ROOT (SHR) transcription factors act in a parallel pathway to maintain the QC identity via WOX5 and stem cell division. By contrast, CLE40 peptide signaling restricts *WOX5* expression in QC, while the root growth factor (RGF) peptides positively regulate the auxin accumulation in RAM. The cytokinin signaling (pink) acts antagonistically with auxin to promote cell division and differentiation through IAA3/SHORT HYPOCOTYL2 (SHY2) transcriptional repressor to restrain the auxin transport and thus the meristem size, and auxin-regulated ARR1 also serves as a feedback regulatory mechanism for RAM maintenance.

is required to maintain QC and stem cell identity by repressing the expression of *WOX5* (Zhang et al., 2015b). By contrast, a CLAVATA3/EMBRYO SURROUNDING REGION (CLE) peptide is also involved in the regulation of QC identity. CLE40 is expressed in the differentiating descendants of distal root stem cells, and it constrains the *WOX5* expression within the QC by binding to the receptor-like kinase ARABIDOPSIS CRINKLY 4 (ACR4). This determines the position and size of the root stem cells. The elevation of CLE40 level reduces the expression domain of *WOX5* and promotes stem cell differentiation, while reduction of CLE40 levels delays stem cell differentiation and allows stem cells to proliferate (Stahl et al., 2009).

12.2.2.3 Two parallel pathways in the maintenance of root apical meristem

Root growth is achieved by the balance of the maintenance of stem cells and the specification of these cells into differentiated cell types. The maintenance or specification of stem cells in RAM involves two parallel pathways. One involves the plant-specific AP2-type transcription factors of the PLETHORA (PLT) family. *PLT* genes are expressed in the root stem cell niches. Mutations in *PLT* genes cause subtle defects in distal stem cell division and root growth, while embryonic-specific expression of *PLT* genes induces root meristem formation in the hypocotyl and shoot (Aida et al., 2004). Interestingly, high levels of PLT activity promotes stem cell identity and maintenance, and medium levels promote mitotic activity of stem cell daughters, and a low PLT level is required for cell differentiation. Hence, the gradient of expression of the PLTs is likely to balance the maintenance and specification of stem cells of RAM in a dose-dependent manner (Galinha et al., 2007). The other parallel pathway of root stem cell maintenance involves the GRAS family transcription factor SHORT-ROOT (SHR) and SCARECROW (SCR) (Fig. 12.2). Loss of function of *SHR* or *SCR* genes results in a defect in the maintenance of QC identity and termination of root stem cells (Sabatini et al., 2003). The SHR protein is primarily expressed in the stele, and it can move into adjacent cell layers to activate *SCR* transcription and thus maintains QC identity (Cui et al., 2007; Helariutta et al., 2000). Accordingly, *SCR* expression in the QC is required to maintain identity of the QC and stem cells through a cell-autonomous mechanism (Sabatini et al., 2003).

12.2.2.4   Hormonal interactions in the maintenance of the root apical meristem

Plant hormones are pivotal for the maintenance of the RAM, in which auxin and cytokinin act antagonistically and possibly synergistically in the maintenance and differentiation of stem cells in the RAM (Fig. 12.2) (Schaller et al., 2015; Sparks et al., 2013). Typically, an auxin maximum forms within the distal stem cell region, via local biosynthesis and polar auxin transport directed by the PIN efflux facilitators, and such a maximum is required for QC function (Blilou et al., 2005; Grieneisen et al., 2007; Petersson et al., 2009; Zhao, 2010). The auxin-inducible genes encoding PLT1-PLT4 transcription factors are master regulators of root stem cell activity, and their expression gradient is dependent on auxin distribution and is required for the maintenance or differentiation of stem cells in the RAM (Aida et al., 2004; Galinha et al., 2007).

Cytokinins act antagonistically with auxin in controlling root stem cell activity. In the basal hypophysis, auxin signaling induces transcription of genes encoding the A-type ARR7 and ARR15 proteins, which are repressors of cytokinin signaling. Loss of function of *ARR7* and *ARR15* or ectopic expression of B-type ARR10 in the basal cells during early embryogenesis can result in defective root stem cell formation (Müller and Sheen, 2008). Exogenous application of cytokinin leads to a reduction of meristem size, indicating its role to promote cell differentiation (Dello Ioio et al., 2007). Accordingly, mutants defective in cytokinin biosynthesis or signaling, such as *ipt3*, *ahk3*, or *arr1*, display increased size of the RAM (Dello Ioio et al., 2008; Miyawaki et al., 2004). It is also demonstrated that the B-type ARR proteins, such as ARR1, ARR10, and ARR12, could interfere with auxin signaling and thus affect the maintenance of the RAM. For instance, the Aux/IAA protein IAA3, also known as SHORT HYPOCOTYL2 (SHY2), is a direct target of ARR1, and SHY2 is a transcription repressor of auxin signaling that could activate cytokinin biosynthesis via induction of *IPT5* transcription. Moreover, SHY2 can also interfere with the auxin efflux PIN proteins, and thus generate a regulatory circuit of auxin and cytokinin in controlling the root meristem size (Dello Ioio et al., 2008).

Peptide signaling is also critical for root stem cell maintenance. Besides CLE40 peptide signaling, the root growth factors (RGFs) are a group of Tyr-sulfated peptides mainly accumulated in the stem cell area and the innermost layer of central columella cells (see Song et al., Chapter 11). The TYROSYLPROTEIN SULFOTRANSFERASE (TPST) enzyme sulfates RGFs posttranscriptionally and thus upregulate *PLT* genes to define the expression level and patterning of the PLTs. In combination with RGFs, two other tyrosine-sulfated peptides, phytosulfokine (PSK) and plant peptide containing sulfated tyrosine 1 (PSY1), are also involved in the maintenance of stem cell activity, providing evidence for cross talk between auxin and peptide signaling pathways (Matsuzaki et al., 2010; Ou et al., 2016; Zhou et al., 2010).

Other plant hormones or environmental cues are also involved in the regulation of root meristem activity, mainly via cross talk with auxin or cytokinin. BR is involved in the promotion of QC renewal and distal stem cell differentiation. Interestingly, mutation in BR biosynthesis genes or exogenous application of BR all lead to the reduction of root meristem size, suggesting that balanced BR signaling is critical for root meristem maintenance (Gonzalez-Garcia et al., 2011; Hacham et al., 2011; Lee et al., 2015). During the postembryonic development of the root, ethylene could restrain cell division of the QC, and the cells formed through ethylene-induced divisions retain characteristics of the QC (Ortega-Martínez et al., 2007). GA signaling can promote stem cell proliferation because GA biosynthetic mutants display a reduction of root meristem size (Ubeda-Tomás et al., 2009; Ubeda-Tomas et al., 2008). It is reported that high levels of NO reduce auxin transport and response via repression of *PIN1* expression, and thus reduce root meristem activity concomitantly (Fernández-Marcos et al., 2011). In addition, ABA can also regulate root meristem activity via its effect on polar auxin transport under osmotic stress (Rowe et al., 2016).

12.3   Hormones and de novo stem cell niche formation

12.3.1   Hormone-directed somatic cell pluripotency

12.3.1.1   Auxin and cytokinin in the acquisition of cell pluripotency

It is well realized that plant somatic cells retain great potential to acquire pluripotency, either induced by plant hormones or by wounding. In a typical plant in vitro regeneration system, detached organs or differentiated tissues could form a mass of pluripotent cells termed callus on an auxin-rich callus-inducing medium (CIM). Subsequent cultures of callus on shoot-inducing medium (SIM) or root-inducing medium (RIM) that contain different ratios of cytokinin and auxin enable the de novo formation of shoots or roots, respectively (Skoog and Miller, 1957). Building on this concept, a wide range of plant regeneration systems have been established in a variety of plant species, and plant in vitro regeneration has become a key platform for plant biotechnology.

It is only after 2007 the origin and molecular control of callus formation have been described. In Arabidopsis, CIM-induced callus formation in multiple organs, including root, hypocotyl, cotyledon, and petal, actually occurs from the pericycle or pericycle-like cells, and the derived callus tissues have characteristics of the root meristem as revealed by expression of root meristem genes (Atta et al., 2009; Che et al., 2007; Sugimoto et al., 2010). Consistent with this, disruption of either the pericycle cell identity or its competence leads to a callus-forming defect in multiple organs (Atta et al., 2009; Che et al., 2007; Sugimoto et al., 2010). It still remains unclear whether callus formation in other somatic cell types or other plant species and tissues also follows such a root developmental pathway. For example, the early practice of tissue culture was performed using the phloem cells of carrot roots to successfully prove the hypothesis of cell totipotency in plants (Steward et al., 1958).

Auxin and cytokinin have long been considered to be pivotal in the acquisition of pluripotency in somatic cells, and discovery of auxin and cytokinin as growth regulators also coincided with the history of tissue culture (Gautiteret, 1983). In classical tissue culture conditions, the pluripotent callus cells are induced on auxin-rich CIM (Skoog and Miller, 1957; Valvekens et al., 1988), indicating that auxin is critical for somatic cells to acquire pluripotency. Indeed, mutants defective in auxin signaling exhibit a compromised callus-forming capacity, such as the *auxin resistant 1* (*axr1*) mutant which is defective in auxin response, and mutant of the PROPORZ1 adaptor protein involved in mediating auxin response (Anzola et al., 2010; Sieberer et al., 2003). It was shown that the four auxin-inducible LATERAL ORGAN BOUNDARIES DOMAIN (LBD) transcription factors, which act downstream of ARF7 and ARF19 to mediate auxin signaling in lateral root formation, play a key role in directing CIM-induced callus formation (Fig. 12.3) (Fan et al., 2012). CIM-induced *LBD* expression occurs in pericycle or pericycle-like cells where the callus formation initiates, and ectopic expression of the *LBD*s is sufficient to trigger autonomous callus formation without plant hormone, while suppression of LBD function inhibits the CIM-induced callus formation (Fan et al., 2012). This finding provides the molecular link between auxin signaling and callus induction.

On the other hand, cytokinin has been found to be involved in wound-induced callus formation. Interestingly, wound-induced callus originates from cells of the wounded sites but not from the pericycle cells, and such callus does not exhibit root meristem characteristics, suggesting that wound-induced callus formation is likely to be a process of cell dedifferentiation (Iwase et al., 2011). The wound-induced callus formation is directed by the AP2 transcription factors WOUND INDUCED DEDIFFERENTIATIONs (WINDs), mainly via manipulation of the cytokinin signaling mediated by ARR1 and ARR12 (Iwase et al., 2011), suggesting that cytokinin signaling may be required for wound-induced callus formation (Fig. 12.3). Consistent with this suggestion, when plants are attacked by a bacterial pathogen such as *Agrobacterium tumefaciens*, the formation of crown gall tissue is correlated with elevation of endogenous cytokinin level (Ito and Machida, 2015). Thus, there are at least two pathways in mediating acquisition of pluripotency in plant somatic cells, which may act synergistically in culture systems in vitro, and in plants. Moreover, a study shows that ARR15 could repress callus formation downstream of ARF10,

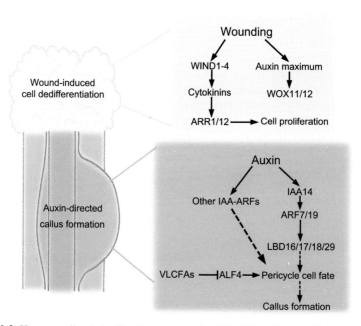

Figure 12.3 Hormone-directed callus formation and cell dedifferentiation. High auxin ectopically induces the expression of LATERAL ORGAN BOUNDARIES DOMAIN (LBD) transcription factors downstream of IAA14-ARF7 and IAA14-ARF19, and thus directs the pericycle or pericycle-like cells to form callus with root meristem properties. Other IAA-ARF modules may be involved in this process. The pericycle competence for callus formation is restricted by VLCFA-regulated *ABERRANT LATERAL ROOT FORMATION 4* (*ALF4*) expression. Cytokinin signaling participates in the wound-induced callus formation, a possible cell differentiation process directed by WOUND INDUCED DEDIFFERENTIATION (WIND) transcription factors. Wound-induced accumulation of auxin and its downstream targets *WOX11* and *WOX12* in the wounding site are also required for de novo organogenesis.

indicating the mutual interactions of cytokinin and auxin in the acquisition of pluripotency (Liu et al., 2016).

12.3.1.2 Mechanisms restricting the fate of somatic cells

Although the above advances have been made in understanding how plant cells acquire pluripotency directed by hormones and wounding, it remains largely unclear about how different cell fates are strictly maintained or restricted under normal development conditions. Findings might provide some thought-provoking clues in this critical issue. In Arabidopsis, the plant-specific nuclear protein Aberrant Lateral Root Formation 4 (ALF4) appears to play an important role in maintenance of pluripotency of pericycle cells and possibly other cell types. Loss of ALF4 function results in the failure of lateral root and callus formation (DiDonato et al., 2004; Sugimoto et al., 2010), and the protoplasts of *alf4-1* mutant are incapable of initiating cell proliferation

(Chupeau et al., 2013). On the other hand, the endodermis-derived very-long-chain fatty acids (VLCFAs) or their derivatives have recently been reported to act as cell-layer signaling components to restrict pericycle cell competence for callus formation and thus the regeneration capacity via suppression of *ALF4* transcription (Fig. 12.3). This finding suggests that VLCFAs or their derivatives serve as inhibitory signals to restrain the callus-forming capacity of pericycle cells or the possible pluripotency of other cell types during normal growth and development (Shang et al., 2016). This also opens a door to further explore the signaling and molecular mechanism of how cell fate and pluripotency are maintained in different cell types in plants.

12.3.2 Hormone-directed de novo stem cell niche formation

12.3.2.1 Formation of stem cell niches de novo

During postembryonic development, certain types of somatic cells can be switched into stem cell niches resembling those in the SAM and the RAM (Duclercq et al., 2011; Ikeuchi et al., 2016). With de novo stem cell niche formation, plants continuously regenerate axillary buds and lateral roots to shape their architecture (Ikeuchi et al., 2016; Kerstetter et al., 1997; Steeves and Sussex, 1989). Unlike their animal counterparts, plant organs or tissues exhibit a remarkable regenerative ability under in vitro culture conditions. In a wide variety of species, a piece of adult tissue is capable of regenerating the entire plant body through de novo specification of stem cell niche. The classic study by Skoog and Miller (1957) six decades ago demonstrated the critical roles of auxin and cytokinin in these processes. Incubation of explants in medium with high cytokinin/auxin ratios induces the formation of shoots, whereas incubation in medium with high auxin/cytokinin ratios induces root formation. A large number of plant species have been successfully propagated through successive regeneration of shoots and roots. Therefore, in vitro regeneration not only lays foundations for plant propagation and genetic transformation but also provides ideal systems for investigating the mechanisms underlying de novo stem cell niche specification.

12.3.2.2 Regulating events for the de novo formation of stem cell niches

Regenerated shoot meristems are histologically and functionally indistinguishable from the embryonic SAM. In Arabidopsis, a number of key factors regulating the initiation and maintenance of the SAM also play essential roles in de novo formation of the shoot stem cell niche and subsequent shoot regeneration (Duclercq et al., 2011; Gordon et al., 2007). For instance, *WUS* expression marks the onset of shoot meristem formation, and promotes cell fate transition from callus cells into the OC, which is critical for the specification of the stem cell niche (Aichinger et al., 2012; Duclercq et al., 2011). Shoot regeneration can be significantly reduced in *wus* mutants but enhanced when *WUS* expression is specifically increased (Cheng et al., 2013; Li et al., 2011). Loss of function of *STM* completely suppresses shoot regeneration (Daimon et al., 2003). Single and double mutation of *CUC1* and *CUC2* impairs shoot formation

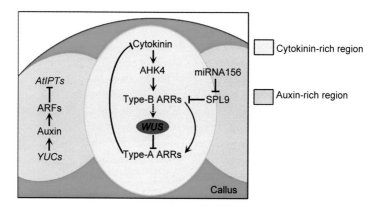

Figure 12.4 Hormone signaling in de novo shoot-stem cell niche formation. During shoot regeneration, in the central region of the pre-meristem, CK signaling mediated by AHK4 activates type-B ARRs, which in turn initiates *WUS* expression. WUS directly represses the expression of type-A *ARR* genes, the negative regulators of cytokinin signaling. Thus, cytokinin signaling and WUS regulation constitute a positive feedback loop, which initiates and maintains the specification of the OC. In addition, type-B *ARR* genes are negatively regulated by a miRNA156-target gene, *SPL9*. In the surrounding region, spatially expressed *YUC* genes are required for local biosynthesis of auxin. Auxin signaling inhibits cytokinin accumulation through direct repression of *AtIPTs* by ARFs. This mutually exclusive distribution pattern of cytokinin and auxin signaling is critical for de novo shoot-stem cell niche formation.

while overexpression of these genes promotes it (Aida et al., 1997; Daimon et al., 2003). Studies indicate that cytokinin and auxin determine shoot regeneration mainly through regulating these meristem-related key factors (Fig. 12.4).

12.3.2.3 Auxin-regulated regeneration programs

Spatial biosynthesis, polar transport, and signaling transduction of auxin are required for shoot regeneration (Gordon et al., 2007; Kareem et al., 2015). Although it is currently unknown how exogenous auxin regulates de novo shoot formation, the spatial biosynthesis of endogenous IAA is involved in this process. The *YUC* genes encode a family of enzymes catalyzing a rate-limiting step in the tryptophan-dependent auxin biosynthesis pathway. The transcript levels of two members of this family, *YUC1* and *YUC4*, are significantly upregulated during shoot induction. Furthermore, expression analyses of *YUC4* demonstrate that a dynamic pattern is gradually restricted to the regions apical and peripheral to the *WUS*-expressing domain. The double mutant *yuc1 yuc4* and overexpressing *YUC1* or *YUC4*, both repress shoot regeneration (Cheng et al., 2013).

Polar transport of auxin regulates shoot regeneration mainly through the function of PIN efflux carriers. This regulation starts at early stages because PIN1 is the earliest marker for regenerating shoot progenitor cells (Gordon et al., 2007; Kareem et al., 2015). Incubation of callus in SIM for 1 day induces the polarized membrane localization of PIN1, while prolonged incubation restricts PIN1 to the outermost cell layer of the region apical to the *WUS*-expressing domain (Cheng et al., 2013). The PLT3, PLT5, and PLT7 proteins, which establish callus pluripotency, are responsible for PIN1 accumulation in shoot primordia (Kareem et al., 2015). Both pharmacological and genetic disruptions of PIN1 repress shoot formation (Cheng et al., 2013; Gordon et al., 2007). Although both *WUS* and *PIN1* are early regulators of shoot regeneration, the phenotype of the *pin1-4* mutant was not as severe as that of *wus-1* (Gordon et al., 2007). This is possibly due to the functional redundancy of PIN family members.

Transduction of auxin signaling depends on ARF-mediated transcriptional regulation (Salehin et al., 2015). The ARF5/MP protein, a well-characterized member of ARF family, has also been positioned in the genetic network underlying shoot regeneration (Ckurshumova et al., 2014). When explants are cultured on CIM, *ARF5/MP* is induced in pericycle and pericycle-derived cells and subsequently switches to the edge of callus. After the callus is transferred onto SIM, *ARF5/MP* expression is detected in discrete spots where the shoot meristem will occur. Moreover, the *ARF5/MP* expression signals exhibit a ring-shaped pattern, with weaker expression in the center. When *ARF5/MP* is disrupted, the ability for de novo shoot formation dramatically diminishes. Interestingly, expression from the native promoter of a truncated variant of ARF5/MP, which lacks the C-terminal domains required for interactions with Aux/IAAs, results in increased shoot regeneration. Because Aux/IAAs are transcriptional repressors that recruit TOPLESS corepressors (Szemenyei et al., 2008), deletion of the C-terminal domain of ARF5/MP derepresses *ARF5/MP* functions. In this way, shoot regeneration can be promoted by enhancing *ARF5/MP*-mediated auxin signaling.

12.3.2.4   Cytokinin-regulated regeneration programs

Cytokinin signaling is positively correlated with de novo shoot regeneration (Fig. 12.4). The well-characterized model of cytokinin signaling involves the canonical two-component transcription circuitry (Hwang et al., 2012). The receptors AHK2, AHK3, and AHK4 are activated when perceiving cytokinin, and then transfer a phosphoryl group to ARABIDOPSIS HISTIDINE-CONTAINING PHOSPHOTRANSFER (AHP) proteins, and subsequently to type-B ARRs. Defects in *AHK4* cause reduced response to cytokinin and prevent shoot induction (Inoue et al., 2001). During in vitro culture, CIM incubation induces *AHK4* expression in callus cells. After transfer onto SIM, *WUS* expression is induced in the cells expressing *AHK4*, suggesting regulation by cytokinin signaling of *WUS* expression and stem cell niche specification (Gordon et al., 2009). The type-B ARRs are critical transcription factors responsible for regulating the transcription of primary cytokinin response genes, including the type-A *ARRs* that encode negative regulators of cytokinin signaling (Hwang et al., 2012). When the primary members of type-B ARRs are mutated, shoot regeneration is repressed. For example, the *arr1* single mutant displays reduced shoot regeneration capacity (Sakai et al., 2001). In the double mutants *arr1 arr10* or *arr1 arr12*, the callus can become green but shoot formation hardly occurs (Hill et al., 2013; Ishida et al., 2008). In the *arr1 arr10 arr12* triple mutant, shoot regeneration is completely inhibited (Ishida et al., 2008). Furthermore, overexpression of *ARR1* by using 35S promoter enhances ectopic shoot formation and de novo shoot regeneration (Sakai et al., 2001). Thus, *ARR1*, *ARR10*, and *ARR12* are necessary for shoot regeneration. In contrast, shoot formation is reduced when *ARR7* and *ARR15*, the type-A *ARRs*, are overexpressed (Buechel et al., 2010).

Regenerative capacity progressively declines as plants age, which is related to the functions of cytokinin signaling. The SQUAMOSA PROMOTER BINDING PROTEIN-LIKE9 (SPL9) transcription factor, interacts with ARR2 and represses ARR-mediated transcriptional activity. The *SPL9* gene is targeted by miRNA156 and as the plant ages, miRNA156 level diminishes while *SPL9* transcript level increases accordingly. As a result, the regenerative capacity that depends on ARR-mediated transcription gradually declines (Zhang et al., 2015a).

12.3.2.5 Auxin–cytokinin cross talk regulates de novo specification of shoot stem cell niche

Auxin and cytokinin are known to act synergistically in controlling meristem initiation and development (Fig. 12.4) (Moubayidin et al., 2009). The molecular mechanisms through which these two hormones interact to regulate de novo formation of shoot stem cell niche have begun to be understood.

During the regeneration of shoot meristems, auxin and cytokinin signals show dynamic spatiotemporal patterns and form a mutually exclusive distribution before the appearance of a promeristem. At the beginning of shoot induction, both auxin and cytokinin response signals distribute evenly at the edge of the callus. Subsequently, auxin response signals are progressively switched to restrictive regions in the outermost cell layers. When *WUS* expression occurs, auxin response is located in a ring-like region, apical and peripheral to the *WUS*-expressing domain. After the formation of a meristem, auxin signals are restricted to the L1 cell layer. Cytokinin response signals translocate to the region of future *WUS* expression and overlap with the *WUS*-expressing domain thereafter (Fig. 12.4). Therefore, at the appearance of *WUS* expression, auxin response signals form a ring-like pattern while the cytokinin signal is restricted to the center of the "auxin ring" (Cheng et al., 2013).

The mutually exclusive distribution pattern results from antagonistic regulation between auxin and cytokinin. Spatiotemporal biosynthesis and polar transport lead to specific auxin distribution. ARF3 binds to the promoter of the *ATP/ADP ISOPENTENYLTRANSFERASE5* (*AtIPT5*) gene and represses its expression, thus inhibiting cytokinin biosynthesis and giving rise to an auxin signaling-rich (ASR) region in a radial pattern. In the center of the ASR, a cytokinin signaling-rich (CSR) region is observed. The remaining question is how auxin accumulation is suppressed in the central CSR (Cheng et al., 2013). A possible answer is that cytokinin signaling represses auxin biosynthesis-related genes directly or indirectly. The specific auxin–cytokinin distribution pattern is critical for de novo specification of shoot stem cell niche. Mutation in *ARF3* or pharmacological disruption of auxin polar transport causes ectopic expression of *AtIPT5* and abolishes shoot regeneration. However, the underlying mechanisms remain to be elucidated. The *WUS* and *CUC2* genes are expressed in non-overlapping domains within the callus (Gordon et al., 2007). During shoot primordia formation, *CUC2* is expressed in a radial region, surrounding the *WUS*-expressing domain, resembling the auxin–cytokinin pattern. *WUS* can be induced by cytokinin whereas *CUC2* responds to auxin (Birnbaum and Sánchez Alvarado, 2008; Gordon et al., 2009). Thus the significance of forming the mutually exclusive auxin–cytokinin pattern might be critical to induce key regulatory genes for stem cell niche regeneration.

12.3.2.6   Wound-induced de novo regeneration

Apical meristems have the ability to recover from injury or regeneration after excision. After laser ablation of the CZ and *WUS*-expressing region in tomato SAM, the lesions can be gradually displaced within 2 days. This displacement is suggested to be a result of activation of new meristem center at the flank, because *LeWUS* expression is induced in the PZ and confined to one side of the lesion before the appearance of the new center. In some cases, two new centers are generated on opposite sides of the lesion and give rise to a split meristem (Reinhardt et al., 2003). Similarly, after the laser ablation of the QC in Arabidopsis, as well as excision of root tips in pea and maize, a new QC could be reformed within a short time. In Arabidopsis, excision at 130 μm from the root tip, removing the QC and surrounding stem cells, results in a high regeneration percentage of root-stem cell niche. Mutations that fail to maintain the stem cell niche do not impair root regeneration, indicating that a functional stem cell niche is not required for root regeneration. The reestablishment of root-stem cell niche depends on cell fate transition. However, excision at 200 μm from the root tip dramatically reduces root regeneration, suggesting that not all cell types possess the competency for fate-reprogramming (Sena et al., 2009; Xu et al., 2006).

In many plant species, organs such as stems and leaves can generate adventitious roots after wounding or detachment depending on the de novo regeneration of stem cell niches (Chen et al., 2014). Root organogenesis de novo can be mimicked in tissue culture systems. By culturing Arabidopsis leaves detached from the seedling on B5 medium without exogenous hormones, adventitious roots can be obtained from the mid-vein near the wound (Liu et al., 2014). Using this in vitro system, molecular mechanisms underlying root organogenesis, especially the regulatory roles of auxin, have been elucidated.

Adventitious roots initiate through cell fate transition from the procambium or cambium regions in two steps (Liu et al., 2014; Xu and Huang, 2014). Firstly, *WOX11* and *WOX12* act redundantly to regulate the transition from procambium cells into root founder cells. Secondly, root founder cells are further reprogrammed into root primordium cells. Endogenous auxin plays critical roles in the cell fate reprogramming process by inducing the expression of *WOX11*. After detachment of the leaf explant, auxin response signals accumulate in procambium cells within the vascular tissues. *WOX11* expression is induced in the overlapping regions. Mutations of the auxin response elements within its promoter or NPA treatment disrupt the expression pattern of *WOX11*, indicating that *WOX11* is directly induced by auxin signaling (Liu et al., 2014).

Endogenous auxin accumulation that promotes *WOX11* expression is triggered by wounding. The detachment of leaf explant leads to responses of *YUC* genes. Different *YUC* genes exhibit division of labor and orchestrate auxin biosynthesis. Of these *YUC1* and *YUC4* quickly respond to wounding signal under both light and dark conditions, whereas *YUC5*, *YUC8*, and *YUC9* mainly produce auxin in the leaf margin and mesophyll cells in response to darkness. *YUC2* and *YUC6* contribute to the basal auxin level (Chen et al., 2016). Local biosynthesis and polar transport result in the accumulation of auxin in competent cells and start the cell fate reprogramming. How wounding signals trigger the spatial expression of *YUC* genes is currently unknown.

12.4 Summary points

1. Auxin and cytokinin signaling are spatially and functionally antagonistic in the SAM. Cytokinin signaling distributes in the central region, maintains the undifferentiated state and promotes stem cell activity. Auxin signaling inhibits meristematic identity and starts differentiation at the peripheral region.
2. Auxin gradients and signaling play a key role in maintenance of stem cell activity in the RAM, while cytokinin and CLE peptide act antagonistically with auxin to restrict stem cell activity by promoting cell differentiation. Other plant hormones are involved in the regulation of RAM activity mainly via interfering with auxin and/or cytokinin signaling.
3. There are at least two pathways that direct plant somatic cells to acquire pluripotency. Auxin-induced callus formation from pericycle or pericycle-like cells follows a root developmental program, while cytokinin is involved in wound-induced callus formation possibly by inducing cell dedifferentiation.
4. During de novo regeneration of new stem cell niches, a mutually exclusive distribution patterning between auxin and cytokinin provides the positional information inducing the expression of key SAM regulators, and thus reestablishes the proper pattern and function of the shoot meristem.
5. Wound-induced hormone concentrations promote fate transition of competent cells, which allows the recovery or regeneration of new apical meristems.

12.5 Future perspectives

1. A refined picture of the regulation of meristem activity has started to emerge in recent years, in which the interplay among plant hormonal signals, transcriptional regulatory networks, and chromatin remodeling factors are outlined. However, how this network responds to environmental signals and how these inputs are translated into cellular behavior are yet to be elucidated.
2. Because SAM and RAM are inaccessible, and observation often requires invasive measures, the dynamic regulation of stem cell niches in SAM and RAM is still challenging. It is necessary to further study hormone responses dynamically and noninvasively to reveal their effects on stem cell activities.

3. Hormones, especially auxin and cytokinin, have been shown to regulate stem cell niche maintenance and meristem activity through controlling the expression of genes encoding the key factors such as WUS, CLV3, STM, WOX5, and PLT. However, the molecular mechanisms underlying hormonal regulation of these key factors remains to be revealed.
4. It is largely unknown how cell pluripotency is established and dynamically maintained during development and in response to environmental signaling. Moreover, definitions of varied states or pluripotency of cells with molecular markers are still challenging in plants.
5. During de novo regeneration of shoot or root meristems, the effects of exogenous hormones and their relationship with endogenous hormones need to be further explored.
6. Our knowledge about plant regeneration is mainly derived from a few model plant species. Whether these regulatory mechanisms are widely applicable in crops and other plant species should be addressed in the future.

Abbreviations

ACR4	ARABIDOPSIS CRINKLY 4
AHK	*ARABIDOPSIS HISTIDINE KINASE*
AHP	HISTIDINE PHOSPHOTRANSFER PROTEINS
ALF4	ABERRANT LATERAL ROOT FORMATION 4
ARF	AUXIN RESPONSE FACTOR
ARR	*ARABIDOPSIS RESPONSE REGULATOR*
ASR	Auxin signaling-rich region
BR	Brassinosteroid
CIM	Callus-inducing medium
CKX	CYTOKININ OXIDASE
CLV3	*CLAVATA3*
CSR	Cytokinin signaling-rich region
CTR1	*CONSTITUTIVE TRIPLE RESPONSE 1*
CUC	*CUP-SHAPED COTYLEDON*
CZ	Central zone
GA	Gibberellic acid
IAA	Indole-3-acetic acid
IPT	*ISOPENTENYLTRANSFERASE*
LBD	LATERAL ORGAN BOUNDARIES DOMAIN
LOG	LONELYGUY
NPA	N-1-naphthylphthalamic acid
OC	Organizing center
PHD	Plant homeodomain
PIN	PIN-FORMED
PLT	PLETHORA
PSK	Phytosulfokine
PSY1	Plant peptide containing sulfated tyrosine 1
PZ	Peripheral zone
QC	Quiescent center
RAM	Root apical meristem
RGF	Root growth factor

ROW1	REPRESSOR OF WUSCHEL 1
RZ	Rib zone
SAM	Shoot apical meristem
SCR	SCARECROW
SHR	SHORT-ROOT
SIM	Shoot-inducing medium
SPL9	SQUAMOSA PROMOTER BINDING PROTEIN-LIKE 9
STM	SHOOT MERISTEMLESS
taa1	*Tryptophan aminotransferase 1*
tir1	*Transport inhibitor response 1*
VLCFA	Very-long-chain fatty acid
WIND	WOUND INDUCED DEDIFFERENTIATION
WOX	WUSCHEL-RELATED HOMEOBOX
WUS	*WUSCHEL*
YUC	*YUCCA*

Acknowledgments

The research on hormones and stem cells was supported by grants from the National Natural Science Foundation (NNSF) of China (31230009 and 91217308) and the Ministry of Science and Technology (MOST) of China (2013CB967300). No conflicts of interest declared.

References

Aichinger, E., Kornet, N., Friedrich, T., et al., 2012. Plant stem cell niches. Annu. Rev. Plant Biol. 63, 615–636.

Aida, M., Beis, D., Heidstra, R., et al., 2004. The PLETHORA genes mediate patterning of the *Arabidopsis* root stem cell niche. Cell 119, 109–120.

Aida, M., Ishida, T., Fukaki, H., et al., 1997. Genes involved in organ separation in *Arabidopsis*: an analysis of the cup-shaped cotyledon mutant. Plant Cell 9, 841–857.

Anzola, J.M., Sieberer, T., Ortbauer, M., et al., 2010. Putative *Arabidopsis* transcriptional adaptor protein (PROPORZ1) is required to modulate histone acetylation in response to auxin. Proc. Natl. Acad. Sci. U.S.A. 107, 10308–10313.

Atta, R., Laurens, L., Boucheron-Dubuisson, E., et al., 2009. Pluripotency of *Arabidopsis* xylem pericycle underlies shoot regeneration from root and hypocotyl explants grown in vitro. Plant J. 57, 626–644.

Bartrina, I., Otto, E., Strnad, M., et al., 2011. Cytokinin regulates the activity of reproductive meristems, flower organ size, ovule formation, and thus seed yield in *Arabidopsis thaliana*. Plant Cell 23, 69–80.

Bell, E.M., Lin, W.C., Husbands, A.Y., et al., 2012. *Arabidopsis* lateral organ boundaries negatively regulates brassinosteroid accumulation to limit growth in organ boundaries. Proc. Natl. Acad. Sci. U.S.A. 109, 21146–21151.

Benková, E., Michniewicz, M., Sauer, M., et al., 2003. Local, efflux-dependent auxin gradients as a common module for plant organ formation. Cell 115, 591–602.

Birnbaum, K.D., Sánchez Alvarado, A., 2008. Slicing across kingdoms: regeneration in plants and animals. Cell 132, 697–710.

Blilou, I., Xu, J., Wildwater, M., et al., 2005. The PIN auxin efflux facilitator network controls growth and patterning in *Arabidopsis* roots. Nature 433, 39–44.

Brand, U., Fletcher, J.C., Hobe, M., et al., 2000. Dependence of stem cell fate in *Arabidopsis* on a feedback loop regulated by CLV3 activity. Science 289, 617–619.

Buechel, S., Leibfried, A., To, J.P., et al., 2010. Role of A-type *ARABIDOPSIS RESPONSE REGULATORS* in meristem maintenance and regeneration. Eur. J. Cell Biol. 89, 279–284.

Byrne, M.E., Simorowski, J., Martienssen, R.A., 2002. *ASYMMETRIC LEAVES*1 reveals *knox* gene redundancy in *Arabidopsis*. Development 129, 1957–1965.

Che, P., Lall, S., Howell, S.H., 2007. Developmental steps in acquiring competence for shoot development in *Arabidopsis* tissue culture. Planta 226, 1183–1194.

Chen, H., Banerjee, A.K., Hannapel, D.J., 2004. The tandem complex of BEL and KNOX partners is required for transcriptional repression of *ga20ox1*. Plant J. 38, 276–284.

Chen, L., Tong, J., Xiao, L., et al., 2016. *YUCCA*-mediated auxin biogenesis is required for cell fate transition occurring during *de novo* root organogenesis in *Arabidopsis*. J. Exp. Bot. 67, 4273–4284.

Chen, X., Qu, Y., Sheng, L., et al., 2014. A simple method suitable to study *de novo* root organogenesis. Front. Plant Sci. 5, 208.

Cheng, Y., Dai, X., Zhao, Y., 2007. Auxin synthesized by the YUCCA flavin monooxygenases is essential for embryogenesis and leaf formation in *Arabidopsis*. Plant Cell 19, 2430–2439.

Cheng, Z.J., Wang, L., Sun, W., et al., 2013. Pattern of auxin and cytokinin responses for shoot meristem induction results from the regulation of cytokinin biosynthesis by AUXIN RESPONSE FACTOR3. Plant Physiol. 161, 240–251.

Chupeau, M.C., Granier, F., Pichon, O., et al., 2013. Characterization of the early events leading to totipotency in an *Arabidopsis* protoplast liquid culture by temporal transcript profiling. Plant Cell 25, 2444–2463.

Ckurshumova, W., Smirnova, T., Marcos, D., et al., 2014. Irrepressible *MONOPTEROS/ARF5* promotes *de novo* shoot formation. New Phytol. 204, 556–566.

Clark, S.E., 2001. Meristems: start your signaling. Curr. Opin. Plant Biol. 4, 28–32.

Crawford, B.C., Sewell, J., Golembeski, G., et al., 2015. Genetic control of distal stem cell fate within root and embryonic meristems. Science 347, 655–659.

Cui, H., Levesque, M.P., Vernoux, T., et al., 2007. An evolutionarily conserved mechanism delimiting SHR movement defines a single layer of endodermis in plants. Science 316, 421–425.

Daimon, Y., Takabe, K., Tasaka, M., 2003. The *CUP-SHAPED COTYLEDON* genes promote adventitious shoot formation on calli. Plant Cell Physiol. 44, 113–121.

Daum, G., Medzihradszky, A., Suzaki, T., et al., 2014. A mechanistic framework for non-cell autonomous stem cell induction in *Arabidopsis*. Proc. Natl. Acad. Sci. U.S.A. 111, 14619–14624.

De Rybel, B., Mahonen, A.P., Helariutta, Y., et al., 2016. Plant vascular development: from early specification to differentiation. Nat. Rev. Mol. Cell Biol. 17, 30–40.

Dello Ioio, R., Linhares, F.S., Scacchi, E., et al., 2007. Cytokinins determine *Arabidopsis* root-meristem size by controlling cell differentiation. Curr. Biol. 17, 678–682.

Dello Ioio, R., Nakamura, K., Moubayidin, L., et al., 2008. A genetic framework for the control of cell division and differentiation in the root meristem. Science 322, 1380–1384.

DiDonato, R.J., Arbuckle, E., Buker, S., et al., 2004. *Arabidopsis ALF4* encodes a nuclear-localized protein required for lateral root formation. Plant J. 37, 340–353.

Ding, Z., Friml, J., 2010. Auxin regulates distal stem cell differentiation in *Arabidopsis* roots. Proc. Natl. Acad. Sci. U.S.A. 107, 12046–12051.

Dinneny, J.R., Benfey, P.N., 2008. Plant stem cell niches: standing the test of time. Cell 132, 553–557.

Domagalska, M.A., Leyser, O., 2011. Signal integration in the control of shoot branching. Nat. Rev. Mol. Cell Biol. 12, 211–221.

Duclercq, J., Sangwan-Norreel, B., Catterou, M., et al., 2011. *De novo* shoot organogenesis: from art to science. Trends Plant Sci. 16, 597–606.

Fan, M., Xu, C., Xu, K., et al., 2012. LATERAL ORGAN BOUNDARIES DOMAIN transcription factors direct callus formation in *Arabidopsis* regeneration. Cell Res. 22, 1169–1180.

Fernández-Marcos, M., Sanz, L., Lewis, D.R., et al., 2011. Nitric oxide causes root apical meristem defects and growth inhibition while reducing PIN-FORMED 1 (PIN1)-dependent acropetal auxin transport. Proc. Natl. Acad. Sci. U.S.A. 108, 18506–18511.

Fisher, A.P., Sozzani, R., 2016. Uncovering the networks involved in stem cell maintenance and asymmetric cell division in the *Arabidopsis* root. Curr. Opin. Plant Biol. 29, 38–43.

Fleet, C.M., Sun, T.P., 2005. A DELLAcate balance: the role of gibberellin in plant morphogenesis. Curr. Opin. Plant Biol. 8, 77–85.

Fletcher, J.C., Brand, U., Running, M.P., et al., 1999. Signaling of cell fate decisions by *CLAVATA3* in *Arabidopsis* shoot meristems. Science 283, 1911–1914.

Gaillochet, C., Daum, G., Lohmann, J.U., 2015. O Cell, Where Art Thou? The mechanisms of shoot meristem patterning. Curr. Opin. Plant Biol. 23, 91–97.

Gaillochet, C., Lohmann, J.U., 2015. The never-ending story: from pluripotency to plant developmental plasticity. Development 142, 2237–2249.

Galinha, C., Bilsborough, G., Tsiantis, M., 2009. Hormonal input in plant meristems: a balancing act. Semin. Cell Dev. Biol. 20, 1149–1156.

Galinha, C., Hofhuis, H., Luijten, M., et al., 2007. PLETHORA proteins as dose-dependent master regulators of *Arabidopsis* root development. Nature 449, 1053–1057.

Gautiteret, R.J., 1983. Plant tissue culture: a history. Bot. Mat Tokyo 96, 393–410.

Gendron, J.M., Wang, Z.Y., 2007. Multiple mechanisms modulate brassinosteroid signaling. Curr. Opin. Plant Biol. 10, 436–441.

Gonzalez-Garcia, M.P., Vilarrasa-Blasi, J., Zhiponova, M., et al., 2011. Brassinosteroids control meristem size by promoting cell cycle progression in *Arabidopsis* roots. Development 138, 849–859.

Gordon, S.P., Chickarmane, V.S., Ohno, C., et al., 2009. Multiple feedback loops through cytokinin signaling control stem cell number within the *Arabidopsis* shoot meristem. Proc. Natl. Acad. Sci. U.S.A. 106, 16529–16534.

Gordon, S.P., Heisler, M.G., Reddy, G.V., et al., 2007. Pattern formation during de novo assembly of the *Arabidopsis* shoot meristem. Development 134, 3539–3548.

Grieneisen, V.A., Xu, J., Marée, A.F., et al., 2007. Auxin transport is sufficient to generate a maximum and gradient guiding root growth. Nature 449, 1008–1013.

Hacham, Y., Holland, N., Butterfield, C., et al., 2011. Brassinosteroid perception in the epidermis controls root meristem size. Development 138, 839–848.

Haecker, A., Groß-Hardt, R., Geiges, B., et al., 2004. Expression dynamics of *WOX* genes mark cell fate decisions during early embryonic patterning in *Arabidopsis thaliana*. Development 131, 657–668.

Hamann, T., Benkova, E., Bäurle, I., et al., 2002. The *Arabidopsis BODENLOS* gene encodes an auxin response protein inhibiting MONOPTEROS-mediated embryo patterning. Genes Dev. 16, 1610–1615.

Hamant, O., Nogué, F., Belles-Boix, E., et al., 2002. The KNAT2 homeodomain protein interacts with ethylene and cytokinin signaling. Plant Physiol. 130, 657–665.

Hardtke, C.S., Berleth, T., 1998. The *Arabidopsis* gene *MONOPTEROS* encodes a transcription factor mediating embryo axis formation and vascular development. EMBO J. 17, 1405–1411.

Hay, A., Kaur, H., Phillips, A., et al., 2002. The gibberellin pathway mediates KNOTTED1-type homeobox function in plants with different body plans. Curr. Biol. 12, 1557–1565.

Heisler, M.G., Ohno, C., Das, P., et al., 2005. Patterns of auxin transport and gene expression during primordium development revealed by live imaging of the *Arabidopsis* inflorescence meristem. Curr. Biol. 15, 1899–1911.

Helariutta, Y., Fukaki, H., Wysocka-Diller, J., et al., 2000. The *SHORT-ROOT* gene controls radial patterning of the *Arabidopsis* root through radial signaling. Cell 101, 555–567.

Hill, K., Mathews, D.E., Kim, H.J., et al., 2013. Functional characterization of type-B response regulators in the *Arabidopsis* cytokinin response. Plant Physiol. 162, 212–224.

Hwang, I., Sheen, J., Müller, B., 2012. Cytokinin signaling networks. Annu. Rev. Plant Biol. 63, 353–380.

Ikeuchi, M., Ogawa, Y., Iwase, A., et al., 2016. Plant regeneration: cellular origins and molecular mechanisms. Development 143, 1442–1451.

Inoue, T., Higuchi, M., Hashimoto, Y., et al., 2001. Identification of CRE1 as a cytokinin receptor from *Arabidopsis*. Nature 409, 1060–1063.

Ishida, K., Yamashino, T., Yokoyama, A., et al., 2008. Three type-B response regulators, ARR1, ARR10 and ARR12, play essential but redundant roles in cytokinin signal transduction throughout the life cycle of *Arabidopsis thaliana*. Plant Cell Physiol. 49, 47–57.

Ito, M., Machida, Y., 2015. Reprogramming of plant cells induced by 6b oncoproteins from the plant pathogen *Agrobacterium*. J. Plant Res. 128, 423–435.

Iwase, A., Mitsuda, N., Koyama, T., et al., 2011. The AP2/ERF transcription factor WIND1 controls cell dedifferentiation in *Arabidopsis*. Curr. Biol. 21, 508–514.

Jasinski, S., Piazza, P., Craft, J., et al., 2005. KNOX action in *Arabidopsis* is mediated by coordinate regulation of cytokinin and gibberellin activities. Curr. Biol. 15, 1560–1565.

Jeong, S., Trotochaud, A.E., Clark, S.E., 1999. The *Arabidopsis CLAVATA2* gene encodes a receptor-like protein required for the stability of the CLAVATA1 receptor-like kinase. Plant Cell 11, 1925–1934.

Kareem, A., Durgaprasad, K., Sugimoto, K., et al., 2015. *PLETHORA* genes control regeneration by a two-step mechanism. Curr. Biol. 25, 1017–1030.

Kerstetter, R.A., Laudencia-Chingcuanco, D., Smith, L.G., et al., 1997. Loss-of-function mutations in the maize homeobox gene, *knotted1*, are defective in shoot meristem maintenance. Development 124, 3045–3054.

Kurakawa, T., Ueda, N., Maekawa, M., et al., 2007. Direct control of shoot meristem activity by a cytokinin-activating enzyme. Nature 445, 652–655.

Laux, T., 2003. The stem cell concept in plants: a matter of debate. Cell 113, 281–283.

Laux, T., Würschum, T., Breuninger, H., 2004. Genetic regulation of embryonic pattern formation. Plant Cell 16, S190–S202.

Lee, H.S., Kim, Y., Pham, G., et al., 2015. Brassinazole resistant 1 (BZR1)-dependent brassinosteroid signalling pathway leads to ectopic activation of quiescent cell division and suppresses columella stem cell differentiation. J. Exp. Bot. 66, 4835–4849.

Leibfried, A., To, J.P., Busch, W., et al., 2005. WUSCHEL controls meristem function by direct regulation of cytokinin-inducible response regulators. Nature 438, 1172–1175.

Lenhard, M., Jürgens, G., Laux, T., 2002. The *WUSCHEL* and *SHOOTMERISTEMLESS* genes fulfil complementary roles in *Arabidopsis* shoot meristem regulation. Development 129, 3195–3206.

Li, W., Liu, H., Cheng, Z.J., et al., 2011. DNA methylation and histone modifications regulate *de novo* shoot regeneration in *Arabidopsis* by modulating *WUSCHEL* expression and auxin signaling. PLoS Genet. 7, e1002243.

Liu, J., Sheng, L., Xu, Y., et al., 2014. *WOX11* and *12* are involved in the first-step cell fate transition during *de novo* root organogenesis in *Arabidopsis*. Plant Cell 26, 1081–1093.

Liu, Z., Li, J., Wang, L., et al., 2016. Repression of callus initiation by the miRNA-directed interaction of auxin-cytokinin in *Arabidopsis thaliana*. Plant J. 87, 391–402.

Long, J.A., Moan, E.I., Medford, J.I., et al., 1996. A member of the KNOTTED class of homeodomain proteins encoded by the *STM* gene of *Arabidopsis*. Nature 379, 66–69.

Müller, B., Sheen, J., 2008. Cytokinin and auxin interaction in root stem-cell specification during early embryogenesis. Nature 453, 1094–1097.

Möller, B., Weijers, D., 2009. Auxin control of embryo patterning. Cold Spring Harb Perspect. Biol. 1, a001545.

Matsuoka, M., Ichikawa, H., Saito, A., et al., 1993. Expression of a rice homeobox gene causes altered morphology of transgenic plants. Plant Cell 5, 1039–1048.

Matsuzaki, Y., Ogawa-Ohnishi, M., Mori, A., et al., 2010. Secreted peptide signals required for maintenance of root stem cell niche in *Arabidopsis*. Science 329, 1065–1067.

Mayer, K.F., Schoof, H., Haecker, A., et al., 1998. Role of WUSCHEL in regulating stem cell fate in the *Arabidopsis* shoot meristem. Cell 95, 805–815.

Miyawaki, K., Matsumoto-Kitano, M., Kakimoto, T., 2004. Expression of cytokinin biosynthetic isopentenyltransferase genes in *Arabidopsis*: tissue specificity and regulation by auxin, cytokinin, and nitrate. Plant J. 37, 128–138.

Mockaitis, K., Estelle, M., 2008. Auxin receptors and plant development: a new signaling paradigm. Annu. Rev. Cell Dev. Biol. 24, 55–80.

Moubayidin, L., Di Mambro, R., Sabatini, S., 2009. Cytokinin-auxin crosstalk. Trends Plant Sci. 14, 557–562.

Murray, J.A., Jones, A., Godin, C., et al., 2012. Systems analysis of shoot apical meristem growth and development: integrating hormonal and mechanical signaling. Plant Cell 24, 3907–3919.

Ortega-Martínez, O., Pernas, M., Carol, R.J., et al., 2007. Ethylene modulates stem cell division in the *Arabidopsis thaliana* root. Science 317, 507–510.

Ou, Y., Lu, X., Zi, Q., et al., 2016. RGF1 INSENSITIVE 1 to 5, a group of LRR receptor-like kinases, are essential for the perception of root meristem growth factor 1 in *Arabidopsis thaliana*. Cell Res. 26, 686–698.

Perilli, S., Di Mambro, R., Sabatini, S., 2012. Growth and development of the root apical meristem. Curr. Opin. Plant Biol. 15, 17–23.

Petersson, S.V., Johansson, A.I., Kowalczyk, M., et al., 2009. An auxin gradient and maximum in the *Arabidopsis* root apex shown by high-resolution cell-specific analysis of IAA distribution and synthesis. Plant Cell 21, 1659–1668.

Petricka, J.J., Winter, C.M., Benfey, P.N., 2012. Control of *Arabidopsis* root development. Annu. Rev. Plant Biol. 63, 563–590.

Reinhardt, D., Frenz, M., Mandel, T., et al., 2003. Microsurgical and laser ablation analysis of interactions between the zones and layers of the tomato shoot apical meristem. Development 130, 4073–4083.

Rowe, J.H., Topping, J.F., Liu, J., et al., 2016. Abscisic acid regulates root growth under osmotic stress conditions via an interacting hormonal network with cytokinin, ethylene and auxin. New Phytol. 211, 225–239.

Sabatini, S., Heidstra, R., Wildwater, M., et al., 2003. SCARECROW is involved in positioning the stem cell niche in the *Arabidopsis* root meristem. Genes Dev. 17, 354–358.

Sablowski, R., 2007. The dynamic plant stem cell niches. Curr. Opin. Plant Biol. 10, 639–644.

Sakai, H., Honma, T., Aoyama, T., et al., 2001. ARR1, a transcription factor for genes immediately responsive to cytokinins. Science 294, 1519–1521.

Sakamoto, T., Kamiya, N., Ueguchi-Tanaka, M., et al., 2001. KNOX homeodomain protein directly suppresses the expression of a gibberellin biosynthetic gene in the tobacco shoot apical meristem. Genes Dev. 15, 581–590.

Salehin, M., Bagchi, R., Estelle, M., 2015. SCF$^{TIR1/AFB}$-based auxin perception: mechanism and role in plant growth and development. Plant Cell 27, 9–19.

Sarkar, A.K., Luijten, M., Miyashima, S., et al., 2007. Conserved factors regulate signalling in *Arabidopsis thaliana* shoot and root stem cell organizers. Nature 446, 811–814.

Schaller, G.E., Bishopp, A., Kieber, J.J., 2015. The yin-yang of hormones: cytokinin and auxin interactions in plant development. Plant Cell 27, 44–63.

Scheres, B., 2007. Stem-cell niches: nursery rhymes across kingdoms. Nat. Rev. Mol. Cell Biol. 8, 345–354.

Schlereth, A., Möller, B., Liu, W., et al., 2010. MONOPTEROS controls embryonic root initiation by regulating a mobile transcription factor. Nature 464, 913–916.

Schoof, H., Lenhard, M., Haecker, A., et al., 2000. The stem cell population of *Arabidopsis* shoot meristems is maintained by a regulatory loop between the *CLAVATA* and *WUSCHEL* genes. Cell 100, 635–644.

Sena, G., Wang, X., Liu, H., et al., 2009. Organ regeneration does not require a functional stem cell niche in plants. Nature 457, 1150–1153.

Shang, B., Xu, C., Zhang, X., et al., 2016. Very-long-chain fatty acids restrict regeneration capacity by confining pericycle competence for callus formation in *Arabidopsis*. Proc. Natl. Acad. Sci. U.S.A. 113, 5101–5106.

Sieberer, T., Hauser, M.T., Seifert, G.J., et al., 2003. PROPORZ1, a putative *Arabidopsis* transcriptional adaptor protein, mediates auxin and cytokinin signals in the control of cell proliferation. Curr. Biol. 13, 837–842.

Skoog, F., Miller, C.O., 1957. Chemical regulation of growth and organ formation in plant tissues cultured in vitro. Symp. Soc. Exp. Biol. 11, 118–130.

Skylar, A., Hong, F., Chory, J., et al., 2010. STIMPY mediates cytokinin signaling during shoot meristem establishment in *Arabidopsis* seedlings. Development 137, 541–549.

Smit, M.E., Weijers, D., 2015. The role of auxin signaling in early embryo pattern formation. Curr. Opin. Plant Biol. 28, 99–105.

Soyars, C.L., James, S.R., Nimchuk, Z.L., 2016. Ready, aim, shoot: stem cell regulation of the shoot apical meristem. Curr. Opin. Plant Biol. 29, 163–168.

Sparks, E., Wachsman, G., Benfey, P.N., 2013. Spatiotemporal signalling in plant development. Nat. Rev. Genet. 14, 631–644.

Stahl, Y., Wink, R.H., Ingram, G.C., et al., 2009. A signaling module controlling the stem cell niche in *Arabidopsis* root meristems. Curr. Biol. 19, 909–914.

Steeves, T.A., Sussex, I.M., 1989. Patterns in Plant Development, second ed. Cambridge University Press, Cambridge.

Stepanova, A.N., Robertson-Hoyt, J., Yun, J., et al., 2008. TAA1-mediated auxin biosynthesis is essential for hormone crosstalk and plant development. Cell 133, 177–191.

Steward, F.C., Marion, O.M., Mears, K., 1958. Growth and organized development of cultured cells. II. Organization in cultures grown from freely suspended cells. Am. J. Bot. 45, 705–708.

Sugimoto, K., Jiao, Y., Meyerowitz, E.M., 2010. *Arabidopsis* regeneration from multiple tissues occurs via a root development pathway. Dev. Cell 18, 463–471.

Szemenyei, H., Hannon, M., Long, J.A., 2008. TOPLESS mediates auxin-dependent transcriptional repression during *Arabidopsis* embryogenesis. Science 319, 1384–1386.

ten Hove, C.A., Lu, K.J., Weijers, D., 2015. Building a plant: cell fate specification in the early *Arabidopsis* embryo. Development 142, 420–430.

Tian, H., De Smet, I., Ding, Z., 2014. Shaping a root system: regulating lateral versus primary root growth. Trends Plant Sci. 19, 426–431.

Tokunaga, H., Kojima, M., Kuroha, T., et al., 2012. *Arabidopsis* lonely guy (LOG) multiple mutants reveal a central role of the LOG-dependent pathway in cytokinin activation. Plant J. 69, 355–365.

Tsuda, K., Kurata, N., Ohyanagi, H., et al., 2014. Genome-wide study of *KNOX* regulatory network reveals brassinosteroid catabolic genes important for shoot meristem function in rice. Plant Cell 26, 3488–3500.

Tucker, M.R., Laux, T., 2007. Connecting the paths in plant stem cell regulation. Trends Cell Biol. 17, 403–410.

Ubeda-Tomás, S., Federici, F., Casimiro, I., et al., 2009. Gibberellin signaling in the endodermis controls *Arabidopsis* root meristem size. Curr. Biol. 19, 1194–1199.

Ubeda-Tomas, S., Swarup, R., Coates, J., et al., 2008. Root growth in *Arabidopsis* requires gibberellin/DELLA signalling in the endodermis. Nat. Cell Biol. 10, 625–628.

Valvekens, D., Van Montagu, M., Van Lijsebettens, M., 1988. *Agrobacterium tumefaciens*-mediated transformation of *Arabidopsis thaliana* root explants by using kanamycin selection. Proc. Natl. Acad. Sci. U.S.A. 85, 5536–5540.

Veit, B., 2009. Hormone mediated regulation of the shoot apical meristem. Plant Mol. Biol. 69, 397–408.

Wang, B., Chu, J., Yu, T., et al., 2015. Tryptophan-independent auxin biosynthesis contributes to early embryogenesis in *Arabidopsis*. Proc. Natl. Acad. Sci. U.S.A. 112, 4821–4826.

Wang, Y., Li, J., 2008. Molecular basis of plant architecture. Annu. Rev. Plant Biol. 59, 253–279.

Werner, T., Motyka, V., Laucou, V., et al., 2003. Cytokinin-deficient transgenic *Arabidopsis* plants show multiple developmental alterations indicating opposite functions of cytokinins in the regulation of shoot and root meristem activity. Plant Cell 15, 2532–2550.

Wu, X., Dabi, T., Weigel, D., 2005. Requirement of homeobox gene *STIMPY/WOX9* for *Arabidopsis* meristem growth and maintenance. Curr. Biol. 15, 436–440.

Xie, M., Tataw, M., Venugopala Reddy, G., 2009. Towards a functional understanding of cell growth dynamics in shoot meristem stem-cell niche. Semin. Cell Dev. Biol. 20, 1126–1133.

Xu, J., Hofhuis, H., Heidstra, R., et al., 2006. A molecular framework for plant regeneration. Science 311, 385–388.

Xu, L., Huang, H., 2014. Genetic and epigenetic controls of plant regeneration. Curr. Top. Dev. Biol. 108, 1–33.

Yadav, R.K., Perales, M., Gruel, J., et al., 2011. WUSCHEL protein movement mediates stem cell homeostasis in the *Arabidopsis* shoot apex. Genes Dev. 25, 2025–2030.

Yanai, O., Shani, E., Dolezal, K., et al., 2005. *Arabidopsis* KNOXI proteins activate cytokinin biosynthesis. Curr. Biol. 15, 1566–1571.

Zhang, T., Lian, H., Tang, H., et al., 2015a. An intrinsic microRNA timer regulates progressive decline in shoot regenerative capacity in plants. Plant Cell 27, 349–360.

Zhang, Y., Jiao, Y., Liu, Z., et al., 2015b. ROW1 maintains quiescent centre identity by confining *WOX5* expression to specific cells. Nat. Commun. 6, 6003.

Zhao, Y., 2010. Auxin biosynthesis and its role in plant development. Annu. Rev. Plant Biol. 61, 49–64.

Zhao, Z., Andersen, S.U., Ljung, K., et al., 2010. Hormonal control of the shoot stem-cell niche. Nature 465, 1089–1092.

Zhou, W., Wei, L., Xu, J., et al., 2010. *Arabidopsis* tyrosylprotein sulfotransferase acts in the auxin/PLETHORA pathway in regulating postembryonic maintenance of the root stem cell niche. Plant Cell 22, 3692–3709.

Phytohormonal quantification based on biological principles

13

Yi Su, Shitou Xia, Ruozhong Wang, Langtao Xiao
Hunan Agricultural University, Changsha, China

Summary

Accurate quantification of phytohormones in specific target cells will greatly facilitate the elucidation of the molecular mechanisms of their actions. Since the *Avena* coleoptile curvature test for auxin was established in the 1920s, phytohormonal assays have undergone a long march involving several technological generations. Biological methods such as traditional bioassays and immunoassays dominated phytohormonal quantification from the 1920s to the 1980s. In the past three decades, physical methods based on chromatography and mass spectrometry have largely replaced biological methods and have become the mainstream methods for phytohormonal quantification. However, even the most advanced tandem mass spectrometry today is still unable to perform quantification or to reveal hormone dynamics at the single-cell level. Some molecular phytohormonal quantification methods such as phytohormone-inducible promoter-reporter systems and degrons (phytohormone-inducible signaling repressor protein degradation) have shown the potential for single-cell-level quantification. Furthermore, biosensors offer the ability for fast determination while "nucleic acid terminal protection" methods provide the possibility for improvement of sensitivity. Therefore, phytohormonal quantification based on biological principles has regained the attention from plant biologists and will become increasingly important in future phytohormonal quantification practice. In this chapter, phytohormonal quantification based on biological principles, as well as the sample preparation, localization, and profiling techniques have been reviewed. Moreover, perspectives of ongoing approaches with single-cell dynamics are also discussed.

13.1 Phytohormones and their quantification

13.1.1 Identified phytohormone categories, chemical structures, and properties

Phytohormones or plant hormones, are naturally occurring small organic molecules or substances which influence physiological processes in plants at very low concentrations (Davies, 2004). In other words, phytohormones are chemical messengers that coordinate cellular activities of plants (Fleet and Williams, 2011). From the early discovery of auxin as the first phytohormone (Went, 1935) to the most recent identification of strigolactones (SL) (Gomez-Roldan et al., 2008), nine categories of phytohormones, that is, auxins, cytokinins (CK), gibberellins (GA), abscisic acid

Hormone Metabolism and Signaling in Plants. http://dx.doi.org/10.1016/B978-0-12-811562-6.00013-X

(ABA), ethylene (ETH), brassinosteroids (BR), salicylates (SA), jasmonates (JA) and strigolactones (SL), have been identified so far. The first five (auxin, CK, GA, ABA, and ETH) are sometimes referred to as the "classical" phytohormones, while the latter four are more later additions to the growing phytohormonal family (Fleet and Williams, 2011). Although nitric oxide (NO) and reactive oxygen species are important signaling molecules in plants, they are not widely recognized as phytohormones among plant biologists mainly because they are inorganic chemicals. Signaling peptides are increasingly important in plant biology but they are macromolecules rather than organic chemicals, and are discussed in a separate chapter in this book (Song et al., 2017). Related to the term "phytohormone," the term "plant growth regulator" (PGR) refers to non-naturally occurring synthetic compounds with phytohormone-like activities while the term "plant growth substance" (PGS) includes both phytohormones and PGRs.

Chemical structures and physicochemical properties of phytohormones are critical for the establishment of phytohormonal quantification methodology. Regarding the chemical structures, auxins are indole derivatives; ABA is a sesquiterpene; ETH is the simplest alkene; CKs are adenine derivatives; GAs are tetracyclic diterpenoid acids; BRs are polyhydroxysteroids; JA is a fatty acid derivative from linolenic acid; SLs are terpenoid lactones derived from carotenoids (Fleet and Williams, 2011). Generally, auxins, GAs, ABA, JA, and SA are acidic and CKs are alkaline; thus, the diversity of chemical structures and properties of different phytohormones make the specialized extraction, purification, and quantification, as well as localization, procedures necessary (Han et al., 2012). Consideration of the diverse physicochemical properties of phytohormones is discussed in a separate chapter in this book which describes the physical methods of liquid chromatography (LC) and mass spectrometry (MS) for the identification and quantification of plant hormones (Fang et al., 2017).

13.1.2   Significance of phytohormonal quantification in research

Phytohormonal quantification consists of several necessary steps, including sampling, extraction, purification, and quantification. This chapter mainly focuses on the development of the representative methods belonging to different technical generations in phytohormonal biological quantification methodology over the past 90 years. The future goals and perspectives of novel techniques for analyzing single-cell dynamics, such as phytohormonal degron-based molecular methods, are also discussed.

Advances in phytohormonal research have been regarded as one of the most important driving forces for the "Green Revolution" and the progress of agricultural science and technology because almost all known "Green Revolution" genes, that is, the genes responsible for important agronomic traits such as plant height, tiller number, seed development, and yield, are found to be phytohormone-related genes (Peng et al., 1999; Sasaki et al., 2002; Dockter et al., 2014; Qian et al., 2016). High sensitivity and accuracy together with high-throughput quantification of phytohormones is crucial for frontier studies to reveal phytohormonal metabolism, transport, and molecular mechanisms in plants. Therefore, driven by the urgent needs of modern plant biologists and boosted by the intensive investment of research funding from those countries that value basic research in plant sciences (Chong and Xu, 2014), great efforts have been invested in the field of phytohormonal assay methodology and notable advances have been achieved now.

13.1.3 Significance of biological methods in phytohormonal quantification

The *Avena* (oat) coleoptile curvature test is the first phytohormonal quantification method established based on the physiological activity of phytohormones, and quantifies through the specificity of phytohormonal responses in certain plant tissues and organs (Went, 1926). Although early biological methods had limited sensitivity and selectivity, they have made great contributions to technical progress in the early stages of phytohormone assay methodology. Generally, biological quantification methods are simple and easy to establish, and the results can accurately reflect the physiological activity of phytohormones. Along with the development of modern molecular biology, some promising phytohormonal quantification approaches based on phytohormonal signaling have shown the potential for highly sensitive in vivo and real-time measurement of phytohormones (Brunoud et al., 2012; Wells et al., 2013; Larrieu et al., 2015).

13.2 Sample preparation for phytohormonal assay

13.2.1 Sampling

13.2.1.1 Sample collection and treatment

Sampling should provide material with an adequate amount of phytohormones detectable by the final quantification measures, as well as provide a sample that will help to answer the questions being examined (Brenner, 1981). Since phytohormones are changing dynamically in plants, fresh materials for phytohormonal analysis need to be properly collected according to certain quantification requirements. When the plant materials are plentiful, the representativeness of sample collection is critical; when the plant materials are available only in trace amounts, accuracy is more important. Several new sample collection methods have been adopted in phytohormonal analysis. Laser capture microdissection (LCM) has been employed to collect target cells or cell groups under a microscope (Böhm et al., 1997; Suwabe et al., 2008). Fluorescence activated cell sorter (FACS) technology has been used to sort, collect, and count the sampled cells or protoplasts when different types of cells are molecularly labeled by fluorescent proteins (Brunoud et al., 2012; Wells et al., 2013).

After sample collection from plants or cell cultures, careful operations with the minimum of incision or other wounding is required in order to avoid enzymatic degradation. Measures to minimize unnecessary exposure to light, heat, and oxygen during the sampling process are also important. Except for in vivo analysis, fast freezing of the target tissue in liquid nitrogen is necessary for the newly collected samples. This applies to samples for instant extraction, for short-term storage in the freezer, or for samples to be freeze-dried for long-term storage. Technical details for sample preparation have been discussed and improvement suggestions have been proposed by some researchers (reviewed in Du et al., 2012; Fang et al., 2017).

13.2.1.2   Recovery analysis

Recovery analysis calculates the recovery rate of a target analyte after a certain process, usually by adding a fixed amount of a standard prior to the process and analyzing the amount of standard recovered after the process. Since extensive manual operations are involved in sample preparation, such as, freezing in liquid nitrogen, lyophilization, grinding and homogenization, separation, and sample transfer, all of which will result in degradation of phytohormones caused by light, heat, hydrolysis, and oxidation, phytohormone loss must be considered. Recovery rate varies with the stability of the particular phytohormone during sample preparation. For example, Indole-3-acetic acid (IAA) is more sensitive to environmental factors than many other phytohormones, so substantial loss of IAA could be found during sample preparation. Generally, the loss can be estimated and corrected by recovery analysis using radioactively labeled phytohormones as tracers (Ljung et al., 2004).

13.2.2   Extraction and purification
13.2.2.1   Solvent extraction

Extraction and purification are critical steps in sample preparation for phytohormonal quantification because phytohormones need to be extracted from plant samples and the crude plant extracts with co-extracted metabolites are very complex. A number of methods based on different principles for plant extraction and purification have been reported since the 1960s. Solvent extraction is a common method for the extraction of most phytohormones. Starting from some early studies in the 1960s, the influence of solvents on the enzymic action of plant tissue during extraction has been examined (Bieleski, 1964). Although a few studies used distilled water to extract phytohormones from plant samples, organic solvents with polarity close to the target phytohormones are strongly suggested (Loveys and Van Dijk, 1988). Different extraction protocols have been reported in previous studies and various organic solvents such as methanol, ethanol, acetonitrile, chloroform, acetone, propanol, ethylacetate, and acetic acid have been used. Among the organic solvents tested, 80% methanol has become a preferred solvent for its relatively good extraction efficiency for most phytohormones. In addition to a single solvent, mixtures comprising different solvents have been frequently used. For example, the modified Bieleski's solvent (methanol/formic acid/water

15:1:4) showed better extraction efficiency than a single solvent for simultaneous extraction of multiple phytohormone categories with inhibition of enzymatic degradation and reduced co-extraction of lipids which can be abundant in plant samples (Kojima et al., 2009; Giannarelli et al., 2010).

13.2.2.2 Ultrasound-assisted extraction

Ultrasonication is the application of high-intensity and high-frequency sound waves, which through interaction with the sample material, disrupt the physicochemical properties of the sample (Mason and Lorimer, 1988). In the case of raw plant tissues, ultrasound has been found to disrupt plant cell walls thereby facilitating the release of extractable compounds and enhancing mass transfer of solvent from the continuous phase into plant cells (reviewed in Vinatoru, 2001). Extraction processes have greatly benefited from the use of ultrasound, since it enhances the mass transfer efficiency, reducing the extraction time, lowering loss due to thermal degradation, and reducing consumption of water and energy. Ultrasound-assisted extraction (UAE) has been applied to improve the extraction for phytohormones, either alone or in association with other non-thermal technologies (Zhang et al., 2015).

13.2.2.3 Vacuum microwave-assisted extraction

Microwave-assisted extraction (MAE) has been widely used as a sample preparation technique for different analytical purposes such as environmental detection, food inspection, and agricultural sample analysis, because of low solvent consumption, short extraction time, and high extraction efficiency. However when MAE is performed in open mode, exposed to the atmosphere (air-MAE), the operating temperature which is near to the boiling point of the solvent is usually too high for thermo-sensitive and oxygen-sensitive phytohormones. To reduce degradation at high operating temperature, vacuum microwave-assisted extraction (VMAE) has been developed to operate at low oxygen subpressure in vacuo (Wang et al., 2008). When conventional solvent extraction was replaced by VMAE to extract IAA from plant tissues, improved extraction efficiency and decreased extraction time was obtained (Hu et al., 2011).

13.2.2.4 Liquid–liquid extraction

Liquid–liquid extraction (LLE) is based on the principle that a solute or an analyte can distribute itself in a certain ratio between two immiscible solvents, usually water (aqueous phase) and organic solvent (organic phase). LLE is widely used in sample preparation for cleanup and enrichment, which results in signal enhancement. In the early 1970s, classical LLE was first applied to purify cytokinins (Hemberg and Westlin, 1973; Dekhuijzen and Gevers, 1975). It was found that when kinetin in acidic aqueous solution was partitioned three times with equal volumes of ethyl ether, only about 50% of the kinetin went into the ether phase. However when partitioned with ethyl acetate, more than 90% of the kinetin partitioned into the ethyl acetate phase

(Dekhuijzen and Gevers, 1975). Subsequently, more categories of phytohormones, such as IAA, ABA, GA, SA, and JA have been purified using LLE (Schmelz et al., 2003; Durgbanshi et al., 2005; Xie et al., 2011).

All naturally occurring phytohormones can be isolated from plant extracts with 50% or greater recovery rate using LLE. However, its shortcomings are apparent because it is unsuitable for ultra-trace determinations, it is time-consuming, and the solvents are environmentally unsafe. Therefore, some improved LLE methods have been developed to simplify sample preparation procedures, including single drop microextraction (SDME), wetting film extraction (WFE), cloud point extraction (CPE), homogeneous liquid–liquid extraction (HLLE), dispersive liquid–liquid microextraction (DLLME), and dispersive liquid–liquid microextraction based on solidification of a floating organic drop (DLLME-SFO) (Paleologos et al., 2005; Melchert et al., 2012; reviewed in Kokosa, 2015). Hollow fiber-based liquid–liquid–liquid microextraction (HF–LLLME) and DLLME have already been applied for the purification of phytohormones, especially for the simultaneous extraction and enrichment of multiple phytohormones by providing improved enrichment factors and recoveries (Gupta et al., 2011; Wu and Hu, 2009).

13.2.2.5 Solid phase extraction

Solid phase extraction (SPE) is a sample preparation process by which compounds that are dissolved or suspended in a liquid mixture are separated from other compounds in the mixture according to their physicochemical properties. SPE can be used to isolate analytes of interest from a wide variety of matrices and shows higher efficiency and flexibility than LLE. Thus, SPE methods using cartridges filled with matrix compounds which extract compounds based on diverse interactions, including adsorption, hydrogen bonding, polar and nonpolar interactions, cation or anion exchange or size exclusion, have been widely used for high-throughput phytohormone extraction and purification (Poole, 2003). Many SPE columns, such as Sep-Pak C18, Oasis HLB, Oasis MCX, and Oasis MAX containing different sorbents, are now commercially available for phytohormonal purification from crude extracts, so that SPE has become the mainstream method in purification of phytohormones.

A more recently developed SPE using fibers coated with an extracting phase of either a liquid (polymer) or a solid (sorbent) is solid phase microextraction (SPME) which can extract both volatile and nonvolatile analytes from different media (liquid or gas). In the extraction and purification of phytohormones, SPME application has the advantage of high recovery rate (>80%) and easy operation (Liu et al., 2007; Ding et al., 2013).

Magnetic solid phase extraction (MSPE) is an improved SPE for the enrichment and purification of target analytes from large volumes based on the use of magnetic or magnetizable adsorbents. In the MSPE procedure, magnetic adsorbent is added to a solution or suspension containing the target analyte. The analyte will be adsorbed on to the magnetic adsorbent and then the adsorbent with the adsorbed analyte is recovered from the suspension using a magnetic separator. Subsequently, the analyte is eluted from the recovered adsorbent. MSPE has drawn much attention for sample preparation because of its rapid and easy operation (Fig. 13.1). Compared with traditional

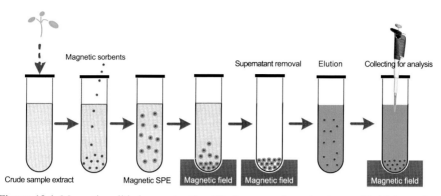

Figure 13.1 Magnetic solid phase extraction. Firstly, magnetic sorbents are added into crude sample extract and the phytohormone molecules are absorbed onto the sorbents. Secondly, the magnetic sorbents with phytohormone molecules are precipitated at the bottom of the container through the magnetic field and then the supernatant is removed. Finally, the phytohomone is eluted from magnetic sorbents and then quantified.

SPE, the magnetic adsorbents are dispersed in solution without being packed into a SPE cartridge, so that the column blocking frequently seen in SPE can be avoided. Moreover, mass transfer of analytes can be facilitated by drastically increasing the interfacial area between adsorbent and sample solution, resulting in improved sample purification efficiency (Gao et al., 2010). In recent years, MSPE has been widely used for the enrichment and purification of phytohormones and this procedure has been greatly simplified. For example, one-step MSPE purification has been established using Fe_3O_4 @TiO_2 (a nanoparticle with Fe_3O_4 as the core and TiO_2 as the shell) magnetic nanoparticles prepared by liquid phase desorption (LPD) to extract and purify CKs, IAA, ABA, and GAs in rice (Cai et al., 2015). Nowadays, an MSPE method using β-cyclodextrin (β-CD)-functionalized magnetic reduced graphene oxide (RGO) composite (Fe_3O_4/RGO@β-CD) as a selective adsorbent has been applied for the determination of NAA and 2- naphthoxyacetic acid (NOA). In this method, high recovery rates ranging from 90%–95% have been obtained (Li et al., 2016).

13.2.2.6 Chromatographic purification

Chromatographic purification methods, including thin layer chromatography (TLC), immunoaffinity chromatography (IAC), and high-performance liquid chromatography (HPLC) have been widely used in the purification of phytohormones. As an easy purification method with low cost, TLC was widely used in early purification practices. It is usually performed on a plate coated with a thin layer of adsorbent such as silica gel, aluminum oxide, or cellulose. The sample will be driven to ascend the plate by the mobile phase (developing solvent) at a distance according to the retention factor (R_f) of the target compound. The separated compound can be then visualized and collected under UV light. TLC has been applied for zeatin and auxin purification since the 1960s, with a recovery rate of about 60% (Sagi, 1969;

Dekhuijzen and Gevers, 1975). In addition, high-performance thin layer chromatography (HPTLC), an upgraded TLC, has been applied in phytohormonal purification (Goswami et al., 2015).

Immunoaffinity chromatography (IAC) is a highly selective purification technique with high affinity based upon antigen–antibody recognition, which enables "all-in-one" extraction, purification, and enrichment of target compound, providing high sensitivity in the final determinations using physical methods such as MS. Application of IAC in purifying phytohormones has been frequently reported with highly selective enrichment (Novák et al., 2003; Hauserová et al., 2005; Tarkowski et al., 2009). For example, an IAC method has been developed using a polyclonal ABA antibody coupled to N-hydroxysuccinimide-ester-activated agarose as the immunoaffinity gel. The reusable IAC column was created by dispensing the immunoaffinity gel into a syringe (Hradecká et al., 2007). Moreover, IAC has been employed to purify trace amounts of CKs (Hauserová et al., 2005). A novel category-specific immunoaffinity column for the natural isoprenoid CKs has been developed by using *trans*-zeatin riboside as the hapten to generate a complete antigen. This purification method allowed simultaneous analysis of eight isoprenoid CKs in Arabidopsis leaves with a wide linear range from 25 to 500 pg/g fresh weight, a detection limit of 12.5 pg/g fresh weight, and a recovery rate higher than 70% (Liang et al., 2012).

High-performance liquid chromatography (HPLC) is a common technique which is able to separate and quantify compounds in a mixture, usually using operational pressures significantly higher than ordinary liquid chromatography. Although HPLC has been a phytohormonal quantification tool since the 1970s, it has become a tool for purification rather than quantification now because of the limited sensitivity of detectors (Pool and Powell, 1972). Generally, over 80% recovery rate can be obtained for purification of phytohormone using HPLC (Pool and Powell, 1972). Recently, several two-dimensional HPLC (2D-HPLC) purification systems have been developed, which consist of two different chromatographic columns to purify the analyte. Eluent from the first-dimension column is operated under gradient conditions and the target analyte is collected in a loop by means of a computer-controlled six-port valve. The second pump then flushes this loop to the second-dimension column for further purification, after which the doubly purified target analyte can be collected for final determination. In phytohormone assays, 2D-HPLC has been explored to purify auxin and other PGSs with better purification efficiency and higher recovery rate (Dobrev et al., 2005; Han et al., 2011; Hu et al., 2013). The online 2D-HPLC has been configured for the purification of phytohormones (Zhong et al., 2014). Compared with the offline UHPLC system, this online system has shown significant advantages such as higher sensitivity, shorter analysis time, and better repeatability. The feasibility of the system was demonstrated by the direct analysis of three auxins from different plant tissues, including leaves, buds, and petals. Under the optimized conditions, the whole analysis procedure took only 7 min. All the correlation coefficients were greater than 0.9987, while the limits of detection and quantification were in the range of 0.560–0.800 ng/g and 1.80–2.60 ng/g, respectively. The recovery rates of the authentic samples ranged from 61.0%–117% (Zhong et al., 2014).

13.2.2.7 Molecularly imprinted polymers

Molecularly imprinted polymer (MIP) is a polymer produced through molecular imprinting, which leaves cavities in the polymer matrix with affinity to a chosen "template" target molecule. These MIPs have affinity for the target molecule and can be used in affinity-related applications, such as immunoassays and affinity purification. Although the first appearance of molecular imprinting dates back to the early 1930s and has been widely applied in detection of environmental analytes, IAA-MIPs with significant selectivity for IAA have been applied in phytohormonal quantification now (Kugimiya and Takeuchi, 1999). The affinity of MIPs is obviously influenced by the functional monomer as a result of electrostatic and hydrogen bonding interactions with the template molecule. It has been found that the MIP prepared with N, N-dimethylaminoethyl methacrylate showed higher affinity and selectivity for IAA than the polymer prepared with methacrylic acid (Porobić et al., 2013). Further investigations indicated that the MIP membrane prepared with 9-vinyladenine had higher permeability-selectivity (permselectivity) for IAA in comparison with the imprinted membrane made with methacrylic acid, and exhibited higher selectivity for the template molecule IAA than IBA or kinetin (Chen et al., 2006). A new purification strategy has been reported for isolation and enrichment of auxins using magnetic molecularly imprinted polymer (mag-MIP) beads, which were developed by microwave heating initiated suspension polymerization using IAA as template. The mag-MIP beads fabricated with 4-vinyl pyridine and β-CD as binary functional monomers exhibited improved recognition ability to IAA and higher specific recognition for the template than the non-imprinted polymer (mag-NIP) beads. This method achieved purification recovery rates in the range of 70.1–93.5% (Zhang et al., 2010).

13.2.2.8 Other extraction and purification methods

In addition to the above methods, more methods such as "quick easy cheap effective rugged and safe" (QuEChERS) and "dual-cloud point extraction" (dCPE) have been also explored in phytohormonal extraction and purification (Flores et al., 2011; Yin et al., 2010). Moreover, assembling multiple methods may further increase the extraction and purification efficiency; thus, the combination of two or more methods (combo-method) has been frequently applied in practice. For example, HPLC plus SPE (Hou et al., 2008; Han et al., 2012; Novák et al., 2012; Cui et al., 2015) or MIP plus SPE (Porobić et al., 2013) has become more and more popular in phytohormonal extraction and purification.

13.2.3 Derivatization

13.2.3.1 The need for derivatization

Derivatization, the transformation of a compound into a derivative product, is often needed prior to the final determination in order to increase the stability and detectability of the extracted and purified phytohormones. This approach is mainly used when phytohormones are to be characterized and quantified by physical methods such as chromatography and MS and also used for bioassay methods such as immunoassay. Generally, a specific functional group of the compound participates in the derivatization reaction and transforms the compound to a derivate with improved physicochemical properties, which can be used for the quantification or separation of the original compound. Derivatization techniques, including methylation, silylation, and others, are frequently employed in phytohormonal analysis in order to increase the stability, volatility, and other properties as required by successive determination methods.

13.2.3.2   Methylation

Methylation has been applied to acidic phytohormones, and is usually performed using chemicals such as diazomethane, 1,1-Dimethoxytrimethylamine (DMF-DMA), tetramethylammonium hydroxide (TMAH), MeI–potassium and N-methyl-N-(*tert.*-butyldimethylsilyl) trifluoroacetamide (MTBSTFA) (Schlenk et al., 1960; Birkemeyer et al., 2003). Methylation with diazomethane was reported in the quantification of jasmonic acid (Miersch et al., 1999), auxin, salicylic acid, and abscisic acid (Müller et al., 2002).

13.2.3.3   Silylation

Silylation has been widely used especially in the determination of low-volatility polar compounds which show low detection sensitivity. Thermally stable and highly volatile derivatives can be easily obtained by the silylation reaction. Various silylation reagents for the derivatization of the hydroxyl group, such as nitrogen-containing silyl ethers, trimethylsilyl ether, *bis*(trimethylsilyl)trifluoroacetamide (BSTFA), *bis*(trimethylsilyl)acetamide, and pentafluorophenylsilyl ether, have been widely used because of their fast reactions with various hydroxyl compounds at moderate conditions (Li et al., 2001). For example, trimethylsilylation of cytokinin and auxin, and *tert.*-butyldimethylsilylation of cytokinins have been reported (Palni et al., 1983; Badenoch-Jones et al., 1984; Hocart et al., 1986; reviewed in Pan et al., 2011).

13.2.3.4   Other derivatization methods

Other derivatization methods for phytohormone analysis have also been employed. Since acylation and alkylation (or arylation) can protect the —OH, —SH, and —NH groups, tri-fluoroacetylation has been used in cytokinin analysis (Ludewig et al., 1982) while alkylation with pentafluorbenzylbromide has been applied to both cytokinin (Letham et al., 1991) and auxin (Prinsen et al., 2000). Two-step procedures consisting of alkylation with diazomethane and subsequent trimethylsilylation were described for auxin (Edlund et al., 1995) and gibbberellins (Croker et al., 1994). Now more derivatization reactions have been applied in phytohormonal quantification in order to increase their stability. For example, 4-(dimethyl-amino) phenylboronic acid (4-DMAPBA) (Ding et al., 2014) and aminophenylboronic acid (Wu et al., 2013) have been used for the derivatization of BRs.

13.3 Biological methods for phytohormonal quantification

13.3.1 Traditional growth response bioassays

13.3.1.1 The value of growth response assays

Traditional growth response bioassays for phytohormones, which are based on easily measurable morphological responses of certain plant organs or tissue sections sensitive to a target phytohormone, used to be the only available tool for phytohormone determination, yet greatly contributed to the identification of the first phytohormone auxin (Went, 1935). To date, various bioassays based on different growth responses have been established, such as coleoptile curvature test and hypocotyl fresh weight increase for auxin, epidermal strip stomatal opening for ABA, and leaf epinasty for ETH. Traditional bioassays have contributed to the phytohormonal research field for more than half a century and still can occasionally be seen today especially in qualitative studies (Bose et al., 2013).

13.3.1.2 Coleoptile curvature test for auxin

The *Avena* (oat) coleoptile curvature test for auxin was established by pioneers in phytohormone determination. In this test, the coleoptile stump of a decapitated seedling will curve in darkness after a small gelatin block containing auxin is placed on one side of a straight coleoptile stump. Agar blocks can take up auxin by diffusion from plant material such as coleoptile tips or can have plant extract incorporated. Generally, the curvature angle which is positively correlated to the dose is used to quantify the auxin in gelatin (Fig. 13.2). The test has proved to be specific for PGSs of the auxin category with a limit of detection around 1.5×10^{-7} mol/L and a test efficiency of 20–30 fractions or samples per day (Went, 1926). Subsequently, the test has undergone a series of improvements and modifications. For example, a simplified curvature test which can be performed under diffuse daylight has been reported (Söding, 1952; Kaldewey and Stahl, 1964). Moreover, another much faster test able to test 30–80 fractions per day has been developed by using etiolated *Avena* coleoptiles and measuring the coleoptile based on projection and shadowgraph using photographic film (Kaldewey et al., 1968).

13.3.1.3 Hypocotyl fresh weight increase bioassay for auxin

The hypocotyl fresh weight increase test uses hypocotyl sections from etiolated cucumber seedlings and responds to IAA when incubated in a simple medium, usually containing 2 mmol/L KCl, 0.1 mmol/L $CaCl_2$, and 10 mg/L chloramphenicol. The bioassay has been found to be easy and fast for the detection of IAA (Epel et al., 1987). The growth of the etiolated hypocotyl sections is relatively specific for auxin IAA, while being insensitive to GA and CK. This easily performed test exhibits a significant response and requires only 3 h incubation, by generating a steep slope log-linear concentration-response standard curve with good accuracy.

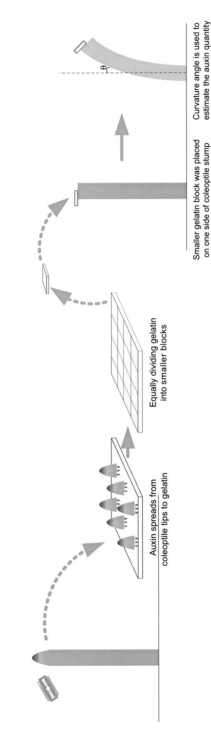

Figure 13.2 Traditional growth response bioassays. Firstly, the coleoptile tips are cut off and placed on gelatin. Then auxin in coleoptile tips spreads into gelatin. After equally dividing gelatin into smaller blocks, the gelatin block is placed on one side of the coleoptile stump which will curve in darkness because of the auxin-caused cell elongation. Finally, the angle of curvature is measured to estimate the auxin quantity.

13.3.1.4   Adventitious root formation bioassay for auxin

The adventitious root formation test for determining auxin activity was developed using the easily obtainable tropical plant *Ageratum houstonianum* (Van Raalte, 1950). Auxin activity is determined by the number of roots formed by the treated petioles of excised leaves, and the total number of roots formed in each group is compared. Compared to the adjacent petioles on the same plant, the opposite petioles of the same leaf pair are preferred because of more uniform responses. The root formation response is independent of the solution pH within the range of 4.5 to 7.0. The standard curve is established based on the linear relationship between the logarithm of the number of roots formed and the logarithm of the auxin concentration used.

13.3.1.5   Hypocotyl extension bioassay for cytokinin

The hypocotyl extension test is based on the extensions of hypocotyl segments cut from etiolated tomato seedlings which show an increase in length when incubated in deionized water in total darkness on a slowly revolving clinostat, but no increase in diameter occurs over the period (Pegg, 1962). The soybean hypocotyl extension test was developed in the 1970s (Manos and Goldthwaite, 1976; Newton et al., 1980). The researchers found that hypocotyl sections, grown under sterile conditions and cultured on a medium containing zeatin, showed a measurable response after 9–13 days. There was an approximately linear response with low variability over the concentration range 5×10^{-10} to 10^{-6} mol/L zeatin.

13.3.1.6   Leaf epinasty bioassay for ethylene

Leaf epinasty is caused by ETH at mg/L concentrations, both in vitro and in vivo (Crocker et al., 1935), and therefore provides a test to quantify the amount of ETH. The response curve for epinasty is dose dependent, similar to other ETH-mediated processes such as organ abscission and growth reduction. Levels of ETH as low as 17 µL/L can cause some curvature (Crocker et al., 1935), while high ETH concentration at 100 µL/L shows no additional effect (Lyon, 1970). Using leaf epinasty as a qualitative bioassay, the ETH released from detached leaves of potato, dandelion, rhubarb, calla, hollyhock, and leafy stalks of peony has been determined (Denny, 1935). Some extended ETH quantification studies have been performed using chopped leaves of rhubarb, lettuce, potato, tomato, onion, Virginia creeper, rose, and cotton as the test materials. The reported detection limits of the potato leaf epinasty test range from 0.5 to 10.0 ppm (Denny, 1935; Nelson and Harvey, 1935; Hall, 1951).

13.3.1.7   Epidermal strip stomatal opening bioassay for abscisic acid

Based on the effects of ABA inhibition of the light-induced opening of stomata, the epidermal strip stomatal opening test has been developed as the bioassay for ABA (Tucker and Mansfield, 1971). In this test, isolated epidermal strips of *Commelina communis* L. show progressively smaller stomatal openings when incubated in ABA solutions at concentrations ranging from 10^{-8} to 10^{-4} mol/L. The test is simple and specific because the ABA effects on epidermal strip stomatal opening are reproducible and not influenced by the presence of other phytohormones.

13.3.1.8   Lamina inclination bioassay for brassinosteroids

The rice lamina inclination test has been developed as a bioassay for BRs because they are able to strongly stimulate lamina inclination of rice at μg/L levels (Wada et al., 1981). The bioassay has been found to be highly sensitive and specific for BRs and related compounds with a linearity in the concentration range of 5×10^{-5} to 5×10^{-3} μg/mL in uniform seedlings of the rice cultivars Arborio J-l and Nipponbare (Wada et al., 1981). Moreover, the rice lamina inclination test can be used both as a micro-quantitative bioassay for BRs and as a method for detecting anti-BR compounds (Wada et al., 1984).

13.3.1.9   Amylase activity bioassay for gibberellins

The α-amylase activity test for GAs has been developed based on the release of α-amylase by the aleurone layers of barley seeds induced by GA treatment (Yomo, 1960; Varner and Chandra, 1964; Jones and Varner, 1966). The highly specific bioassay has proven to be reproducible and insensitive to other soluble residues or substances present in crude plant extracts and has been successfully applied to the quantification of gibberellins. The amount of α-amylase released from embryoless half-seeds of barley in response to GA_3 application is proportional in the range from 0.0005 μg/mL to 0.05 μg/mL. GA_1, GA_4, and GA_7 were found to be comparable to GA_3 with respect to their activity in α-amylase release (Jones and Varner, 1966).

13.3.2   Immunological methods

13.3.2.1   The principle

Immunoassay is a method based on the specific binding between antibody and its corresponding antigen(s). Immunological methods, including enzyme-linked immunosorbent assay (ELISA) and radioimmunoassay (RIA), have been introduced for phytohormonal analysis since the late 1960s (Fuchs and Fuchs, 1969) and have widely served in the quantification of major phytohormone categories (De Diego et al., 2013; Guan et al., 2015; Wang et al., 2015).

13.3.2.2   Enzyme-linked immunosorbent assay

The ELISA system uses the specificity of antibodies to identify a substance, combined with the sensitivity of a simple enzyme assay involving a color change. Typically this is carried out using a solid phase enzyme immunoassay to detect the presence of a substance, usually an antigen, in a liquid sample. Antigens from the sample are attached to a surface, and then a specific antibody is applied over the surface so it can bind to the antigen. This antibody is linked to an enzyme which catalyzes a chromogenic reaction for accurate signal quantification. Since the antigen can be labeled with easily detectable enzymes, several types of ELISA have been established, including direct, indirect, solid phase, and other types (Weiler, 1982; Cahill and Ward, 1989; Blintsov and Gusakovskaya, 2006). The sensitivity and accuracy of enzyme immunoassays depend upon the amplification of the final enzyme reaction. ELISA has been employed to detect and quantify several PGSs since the late 1960s (Fuchs and Fuchs, 1969), yielding good sensitivity for major phytohormones in crude plant extracts. Now, different phytohormonal ELISA kits have become commercially available (Tarkowski et al., 2009; Guan et al., 2015), and the lowest detection limit is around the picogram level (Su et al., 2013; Wang et al., 2015).

13.3.2.3   Radioimmunoassay

In RIA, antigen is labeled with radioactive tracers such as tritium, and reacted with a specific antibody. Addition of a sample containing the analyte requiring quantification leads to competition between labeled and unlabeled analyte. The amount of analyte is quantified by measuring the radioactivity in the antigen–antibody complex, following precipitation with saturated ammonium sulfate. RIA is a very sensitive in vitro assay because it combines the unique specificity of the antigen–antibody reaction with the sensitivity of determination of trace amounts of radioactivity. For both low- and high-molecular weight analytes, it usually provides detection limits from 10^{-12} to 10^{-15} mol/L. RIA has been employed to quantify ABA, IAA, CK, and GA in crude plant extracts with nmol/L detection limits (Pengelly and Meins, 1977; Weiler and Ziegler, 1981; Tarkowski et al., 2009). However, RIA requires instruments to measure radioactivity, as well as training and a licence for researchers to deal with isotopes.

13.3.3   Biosensor methods

13.3.3.1   Biosensors explained

Biosensors are devices that record a physical, chemical, or biological change and convert that into a measurable signal. A biosensor usually consists of a biologically sensitive unit, a transducer unit, and a signal reader unit (Fig. 13.3). Phytohormonal biosensors, in which phytohormone-sensitive units are used, have been employed to detect phytohormones for more than a decade (Kugimiya and Takeuchi, 1999; Li et al., 2002). Regarding the term "biosensor" in phytohormonal quantification, the phytohormonal "degrons" (Nishimura et al., 2009) are sometimes also referred to as genetically encoded biosensors, but they are based on totally different principles from the above biosensor definition and thus will be reviewed in the "molecular methods" section.

Within the past decade, several types of phytohormonal biosensors with different sensing principles have been constructed. Among them, quartz crystal microbalance (QCM) biosensors measure a frequency shift (Li et al., 2002); electrochemical biosensors measure current (amperometric) or impedance changes (Wang et al., 2006; Li et al., 2003, 2010); optical biosensors measure luminescence (Liu et al., 2010; Hun et al., 2012; Li et al., 2012); the surface plasmon resonance (SPR) (Wei et al., 2011)

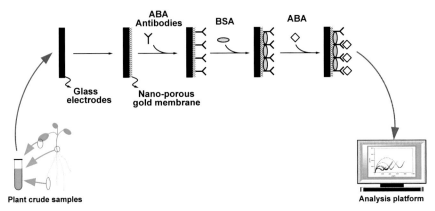

Figure 13.3 An electrochemical biosensor configuration. The glass electrode modified with nano-porous gold absorbs Abcisic acid (ABA) antibodies and the remaining active surface is blocked by bovine-serum-albumin (BSA). The added ABA will only bind with its antibody and then causes a change in electrical impedance. Thus, the analysis platform can record the impedance change signals and compute the content of ABA.

or the Förster resonance energy transfer (FRET) (Jones et al., 2014) measures the fluorescence thus produced; microbial biosensors measure microorganism-catalyzed products (Defraia et al., 2008; Shimojo et al., 2011). For the phytohormone-sensitive units, phytohormonal antibodies were often used and other sensitive molecules such as phytohormone binding proteins (Nikolelis et al., 2008), MIPs (Kugimiya and Takeuchi, 1999), and DNA aptamers were also explored (Qi et al., 2013). Materials such as self-assembled membrane (Li et al., 2010), nanogold particles (Hun et al., 2012), graphene (Wen et al., 2012), and many others have been used for the fabrication of the sensing interface.

Up to now, biosensors designed for the detection of IAA (Liu et al., 2010; Zhou et al., 2013), ABA (Li et al., 2008), CK (Kowalska et al., 2011), GA (Li et al., 2012), JA (Gan et al., 2010), Me-JA (Sun et al., 2015), SA (Defraia et al., 2008), ETH (Kathirvelan and Vijayarahavan, 2014), as well as PGSs such as 2,4-D (Shimojo et al., 2011), NAA (Nikolelis et al., 2008), and NO (Wen et al., 2012), have been reported, and some of them presented relatively high specificity and sensitivity by achieving sub-nanogram level limits of detection (LOD) (Li et al., 2008; Hun et al., 2012). In addition, phytohormonal biosensors have shown the advantages of detecting gaseous phytohormones (Wen et al., 2012; Kathirvelan and Vijayarahavan, 2014) and monitoring tissue phytohormonal dynamics based on their easy sample preparation and real-time detection potential (Jones et al., 2014).

13.3.3.2 Electrochemical biosensors

Electrochemical biosensors combine the sensitivity of electrochemical transducers, as indicated by low detection limits, with the high specificity of biological recognition processes by converting a biological event into an electronic signal.

These devices contain a biological recognition element that selectively reacts with the target analyte and produces an electrical signal that is related to analyte concentration. Electrochemical biosensors can be divided into two main categories based on the nature of the biological recognition process, that is, biocatalytic and affinitive sensors (Wiley and Hoboken, 2006).

The direct electrochemical biosensor for phytohormones utilizes the bioactive substances, such as enzyme, antibody, receptor, and binding protein as the recognition element. The recognition molecules are immobilized on an electrode surface through a physical or chemical method which is similar to enzyme-immobilization and immune-immobilization techniques, and thus can also be classified into enzyme electrode and immuno-electrode methods. In practice, a phytohormone at µg/mL levels can be quantified through direct electrochemical biosensors. The direct electrochemical sensors have been developed to quickly detect several phytohormones (IAA, ABA, JA, and SA) based on their electrochemical activities. This biosensor was firstly applied in IAA and ABA determination (Hernández et al., 1996), and then more advanced direct electrochemical biosensors have been developed with the reported LODs at µg/mL level because of limited electrochemical activities (Li et al., 2014; Wang et al., 2015). Now more direct electrochemical biosensors are being applied in trace phytohormonal determination because of the progresses of nanophase materials and signal amplification techniques. Nowadays, the LOD of direct electrochemical biosensors for some phytohormones has reached ng/mL levels, showing good potential in phytohormonal quantification (Yin et al., 2010; Li et al., 2010; Wang et al., 2009).

The photoelectrochemical (PEC) biosensor is a novel and promising analytical method based on electron transfer among analyte, photoactive species, and electrode upon photoirradiation (Sun et al., 2013). Benefiting from using photoirradiation as the excitation source coupled with electrochemical detection, high sensitivity can be obtained. Furthermore, the inexpensive photoelectric devices and rapid high-throughput assay procedure makes PEC analysis more attractive (Long et al., 2011; Zhang and Guo, 2012). As a key unit of PEC sensors, a variety of photoactive materials have been utilized, including TiO_2, ZnO, CdS, and CdSe. For example, novel mercaptopropionic acid (MPA)-CdS/RGO nanocomposites were synthesized through a facile solventhermal procedure by assembling MPA-decorated CdS nanoparticles onto the RGO layers. The MPA-CdS/RGO nanocomposites with good crystalline structure produced enhanced photocurrent response benefiting from the synergistic effect of RGO and MPA-functionalized CdS nanoparticles. MPA improved the uniform distribution of CdS nanoparticles onto the RGO sheets and acted as a bridge to immobilize the biosensing unit. Due to improved PEC performance of the MPA-CdS/RGO nanocomposites, a sensitive PEC immunosensor was fabricated for IAA analysis. The proposed photoelectrochemical immunosensor for IAA operated in a linear range from 0.1 to 1000 ng/mL with 0.05 ng/mL LOD based on the decline of the photocurrent, dependent on the IAA concentration (Sun et al., 2014). Therefore, PEC sensors show the potential to offer a highly sensitive, reproducible, and specific resolution for trace phytohormonal quantification.

At present, few applications in phytohormonal determination based on PEC sensors have been reported (Sun et al., 2014). Thus, systemic theory research, new material application, and new signal recognition models are urgently needed in future studies as most phytohormones do not possess high electro-activities. Therefore, PEC sensors coupled with MIPs can compensate for this shortage: a MIP electrochemical biosensor containing a selective MIP electrode for a certain phytohormone has shown a lower LOD at pmol/L level (Zhang et al., 2014).

13.3.3.3 Piezoelectric biosensors or quartz crystal microbalance

A piezoelectric quartz crystal resonator is a precisely cut slab from a natural or synthetic crystal of quartz. A QCM consists of a thin quartz disk with electrodes placed on it. As the QCM is piezoelectric, when a crystal is dipped into a solution, the oscillating frequency depends on the solvent used. QCM systems have been designed to reliably measure mass changes up to 100 mg, whereas the minimum detectable mass change is typically $1 \, ng \, cm^2$ (O'sullivan and Guilbault, 1999). QCM systems have traditionally been widely used in analysis related to mass measurement. Various approaches can be taken once a suitable recognition layer has been coated on the crystal, such as to use a single sensor or an array of sensors with different coatings combined with pattern recognition. The options are many but the primary goal is to find a suitable coating layer and a method of reproducible application.

The piezoelectric crystal was first employed to develop and commercialize a piezoelectric detector, which could detect moisture of 0.1 ppm and hydrocarbons such as xylene to 1 ppm (King, 1964). An area which has captured a lot of attention has been the use of antibodies as the crystal coating, offering inherent bioselectivity. These antibody-coated crystals are referred to as QCM-based immunosensors. The first piezoelectric immunosensor was developed several decades ago (Shons et al., 1972). A plethora of methods for antibody immobilization on the crystal surface has been investigated. Adsorption of antibodies is a method widely used in the detection of various analytes including DNA and RNA (Sakai et al., 1995; Fawcett et al., 1988). A novel piezoelectric immunosensor was developed for the determination of β-indole acetic acid (IAA) in dilute solutions. The detection is based on competitive immunoreaction between a hapten (i.e., IAA) and an antigen (i.e., IAA-bovine serum albumin) bound to an anti-IAA antibody, immobilized on a QCM. The frequency change of the sensor caused by antigen binding is linearly related to the logarithm of the concentration of IAA in the range of 0.5 ng/mL to 5.0 μg/mL (Li et al., 2002).

13.3.3.4 Förster resonance energy transfer biosensors

FRET is an energy transfer mechanism between two fluorophores, and is widely used in life sciences for determining small molecular distances (Roy et al., 2008). This energy transfer is crucially dependent on the separation and orientation of two coupled chromophores and is mediated by near-field dipole–dipole coupling (Lakowicz, 2006). FRET tools were developed to measure the levels of ABA within defined compartments of individual cells in living plants and in real time (Jones et al., 2014; Waadt et al., 2014). For example, to directly monitor the cellular ABA dynamics in response to environmental cues, ABA-specific optogenetic reporters (ABA-inducible genetically encoded fluorescent proteins) were employed to instantaneously convert the phytohormone-triggered interaction of ABA receptors with PP2C-type phosphatases to emit a FRET signal in response to ABA. This FRET sensor with ABA affinities in the range of 100–600 nmol/L was used to map tissue ABA dynamics with high spatial and temporal resolution (Waadt et al., 2014).

13.3.3.5 Surface plasmon resonance and other optical biosensors

Optical biosensors are based on the detection mechanism of optical signals produced by the biological recognition with the analyte either through biocatalytic or bio-affinitive reactions. The surface plasmon resonance or surface plasma resonance (SPR) biosensor is based on the exploitation of evanescent waves and is one of the more widely used optical biosensors (Liedberg et al., 1983). The technique can be used to study the interaction between immobilized receptors and analytes in real time without labeling the analyte. Observed binding can be interpreted in different ways to provide information on analyte concentration or on the specificity, kinetics, and affinity of the interaction. SPR is responsible for a dip in reflectance at a specific wavelength, resulting from the absorption of optical energy in the metal, typically gold or silver. The surface wave is highly sensitive to changes in the refractive index near the metal surface within the range of the surface plasmon field. Such a change may result in a shift in the resonant wavelength of the incident light (Homola et al., 1995), a change in the intensity of the reflected light (Manuel et al., 1993), or a change in the resonant angle of the incident light (Liedberg et al., 1993). The magnitude of such shifts is quantitatively related to the magnitude of the change in the refractive index of the medium in contact with the metal surface.

SPR biosensors have been used for the detection of different sample types (gaseous, liquid, or solid). An optical SPR immunosensor was reported for direct, label-free analyte detection (Gobi et al., 2005). The analyte was conjugated with bovine serum albumin (BSA) protein, and the conjugate was immobilized onto a thin gold-film SPR sensor chip mounted in a continuous flow system by simple physical adsorption. Functionalization of sensor surface by physical adsorption of analyte-BSA conjugate and subsequent immunoreaction of surface-bound analyte-BSA with a monoclonal anti-conjugate antibody were studied by the SPR angle interrogation method. The quantification of free analyte was based on competitive immunoassay, which was performed by co-injecting the sample and the monoclonal antibody over the physically absorbed thin gold-film SPR chip and measuring the amount of specifically bound monoclonal antibodies. Such an SPR system has been shown to be capable of detecting concentrations of 2,4-dichlorophenoxyacetic acid (2,4-D) ranging from 0.5 ng/mL to 1 μg/mL with an LOD at 0.5 ng/mL and a response time of 20 min (Gobi et al., 2005). Moreover, SPR biosensors have been applied in the identification and quantification of IAA in different plant samples. To achieve high selectivity, an MIP monolayer (MIM) was prepared from alkanethiols self-assembled on an SPR sensor chip with pre-adsorbed template IAA. Satisfactory results have been obtained using this MIM-SPR sensor to determine IAA concentrations in different tissue extracts. The LODs obtained ranged from 0.20 to 0.32 pmol/L (Wei et al., 2011).

Furthermore, the SPR biosensors were used to identify and detect bioactive GAs accurately and rapidly based on bioactive GA-dependent interaction between AtGID1a and DELLA protein (Zhao et al., 2015).

In addition to SPR biosensors, other optical biosensors have been also developed for the detection of phytohormones. For example, a colorimetric luminescence biosensor for IAA was fabricated by using green emissive quantum dots (QDs) of cadmium telluride (CdTe QDs) as a background layer and a red emissive europium chelate as a specific sensing layer coated on the surface of glass slides. The luminescence response of the sensor is given by the dramatic changes in emission colors from green to red at different IAA concentrations. This approach provides a rapid and sensitive method for the detection of IAA (Liu et al., 2010). Another highly sensitive chemiluminescence (CL) biosensor for IAA has been developed by using G-rich DNA labeled gold nanoparticles (AuNPs) as a CL probe, which offers a significant amplification for the detection of target molecules. The IAA antibody was immobilized on carboxyl groups terminated magnetic beads (MBs). In the presence of IAA, antibody-labeled AuNPs are captured by these antibody-functionalized MBs. The DNA on the AuNPs is released by a ligand exchange process induced by the addition of DTT. The released DNA hybridized with the captured DNA on MBs, and then the DNA-MBs complex hybridized to AuNPs and probe DNA. The CL signal is obtained via the instantaneous derivatization reaction between specific CL reagents, such as 3,4,5-trimethoxyl-phenylglyoxal, and the G-rich DNA on the AuNP's CL probe. IAA can be detected in the concentration range of 0.02–30 ng/mL, and LOD is 0.01 ng/mL (Hun et al., 2012).

13.3.4   Molecular methods

13.3.4.1   The principles

Molecular methods employed in phytohormonal quantification include assays based on phytohormonal action and phytohormonal signal transduction pathways, as well as other assays using molecular techniques. Molecular methods have been employed in phytohormonal assays since the early 1990s based on the inducible promoter of *Pg5* (the promoter of the gene 5 of the *Agrobacterium tumefaciens* octopine plasmid pTiAch5) (Boerjan et al., 1992). Subsequently, other phytohormone-inducible promoter-reporter systems have been explored to monitor phytohormonal distribution in plants (Sabatini et al., 1999; Ulmasov et al., 1997). Since phytohormone perception often triggers the degradation of the signaling repressor proteins, the phytohormone-responsive degradation of repressor proteins, referred to as phytohormonal degrons, have been explored to monitor cellular phytohormone levels (Brunoud et al., 2012). Now, some nucleic acid-based methods have been established for phytohormonal quantification, such as the terminal protection of small-molecule-linked DNA (Wu et al., 2009) based on the stereo-hindrance effect caused by molecule recognition (Hu et al., 2014). It is worth emphasizing that molecular methods have quickly emerged as new phytohormonal quantification tools with in vivo, dynamic, and cell-specific potential.

13.3.4.2 Phytohormone-inducible promoter-reporter systems

Phytohormone-inducible promoters driving a certain reporter are powerful tools for relative quantification of phytohormones although the output signals are indirect, rather than the actual phytohormone content in the cell (Fig. 13.4). The auxin-inducible promoter *DR5* consists of seven to nine copies of an auxin response element (AuxRE) fused upstream of a minimal promoter derived from the cauliflower mosaic virus 35S promoter (Ulmasov et al., 1997; Benková et al., 2003), and has been frequently applied in auxin visualization and quantification. The main locations of auxin distribution are visualized in different plant tissues using DR5rev:GFP and DR5:GUS reporter gene systems (Friml et al., 2002, 2003). Furthermore, through modeling the relation of fluorescence intensity and IAA concentration in sorted protoplast cells from 14 GFP-expressing Arabidopsis lines covering different types of cells in the Arabidopsis root apex, an IAA distribution map has been constructed (Petersson et al., 2009).

Similarly, more phytohormone-inducible promoter-reporter systems have been constructed and applied in phytohormonal quantification to monitor the tissue phytohormonal levels, including a transcriptional response marker TCS::GFP (two-component output sensor) for CKs (Müller and Sheen, 2008), and two ABA-responsive promoters driving reporters, namely *pAtHB6T::LUC* (Christmann et al., 2005) and *ProRAB18::GFP* (Duan et al., 2013). The transcriptional output of a given signaling pathway in combination with a phytohormone-inducible promoter-reporter system can quantitatively monitor a given signal in a cell (Wells et al., 2013).

13.3.4.3 Phytohormonal degrons

A phytohormonal degron, which generally contains a constitutive promoter and a transcriptional inhibitor fused with a reporter, is based on the phytohormone-triggered proteasome-dependent degradation of transcriptional inhibitors. Recently, phytohormonal

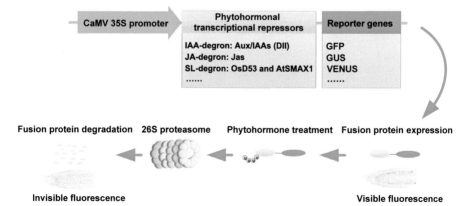

Figure 13.4 Visualized localization by a phytohormonal degron. Recombinant plasmids are constructed and the expression of phytohormonal transcriptional repressor (or its functional domain) fused fluorescence reporter is driven by a 35S promoter. The visible fluorescence will be detected in lower phytohormone level cells. After treating with phytohormone, the fusion proteins will be degraded by 26S proteasome and the fluorescence will disappear.

degrons have been used successfully in detecting different signaling molecules in living cells. A successful example is the auxin degron system. In the auxin signaling pathway, auxin perception triggers rapid degradation of the Aux/IAA transcriptional repressors (Chapman and Estelle, 2009; Qu et al., 2017), and the Aux/IAA proteins also function as auxin co-receptors by binding to Transport Inhibitor Response1 (TIR1)/Auxin Signaling F-Box (AFB) proteins (Dharmasiri et al., 2005; Kepinski and Leyser, 2005). Based on the above mechanism, an auxin-inducible degron named DII-VENUS (domain II of Auxin resistant/indole-3-acetic acid inducible fused VENUS) has been developed for screening auxin distribution in transgenic plants (Brunoud et al., 2012). In this auxin degron system, a chimeric gene was constructed which encoded the auxin-binding domain of IAA28 fused in frame to the VENUS protein and expressed using the 35S promoter. The use of the fast-maturating YFP variant VENUS (with faster oxidation of chromophore) was essential because Aux/IAAs exhibit rapid turnover and have proved to be very difficult to visualize using GFP tagging (Zhou et al., 2011). Extensive characterization of the DII-VENUS degron system indicates that fluorescence levels are inversely correlated to endogenous auxin levels (Brunoud et al., 2012). The DII-VENUS degron system could be used not only to map auxin distribution with cellular resolution in the root tip but also to follow auxin dynamics in several independent developmental processes. Through computer modeling for fluorescence intensity and cellular IAA concentration with DII-VENUS, a system for the analysis of auxin transport in the Arabidopsis root apex was provided (Band et al., 2014). In addition, an improved auxin-inducible degron system also worked in non-plant eukaryotic cells although other eukaryotes lack the auxin response but share the SCF degradation pathway. In this system, the fluorescence intensity showed negative correlation with auxin concentration in the range of 0.0–20.0 μmol/L and treatment time in the range of 0–90 min (Nishimura et al., 2009).

Using similar approaches, researchers have exploited knowledge gained about how other signaling molecules are perceived in plant cells. In addition to auxin, GAs, Jas, and SLs employ similar signaling pathways in the degradation of transcriptional regulators (Shan et al., 2012; Zhou et al., 2013; Soundappan et al., 2015; Liang et al., 2016; Smith et al., 2017). Visualizing the degradation of these transcriptional regulators thus provides a measure to estimate the cellular concentration and activity of the corresponding phytohormones. Since GA perception by the GID1 receptors rapidly leads to the degradation of the DELLA transcriptional repressors, A gene encoding a GFP translational fusion to the DELLA protein (GFP-RGA) expressed under the RGA endogenous promoter has been used to monitor cellular GA changes in Arabidopsis (Silverstone et al., 2001; Achard et al., 2006; reviewed in Sun, 2011). Recently, a JA degron Jas9-VENUS has been constructed and used to quantify JA dynamics in response to stress with high spatiotemporal sensitivity (Larrieu et al., 2015). The abundance of Jas9-VENUS is dependent on bioactive JA isoforms, the COI1 co-receptor, and a functional Jas motif and proteasome activity (Zhai et al., 2017). The results also demonstrate the value of developing quantitative degron systems such as Jas9-VENUS to provide high-resolution spatiotemporal data about phytohormone distribution in response to plant abiotic and biotic stresses.

13.3.4.4 Nucleic acid terminal protection

Nucleic acid terminal protection is based on the protection of a nucleic acid from degradation or extension by tethering or modification of the nucleic acid terminus with a small molecule. Nucleic acid terminal protection provides a promising strategy for specific, sensitive detection of the binding events between small molecules and their protein targets, and has been applied in quantification of small molecules (Fig. 13.5). When a single-stranded DNA (ssDNA) is terminally tethered to a small molecule, which in turn is bound to a protein, it can be protected from degradation by exonuclease I (*Exo* I). Presumably, the terminal protection is attributed to steric hindrance of the bound protein molecule, which prevents *Exo* I from approaching and cleaving the phosphodiester bond adjacent to the 3' terminus. The nucleic acid terminal protection approach was first applied in the determination of the interaction between small molecules and proteins (folate and its receptor FR) (Wu et al., 2009; Cao et al., 2012). These small molecule analysis platforms based on nucleic acid terminal protection can also be applied in phytohormonal

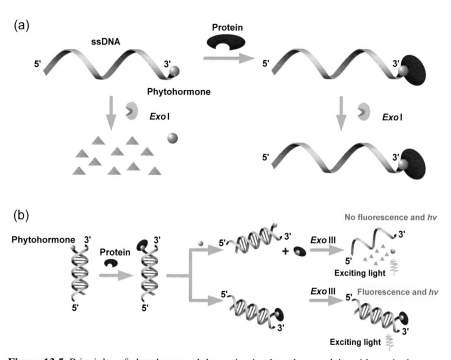

Figure 13.5 Principles of phytohormonal determination based on nucleic acid terminal protection. (a) ssDNA terminal protection. ssDNA bound with a certain phytohormone in 3' terminal can be fragmented by *Exo* I, but if the 3' terminal binds a protein or antibody, *Exo* I will not cut the ssDNA. (b) dsDNA terminal protection. 5' terminals of dsDNA are modified with phytohormone and then phytohormone affinitive protein or antibody can bind to 5' terminals and protect the terminal from *Exo* III degradation. If the phytohormonal affinitive protein is competitively bound by added phytohormone molecules, dsDNA will be degraded by *Exo* III (Wang et al., 2013; Hu et al., 2014).

quantification. Based on a non-nuclease-assisted terminal protection assay, the quantitative detection of IAA was accomplished on a platform of quartz-crystal-microbalance (Fei et al., 2011). In another study using a rolling circle amplification (RCA)-integrated strategy, based on the principle of terminal protection of small molecule-linked DNA, SA was sensitively detected with a detection limit as low as 0.4 pmol/L with high selectivity (Wang et al., 2013; Hu et al., 2014).

13.4   Biological methods for phytohormonal localization and profiling

13.4.1   Development of methods for phytohormone localization

Phytohormonal localization can be traced back to the late 1960s when early radioautographic analysis, based on radioactively labeled phytohormones as the tracers, was employed (Liao and Hamilton, 1966). Subsequently, other techniques have also been developed in phytohormonal localization, including immunocytochemical methods and the immunocolloidal gold methods. Both are based on the specificity of antigen–antibody recognition including the binding between phytohormone-bound primary antibodies with secondary antibodies labeled by observable marker enzymes, fluorescent molecules, or colloidal gold particles (Driss-Ecole and Perbal, 1987). These methods have been widely used for describing the in situ localization of several phytohormones (Moncaleán et al., 2001; Nakayama et al., 2002; Dong et al., 2012; De Diego et al., 2013; Pérez et al., 2015). Now, new methods have been introduced for phytohormonal localization. For example, QD, a nanocrystal small enough to exhibit quantum mechanical properties and fluorescence excitation, has been explored as an inorganic fluorescence label in the localization of phytohormone binding proteins and receptors (Liu et al., 2014). Besides the use in phytohormonal quantification, phytohormonal degrons can be also used to visualize the phytohormonal localization in plant tissues, as can the phytohormone-inducible promoter-reporter systems, as reviewed above (Brunoud et al., 2012).

13.4.2   Histochemical methods

13.4.2.1   Immunocytochemical methods

The immunocytological techniques are based on the principles of cytology with the specificity of the immunological reaction to reveal the localization and the relative abundance of cellular antigenic constituents (Avrameas, 1982). Immunocytochemical techniques together with low-temperature histological techniques have been employed for in situ localization of phytohormones since the late 1980s (Driss-Ecole and Perbal, 1987). For example, in situ immunocytochemical localization methods have been established for auxin (Dewitte and Van Onckelen, 2001) and ABA (Peng et al., 2006). The localization of CKs in corn root tips was investigated using antibodies or antibody fragments directed against dihydrozeatin riboside and labeled with rhodamine or colloidal gold (Zavala and Brandon, 1983). Immunolocalization of IAA and ABA was carried out in apical needles and root apex of radiate pine (*Pinus radiata*) (De Diego et al., 2013). Immunolocalization of free ABA was also performed on the axis of embryos at different developmental stages (Pérez et al., 2015).

The localization of phytohormones is difficult because they occur at low concentrations and they are highly soluble in solvents used in preparing samples for microscopy. Generally, factors such as phytohormone diffusion and detection specificity, differences in organelle and cell volumes, permeability for the antibodies, and variations in the efficiency of fixation must be taken into account when carrying out immunocytochemical localization.

13.4.2.2 Immunocolloidal gold method

The immune colloidal gold technique is an immune diagnostic technique based on labeling the second antibody with colloidal gold rather than with other conventional markers that are widely applied for the detection of antibody and antigen (such as luciferin, radioactive isotope, and enzyme). Antibodies can in principle be labeled with several metals but electron-dense colloidal gold is ideal for electron microscopic observation. Particles of colloidal gold carry a net negative charge in water, and their stability is maintained by electrostatic repulsion. However the addition of strong electrolytes alters the electrochemical environment and allows them to adhere and flocculate. This can be prevented by adding a "protective" colloid, such as protein. The proteins are adsorbed onto particles of colloidal gold, and the gold is stabilized against subsequent flocculation by electrolytes. It is therefore reasonable to stabilize the gold sol with antibodies, which can hence be observed through electron microscopy. As an immuno-method, the immunocolloidal gold method is based on specific antibody–antigen binding. Since colloidal gold is commercially available, it is quite convenient to detect bound primary antibodies in a two-step protocol in which the binding of the primary antibody is revealed by an electron microscope. The immunocolloidal gold method has been widely applied to localize different phytohormones since the 1990s (Ohmiya et al., 1990).

13.4.2.3 In vivo cytochemical imaging

Along with the development of fluoresceins and confocal laser scanning microscope (CLSM) techniques, phytohormones can now be accurately imaged in vivo. For example, the fluorescent dye fluorescein has been used to visualize GA distribution. The green fluorescein (FI) molecule can form a red complex with GA_3 (FI-GA_3); thus, GA distribution can be visualized by labeling the GA with the fluorescein tag. This method is capable of tissue and subcellular resolution and has been successfully used in GA localization (Shani et al., 2013). Similarly, auxin can form a conjugate with two specific fluorescent-dye tracers, fluorescein isothiocyanate (FITC) and rhodamine isothiocyanate (RITC), which allows direct visualization of auxin localization and transport in plant tissues (Sokołowska et al., 2014).

13.4.3 Molecular probes for maximum, minimum, and flux

Phytohormonal maximum or minimum refers to local accumulation or depletion of a certain phytohormone in a cell group compared to their neighboring cells. Local concentration gradients of phytohormones regulate the positioning of organ primordia and stem cell niches and direct cell division and differentiation. It has long been believed that auxin acts through the development of gradients and concentration maxima within developing tissues, and these gradients provide positional cues for diverse developmental processes (Bhalerao and Bennett, 2003; Benjamins and Scheres, 2008; Ikeda et al., 2009). Phytohormonal flux refers to the direction and quantity of phytohormonal transport in tissues. In the case of auxin, IAA fluxes mainly depend on the PIN-FORMED (PIN) family of auxin efflux transporters (reviewed in Adamowski and Friml, 2015) and the AUX/LAX family of influx carriers (reviewed in Swarup and Péret, 2012). The auxin fluxes driven by these carriers will generate auxin asymmetries that regulate cell differentiation and division (reviewed in Schaller et al., 2015). Physiological and molecular evidence for long-distance transport has been also obtained for ABA, CKs, and SLs (Boursiac et al., 2013; Zhang et al., 2014; Kohlen et al., 2011).

Generally, phytohormonal maximum, minimum, and fluxes can be visualized and quantified in relative terms by related molecular probes, such as phytohormonal signaling repressors and transporters fused to reporters (Grieneisen et al., 2007; Vernoux et al., 2011; Brunoud et al., 2012). Systems such as DII-VENUS, Jas9-VENUS, and D53-GFP based on transcriptional repressors have been used as molecular probes for in vivo phytohormonal maximum and minimum visualization (as described above). For auxin fluxes, auxin transporter proteins such as PINs (responsible for auxin efflux) (Heisler et al., 2005), AUXs (responsible for auxin influx) (Dharmasiri et al., 2006), and PILSs (responsible for auxin transport across the endoplasmic reticulum) (Barbez et al., 2012) have been also used.

13.4.4 Phytohormonal profiling-based metabolomics

Phytohormonal profiling, establishing a chemical fingerprint or profile through the simultaneous determination of free phytohormones and phytohormonal metabolites, is an indispensable research tool for the study of biosynthesis, metabolism, homeostasis, transport, and cross talk of phytohormones (Anegawa et al., 2015). Phytohormones interact by mutually affecting (Tarkowská et al., 2014). Metabolomics methods are mainly based on tandem MS and data modeling (Castillo et al., 2014; Anegawa et al., 2015). Starting from the late 1990s, tandem MS has been introduced into the analysis of multiple phytohormones (Laureys et al., 1998; Castillo et al., 2014), and has become a mainstream tool for phytohormonal profiling (Chiwocha et al., 2003; Böttcher et al., 2007; Lin et al., 2015). For example, more than 40 phytohormones, conjugates, and derivatives in different parts of a single rice plant have been simultaneously identified and quantified by using a tandem MS-based method with a sub-fmol LOD (Kojima et al., 2009); ABA and over 20 CKs have been identified by tandem MS in fungi

(Morrison et al., 2015). Homeostasis of ABA and IAA has also been monitored using a tandem MS system (Boycheva et al., 2015). Concurrently, nuclear magnetic resonance (NMR) spectroscopy, which has an advantage of requiring less analyte separation, has been applied in phytohormonal metabolomics to identify the compounds involved (Glauser et al., 2010; Kim et al., 2013), although NMR is less sensitive than MS-based methods. In addition to the above detection methods, other supporting techniques for phytohormonal profiling, such as the statistical methods for large-scale NMR and MS data (Glauser et al., 2010), the separation techniques prior to the detection methods have also been examined (Chen et al., 2011; Van Meulebroek et al., 2012).

13.5   Summary points

- Since the first *Avena* coleoptile curvature bioassay was established in the 1920s, phytohormonal assays have undergone a long march of several technological generations, including traditional bioassays, immunoassays, GC-MS, LC-MS, and tandem mass spectrometry.
- Biological methods for phytohormonal quantification include traditional growth response bioassays, immunological methods, biosensor methods, and molecular methods.
- Traditional bioassays and immunoassays dominated phytohormonal quantification from the 1920s to the 1980s, but biological methods have been replaced by MS as the mainstream quantification methods in the past three decades.
- Single-cell visualization and dynamics still remains a huge challenge today.
- Molecular methods such as phytohormone inducible promoter-reporter systems and phytohormonal degrons have shown the potential for single-cell quantification. Phytohormonal biosensors offer the ability for fast determination while nucleic acid terminal protection methods provide the possibility for improvement of sensitivity.
- Phytohormonal quantification based on biological principles has regained the attention of plant biologists and should become increasingly important in future research practice.

13.6   Future perspectives

- To further improve quantification sensitivity with single-cell quantification potential, through the exploration of new biological methods or combinations of biological methods with other methods, in order to reveal single-cell phytohormonal dynamics.
- To develop in situ, in vivo spatiotemporal quantification and localization methods based on molecular methods such as phytohormone-inducible promoter-reporter systems and phytohormonal degrons.
- To establish profiling methods at the whole phytohormone level to reveal the interactions between different phytohormones based on the approach of metabolomics, phytohormonal biosensor chips, and modeling.
- To develop other supporting technologies as required by single-cell phytohormonal dynamics studies, such as new sampling methods able to pick up specific target cells and purification methods such as 2D-HPLC capable of online target capture.

Abbreviations

2D-HPLC	two-dimensional HPLC
4-DMAPBA	4-dimethyl-amino phenylboronic acid enzyme
ABA	abscisic acid
AuNPs	gold nanoparticles
AuxRE	auxin response element
BRs	brassinosteroids
BSTFA	bistrimethylsilyltrifluoroacetamide
CdTe QDs	quantum dots of cadmium telluride
CL	chemiluminescence
CLSM	confocal laser scanning microscope
CPE	cloud point extraction
CTKs	cytokinins
dCPE	dual-cloud point extraction
DLLME	dispersive liquid–liquid microextraction
DLLME-SFO	dispersive liquid–liquid microextraction based on solidification of a floating organic drop
DMF-DMA	diazomethane, 1,1-Dimethoxytrimethylamine
ELISA	enzyme-linked immunosorbent assay
ETH	ethylene
FACS	fluorescence activated cell sorter
FITC	fluorescein isothiocyanate
FR	folate receptor
FRET	Förster resonance energy transfer
GAs	gibberellins
HF–LLLME	hollow fiber-based liquid–liquid–liquid microextraction
HLLE	homogeneous liquid–liquid extraction
HPLC	high-performance liquid chromatography
HPTLC	high-performance thin layer chromatography
IAA	indole-3-acetic acid
IAC	immunoaffinity chromatography
JA	jasmonates
LCM	laser capture microdissection
LLE	liquid–liquid extraction
LOD	limit of detection
LPD	liquid-phase desorption
MAE	microwave-assisted extraction
mag-MIP	magnetic molecularly imprinted polymer
MBs	magnetic beads
MIP	molecular imprinted polymer
MSPE	magnetic solid phase extraction
MTBSTFA	N-methyl -N-tert–butyldimethylsilyltrifluoroacetamide
NMR	nuclear magnetic resonance
NO	nitric oxide
PEC	photoelectrochemical
PGR	plant growth regulator

PGS	plant growth substance
QCM	quartz crystal microbalance
QD	quantum dot
QuEChERS	quick easy cheap effective rugged and safe
RCA	rolling circle amplification
RIA	radioimmunoassay
SA	salicylic acid
SDME	single drop microextraction
SL	strigolactones
SPE	solid phase extraction
SPME	solid phase microextraction
SPR	surface plasmon resonance
ssDNA	single-stranded DNA
TLC	Thin layer chromatography
TMAH	tetramethylammonium hydroxide
UAE	ultrasound-assisted extraction
VMAE	vacuum microwave-assisted extraction
WFE	wetting film extraction
β-CD	β-cyclodextrin

Acknowledgment

We thank National Natural Science Foundation of China (Grants 90817101, 91117006, 91317312, 31570372, 31671777 and 30970247) to support our research in phytohormonal quantification.

References

Achard, P., Cheng, H., De Grauwe, L., et al., 2006. Integration of plant responses to environmentally activated phytohormonal signals. Science 311, 91–94.

Adamowski, M., Friml, J., 2015. PIN-dependent auxin transport: action, regulation, and evolution. Plant Cell 27, 20–32.

Anegawa, A., Ohnishi, M., Takagi, D., et al., 2015. Altered levels of primary metabolites in response to exogenous indole-3-acetic acid in wild type and auxin signaling mutants of *Arabidopsis thaliana*: a capillary electrophoresis-mass spectrometry analysis. Plant Biotechnol. 32, 65–79.

Avrameas, S., 1982. Immunocytological methods. In: Nicolini, C. (Ed.), Cell Growth. Springer Press, New York, pp. 51–59.

Badenoch-Jones, J., Summons, R., Rolfe, B., et al., 1984. Phytohormones, *Rhizobium* mutants, and nodulation in legumes. IV. Auxin metabolites in pea root nodules. J. Plant Growth Regul. 3, 23–39.

Band, L., Wells, D., Fozard, J., et al., 2014. Systems analysis of auxin transport in the *Arabidopsis* root apex. Plant Cell 26, 862–875.

Barbez, E., Kubeš, M., Rolčík, J., et al., 2012. A novel putative auxin carrier family regulates intracellular auxin homeostasis in plants. Nature 485, 119–122.

Benjamins, R., Scheres, B., 2008. Auxin: the looping star in plant development. Annu. Rev. Plant Biol. 59, 443–465.

Benková, E., Michniewicz, M., Sauer, M., et al., 2003. Local, efflux-dependent auxin gradients as a common module for plant organ formation. Cell 115, 591–602.

Bhalerao, R., Bennett, M., 2003. The case for morphogens in plants. Nat. Cell Biol. 5, 939–943.

Bieleski, R., 1964. The problem of halting enzyme action when extracting plant tissues. Anal. Biochem. 9, 431–442.

Birkemeyer, C., Kolasa, A., Kopka, J., 2003. Comprehensive chemical derivatization for gas chromatography–mass spectrometry-based multi-targeted profiling of the major phytohormones. J. Chromatogr. 993, 89–102.

Blintsov, A., Gusakovskaya, M., 2006. Immunoenzyme method for differential assay of free and bound ABA. Russ. J. Plant Physiol. 53, 407–412.

Boerjan, W., Genetello, C., Montagu, M., et al., 1992. A new bioassay for auxins and cytokinins. Plant Physiol. 99, 1090–1098.

Böhm, M., Wieland, I., Schütze, K., et al., 1997. Microbeam MOMeNT: non-contact laser microdissection of membrane-mounted native tissue. Am. J. Pathol. 151, 63–67.

Bose, A., Shah, D., Keharia, H., 2013. Production of indole-3-acetic-acid (IAA) by the white rot fungus *Pleurotus ostreatus* under submerged condition of *Jatropha* seedcake. Mycology 4, 103–111.

Böttcher, C., Roepenack-Lahaye, E., Willscher, E., et al., 2007. Evaluation of matrix effects in metabolite profiling based on capillary liquid chromatography electrospray ionization quadrupole time-of-flight mass spectrometry. Anal. Chem. 79, 1507–1513.

Boursiac, Y., Léran, S., Corratgé-Faillie, C., et al., 2013. ABA transport and transporters. Trends Plant Sci. 18, 325–333.

Boycheva, S., Dominguez, A., Rolcik, J., et al., 2015. Consequences of a deficit in vitamin B6 biosynthesis *de Novo* for hormone homeostasis and root development in *Arabidopsis*. Plant Physiol. 167, 102–117.

Brenner, M., 1981. Modern methods for plant growth substance analysis. Ann. Rev. Plant Physiol. 32, 511–538.

Brunoud, G., Wells, D., Oliva, M., et al., 2012. A novel sensor to map auxin response and distribution at high spatio-temporal resolution. Nature 482, 103–106.

Cahill, D., Ward, E., 1989. An indirect enzyme-linked immunosorbent assay for measurement of abscisic acid in soybean inoculated with *Phytophthora megasperma* f. sp. *glycinea*. Phytopathology 79, 1238–1242.

Cai, B., Yin, J., Hao, Y., et al., 2015. Profiling of phytohormones in rice under elevated cadmium concentration levels by magnetic solid-phase extraction coupled with liquid chromatography tandem mass spectrometry. J. Chromatogr. A 1406, 78–86.

Cao, Y., Sha, Z., Yu, J., et al., 2012. Protein detection based on small molecule-linked DNA. Anal. Chem. 84, 4314–4320.

Castillo, G., Torrecillas, A., Nogueiras, C., et al., 2014. Simultaneous quantification of phytohormones in fermentation extracts of *Botryodiplodia theobromae* by liquid chromatography–electrospray tandem mass spectrometry. World J. Microbiol. Biotechnol. 30, 1937–1946.

Chapman, E., Estelle, M., 2009. Mechanism of auxin-regulated gene expression in plants. Annu. Rev. Genet. 43, 265–285.

Chen, C., Chen, Y., Zhou, J., et al., 2006. A 9-vinyladenine-based molecularly imprinted polymeric membrane for the efficient recognition of plant hormone ^{1}H-indole-3-acetic acid. Anal. Chim. Acta 569, 58–65.

Chen, M., Huang, Y., Liu, J., et al., 2011. Highly sensitive profiling assay of acidic plant hormones using a novel mass probe by capillary electrophoresis-time of flight-mass spectrometry. J. Chromatogr. B 879, 938–944.

Chiwocha, S., Abrams, S., Ambrose, S., et al., 2003. A method for profiling classes of plant hormones and their metabolites using liquid chromatography-electrospray ionization tandem mass spectrometry: an analysis of hormone regulation of thermodormancy of lettuce (*Lactuca sativa* L.) seeds. Plant J. 35, 405–417.

Chong, K., Xu, Z., 2014. Investment in plant research and development bears fruit in China. Plant Cell Rep. 33, 541–550.

Christmann, A., Hoffmann, T., Teplova, I., et al., 2005. Generation of active pools of abscisic acid revealed by *in vivo* imaging of water-stressed *Arabidopsis*. Plant Physiol. 137, 209–219.

Crocker, W., Hitchcock, A., Zimmerman, P., 1935. Similarities in the effects of ethylene and the plant auxins. Contrib. Boyce Thompson Inst. 7, 231–248.

Croker, S.J., Gaskin, P., Hedden, P., et al., 1994. Quantitative analysis of gibberellins by isotope dilution mass spectrometry: a comparison of the use of calibration curves, an isotope dilution fit program and arithmetical correction of isotope ratios. Phytochem. Anal. 5, 74–80.

Cui, K., Lin, Y., Zhou, X., et al., 2015. Comparison of sample pretreatment methods for the determination of multiple phytohormones in plant samples by liquid chromatography–electrospray ionization-tandem mass spectrometry. Microchem. J. 121, 25–31.

Davies, P. (Ed.), 2004. Plant Hormones: Biosynthesis, Signal Transduction, Action!. Springer Science & Business Media Press, New York.

De Diego, N., Rodríguez, J., Dodd, I., et al., 2013. Immunolocalization of IAA and ABA in roots and needles of radiata pine (*Pinus radiata*) during drought and rewatering. Tree Physiol. 33, 537–549.

Defraia, C., Schmelz, E., Mou, Z., 2008. A rapid biosensor-based method for quantification of free and glucose-conjugated salicylic acid. Plant Methods 4, 28.

Dekhuijzen, H., Gevers, C., 1975. The recovery of cytokinins during extraction and purification of clubroot tissue. Physiol. Plant. 35, 297–302.

Denny, F., 1935. Testing plant tissues for emanations causing leaf epinasty. Contrib. Boyce Thompson Inst. 3, 341.

Dewitte, W., Van Onckelen, H., 2001. Probing the distribution of plant hormones by immunocytochemistry. Plant Growth Regul. 33, 67–74.

Dharmasiri, N., Dharmasiri, S., Estelle, M., 2005. The F-box protein TIR1 is an auxin receptor. Nature 435, 441–445.

Dharmasiri, S., Swarup, R., Mockaitis, K., et al., 2006. AXR4 is required for localization of the auxin influx facilitator AUX1. Science 312, 1218–1220.

Ding, J., Jiang, L., Feng, Y., 2014. An automatic and sensitive method for the determination of endogenous brassinosteroids in plant tissues by an online trapping-*in situ* derivatization-ultra performance liquid chromatography-tandem mass spectrometry system. Chin. J. Chromatogr. 32, 1094–1103.

Ding, J., Mao, L., Wang, S., et al., 2013. Determination of endogenous brassinosteroids in plant tissues using solid-phase extraction with double layered cartridge followed by high-performance liquid chromatography-tandem mass spectrometry. Phytochem. Anal. 24, 386–394.

Dobrev, P., Havlíček, L., Vágner, M., et al., 2005. Purification and determination of plant hormones auxin and abscisic acid using solid phase extraction and two-dimensional high performance liquid chromatography. J. Chromatogr. 1075, 159–166.

Dockter, C., Gruszka, D., Braumann, I., et al., 2014. Induced variations in brassinosteroid genes define barley height and sturdiness, and expand the Green Revolution genetic toolkit. Plant Physiol. 166, 1912–1927.

Dong, N., Pei, D., Yin, W., 2012. Tissue-specific localization and dynamic changes of endogenous IAA during poplar leaf rhizogenesis revealed by in situ immunohistochemistry. Plant Biotechnol. Rep. 6, 165–174.

Driss-Ecole, D., Perbal, G., 1987. Intracellular localization of ^{3}H-IAA in the apical bud of *Lycopersicon esculentum*. J. Exp. Bot. 38, 1362–1372.

Du, F., Ruan, G., Liu, H., 2012. Analytical methods for tracing plant hormones. Anal. Bioanal. Chem. 403, 55–74.

Duan, L., Dietrich, D., Ng, C., et al., 2013. Endodermal ABA signaling promotes lateral root quiescence during salt stress in *Arabidopsis* seedlings. Plant Cell 25, 324–341.

Durgbanshi, A., Arbona, V., Pozo, O., et al., 2005. Simultaneous determination of multiple phytohormones in plant extracts by liquid chromatography-electrospray tandem mass spectrometry. J. Agric. Food Chem. 53, 8437–8442.

Edlund, A., Eklof, S., Sundberg, B., et al., 1995. A microscale technique for gas chromatography-mass spectrometry measurements of picogram amounts of indole-3-acetic acid in plant tissues. Plant Physiol. 108, 1043–1047.

Epel, B., Erlanger, M., Yahalom, A., 1987. The etiolated cucumber hypocotyl weight-growth test: a sensitive, easy and ultrafast bioassay for IAA. Plant Growth Regul. 5, 3–14.

Fang, S., Xin, P., Guo, Z., Chen, Y., Chu, J., 2017. Quantitative Analysis of Plant Hormones Based on LC-MS/MS.

Fawcett, N., Evans, J., Chien, L., et al., 1988. Nucleic acid hybridization detected by piezoelectric resonance. Anal. Lett. 21, 1099–1114.

Fei, Y., Liu, D., Wu, Z., et al., 2011. DNA-encoded signal conversion for sensitive microgravimetric detection of small molecule-protein interaction. Bioconjug. Chem. 22, 2369–2376.

Fleet, C., Williams, M., 2011. Gibberellins. Teaching tools in plant biology: lecture notes. Plant Cell. 110. www.plantcell.org/cgi/doi/101105/tpc.

Flores, M., Romero-González, R., Frenich, A., et al., 2011. QuEChERS-based extraction procedure for multifamily analysis of phytohormones in vegetables by UHPLC-MS/MS. J. Sep. Sci. 34, 1517–1524.

Friml, J., Vieten, A., Sauer, M., et al., 2003. Efflux-dependent auxin gradients establish the apical–basal axis of *Arabidopsis*. Nature 426, 147–153.

Friml, J., Wiśniewska, J., Benková, E., et al., 2002. Lateral relocation of auxin efflux regulator PIN3 mediates tropism in *Arabidopsis*. Nature 415, 806–809.

Fuchs, S., Fuchs, Y., 1969. Immunological assay for plant hormones using specific antibodies to indoleacetic acid and gibberellic acid. Bba-general Sub 192, 528–530.

Gan, T., Hu, C., Chen, Z., et al., 2010. Fabrication and application of a novel plant hormone sensor for the determination of methyl jasmonate based on self-assembling of phosphotungstic acid–graphene oxide nanohybrid on graphite electrode. Sens. Actuators B Chem. 151, 8–14.

Gao, Q., Luo, D., Ding, J., et al., 2010. Rapid magnetic solid-phase extraction based on magnetite/silica/poly (methacrylic acid–co–ethylene glycol dimethacrylate) composite microspheres for the determination of sulfonamide in milk samples. J. Chromatogr. A 1217, 5602–5609.

Giannarelli, S., Muscatello, B., Bogani, P., et al., 2010. Comparative determination of some phytohormones in wild-type and genetically modified plants by gas chromatography–mass spectrometry and high-performance liquid chromatography–tandem mass spectrometry. Anal. Biochem. 398, 60–68.

Glauser, G., Boccard, J., Rudaz, S., et al., 2010. Mass spectrometry-based metabolomics oriented by correlation analysis for wound-induced molecule discovery: identification of a novel jasmonate glucoside. Phytochem. Anal. 21, 95–101.

Gobi, K., Tanaka, H., Shoyama, Y., et al., 2005. Highly sensitive regenerable immunosensor for label-free detection of 2,4-dichlorophenoxyacetic acid at ppb levels by using surface plasmon resonance imaging. Sens. Actuators B Chem. 111, 562–571.

Gomez-Roldan, V., Fermas, S., Brewer, P., et al., 2008. Strigolactone inhibition of shoot branching. Nature 455, 189–194.

Goswami, D., Thakker, J., Dhandhukia, P., 2015. Simultaneous detection and quantification of indole-3-acetic acid (IAA) and indole-3-butyric acid (IBA) produced by rhizobacteria from Ltryptophan (Trp) using HPTLC. J. Microbiol. Methods 110, 7–14.

Grieneisen, V., Xu, J., Marée, A., et al., 2007. Auxin transport is sufficient to generate a maximum and gradient guiding root growth. Nature 449, 1008–1013.

Guan, C., Ji, J., Zhang, X., et al., 2015. Positive feedback regulation of a *Lycium chinense*-derived *VDE* gene by drought-induced endogenous ABA, and over-expression of this *VDE* gene improve drought-induced photo-damage in *Arabidopsis*. J. Plant Physiol. 175, 26–36.

Gupta, V., Kumar, M., Brahmbhatt, H., et al., 2011. Simultaneous determination of different endogenetic plant growth regulators in common green seaweeds using dispersive liquid–liquid microextraction method. Plant Physiol. Biochem. 49, 1259–1263.

Hall, W., 1951. Studies on the origin of ethylene from plant tissues. Bot. Gaz. 113, 55–65.

Han, Y., Bai, Y., Xiao, Y., et al., 2011. Simultaneous discrimination of jasmonic acid stereoisomers by CE-QTOF-MS employing the partial filling technique. Electrophoresis 32, 2693–2699.

Han, Z., Liu, G., Rao, Q., et al., 2012. A liquid chromatography tandem mass spectrometry method for simultaneous determination of acid/alkaline phytohormones in grapes. J. Chromatogr. B 881, 83–89.

Hauserová, E., Swaczynová, J., Doležal, K., et al., 2005. Batch immunoextraction method for efficient purification of aromatic cytokinins. J. Chromatogr. A 1100, 116–125.

Heisler, M., Ohno, C., Das, P., et al., 2005. Patterns of auxin transport and gene expression during primordium development revealed by live imaging of the *Arabidopsis* inflorescence meristem. Curr. Biol. 15, 1899–1911.

Hemberg, T., Westlin, P.E., 1973. The quantitative yield in purification of cytokinins. model-experiments with kinetin, 6-furfuryl-amino-purine. Physiol. Plant 28, 228–231.

Hernández, L., Hernández, P., Patón, F., 1996. Adsorptive stripping determination of indole-3-acetic acid at a carbon fiber ultramicroelectrode. Anal. Chim. Acta 327, 117–123.

Hocart, C., Wong, O., Letham, D., et al., 1986. Mass spectrometry and chromatography of t-butyldimethylsilyl derivatives of cytokinin bases. Anal. Biochem. 153, 85–96.

Homola, J., Schwotzer, G., Lehmann, H., et al., 1995. Fiber optic sensor for adsorption studies using surface plasmon resonance. In: European Symposium on Optics for Environmental and Public Safety, 2508. International Society for Optics and Photonics, pp. 324–333.

Hou, S., Zhu, J., Ding, M., et al., 2008. Simultaneous determination of gibberellic acid, indole-3-acetic acid and abscisic acid in wheat extracts by solid-phase extraction and liquid chromatography-electrospray tandem mass pectrometry. Talanta 76, 798–802.

Hradecká, V., Novák, O., Havlíček, L., et al., 2007. Immunoaffinity chromatography of abscisic acid combined with electrospray liquid chromatography–mass spectrometry. J. Chromatogr. B 847, 162–173.

Hu, C., Wu, Z., Tang, H., et al., 2014. Terminal protection of small molecule-linked DNA for small molecule–protein interaction assays. Inter J. Mol. Sci. 15, 5221–5232.

Hu, J., Wei, F., Dong, X., et al., 2013. Characterization and quantification of triacylglycerols in peanut oil by off-line comprehensive two-dimensional liquid chromatography coupled with atmospheric pressure chemical ionization mass spectrometry. J. Sep Sci. 36, 288–300.

Hu, Y., Li, Y., Zhang, Y., et al., 2011. Development of sample preparation method for auxin analysis in plants by vacuum microwave-assisted extraction combined with molecularly imprinted clean-up procedure. Anal. Bioanal. Chem. 399, 3367–3374.

Hun, X., Mei, Z., Wang, Z., et al., 2012. Indole-3-acetic acid biosensor based on G-rich DNA labeled AuNPs as chemiluminescence probe coupling the DNA signal amplification. Spectrochim. Acta A Mol. Biomol. Spectrosc. 95, 114–119.

Ikeda, Y., Men, S., Fischer, U., et al., 2009. Local auxin biosynthesis modulates gradient-directed planar polarity in *Arabidopsis*. Nat. Cell Biol. 11, 731–738.

Jones, A., Danielson, J., Manojkumar, S., et al., 2014. Abscisic acid dynamics in roots detected with genetically encoded FRET sensors. Elife Sci. 3, e01741.

Jones, R., Varner, J., 1966. The bioassay of gibberellins. Planta 72, 155–161.

Kaldewey, H., Wakhloo, J., Weis, A., et al., 1968. The Avena geo-curvature test: a quick and simple bioassay for auxins. Planta 84, 1–10.

Kaldewey, P., Stahl, E., 1964. Die quantitative Auswertung dünnschichtchromatographisch getrennter auxine im *Avena*-tageslichttest nach söding (*Avena*-silicagel-krümmungstest). Planta 62, 22–38.

Kathirvelan, J., Vijayaraghavan, R., 2014. Development of prototype laboratory setup for selective detection of ethylene based on multiwalled carbon nanotubes. J. Sens. 3, 1–6.

Kepinski, S., Leyser, O., 2005. The *Arabidopsis* F-box protein TIR1 is an auxin receptor. Nature 435, 446–451.

Kim, J., Jung, Y., Song, B., et al., 2013. Discrimination of cabbage (*Brassica rapa* ssp. pekinensis) cultivars grown in different geographical areas using 1 H NMR-based metabolomics. Food Chem. 137, 68–75.

King, W., 1964. Correction: piezoelectric sorption detector. Anal. Chem. 36, 1735–1739.

Kohlen, W., Charnikhova, T., Liu, Q., et al., 2011. Strigolactones are transported through the xylem and play a key role in shoot architectural response to phosphate deficiency in nonarbuscular mycorrhizal host *Arabidopsis*. Plant Physiol. 155, 974–987.

Kojima, M., Kamada-Nobusada, T., Komatsu, H., et al., 2009. Highly sensitive and high-throughput analysis of plant hormones using MS-probe modification and liquid chromatography–tandem mass spectrometry: an application for hormone profiling in *Oryza sativa*. Plant Cell Physiol. 50, 1201–1214.

Kokosa, J., 2015. Recent trends in using single-drop microextraction and related techniques in green analytical methods. TrAC Trends Anal. Chem. 71, 194–204.

Kowalska, M., Tian, F., Šmehilová, M., et al., 2011. Prussian Blue acts as a mediator in a reagentless cytokinin biosensor. Anal. Chim. Acta 701, 218–223.

Kugimiya, A., Takeuchi, T., 1999. Molecularly imprinted polymer-coated quartz crystal microbalance for detection of biological hormone. Electroanalysis 11, 1158–1160.

Lakowicz, J., 2006. Plasmonics in biology and plasmon-controlled fluorescence. Plasmonics 1, 5–33.

Larrieu, A., Champion, A., Legrand, J., et al., 2015. A fluorescent hormone biosensor reveals the dynamics of jasmonate signalling in plants. Nat. Commun. 6, 1–8.

Laureys, F., Dewitte, W., Witters, E., et al., 1998. Zeatin is indispensable for the G 2-M transition in tobacco BY-2 cells. FEBS Lett. 426, 29–32.

Letham, D., Singh, S., Wong, O., 1991. Mass spectrometric analysis of cytokinins in plant tissue VII. Quantification of cytokinin bases by negative ion mass spectrometry. J. Plant Growth Regul. 10, 107–113.

Li, D., Park, J., Oh, J., 2001. Silyl derivatization of alkylphenols, chlorophenols, and bisphenol A for simultaneous GC/MS determination. Anal. Chem. 73, 3089–3095.

Li, J., Li, S., Wei, X., et al., 2012. Molecularly imprinted electrochemical luminescence sensor based on signal amplification for selective determination of trace gibberellin A3. Anal. Chem. 84, 9951–9955.

Li, J., Wu, Z., Xiao, L., et al., 2002. A novel piezoelectric biosensor for the detection of phytohormone beta-indole acetic acid. Anal. Sci. 18, 403–407.

Li, J., Xiao, L., Zeng, G., et al., 2003. Amperometric immunosensor based on polypyrrole/ poly (m-pheylenediamine) multilayer on glassy carbon electrode for cytokinin N6-(Δ2-isopentenyl) adenosine assay. Anal. Biochem. 321, 89–95.

Li, N., Chen, J., Shi, Y., 2016. Magnetic reduced graphene oxide functionalized with β-cyclodextrin as magnetic solid-phase extraction adsorbents for the determination of phytohormones in tomatoes coupled with high performance liquid chromatography. J. Chromatogr. A 1441, 24–33.

Li, Q., Wang, R., Huang, Z., et al., 2010. A novel impedance immunosensor based on O-phenylenediamine modified gold electrode to analyze abscisic acid. Chin. Chem. Lett. 21, 472–475.

Li, R., Wang, C., Hu, Y., et al., 2014. Electrochemiluminescence biosensor for folate receptor based on terminal protection of small-molecule-linked DNA. Biosens. Bioelectron. 58, 226–231.

Li, Y., Xia, K., Wang, R., et al., 2008. An impedance immunosensor for the detection of the phytohormone abscisic acid. Anal. Bioanal. Chem. 391, 2869–2874.

Liang, Y., Zhu, X., Zhao, M., et al., 2012. Sensitive quantification of isoprenoid cytokinins in plants by selective immunoaffinity purification and high performance liquid chromatography-quadrupole-time of flight mass spectrometry. Methods 56, 174–179.

Liang, Y., Ward, S., Li, P., et al., 2016. SMAX1-LIKE7 signals from the nucleus to regulate shoot development in *Arabidopsis* via partially EAR motif-independent mechanisms. Plant Cell 28, 1581–1601.

Liao, S., Hamilton, R., 1966. Intracellular localization of growth hormones in plants. Science 151, 822–823.

Liedberg, B., Lundström, I., Stenberg, E., 1993. Principles of biosensing with an extended coupling matrix and surface plasmon resonance. Sens. Actuators B Chem. 11, 63–72.

Liedberg, B., Nylander, C., Lunström, I., 1983. Surface plasmon resonance for gas detection and biosensing. Sens. Actuators 4, 299–304.

Lin, G., Chang, C., Lin, H., 2015. Systematic profiling of indole-3-acetic acid biosynthesis in bacteria using LC-MS/MS. J. Chromatogr. B 988, 53–58.

Liu, F., Yu, Y., Lin, B., et al., 2014. Visualization of hormone binding proteins in vivo based on Mn-doped CdTe QDs. Spectrochim. Acta A Mol. Biomol. Spectrosc. 131, 9–16.

Liu, H., Li, Y., Luan, T., et al., 2007. Simultaneous determination of phytohormones in plant extracts using SPME and HPLC. Chromatographia 66, 515–520.

Liu, Y., Dong, H., Zhang, W., et al., 2010. Preparation of a novel colorimetric luminescence sensor strip for the detection of indole-3-acetic acid. Biosens. Bioelectron. 25, 2375–2378.

Ljung, K., Sandberg, G., Moritz, T., 2004. Hormone analysis. In: Davies, P.J. (Ed.), Plant Hormones: Biosynthesis, Signal Transduction, Action, third ed. Kluwer Academic Publishers, Dordrecht, Boston, London, pp. 1–15.

Long, M., Jiang, J., Li, Y., et al., 2011. Effect of gold nanoparticles on the photocatalytic and photoelectrochemical performance of Au modified $BiVO_4$. Nano-micro Lett. 3, 171–177.

Loveys, B., Van Dijk, H., 1988. Improved extraction of abscisic acid from plant tissue. Funct. Plant Biol. 15, 421–427.

Ludewig, M., Dörffling, K., König, W., 1982. Electron-capture capillary gas chromatography and mass spectrometry of trifluoroacetylated cytokinins. J. Chromatogr. A 243, 93–98.

Lyon, C., 1970. Ethylene inhibition of auxin transport by gravity in leaves. Plant Physiol. 45, 644.

Manos, P., Goldthwaite, J., 1976. An improved cytokinin bioassay using cultured soybean hypocotyl sections. Plant Physiol. 57, 894–897.

Manuel, M., Vidal, B., López, R., et al., 1993. Determination of probable alcohol yield in musts by means of an SPR optical sensor. Sens. Actuators B Chem. 11, 455–459.

Mason, T., Lorimer, J., 1988. Theory, Applications and Uses of Ultrasound in Chemistry. Ellis Harwood Limited, John Wiley, New York.

Melchert, W.R., Reis, B.F., Rocha, F.R.P., 2012. Green chemistry and the evolution of flow analysis. A review. Anal. Chim. Acta 714, 8–19.

Miersch, O., Bohlmann, H., Wasternack, C., 1999. Jasmonates and related compounds from *Fusarium oxysporum*. Phytochemistry 50, 517–523.

Moncaleán, P., López-Iglesias, C., Fernández, B., et al., 2001. Immunocytochemical location of endogenous cytokinins in buds of kiwifruit (*Actinidia deliciosa*) during the first hours of *in vitro* culture. Histochem. J. 33, 403–411.

Morrison, E., Knowles, S., Hayward, A., et al., 2015. Detection of phytohormones in temperate forest fungi predicts consistent abscisic acid production and a common pathway for cytokinin biosynthesis. Mycologia 107, 667–692.

Müller, A., Düchting, P., Weiler, E., 2002. A multiplex GC-MS/MS technique for the sensitive and quantitative single-run analysis of acidic phytohormones and related compounds, and its application to *Arabidopsis thaliana*. Planta 216, 44–56.

Müller, B., Sheen, J., 2008. Cytokinin and auxin interaction in root stem-cell specification during early embryogenesis. Nature 453, 1094–1097.

Nakayama, A., Park, S., Zhengjun, X., et al., 2002. Immunohistochemistry of active gibberellins and gibberellin-inducible alpha-amylase in developing seeds of morning glory. Plant Physiol. 129, 1045–1053.

Nelson, R., Harvey, R., 1935. The presence in self-blanching celery of unsaturated compounds with physiological action similar to ethylene. Science 82, 133–134.

Newton, C., Morgan, C., Morgan, D., 1980. Evaluation of a bioassay for cytokinins using soybean hypocotyl sections. J. Exp. Bot. 31, 721–729.

Nikolelis, D., Chaloulakos, T., Nikoleli, G., et al., 2008. A portable sensor for the rapid detection of naphthalene acetic acid in fruits and vegetables using stabilized in air lipid films with incorporated auxin-binding protein 1 receptor. Talanta 77, 786–792.

Nishimura, K., Fukagawa, T., Takisawa, H., et al., 2009. An auxin-based degron system for the rapid depletion of proteins in nonplant cells. Nat. Methods 6, 917–922.

Novák, O., Hényková, E., Sairanen, I., et al., 2012. Tissue-specific profiling of the *Arabidopsis thaliana* auxin metabolome. Plant J. 72, 523–536.

Novák, O., Tarkowski, P., Tarkowská, D., et al., 2003. Quantitative analysis of cytokinins in plants by liquid chromatography-single-quadrupole mass spectrometry. Anal. Chim. Acta 480, 207–218.

O'sullivan, C., Guilbault, G., 1999. Commercial quartz crystal microbalances–theory and applications. Biosens. Bioelectron. 14, 663–670.

Ohmiya, A., Hayashi, T., Kakiuchi, N., 1990. Immuno-gold localization of indole-3-acetic acid in peach seedlings. Plant Cell Physiol. 31, 711–715.

Paleologos, E., Giokas, D., Karayannis, M., 2005. Micelle-mediated separation and cloud-point extraction. TrAC Trends Anal. Chem. 24, 426–436.

Palni, L., Summons, R., Letham, D., 1983. Mass spectrometric analysis of cytokinins in plant tissues: v. Identification of the cytokinin complex of datura innoxia crown gall tissue. Plant Physiol. 72, 858–863.

Pan, J., Tan, W., Gongke, L., et al., 2011. Progress in the analysis of brassinosteroids. Chin. J. Chromatogr. 29, 105–110.

Pegg, G., 1962. A new method for growth assays: the extension growth of segments of etiolated tomato seedling hypocotyls. Ann. Bot. 26, 207–218.

Peng, J., Richards, D., Hartley, N., et al., 1999. 'Green revolution' genes encode mutant gibberellin response modulators. Nature 400, 256–261.

Peng, Y., Zou, C., Wang, D., et al., 2006. Preferential localization of abscisic acid in primordial and nursing cells of reproductive organs of *Arabidopsis* and cucumber. New Phytol. 170, 459–466.

Pengelly, W., Meins Jr., F., 1977. A specific radioimmunoassay for nanogram quantities of the auxin, indole-3-acetic acid. Planta 136, 173–180.

Pérez, M., Viejo, M., LaCuesta, M., et al., 2015. Epigenetic and hormonal profile during maturation of *Quercus suber* L. somatic embryos. J. Plant Physiol. 173, 51–61.

Petersson, S., Johansson, A., Kowalczyk, M., et al., 2009. An auxin gradient and maximum in the *Arabidopsis* root apex shown by high-resolution cell-specific analysis of IAA distribution and synthesis. Plant Cell 21, 1659–1668.

Pool, R., Powell, L., 1972. The use of pellicular ion-exchange resins to separate plant cytokinins by high-pressure liquid chromatography. HortScience 7, 330.

Poole, C., 2003. New trends in solid-phase extraction. TrAC Trend Anal. Chem. 22, 362–373.

Porobić, I., Kontrec, D., Šoškić, M., 2013. Molecular recognition of indole derivatives by polymers imprinted with indole-3-acetic acid: a QSPR study. Bioorg. Med. Chem. 21, 653–659.

Prinsen, E., Van, L., Oden, S., et al., 2000. Auxin analysis. Methods Mol. Biol. 141, 49–65.

Qi, C., Bing, T., Mei, H., et al., 2013. G-quadruplex DNA aptamers for zeatin recognizing. Biosens. Bioelectron. 41, 157–162.

Qian, Q., Guo, L., Smith, S., et al., 2016. Breeding high-yield superior-quality hybrid super-rice by rational design. Natl. Sci. Rev. 3 , 283–294.

Qu, L., Jiang, Z., Li, J., 2017. AUXINS (This Book).

Roy, R., Hohng, S., Ha, T., 2008. A practical guide to single-molecule FRET. Nat. Methods 5, 507–516.

Sabatini, S., Beis, D., Wolkenfelt, H., et al., 1999. An auxin-dependent distal organizer of pattern and polarity in the *Arabidopsis* root. Cell 99, 463–472.

Sagi, F., 1969. Silica gel or cellulose for the thin-layer chromatography of indole-3-acetic acid? J. Chromatogr. A 39, 334–335.

Sakai, G., Saiki, T., Uda, T., et al., 1995. Selective and repeatable detection of human serum albumin by using piezoelectric immunosensor. Sens. Actuators B Chem. 24, 134–137.

Sasaki, A., Ashikari, M., Ueguchi-Tanaka, M., et al., 2002. Green revolution: a mutant gibberellin-synthesis gene in rice. Nature 416, 701–702.

Schaller, G., Bishopp, A., Kieber, J., 2015. The yin-yang of hormones: cytokinin and auxin interactions in plant development. Plant Cell 27, 44–63.

Schlenk, H., Gellerman, J., Chem, A., 1960. Esterification of fatty acids with diazomethane on a small scale. Anal. Chem. 32, 1412–1414.

Schmelz, E., Engelberth, J., Alborn, H., et al., 2003. Simultaneous analysis of phytohormones, phytotoxins, and volatile organic compounds in plants. Proc. Natl. Acad. Sci. U.S.A. 100, 10552–10557.

Shan, X., Yan, J., Xie, D., 2012. Comparison of phytohormone signaling mechanisms. Curr. Opin. Plant Biol. 15, 84–91.

Shani, E., Weinstain, R., Zhang, Y., et al., 2013. Gibberellins accumulate in the elongating endodermal cells of *Arabidopsis* root. Proc. Natl. Acad. Sci. U.S.A. 110, 4834–4839.

Shimojo, M., Amada, K., Koya, H., et al., 2011. Flow Injection Biosensor System for 2,4-Dichlorophenoxyacetate Based on a Microbial Reactor and Tyrosinase-modified Electrode. In: Somerset V. (Ed.), Environmental Biosensors InTechOpen Publisher; Rijeka, pp. 341–356.

Shons, A., Dorman, F., Najarian, J., 1972. The piezoelectric quartz immunosensor. J. Biomed. Mater. Res. 6, 565–570.

Silverstone, A., Jung, H., Dill, A., et al., 2001. Repressing a repressor gibberellin-induced rapid reduction of the RGA protein in *Arabidopsis*. Plant Cell 13, 1555–1566.

Smith, S.M., Li, C., Li, J., 2017. Hormone function in plants. In: Li, J., Li, C., Smith, S.M. (Eds.), Hormone metabolism and signalling in plants. Academic Press, United States of America, 1–32.

Söding, H., 1952. Die Wuchsstofflehre. Georg Thieme, Stuttgart.

Sokołowska, K., Kizińska, J., Szewczuk, Z., et al., 2014. Auxin conjugated to fluorescent dyes - a tool for the analysis of auxin transport pathways. Plant Biol. 16, 866–877.

Song, X., Ren, S., Liu, C., 2017. Peptide hormones. In: Li, J., Li, C., Smith, S.M. (Eds.), Hormone Metabolism and Signalling in Plants. Academic Press, United States of America, 361–395.

Soundappan, I., Bennett, T., Morffy, N., et al., 2015. SMAX1-LIKE/D53 family members enable distinct MAX2-dependent responses to strigolactones and karrikins in *Arabidopsis*. Plant Cell 27, 1898–1899.

Su, Y., Su, Y., Liu, Y., et al., 2013. Abscisic acid is required for somatic embryo initiation through mediating spatial auxin response in *Arabidopsis*. Plant Growth Regul. 69, 167–176.

Sun, B., Chen, L., Xu, Y., et al., 2014. Ultrasensitive photoelectrochemical immunoassay of indole-3-acetic acid based on the MPA modified CdS/RGO nanocomposites decorated ITO electrode. Biosens. Bioelectron. 51, 164–169.

Sun, B., Zhang, K., Chen, L., et al., 2013. A novel photoelectrochemical sensor based on PPIX-functionalized WO$_3$-rGO nanohybrid-decorated ITO electrode for detecting cysteine. Biosens. Bioelectron. 44, 48–51.

Sun, J., Wu, Z., Hu, D., et al., 2015. Preparation of reduced graphene oxide–poly (safranine T) film via one-step polymerization for electrochemical determination of methyl jasmonate. Fuller Nanotubes Carbon Nanostruct. 23, 701–708.

Sun, T., 2011. The molecular mechanism and evolution of the GA–GID1–DELLA signaling module in plants. Curr. Biol. 21, R338–R345.

Suwabe, K., Suzuki, G., Takahashi, H., et al., 2008. Separated transcriptomes of male gametophyte and tapetum in rice: validity of a laser microdissection (LM) microarray. Plant Cell Physiol. 49, 1407–1416.

Swarup, R., Peret, B., 2012. AUX/LAX family of auxin influx carriers-an overview. Front Plant Sci. 3, 225.

Tarkowská, D., Novák, O., Floková, K., et al., 2014. Quo vadis plant hormone analysis? Planta 1–22.

Tarkowski, P., Ge, L., Yong, J., et al., 2009. Analytical methods for cytokinins. TrAC Trend Anal. Chem. 28, 323–335.

Tucker, D., Mansfield, T., Hagen, G., 1971. A simple bioassay for detecting "antitranspirant" activity of naturally occurring compounds such as abscisic acid. Planta 98, 157–163.

Ulmasov, T., Murfett, J., Hagen, G., et al., 1997. Aux/IAA proteins repress expression of reporter genes containing natural and highly active synthetic auxin response elements. Plant Cell 9, 1963–1971.

Van Meulebroek, L., Bussche, J., Steppe, K., et al., 2012. Ultra-high performance liquid chromatography coupled to high resolution Orbitrap mass spectrometry for metabolomic profiling of the endogenous phytohormonal status of the tomato plantAgeratum houstonianum. J. Chromatogr. A 1260, 67–80.

Van Raalte, M., Chandra, G., 1950. Root formation by the petioles of *Ageratum houstonianum* Mill. as a test for auxin activity in tropical countries. Ann. Bogor. 1, 13–26.

Varner, J., Chandra, G., Farcot, E., 1964. Hormonal control of enzyme synthesis in barley endosperm. Proc. Natl. Acad. Sci. U.S.A. 52, 100–106.

Vernoux, T., Brunoud, G., Farcot, E., et al., 2011. The auxin signalling network translates dynamic input into robust patterning at the shoot apex. Mol. Syst. Biol. 7, 508–313.

Vinatoru, M., Hitomi, K., Nishimura, N., 2001. An overview of the ultrasonically assisted extraction of bioactive principles from herbsArabidopsis. Ultrason. Sonochem. 8, 303–313.

Waadt, R., Hitomi, K., Nishimura, N., et al., 2014. FRET-based reporters for the direct visualization of abscisic acid concentration changes and distribution in *Arabidopsis*. Elife Sci. 3, 1780–1788.

Wada, K., Marumo, S., Abe, H., et al., 1984. A rice lamina inclination test—a micro-quantitative bioassay for brassinosteroids. Agric. Biol. Chem. 48, 719–726.

Wada, K., Marumo, S., Ikekawa, N., et al., 1981. Brassinolide and homobrassinolide promotion of lamina inclination of rice seedlings. Plant Cell Physiol. 22, 323–325.

Wang, J., Xiao, X., Li, G., 2008. Study of vacuum microwave-assisted extraction of polyphenolic compounds and pigment from Chinese herbs. J. Chromatogr. A 1198–1199, 45–53.

Wang, Q., Jiang, B., Xie, J., et al., 2013. Coupling of background reduction with rolling circle amplification for highly sensitive protein detection via terminal protection of small molecule-linked DNA. Analyst 138, 5751–5756.

Wang, R., Xiao, L., Yang, M., et al., 2006. Amperometric determination of indole-3-acetic acid based on platinum nanowires and carbon nanotubesArabidopsis. Chin. Chem. Lett. 17, 1585–1588.

Wang, R., Li, Y., Li, Q., et al., 2009. A novel amperometric immunosensor for phytohormone abscisic acid based on in situ chemical reductive growth of gold nanoparticles on glassy carbon electrode. Anal. Lett. 42, 2893–2904.

Wang, Z., Gehring, C., Zhu, J., et al., 2015. The Arabidopsis vacuolar sorting receptor1 is required for osmotic stress-induced abscisic acid biosynthesis. Plant Physiol. 167, 137–152.

Wei, C., Zhou, H., Chen, C., et al., 2011. On-line monitoring ^{1}H-indole-3-acetic acid in plant tissues using molecular imprinting monolayer techniques on a surface plasmon resonance sensor. Anal. Lett. 44, 2911–2921.

Weiler, E., Ziegler, H., 1981. Determination of phytohormones in phloem exudate from tree species by radioimmunoassay. Planta 152, 168–170.

Weiler, E., Laplaze, L., Bennett, M., 1982. An enzyme-immunoassay for cis-(+)-abscisic acid. Physiol. Plant 54, 510–514.

Wells, D., Laplaze, L., Bennett, M., et al., 2013. Biosensors for phytohormone quantification: challenges, solutions, and opportunities. Trends Plant Sci. 18, 244–249.

Wen, W., Chen, W., Ren, Q., et al., 2012. A highly sensitive nitric oxide biosensor based on hemoglobin–chitosan/graphene–hexadecyltrimethylammonium bromide nanomatrix. Sens. Actuators B Chem. 166, 444–450.

Went, F., 1935. Auxin, the plant growth-hormoneAvena sativa. Bot. Rev. 1, 162–182.

Went, F., Hoboken, S., 1926. On growth-accelerating substances in the coleoptile of Avena sativa. Proc. K. Ned. Akad. Wet. 30, 10–19. New Jersey, USA.

Wiley, J., Hoboken, S., 2006. Analytical Electrochemistry. New Jersey, USA.

Wu, Q., Wu, D., Shen, Z., et al., 2013. Quantification of endogenous brassinosteroids in plant by on-line two-dimensional microscale solid phase extraction-on column derivatization coupled with high performance liquid chromatography-tandem mass spectrometry. J. Chromatogr. A 1297, 56–63.

Wu, Y., Hu, B., Jiang, J., 2009. Simultaneous determination of several phytohormones in natural coconut juice by hollow fiber-based liquid–liquid–liquid microextraction-high performance liquid chromatography. J. Chromatogr. A 1216, 7657–7663.

Wu, Z., Zhen, Z., Jiang, J., et al., 2009. Terminal protection of small-molecule-linked DNA for sensitive electrochemical detection of protein binding via selective carbon nanotube assembly. J. Am. Chem. Soc. 131, 12325–12332.

Xie, W., Han, C., Zheng, Z., et al., 2011. Determination of gibberellin A3 residue in fruit samples by liquid chromatography–tandem mass spectrometry. Food Chem. 127, 890–892.

Yin, X., Guo, J., Wei, W., 2010. Dual-cloud point extraction and tertiary amine labeling for selective and sensitive capillary electrophoresis-electrochemiluminescent detection of auxinsα. J. Chromatogr. 1217, 1399–1406.

Yomo, H., Brandon, D., 1960. Studies on the α-amylase activating substance. IV, on the amylase activating action of gibberellin. Hakko Kyokaishi 18, 600–602.

Zavala, M., Brandon, D., Li, L., 1983. Localization of a phytohormone using immunocytochemistry. J. Cell Biol. 97, 1235–1239.

Zhai, Q., Yan, C., Li, L., et al., 2017. Metabolism and Signaling of Jasmonate[2+]. Biosens. Bioelectron. 37, 112–115.

Zhang, B., Guo, L.H., Wong, W., 2012. Highly sensitive and selective photoelectrochemical DNA sensor for the detection of Hg^{2+} in aqueous solutions. Biosens. Bioelectron. 37, 112–115.

Zhang, H., Tan, S., Wong, W., et al., 2014. Mass spectrometric evidence for the occurrence of plant growth promoting cytokinins in vermicompost tea. Biol. Fertil. Soils 50, 401–403.

Zhang, H., Tan, S., Teo, C., et al., 2015. Analysis of phytohormones in vermicompost using a novel combinative sample preparation strategy of ultrasound-assisted extraction and solid-phase extraction coupled with liquid chromatography–tandem mass spectrometryβ. Talanta 139, 189–197.

Zhang, Y., Li, Y., Hu, Y., et al., 2010. Preparation of magnetic indole-3-acetic acid imprinted polymer beads with 4-vinylpyridine and β-cyclodextrin as binary monomer via microwave heating initiated polymerization and their application to trace analysis of auxins in plant tissues. J. Chromatogr. A 1217, 7337–7344.

Zhao, Z., Xing, Z., Zhou, M., et al., 2015. Functional analysis of synthetic DELLA domain peptides and bioactive gibberellin assay using surface plasmon resonance technology. Talanta 144, 502–509.

Zhong, Q., Qiu, X., Lin, C., et al., 2014. An automatic versatile system integrating solid-phase extraction with ultra-high performance ldchromatography–tandem mass spectrometry using a dual-dilution strategy for direct analysis of auxins in plant extracts. J. Chromatogr. 1359, 131–139.

Zhou, F., Lin, Q., Zhu, L., et al., 2013. D14-SCFD3-dependent degradation of D53 regulates strigolactone signallingArabidopsis. Nature 504, 406–410.

Zhou, R., Benavente, L., Stepanova, A., et al., 2011. A recombineering-based gene tagging system for *Arabidopsis*. Plant J. 66, 712–723.

Quantitative analysis of plant hormones based on LC-MS/MS

14

Jinfang Chu[1], Shuang Fang[1], Peiyong Xin[1], Zhenpeng Guo[2], Yi Chen[2]
[1]National Center for Plant Gene Research (Beijing), Institute of Genetics and Developmental Biology, Chinese Academy of Sciences, Beijing, China; [2]Key Laboratory of Analytical Chemistry for Living Biosystems, Institute of Chemistry, Chinese Academy of Sciences, Beijing, China

Summary

Endogenous plant hormones are naturally occurring low molecular weight compounds that have various physiological functions in the regulation of plant growth and development. There are several groups or families of hormones each with different physical and chemical properties. The ability of sensitive, fast, accurate and quantitative analysis of plant hormones is very important to elucidate metabolic and signaling processes in plants. Different analytical methods, including immunoassays, electrochemistry, chromatography, and mass spectrometry, have been used over the years, to analyze plant hormones quantitatively. The coupling of liquid chromatography (LC) to fractionate compounds, together with tandem mass spectrometry (MS/MS) to identify compounds, has established LC-MS/MS as the most powerful tool for quantitative analysis of plant hormones due to its high sensitivity, selectivity, and ease of manipulation. In this chapter, we review and introduce the current advances of plant hormones quantitative analysis methods based on LC-MS/MS, including sample preparation procedures and LC-MS/MS detection. Easy-to-follow quantification methods for different plant hormones based on LC-MS/MS are also presented in this chapter to facilitate the plant hormone community.

14.1 Introduction to the history of plant hormone analysis

Plant hormones are naturally occurring small molecule compounds and play particularly important roles in the normal functioning of plant. According to structural and chemical diversity, plant hormones are grouped into several classes, including auxins, cytokinins (CKs), abscisic acid (ABA), gibberellins (GAs), ethylene, jasmonic acid (JA), salicylic acid (SA), brassinosteroids (BRs), and strigolactones (SLs). The elucidation of the molecular mechanisms of plant hormone action contributes to understanding of plant biology and improving crops because of the particularly important roles of hormones in the regulation of plant growth and development (Li et al., 2017; Davies, 2013; Davies, 2010; Hooykaas et al., 1999). Progress toward a better understanding of the biosynthesis, metabolism, and mechanisms of plant hormone function requires detailed information on plant hormone structure, plant hormone distribution, and changing levels in diverse plant organs and tissues. The biological activities of plant hormones depend

Hormone Metabolism and Signaling in Plants. http://dx.doi.org/10.1016/B978-0-12-811562-6.00014-1

on their cellular concentration in the plant, so the availability of high-performance and high-sensitivity methods of analysis, especially sensitive and accurate quantification methods, is essential for facilitating our understanding of biosynthesis and mechanisms of action of plant hormones.

In the analytical history of plant hormones, various kinds of strategies have been used to quantify plant hormones, including biological approaches such as bioassays and immunoassays (Su et al., 2017), or physical and chemical analysis methods including electrochemical analysis, chromatography, and capillary electrophoresis (CE) with UV or fluorescence (FLR) detectors. All of these methods have played important roles in the quantification of plant hormones, although there are different problems or limitations during the application of these methods that have been reviewed (Davies, 2010; Du et al., 2012b; Fu et al., 2011; Porfirio et al., 2016). In the past decades, mass spectrometry (MS) has undergone spectacular development in the extent of its application in analytical chemistry, proteomics, metabolomics, and other fields. As MS instrumentation became more widely available, it was quickly adopted for use in the identification, characterization, and quantitative determination of plant hormones. The development of MS-based analytical methods made quantification of plant hormones more sensitive, accurate, and convenient.

Since the 1980s, gas chromatography (GC) coupled to MS (GC-MS) has become a powerful tool for determination of some plant hormones and their metabolites. Many methods of analysis of plant hormones based on GC-MS have been published (Barkawi et al., 2010; Birkemeyer et al., 2003; Björkman and Tillberg, 1996; Ikekawa et al., 1984; Koek et al., 2011; Liu et al., 2012b; Müller et al., 2002; Schmelz et al., 2003, 2004). The application of GC-MS helped further development of analysis techniques for plant hormones, but there is a drawback of the GC-MS-based method for quantification of nonvolatile plant hormones. Analytes must be separated in the gas phase at high temperature, but many compounds are not volatile or are unstable at high temperature. Derivatization must be carried out to convert polar and thermo-labile plant hormones into corresponding derivatives that are thermo stable and volatile, prior to being analyzed by GC-MS.

During the last 10 years, liquid chromatography-mass spectrometry (LC-MS), especially liquid chromatography coupled to tandem mass spectrometry (LC-MS/MS) has become the most widely used approach to plant hormone analysis (Alarcon Flores et al., 2011; Balcke et al., 2012; Flokova et al., 2014; Fu et al., 2012; Kojima et al., 2009; Li et al., 2016; Liu et al., 2012c; Matsuura et al., 2009; Pan et al., 2010; Prinsen et al., 1997; Xin et al., 2013; Yoneyama et al., 2008). This overcame some limitations of GC-MS for analyzing trace level plant hormones in complex plant samples, due to the good separation performance of LC and its versatility, without the need for derivatization. The use of tandem MS provides a means to characterize molecular ions at each of two cycles of detection, which thus provides detailed structural information to aid the identification of the analytes.

In this chapter, different plant hormone quantification methods based on LC-MS/MS will be discussed, including auxins, CKs, ABA, GAs, JA, SA, BRs, and SLs. The aim of this chapter is to provide a source of information to plant researchers for understanding the principles and methodology of LC-MS/MS quantification methods for detecting endogenous plant hormone levels accurately. As for quantitative analysis method of BRs, GC-MS methods are also discussed together with LC-MS/MS because both of them are still used widely in plant science research.

14.2   The analytical principle and problems

Plant extracts are extremely complex, multicomponent mixtures and the degree of difficulty in accurate quantification of plant hormones is determined mainly by their low concentration in extracts (Hooykaas et al., 1999). As we know, endogenous plant hormones are present at trace amounts in plants, usually at the level of 0.1–50 ng/g fresh weight (FW). However, hundreds of primary and second metabolites are present at much higher amounts in plants (Fig. 14.1), which makes analysis of plant hormones challenging to both biologists and analytical chemists. An ideal analytical method must be extremely highly selective and sensitive to quantify the relatively much lower content of plant hormones in the presence of many kinds of more abundant compounds in the complex plant extracts.

The whole analysis procedure of plant hormones based on LC-MS/MS recommended for typical plant material can be divided into five stages: (1) Sampling, (2) Grinding, (3) Extraction, (4) Purification, and (5) LC-MS/MS detection (Fig. 14.2(a)).

The former four stages are generally considered to be the part of the sample preparation or pretreament. The nature and goal of sample preparation operation is to decompose the matrix structure, to isolate the target analytes from potential interference substances, and to make the tested compounds detectable. It is evident that the sample preparation part has a profound influence on both the time cost of completing

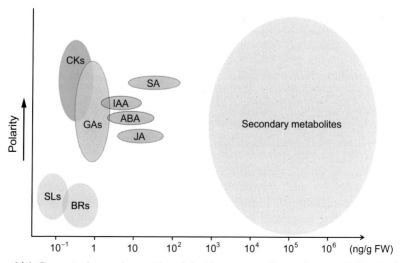

Figure 14.1 Concentrations and properties of plant hormones and secondary metabolites in plants.

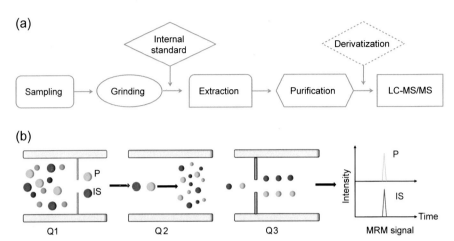

Figure 14.2 (a) Flow scheme for plant hormone quantification based on LC-MS/MS. (b) The principle of plant hormone quantification using a triple quadrupole MS in multiple reaction monitoring (MRM) mode. IS, internal standard; P, plant hormone.

the analysis process and the quality of analytical results. Therefore sample preparation is undoubtedly the most vital part, which usually takes up to 70%–80% of the total analysis work. Higher demands are being placed on selectivity, enrichment, and operability in sample preparation because of extremely low concentrations, complicated matrixes, and instability of plant hormones. The sample preparation procedures usually include sample homogenization, extraction, purification, and clean-up, and might even require a derivatization or labeling step depending on different strategies used in different laboratories or the requirements of particular experiments. Different methods for further purification of the crude extracts have been reported. Compared to the traditional liquid-liquid extraction (LLE) methods in which analytes are partitioned between different solvents, solid-phase extraction (SPE) is becoming the preferred purification technique used in plant hormone quantification for its advantages of saving time and solvents, and its higher recovery and throughput. In this article, sample preparation will be described as one part of the whole analysis strategy according to the particular requirements of different plant hormones.

Electrospray ionization (ESI) ion source and multiple reaction monitoring (MRM) mode based on the MS/MS technique are most commonly used for quantification of auxins, ABA, JA, SA, CKs, BRs, SLs, and their metabolites. The ESI source subjects the liquid phase to a very high voltage which creates an aerosol and ionizes the analytes. MRM mode can be used in different MS instruments classified according to different mass analyzers, such as quadrupole time-of-flight (Q-TOF), triple quadrupole (QQQ), and triple quadrupole linear ion trap (Q-Trap). In MRM, a precursor mass ion of a targeted analyte is selected in the first quadrupole (Q1) and then fragmented in the collision cell (Q2), and finally the diagnostic product ion is filtered and monitored by the third quadrupole (Q3). In this way the only signal that is detected is the diagnostic ion filtered by Q3, whose precursor ion was first selected by Q1. Since every compound has a distinct precursor-to-product ion transition in MRM mode, it is diagnostic for a particular substance in a complicated plant extract. The principle of the strategy of MRM mode of LC-MS/MS together with a stable-isotope-labeled internal standard (IS) is shown in Fig. 14.2(b). Now LC-MS/MS has become the most popular means for quantitative measurements of plant hormones and their metabolites because of its high specificity, sensitivity, ease of operation, and reproducibility.

For the analysis of endogenous plant hormones, LC-MS/MS is a powerful tool, but most methods involve multi-steps in the handling and preparation of samples. In addition, ion enhancement or suppression of the analyte signal during the ionization process is always subject to matrix effects, which may hamper the accuracy and precision. Therefore, a stable-isotope-labeled IS strategy is widely used to ensure highly accurate quantification results in such plant hormone analysis. Stable-isotope-labeled chemical standards of the target plant hormones under investigation share the same chemical and physical properties with the analytes, but they are distinct in molecular mass, which allows them to be distinguished by MS. Thus stable-isotope-labeled compounds added in a defined amount to the plant extract can correct for analyte loss and matrix effects, and so are used as IS to quantify trace levels of plant hormones with LC-MS/MS. In general, a known amount of stable-isotope-labeled IS is added before the plant material is fractionated. The ratio of analyte/IS is unchanged during the preparation and detection procedures and the intensity ratio of analyte/IS produced from mass spectrometry is used to quantify the level of target endogenous plant hormone in the sample. Deuterium (^{2}H), ^{13}C, or sometimes ^{15}N-labeled plant hormone compounds are often used as IS, which can be easily distinguished from endogenous plant hormones. If possible, IS labeled with more than three heavy-isotope atoms are preferred to monitor endogenous plant hormones. To ensure the accuracy of quantification, equations of linear regression related to the amount ratios to area ratios of unlabeled and labeled plant hormone compounds should be investigated. Only when the value of amount ratios of endogenous plant hormones and IS are covered by the standard curve, can the concentration of endogenous plant hormone in plant material be calculated with the following formula:

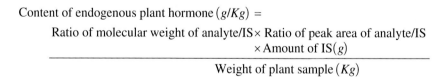

$$\text{Content of endogenous plant hormone } (g/Kg) =$$
$$\frac{\text{Ratio of molecular weight of analyte/IS} \times \text{Ratio of peak area of analyte/IS} \times \text{Amount of IS}(g)}{\text{Weight of plant sample } (Kg)}$$

It should be noted that when a QQQ or a Q-Trap is used to detect the concentration of plant hormones, there are some factors to which particular attention should be paid. Since the resolution of these instruments is only 1 Dalton, it is not enough to distinguish interfering peaks from the natural isotope cluster. One circumstance is that if the amount of IS added is too much or too little, it will result in a ratio value which is beyond the linear range of the standard curve, so the quantification result maybe inaccurate. Another circumstance is that if an IS labeled with one or two ^{2}H or ^{13}C atoms is used, there is only one or two mass units difference from the endogenous analyte, so a correction factor is needed to correct the analysis result.

14.3 Indole-3-acetic acid, abscisic acid, jasmonic acid and salicylic acid

14.3.1 Structures and physiochemical properties of indole-3-acetic acid, abscisic acid, jasmonic acid, and salicylic acid

All plant hormones can be divided into neutral, acidic, and basic chemical compounds, according to their molecular structures (Chiwocha et al., 2003). The main bioactive auxin, indole-3-acetic acid (IAA), together with ABA, GAs, JA, and SA are acidic plant hormones since they contain carboxyl groups (Fig. 14.3), and CKs are classified as basic compounds. Neutral hormones include SLs and BRs which are low-polar compounds and neither of them is easily ionized (Wu et al., 2014a).

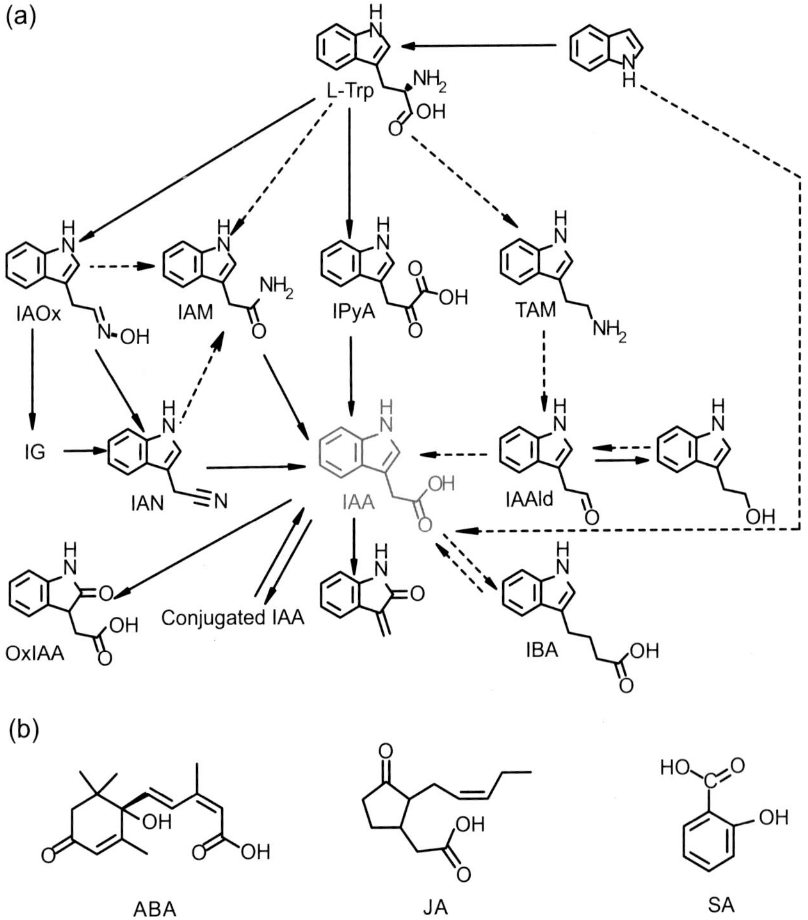

Figure 14.3 Structures of four acidic plant hormones. (a) Indole-3-acetic acid (IAA) and key metabolites (b) Abscisic acid (ABA), Jasmonic acid (JA), and Salicylic acid (SA).

Auxins are the first hormones discovered (Darwin and Darwin, 1880; Went, 1935) and are by far the most studied group of plant hormones for their biological functions in regulating plant growth and development (Zhao, 2010). Auxins are a series of indolic small molecule compounds and IAA is the most physiologically active substance in almost all plants species. In addition to IAA, quantification methods for precursors, conjugates, and metabolites of IAA, including tryptophan, indole-3-butyric acid (IBA), indole-3-pyruvic acid, and so on (Fig. 14.3(a)), are all essential to clarify the biosynthesis and mechanisms of auxin action.

ABA is an optically active sesquiterpenoid (15-carbon) that plays a very important regulatory role in many plant physiological processes, such as embryo development, seed germination and maturation, flowering, and response to drought, cold, and salt stress (Cutler et al., 2010; Nambara and Marion-Poll, 2005). ABA is a relatively hydrophobic compound containing a carboxylic acid (Fig. 14.3(b)). According to the position of the carboxyl group on the side chain, *cis-* and *trans-* isomers can be distinguished, and the former one exhibits the biological activity. ABA can be conjugated through a multitude of pathways (Feurtado et al., 2007), and ABA glucosyl ester is considered as one of the major inactive forms of ABA.

JA is a cyclopentanone carboxylic acid (Fig. 14.3(b)) and is well-known for its function as a signal in the integration of biotic and abiotic stresses with endogenous responses in plant persistence and survival responses. Endogenous JA levels can vary over several orders of magnitude ranging from ng/g to μg/g FW, depending on the environmental conditions and the specific tissues. In particular, JA levels in healthy leaf tissues can increase transiently in response to wounding, herbivory, elicitor treatments, and other biotic and abiotic stimuli (Browse, 2009; Robert-Seilaniantz et al., 2011). Methyl jasmonate, (+)-7-*iso*-jasmonoyl-L-isoleucine and 12-oxo-phytodienoic acid are major metabolites of JA which are important in the biosynthesis and signaling pathways of JA (Zhai et al., 2017).

SA is considered to be an important signaling molecule which is involved in local and systemic disease resistance in plants in response to various pathogenic attacks (Hayat et al., 2010). Chemically, SA is a phenolic derivative that possesses an aromatic ring with a hydroxyl group (Fig. 14.3(b)). It is a natural product of phenylpropanoid metabolism. Decarboxylation of *trans*-cinnamic acid to form benzoic acid and its subsequent 2-hydroxylation results in SA (Hayat et al., 2007). The SA compound is a crystalline powder having a melting point of 157–159°C and a pK_a of 2.98 (Raskin, 1992b). SA is distributed in a wide range of plant species. It undergoes metabolism by conjugating with glucose to form SA glucoside. The changes in level of SA in plant cells will dramatically influence immune responses to environmental factors and developmental processes (Raskin, 1992a; Vlot et al., 2009), so precise quantification of SA in plants is essential to clarify mechanisms of SA action.

Because ABA, JA, and SA exhibit many similar chemical properties with IAA, they can be analyzed together with IAA using a single-run purification and quantification method using LC-MS/MS. Compared with these four acidic plant hormones, the levels of GAs are much lower in plants, which makes it more difficult to analyze and quantify them, so special preparation procedures and analysis are required. Quantification methods of endogenous multicomponent GAs based on LC-MS/MS will be explained in a subsequent subsection in this chapter. Quantification methods of IAA, ABA, JA, and SA using LC-MS/MS, including sampling, extraction, purification, and detection procedures, will be introduced in detail here.

14.3.2   Sample preparation

14.3.2.1   Sampling

Sample collection is a vital step in the procedures of sample preparation. Since distribution and levels of endogenous plant hormones in plants are very easily affected by slight changes in environmental factors such as temperature, light intensity, and humidity, conditions for both plant growth and sample collection should be rigidly controlled among different groups (Chiwocha et al., 2003; Hayat et al., 2010). If not, the quantification results of plant hormones cannot represent the real concentrations in different sample groups, even though the following analytical procedures are completed perfectly. Biological variability is usually much greater than variability in analytical procedures. Generally, sample tissues need to be cut from the plant or collected from culture vessels, so rapid manipulation with the minimum of disturbance is very important to avoid changes in the level of plant hormones induced by environmental factors or plant wounding (Chiwocha et al., 2003). After harvesting plant samples, the frozen plant material can be kept at −80°C before analysis.

In order to calculate plant hormone concentration accurately, the weight of plant materials need to be recorded exactly. In different methods of quantification of IAA, ABA, JA, and SA, both dry weight (DW) and FW are used to express plant hormone concentrations in plant tissues (Chiwocha et al., 2003; Kowalczyk and Sandberg, 2001; Müller and Munné-Bosch, 2011). For FW values, frozen fresh tissue can be weighed while for DW values the tissue can be lyophilized and weighed. Levels of plant hormones in lyophilized samples are stable at room temperature under dry conditions.

Homogenization is another key step after sample collection, and aims at breaking the cell walls and allowing plant hormones to be dissolved in extraction solvent (Tarkowská et al., 2014). Well-homogenized plant samples will improve the extraction efficiency of plant hormones from plant tissues. For large amounts of plant material, samples are homogenized to a fine powder using a mortar and pestle or tissue homogenizer with steel ball under liquid nitrogen (Davies, 2013; Fu et al., 2012; Sun et al., 2014). If only small amounts of tissue (mg) are available to quantify plant hormone content, the sample can be homogenized directly in 1.5 mL plastic micro tubes using a vibrating-ball micro mill. To the tube containing plant tissue, tungsten carbide beads and extraction buffer are added and the tubes vibrated at the appropriate frequency for the required time (Wang et al., 2015). No matter what grinding method is used, the plant material must always be kept cold to avoid enzymatic or chemical degradation of plant hormones.

14.3.2.2 Extraction

Most (if not all) plant tissues are solid or semisolid samples, and the next step after sampling in analysis methods of plant hormones is usually the exhaustive extraction of target hormones from the matrix and their transfer into liquid solvent, which makes it convenient for the following operations. The extraction efficiency depends not only on the physicochemical properties of target compounds but also on their subcellular localization and their association with others compounds such as proteins, pigments, lipids, and phenolics (Hillman, 1978). An ideal extraction solvent should meet two criteria. First, the solvent must be capable of extracting the target plant hormone efficiently from plant tissues. Second, the selected solvent should minimize extracting interfering substances. It is hard to find such an ideal extraction solvent because the interfering compounds are in large excess over plant hormones in plant tissues.

Solvent extraction is the most commonly used method for plant hormone extraction. IAA, ABA, JA, and SA, as acidic plant hormones, are freely soluble in methanol (MeOH) and ethanol (EtOH), soluble in acetone, ethyl acetate and diethyl ether, and slightly soluble in water and chloroform (Hillman, 1978). Depending on different plant materials and target plant hormones, many different solvents have been used, such as MeOH (Fu et al., 2012), MeOH: water (Cui et al., 2015), MeOH: water: acetic acid (Lu et al., 2015), MeOH: water: formic acid (FA) (Kojima et al., 2009), acetone (Wu et al., 2014c), acetone: water (Matsuda et al., 2005), isopropanol: water: HCl (Pan et al., 2008, 2010), and isopropanol: imidazole buffer (Strader et al., 2010). Several extraction solvents were compared and the results showed that mixtures of MeOH and water provided higher extraction efficiency for IAA, ABA, JA, and SA because of its good penetrability into plant cells (Trapp et al., 2014). The optimal solvent depends not only on the extraction efficiency but also on the following purification method. It should be mentioned that EtOH used as extraction solvent can possibly result in the esterification of acidic plant hormones (Barkawi et al., 2010). IAA and ABA are easily oxidized when exposed to oxygen and light, so adding a small amount of antioxidant to the extraction solvent is suggested before extracting plant tissues, to protect the analyte against degradation. Diethyldithiocarbamic acid sodium salt (Novak et al., 2012) and butylated hydroxytoluene (Karadeniz et al., 2006) are the most widely used antioxidants during extraction. Moreover, the extraction is normally carried out at low temperature to reduce plant hormone degradation. As for extraction time, it ranges from 15 min to 24 h depending on different methods. The volume of extraction solvent must be enough to extract most target plant hormones into the solvent. A ratio of sample: extraction solvent of 1:10 (mg/μL) is typical (Pan et al., 2010). Other techniques, such as vacuum microwave-assisted extraction (Hu et al., 2011), magnetic phase extraction (MSPE) (Cai et al., 2015b), and polyaniline-sheathed electrospun nanofiber bar extraction (Wu et al., 2014a), have also been established for extraction of IAA, ABA, JA, and SA, aiming to improve the extraction efficiency. Stable-isotope-labeled acidic plant hormones, such as $^{13}C_6$-IAA and $^2H_2$-IAA for endogenous IAA, $^2H_6$-ABA for ABA, $^2H_6$-JA and $^2H_5$-JA for JA, and $^2H_4$-SA for SA, are commonly added into the sample before or after extraction solvent is added (Balcke et al., 2012; Chen et al., 2011; Fu et al., 2012; Jikumaru et al., 2013; Pan et al., 2010).

14.3.2.3   Purification and clean-up

Not only are the target compounds IAA, ABA, JA, and SA extracted in trace amounts, but also a large quantity of non-target compounds in much higher concentration are extracted during extraction step. Since the non-target compounds will probably result in serious interference, the crude extract may not be injected directly into the LC-MS instrument for analysis without further purification and clean-up steps. Purification and clean-up of crude extracts from plant samples can remove the bulk of interfering compounds and improve separation by high performance liquid chromatography (HPLC) and detection by MS. The development of a high efficiency purification method for plant hormones has been considered to be a bottleneck for plant hormone analytical procedures for a long time.

The choice of purification methods depends on not only the target analytes but also on the type of analysis to be performed, as well as the available analytical instruments (Pan and Wang, 2009). In the LC-MS/MS strategy for IAA, ABA, JA and SA, several techniques have been adapted for sample purification. LLE is used widely to purify acidic plant hormones in the early steps of application of LC-MS/MS. Different organic solvents were chosen for IAA, ABA, JA, SA, and related metabolites, such as dichloromethane (Pan et al., 2010) or methylene chloride (Pan et al., 2008). Ethyl acetate has been used for IAA, ABA, and JA (Durgbanshi et al., 2005), and diethyl ether for JA stereoisomers (Han et al., 2012a). Traditional LLE involves many extraction steps with different solvents involving long handling time and leading to the consumption of solvents and generation of waste solvents. In addition, for the water-soluble plant hormones IAA, ABA, JA, and SA, it is hard to improve extraction efficiency of LLE due to emulsification (Fu et al., 2011). To resolve the disadvantage of traditional LLE, liquid-liquid micro-extraction (LLME) methods were applied to extract and enrich plant hormones. The hollow fiber-based liquid-liquid extraction (HF-LPME) method can efficiently extract and enrich IAA, ABA, JA, and SA, as demonstrated in natural coconut juice samples (Wu and Hu, 2009). In contrast to HF-LPME, dispersive liquid–liquid micro-extraction (DLLME) requires no special interface when used in conjunction with HPLC, and its simple operation makes it a good technique in cleaning extracts containing IAA, ABA, and JA, as demonstrated in peach samples (Lu et al., 2015). A sequential solvent-induced phase transition extraction (SIPTE) method was successfully used to purify and enrich IAA, ABA, JA, and SA in 5-mg plant sample, prior to UPLC-MS/MS detection (Cai et al., 2015a).

Although it was reported that IAA, ABA, JA, and SA and other plant hormones in crude plant extracts could be quantified directly with LC-MS/MS (Pan et al., 2008) without further clean-up or purification, the purification procedure is still preferred before the sample is injected. It is known that co-extracted substances in the crude extract can interfere with the analysis and cause ionization enhancement or suppression when LC-MS/MS is used. Therefore, the complexity of the matrix together with the small amounts of analytes present, require that the study of matrix effects is an important step in order to optimize procedures to obtain reliable data (Raskin, 1992a; Silva et al., 2012). Besides purification as a means to reduce matrix effects for plant hormone quantification, a clean sample is very helpful to prolong the lifetime of the column in the LC-ESI-MS/MS system and to maintain the MS instrument in good condition and thus deliver reproducible results.

SPE is probably the most widely accepted technique for pre-concentration and purification of analytes from fluids and aqueous samples (Ramos, 2012). The large variety of sorbents available, either commercially or home-synthesized, makes this technique suitable for separation and clean-up of target compounds with different chemical structures and polarities. Because of its higher efficiency than LLE, SPE is by far the most popular purification technique for LC-MS/MS quantification of IAA, ABA, JA, and SA. Generally, the SPE solid sorbent binds or traps target plant hormones on the basis of different interactions, such as hydrogen bonding, polar or non-polar interaction, and cation or anion interaction depending on the property of the selected SPE cartridge (Buchanan et al., 2015). Therefore, selection of the appropriate SPE sorbent is critical for developing a good purification method, and the researcher must have a good knowledge of the mechanism of interactions between the SPE sorbents and the target compounds.

The hormones IAA, ABA, JA, and SA belong to a series of weak acidic small molecule compounds, since they all contain a carboxyl group, and diverse SPE cartridges have been applied to purify and concentrate them individually or simultaneously in the whole quantification strategy using LC-MS/MS. The most widely used sorbent when the SPE technique was firstly used in purifying the weakly acidic plant hormones was C18. Based on hydrophobic interactions, the C18 sorbent of the "Sep-Pak" C18 SPE cartridge was successfully used to purify 15 plant hormones including IAA and ABA from lettuce seed extract (Chiwocha et al., 2003). Now, due to their chemical composition, the lipophilic divinylbenzene and the hydrophilic N-vinylpyrrolidone, Oasis HLB SPE sorbents have been widely used to quantitatively extract acidic plant hormones including IAA, ABA, JA, SA, and related metabolites from plant extracts (Flokova et al., 2014; Gouthu et al., 2013; Han et al., 2012b; Kojima et al., 2009; Matsuura et al., 2009; Novak et al., 2012). The separation mode of C18 and HLB sorbents is based on a reversed phase (RP) retention mechanism. Mixed-mode, reversed-phase cationic-exchange cartridges or anionic-exchange cartridges, were found to be highly suitable to purify IAA and similar acidic hormones due to the two kinds of interaction involved, and improved the sensitivity of LC-MS/MS quantitation of these acidic plant hormones (Bosco et al., 2014; Fu et al., 2012). More and more purification methods with both C18 and mixed-mode sorbents have been reported for simultaneous quantification of multiple plant hormones, including IAA, JA, SA, ABA, GAs, CKs, BRs, and their metabolites (Izumi et al., 2009; Jikumaru et al., 2013; Kojima et al., 2009).

Improved SPE techniques, including solid-phase micro-extraction (SPME), MSPE (Cai et al., 2015b), molecular imprinted polymers (Hu et al., 2011), and polyaniline-sheathed electrospun nanofiber bar extraction in vivo (Wu et al., 2014b), were also introduced to improve purification efficiency for IAA, ABA, JA, and SA. Sample preparation and LC-MS/MS detection methods for IAA, ABA, JA, and SA are summarized in Table 14.1.

Table 14.1 Summary of sample preparation and quantification methods for IAA, ABA, JA, and SA based on LC-MS/MS

Plant hormones	Extraction solvent	Purification and materials	Plant sample	Quantification instrument	References
ABA, SA, JA	MeOH:acetic acid (99/1, v/v, %)	No purification	Arabidopsis, 50 mg FW	MRM, UFLC-MS/MS (QQQ MS)	Kasote et al. (2016)
ABA, SA	MeOH:water:acetic acid (80/19/1, v/v/v, %)	DLLME	Peach fruit, 250 mg FW	MRM, LC-ITMS (IT MS)	Lu et al. (2015)
IAA, ABA, SA, JA	Acetonitrile:water:FA (80/19/1, v/v/v, %)	SPE-LLE, Oasis MCX cartridge and EtOAc	Leaves of oilseed rape, 1 g FW	MRM, LC-MS/MS (Q-Trap MS)	Cui et al. (2015)
IAA, ABA, JA	Acetonitrile	Magnetic phase extraction (MSPE)	Rice seedlings, 100 mg FW	MRM, LC-MS/MS (Q-Trap MS)	Cai et al. (2015b)
IAA, ABA, JA, SA	Acetonitrile	Sequential phase transition extraction (SIPTE)	Rice and Arabidopsis, 5 mg FW	MRM, LC-MS/MS (Q-Trap MS)	Cai et al. (2015a)
IAA, ABA, JAs	MeOH	SPE, DEAE-Sephadex	Arabidopsis leaves, 20–80 mg FW	MRM, LC-MS/MS (Q-Trap MS)	Ziegler et al. (2014)
IAA, ABA, JA, SA	MeOH	SPE, Bond Elut CBA cartridge	Arabidopsis seed, 0.8 mg DW	MRM, LC-MS/MS (Q-Trap MS)	Sun et al. (2014)
IAA, ABA, JA	Acetonitrile	MSPE, TiO₂/magnetic hollow mesoporous silica sphere (MHMSS)	Arabidopsis, 100 mg	MRM, UPLC-MS/MS (QQQ MS)	Liu et al. (2014)
IAA, ABA, JA, SA	MeOH:water (10/90, v/v, %)	SPE, Oasis HLB cartridge	Arabidopsis, 20–25 mg FW	MRM, UHPLC-MS/MS (QQQ MS)	Flokova et al. (2014)
IAA, ABA, JA, SA	EtOH:water (70/30, v/v, %)	LLE, ethyl acetate	Fermentation broths of *Botryodiplodia theobromae*, 3 mL	MRM, RPLC-MS/MS (IT MS)	Castillo et al. (2014)

Hormones	Mobile phase	Purification	Sample	MS method	Reference
ABA, JA, SA	MeOH:water:acetic acid (10/89/1, v/v/v, %)	SPE, Oasis MCX cartridge	Rose leaves, 20 mg	MRM, UHPLC-MS/MS (QQQ MS)	Bosco et al. (2014)
JA, ABA	Electrospun fibers	In vivo solid-phase micro-extraction, polyaniline-sheathed electrospun nanofiber bar	Aloe leaf tissues, in vivo	LC-MS/MS (QQQ MS)	Wu et al. (2014)
IAA, ABA	MeOH:acetonitrile: water:acetic acid (40/40/20/1, v/v/v/v, %).	No purification	Bean sprout and tomato, 10 g	MRM,HPLC-MS/ MS (QQQ MS)	Ma et al. (2013)
IAA, ABA, JA, SA	Acetonitrile:water:acetic acid (80/19/1, v/v/v, %)	SPE, Oasis HLB Oasis MAX	Arabidopsis, 5 mg DW	MRM, LC-MS/MS (QQQ MS)	Jikumaru et al. (2013)
IAA, ABA	MeOH:FA:water (15/1/4,v/v/v)	SPE, Oasis HLB Oasis MCX	Grape berry, 50 mg FW	MRM, LC-MS/MS (Q-Trap MS)	Gouthu et al. (2013)
IAA, ABA, JA, SA	MeOH:FA:water (15/1/4, v/v/v)	Filtered with 30kDa Amicon Ultra centrifugal filter unit	Tomato leaf and fruit tissue, 100 mg	U-HPLC-MS (Orbitrap MS)	Van Meulebroek et al. (2012)
IAA	Sodium phosphate buffer (50 mM, pH 7.0)	SPE, Oasis HLB	Arabidopsis seedlings, 20 mg FW	MRM, LC-MS/MS (QQQ MS)	Novak et al. (2012)
IAA, ABA, JA, SA	MeOH:water: acetic acid (80/19/1, v/v/v, %)	No purification	Rice, 100 mg	MRM, UFLC-MS/ MS (Q-Trap MS)	Liu et al. (2012d)
IAA, ABA, JA, SA	MeOH	SPE, Oasis MAX	Arabidopsis, 20–200 mg FW	MRM, UPLC-MS/ MS (QQQ MS)	Fu et al. (2012)
IAA,ABA, JA,SA	MeOH:water (80/20, v/v, %)	SPE, cartridges containing C18 adsorbent and SAX adsorbent	Rice, 5 mg	Nano-LC-MS (Q-TOF MS)	Chen et al. (2012)
IAA, IBA	MeOH:FA (99/1, v/v, %)	SPE, Oasis HLB	Grapes, 2.0 g	SRM, HPLC-MS/ MS (QQQ MS)	Han et al. (2012b)

continued

Table 14.1 Summary of sample preparation and quantification methods for IAA, ABA, JA, and SA based on LC-MS/MS—cont'd

Plant hormones	Extraction solvent	Purification and materials	Plant sample	Quantification instrument	References
SA, ABA, JA	MeOH	SPE, a strong cation-exchange HR-XC material (96-well filter plates)	Tomato, 20–50 mg FW	MRM, UPLC-MS/MS (Q-Trap MS)	Balcke et al. (2012)
IAA, ABA, JA, SA, IBA	MeOH:water (80/20, v/v, %)	SPE, C18 cartridges	Rice leaves, 3 g	Self-made CE-TOF-MS	Chen et al. (2011)
IAA, ABA, JA, SA	2-Propanol:water:concentrated HCl (2/1/0.002, v/v/v)	LLE, dichloromethane	Arabidopsis, 50 mg FW	MRM, LC-MS/MS (Q-Trap MS)	Pan et al. (2008, 2010)
IAA, ABA, JA, SA	MeOH:water:FA (15/4/1, v/v/v)	SPE, Oasis HLB 96-well plate, Oasis MCX 96-well plate	Rice, 100 mg FW	MRM, UPLC-MS/MS (QQQ MS)	Kojima et al. (2009)
JA, SA	80% MeOH with 2% acetic acid (1:30, w/v)	SPE, Bond Elut C18 cartridges	Tobacco, 2–3 g	MRM, UPLC-MS/MS (QQQ MS)	Matsuura et al. (2009)
IAA, ABA	MeOH:water:FA (15/4/1, v/v/v)	SPE, Oasis HLB and Oasis MCX	Arabidopsis and tobacco, 100 mg FW	MRM, nano-LC-MS/MS (IT MS)	Izumi et al. (2009)
IAA, ABA	MeOH:water (80/20, v/v, %)	SPE, C18	Wheat tissues, 2 g	SRM, LC-MS/MS (LTQ IT MS)	Hou et al. (2008)
IAA, ABA, IPA	Isopropanol:glacial acetic acid (99:1, v/v, %)	SPE, Sep-Pak C18	Lettuce seeds, 50–100 mg	HPLC-MS/MS (QQQ MS)	Chiwocha et al. (2003)

14.3.3   LC-MS/MS detection

Because no derivatization is required, LC is a more suitable technique, compared with GC, to separate and analyze acid plant hormones coupled with different detectors, including UV-detector (Absalan et al., 2008; Liu et al., 2007; Wu and Hu, 2009; Yan et al., 2012), FLR detector (Lu et al., 2010; Zhang et al., 2010a), and MS (Table 14.1). Given the sensitivity, selectivity, and the range of adaptability, MS is by far the optimum choice and the most popular detector combined with LC for quantification of IAA, ABA, JA, SA, and other plant hormones.

The analytical technique of MS detects the mass-to-charge ratio (m/z) of charged particles of analytes ionized by the ion source. Various types of mass analyzers combined with LC, including ion trap mass, Q-TOF, QQQ, and Q-Trap have been reported in different methods of plant hormones analysis (Table 14.1).

The m/z separation capability of the mass analyzer improves the selectivity of target compounds, which is very helpful for quantification of the trace levels of plant hormones in plant extract after purification by SPE (Fu et al., 2011). Compared with full scan MS mode in which all ions are monitored simultaneously, MRM mode, sometimes called selected reaction monitoring mode, provides higher selectivity and specificity when MS/MS is used to analyze multiple plant hormones. In particular, the MRM mode is the most commonly used MS scan mode to quantify IAA, ABA, SA, JA, and related plant hormones due to its higher selectivity and higher sensitivity. The LC-MS/MS strategy has been performed successfully with extracts from various plants such as Arabidopsis, rice, wheat, tomato, and lettuce. The amount of plant tissues required has been reduced significantly from g (FW) amounts to mg (FW) amounts, as a result of the application of LC-MRM-MS/MS (Table 14.1). This has made quantification of the acidic plant hormones IAA, ABA, SA, and JA in specific tissue, or organs, readily achievable. Monitoring changes in plant hormone levels is very important to understand the functions and molecular mechanisms of different plant hormones. It is also particularly valuable when applied to the analysis of mutants for which amounts of material may be limited.

The negative ion mode of the mass spectrometer is generally much less sensitive than the positive ion mode. Therefore, derivatization of negatively charged compounds such as IAA, ABA, JA, and SA by incorporating a positively charged quaternary ammonium group is an effective way to improve the MS signal response. Such derivatization has been successfully used to further improve the sensitivity of detection of IAA, ABA, JA, SA, and other acidic plant hormone compounds. Such compounds can be accurately quantified in mg or even sub-milligram amounts of Arabidopsis seedlings, and in a single seed which is typically about 25 µg (FW). The auxin IBA is difficult to quantify in plants using LC-MS/MS methods due to its much lower level than IAA, but was successfully detected in a single Arabidopsis seed following derivatization (Sun et al., 2014).

Selection of the appropriate IS is a critical factor for quantitative analysis, and stable-isotope-labeled IS is commonly used in quantification of plant hormones based on LC-MS/MS. This cannot only correct for loss of sample during preparation but can also eliminate matrix effects from interfering substances. IS labeled with ^{13}C atoms is the best choice, but ^2H, ^{15}N, or ^{18}O labeled compounds can be used. $^{13}C_6$-IAA, $^2H_2$-IAA, $^2H_4$-IAA, $^2H_5$-IAA, $^2H_6$-ABA, $^2H_4$-SA, $^2H_5$-JA, and $^2H_6$-JA can be obtained commercially or can be synthesized. It should be noted that not all stable-isotope-labeled plant hormone compound can be used as IS to quantify endogenous plant hormone levels. The whole analysis strategy and validation method must be considered. If it is difficult to obtain a stable-isotope-labeled IS, a stable-isotope-labeled derivatization agent can be used to modify the plant hormone extract with a chemical tag. In this way the relative quantification of IAA, ABA, JA, SA, and other plant hormones can be accomplished, which is especially valuable when stable-isotope-labeled IS is unavailable.

14.3.4 An easy-to-follow quantification method for indole-3-acetic acid, abscisic acid, jasmonic acid, and salicylic acid

The whole quantification strategy is summarized in Fig. 14.4(a). Plant material is collected and ground in liquid nitrogen with a mortar and pestle. About 50–100 mg (FW) powdered plant sample is weighed accurately, then MeOH added together with $^{13}C_6$-IAA, $^2H_6$-ABA, $^2H_4$-SA, and $^2H_5$-JA as ISs. The sample is extracted overnight at −20°C and then centrifuged for 15 min at 15,000 g at 4°C. The supernatant is then collected and evaporated under nitrogen. The residue is dissolved in ammonia solution (5%,v/v).

To clean plant extracts Oasis MAX SPE cartridges are used. Before loading the sample, the SPE cartridge needs to be preconditioned sequentially with MeOH, water, and ammonia solution. Then the sample is loaded onto a cartridge and washed sequentially with ammonia solution, water and MeOH. Finally, MeOH containing 10% (v/v) FA is used to elute the target plant hormones, IAA, ABA, JA, and SA. The eluted solution is dried and reconstituted in 80% (v/v) MeOH for LC-MS/MS analysis.

The detection system in the present example uses a UPLC-MRM-MS/MS strategy. An ACQUITY UPLC (Waters) system combined with 5500 Q-Trap MS equipped with an ESI source (AB SCIEX) is used to separate and detect the four acidic plant hormones of IAA, ABA, JA, and SA. Five to 10 μL of sample is injected onto a BEH C18 column (2.1 mm × 50 mm i.d., 1.7 μm) and the column temperature is kept at 35°C. Analytes of IAA, ABA, JA, and SA are separated with two mobile phases comprising (A) 0.05% acetic acid in water, and (B) 0.05% acetic acid in acetonitrile (ACN). The gradient runs initially with 15% B, then increased to 40% B in 5 min and further increased to 80% B in the next 0.5 min. The flow rate is 0.5 mL/min. The optimized mass spectrometer parameters are set as follows: curtain gas at 40 psi, collision gas at 6 psi, ion spray voltage at −4300 V, and temperature at 550°C. The optimized MRM parameters are listed in Table 14.2. Example chromatograms are shown in Fig. 14.4 (b).

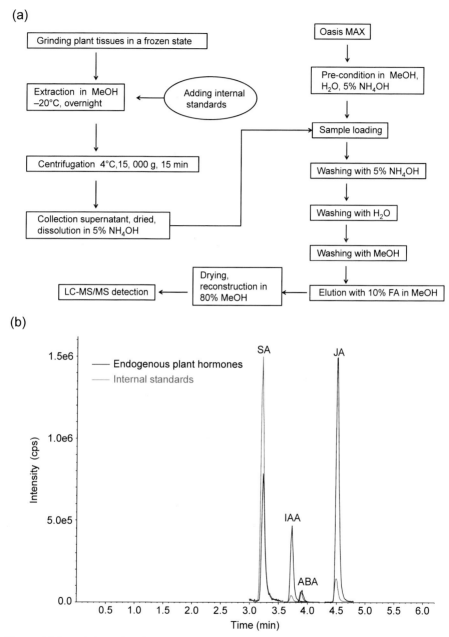

Figure 14.4 (a) Schematic of LC-MS/MS quantification strategy of IAA, ABA, JA, and SA. (b) UPLC-MS/MS MRM chromatograms of quantification of IAA, ABA, JA, and SA in Arabidopsis seedling.

Table 14.2 **Optimized MRM parameters for target plant hormones**

Compounds	Q1/Q3	Collision energy/V	Declustering potential/V	Entrance potential/V	Collision cell exit potential/V
IAA	174/130.1	−13	−85	−9	−8
$^{13}C_6$-IAA	180/136.1	−13	−85	−9	−8
ABA	263/153.1	−15	−85	−10	−9
$^2H_6$-ABA	269.1/159.3	−15	−85	−10	−9
JA	209/59	−16	−80	−9	−6
$^2H_5$-JA	214.1/59	−16	−80	−9	−6
SA	137/93	−22	−70	−9	−10
$^2H_4$-SA	141/9	−22	−70	−9	10

14.4 Gibberellins

14.4.1 Structures and analysis of gibberellins

GAs are a major set of endogenous plant hormones, widely distributed throughout the plant kingdom and playing many key roles in higher plant growth and development, such as seed germination, stem elongation, the transition to flowering, pollen development, and fruit growth (Bethke and Jones, 1998; Hooley, 1994; Olszewski et al., 2002; Segarra et al., 2006; Shani et al., 2013; Tyler et al., 2004). As a class of tetracyclic diterpenoid carboxylic acids with either *ent*-gibberellane or 20-*nor-ent*-gibberellane carbon skeletons, GAs comprise at least 136 highly similar endogenous analogs. Only a few of them have intrinsic biological activity, including GA_1, GA_3, GA_4, GA_5, GA_6, and GA_7, while GA_1, GA_3, and GA_4 are the predominant bioactive GAs, but appear only at an ultra-trace level in the range of fmol to pmol per gram (FW) in plants, depending on the tissues and species (Hedden and Thomas, 2012; Mander, 1992). It is essential to know not only the concentrations of the bioactive forms of GAs, but also their precursors and metabolites, to provide important information of GA metabolism and function. However, it is a serious challenge to analyze, and especially to quantify, these complex and chemically similar ultra-trace GAs. The challenge is further increased by the poor detectability of GAs, which have no FLR and only weak UV absorbance, and in some cases by the limited availability of specific biological samples.

Various potential methods have been tried mainly based on different separation principles and spectrometry. Separation methods have the power to resolve complex GAs with various choices such as GC (Lange et al., 2005; Moritz and Olsen, 1995), LC (Barendse et al., 1980; Koshioka et al., 1983; Lu et al., 2016; Urbanova et al., 2013), CE, and electrochromatography (Ge et al., 2007, 2008; Zhu et al., 2013). However, their applicability depends much on the detectability of trace amounts of GAs. The most advanced separation tools are now coupled with MS to enhance not only the detection sensitivity but also the ability to identify separated peaks (Pan and Wang, 2009). LC-MS/MS is now becoming the most widely employed method for the simultaneous qualitative and quantitative detection of multiple GAs.

A complete LC-MS/MS analytical process for GA analysis typically includes several steps that are shown in Fig. 14.2. The strategies and technical developments of each step will be illustrated here, with emphasis on some promising methods that have been developed.

14.4.2   Sample preparation

Although an in vivo method for extracting analytes directly from living plants is in principle the ideal approach, the limited development of such methods and the very low levels of GAs in vivo mean that the harvesting and homogenization of tissue in vitro is still the best approach. Since wounding can induce changes in the level of some hormones (Rappaport and Sachs, 1967), rapid harvesting and immediate freezing in liquid nitrogen are accepted strategies to avoid chemical or enzymatic degradation of the GAs.

The sample matrix remains a serious source of interference in LC-MS/MS, not only reducing the resolution of LC and the ionization efficiency of MS, but also increasing the detector "noise" and ultimately reducing the limits of detection (Chen et al., 2008). Therefore, it is crucial to choose the right sample preparation approaches.

Prior to the extraction of GAs from the plant samples, isotope-labeled GAs should be added as IS, and frozen plant samples pulverized or homogenized directly with extraction solution at low temperature. Sample loss should be minimized during this procedure, especially when dealing with very small quantities of plant material (mg and sub-mg amounts).

Effective extraction of GAs from solid plant materials is critical and therefore can be time-consuming, usually needing several hours at low temperature. Many different protocols have been used for extraction, including the use of polar organic solvents such as MeOH and ACN usually mixed with water, and FA can also be added to help extraction (Tarkowska et al., 2014). The extraction protocol should be optimized depending on the type and the amount of plant material to achieve high efficiency, good recovery, and high reproducibility.

Purification or clean-up of the crude extracts before analysis commonly employs LLE or SPE. As weak organic acids, with dissociation constants (pK_a) around 4.0, GAs can be extracted into organic solvent (ethyl acetate or dichloromethane) from acidic aqueous solution (pH 2.5–3.0), so LLE is an effective way to purify the plant extracts (Li et al., 2016; Pan et al., 2010). SPE with C18, ion-exchange and mixed-mode (reversed-phase retention and ion-exchange) sorbents are also widely used to extract and enrich GAs from the crude extracts (Chiwocha et al., 2003; Izumi et al., 2009). LLE, coupled to SPE, has been exploited as an effective purification procedure (Chen et al., 2012). Now, to improve the extraction efficiency and selectivity, several new approaches have been used including magnetic molecularly imprinted polymer beads (Zhang et al., 2012), hybrid magnetic metal-organic frameworks (Hu et al., 2013), and TiO_2/magnetic hollow mesoporous silica sphere sorbents (hydrophilic SPE) (Liu et al., 2014). MSPE procedures have also been established for the removal of the sample matrix and the enrichment of some GAs (Liu et al., 2014). Some micro-extraction methods have also been applied in the determination of GAs, such as dispersive DLLME and single-drop liquid-liquid-liquid micro-extraction for GA_3 (Bai et al., 2012; Gupta et al., 2011) and hollow fiber-based liquid-liquid-liquid micro-extraction for eight GAs (Wu et al., 2012), with the results showing good enrichment factors.

14.4.3 Derivatization

An important issue faced in the determination of GAs by ESI-MS is the poor sensitivity resulting from the low ionization efficiency in negative ion mode. Chemical labeling is a straightforward strategy to improve the sensitivity of GA analysis using ESI-MS, by converting carboxyl groups to positively charged moieties, permitting measurement in the positive ion mode to enhance ionization efficiency and thus improving detection sensitivity (Qi et al., 2014). The derivatization reaction should be conducted under very mild conditions, since some GAs are labile. For example, lactone rearrangement in GA_3 and GA_7 under alkaline conditions (pH > 8.0), and structural rearrangement of the C/D rings and hydration of the 16,17-double bond in GA_1 under strongly acidic conditions (pH < 1.0), result in identification difficulties. However the labeling reaction(s) should be highly reactive to enable the direct labeling of GAs present in trace amounts. Bromocholine bromide has been used to derivatize acidic plant hormones including 12 GAs, but the conditions of the labeling reaction were somewhat harsh involving treatment with triethylamine (TEA) at 80°C (Kojima et al., 2009). Subsequently, 3-bromoactonyltrimethylammonium bromide was synthesized, which can react efficiently with GAs in ACN containing TEA at room temperature (Chen et al., 2012). A one-reagent labeling technique that used only N-(3-dimethylaminopropyl)-N-ethylcarbodiimide (EDC) was reported to directly react with ultra-trace monocarboxylic acid GAs under very mild conditions, typically at pH 4.5 and 35°C. This innovative labeling reaction is able to directly derivatize ultra-trace GAs at a concentration down to 0.1 pM, and allows establishment of an ultrasensitive LC-ESI-MS/MS procedure for GA determination (Li et al., 2016).

14.4.4 LC-MS/MS detection

Combining RP-LC with different types of MS instruments, such as single quadrupole or QQQ (Mueller and Munne-Bosch, 2011; Susawaengsup et al., 2011; Zhang et al., 2015a), ion trap (Chiwocha et al., 2003), and Q-TOF (Ayele et al., 2010; Fletcher and Mader, 2007; Varbanova et al., 2007) have been used for GA analysis. The preferential approach may be LC-ESI-MS/MS with MRM mode because it provides better selectivity, specificity, and sensitivity for the simultaneous analysis and quantification of multiple GAs. In order to rapidly and effectively separate multiple GAs, UPLC provides high chromatographic resolution, high peak capacity, and rapid speed of analysis. For example, a UPLC-ESI-MS/MS method was established for the determination of 20 GAs as free acids, including all the bioactive GAs and their biosynthetic precursors and metabolic products, with limits of detection ranging between 0.08 and 10 fmol (Urbanova et al., 2013). Coupled with two-step SPE extraction and pre-concentration procedures, the method was successfully applied for the quantitation of GAs in 100 mg (FW) plant tissue using isotope dilution analysis. The chemical label-based LC-ESI-MS/MS methods can potentially improve the detection sensitivity of GAs, so that the sample size can be reduced further. The quantification limits of GAs increased up to 50-fold after chemical derivatization with bromocholine bromide, and the developed method needed less than 100 mg (FW) of plant tissues to determine plant hormone profiles (Kojima et al., 2009). The required sample size could be scaled down to 5 mg (FW) using laboratory-synthesized 3-bromoactonyltrimethylammonium bromide (Chen et al., 2012) as labeling reagent. Nevertheless, such LC-ESI-MS/MS strategies remain unable to directly measure the ultra-trace GAs in fresh tissue less than 5 mg. To further lower the amount of sample required, an analytical method was described that combined a newly developed ultra-trace labeling technique with UPLC-ESI-MS/MS, with a lowest limit of detection (LOD) down to 0.327 pM and limit of quantitation (LOQ) of 1.09 pM, enabling quantitative analysis of ultra-trace GAs in tiny samples of fresh plant material such as a single flower and even a single stamen (c.0.020 mg). With this organ-based assay, GA_1, GA_4, GA_5, GA_8, GA_{29}, $GA_{34,}$ and GA_{51} were successfully quantified using only one flower of Arabidopsis, and a high level of GA_1 was revealed in stamens (Li et al., 2016).

In quantitative analysis, deuterium isotopic-labeled GAs have been used as IS to enhance the accuracy and precision of the analytical method. However, accurate quantification of GAs will not be achieved without eliminating the isotopic interference, especially when the amount of endogenous GAs is much higher than spiked IS, since commercially obtained [$^2H_2$] GAs have the same mass and chromatographic peak position as the natural isotopic GAs at (M + 2). The interference is related to the concentrations of the analytes and IS, when the added amounts of IS are very close to the contents of the analytes, and quantitation can be simply achieved based on the peak area ratio of the analytes and IS. For other situations, a method to eliminate the isotopic interference between the analytes and IS was reported, which employs a mathematical approach using some constants with an experimentally determined adjustable parameter (Rule et al., 2013).

14.4.5   An easy-to-follow quantification method for gibberellins

14.4.5.1   Sample preparation

IS ($^2H_2$-GAs) are added to 1 mg (FW) frozen plant sample, ground into a fine powder in the frozen state, 100 μL extraction solvent composed of 75% MeOH, 20% H_2O, and 5% FA (v/v/v) are added and extracted at 4°C overnight. The extract is centrifuged at 10,000 g for 10 min at 4°C, the lower pellet re-extracted twice with 50 μL MeOH for 10 min each, the combined supernatant evaporated and then dissolved in 20 μL H_2O acidified to pH 2.5 with 1 M HCl, further extracted with ethyl acetate (3 × 60 μL), and the combined ethyl acetate phase evaporated to dryness.

14.4.5.2   Derivatization with N-(3-dimethylaminopropyl)-N-ethylcarbodiimide

To the sample, 50 µL of 20 mM EDC aqueous solution (freshly prepared) is added, ultrasonicated for 5 min, and shaken at 35°C, 1200 g, and 750 rpm for 1.5 h. After centrifugation, supernatant is subjected to UPLC-ESI-MS/MS analysis using the positive ion mode, or otherwise stored at -20°C for later analysis.

14.4.5.3   UPLC-MS/MS conditions

The LC-MS/MS (model LCMS-8050, Shimadzu) consists of CBM-20A system controller, LC-30AD pump, DGU-20A5R degasser, SIL-30AC autosampler, and CTO-30A column oven, coupled to a quadrupled type tandem mass spectrometer via an ESI interface, controlled by LabSolutions LC-MS Ver.5.6.

The LC conditions are as follows: injection, 10 µL; column, RP packed column (XR-ODS, 50 mm X 3.0 mm I.D., 2.2 µm, Shimadzu); column temperature, 40°C; flow rate, 0.3 mL/min. Typical gradients of binary solvents (mobile A, 0.1% FA in water; mobile B, 0.1% FA in ACN) are set as follows: the gradient is run with initial 1% B, then increased to 4% B over 1 min, to 6% over the next 1 min, to 9% over the next 3 min, to 12% over the next 4 min, to 34% over the next 12 min, to 50% B over the next 5 min sequentially, and further to 100% over the next 2 min.

The MS/MS conditions are set as follows: nebulizing gas flow at 3 L/min, drying gas flow at 15 L/min, heating gas flow at 10 L/min and interface temperature at 300°C, desolvation lines temperature at 300°C, heat block temperature at 450°C, and collision-induced dissociation (CID) gas at 230 kPa. Positive MRM mode, typical MS/MS transitions, collision energy, and retention time of each labeled GA are summarized in Table 14.3. Representative multiple-reaction monitoring chromatograph of quantified GAs from six stamens of fresh Arabidopsis flowers is shown in Fig. 14.5(b).

14.5   Cytokinins

14.5.1   Structures and properties of cytokinins

Natural CKs are adenine-derived compounds with isoprenoid or aromatic substituents at the N6 position of the adenine ring. The number of identified CKs so far exceeds 40, among which are free bases, nucleosides (ribosides), nucleotides, and glycosides (O- and N-glycosides) (Table 14.4) (Liang et al., 2009; Feng et al., 2017). Although many researchers believe that CK free bases are the bioactive CKs in plants (Schmitz et al., 1972; Spiess, 1975), some evidence indicates that other types of CKs, such as nucleosides and glycosides, also display bioactivity (Romanov et al., 2006; Sakakibara, 2006). Furthermore, CKs can be classified into isopentenyladenine (iP), *trans*-zeatin (tZ), dihydrozeatin (DHZ), and *cis*-zeatin (cZ) groups based on their side chain. The iP and tZ types are considered as the most abundant and important compounds (Sakakibara, 2006). To better understand CK bioactivity, biosynthesis, metabolism, and signaling, it is of importance to be able to measure the various CKs within plant tissues.

Table 14.3 Selected reaction monitoring conditions for EDC-GAs and EDC-^{2}H$_2$-GAs

Analytes	Quasi-molecular ions	Fragment ions	Collision energy	Internal standard	Quasi-molecular ions	Fragment ions	Collision energy	RT (min)
EDC-GA$_1$	504.30	388.20	−34.0	EDC-^{2}H$_2$-GA$_1$	506.30	390.20	−34.0	12.5
EDC-GA$_3$	502.25	431.30	−24.0	EDC-^{2}H$_2$-GA$_3$	504.25	433.30	−24.0	12.2
EDC-GA$_4$	488.30	372.25	−35.0	EDC-^{2}H$_2$-GA$_4$	490.30	374.25	−35.0	19.7
EDC-GA$_5$	486.30	370.20	−35.0	EDC-^{2}H$_2$-GA$_5$	488.30	372.20	−35.0	16.2
EDC-GA$_6$	502.30	386.30	−36.0	EDC-^{2}H$_2$-GA$_6$	504.30	388.30	−36.0	14.3
EDC-GA$_7$	486.30	370.20	−32.0	EDC-^{2}H$_2$-GA$_7$	488.30	372.20	−32.0	19.5
EDC-GA$_8$	520.30	404.15	−37.0	EDC-^{2}H$_2$-GA$_8$	522.30	406.15	−37.0	7.7
EDC-GA$_9$	472.30	356.15	−34.0	EDC-^{2}H$_2$-GA$_9$	474.30	358.15	−34.0	21.9
EDC-GA$_{15}$	486.30	370.25	−35.0	EDC-^{2}H$_2$-GA$_{15}$	488.30	372.25	−35.0	23.2
EDC-GA$_{20}$	488.30	372.15	−35.0	EDC-^{2}H$_2$-GA$_{20}$	490.30	374.15	−35.0	16.4
EDC-GA$_{29}$	504.30	388.20	−35.0	EDC-^{2}H$_2$-GA$_{29}$	506.30	390.20	−35.0	8.5
EDC-GA$_{34}$	504.30	388.20	−35.0	EDC-^{2}H$_2$-GA$_{34}$	506.30	390.20	−35.0	17.6
EDC-GA$_{44}$	502.30	386.20	−36.0	EDC-^{2}H$_2$-GA$_{44}$	504.30	388.20	−36.0	18.2
EDC-GA$_{51}$	488.30	372.15	−35.0	EDC-^{2}H$_2$-GA$_{51}$	490.30	374.15	−35.0	17.8

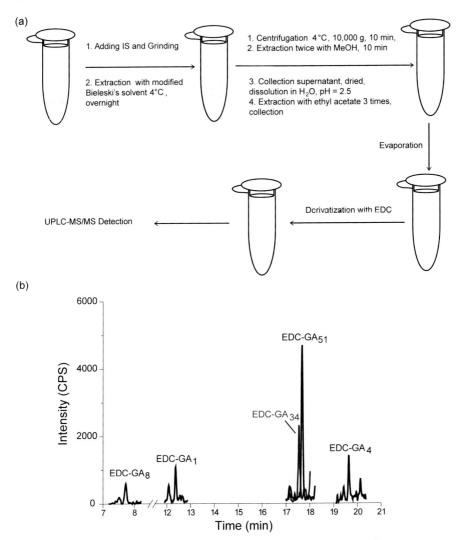

Figure 14.5 (a) Scheme of quantitative analysis for GAs from plant sample. (b) Representative multiple-reaction monitoring chromatographs of quantified GAs from six stamens of fresh Arabidopsis flowers.

Compared with acidic plant hormones, such as IAA, JA, SA, and ABA, CKs are typically present at extremely low concentrations in plants (Saenz et al., 2010; Svačinová et al., 2012; von Schwartzenberg et al., 2007), in the pmol/g (FW) range. Therefore, CK analysis can suffer seriously from interferences from the numerous metabolic components in plants. Additionally, the physiochemical properties of CKs vary due to the different substituents, and most of these compounds are amphoteric, potentially with several pK_a or pK_b values. Therefore, it is a really challenging work to build highly sensitive, selective, and reliable determination methods for CK quantification. Various

Table 14.4 Structures, names, and abbreviations of CKs

Main structure	R1	R2	R3	Name	Abbreviations
	CH₃ / CH₃ / CH₂	H	–	N⁶-Isopentenyladenosine	iP
		R	–	N⁶-Isopentenyladenosine	iPR
		RP	–	N⁶-Isopentenyladenosine-5′-monophosphate	iPMP
		G	–	N⁶-Isopentenyladenosine-9-glucoside	iP9G
	CH₂OR₃ / CH₃ / CH₂	H	H	trans-Zeatin	tZ
		H	G	trans-Zeatin O-glucoside	tZOG
		R	H	trans-Zeatin riboside	tZR
		R	G	trans-Zeatin riboside O-glucoside	tZROG
	CH₃ / CH₂OR₃ / CH₂	H	H	cis-Zeatin	cZ
		R	H	cis-Zeatin riboside	cZR
	CH₂OR₃ / CH₃ / CH₂	H	H	Dihydrozeatin	DHZ
		R	H	Dihydrozeatin riboside	DHZR
	(phenyl)-CH₂	H	–	N⁶-Benzylaminopurine	BAP
		R	–	N⁶-Benzylaminopurine riboside	BAPR

Continued

Table 14.4 Structures, names, and abbreviations of CKs—cont'd

Main structure	R1	R2	R3	Name	Abbreviations
		H	–	*meta*-Topolin	mT
		R	–	*meta*-Topolin riboside	mTR
		H	–	*para*-Topolin	pT
		R	–	*para*-Topolin riboside	pTR
		H	–	Kinetin	K
		R	–	Kinetin riboside	KR

Main structure	R1	R2	R3	Name	Abbreviations
		G	–	Dihydrozeatin-3-glucoside	DHZ3G
		G	–	trans-Zeatin-7-glucoside	tZ7G

analytical methods have been employed, such as ELISA (Sousa et al., 2011), GC (Björkman and Tillberg, 1996), HPLC (Prinsen et al., 1995), and CE (Ge et al., 2006). Among these techniques, LC-MS is the most commonly used technique because of its excellent performance in separation and unequivocal identification.

14.5.2   Sample preparation

14.5.2.1   Sampling and extraction

Sample collection, storage, homogenization and weighing should be done as previously described in subsection 14.3.2.1. Several solvents have been used for CK extraction, such as MeOH, ACN, EtOH, and mixture of these with water, chloroform, or FA (Ai et al., 2015; Chen et al., 2010; Du et al., 2015; Liu et al., 2012a). In addition to extraction efficiency, extra attention is required during CK extraction to suppress the dephosphorylation of CK nucleotides catalyzed by the phosphatases. Many researchers use Bieleski's solvent, comprising MeOH/chloroform/water/FA (12/5/2/1, v/v/v/v), to extract samples at −20°C to inactivate the phosphatases (Du et al., 2015; Hussain et al., 2010; Svačinová et al., 2012). However, there have been reports that the presence of chloroform in solvents may increase the extraction of lipophilic components that complicate further purification (Horgan and Scott, 1987; Laloue et al., 1974). Comparison of three different solvents: 80% (v/v) MeOH, Bieleski's solvent, and modified Bieleski's solvent (MeOH/water/FA, 15/4/1, v/v/v), found that the modified Bieleski's solvent was better for subsequent purification and gave better results on LC-MS (Hoyerova et al., 2006).

14.5.2.2   Purification

Many purification and enrichment methods of CKs have been developed, thus giving researchers several choices. Typically used methods include LLE, SPE, and immunoaffinity purification (IAP), and there have also been trials of the application of polymer monolith extraction (PME) and MSPE. For CK enrichment in the past, LLE was used, but the low efficiency and recovery of this method are weaknesses. Based on antibody–antigen interactions, IAP can provide selective CK enrichment, but suffers from being slow, expensive, and with poor reusability.

PME and MSPE have become alternatives for CK purification. Various monolith polymer techniques and custom-made monolith polymer materials have been employed, such as molecular imprinted poly (methacrylic acid-*co*-ethylene glycol dimethacrylate) monolith (Du et al., 2012a), styrene, and divinylbenzene containing "high internal phase emulsion" polymers (Du et al., 2015), and poly (2-acrylamido-2-methyl-1-propanesulfonic acid-*co*-ethylene dimethacrylate) monolith (Liu et al., 2010). These materials can provide excellent selectivity, stability, and reusability. The MSPE technique disperses magnetic or magnetizable absorbents in the sample solution, which can then be isolated with a magnetic lure, thus simplifying sample pretreatment. Fe_3O_4/SiO_2 magnetic nanoparticles (Cai et al., 2014) and $Fe_3O_4/SiO_2/poly$ (2-acrylamido-2-methyl-1-propanesulfonic acid-*co*-ethylene glycol dimethacrylate) magnetic porous polymer (Liu et al., 2012c) are used for CK purification. Both PME and MSPE methods generally require self-prepared extraction medium.

The SPE approach has been widely applied for the purification of many compounds, especially trace metabolites in complex biological samples. There are several advantages of SPE, such as low solvent consumption, simple operations, high extraction recovery, and availability of various sorbents. It is probable that SPE is the most widely used method for CK purification. Commercially available sorbents for SPE vary, such as hydrophilic sorbents, hydrophobic sorbents, cation/anion exchange sorbents and mixed-mode sorbents.

Purification and separation with SPE generally take advantage of polarity and ionization properties of CKs. RP SPE, such as C18 and HLB columns have been used for pre-concentration of CKs. When passing plant extracts through a RP SPE column, compounds with low polarity, such as lipids and plant pigments, are bound to the sorbent due to their high hydrophobicity. Other metabolites with relatively higher polarity including CKs are more concentrated in the flow-through. On the other hand, cation-exchange SPE cartridges, such as MCX columns, are generally used for further purification considering the alkaline properties of the adenine structure. Notably, the MCX SPE cartridges contain sorbent of a sulfonated poly (divinylbenzene-*co*-N-vinylpyrrolidone) copolymer, which gives the sorbent reversed-phase character and increased water wettability. The sulfonated part has cation-exchange capacity of 1 meq/g at pH range of 0–14. Therefore, the MCX columns show mixed-mode, reversed-phase and cation-exchange characteristics. Predominately positively charged CK molecules can be absorbed on MCX sorbent, leading to a more purified residue. Other SPE methods have also been tried for CK sample preparation, such as hydrophilic interaction chromatography (HILIC) SPE that takes advantage of the hydrophilic character of CKs (Cai et al., 2013).

It is a common understanding that combination of different SPE columns, or even different purification methods, would make possible the elimination of different interferences and provide efficient extraction. This would facilitate the subsequent LC-MS/MS analysis to gain accurate analytical results. Sample pretreatment and quantification methods for CKs based on LC-MS/MS are summarized in Table 14.5.

14.5.3 LC-MS/MS detection

14.5.3.1 LC separation

RP columns (Du et al., 2015; Takei et al., 2003; von Schwartzenberg et al., 2007) are most commonly used in CK separation. The free bases of CKs exhibit gradation in polarity due to the different substituents at the N6 position, which is particularly suitable for LC separation prior to ESI-MS analysis. Moreover, substituents at the N9 or O position also affect the retention of CKs on RP columns. Considering that CK free bases, glucosides, and nucleosides generally behave as weak bases, acetic acid is sometime added to the mobile phase to improve separation and chromatographic peak shape, as well as to enhance ESI signal.

Among the four classes of CKs, CK nucleotides exhibit different characteristics in retention due to the high polarity of the molecules. This needs extra attention in LC separation. The main problems in LC-ESI-MS analysis of CK nucleotides are

Table 14.5 Sample pretreatment and quantification methods for CKs based on LC-MS/MS

Analyte	Extraction solvent	Purification method	Plant matrix	Instrumental method	References
K, KR, tZ, and mT	Modified Bieleski's solvent	PME	Tobacco leaves, 5 g	HPLC-Q-Trap MRM	Du et al. (2012a)
iP, iPR,iP9G, tZ, tZR,tZ9G,DHZ, DHZR,	Modified Bieleski's solvent	C18 SPE-PME	Arabidopsis and rice seedlings, 1 g	HPLC-Q-Trap MRM	Liu et al. (2010)
tZ, mT, K, KR	Modified Bieleski's solvent	PME	Tobacco leaves and bean leaves, 100 mg	UHPLC-QQQ SRM	Du et al. (2015)
iP;iPR,iP9G,tZ,tZR,tZ9G,cZ, cZR,cZOG,DHZ,	ACN	MSPE	Rice, 50 mg	UPLC-QQQ MRM	Cai et al. (2014)
iP,iPR,iP9G,tZ,tZR,tZ9G DHZ,DHZR	Modified Bieleski's solvent	C18 SPE-MSPE	Rice root, leaves and Arabidopsis seedlings, 200 mg	HPLC-Q-Trap MRM	Liu et al. (2012c)
iP;iPR,iP7G,iPMP,tZ,tZR,tZ7G,tZ9G tZOG,tZMP,cZ,cZ9G,cZR,cZOG, DHZR,DHZ9G,DHZ7G,DHZOG	Bieleski's solvent	In-tip microSPE	Arabidopsis, 1–5 mg	UPLC-QQQ MRM	Antoniadi et al. (2015), Svačinová et al. (2012)
iP;iPR,iP9G,iPRPs,tZ,tZR,tZ7G,tZ9G tZROG,tZOG,tZRPs,tZRPOG,cZ, cZRPOG,cZR,cZOG,cZROG,cZRPs, DHZR,DHZ9G	Modified Bieleski's solvent	MCX SPE	Rice roots and maize roots, 100 mg-1 g	UPLC-QQQ MRM	Chen et al. (2010), Kojima et al., 2009 and Liu et al. (2012a)
iP;iPR, iPRMP,tZ,tZR,tZROG,tZOG, tZRMP,cZ,cZR,cZOG,cZROG, cZRMP, DHZ,DHZR, DHZROG,BAR,BA,mT, oT	70% EtOH, 80% MeOH	IAP	*Physcomitrella patens*	UPLC-QQQ MRM	Sousa et al. (2011), Strnad (1996), Takei et al. (2003) and von Schwartzenberg et al. (2007)

firstly that CK nucleotides have shorter retention time and are poorly separated in the RP column, and secondly that the absolute ESI ionization of CK nucleotides is relatively low. To solve these problems acetate/formate buffers or ion pairing reagents in the LC mobile phase can be used. Additionally, there have been two methods to improve analysis of CK nucleotides. The first approach is to convert the nucleotides to nucleosides with alkaline phosphatase (Kojima et al., 2009; Takei et al., 2003). This method needs CK nucleotides to be separated from others in SPE purification. The other approach is derivatization, making the analytes more hydrophobic, thus improving both the chromatographic retention properties of CK nucleotides and the ESI process. Propionyl and benzoyl ester formation as the mode of derivatization achieved excellent separation and higher detection signal (Nordström et al., 2004).

As a variant of normal phase chromatography, HILIC is a new technique developed for the separation of polar and hydrophilic compounds. Since some CKs, especially CK nucleotides, exhibit high polarity, HILIC LC was also applied in CK separation. The elution order of CKs in HILIC is more or less the opposite of that seen in RP-LC (Liu et al., 2010).

14.5.3.2 MS/MS detection

For quantitative analysis of CKs, MRM is the most commonly used MS operation mode, and stable-isotope dilution is the most accurate method, as previously described for other plant hormones.

Protonated CKs ($[M+H]^+$) produce a typical fragmentation pattern after CID in mass spectrometry, characterized by the loss of the nucleosyl, glycosyl, or nucleotidyl group and the fragmentation of the N6-substituent for the aglycons (Prinsen et al., 1995; Svačinová et al., 2012). The corresponding product ions are among the most abundant ions in the MS/MS spectra, so the MRM transitions are used for identification and quantification of CK metabolites. Based on the fragmentation pattern of different substituent types, a method to characterize types of unknown CKs has been introduced (Liu et al., 2012a). A simple and fast quantification method based on experience in our lab is recommended and shown as Fig. 14.6.

14.5.4 An easy-to-follow quantification method for cytokinins

14.5.4.1 Sample preparation

Plant material (about 200 mg, FW) is frozen in liquid nitrogen and then homogenized to a fine powder. The powder is soaked in 2 mL 90% (v/v) MeOH and the stable-isotope-labeled CKs added as ISs. The homogenate is kept at -20°C overnight. After centrifugation at 15,000 g for 15 min, the supernatant is collected and loaded onto a MCX cartridge that has been pre-equilibrated with MeOH and water. The cartridge is sequentially washed with 2% FA and MeOH. Then CKs are eluted with 5% (v/v) NH_4OH in MeOH. Eluate is evaporated to dryness under vacuum and the residue is reconstituted in 20% (v/v) MeOH and passed through a micro-filter for LC-MS/MS analysis.

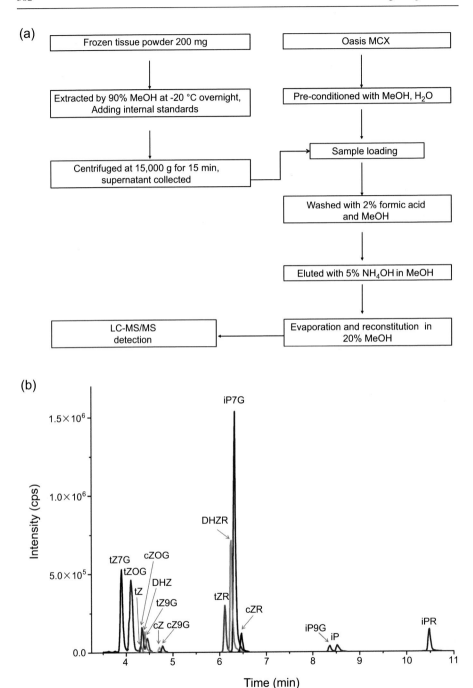

Figure 14.6 (a) Flow scheme for quantitative analysis of CKs in plant tissues. (b) Chromatograms of 15 CKs in a Col-0 Arabidopsis stem sample detected by LC-MS/MS method.

14.5.4.2   LC-MS/MS detection

LC-MS/MS analysis of samples is performed on an ACQUITY UPLC System (Waters) coupled to a 5500 Q-Trap MS System (AB SCIEX). The samples are injected onto an RP column (Acquity UPLC BEH C18 1.7 mm, 2.1 × 100 mm; Waters) and separated using a gradient of 0.01% acetic acid (A) and 0.01% acetic acid in ACN (B) at a flow rate of 0.30 mL/min. The gradient used is: 0 min, 98:2 (A:B) to 12.0 min, 70:30 (A:B) to 13.0 min, 15:85 (A:B). Column temperature is set to 30°C and sample temperature to 10°C. The MRM transitions are selected based on the fragment pattern of CKs (Table 14.6). Conditions in positive ion mode are as follows: ion spray voltage, 4500 V; desolvation temperature, 600°C; nebulizing gas 1, 50 psi; desolvation gas 2, 50 psi; and curtain gas, 30 psi. This method can be employed to determine CK contents of various plant materials. Fig. 14.6(b) shows the results of 15 CKs in a sample of Arabidopsis stem.

Table 14.6 **Optimized MRM parameters for CK detection**

Analyte	Q1 mass/ Q3 mass	Collision energy/V	Declustering potential/V	Entrance potential/V	Collision cell exit potential/V
tZ7G	382.2/220.1	30	130	10	12
$^2H_5$-tZ7G	387.2/225.1	30	130	10	12
tZOG	382.2/220.1	25	100	10	12
$^2H_5$-tZOG	387.2/225.1	25	100	10	12
tZ	220.1/136.0	24	100	11	12
$^2H_5$-tZ	225.1/137.1	24	100	11	12
cZOG	382.2/220.1	25	110	10	12
DHZ	222.1/136.0	29	105	9	12
$^2H_3$-DHZ	225.1/136.1	29	105	9	12
tZ9G	382.2/220.1	29	130	10	12
$^2H_5$-tZ9G	387.2/225.1	29	130	10	12
cZ	220.1/136.1	24	105	10	12
cZ9G	382.2/220.1	25	120	10	12
tZR	352.2/220.1	27	110	10	12
$^2H_5$-tZR	357.2/225.1	27	110	10	12
DHZR	354.2/222.1	29	110	10	12
$^2H_3$-DHZR	357.2/225.1	29	110	10	12
iP7G	366.2/204.1	28	150	8	12
$^2H_6$-iP7G	372.2/210.2	28	150	8	12
cZR	352.2/220.1	24	120	10	12
iP9G	366.2/136.1	45	140	9	12
$^2H_6$-iP9G	372.2/142.1	45	140	9	12
iP	204.1/136.0	22	105	8	12
$^2H_6$-iP	210.2/137.1	22	105	8	12
iPR	336.1/204.1	25	150	11	12
$^2H_6$-iPR	342.2/210.2	25	150	11	12

14.6   Brassinosteroids

14.6.1   Structures, properties, and analysis of brassinosteroids

All BRs are structurally featured with a four-ring 5α-cholestan skeleton connected to a side chain (Wang et al., 2017b). The diversity of BR structures mainly results from variations in the A ring, B ring, and the side chain substituent groups. Alkyl-substitution on the C_{24} site of the side chain determines the number of carbons and their classification into C_{27}, C_{28}, and C_{29} BRs. Different oxidation forms of the B ring classifies BRs as B-ring lactone-type, B-ring cycloketone-type, and 6-deoxo-type BRs. Another difference in BR structure is the hydroxyl substituents at C_2, C_3 sites on the A ring and C_{22}, C_{23} sites on the side chain.

Dependent on their structures, BRs are all neutral compounds containing no ionizable groups and they exhibit stronger hydrophobicity than other plant hormones such as IAA, GAs, CKs, ABA, and JA. When developing analysis methods, the high hydrophobicity and neutral property of BRs should be considered and utilized. The hydrophobicity orders of BRs are: C_{27} BRs < C_{28} BRs < C_{29} BRs, B-ring lactone-type < B-ring cycloketone-type < 6-deoxo-type BRs. In addition, the more hydroxyl groups present the less hydrophobic the BRs are. Another structural feature of major BRs is the occurrence of vicinal diols on the A ring and/or the side chain, which is frequently exploited in BR enrichment and derivatization via boronate affinity interaction.

Compared with other plant hormones, there remain some unique difficulties in BR analysis. First, BRs are present in plants at extremely low concentrations ranging from 0.01 to 100 ng/g (FW) (Hayat and Ahamad, 2011). Second, BRs are neutral hydrophobic steroids lacking photosensitive, electrosensitive, or ionizable groups (Hayat and Ahamad, 2011). Therefore BRs are difficult to detect with common detectors directly, such as UV, FLR detectors or MS. These difficulties mean that BR detection is more likely to be influenced by co-existent interference from plant matrix than other acidic and basic plant hormones. Therefore, BR analysis often requires sufficient sample pretreatment to remove interfering substances, derivatization to enhance detection responses, and high-selectivity detectors to improve detection specificity. In the early stages of BR research, HPLC-UV, HPLC-FLR detection after derivatization or HPLC-ELISA were used (Gamoh et al., 1988, 1994; Lv et al., 2014; Pan et al., 2013; Swaczynova et al., 2007; Zhang et al., 2010b), but these methods have disadvantages in sensitivity and specificity compared with MS-based technologies and have been rarely adopted by plant physiologists. The most widely used and commonly accepted approaches are now MS-based methods.

14.6.2   Brassinosteroids quantitative analysis based on GC-MS

14.6.2.1   Sample preparation

GC-MS has always been a widely used method for BR analysis since the discovery of BRs in plants. This method has provided most of the BR quantitation data for plant physiological studies in the past (Joo et al., 2015; Kim et al., 2014, 2006; Noguchi et al., 2000; Ohnishi et al., 2012; Sakamoto et al., 2006). For sample preparation the harvested plant material is typically lyophilized and then ground to a fine powder, followed by MeOH/chloroform extraction. The concentrate is then partitioned three times with chloroform/water after which the chloroform fraction is subjected to silica gel column for BRs enrichment, and the BR containing fraction collected is purified with Sephadex LH-20 column and Sep-Pak Plus C18 cartridge, in sequence. Finally, eluates are subject to preparative HPLC and each fraction is derivatized for GC/MS analysis in selected ion monitoring (SIM) mode.

14.6.2.2   Derivatization

The occurrence of several hydroxyl groups in most BR structures makes them polar and difficult to vaporize in GC, so derivatization is employed mainly to improve vaporization rather than sensitivity. The derivatization reaction can be divided into three types. For BRs containing vicinal diol groups, derivatization is often achieved through the formation of a boronate ester. The frequently used derivatization reagents are methaneboronic acid and similar compounds. For some BR precursors containing no vicinal hydroxyls, BRs are often derivatized by trimethylsilylation with N-Methyl-N-trimethylsilyl-trifluoroacetamide, or N,O-*bis*(trimethylsilyl) acetamide. Finally, if BRs such as teasterone (TE) and typhasterol (TY) contain both vicinal and individual hydroxyls, both types of derivatization reactions can be used.

14.6.2.3   GC-MS detection

When performing GC-MS analysis, derivatized BRs are separated by GC and transferred into the EI source, where they are ionized and fragmented for detection. The scan modes of GC-MS include full scan and SIM, depending on requirements. Full scan is often used for identification of BRs, while SIM mode for quantitation. In addition, SIM mode can provide better sensitivity than full scan because the isolation of target ions by the quadrupole in SIM mode lowers the background.

14.6.3   Brassinosteroids quantitative analysis based on liquid chromatography coupled to tandem mass spectrometry

14.6.3.1   Advantages of liquid chromatography coupled to tandem mass spectrometry quantification for brassinosteroids

The amount of plant material available is an important factor because some tissues or mutant materials are difficult to acquire and so a highly sensitive method is required. Generally, GC-MS methods consume 20–40 g fresh plant material, which is a rather large amount, especially for some small plant seedlings such as Arabidopsis. The sample pretreatment process, including freeze drying, LLE, and multi-step SPE and HPLC fractionation, is obviously time-consuming and labor-intensive. It can take 1 or 2 days to prepare one batch of samples. The limitations of GC-MS-based methods in terms of sensitivity and throughput restrict the number of samples and replicates that are analyzed, so the accuracy and reliability of the resulting data can be compromised.

   With the development of (atmospheric pressure ionization) API sources and triple quadrupole MS, LC-MS/MS has become more powerful and hence more popular. LC-MS/MS can simultaneously provide efficient separation, high detection sensitivity, and excellent peak specificity. Considering all these advantages, BR analysis methods have been increasingly developed based on LC-MS/MS techniques (Table 14.7).

Table 14.7 BR quantification methods based on LC-MS/MS

BRs	Sample pretreatment procedure	Derivatization	Plant matrix	Detection	References
24-epiBL, 28-homoBL, 28-epihomoBL	Extraction, Bond Elut Plexa SPE, CENTRICON membrane, Strata RP column	BPyBA	Arabidopsis, leaves, 2–10 g	UPLC-ESI-MS/MS	Huo et al. (2012)
28-epihomoBL	Extraction, Bond Elut Plexa SPE, CENTRICON membrane, Strata RP column, derivatization, online polymer monolith micro-extraction (home-made)	BPyBA	Arabidopsis, 400 mg	LC-QTof-MS/MS	Wang et al. (2013)
24-epiBL, 24-epiCS, 6-deoxo-24-epiCS, DS, TE, TY	In-line MSPD (self-packing)-MAX-MCX SPE	m-APBA	Rice, 0.2 g	HPLC-ESI-MS/MS	Wang et al. (2014a)
24-epiBL, 24-epiCS, 6-deoxo-24-epiCS, TE, TY	Extraction, online 2D µSPE (PBA silica-C18 silica, self-packing)-on column derivatization (OCD)	m-APBA	Tomato leaves, 225 mg DW	HPLC-MS/MS	Wu et al. (2013)
BL, CS, epiBL, epiCS	Three liquid-liquid distribution steps, TLC and reversed-phase HPLC fractioning	Dansyl-3-aminophenylboronic acid	Arabidopsis, 26–57 g	HPLC-ESI-MS/MS	Svatos et al. (2004)
24-epiBL	Integrated MIP monolithic column based in-tube SPME (home-made), derivatization and FSS-UPLC, termed in-tube SPME-DE-FSS-UPLC	9-Phenanthreneboronic acid	Broad bean pollens, flowering Chinese cabbage flowers and corn seeds, 5 g	UPLC-FLR	Pan et al. (2013)
24-epiBL, homoBL	Microwave-assisted extraction, vacuum distillation, LLE*3 times, vacuum distillation, styrene-co-4-vinylpyridine magnetic polymer beads (Homemade)	9-Phenanthreneboronic acid	Breaking-wall rape pollen, fresh cole, 25–50 g	HPLC-FLR	Zhang et al. (2010b)

BRs	Sample pretreatment procedure	Derivatization	Plant matrix	Detection	References
28-norBL, 28-norCS, 24-epiBL, BL, CS, 24-epiCS, homoBL	Extraction, 2-step dehydration, GCB/PSA double layered SPE (self-packing), drying, reconstruction, LLE*3 times, drying, reconstruction	NA	Rice shoots, rape leaves and flowers, 0.5–1 g	HPLC-ESI-MS/MS	Ding et al. (2013a)
28-norBL, 28-norCS, BL, CS, homoBL	Extraction, 2-step dehydration, GCB/PSA double layered SPE (self-packing), drying, reconstruction, BA/PMME, drying, reconstruction	NA	Rice shoots, rape leaves and flowers, 0.5–1 g	HPLC-ESI-MS/MS	Ding et al. (2013b)
28-norBL, 28-norCS, BL, CS, homoBL	Extraction, drying, reconstruction, TiO$_2$-coated magnetic hollow mesoporous silica sphere (TiO$_2$/MHMSS), in situ derivatization (ISD)	4-DMAPBA	Rice shoots, rape, 100 mg	UPLC-ESI-MS/MS	Ding et al. (2014)
BL, CS	Extraction, loading, packing, SPE fractioning	Naphthaleneboronic acid	Bee-collected pollen of broad bean extracts	HPLC-UV	Gamoh et al. (1994)
BL, CS, TE, TY	NA	Naphthaleneboronic acid	Active fraction from the seeds of Cannabis sativa L	LC-APCI-MS	Gamoh et al. (1996)
BL	Ultra-sonic-assisted dispersive liquid-liquid micro-extraction (UA-DLLME) coupled with derivatization	9-Phenanthreneboronic acid	Leaves of Arabidopsis, Daucus carota, and B. campestris, 2 g	HPLC-FLR	Lv et al. (2014)
BL, CS, epiBL, epiCS, 28-homoBL, 28-homoCS	Extraction, C18 SPE, Tandem DEAE-Sephadex A-25-Strata X reversed-phase column SPE	NA	Phaseolus vulgaris, Arabidopsis, Rape pollen, D. carota, 1 g	HPLC-ELISA, HPLC-MS	Swaczynova et al. (2007)

NA, not available.

14.6.3.2 Sample preparation

When performing quantitative analysis by LC-MS/MS, the most frequently used scan mode is MRM which can efficiently eliminate the co-eluting interferences and lower the baseline because of simultaneously isolating precursor ions and product ions in Q1 and Q3. This means that the sample cleanliness after pretreatment may be relative low and the sample pretreatment can be somewhat simplified. However, considering the extremely low content of endogenous BRs, care and attention should be paid to sample processing.

Some methods have been presented using custom-made sample pretreatment materials, which include boronate affinity magnetic nanoparticles, TiO_2-coated magnetic hollow mesoporous silica spheres, boronate affinity polymer monoliths, and styrene-*co*-4-vinylpyridine porous magnetic polymer beads (Ding et al., 2013, 2014; Xin et al., 2013a; Zhang et al., 2010b). Using these materials combined with other commercial SPE sorbents makes the sample clean-up efficient and easy. In addition, some methods based on new instrumentation or utilizing new pretreatment techniques have also been presented, such as an in-line coupled MSPD-MAX-MCX system, an online two-dimensional microscale SPE-on-column derivatization system (online 2D µSPE-OCD), and online polymer monolith micro-extraction (Wang et al., 2014a, 2013; Wu et al., 2013).

There are also some efficient designs enabling successful quantification of plant endogenous BRs. The combination of MCX and MAX SPE columns can eliminate all ionizable interferences and polar neutral compounds from the plant matrix and make the detection of BRs easy after derivatization (Xin et al., 2013b). All the materials used in this method are commercially available and the sample pretreatment is relatively simple, which makes it easy to use in many laboratories.

14.6.3.3 Derivatization

In general, an atmospheric pressure chemical ionization (APCI) source may be highly suitable for detection of free BRs because most BRs have medium to high hydrophobicity (Gamoh et al., 1996). However, the LOD of 24-epibrassinolide (24-epiBL) with APCI was found to be 10 times higher than with the ESI source (Xin et al., 2013b). Thus, derivatization combined with ESI-MS detection is a good option.

In contrast to GC-MS, derivatization in LC-MS-based methods only aims at enhancing ionization efficiency and MS response of BRs. The most common method uses an N-containing organic boronic acid compound to anchor an easy-ionization N-containing group onto BRs via a boronate esterification reaction. Some derivatization reagents often used include 2-bromopyridine-5-boronic acid, 2-methoxy-5-pyridineboronic acid, 4-(dimethylamino)-phenylboronic acid, 3-(dimethylamino)-phenylboronic acid (DMAPBA), 3-aminophenylboronic acid, 2-chloropyridine-5-boronic acid, and dansyl-3-aminophenylboronates.When choosing derivatization reagents, the most important considerations include the degree of the signal improvement, whether the reaction condition is harsh or mild and the reaction speed.

14.6.3.4   LC-MS/MS detection

When performing quantitative analysis with triple quadrupole MS, MRM is the most often used operation mode. Generally, one MRM transition is used for quantification and an additional one for identification of the target compound. If the intensity ratio of these two channels does not equal that of the authentic standards, the specificity of the target compound cannot be assured.

The most widely used method for quantification of BRs with LC-MS/MS is stable-isotope dilution, similar to other plant hormones, and can provide high accuracy and precision. The main drawback is the difficulty in obtaining commercially available isotope-labeled BRs, which limits the development of MS-based BR analysis methods. To address this issue, IS quantitation and relative quantitation may be the next-best option.

In summary, the advances in instrumentation and sample preparation have facilitated improvements in sensitivity for BR detection. The sample weight required has been reduced to the sub-gram level, and the sample pretreatment has been much simplified for LC-MS/MS analysis. The usual process including only a few SPE steps and subsequent derivatization has shortened the time for overall analysis to no more than one day. The throughput of these methods have been greatly improved. A challenge for these newly developed methods is their repeatability because most of these methods are based on custom-made sample pretreatment materials or self-set-up pretreatment instruments. The consistency of such materials and instruments is usually not comparable to commercial ones, and the reproducibility of different batches of materials has not been investigated in these studies. Based on these considerations discussed above, an easy-to-follow quantification method for BRs reported by Xin et al. (2013b) is recommended.

14.6.4   An easy-to-follow quantification method for brassinosteroids

14.6.4.1   Sample preparation

The collected plant tissues are first ground to a fine powder under liquid nitrogen. The plant material powder (0.5–1 g, FW) is extracted with 6 mL of 95% (v/v) aqueous MeOH twice at 4°C overnight. For quantification of BRs, $^2H_3$ labeled brassinolide (BL) and $^2H_3$ labeled castasterone (CS) are added to the extract as ISs. Then the sample is centrifuged for 15 min at 30,000 g at 4°C and the supernatant is collected for SPE.

A MAX cartridge is activated and equilibrated with MeOH, water, 1 M KOH, and 95% (v/v) MeOH in turn, and a MCX cartridge with MeOH, water, 5% (v/v) FA, and 10% (v/v) MeOH. Then, the supernatant is passed through the MAX column and collected to be dried by nitrogen stream. The samples are then dissolved in 10% (v/v) MeOH, and loaded onto the equilibrated MCX cartridge. After sequential washing with 5% (v/v) FA in 5% (v/v) MeOH, 5% (v/v) MeOH, 5% (v/v) NH_4OH in 5% (v/v) MeOH, and 5% (v/v) MeOH, BRs are eluted with MeOH. If the BR synthetic precursors, including 6-deoxo-CS, 6-deoxo-TY and so on, are also considered for detection, the MeOH percentage should be increased appropriately. The eluate is dried and then dissolved in of ACN and DMAPBA is added to undergo the derivatization reaction at 40°C for 1 h.

14.6.4.2 UPLC-MS/MS detection

The BR-DMAPBA detection is performed on a UPLC (Waters) instrument combined with a 5500 Q-Trap MS equipped with an ESI source (AB SCIEX). Five microliters of each sample are injected onto a BEH C18 column (100 mm × 2.1 mm, 1.7 μm). The inlet method was set as follows: mobile phase A, 0.05% acetic acid in water and B, 0.05% acetic acid in ACN. Gradient: 0–3 min, 65% B to 75% B; 3–11 min, 75% B to 95% B; 11–12 min, 95% B; 12–13.5 min, 95% B to 65% B; and 13.5–16 min, 65% B. BR-DMAPBA is detected in positive MRM mode. The ESI source parameters are set as: ion spray voltage, 5500 V; desolvation temperature, 550°C; nebulizing gas 1, 45 psi; desolvation gas 2, 45 psi; and curtain gas, 28 psi. The optimized MRM parameters are listed in Table 14.8.

The LOQ (signal to noise ratio = 10) of BL and CS are 1.7 pg/g FW and 3.9 pg/g FW respectively (in a rice leaf matrix), which are suitable for quantitative analysis of BL and CS under most circumstances. Relative standard deviation of the peak areas of endogenous CS in 18 duplicates of the same plant sample has been determined to be 7.03%, indicating good reproducibility of the method. Representative MRM detection results are depicted in Fig. 14.7. Furthermore, the wide applicability of the method has been verified in different plant matrixes such as rosettes, stems, and siliques of Arabidopsis and leaves, stems, panicles of rice.

Table 14.8 Optimized MRM parameters for BRs-DMAPBA

Compounds	Q1 mass/Q3 mass	Collision energy/V	Declustering potential/V	Entrance potential/V	Collision cell exit potential/V
BL-DMAPBA	610.40/190.1	55	35	10	12
24-epiBL-DMAPBA	610.40/176.1	73	35	10	12
^{2}H$_3$-BL-DMAPBA	613.40/190.1	55	35	10	12
	613.40/176.1	73	35	10	12
CS-DMAPBA	594.50/190.1	60	15	10	12
	594.50/176.1	70	15	10	12
^{2}H$_3$-CS-DMAPBA	597.50/190.1	60	15	10	12
	597.50/176.1	70	15	10	12
TE-DMAPBA	578.45/190.1	56	10	10	12
	578.45/176.1	69	10	10	12
TY-DMAPBA	578.40/190.1	62	15	10	12
	578.40/176.1	69	15	10	12

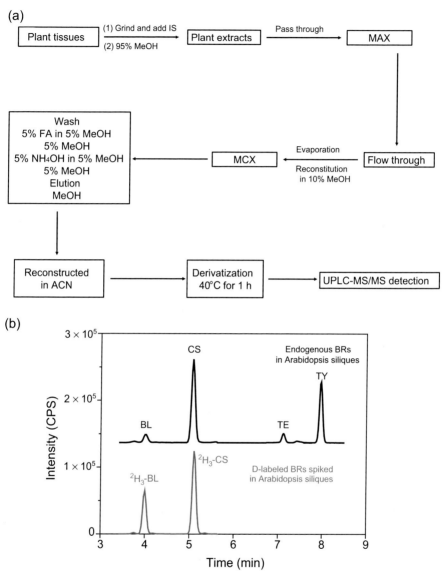

Figure 14.7 (a) MAX-MCX SPE procedure for BR analysis. (b) UPLC-MRM-MS detection of DMAPBA-derivatized BRs (upper trace) and $^2H_3$-BRs (lower trace) from 1 g of Arabidopsis siliques.

14.7 Strigolactones

14.7.1 Structures and analysis of strigolactones

SLs are the latest identified plant hormones and are derived from carotenoids (Alder et al., 2012). First known for their functions in promoting parasitic plant germination (Cook et al., 1966) and arbuscular mycorrhizal symbiosis establishment (Akiyama et al., 2005), SLs have been implicated in many biological processes in plants (Wang et al., 2017a,b), such as shoot branching (Gomez-Roldan et al., 2008; Umehara et al., 2008), root growth (Kapulnik et al., 2011), seed germination (Toh et al., 2012), hypocotyl elongation (Jia et al., 2014), and plant defense (Torres-Vera et al., 2014). It is hence important to quantitatively analyze SL levels and distribution of plants.

The structures of natural SLs identified so far typically present a common structure of a tetracyclic skeleton (A, B, C, and D rings), with a tricyclic lactone (ABC rings) connected by an enol ether linkage to an α,β-unsaturated furanone moiety (D ring) (Wang et al., 2017a). It is noteworthy that the D ring and the stereochemistry at the C-2' position are the same in all natural SLs to date, while A and B rings bear various oxygen-containing functions. Such SLs are typically found in root exudates and are referred to as canonical SLs. Evidence has emerged that SLs that function as endogenous plant hormones may not contain B and C rings (Al-Babili and Bouwmeester, 2015; Brewer et al., 2016; Seto et al., 2014).

The common structure of canonical SLs may result in two problems. First, SLs are neutral compounds (Wang et al., 2017a) with relatively low ionization efficiency compared to acidic or alkaline compounds, and thus may be subject to interference in MS detection. Second, the lactone structure of C and D rings in SL molecules exhibit low stability, and may hydrolyze in acidic or alkaline conditions. SL standard hydrolyzed significantly in a solution of EtOH:water (1:4, v/v) at room temperature (Boyer et al., 2012), thus minimizing SL hydrolysis during analysis procedures is important. In addition, many researches have shown that SL levels in plants are dramatically influenced by plant growth conditions (Jamil et al., 2011; Umehara et al., 2010), which makes it difficult to obtain stable and reliable results. Considering all these factors, extreme care is required to perform SL determination successfully.

14.7.2   Plant growth conditions

Originally, SLs were known as rhizosphere signaling molecules for parasitic plant germination and establishment of arbuscular mycorrhizal symbiosis. For both of these biological roles, SLs function in the rhizosphere at very low concentrations after they are secreted from the plant root system (Akiyama et al., 2005; Yokota et al., 1998). Therefore, researchers have focused particularly on analysis of SLs in root exudates for several years. Hydroponic culture was hence employed, since root exudate collection is very convenient compared to plants grown in soil culture. After SLs were classified as plant hormones, endogenous SL analysis became much more important. Several researchers have revealed that the levels of SLs vary tremendously when plant growth conditions change.

It is recognized that macronutrients, such as nitrogen and phosphorus, negatively affect SL production and exudation. When rice was grown in different phosphorus levels, it was found that concentrations of *ent-2'-epi*-5DS (now also referred to as 4-deoxyorobanchol, 4DO) in both root and root exudates were greatest without phosphorus in the hydroponic culture medium (Umehara et al., 2010). When the phosphorus level was increased to the level of full nutrient medium, *ent-2'-epi*-5DS (4DO) was not detected. Although the SL content varies with phosphorus levels, the relationship between SL content of WT and mutants at a specific phosphorus level remains constant. For example, SL content of the SL-insensitive mutant *d3-1* was higher than WT, while that of SL biosynthesis mutant *d10-1* was much lower. Similarly, it is found that the content of sorgomol and 5-DS in sorghum were higher in low nitrogen or low phosphorus conditions (Yoneyama et al., 2013).

Therefore, hydroponic culture combining phosphorus deficiency has been commonly used as plant growth conditions for SL analysis. A typical plant culture cycle for SL determination is as follows (Yoneyama et al., 2013). During seed germination and the seedling early growth stage, growth conditions include culture medium with full nutrients. Subsequently, plants are transferred to low phosphorus medium and maintained for an extended period. Without the low phosphorus period, SL levels would be so low that even highly sensitive LC-MS/MS may miss them (Umehara et al., 2010). This highlights the challenge of detecting and quantifying SLs under normal growth conditions. SL quantitative analytical methods based on LC-MS/MS are summarized in Table 14.9.

14.7.3   Sample preparation

14.7.3.1   Extraction and purification of strigolactones of root exudates

There are typically two methods to extract SL compounds from the hydroponic medium. The first one (Akiyama et al., 2005) involves placing activated charcoal cartridges in the water intake port of the hydroponic system to extract SLs continuously from the solution. Another method, which is used with more frequency, is to directly extract SLs from the hydroponic culture medium by LLE with ethyl acetate (Yoneyama et al., 2007a,b, 2012). The LLE method requires intermittent removal of sample for extraction. Typically 0.5–5 L is extracted with benzene or ethyl acetate (Cook et al., 1966; Yoneyama et al., 2008). Compared with plant tissues, hydroponic culture medium contains relatively simple components, making LLE an acceptable purification method, and LLE has also been used to further purify the fraction eluted from the activated charcoal cartridges in the first method (Akiyama et al., 2005; Yoneyama et al., 2008). Although LLE has been performed successfully for SL enrichment from root exudates, new methods which employ the SPE typically used for roots samples, have also been used for root exudate purification (Gomez-Roldan et al., 2008; Jamil et al., 2010; López-Ráez et al., 2008a,b).

14.7.3.2   Sampling, extraction, and purification of plant tissues

Collection, storage, homogenization and weighing of plant samples are carried out as previously described above for other plant hormones. Root samples contain many more metabolites that may interfere with SL analysis, thus making more elaborate extraction and purification methods necessary (Table 14.9). Considering the instability of SLs in aqueous solution, most of these protocols concentrate SLs based on their polarity properties. Ethyl acetate (Jamil et al., 2012; Liu et al., 2013a) or acetone

Table 14.9 Sample preparation and quantification methods based on LC-MS/MS for SLs

Analyte	Extraction	Purification method	Plant matrix	Instrumental method	References
5-DS, orobanchol, orobanchyl acetate	NA	Activated charcoal	*Lotus. japonicus*, fabaceae plants root exudates	HPLC-QQQ MRM	Akiyama et al. (2005), Yoneyama et al. (2008)
5-DS, orobanchol, orobanchyl acetate, sorgomol,	NA	LLE by ethyl acetate	Red clover, sorghum, Chinese milk vech, alfalfa, lettuce, tomato, wheat and marigold root exudate, 0.45–1.2 L	HPLC-QQQ MRM	Yoneyama et al. (2007a,b, 2012)
5-DS, orobanchol, orobanchyl acetate, sorgomol, solanacol, didehydro-orobanchol isomers, *ent-2'-epi*-5DS, fabacyl acetate	NA	SPE-C18	Pea, tomato, rice, Arabidopsis and *L. Japonicus* root exudates, 0.05–1 L	HPLC/UPLC-QQQ/ Q-Trap MRM	Foo and Davies (2011), Gomez-Roldan et al. (2008), Jamil et al. (2010, 2012), López-Ráez (2008a,b), Kohlen et al. (2011) and Liu et al. (2013a)
5-DS *ent-2'-epi*-5DS, orobanchol, orobanchyl acetate, fabacyl acetate	Ethyl acetate Ethyl acetate	LLE SPE	Sorghum root Rice, Arabidopsis and pea root	HPLC-QQQ MRM HPLC/UPLC-Q-TOF/ QQQ	Yoneyama et al. (2007a) Foo and Davies (2011), Kohlen et al. (2011) and Umehara et al. (2008)

(Umehara et al., 2008, 2010) are frequently used as extraction solvent extracting SLs from plant tissues. Although some researchers perform the extraction step at 4°C for 3 days (Foo and Davies, 2011; Yoneyama et al., 2007a), sonication could reduce this period to about 10 min (Kohlen et al., 2011; Liu et al., 2013a). For further enrichment, SPE has been the most popular purification method, and some researchers perform an LLE step prior to SPE (Umehara et al., 2010). Polarity properties are usually the basis for SL concentration by SPE. RP SPE sorbents, such as HLB and C18, as well as normal phase sorbents such as silica have been applied in SL purification (Foo and Davies, 2011; Kohlen et al., 2011). Methods combining RP SPE with normal phase SPE have also been developed (Kohlen et al., 2011; Umehara et al., 2008). Some of the SPE methods have also been applied in sample preparation of root exudates (Gomez-Roldan et al., 2008; Jamil et al., 2010, 2012; López-Ráez et al., 2008a,b; Kohlen et al., 2011).

14.7.4   LC-MS/MS detection

Seed germination and hyphal branching bioassays have been useful to estimate the amount of SL compounds in a specific sample (López-Ráez et al., 2008a,b), because SLs can promote parasitic plants seed germination and arbuscular mycorrhizal hyphal branching at extremely low concentrations (10^{-13}–10^{-9} M) (Akiyama et al., 2005; Kim et al., 2010). However, absolute quantitation and identification require LC-MS/MS methods, which now provide high-throughput, selectivity, and sensitivity, especially in MRM mode. Considering the terpenoid lactone structure and the relatively low hydrophilicity of SL molecules, the chromatographic separation of SLs has generally been performed on a RP column in a gradient elution mode (Kohlen et al., 2011; Yoneyama et al., 2007b). There are some specific fragmentation patterns of SL molecules that have been widely applied in their identification and quantification by MS. The structure and stereochemistry of the D ring are exactly the same in all natural SL molecules. Therefore, the neutral loss of the D ring is the most characteristic fragmentation pattern for all SLs, and the MRM transitions $[M+H/Na]^+ > [M+H/Na\text{-D ring}]^+$ and $[M+H/Na]^+ > [D\text{ ring}]^+$ have been selected for SL detection (Xie and Yoneyama, 2010; Yoneyama et al., 2012).

14.7.5   An easy-to-follow quantification method for strigolactones

14.7.5.1   Rice growth conditions

This quantification method for SLs in rice has been used in several key SL studies (Jiang et al., 2013; Zhou et al., 2013). Rice seedlings are grown hydroponically (Fig. 14.8). Surface-sterilized rice seeds are incubated in sterile water at 37°C in the dark for 2–3 days. Germinated seeds are transferred to hydroponic culture media (Kamachi et al., 1991) and cultured at 28°C with a 16 h light/8 h dark photoperiod for about 1 week. The seedlings are then transferred to phosphorus-deficient hydroponic culture

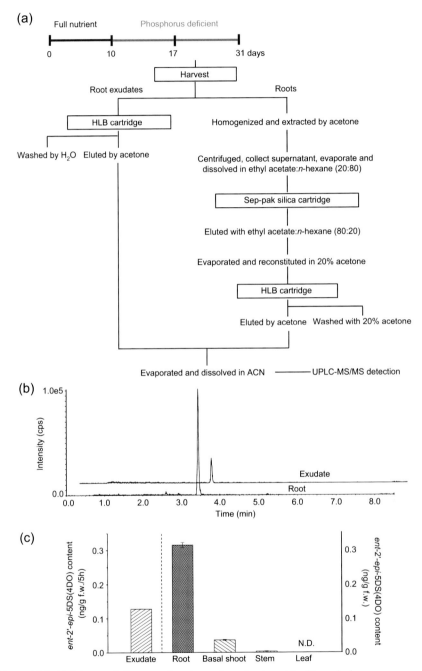

Figure 14.8 (a) Scheme of plant growth and sample pretreatment for SL analysis. (b) Chromatography of *ent*-2'-*epi*-5DS (4DO) in rice root and exudates. (c) Content of *ent*-2'-*epi*-5DS (4DO) in rice.

media and grown. Harvest of root exudates could be carried out within another 1–3 weeks.

14.7.5.2 Strigolactone extraction from root exudates

After collection of hydroponic culture media, isotope-labeled IS are added. Hydroponic culture media are then loaded onto HLB cartridges (Waters), then washed with water, and eluted with acetone.

14.7.5.3 Strigolactone extraction from plant tissues

Plant tissues (0.2 g, FW) are homogenized to a fine powder under liquid nitrogen. Acetone is used for SL extraction. After extraction at -20°C overnight, the homogenates are centrifuged at 10,000 g for 15 min. Supernatant is collected and evaporated to dryness under nitrogen gas. The residue is dissolved in ethyl acetate:n-hexane (20:80,v/v), and applied onto a Sep-pak Silica cartridge. The cartridge is eluted with ethyl acetate:n-hexane (80:20,v/v), and the eluate is evaporated to dryness. After reconstitution in 20% (v/v) acetone, the residue is loaded onto an HLB cartridge and eluted with 80% (v/v) acetone. The eluate is dried and re-dissolved in ACN for LC-MS/MS analysis.

14.7.5.4 LC-MS/MS detection

The purified SL-containing fractions are subjected to LC-MS/MS analysis using a system consisting of a 5500 Q-Trap mass spectrometer (AB SCIEX) and a UPLC (Waters) equipped with a RP column (Acquity UPLC BEH-C18, 2.1 × 100 mm, 1.7 μm; Waters). The LC conditions are set as follows: mobile phase A, 0.05% acetic acid; mobile phase B, 0.05% acetic acid in ACN; linear gradients: 0 min, 30% B; 1 min, 30% B; 7 min, 60% B; 10 min, 80% B; wash with 95% B for 2 min. SLs are detected in positive MRM mode. MS parameters are: Ion spray voltage, 4500 V; desolvation temperature, 600°C; nebulizing gas 1, 50 psi; desolvation gas 2, 55 psi; and curtain gas, 40 psi. Measurement of ent-2'-epi-5DS (4DO) is with MRM transition 331.1/97.0; collision energy, 27 V; declustering potential, 80 V; entrance potential, 8 V; and collision cell exit potential, 10 V. Illustration of this method applied to determine the content of ent-2'-epi-5DS (4DO) in rice tissues and exudates is shown in Fig. 14.8(c).

14.8 Multiple plant hormones

14.8.1 Classification of plant hormones

Analytical techniques for each class of plant hormone have helped in understanding of plant growth and development. Increasing evidence indicates that multiple plant hormones act through signaling networks and often impact the same biological process by additive, synergistic or antagonistic actions. Reported hormonal interactions include IAA and GAs in growth regulation (Tanimoto, 2005; Zhang et al., 2005); CKs, IAA, GAs, ABA and SLs in apical dominance (Cline and Oh, 2006; Luisi et al., 2011; Wang et al., 2006); and SA, JA, ABA, IAA, GAs, CKs, and BRs in plant defense (Bari and Jones, 2009). To better understand the molecular mechanisms and interactions of plant hormones in complex signaling networks, it is important to investigate the concentration, diversification, spatial and temporal distribution of multiple plant hormones simultaneously. Consequently, accurate and efficient quantitative measurement of multiple major plant hormones is required.

In the case of simultaneous determination of multiple classes of plant hormones, it is common to classify plant hormones based on their physicochemical properties. As mentioned earlier, IAA, ABA, JA, SA, and GAs are acidic compounds; CKs are alkaline compounds; SLs and BRs are neutral compounds. In a background of a wide range of more abundant primary and secondary metabolites, the ultra-trace amounts of target analytes present a major challenge for analytical methods to simultaneously profile multiple plant hormones. To efficiently separate and accurately quantitate multiple plant hormones, extra consideration needs to be taken in sample preparation, so as to meet the needs of highly sensitive, high-throughput, and comprehensive analytical approaches. Sample preparation needs to enrich a relatively wide range of compounds that may have various physicochemical properties, while also eliminating interfering substances.

14.8.2 Sample preparation

14.8.2.1 Sample extraction

Sample collection, storage, homogenization, and weighing should be conducted as previously described. The extraction solvent should provide high extraction efficiency for target plant hormones. Diverse solvents have been used for different plant hormones. For the determination of all acidic plant hormones, including IAA, ABA, SA, JA, GAs, and some of their derivatives, 80% (v/v) MeOH (Chen et al., 2011, 2012), 1-propanol/water/HCl (12/1/0.002, v/v/v) (Li et al., 2011) are commonly used for extraction. For extraction of both acidic and alkaline plant hormones, modified Bieleski's solvent (MeOH/water/FA, 15/4/1, v/v/v), which is a typical extraction solvent for CKs, has been widely used (Cui et al., 2015; Gupta et al., 2011; Izumi et al., 2009; Kojima et al., 2009; Wang et al., 2014b). In addition, MeOH/isopropanol (20/80, v/v) (Mueller and Munne-Bosch, 2011) and 1-propanol/water/concentrated HCl (12/1/0.002, v/v/v) (Pan et al., 2008, 2010) are also used. When extraction needs to cover acidic and alkaline plant hormones, as well as BRs, several solvents have been applied, such as MeOH/water/FA (15/10/5, v/v/v) (Zaveska Drabkova et al., 2015), 80%(v/v) MeOH containing 0.1% (v/v) FA (Van Meulebroek et al., 2012) and modified Bieleski's solvent (Fan et al., 2011). Solvents such as MeOH, ACN, and propanol, containing different proportions of water or formic/acetic acid have frequently been tested in plant hormone extractions. The optimal solvent should be selected based on both the physicochemical properties of the target analytes and the types of plant tissues.

14.8.2.2 Sample purification

Various purification techniques have been applied to further enrich trace plant hormones from the crude plant extracts, such as LLE, SPE, DLLME, and MSPE. Some of the purification methods work well for individual class of plant hormones, but employing some of those methods to analyze multiple classes of plant hormones is more challenging. A protocol has been reported using LLE to purify SA, IAA, ABA, JA, GA_3, GA_4 and several related metabolites from crude plant extracts (Pan et al., 2010). DLLME was employed to concentrate ABA, GA_3, IAA, SA, and KR from green seaweeds (Gupta et al., 2011). A method using Fe_3O_4/TiO_2-based MSPE was developed to extract 17 plant hormones, including 7 CKs, IAA, JA, ABA, and 7 GAs in rice (Cai et al., 2015b). Compared to analysis of individual class of plant hormones, sample purification steps for simultaneous analysis of multiple plant hormones should pay more attention to being rapid and inclusive, as well as not losing sensitivity and selectivity.

Considering the diversity in physicochemical properties of multiple classes of plant hormones, quantification of multiple plant hormones by a single-run LC-MS/MS method is not recommended unless it is demanded in particular experiment. One reason is that it will inevitably include more interfering compounds in the sample when various plant hormones should be covered. Another reason is that acidic and alkaline plant hormones need to be detected in different MS ionization modes, which would be impractical in most mass spectrometers. Therefore, it is preferable to separate plant hormones into several fractions based on their physicochemical properties during the purification steps, thus to reduce interferences and facilitate LC-MS/MS analysis. Accordingly, purification methods based on a single protocol are not yet very efficient. SPE with various kinds of sorbents is the most commonly used procedure (Table 14.10).

14.8.3 LC-MS/MS detections

Considering the physicochemical properties of various plant hormones, they need to be detected using different ionization modes of the mass spectrometer. Generally, acidic plant hormones are detected under negative-ion mode, while basic and neutral plant hormones are detected under positive ion mode. Thus simultaneous detection under different ionization modes in a single LC-MS/MS run is not possible unless the targeted compounds are relatively few and the mass spectrometer is an advanced model. A Waters Quattro Premier XE tandem mass spectrometer (Waters), which can switch mode between detection functions separated by different time windows, could detect GA_4, ABA, IAA, tZR (Niu et al., 2014). Using an ABI 4000 Q-Trap (AB SCIEX), which pauses for only 5 ms between MRM transitions and switches polarities within 700 ms, it was possible to analyze SA, IAA, ABA, JA, GA_3, GA_4 and several related metabolites (Pan et al., 2008).

In most cases, determination of acidic and alkaline plant hormones in separate LC-MS/MS runs is a more reasonable option (Liu et al., 2013b; Mueller and

Table 14.10 Sample pretreatment and quantification methods for multiple classes of plant hormones

Plant matrix	Analyte	Extraction solvent	Purification method	Instrumental method	References
Barley, citrus, papaya, pear, Arabidopsis	Auxins, ABA, SA, JA, GAs, CKs	80% MeOH; 1-propanol/water/concentrated HCl(2:1:0.002, v/v/v)	LLE	HPLC/UPLC-QQQ/Q-Trap MRM	Durgbanshi et al. (2005), Niu et al. (2014) and Pan et al. (2008)
Nicotiana tabacum, Arabidopsis, oilseed rape, grape, coconut water, lettuce seeds	Auxins, ABA, SA, JA, GAs, CKs	80% MeOH; 1-propanol/water/concentrated HCl(2:1:0.002, v/v/v); 1% FA in MeOH; 1% acetic acid in 2-propanol	C18/HLB-SPE	HPLC/UPLC-QQQ/Q-Trap MRM	Flokova et al. (2014), Grokinsky et al. (2014), Han et al. (2012a,b), Li et al. (2011), Ma et al. (2008) and Sheila D. S. Chiwacha et al. (2003)
Vermicompost, oilseed rape, rice, Arabidopsis, bryophytes, maize	Auxins, ABA, SA, JA, GAs, CKs, BRs	1% acetic acid in 2-propanol; modified Bieleski's solvent; 80% MeOH: MeOH:water: FA (15:10:5, v/v/v)	MCX-based-SPE	HPLC/UPLC/nano-LC-QQQ/Q-Trap MRM	Cui et al. (2015), Izumi et al. (2009), Kojima et al. (2009), Liu et al. (2013a,b), Zaveska Drabkova et al. (2015) and Zhang et al. (2015b)
Rice	IAA, ABA, SA, JA, GAs	80% MeOH	C18 SPE-derivatization	Nano-LC-Q-TOF CE-Q-TOF	Chen et al. (2011, 2012)
Green seaweeds	IAA, ABA, SA, GA₃, KR	Modified Bieleski's solvent	DLLME	HPLC-Q-TOF	Gupta et al. (2011)
Rice	IAA, ABA, JA, GAs, CKs	Acetonitrile	MSPE	HPLC-Q-Trap MRM	Cai et al. (2015b)

Munne-Bosch, 2011; Zaveska Drabkova et al., 2015). The LC-MS/MS conditions can be optimized as described in previous sections of this chapter. Compared to LC-MS/MS analysis of a specific class of plant hormone, there is little change in the parameters of MS in simultaneous detection of multiple classes of plant hormones. The MRM transitions described in previous sections are also available here. Although multiple plant hormone analysis requires more peaks be separated in a single LC separation run, highly selective and sensitive MS detectors reduce the necessity of baseline separation of every compound. Therefore, LC conditions are similar to those used for the separation of individual class of plant hormones.

14.8.4   Multiple plant hormone profiling based on LC-MS/MS

Analysis methods that can cover global plant hormones, including acidic, alkaline, and neutral molecules, have always been a challengeable problem. A high-through-put and highly sensitive LC-MS/MS method was reported, that could quantitatively analyze 43 molecular species of CKs, auxins, ABA, and GAs (Kojima et al., 2009). This number of simultaneously analyzed plant hormones is the highest to date. However, some plant hormones, such as JA, SA, SLs, and BRs, are not included in this method. Another analysis method carried out the profiling of plant hormones in bryophytes via a single MCX SPE cartridge purification (Zaveska Drabkova et al., 2015). Data for only some classes of hormones were presented, including IAA, SA, ABA, JA, CKs, and their metabolites, while GAs and BRs were not reported. Thus, the development of quantitative analysis methods of multiple plant hormones with high sensitivity is still lacking. Additionally, the gaseous plant hormone ethylene cannot be detected with LC-MS/MS. However, the amino acid precursor 1-amino-cyclopropane 1-carboxylate (ACC) can be detected and used as a marker of ethylene (Birkemeyer et al., 2003; Schmelz et al., 2004), thus allowing more complete plant hormone coverage using LC-MS/MS techniques. It is a very important goal to develop a simple and highly-sensitive quantitative analysis method using LC-MS/MS, which enables profiling of all the active plant hormones that will be used in research of signaling networks and cross talk of plant hormones. An easy-to-follow quantification method for acidic and alkaline plant hormones (Kojima et al., 2009) is summarized in Fig. 14.9 and described below.

14.8.5   An easy-to-follow quantification methods for multiple plant hormones

14.8.5.1   Sample preparation

About 100 mg (FW) plant material is frozen, and crushed to a fine powder, then soaked in 1 mL of modified Bieleski's solvent. A mixture of stable-isotope-labeled ISs is added to the homogenate. After extraction at $-30°C$ for at least 16 h, the homogenate is centrifuged at 10,000 g for 15 min. Supernatant is collected, and the pellet is re-extracted with 0.2 mL of extraction solvent. Combined supernatant fractions are applied to an HLB cartridge, which has been equilibrated with 1 M FA. Flow-through of the

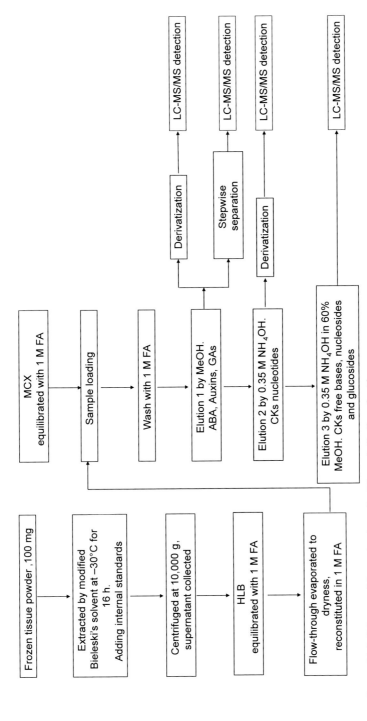

Figure 14.9 Scheme of determination of acidic and alkaline plant hormones.

HLB cartridge is collected. The eluate is evaporated to dryness, and then re-dissolved in 1 mL of 1 M FA for further MCX purification. An Oasis MCX (30 mg) is pre-conditioned with 1 M FA prior to sample loading. After washing with 1 M FA, ABA, auxins, and GAs are eluted with MeOH (Elution 1). CK nucleotides are eluted with 0.35 M NH$_4$OH (Elution 2), and CK free bases, nucleosides and glucosides are eluted with 0.35 M NH$_4$OH in 60% (v/v) MeOH (Elution 3). Each fraction is evaporated to dryness.

Elution 3, containing CK free bases, nucleosides and glucosides, is reconstituted with 0.1% (v/v) acetic acid for LC-MS analysis. Derivatization is performed for analysis of CK nucleotides in Elution 2, which converts CK nucleotides to nucleosides with alkaline phosphates. ABA, auxins, and GAs in Elution 1 are further prepared either by a stepwise separation method or an MS-probe method. In the stepwise separation method, the residue of Elution 1 is reconstituted with 0.2 mL H$_2$O and then passed through a DEAE-cellulose column, and eluted with a series of FA aqueous solutions with increasing concentration of acid. In the MS-probe modification method, samples are passed through a DEAE-cellulose column, then acidic plant hormones are derivatized with bromocholine (MS-probe).

14.8.5.2 LC-MS/MS detection

CKs and acidic plant hormones are measured with a UPLC-MS/MS system with a BEH C18 column (100 mm × 2.1 mm, 1.7 mm). For CK separation, linear gradient of solvent A (0.06% acetic acid) and solvent B (0.06% acetic acid in MeOH) are set as follows: 0 min, 1% B; 4 min, 45% B; 7 min, 70% B; then wash with 99% B. Capillary voltage is 3.13 kV. Cone voltage (V) and collision energy (eV) are as follows: tZR, cZ riboside, DZR: 52 V, 15 eV; iP: 53 V, 14 eV; iPR: 52 V, 18 eV; tZ7G, tZOG, tZ9G, DZ-9-N-glucoside: 55 V, 20 eV; tZROG: 65 V, 28 eV; iP7G: 70 V, 21 eV; iP9G: 52 V, 20 eV. For acidic plant hormone derivatives, gradient of solvent A (0.05% FA) and solvent B (0.05% FA in ACN) is set as follows: 0 min, 1% B; 2 min, 14% B; 6 min, 16% B; 13 min, 55% B; 14 min, 70% B. Capillary voltage is 3.2 kV. Cone voltage (V) and collision energy (eV) are as follows: IAA, 30 V, 20 eV; IA-Ala, 35 V, 20 eV; IA-Trp, IA-Phe: 40 V, 20 eV; IA-Asp: 40 V, 15 eV; IA-Ile, IA-Leu: 35 V, 20 eV; GA1, GA$_3$, GA$_{24}$, GA$_{44}$, GA$_{53}$: 50 V, 30 eV; GA$_4$, GA$_7$, GA$_{12}$, GA$_{19}$, GA$_{20}$: 40 V, 20 eV; GA$_8$, GA$_9$: 40 V, 30 eV; and ABA: 40 V, 20 eV.

14.9   Summary points

- Analytical methods for endogenous plant hormones are essential to study the plant hormone metabolism and signaling processes. Approaches based on LC-MS/MS have become increasingly important quantification methods. Plant extracts are extremely complex and plant hormones are present in trace amounts, so the preparation procedures including sampling, extraction, purification and derivatization are very important to build high-sensitive LC-MS/MS methods.

- Solutions of MeOH and isopropanol with a little water and FA are used to extract most plant hormones, while acetone is suitable to SLs. SPE is then widely used to purify extracts.
- Different mass spectrometers coupled with LC are used to develop plant hormone quantification method, in which MRM mode using QQQ or Q-Trap has much higher sensitivity. The simultaneous quantification of all plant hormones (plant hormone profiling) and related metabolites can help understand the crosstalk within whole plant hormone network. Simple and efficient methods determining multiple classes of plant hormones based on LC-MS/MS is becoming possible.
- Various quantification methods based on LC-MS/MS for most plant hormones in authors' lab and easy-to-follow analytical protocols are presented.

14.10   Future perspectives

- Improving sensitivity is an eternal pursuit and depends on the enhancement of every step from sample preparation through to detection. Selective enrichment methods and novel derivatization reagents providing higher MS responses offer particular promises.
- High-throughput methods can improve the efficiency of plant hormone analysis. Increasing the level of automation of sample pretreatment is a promising way. Simultaneous multi-hormone sample pretreatment via combination of multiple SPE mechanisms can present all hormonal information from one sample. Improvements in MS techniques enabling simultaneous monitoring of hundreds of MRM transitions and fast positive/negative switch, will make simultaneous multi-plant hormone detection attainable.
- The specificity of target compounds need to be improved sometimes when quantified with LC-MS/MS methods because the matrix is complicated so that the co-eluting interferences are unavoidable. MRM^3 quantitation approach can further filter the interferences from matrix when performing analysis with a Q-Trap MS system. Differential mobility spectrometry (DMS) is an another alternative technique to increase the peak specificity because it brings ion mobility separation to LC-MS/MS system, enabling further eliminating co-eluting contaminant from LC and reducing the high background noise.
- In situ detection with MS imaging can provide spatial distributions of metabolites and their relative abundances in plants, which has been demonstrated by a range of pioneering studies. However, it still remains far from routine for plant hormones beacause the endogenous levels of plant hormones are so low that they cannot be detected without sample cleaning. Particular sample pretreatment such as in situ selective chemical derivatization of target molecules may be required for in situ monitoring of plant hormones.

Abbreviations

ABA	abscisic acid
ACN	acetonitrile
APCI	atmospheric pressure chemical ionization
BL	brassinolide
BRs	brassinosteroids
CE	capillary electrophoresis

CID	collision-induced dissociation
CKs	cytokinins
CS	castasterone
cZ	*cis*-zeatin
DHZ	dihydrozeatin
DLLME	dispersive liquid-liquid micro-extraction
DMAPBA	3-(dimethylamino)-phenylboronic acid
DMS	differential mobility spectrometry
DW	dry weight
EDC	N-(3-dimethylaminopropyl)-N-ethylcarbodiimide
ESI	electrospray ionization
EtOH	ethanol
FA	formic acid
FLR	fluorescence
FW	fresh weight
GAs	gibberellins
GC	gas chromatography
HILIC	hydrophilic interaction chromatography
HPLC	high performance liquid chromatography
IAA	indole-3-acetic acid
IAP	immunoaffinity purification
IBA	indole-3-butyric
iP	isopentenyladenine
IS	internal standard
JA	jasmonic acid
LC	liquid chromatography
LC-MS/MS	liquid chromatography coupled to tandem mass spectrometry
LLE	liquid-liquid extraction
LLME	liquid-liquid micro-extraction
LOD/LOQ	limit of detection/limit of quantification
m/z	mass-to-charge ratio
MeOH	methanol
MRM	multiple reaction monitoring
MS/MS	tandem mass spectrometry
MSPE	magnetic solid phase extraction
PME	polymer monolith extraction
QQQ	triple quadrupole
Q-TOF	quadrupole time-of-flight
Q-Trap	triple quadrupole linear ion trap
RP	reversed phase
SA	salicylic acid
SIM	selected ion monitoring
SLs	strigolactones
SPE	solid-phase extraction
SPME	solid-phase micro-extraction
TE	teasterone
TEA	triethylamine
TY	typhasterol
tZ	*trans*-zeatin

Acknowledgments

We thank Prof. Steven Smith for discussion and revision about the manuscript. We thank Dr. Lin Li and Ms Dongmei Li for preparing figures. This work is supported by the National Natural Science Foundation of China (grant nos. 90817008, 91117016, 91417314 and 31470433) and CAS Key Technology Talent Program to Jinfang Chu, and National Natural Science Foundation of China to Peiyong Xin (grant no. 31500299) and Yi Chen (grant nos. 90717120 and 91117010).

References

Absalan, G., Akhond, M., Sheikhian, L., 2008. Extraction and high performance liquid chromatographic determination of 3-indole butyric acid in pea plants by using imidazolium-based ionic liquids as extractant. Talanta 77, 407–411.

Ai, L., Yue, Z., Ma, Y., et al., 2015. Simple and rapid determination of N(6)-(Delta(2)-isopentenyl)adenine, zeatin, and dihydrozeatin in plants using on-line cleanup liquid chromatography coupled with hybrid quadrupole-Orbitrap high-resolution mass spectrometry. J. Sep. Sci. 38, 1858–1865.

Akiyama, K., Matsuzaki, K., Hayashi, H., 2005. Plant sesquiterpenes induce hyphal branching in arbuscular mycorrhizal fungi. Nature 435, 824–827.

Alarcon Flores, M.I., Romero-Gonzalez, R., Garrido Frenich, A., et al., 2011. QuEChERS-based extraction procedure for multifamily analysis of phytohormones in vegetables by UHPLC-MS/MS. J. Sep. Sci. 34, 1517–1524.

Al-Babili, S., Bouwmeester, H.J., 2015. Strigolactones, a novel carotenoid-derived plant hormone. Ann. Rev. Plant Biol. 66, 161–186.

Alder, A., Jamil, M., Marzorati, M., et al., 2012. The path from beta-carotene to carlactone, a strigolactone-like plant hormone. Science 335, 1348–1351.

Antoniadi, I., Plackova, L., Simonovik, B., Dolezal, K., Turnbull, C., Ljung, K., Novak, O., 2015. Cell-type-specific cytokinin distribution within the Arabidopsis primary root apex. Plant Cell 27, 1955–1967.

Ayele, B.T., Magnus, V., Mihaljevic, S., et al., 2010. Endogenous gibberellin profile during Christmas rose (*Helleborus niger* L.) flower and fruit Development. J. Plant Growth Regul. 29, 194–209.

Bai, Y., Zhang, J., Bai, Y., et al., 2012. Direct analysis in real time mass spectrometry combined with single-drop liquid-liquid-liquid microextraction for the rapid analysis of multiple phytohormones in fruit juice. Anal. Bioanal. Chem. 403, 2307–2314.

Balcke, G.U., Handrick, V., Bergau, N., et al., 2012. An UPLC-MS/MS method for highly sensitive high-throughput analysis of phytohormones in plant tissues. Plant Methods 8, 1–11.

Barendse, G.W.M., Vandewerken, P.H., Takahashi, N., 1980. High-performance liquid-chromatography of gibberellins. J. Chromatogr. A 198, 449–455.

Bari, R., Jones, J.D.G., 2009. Role of plant hormones in plant defence responses. Plant Mol. Biol. 69, 473–488.

Barkawi, L.S., Tam, Y.Y., Tillman, J.A., et al., 2010. A high-throughput method for the quantitative analysis of auxins. Nat. Protoc. 5, 1609–1618.

Bethke, P.C., Jones, R.L., 1998. Gibberellin signaling. Curr. Opin. Plant Biol. 1, 440–446.

Birkemeyer, C., Kolasa, A., Kopka, J., 2003. Comprehensive chemical derivatization for gas chromatography-mass spectrometry-based multi-targeted profiling of the major phytohormones. J. Chromatogr. A 993, 89–102.

Björkman, P.O., Tillberg, E., 1996. Acetylation of cytokinins and modified adenine compounds: a simple and non-destructive derivatization method for gas chromagraphy-mass spectrometric analysis. Phytochem. Anal. 7, 57–68.

Bosco, R., Daeseleire, E., Van Pamel, E., et al., 2014. Development of an ultrahigh-performance liquid chromatography-electrospray ionization-tandem mass spectrometry method for the simultaneous determination of salicylic acid, jasmonic acid, and abscisic acid in rose leaves. J. Agric. Food Chem. 62, 6278–6284.

Boyer, F.D., Germain, A.D., Pillot, J.P., et al., 2012. Structure-activity relationship studies of strigolactone-related molecules for branching inhibition in garden pea: molecule design for shoot branching. Plant Physiol. 159, 1524–1544.

Brewer, P.B., Yoneyama, K., Filardo, F., et al., 2016. *LATERAL BRANCHING OXIDOREDUCTASE* acts in the final stages of strigolactone biosynthesis in Arabidopsis. Proc. Natl. Acad. Sci. U.S.A. 113, 6301–6306.

Browse, J., 2009. Jasmonate passes muster: a receptor and targets for the defense hormone. Ann. Rev. Plant Biol. 60, 183–205.

Buchanan, B.B., Gruissem, W., Vickers, K., et al., 2015. Biochemistry and Molecular Biology of Plants. John Wiley & Sons.

Cai, B., Zhu, J., Shi, Z., et al., 2013. A simple sample preparation approach based on hydrophilic solid-phase extraction coupled with liquid chromatography-tandem mass spectrometry for determination of endogenous cytokinins. J. Chromatogr. B 942–943, 31–36.

Cai, B., Zhu, J., Gao, Q., et al., 2014. Rapid and high-throughput determination of endogenous cytokinins in Oryza sativa by bare Fe_3O_4 nanoparticles-based magnetic solid-phase extraction. J. Chromatogr. A 1340, 146–150.

Cai, B., Ye, E., Yuan, B., et al., 2015a. Sequential solvent induced phase transition extraction for profiling of endogenous phytohormones in plants by liquid chromatography-mass spectrometry. J. Chromatogr. B 1004, 23–29.

Cai, B., Yin, J., Hao, Y., et al., 2015b. Profiling of phytohormones in rice under elevated cadmium concentration levels by magnetic solid-phase extraction coupled with liquid chromatography tandem mass spectrometry. J. Chromatogr. A 1406, 78–86.

Castillo, G., Torrecillas, A., Nogueiras, C., et al., 2014. Simultaneous quantification of phytohormones in fermentation extracts of Botryodiplodia theobromae by liquid chromatography-electrospray tandem mass spectrometry. World J. Microbiol. Biotechnol. 30, 1937–1946.

Chen, Y., Guo, Z., Wang, X., et al., 2008. Sample preparation. J. Chromatogr. A 1184, 191–219.

Chen, W., Gai, Y., Liu, S., et al., 2010. Quantitative analysis of cytokinins in plants by high performance liquid chromatography: electronspray ionization ion trap mass spectrometry. J. Integr. Plant. Biol. 52, 925–932.

Chen, M., Huang, Y., Liu, J., et al., 2011. Highly sensitive profiling assay of acidic plant hormones using a novel mass probe by capillary electrophoresis-time of flight-mass spectrometry. J. Chromatogr. B 879, 938–944.

Chen, M., Fu, X., Liu, J., et al., 2012. Highly sensitive and quantitative profiling of acidic phytohormones using derivatization approach coupled with nano-LC-ESI-Q-TOF-MS analysis. J. Chromatogr. B 905, 67–74.

Chiwocha, S.D.S., Abrams, S.R., Ambrose, S.J., et al., 2003. A method for profiling classes of plant hormones and their metabolites using liquid chromatography-electrospray ionization tandem mass spectrometry: an analysis of hormone regulation of thermodormancy of lettuce (*Lactuca sativa* L.) seeds. Plant J. 35, 405–417.

Cline, M.G., Oh, C., 2006. A reappraisal of the role of abscisic acid and its interaction with auxin in apical dominance. Ann. Bot. 98, 891–897.

Cook, C.E., Whichard, L.P., Turner, B., et al., 1966. Germination of witchweed (*Striga lutea* Lour.): isolation and properties of a potent stimulant. Science 154, 1189–1190.

Cui, K., Lin, Y., Zhou, X., et al., 2015. Comparison of sample pretreatment methods for the determination of multiple phytohormones in plant samples by liquid chromatography-electrospray ionization-tandem mass spectrometry. Microchem. J. 121, 25–31.

Cutler, S.R., Rodriguez, P.L., Finkelstein, R.R., et al., 2010. Abscisic acid: emergence of a core signaling network. Ann. Rev.Plant Biol. 61, 651–679.

Darwin, C., Darwin, F., 1880. The Power of Movement Inplants. John Murray, London.

Davies, P.J., 2010. The Plant Hormones: Their Nature, Occurrence, and Functions. Kluwer Academic Publishers, Netherlands.

Davies, P., 2013. Plant Hormones: Physiology, Biochemistry and Molecular Biology. Springer Science & Business Media.

Ding, J., Mao, L., Wang, S., et al., 2013a. Determination of endogenous brassinosteroids in plant tissues using solid-phase extraction with double layered cartridge followed by high-performance liquid chromatography-tandem mass spectrometry. Phytochem. Anal., 24, 386–394.

Ding, J., Mao, L., Yuan, B., et al., 2013b. A selective pretreatment method for determination of endogenous active brassinosteroids in plant tissues: double layered solid phase extraction combined with boronate affinity polymer monolith microextraction. Plant Methods 9.

Ding, J., Wu, J., Liu, J., et al., 2014. Improved methodology for assaying brassinosteroids in plant tissues using magnetic hydrophilic material for both extraction and derivatization. Plant Methods 10.

Du, F., Ruan, G., Liang, S., et al., 2012a. Monolithic molecularly imprinted solid-phase extraction for the selective determination of trace cytokinins in plant samples with liquid chromatography-electrospray tandem mass spectrometry. Anal. Bioanal. Chem. 404, 489–501.

Du, F., Ruan, G., Liu, H., 2012b. Analytical methods for tracing plant hormones. Anal. Bioanal. Chem. 403, 55–74.

Du, F., Sun, L., Zhen, X., et al., 2015. High-internal-phase-emulsion polymeric monolith coupled with liquid chromatography-electrospray tandem mass spectrometry for enrichment and sensitive detection of trace cytokinins in plant samples. Anal. Bioanal. Chem. 407, 6071–6079.

Durgbanshi, A., Arbona, V., Pozo, O., et al., 2005. Simultaneous determination of multiple phytohormones in plant extracts by liquid chromatography-electrospray tandem mass spectrometry. J. Agric. Food Chem. 53, 8437–8442.

Fan, S., Wang, X., Li, P., et al., 2011. Simultaneous determination of 13 phytohormones in oilseed rape tissues by liquid chromatography-electrospray tandem mass spectrometry and the evaluation of the matrix effect. J. Sep. Sci. 34, 640–650.

Feng, J., Shi, Y., Yang, S., et al., 2017. Cytokinins. In: Hormone Metabolism and Signaling in Plants, Elsevier.

Feurtado, J.A., Yang, J., Ambrose, S.J., et al., 2007. Disrupting abscisic acid homeostasis in western white pine (*Pinus monticola* Dougl. Ex D. Don) seeds induces dormancy termination and changes in abscisic acid catabolites. J. Plant Growth Regul. 26, 46–54.

Fletcher, A.T., Mader, J.C., 2007. Hormone profiling by LC-QToF-MS/MS in dormant *Macadamia integrifolia*: correlations with abnormal vertical growth. J. Plant Growth Regul. 26, 351–361.

Flokova, K., Tarkowska, D., Miersch, O., et al., 2014. UHPLC-MS/MS based target profiling of stress-induced phytohormones. Phytochemistry 105, 147–157.

Foo, E., Davies, N.W., 2011. Strigolactones promote nodulation in pea. Planta 234, 1073–1081.

Fu, J., Sun, X., Wang, J., et al., 2011. Progress in quantitative analysis of plant hormones. Chin. Sci. Bull. 56, 355–366.

Fu, J., Chu, J., Sun, X., et al., 2012. Simple, rapid, and simultaneous assay of multiple carboxyl containing phytohormones in wounded tomatoes by UPLC-MS/MS using single spe purification and isotope dilution. Anal. Sci. 28, 1081–1087.

Gamoh, K., Kitsuwa, T., Takatsuto, S., et al., 1988. Determination of trace brassinosteroids by high-performance liquid-chromatography. Anal. Sci. 4, 533–535.

Gamoh, K., Yamaguchi, I., Takatsuto, S., 1994. Aapid and selective sample preparation for the chromatographic determination of brassinosteroids from plant-material using solid-phase extraction method. Anal. Sci. 10, 913–917.

Gamoh, K., Abe, H., Shimada, K., et al., 1996. Liquid chromatography mass spectrometry with atmospheric pressure chemical ionization of free brassinosteroids. Rapid Commun. Mass Spectrom. 10, 903–906.

Ge, L., Tan, S., Yong, J.W., et al., 2006. CE for cytokinin analyses: a review. Electrophoresis 27, 4779–4791.

Ge, L., Peh, C.Y.C., Yong, J.W.H., et al., 2007. Analyses of gibberellins by capillary electrophoresis-mass spectrometry combined with solid-phase extraction. J. Chromatogr. A 1159, 242–249.

Ge, L., Yong, J.W.H., Tan, S.N., et al., 2008. Analyses of gibberellins in coconut (*Cocos nucifera* L.) water by partial filling-micellar electrokinetic chromatography-mass spectrometry with reversal of electroosmotic flow. Electrophoresis 29, 2126–2134.

Gomez-Roldan, V., Fermas, S., Brewer, P.B., et al., 2008. Strigolactone inhibition of shoot branching. Nature 455, 189–194.

Gouthu, S., Morre, J., Maier, C.S., et al., 2013. An analytical method to quantify three plant hormone families in grape berry using liquid chromatography and multiple reaction monitoring mass spectrometry. In: Gang, R.D. (Ed.), *Phytochemicals, Plant Growth, and the Environment*. Springer New York, New York, NY, pp. 19–36.

Grokinsky, D.K., Albacete, A., Jammer, A., Krbez, P., van der Graaff, E., Pfeifhofer, H., Roitsch, T., 2014. A rapid phytohormone and phytoalexin screening method for physiological phenotyping. Mol. Plant 7, 1053–1056.

Gupta, V., Kumar, M., Brahmbhatt, H., et al., 2011. Simultaneous determination of different endogenetic plant growth regulators in common green seaweeds using dispersive liquid-liquid microextraction method. Plant Physiol. Biochem. 49, 1259–1263.

Han, Y., Zhou, Z., Wu, H., et al., 2012a. Simultaneous determination of jasmonic acid epimers as phytohormones by chiral liquid chromatography–quadrupole time-of-flight mass spectrometry and their epimerization study. J. Chromatogr. A 1235, 125–131.

Han, Z., Liu, G., Rao, Q., et al., 2012b. A liquid chromatography tandem mass spectrometry method for simultaneous determination of acid/alkaline phytohormones in grapes. J. Chromatogr. B 881–882, 83–89.

Hayat, S., Ali, B., Ahmad, A., 2007. Salicylic Acid: Biosynthesis, Metabolism and Physiological Role in Plants. Springer.

Hayat, Q., Hayat, S., Irfan, M., et al., 2010. Effect of exogenous salicylic acid under changing environment: a review. Environ. Exp. Bot. 68, 14–25.

Hayat, S., Ahmad, A., 2011. Brassinosteroids: A Class of Plant Hormone. Springer Science & Business Media.

Hedden, P., Thomas, S.G., 2012. Gibberellin biosynthesis and its regulation. Biochem. J. 444, 11–25.

Hillman, J.R., 1978. Society for Experimental Biology, Seminar Series: Volume 4, Isolation of Plant Growth Substances, vol. 4. CUP Archive.

Hooley, R., 1994. Gibberellins — perception, transduction and responses. Plant Mol. Biol. 26, 1529–1555.

Hooykaas, P.J., Hall, M.A., Libbenga, K.R., 1999. Biochemistry and Molecular Biology of Plant Hormones, vol. 33. Elsevier.

Horgan, R., Scott, I.M., 1987. Cytokinins. In: Rivier, L., Crozier, A. (Eds.), The Principles and Practice of Plant Hormone Analysis. Academic Press, London, pp. 304–365.

Hou, S., Zhu, J., Ding, M., et al., 2008. Simultaneous determination of gibberellic acid, indole-3-acetic acid and abscisic acid in wheat extracts by solid-phase extraction and liquid chromatography-electrospray tandem mass spectrometry. Talanta 76, 798–802.

Hoyerova, K., Gaudinova, A., Malbeck, J., et al., 2006. Efficiency of different methods of extraction and purification of cytokinins. Phytochemistry 67, 1151–1159.

Hu, Y., Li, Y., Zhang, Y., et al., 2011. Development of sample preparation method for auxin analysis in plants by vacuum microwave-assisted extraction combined with molecularly imprinted clean-up procedure. Anal. Bioanal. Chem. 399, 3367–3374.

Hu, Y., Huang, Z., Liao, J., et al., 2013. Chemical bonding approach for fabrication of hybrid magnetic metal-organic framework-5: high efficient adsorbents for magnetic enrichment of trace analytes. Anal. Chem. 85, 6885–6893.

Huo, F., Wang, X., Han, Y., et al., 2012. A new derivatization approach for the rapid and sensitive analysis of brassinosteroids by using ultra high performance liquid chromatography-electrospray ionization triple quadrupole mass spectrometry. Talanta 99, 420–425.

Hussain, A., Krischke, M., Roitsch, T., et al., 2010. Rapid determination of cytokinins and auxin in cyanobacteria. Curr. Microbiol. 61, 361–369.

Ikekawa, N., Takatsuto, S., Kitsuwa, T., et al., 1984. Analysis of natural brassinosteroids by gas-chromatography and gas-chromatography mass-spectrometry. J. Chromatogr. 290, 289–302.

Izumi, Y., Okazawa, A., Bamba, T., et al., 2009. Development of a method for comprehensive and quantitative analysis of plant hormones by highly sensitive nanoflow liquid chromatography-electrospray ionization-ion trap mass spectrometry. Anal. Chim. Acta 648, 215–225.

Jamil, M., Charnikhova, T., Verstappen, F., et al., 2010. Carotenoid inhibitors reduce strigolactone production and *Striga hermonthica* infection in rice. Arch. Biochem. Biophys. 504, 123–131.

Jamil, M., Charnikhova, T., Cardoso, C., et al., 2011. Quantification of the relationship between strigolactones and *Striga hermonthica* infection in rice under varying levels of nitrogen and phosphorus. Weed Res. 51, 373–385.

Jamil, M., Charnikhova, T., Houshyani, B., et al., 2012. Genetic variation in strigolactone production and tillering in rice and its effect on *Striga hermonthica* infection. Planta 235, 473–484.

Jia, K., Luo, Q., He, S., et al., 2014. Strigolactone-regulated hypocotyl elongation is dependent on cryptochrome and phytochrome signaling pathways in Arabidopsis. Mol. Plant 7, 528–540.

Jiang, L., Liu, X., Xiong, G., et al., 2013. DWARF 53 acts as a repressor of strigolactone signalling in rice. Nature 504, 401–405.

Jikumaru, Y., Seo, M., Matsuura, H., et al., 2013. Profiling of jasmonic acid-related metabolites and hormones in wounded leaves. Methods Mol. Biol. 1011, 113–122.

Joo, S., Jang, M., Kim, M.K., et al., 2015. Biosynthetic relationship between C28-brassinosteroids and C29-brassinosteroids in rice (*Oryza sativa*) seedlings. Phytochemistry 111, 84–90.

Kamachi, K., Yamaya, T., Mae, T., et al., 1991. A role for glutamine synthetase in the recombination of leaf nitrogen during natural senescence in rice leaves. Plant Physiol. 96, 411–417.

Kapulnik, Y., Delaux, P.M., Resnick, N., et al., 2011. Strigolactones affect lateral root formation and root-hair elongation in Arabidopsis. Planta 233, 209–216.

Karadeniz, A., Topcuoğlu, Ş., Inan, S., 2006. Auxin, gibberellin, cytokinin and abscisic acid production in some bacteria. World J. Microbiol. Biotechnol. 22, 1061–1064.

Kasote, D.M., Ghosh, R., Chung, J.Y., et al., 2016. Multiple reaction monitoring mode based liquid chromatography-mass spectrometry method for simultaneous quantification of brassinolide and other plant hormones involved in abiotic stresses. Int. J. Anal. Chem. 2016.

Kim, H.B., Kwon, M., Ryu, H., et al., 2006. The regulation of DWARF4 expression is likely a critical mechanism in maintaining the homeostasis of bioactive brassinosteroids in Arabidopsis. Plant Physiol. 140, 548–557.

Kim, H.I., Xie, X., Kim, H.S., et al., 2010. Structure-activity relationship of naturally occurring strigolactones in *Orobanche minor* seed germination stimulation. J. Pest. Sci. 35, 344–347.

Kim, B., Jeong, Y.J., Corvalan, C., et al., 2014. Darkness and gulliver2/phyB mutation decrease the abundance of phosphorylated BZR1 to activate brassinosteroid signaling in Arabidopsis. Plant J. 77, 737–747.

Koek, M.M., Jellema, R.H., van der Greef, J., et al., 2011. Quantitative metabolomics based on gas chromatography mass spectrometry: status and perspectives. Metabolomics 7, 307–328.

Kohlen, W., Charnikhova, T., Liu, Q., et al., 2011. Strigolactones are transported through the xylem and play a key role in shoot architectural response to phosphate deficiency in nonarbuscular mycorrhizal host Arabidopsis. Plant Physiol. 155, 974–987.

Kojima, M., Kamada-Nobusada, T., Komatsu, H., et al., 2009a. Highly sensitive and high-throughput analysis of plant hormones using MS-probe modification and liquid chromatography-tandem mass spectrometry: an application for hormone profiling in *Oryza sativa*. Plant Cell Physiol. 50, 1201–1214.

Koshioka, M., Harada, J., Takeno, K., et al., 1983. Reversed-phase C18 high-performance liquid-chromatography of acidic and conjugated gibberellins. J. Chromatogr. 256, 101–115.

Kowalczyk, M., Sandberg, G., 2001. Quantitative analysis of indole-3-acetic acid metabolites in Arabidopsis. Plant Physiol. 127, 1845–1853.

Laloue, M., Terrrine, C., Cawer, M., 1974. Cytokinins: formation of the nucleoside-5'-triphosphate in tobacco and Acer cells. FEBS Lett. 46, 45–50.

Lange, T., Kappler, J., Fischer, A., et al., 2005. Gibberellin biosynthesis in developing pumpkin seedlings. Plant Physiol. 139, 213–223.

Li, Y., Wei, F., Dong, X., et al., 2011. Simultaneous analysis of multiple endogenous plant hormones in leaf tissue of oilseed rape by solid-phase extraction coupled with high-performance liquid chromatography-electrospray ionisation tandem mass spectrometry. Phytochem. Anal. 22, 442–449.

Li, D., Guo, Z., Chen, Y., 2016. Direct derivatization and quantitation of ultra-trace gibberellins in sub-milligram fresh plant organs. Mol. Plant 9, 175–177.

Li, J., Li, C., Smith, S.M. (Eds.), 2017. Hormone Metabolism and Signaling in Plants. Elsevier.

Liang, Y., Zhao, M., Liu, H., 2009. Research progress in cytokinins analysis. Chin. J. Anal. Chem. 37, 1232–1239.

Liu, H., Li, Y., Luan, T., et al., 2007. Simultaneous determination of phytohormones in plant extracts using SPME and HPLC. Chromatographia 66, 515–520.

Liu, Z., Wei, F., Feng, Y., 2010. Determination of cytokinins in plant samples by polymer monolith microextraction coupled with hydrophilic interaction chromatography-tandem mass spectrometry. Anal. Methods 2, 1676.

Liu, S., Chen, W., Fang, K., et al., 2012a. Classification and characterization of unknown cytokinins into essential types by in-source collision-induced dissociation electrospray ionization ion trap mass spectrometry. Rapid Commun. Mass Spectrom. 26, 2075–2082.

Liu, X., Hegeman, A.D., Gardner, G., et al., 2012b. Protocol: high-throughput and quantitative assays of auxin and auxin precursors from minute tissue samples. Plant Methods 8, 17.

Liu, Z., Cai, B., Feng, Y., 2012c. Rapid determination of endogenous cytokinins in plant samples by combination of magnetic solid phase extraction with hydrophilic interaction chromatography-tandem mass spectrometry. J. Chromatogr. B 891–892, 27–35.

Liu, H., Li, X., Xiao, J., et al., 2012d. A convenient method for simultaneous quantification of multiple phytohormones and metabolites: application in study of rice-bacterium interaction. Plant Methods 8.

Liu, J., Novero, M., Charnikhova, T., et al., 2013a. Carotenoid cleavage dioxygenase 7 modulates plant growth, reproduction, senescence, and determinate nodulation in the model legume Lotus japonicus. J. Exp. Bot. 64, 1967–1981.

Liu, S., Chen, W., Qu, L., et al., 2013b. Simultaneous determination of 24 or more acidic and alkaline phytohormones in femtomole quantities of plant tissues by high-performance liquid chromatography-electrospray ionization-ion trap mass spectrometry. Anal. Bioanal. Chem. 405, 1257–1266.

Liu, J., Ding, J., Yuan, B., et al., 2014. Magnetic solid phase extraction coupled with in situ derivatization for the highly sensitive determination of acidic phytohormones in rice leaves by UPLC-MS/MS. Analyst 139, 5605–5613.

López-Ráez, J.A., Charnikhova, T., Gómez-Roldán, V., et al., 2008a. Tomato strigolactones are derived from carotenoids and their biosynthesis is promoted by phosphate starvation. New Phytol. 178, 863–874.

López-Ráez, J.A., Charnikhova, T., Mulder, P., et al., 2008b. Susceptibility of the tomato mutant high Pigment-2dg (hp-2dg) to *Orobanche* spp. Infect. J. Agric. Food Chem. 56, 6326–6332.

Lu, Q., Chen, L., Lu, M., et al., 2010. Extraction and analysis of auxins in plants using dispersive iiquid-liquid microextraction followed by high-performance liquid chromatography with fluorescence detection. J. Agric. Food Chem. 58, 2763–2770.

Lu, Q., Zhang, W., Gao, J., et al., 2015. Simultaneous determination of plant hormones in peach based on dispersive liquid-liquid microextraction coupled with liquid chromatography-ion trap mass spectrometry. J. Chromatogr. B 992, 8–13.

Lu, Y., Cao, Y., Guo, X., et al., 2016. Determination of gibberellins using HPLC coupled with fluorescence detection. Anal. Methods 8, 1520–1526.

Luisi, A., Lorenzi, R., Sorce, C., 2011. Strigolactone may interact with gibberellin to control apical dominance in pea (*Pisum sativum*). Plant Growth Regul. 65, 415–419.

Lv, T., Zhao, X., Zhu, S., et al., 2014. Development of an efficient hplc fluorescence detection method for brassinolide by ultrasonic-assisted dispersive liquid-liquid microextraction coupled with derivatization. Chromatographia 77, 1653–1660.

Ma, Z., Ge, L., Lee, A.S.Y., Yong, J.W.H., Tan, S.N., Ong, E.S., 2008. Simultaneous analysis of different classes of phytohormones in coconut (*Cocos nucifera* L.) water using high-performance liquid chromatography and liquid chromatography-tandem mass spectrometry after solid-phase extraction. Anal. Chim. Acta 610, 274–281.

Ma, L., Zhang, H., Xu, W., et al., 2013. Simultaneous determination of 15 plant growth regulators in bean sprout and tomato with liquid chromatography-triple quadrupole tandem mass spectrometry. Food Anal. Methods 6, 941–951.

Mander, L.N., 1992. The chemistry of gibberellins – an overview. Chem. Rev. 92, 573–612.

Matsuda, F., Miyazawa, H., Wakasa, K., et al., 2005. Quantification of indole-3-acetic acid and amino acid coniugates in rice by liquid chromatography-electrospray ionization-tandem mass spectrometry. Biosci. Biotechnol. Biochem. 69, 778–783.

Matsuura, H., Aoi, A., Satou, C., et al., 2009. Simultaneous UPLC MS/MS analysis of endogenous jasmonic acid, salicylic acid and their related compounds. Plant Growth Regul. 57.

Moritz, T., Olsen, J.E., 1995. Comparison between high-resolution selected-ion monitoring, selected reaction monitoring, and 4-sector tandem mass-spectrometry in quantitative-analysis of gibberellins in milligram amounts of plant-tissue. Anal. Chem. 67, 1711–1716.

Mueller, M., Munne-Bosch, S., 2011. Rapid and sensitive hormonal profiling of complex plant samples by liquid chromatography coupled to electrospray ionization tandem mass spectrometry. Plant Methods 7, 1–11.

Müller, A., Duchting, P., Weiler, E.W., 2002. A multiplex GC-MS/MS technique for the sensitive and quantitative single-run analysis of acidic phytohormones and related compounds, and its application to *Arabidopsis thaliana*. Planta 216.

Nambara, E., Marion-Poll, A., 2005. Abscisic acid biosynthesis and catabolism. Ann. Rev. Plant Biol. 56, 165–185.

Niu, Q., Zong, Y., Qian, M., et al., 2014. Simultaneous quantitative determination of major plant hormones in pear flowers and fruit by UPLC/ESI-MS/MS. Anal. Methods 6, 1766–1773.

Noguchi, T., Fujioka, S., Choe, S., et al., 2000. Biosynthetic pathways of brassinolide in Arabidopsis. Plant Physiol. 124, 201–209.

Nordström, A., Tarkowski, P., Tarkowska, D., et al., 2004. Derivatization for LC-electrospray ionization-MS: a tool for improving reversed-phase separation and ESI responses of bases, ribosides, and intact nucleotides. Anal. Chem. 76, 2869–2877.

Novak, O., Henykova, E., Sairanen, I., et al., 2012. Tissue-specific profiling of the *Arabidopsis thaliana* auxin metabolome. Plant J. 72, 523–536.

Ohnishi, T., Godza, B., Watanabe, B., et al., 2012. CYP90A1/CPD, a brassinosteroid biosynthetic cytochrome P450 of Arabidopsis, catalyzes C-3 Oxidation. J. Biol. Chem. 287, 31551–31560.

Olszewski, N., Sun, T.P., Gubler, F., 2002. Gibberellin signaling: biosynthesis, catabolism, and response pathways. Plant Cell 14, S61–S80.

Pan, X., Wang, X., 2009. Profiling of plant hormones by mass spectrometry. J. Chromatogr. B 877, 2806–2813.

Pan, X., Welti, R., Wang, X., 2008. Simultaneous quantification of major phytohormones and related compounds in crude plant extracts by liquid chromatography-electrospray tandem mass spectrometry. Phytochemistry 69, 1773–1781.

Pan, X., Welti, R., Wang, X., 2010. Quantitative analysis of major plant hormones in crude plant extracts by high-performance liquid chromatography-mass spectrometry. Nat. Protoc. 5, 986–992.

Pan, J., Huang, Y., Liu, L., et al., 2013. A novel fractionized sampling and stacking strategy for online hyphenation of solid-phase-based extraction to ultra-high performance liquid chromatography for ultrasensitive analysis. J. Chromatogr. A 1316, 29–36.

Porfirio, S., da Silva, M., Peixe, A., et al., 2016. Current analytical methods for plant auxin quantification – a review. Anal. Chim. Acta 902, 8–21.

Prinsen, E., Redig, P., Dongen, W.V., et al., 1995. Quantitative analysis of cytokinins by electrospray tandem mass spectrometry. Rapid Commun. Mass Spectrom. 9, 948–953.

Prinsen, E., VanDongen, W., Esmans, E.L., et al., 1997. HPLC linked electrospray tandem mass spectrometry: a rapid and reliable method to analyse indole-3-acetic acid metabolism in bacteria. J. Mass Spectrom. 32, 12–22.

Qi, B., Liu, P., Wang, Q., et al., 2014. Derivatization for liquid chromatography-mass spectrometry. TrAC Trends Anal. Chem. 59, 121–132.

Ramos, L., 2012. Critical overview of selected contemporary sample preparation techniques. J. Chromatogr. A 1221, 84–98.

Rappapor, L., Sachs, M., 1967. Wound-induced gibberellins. Nature 214, 1149–1150.

Raskin, I., 1992a. Role of salicylic acid in plants. Ann. Rev. Plant Biol. 43, 439–463.

Raskin, I., 1992b. Salicylate, a new plant hormone. Plant Physiol. 99, 799.

Robert-Seilaniantz, A., Grant, M., Jones, J.D.G., 2011. Hormone crosstalk in plant disease and defense: more than just jasmonate-salicylate antagonism. Ann. Rev. Phytopathol. 49, 317–343.

Romanov, G.A., Lomin, S.N., Schmuelling, T., 2006. Biochemical characteristics and ligand-binding properties of Arabidopsis cytokinin receptor AHK3 compared to CRE1/AHK4 as revealed by a direct binding assay. J. Exp. Bot. 57, 4051–4058.

Rule, G.S., Clark, Z.D., Yue, B., et al., 2013. Correction for isotopic interferences between analyte and internal standard in quantitative mass spectrometry by a nonlinear calibration function. Anal. Chem. 85, 3879–3885.

Saenz, L., Azpeitia, A., Oropeza, C., et al., 2010. Endogenous cytokinins in *Cocos nucifera* L. in vitro cultures obtained from plumular explants. Plant Cell Rep. 29, 1227–1234.

Sakakibara, H., 2006. Cytokinins: activity, biosynthesis and translocation. Ann. Rev. Plant Biol. 57, 431–449.

Sakamoto, T., Morinaka, Y., Ohnishi, T., et al., 2006. Erect leaves caused by brassinosteroid deficiency increase biomass production and grain yield in rice. Nat. Biotechnol. 24, 105–109.

Schmelz, E.A., Engelberth, J., Alborn, H.T., et al., 2003. Simultaneous analysis of phytohormones, phytotoxins, and volatile organic compounds in plants. Proc. Natl. Acad. Sci. U.S.A. 100, 10552–10557.

Schmelz, E.A., Engelberth, J., Tumlinson, J.H., et al., 2004. The use of vapor phase extraction in metabolic profiling of phytohormones and other metabolites. Plant J. 39, 790–808.

Schmitz, R.Y., Skoog, F., Playtis, A.J., et al., 1972. Cytokinins: synthesis and biological activity of geometric and position isomers of zeatin. Plant Physiol. 50, 702–705.

Segarra, G., Jáuregui, O., Casanova, E., et al., 2006. Simultaneous quantitative LC-ESI-MS/MS analyses of salicylic acid and jasmonic acid in crude extracts of *Cucumis sativus* under biotic stress. Phytochemistry 67, 395–401.

Seto, Y., Sado, A., Asami, K., et al., 2014. Carlactone is an endogenous biosynthetic precursor for strigolactones. Proc. Natl. Acad. Sci. 111, 1640–1645.

Shani, E., Weinstain, R., Zhang, Y., et al., 2013. Gibberellins accumulate in the elongating endodermal cells of Arabidopsis root. Proc. Natl. Acad. Sci. U.S.A. 110, 4834–4839.

da Silva, C.M.S., Habermann, G., Marchi, M.R.R., et al., 2012. The role of matrix effects on the quantification of abscisic acid and its metabolites in the leaves of *Bauhinia variegata* L. using liquid chromatography combined with tandem mass spectrometry. Braz. J. Plant Physiol. 24, 223–232.

Sousa, C.M., Miranda, R.M., Freire, R.B., 2011. Competitive-IgY-enzyme linked immuno sorbent assay (CIgY-ELISA) to detect the cytokinins in *Gerbera jamesonii* plantlets. Braz. Arch. Biol. Technol. 54, 643–648.

Spiess, L.D., 1975. Comparative activity of isomers of zeatin and ribosyl-zeatin on *Funaria hygrometrica*. Plant Physiol. 55, 583–585.

Strader, L.C., Culler, A.H., Cohen, J.D., et al., 2010. Conversion of endogenous indole-3-butyric acid to indole-3-acetic acid drives cell expansion in Arabidopsis seedlings. Plant Physiol. 153, 1577–1586.

Strnad, M., 1996. Enzyme immunoassays of N 6-benzyladenine and N 6-(meta-hydroxybenzyl) adenine cytokinins. J. Plant Growth Regul. 15, 179–188.

Su, Y., Xia, S., Wang, R., et al., 2017. Phytohormonal quantification based on biological principles. In: Hormone Metabolism and Signaling in Plants. Elsevier.

Sun, X., Ouyang, Y., Chu, J., et al., 2014. An in-advance stable isotope labeling strategy for relative analysis of multiple acidic plant hormones in sub-milligram *Arabidopsis thaliana* seedling and a single seed. J. Chromatogr. A 1338, 67–76.

Susawaengsup, C., Rayanakorn, M., Wongpornchai, S., et al., 2011. Investigation of plant hormone level changes in shoot tips of longan (*Dimocarpus longan* Lour.) treated with potassium chlorate by liquid chromatography-electrospray ionization mass spectrometry. Talanta 85, 897–905.

Svačinová, J., Novák, O., Plačková, L., et al., 2012. A new approach for cytokinin isolation from Arabidopsis tissues using miniaturized purification: pipette tip solid-phase extraction. Plant Methods 8, 1–14.

Svatos, A., Antonchick, A., Schneider, B., et al., 2004. Determination of brassinosteroids in the sub-femtomolar range using dansyl-3-aminophenylboronate derivatization and electrospray mass spectrometry. Rapid Commun Mass Sp 18, 816–821.

Swaczynova, J., Novak, O., Hauserova, E., et al., 2007. New techniques for the estimation of naturally occurring brassinosteroids. J. Plant Growth Regul. 26, 1–14.

Takei, K., Yamaya, T., Sakakibara, H., 2003. A method for separation and determination of cytokinin nucleotides from plant tissues. J. Plant Res. 116, 265–269.

Tanimoto, E., 2005. Regulation of root growth by plant hormones – roles for auxin and gibberellin. Crit. Rev. Plant Sci. 24, 249–265.

Tarkowská, D., Novák, O., Floková, K., et al., 2014. Quo vadis plant hormone analysis? Planta 240, 55–76.

Toh, S., Kamiya, Y., Kawakami, N., et al., 2012. Thermoinhibition uncovers a role for strigolactones in Arabidopsis seed germination. Plant Cell Physiol. 53, 107–117.

Torres-Vera, R., Garcia, J.M., Pozo, M.J., et al., 2014. Do strigolactones contribute to plant defence? Mol. Plant Pathol. 15, 211–216.

Trapp, M.A., De Souza, G.D., Rodrigues-Filho, E., et al., 2014. Validated method for phytohormone quantification in plants. Front. Plant Sci. 5, 417.

Tyler, L., Thomas, S.G., Hu, J.H., et al., 2004. DELLA proteins and gibberellin-regulated seed germination and floral development in Arabidopsis. Plant Physiol. 135, 1008–1019.

Umehara, M., Hanada, A., Yoshida, S., et al., 2008. Inhibition of shoot branching by new terpenoid plant hormones. Nature 455, 195–200.

Umehara, M., Hanada, A., Magome, H., et al., 2010. Contribution of strigolactones to the inhibition of tiller bud outgrowth under phosphate deficiency in rice. Plant Cell Physiol. 51, 1118–1126.

Urbanova, T., Tarkowska, D., Novak, O., et al., 2013. Analysis of gibberellins as free acids by ultra performance liquid chromatography-tandem mass spectrometry. Talanta 112, 85–94.

Van Meulebroek, L., Vanden Bussche, J., Steppe, K., et al., 2012. Ultra-high performance liquid chromatography coupled to high resolution Orbitrap mass spectrometry for metabolomic profiling of the endogenous phytohormonal status of the tomato plant. J. Chromatogr. A 1260, 67–80.

Varbanova, M., Yamaguchi, S., Yang, Y., et al., 2007. Methylation of gibberellins by Arabidopsis GAMT1 and GAMT2. Plant Cell 19, 32–45.

Vlot, A.C., Dempsey, D.M.A., Klessig, D.F., 2009. Salicylic acid, a multifaceted hormone to combat disease. Ann. Rev. Phytopathol. 47, 177–206.

von Schwartzenberg, K., Nunez, M.F., Blaschke, H., et al., 2007. Cytokinins in the bryophyte Physcomitrella patens: analyses of activity, distribution, and cytokinin oxidase/dehydrogenase overexpression reveal the role of extracellular cytokinins. Plant Physiol. 145, 786–800.

Wang, G., Romheld, V., Li, C., et al., 2006. Involvement of auxin and CKs in boron deficiency induced changes in apical dominance of pea plants (Pisum sativum L.). J. Plant Physiol. 163, 591–600.

Wang, X., Ma, Q., Li, M., et al., 2013. Automated and sensitive analysis of 28-epihomobrassinolide in Arabidopsis thaliana by on-line polymer monolith microextraction coupled to liquid chromatography-mass spectrometry. J. Chromatogr. A 1317, 121–128.

Wang, L., Duan, C., Wu, D., et al., 2014a. Quantification of endogenous brassinosteroids in sub-gram plant tissues by in-line matrix solid-phase dispersion-tandem solid phase extraction coupled with high performance liquid chromatography-tandem mass spectrometry. J. Chromatogr. A 1359, 44–51.

Wang, X., Zhao, P., Liu, X., et al., 2014b. Quantitative profiling method for phytohormones and betaines in algae by liquid chromatography electrospray ionization tandem mass spectrometry. Biomed. Chromatogr. 28, 275–280.

Wang, B., Chu, J., Yu, T., et al., 2015. Tryptophan-independent auxin biosynthesis contributes to early embryogenesis in Arabidopsis. Proc. Natil Acad. Sci. 112, 4821–4826.

Wang, B., Wang, Y., Li, J., 2017a. Biosynthesis and mode of action of strigolactones. In: Hormone Metabolism and Signaling in Plants. Elsevier.

Wang, H., Wei, Z., Li, J., et al., 2017b. The synthesis signalling and function of brassinosteroids. In: Hormone Metabolism and Signaling in Plants. Elsevier.

Went, F.W., 1935. Auxin, the plant growth-hormone. Bot. Rev. 1, 162–182.

Wu, Y., Hu, B., 2009. Simultaneous determination of several phytohormones in natural coconut juice by hollow fiber-based liquid-liquid-liquid microextraction-high performance liquid chromatography. J. Chromatogr. A 1216, 7657–7663.

Wu, Q., Wu, D., Duan, C., et al., 2012. Hollow fiber-based liquid-liquid-liquid micro-extraction with osmosis: II. Application to quantification of endogenous gibberellins in rice plant. J. Chromatogr. A 1265, 17–23.

Wu, Q., Wu, D., Shen, Z., et al., 2013. Quantification of endogenous brassinosteroids in plant by on-line two-dimensional microscale solid phase extraction-on column derivatization coupled with high performance liquid chromatography-tandem mass spectrometry. J. Chromatogr. A 1297, 56–63.

Wu, Q., Wang, L., Wu, D., et al., 2014a. Recent advances in sample preparation methods of plant hormones. Chin. J. Chromatogr. 32, 319–329.

Wu, Q., Wu, D., Guan, Y., 2014b. Polyaniline sheathed electrospun nanofiber bar for in vivo extraction of trace acidic phytohormones in plant tissue. J. Chromatogr. A 1342, 16–23.

Wu, T., Liang, Y., Zhu, X., et al., 2014c. Separation and quantification of four isomers of indole-3-acetyl-myo-inositol in plant tissues using high-performance liquid chromatography coupled with quadrupole time-of-flight tandem mass spectrometry. Anal. Bioanal. Chem. 406, 3239–3247.

Xie, X., Yoneyama, K., 2010. The strigolactone story. Ann. Rev. Phytopathol. 48, 93–117.

Xin, P., Yan, J., Fan, J., et al., 2013a. A dual role of boronate affinity in high-sensitivity detection of vicinal diol brassinosteroids from sub-gram plant tissues via UPLC-MS/MS. Analyst 138, 1342–1345.

Xin, P., Yan, J., Fan, J., et al., 2013b. An improved simplified high-sensitivity quantification method for determining brassinosteroids in different tissues of rice and Arabidopsis. Plant Physiol. 162, 2056–2066.

Yan, H., Wang, F., Han, D., et al., 2012. Simultaneous determination of four plant hormones in bananas by molecularly imprinted solid-phase extraction coupled with high performance liquid chromatography. Analyst 137, 2884–2890.

Yokota, T., Sakai, H., Okuno, K., et al., 1998. Alectrol and orobanchol, germination stimulants for *Orobanche minor*, from its host red clover. Phytochemistry 49, 1967–1973.

Yoneyama, K., Xie, X., Kusumoto, D., et al., 2007a. Nitrogen deficiency as well as phosphorus deficiency in *Sorghum* promotes the production and exudation of 5-deoxystrigol, the host recognition signal for arbuscular mycorrhizal fungi and root parasites. Planta 227, 125–132.

Yoneyama, K., Yoneyama, K., Takeuchi, Y., et al., 2007b. Phosphorus deficiency in red clover promotes exudation of orobanchol, the signal for mycorrhizal symbionts and germination stimulant for root parasites. Planta 225, 1031–1038.

Yoneyama, K., Xie, X., Sekimoto, H., et al., 2008. Strigolactones, host recognition signals for root parasitic plants and arbuscular mycorrhizal fungi, from *Fabaceae* plants. New Phytol. 179, 484–494.

Yoneyama, K., Xie, X., Kim, H., et al., 2012. How do nitrogen and phosphorus deficiencies affect strigolactone production and exudation? Planta 235, 1197–1207.

Yoneyama, K., Xie, X., Kisugi, T., et al., 2013. Nitrogen and phosphorus fertilization negatively affects strigolactone production and exudation in *Sorghum*. Planta 238, 885–894.

Zaveska Drabkova, L., Dobrev, P.I., Motyka, V., 2015. Phytohormone profiling across the bryophytes. PLoS One 10.

Zhai, Q., Yan, C., Li, L., et al., 2017. Metabolism and signalling of jasmonate. In: Hormone Metabolism and Signaling in Plants. Elsevier.

Zhang, Z.J., Zhou, W.J., Li, H.Z., 2005. The role of GA, IAA and BAP in the regulation of in vitro shoot growth and microtuberization in potato. Acta Physiol. Plant. 27, 363–369.

Zhang, Y., Li, Y., Hu, Y., et al., 2010a. Preparation of magnetic indole-3-acetic acid imprinted polymer beads with 4-vinylpyridine and beta-cyclodextrin as binary monomer via microwave heating initiated polymerization and their application to trace analysis of auxins in plant tissues. J. Chromatogr. A 1217, 7337–7344.

Zhang, Z., Zhang, Y., Tan, W., et al., 2010b. Preparation of styrene-co-4-vinylpyridine magnetic polymer beads by microwave irradiation for analysis of trace 24-epibrassinolide in plant samples using high performance liquid chromatography. J. Chromatogr. A 1217, 6455–6461.

Zhang, Z., Tan, W., Hu, Y., et al., 2012. Microwave synthesis of gibberellin acid 3 magnetic molecularly imprinted polymer beads for the trace analysis of gibberellin acids in plant samples by liquid chromatography-mass spectrometry detection. Analyst 137, 968–977.

Zhang, Z., Hao, Y., Ding, J., et al., 2015a. One-pot preparation of a mixed-mode organic-silica hybrid monolithic capillary column and its application in determination of endogenous gibberellins in plant tissues. J. Chromatogr. A 1416, 64–73.

Zhang, H., Tan, S.N., Teo, C.H., Yew, Y.R., Ge, L., Chen, X., Yong, J.W., 2015b. Analysis of phytohormones in vermicompost using a novel combinative sample preparation strategy of ultrasound-assisted extraction and solid-phase extraction coupled with liquid chromatography-tandem mass spectrometry. Talanta 139, 189–197.

Zhao, Y., 2010. Auxin biosynthesis and its role in plant development. Ann. Rev. Plant Biol. 61, 49–64.

Zhou, F., Lin, Q., Zhu, L., et al., 2013. D14-SCFD3-dependent degradation of D53 regulates strigolactone signalling. Nature 504, 406–410.

Zhu, G., Long, S., Sun, H., et al., 2013. Determination of gibberellins in soybean using tertiary amine labeling and capillary electrophoresis coupled with electrochemiluminescence detection. J. Chromatogr. B 941, 62–68.

Ziegler, J., Qwegwer, J., Schubert, M., et al., 2014. Simultaneous analysis of apolar phytohormones and 1-aminocyclopropan-1-carboxylic acid by high performance liquid chromatography/electrospray negative ion tandem mass spectrometry via 9-fluorenylmethoxycarbonyl chloride derivatization. J. Chromatogr. A 1362, 102–109.

Author Index

Subject Index

Cytokinins (CKs) (*Continued*)
  response regulators
    type-B ARR function, negative
      regulators, 94
    type-B ARR transcription factors,
      cytokinin-regulated gene expression,
      93–94
  sample preparation
    purification, 498–499, 500t
    sampling and extraction, 498
  structures and types, 78–79, 492–498,
    495t–497t
  transport, 84–85
Cytokinin signaling systems, 85–86, 95
  AHP6 inhibition, 92
  AHP1 to AHP5 function, 91–92
  nitric oxide (NO) regulation, 92–93
  type-B ARR function, negative regulators,
    94
Cytokinin signal transduction, 86–88,
    87f
Cytokinin transport, 84–85

D

DELLA proteins, 26, 94, 130–132, 131f,
    217–218
  antagonism, of hormone signaling,
    136–139
  DELLA protein repression, 133–134,
    134f
  GA-promoted growth, 132–133
  GA signal transduction, 134–136
    DELLA-mediated protein-protein
      interactions, 139–141
    developmental and environmental
      signals, 141–142
    mediators, of GA signaling, 130–132,
      131f
  multiple functions, 136, 138f
De novo stem cell niche formation
  hormone-directed de novo stem cell niche
    formation
    auxin-cytokinin cross talk, 419–420
    auxin-regulated regeneration programs,
      417–418
    cytokinin-regulated regeneration
      programs, 418–419
    regulating events, 416–417, 417f

    stem cell niches de novo, 416
    wound-induced de novo regeneration,
      420–421
  hormone-directed somatic cell
      pluripotency
    auxin and cytokinin, 413–415, 415f
    mechanisms, 415–416
Desiccation tolerance, 168–169
DET2, 294
DFL2, 48–49
Dormancy, 166–169
Dormin, 161
DR5-GFP, 9–10
DR5-GUS, 9–10
Drought rhizogenesis, 170
D53/SMXL proteins, 18
DWARF14, 8

E

EIN2, 16–17
  CEND, 212–213
  cleaved and transport, nucleus, 213–215,
    214f
  translational repression, in cytoplasm,
    215–217
EIN3/EIL1, 217
  F-box proteins EBF1 and EBF2,
    218–220
  regulation, 217–218, 219f
  transcriptional network, 220–221
Electrochemical biosensors, 446–448
Embryogenesis, 20
Enzyme-linked immunosorbent assay
    (ELISA), 444–445
EPFL9, 372
EPF1/2 peptides, 386
ESF1 peptide, 372, 390
Ethylene (ETH), 4, 231
  biology, 203, 204f
  climacteric fruits, 23
  ethylene responses, in *Arabidopsis*, 208
  metabolism
    ACC oxidase, 207–208
    ACC synthase, 205–207
    biosynthesis, in *Arabidopsis*,
      204
Ethylene biology, 203, 204f
Ethylene biosynthesis, 204

Printed in the United States
By Bookmasters